D0606197

Human Physiology

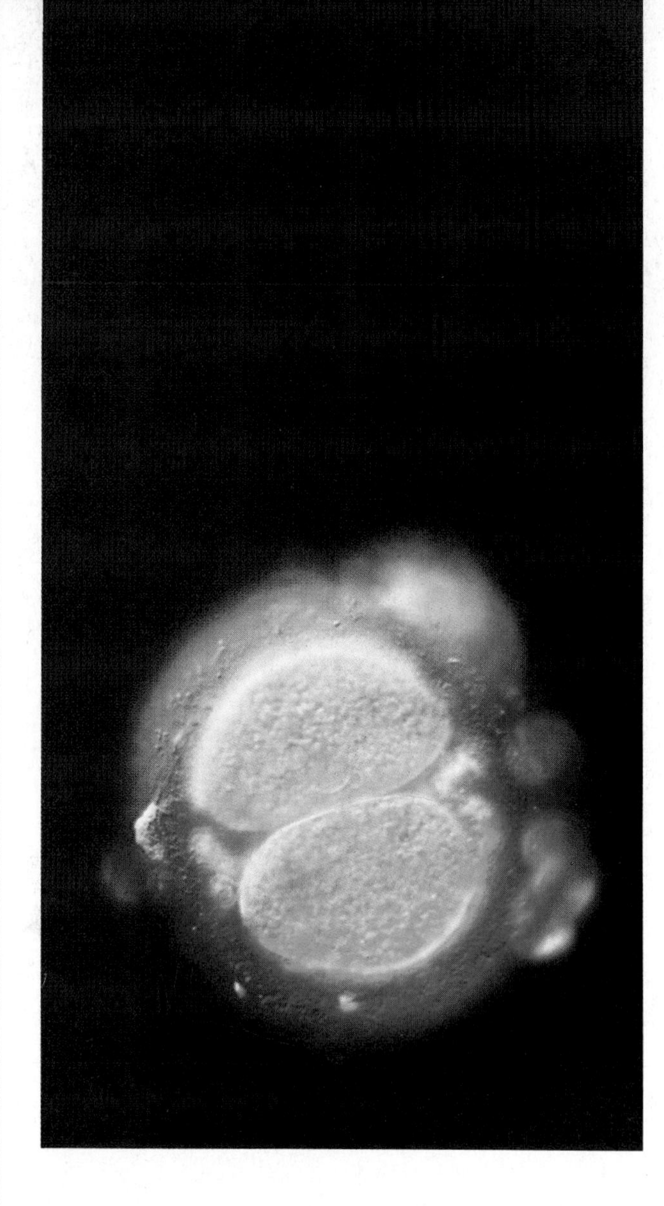

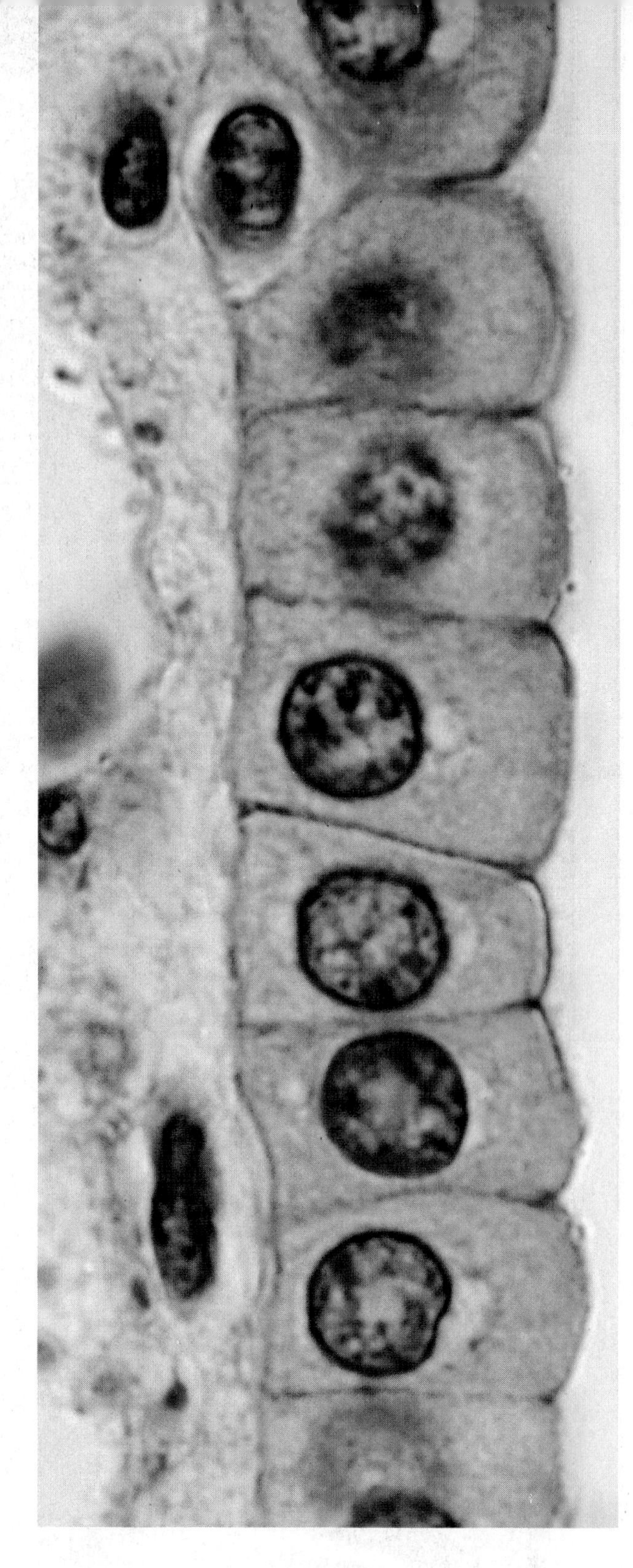

Rodney Rhoades, Ph.D.

Professor and Chairman
Department of Physiology and Biophysics
Indiana University School of Medicine

Richard Pflanzer, Ph.D.

Associate Professor of Biology and Physiology/
Biophysics
Indiana University School of Medicine
Purdue University

Saunders College Publishing

Philadelphia Ft. Worth Chicago San Francisco
Montreal Toronto London Sydney Tokyo

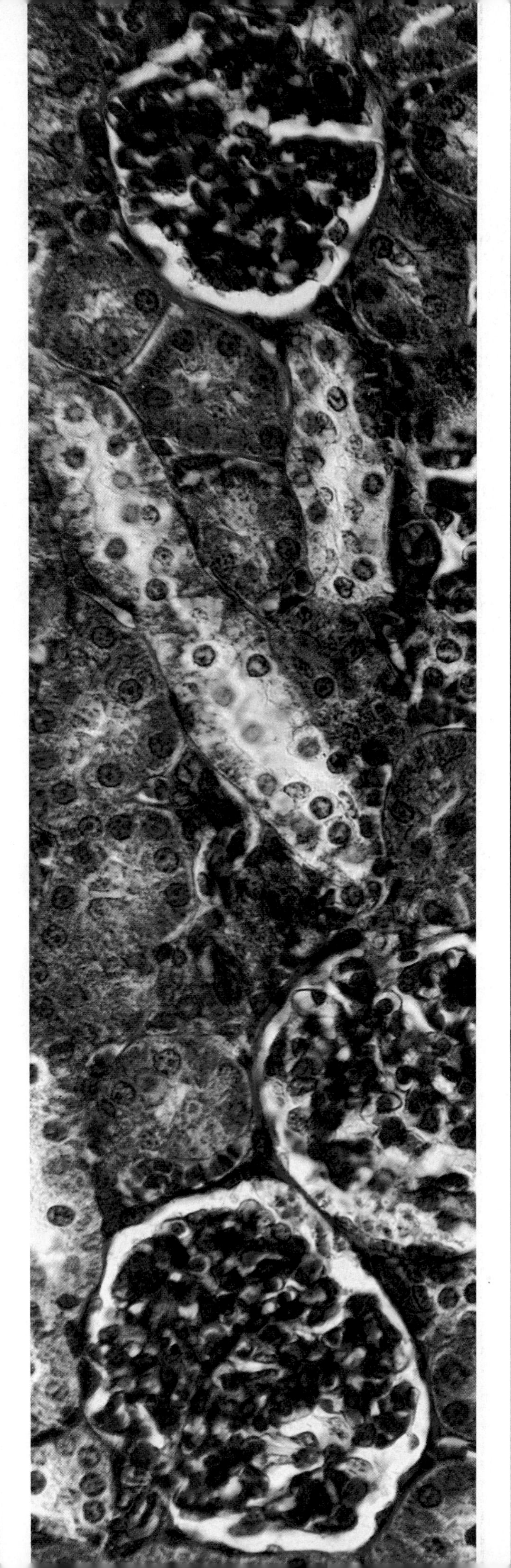

Human Physiology

Text typeface: Palatino
Compositor: York Graphic Services
Acquisitions Editor: Edward F. Murphy
Developmental Editor: Martha Colgan
Production Management: York Production Services
Art Director: Carol C. Bleistine
Art Assistant: Doris Bruey
Text Design: Tracy Baldwin
Text Artwork: York Production Services
Cover Design: Lawrence R. Didona
Cover Credit: Enid V. Hatton
Production Manager: Merry Post

Printed in the United States of America

Library of Congress Catalog Card Number: 88-043545

ISBN 0-03-011254-0

9012 071 987654321

Preface

The explosion of scientific information has given the life and health sciences tremendous opportunities and tools to better understand the world within us. As educators and authors, we face an important challenge to present to students a large quantity of scientific information not as a catalogue of disconnected facts and mysteries, but as a body of unifying principles and concepts. *Human Physiology* explains underlying principles and concepts while acknowledging the excitement of unsolved problems and unexplained facts. We have written this book for students of human biology who, regardless of their science background, have a career interest or personal curiosity about the workings of the human body. Students who plan a career in the life or health sciences will find *Human Physiology* especially beneficial.

Our background material on cellular function and molecular biology will help students who have had little or no college-level science coursework. Students with a stronger background in the sciences will find our basic science coverage a useful review. Underlying principles of human anatomy are also introduced; the emphasis of this text, however, is on function, rather than structure, and students are referred to standard anatomy textbooks for a more thorough treatment of human structure.

The organization and approach of *Human Physiology* have evolved through many years of teaching the subject to students of various backgrounds and abilities. Our research, along with that of many other scientists, helped us a great deal in shaping the content of this text. And our teaching experiences helped us decide how to present it.

Our objectives in writing this text were to

1. develop basic concepts and principles of physiology logically, clearly, and concisely;
2. improve the student's ability to reason scientifically; and
3. develop an understanding and appreciation of normal body functions.

We hope that this book will convey to the student that the human body is a beautifully built machine and that many physiological processes, which we often take for granted, are very complex. For example, at the cellular level, each cell carries all the genetic material needed to create an entire human being; the same genetic pool housed within the nucleus of each cell controls over a thousand chemical reactions. At the organ level, the first breath taken by a newborn triggers a near-miraculous set of physiological events that allow the neonate to separate from a life-supporting placenta and to start air-breathing independence. There are many other examples like these illustrated in this book.

Features

Early coverage of *molecular biology and cell physiology* (Chapters 2 through 6) provides a strong basis for understanding systemic functions. Concepts derived from cellular and molecular biology are applied throughout the text.

Physiological control systems (Section II) are covered before organ functions (Section III) to establish early the concept of *homeostasis.*

Metabolism and energy transformation are covered early (Chapter 6) to present intermediary metabolism as a fundamental concept rather than as a process specific to nutrition.

The artwork has been developed to enhance the use of *flow charts* in human physiology (for example, see Fig. 7–29 on page 242 and Fig. 19–17 on page 591).

End-of-chapter review questions are designed to develop *conceptual understanding.*

Applications include chapter "Focus Boxes" (see Contents) as well as applied topics of chapter length (Chapters 22, 29, 30, and 31).

Organization

The book is organized so that topics progress from the cell, to integrated organ function, to the total body. Chapter 1, designed to develop an appreciation of science, uses a historical perspective to explain the scientific method and introduces basic concepts used throughout the text. The theme of Section I is basic cellular functions, including essential chemistry, biochemistry, and physics required to enhance the student's understanding. Section II focuses on the body's internal environment and analyzes the nature of biologic control systems and the properties of specialized cells—nerves and glands that regulate body function. Section III describes integrated organ function and analyzes the body's coordinated functions (e.g., circulation, digestion, respiration, reproduction) that stem from integration of specialized tissue function.

Throughout this book, we have made every effort to use facts to build on general principles and have used many figures to develop concepts and explanations.

Learning Aids

Many pedagogical aids have been incorporated to help the student learn concepts in physiology. A **chapter outline** is included at the beginning of each chapter, and a **summary outline** and set of **review questions** are given at the end of each chapter. Throughout the text, **boldface** is used to emphasize important new terms. The artwork emphasizes cellular and systemic processes and interactions. Numerous flow charts are used to reinforce the conceptual approach of both text and artwork. We have also introduced **focus boxes** to present subjects in greater depth, to introduce applications of basic concepts, and to illustrate how many human diseases are often linked to altered physiological function. A list of these focus boxes appears on p. **xvi.** At the end of each chapter, **suggested readings** provide greater depth for the biology major.

Supplemental Material

A number of supporting aids accompany the text. These include an **instructor's manual,** a **laboratory manual,** a **study guide,** and **overhead color transparencies** for the classroom. In addition, a set of **lecture outlines** are available that include key figures. These lecture outlines are designed to allow the student to follow the lectures and to encourage better notetaking.

Acknowledgments

We want to express our deepest thanks and appreciation to all the contributors who have written many of the chapters and without whose effort, talent, and insight this book would not have been possible. We would also like to express our gratitude to the editorial staff of Saunders College Publishing for their help and guidance in the development of this book. Most important was our developmental editor Martha Colgan, who strived for the highest possible quality. Martha labored with us over every paragraph and illustration. Her patience in working with us through a maze of details in transforming the initial manuscript to final text is greatly appreciated. She is responsible for initial sketches of most of the artwork that so beautifully illustrates the text. We are indebted to Merry Post, our Production Manager, and to Carol Bleistine, our Art Director, for guiding the book through the production phase. Finally we would like to thank Ed Murphy, our Senior Editor, for believing in this project from the beginning. His support and commitment made this book a reality. He shared with us both his talent and the pressures of meeting a number of impossible deadlines.

We want to thank Ann Hollingsworth, Marilyn Gruenhagen, and Jennifer Conwell for typing portions of the manuscript.

Lastly, we want to thank our wives Diane Pflanzer and Judy Rhoades, who became "university widows" during the final stages of this project. Their love, support, and encouragement carried us through.

We encourage input from our readers and welcome suggestions for future improvements of this textbook.

Reviewers of HUMAN PHYSIOLOGY

We are grateful to a number of faculty members and scientists who reviewed *Human Physiology.* Their suggestions, comments, and insight have been invaluable.

Jerome Yochim, University of Kansas
Arnold J. Sillman, University of California, Davis
Rick Turnquist, Augustana College
Andy Anderson, Utah State University
Byron A. Schottelius, University of Iowa
John T. Fales, University of Missouri, Kansas City
John P. Harley, Eastern Kentucky University
L. Stephen Whitley, Eastern Illinois University
S.J. Coward, University of Georgia
David Noyes, The Ohio State University
Pegge Alciatore, University of Southwest Louisiana
Stephen Williams, Glendale Community College
Kenneth H. Bynum, University of North Carolina, Chapel Hill

Rodney A. Rhoades
Richard G. Pflanzer
November, 1988

Contributing Authors

Reynaldo S. Elizondo, Ph.D.
Dean, College of Science
University of Texas at El Paso
Chapters 28, 29

Janice C. Froehlich, Ph.D.
Assistant Professor of Physiology/Biophysics and Medicine
Indiana University School of Medicine
Chapters 30, 31, 32

Joe R. Haeberle, Ph.D.
Assistant Professor of Physiology/Biophysics and Medicine
Indiana University School of Medicine
Chapters 18, 19

Stephen A. Kempson, Ph.D.
Associate Professor of Physiology and Biophysics
Indiana University School of Medicine
Chapters 1, 2, 3, 4, 5, 6

Leon K. Knoebel, Ph.D.
Professor of Physiology and Biophysics
Indiana University School of Medicine
Chapter 26

Walter C. Low, Ph.D.
Associate Professor of Physiology and Biophysics
Indiana University School of Medicine
Chapters 1, 7, 8, 9, 10, 11

Richard A. Meiss, Ph.D.
Professor of Physiology/Biophysics and Obstetrics/Gynecology
Indiana University School of Medicine
Chapters 1, 16

Daniel E. Peavy, Ph.D.
Associate Professor of Physiology and Biophysics
Indiana University School of Medicine
Research Chemist
R. L. Roudebush VA Medical Center
Chapters 1, 12, 13, 14, 15, 27

Richard G. Pflanzer, Ph.D.
Associate Professor of Biology and Physiology/Biophysics
Indiana University School of Medicine
Purdue University
Chapters 1, 17

Rodney A. Rhoades, Ph.D.
Professor and Chairman
Department of Physiology and Biophysics
Indiana University School of Medicine
Chapters 1, 20, 22

George A. Tanner, Ph.D.
Professor of Physiology and Biophysics
Indiana University School of Medicine
Chapters 1, 23, 24, 25

Wiltz W. Wagner, Ph.D.
Associate Professor of Physiology/Biophysics and Anesthesia
Indiana University School of Medicine
Chapters 1, 21, 22

Contents

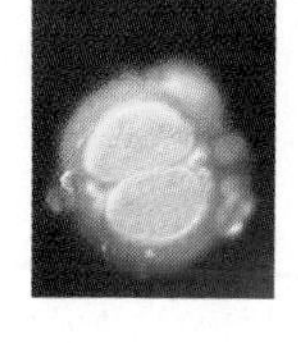

Focus Boxes

These focus boxes are intended to apply basic concepts and illustrate how diseases are characterized by altered functional changes.

The Science of Physiology

The ancient Greeks considered the universe to be composed of four elemental substances including air, water, earth, and fire. The Greeks believed each human being was a miniature universe in whom these substances appeared in the form of four "humors," or fluids: *blood* (corresponding to air); *phlegm* (representing water), *black bile* (corresponding to earth), and *choler* or *yellow bile* (representing fire) (Fig. 1–1).

According to the Greek view, the fluids were not present in each person in equal amounts; one fluid usually predominated and characterized the individual's temperament. For example, one temperament was the "sanguine" type. We still use this word today to describe someone with a cheerful, confident persona. In ancient times, such a personality would have been attributed to a predominance of blood over the other three humors within that person's body, as evidenced perhaps by a ruddy complexion (*sanguine* comes from the Latin *sanguineus,* meaning "bloody"). A "melancholy" person was thought to have more black bile than any of the other humors, whereas a "choleric" person was considered to have a predominance of yellow bile (the humor associated with fire), which produced an excitable, easily angered temperament. A "phlegmatic" person generally was calm and unemotional, even sluggish, supposedly because of a predominance of phlegm, the cold, waterlike humor. The Greeks thought that the unique

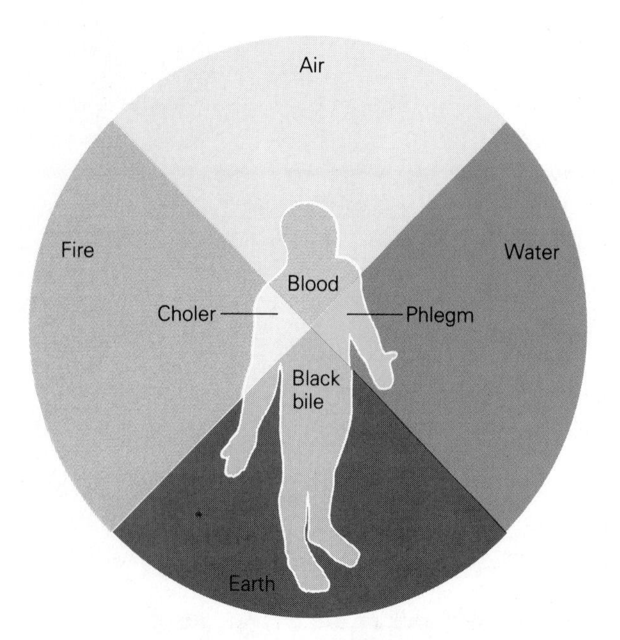

Figure 1–1 The four bodily "humors" of classical and medieval philosophy and physiology. The Greeks thought that the elements of the universe were represented in each human body by blood, phlegm, choler, and black bile. According to this scheme, a predominance of one humor produced each person's temperament.

amounts of each humor and the resulting proportion, or balance, determined a person's physical and psychological makeup.

According to the classical scheme, while a normal predominance of one fluid within each person's unique humoral balance produced a characteristic personality, an abnormal lack or excess of one or more humors resulted in disease. Doctors from ancient times to the Renaissance shaped their therapeutic methods around attempts to restore the proper balance of humors within the sick person's body. Purging, bloodletting, and similar treatments were all intended to adjust humors.

In contrast to the ancient Greek philosophers and physiologists, today we organize the universe of matter into 105, rather than 4, elemental substances, and we attribute each person's uniqueness to a sequence of nucleotides rather than to relative quantities of bodily fluids. Yet despite the revolutions in our conceptions of human physiology, some classical principles influence modern science. One enduring principle is the idea that the human body, if not a world in miniature, is composed of elements from the universe that obey the laws of nature. Another "modern" idea cherished by the ancient philosophers is that of a harmonious balance within the body. While today we speak of homeostasis and recognize it as a self-regulatory process involving not only fluids but dynamic processes, we share the classical view that health is the evidence of balance among bodily functions and substances. Finally, we have inherited the Renaissance attitude—which echoed the ancient Greek belief—that the human body is worthy of the most rigorous study and eloquent description.

The History of Physiology

The story of the development of physiological understanding in modern culture is fascinating. The following brief survey of the last 2000 years of physiological theory and research is meant to provide a context for the study of twentieth-century physiology in all its complexity and sophistication. Further, the discussion communicates a sense of wonder at the vast distances physiologists have come since human beings first began to record theories and observations about the workings of the human body.

Physiological Knowledge in the Classical and Medieval Periods

Since so many Western attitudes about science, philosophy, and the place of human beings in the universe have been influenced by the traditions of the classical Greeks, that period seems a logical starting point to examine the development of physiology as we understand it today. The studies of the Greek philosophers and the investigations of the Greek physiologists at the great Museum in Alexandria influenced European medical and physiological practice until well into the Renaissance because of the mediating influence of Galen. Galen's writings preserved the ancient studies and, in turn, were preserved through the medieval period for examination by Renaissance scholars.

The Observations of Aristotle

Although he is most often thought of as a philosopher, Aristotle (384–322 B.C.) did important work in biology and was one of the first to describe the blood vessels as a system with the heart at the center. Unlike other investigators of his time, he used a bloodless method of killing animals for dissection; Aristotle found that since the blood remained in the vessels during the dissection, they were easier to see and trace.

Some of Aristotle's concepts were corrected through the discoveries of later investigators. Aristotle thought of the heart as both the seat of the intellect and as a furnace that heated the blood to provide needed warmth. (Since warmth disappeared from the body so soon after death, it was thought to be the cause or source of life.) Just as the blood needed a source of heat, so too the furnace needed ventilation. According to Aristotle, this was the purpose of breathing: The lungs were a ventilation system and the air a cooling agent.

Aristotle correctly observed that blood flowed between arteries and veins but he could not see the microscopic capillaries we now know provide the connection between them. He explained the flow by saying, in essence, that the vessels were themselves made of blood and that they disappeared when blood flow stopped.

> . . . the channels of the blood vessels may be compared to the mud which a running stream deposits, they are as it were deposits left by the current of blood in the blood vessels. . . . Thus just as in the irrigation system the biggest channels persist whereas the smallest ones quickly get obliterated by the mud, though when the mud abates they reappear; so in the body the largest blood vessels persist while the smaller ones become flesh in actuality, though potentially they are blood vessels as ever before.

The Theories of Herophilus and Erasistratus

At the Museum at Alexandria, a famous center of Greek culture and learning in what is now Egypt, two men pursued a number of biological investigations in a continuation of Aristotle's search for an understanding of blood flow. Physiological research flourished at Alexandria partly because it was the only place that permitted dissections of human cadavers for the study of anatomy and physiology.

Herophilus (c. 335–280 B.C.), who is considered the first physiologist in the Western tradition by some, identified the brain as the seat of the intellect and demonstrated that the walls of the arteries are thicker than those of the veins. Using the newly invented water clock, he measured the pulse and showed that it varied when disease was present.

Erasistratus (310–240 B.C.) began his physiological studies as an assistant to Herophilus. Erasistratus believed that blood was made in the liver from food and was delivered to the organs of the body by an "ebb and flow" in the veins. He thought of the arteries as air vessels, not blood vessels. The air, called *pneuma,* was thought of as a living force taken in through the lungs to the left side of the heart, where it was transformed into a "vital spirit" and then transported, in "airy" form, through the arteries to the rest of the body. (To explain why blood, not air, flowed from a cut artery, in apparent contradiction with his theory, Erasistratus postulated the existence of tiny connections between veins and arteries that sometimes opened to let blood flow into the arteries.)

The "Pneumatology" of Galen

In a simple experiment with a goose feather, Galen (c. 130–201 A.D.) demonstrated that blood, not air, flowed through the arteries. Galen created a hollow tube by cutting off the ends of the feather. Then he inserted the tube into a tied-off artery, from which blood immediately flowed into the tube. Since, by being tied off, the artery was supposedly separated from the veins, the blood that flowed from it must have been inside the artery from the beginning.

Galen was a Greek who studied physiology both in Greece and at Alexandria. His practice as a surgeon to gladiators in his native city of Pergamum allowed him to observe the internal structure of the body in a time when dissections for the purpose of study were forbidden. Later he was appointed physician to a Roman emperor. Galen is known for his voluminous writings on philosophy, medicine, and physiology, in which he commented and expanded on the work of the investigators who preceded him. Even though many of his ideas have been disproven, we still consider Galen an important figure because of the intellectual sophistication of his physiological schemes, the breadth and scope of his explanations, and the length of time and extent to which his teachings were accepted. His was the prevailing view of human physiology from his own time until the Renaissance.

Like Erasistratus, Galen believed that the blood was produced from food, in the liver (Fig. 1–2). He thought that the blood took on "natural spirits" and carried them through the veins to the bodily organs, which needed the spirits to carry out different functions. After its supply of spirits was depleted, the blood returned along the same venous pathways to the liver to be resupplied. (This idea that the blood flowed both ways in the vessels became one of the most revered of Galen's teachings. When Renaissance anatomists correctly challenged it because of their knowledge of valves, it was very difficult to overthrow.)

According to Galen, some of the blood containing the "natural spirits" went first to the right side of the heart, and then to the left side, where it contacted *pneuma. Pneuma* was a substance produced from air in the lungs and carried into the left side of the heart. When the "natural spirits" contacted *pneuma,* they were transformed into "vital spirits," a higher form of *pneuma.* These "vital spirits" were

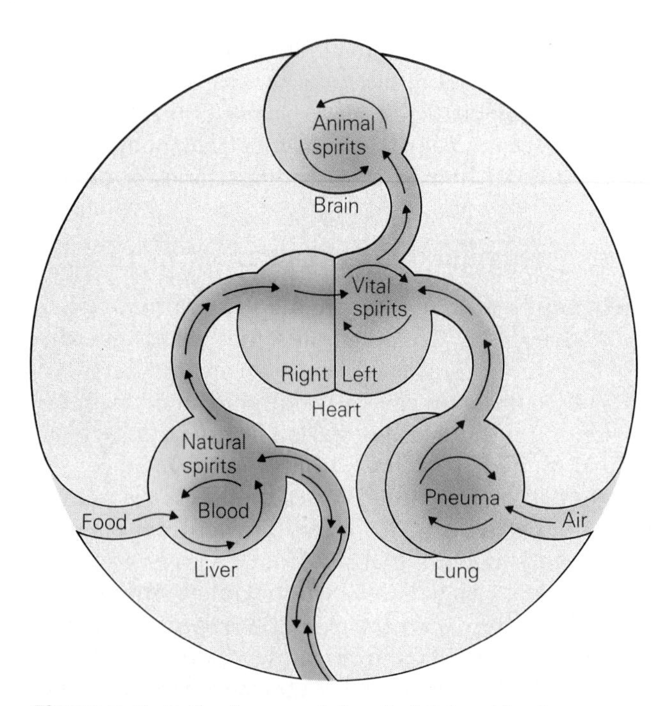

Figure 1–2 In the "pneumatology" of Galen, blood was made in the liver from food and carried "natural spirits" to the bodily organs, traveling both ways in blood vessels. *Pneuma* made from air in the lungs met blood in the heart and was transformed into "vital spirits" and then "animal spirits."

carried on up to the brain, where they were further transformed into yet a higher form of *pneuma* called "animal spirits." These various forms of *pneuma* drove the various processes of the body, according to Galen.

Despite his insistence on the importance of observation in the study of physiology, Galen often made incorrect assumptions about human anatomy based on his dissections of animals. As mentioned previously, however, the logic and elegance of his theories about bodily functions, which unified both the traditions he inherited and the knowledge available to his contemporaries, were convincing. His copious writings further established his authority, and throughout the late classical and medieval periods, Galen's physiology was accepted. It prevailed largely unquestioned until the technological advances and humanistic spirit of the Renaissance permitted a fresh look at the body.

The Discoveries of the European Renaissance Physiologists

In the revival of classical learning that characterized the European Renaissance, Galen's writings about anatomy, physiology, and medicine first began to be widely translated from the Greek and Arabic into Latin. Although the writings were known to exist, and the general outlines of Galen's physiology had dominated medieval and early Renaissance medicine, only in the sixteenth century did the most important writings become available for close scrutiny by European physicians. By this time, the Galenic physiology already had taken hold of the imagination and had come to be regarded almost as religious doctrine. Thus, Renaissance professors of medicine taught the anatomy of Galen, and anyone who departed from Galen's concepts was considered a secular heretic. To explain the discrepancies between the anatomy described by Galen and that visible in a dissected corpse, a professor might claim that the body simply had changed since Galen studied it.

The task of the Renaissance physiologists in the advancement of physiological knowledge thus was first to study Galen's physiology, discover its errors, and then replace it with a modern scheme. Many independent-minded investigators contributed to the revolution in physiological knowledge that occurred in the late Renaissance, although they too defaulted to many of Galen's tenets whenever they could not observe or explain for themselves certain structures and functions. In his teachings, writings, and drawings, Andreas Vesalius corrected many of the inaccuracies of Galen's anatomy while perpetuating others and retaining much of Galen's erroneous physiology. William Harvey, in his discovery of the circulation, finally revised Galen's ancient notion that the air and the blood met in the heart, though he still viewed the lungs, as had Galen, as cooling organs.

The Drawings of Vesalius

Andreas Vesalius (1514–1564) was educated as a doctor and, at the age of 23, became a professor of surgery and anatomy at the great medical school in Padua, in what is now Italy. He had been taught the physiology of Galen and in turn began teaching it to his students. Unlike other professors of his time, however, who taught almost exclusively from Galen's writings, Vesalius centered his lectures not on the ancient texts but on human cadavers that he himself dissected as he lectured. (Previous instructors might not have used a human cadaver or might have lectured from afar while another person conducted the dissection.)

Another pedagogical device that Vesalius introduced into the study of physiology was the anatomical drawing. Galen's texts had been published with few or no illustrations; Vesalius created six anatomical "tables" that showed labelled parts of the body that allowed students to follow his lectures.

1.83 m (equivalent to 6.0 feet) is composed of cells that for the most part have a diameter of about 10 μm. The cells can be discerned only with the aid of a microscope because the human eye cannot see objects smaller than about 100 μm. The cells contain amino acid molecules that are only 1 nm in diameter, and the individual atoms of the amino acids may have diameters that are less than 0.1 nm.

Scientific Notation

In physiology, as in many of the sciences, it is often awkward or inconvenient to express measurements using standard digits (e.g., the diameter of a blood vessel as 10,000 μm, a blood glucose concentration of 0.005 moles per liter). Instead of writing out all the "zeroes," scientists use a system of **scientific notation,** in which quantities are expressed as products of a number plus a power of ten. Powers of ten are expressed with **exponents,** which are small numbers written above and to the right of the base number, ten. The exponent indicates the number of times the base should be used as a factor in the expression. For example, $10^2 = 10 \times 10 = 100$. Thus,

$10,000\ \mu m = 1 \times 10^4\ \mu m$
0.005 moles/liter $= 5 \times 10^{-3}$ moles/liter
0.001 grams $= 1 \times 10^{-3}$ grams

(Note that negative exponents indicate the number of times 10 is used as a *divisor* for quantities less than one. For example, $10^{-3} = 1/10 \times 1/10 \times 1/10$.) Numbers without zeroes can also be expressed in scientific notation (e.g., $46,782 = 4.6782 \times 10^4$).

Following are some simple rules for converting a number expressed in scientific notation back to the original number:

1. For positive exponents: Shift the decimal point to the right by the number of places indicated by the exponent.
2. For negative exponents: Move the decimal point to the left by the number of places indicated by the exponent.
3. When multiplying exponential quantities that have the same base, add the exponents ($5^3 \times 5^2 = 5^5$).

Ratios and Proportions

A **ratio** is an expression that compares two numbers or quantities by division. For example, to compare the numbers 100 and 10, we divide 100 by 10 to get 10/1. If we are describing the number of experimental subjects that were exposed to high concentrations of ozone, we might say that of 100 rats exposed to the ozone, 10 died. The ratio of animals that died to those that survived was 10/100 or 1/10 ("one to ten"). In other words, 1 out of every 10 animals died from the ozone exposure. Ratios can be expressed in several different ways that have the same meaning:

1:250 means 1 part to 250 parts
1/5 means 1 part out of 5 parts
2 cats to 6 dogs means a ratio of 1 cat to every 3 dogs

A **proportion** is a mathematical statement of the equality of two ratios. By arbitrarily using the letters *A*, *B*, *C*, and *D* to express quantities, we can state a proportion in the following way:

A is to *B* as *C* is to *D*.

This is equivalent to saying:

A times *D* equals *B* times *C*.

For example, if $A = 15$, $B = 25$, $C = 3$, and $D = 5$, then

$$\frac{A}{B} = \frac{C}{D} = \frac{15}{25} = \frac{3}{5}$$
$$A \times D = B \times C;\ 15 \times 5 = 25 \times 3.$$

Therefore, it follows that if three of the quantities are known, the value of the fourth can be determined.

For example, assume that an electrocardiogram is being recorded on a moving strip of paper (Fig. 1–9). The speed of the moving paper is 25 mm/sec. If each repeating cycle of the electrocardiogram represents one heartbeat, how many heartbeats are occurring each minute? The problem can be solved as follows:

1. The distance between cycles is 20 mm (as measured from the record).
2. The time interval between cycles is unknown; it equals x sec.
3. The ratios relating distance to time can be expressed as a proportion:

$$\frac{25 \text{ mm}}{1 \text{ sec}} = \frac{20 \text{ mm}}{x \text{ sec}}$$

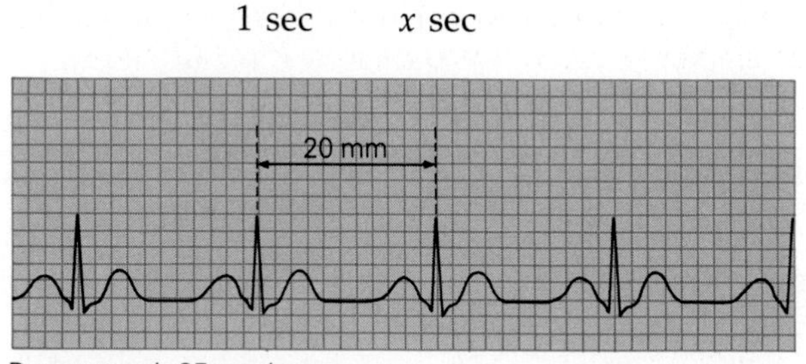

Paper speed: 25 mm/sec

Figure 1–9 From the speed of the paper and the distance between points on this electrocardiogram, a proportion can be used to determine the number of heartbeats per minute.

4. $25x = 20$; $x = 0.8$ The interval between cycles is 0.8 seconds.
5. The number of cycles (beats) occurring each minute = y cycles (beats):

$$\frac{1 \text{ beat}}{0.8 \text{ sec}} = \frac{y \text{ beats}}{60 \text{ sec}} \qquad 0.8y = 60$$

$$y = 75 \text{ beats/minute}$$

Interpretation of Graphic Data

Much of the study of physiology involves learning about relationships between variables. For instance, there is a definite relationship between the heart rate and the rate of exercise. There is a similar relationship between the body's oxygen consumption and the rate of exercise. Often such relationships can be expressed and understood through graphs.

A **graph** is a diagram that expresses a relationship between two or more quantities. In some cases there is a definite cause-and-effect relationship, whereas in others the association is not as direct, but may be due to a third factor. Graphic presentation of data may not explain the reason for the relationship, but it can provide clues by illustrating the shape of it. A graph puts into visual form abstract ideas or experimental data, so that their relationships become apparent.

Variables. The related quantities displayed on a graph are called **variables.** The simplest sort of graph uses a system of coordinates or axes to represent the values that the variables may take on. Usually the relative size of the variable is represented by its position along the axis, and numbers along the axis allow the reader to estimate the values. If the relationship being plotted is one of cause and effect, the variable that expresses the cause is called the **independent variable.** Usually this is represented by the horizontal axis (also sometimes called the **x-axis** or the **abscissa**). The variable that changes as a result of changes in the independent variable is the **dependent variable.** It is usually represented on the vertical axis (also called the **y-axis** or **ordinate**). The two axes are arranged at right angles to each other and cross at a point called the **origin** (Fig. 1–10).

To show the relationship between two variables that are directly related (at some specific value, such as point A in the figure), the value on the x-axis (X_1) is extended vertically, and the corresponding value on the y-axis (Y_1) is extended horizontally. The point A at which these lines cross is determined by their relationship. If another pair of points (X_2 and Y_2) is chosen, their position on the graph also can be plotted; this is point B. A line drawn between points A and B can then give information about how all other x and y values on this graph should relate to each other.

Types of Relationships. As you may have guessed, this explanation represents a very simple case in which some important assumptions were made. We first assumed that for every x-value there was only one y-value, and we further assumed that all of the y-variables were directly related to all of the x-variables. This is the simplest kind of relationship that a graph can represent. It is called a **direct relationship:** the y-values get larger as the x-values get larger. An example of this type of graph is shown in Figure 1–11. The data plotted here could have come from an experiment in which various concentrations of an enzyme were used to study how fast a particular chemical reaction would happen at each concentration.

Some of the imperfections of the "real world" are evident here. The data points show that there is not a perfect relationship between each x-value and its corresponding y-value. We would say that the data show "scatter." (Sometimes this sort of graph

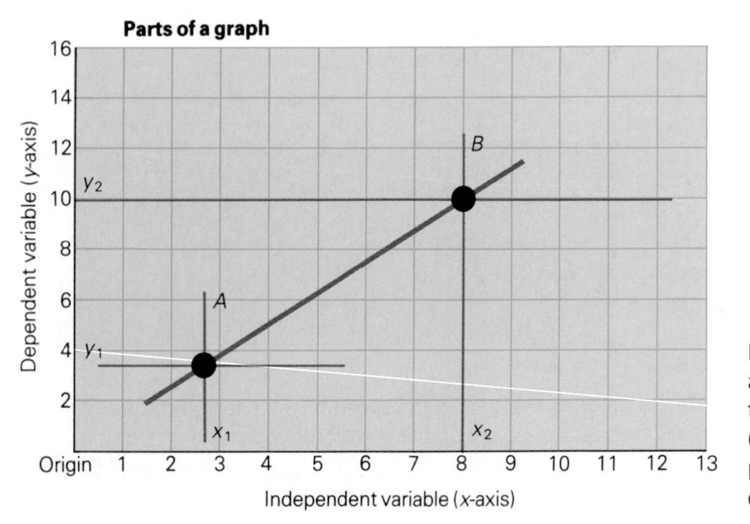

Figure 1–10 In a typical line graph, the vertical axis (the y-axis or ordinate) shows the values for the dependent variable, whereas the horizontal axis (the x-axis or abscissa) shows values for the independent variable. The two axes meet at a point called the origin.

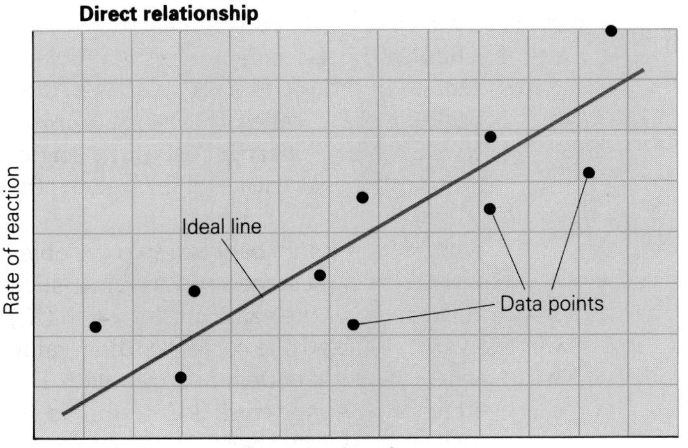

Figure 1–11 In a direct relationship, y-values increase as x-values increase. Data points frequently do not fall exactly on a straight line but are "scattered" about an ideal line, which is drawn to show the general shape of the relationship. A linear relationship can be shown by a straight line.

is called a **scatter diagram.**) In many cases a mathematical procedure may be carried out to determine the "best-fitting" line that would describe the relationship. This is called the **ideal line** in the figure. A relationship that can be described by a straight line is called a **linear relationship.**

Some other relationships that are commonly found are **inverse relationships.** In such relationships, the y-values get *smaller* as the x-values get larger. Such relationships may still be linear, if they are described by straight lines. Other inverse relationships may be curvilinear, as in Figure 1–12. The graph shown here summarizes the experimental finding that a muscle exerting a large force must contract more slowly than one exerting only a small force. A graph of this sort is adequate for the process of interpolation within the range of data, but it is poor for extrapolation outside that range.

Interpolation and Extrapolation. If an experimenter had enough confidence in the reliability of the data, he or she could make two kinds of predictions from such a graph. Predicting data values that fall "between the points" is called **interpolation.** This process is useful if the curve is to be used as a guide for interpreting or testing the reliability of new data that may be obtained. A more risky procedure is **extrapolation,** which involves extending the ideal line into ranges where experimental data are not present. If there is good reason to believe that a direct relationship should hold outside this range, then this prediction could be valid and might permit useful information to be gained.

However, most relationships found in nature are not simple and direct. Over some ranges, a relationship may be direct (and linear, as in the previous example), and then it may change to another form as a wider range of variables is considered. In this case, extrapolation from a limited amount of data could lead to a wrong conclusion.

Take a few moments to flip through the pages of this book. You will see many graphs. Some express simple relationships over their entire range of data, whereas others are more complicated, expressing several relationships at once. Some, especially those that show the way in which an important variable such as blood pressure varies with time during the course of a heartbeat, present a very

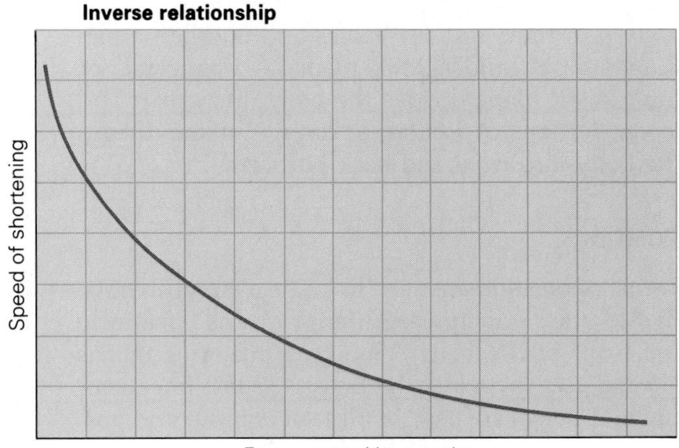

Figure 1–12 In an inverse relationship, y-values decrease as x-values increase, thus producing a downward-sloping line or curve.

complex appearance. In some cases, there are several lines (and possible extra axes) on the graph, each describing some aspect of the idea being presented. Some graphs illustrating other vital processes appear much simpler. Some use bars or parts of a circle instead of lines to illustrate relationships. Whatever their form, all these graphs are designed to present important relationships in the clearest possible way. When you learn to interpret information presented graphically, you are well on your way to understanding the language of physiology.

Anatomical Planes and Positions

In order to study functions of the human body, one must have a basic knowledge of anatomical terminology since function is often described in terms of bodily structures. Conventional lay terms such as *top, bottom, above, below, behind, under,* and *on the side of* can be ambiguous when used in complex descriptions of anatomy. To ensure precise understanding, anatomists have devised a special vocabulary that is part of the language of anatomy and physiology. The following survey of anatomical terms is by no means exhaustive. It is meant to introduce the correct meaning and use of common anatomical terms that appear in subsequent chapters. When encountering new terms, you will find a medical dictionary an invaluable aid for mastering meaning and usage.

Structures of the body are always described with reference to a standard static position of the body called the **anatomical position** (Fig. 1–13). In this position, the body is erect, the face and feet are directed forward, and the arms are straight and directed toward the ground, with the hands rotated so that the palms face forward.

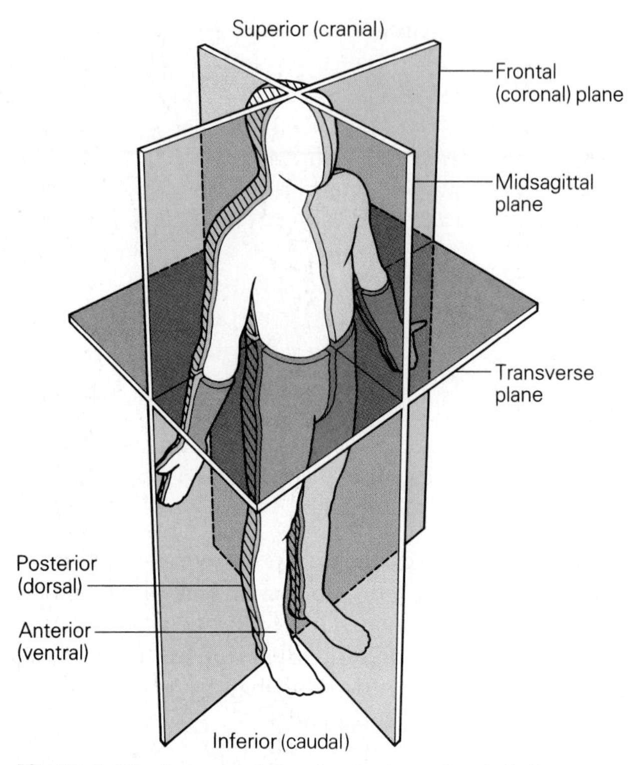

Figure 1–13 A diagram showing basic anatomical planes.

Planes

Imaginary planes are used to divide the body into parts (Fig. 1–13). A **sagittal** plane divides the body into right and left portions. If such a plane passes directly along the midline of the body, it divides the body into right and left halves and is called a **median sagittal (midsagittal) plane.** A **transverse,** or **horizontal, plane** divides the body into upper and lower portions. A **frontal,** or **coronal, plane** divides the body into front and back portions.

Positions

Terms of position are used to locate a structure relative to other structures. **Anterior** means "nearer to the front of the body," whereas **posterior** means "nearer to the back of the body." (Sometimes **ventral** and **dorsal** are used in place of **anterior** and **posterior,** respectively.)

Medial means "nearer to the midsagittal plane," whereas **lateral** means "further from the midsagittal plane." **Superior** means "nearer to the head," whereas **inferior** means "nearer to the lower end of the body." (Sometimes the terms **cranial** and **caudal** are used instead of **superior** and **inferior,** respectively.)

Internal means "nearer the center of an organ, cavity, or part of the body"; **external** means "farther from the center." **Superficial** means "nearer to the surface" of the body; **deep** means "farther from the surface."

Two special terms are used in describing extremities or structures related to their long axes. **Proximal** means "nearer to the origin or point of attachment," whereas **distal** means "farther from the origin or point of attachment." For example, the humerus (the bone in the upper arm) is attached at its proximal end to the shoulder and at its distal end to the elbow.

General Organization of the Body

Living matter is made up of **protoplasm,** a complex mixture of chemicals that displays the attributes of life. These attributes include (1) organization into specific kinds of structural units; (2) the ability to enter into chemical activities that include the transformation of energy and the maintenance or synthesis of protoplasm; (3) the ability to respond to changes in the environment; and (4) the ability to grow and reproduce.

Protoplasm is organized into cells, which are the basic structural and functional units of the human body and of all life. A **cell** can be defined as a microscopic bit of organized protoplasm surrounded by a membrane called the **plasma membrane.** The adult human body comprises about 100 trillion cells, all of which function collectively to maintain an individual's life.

Each human life begins as a single cell, a fertilized ovum, which then divides to form two cells, four cells, eight cells, and so on. In addition to undergoing numerous cell divisions during development, cells also begin to exhibit specialized functions. There are about 200 different cell types in the body, as determined by differences in both structure and function.

Cell Differentiation and Specialization

The process by which a single cell type (the fertilized ovum) develops into many different cell types is known as **cell differentiation.** As you will see, these different cell types are arranged first into tissues, or groups of related cells. Tissues in turn are arranged in groups to form organs, and organs in turn form systems. An analogy can be drawn between the cells of the body and a society of individuals. Just as mail carriers, police, doctors, and teachers each have a specific role for the common good of the entire society, so too do specialized cells such as muscle, nerve, connective, and epithelial cells each serve specific functions to promote the survival of the organism. The body can be thought of as a society or social order of cells.

In a multicellular organism such as a human being, labor is divided among groups of specialized cells, each group performing one principal function such as movement, digestion of food, or reproduction. Specialization of cells has been an essential factor in the development of the large size of multicellular animals. The different groups of cells work in concert to maintain the life of the organism, and considerable interdependence exists among the different cell types. For example, most of the cells of large animals depend upon the red blood cells to obtain their oxygen and transport it to them, while at the same time the red blood cells depend on the pumping action produced by the muscle cells of the heart to be propelled through the body. Such a division and sharing of labor have allowed multicellular organisms to grow to great size, whereas a unicellular organism must carry out for itself all the life processes, and can exist independently only by remaining microscopically small.

Cells and Tissues

Because multicellular organisms require division of labor and specialization among cells, the cells in an individual's body differ in size, shape, internal architecture, and function. Some cells are spherical; others are shaped like cubes, discs, stars, or cylinders. Each cell manifests the form and structure best suited to the function it performs in the body (Fig. 1–14). The neuron (a nerve cell), for example,

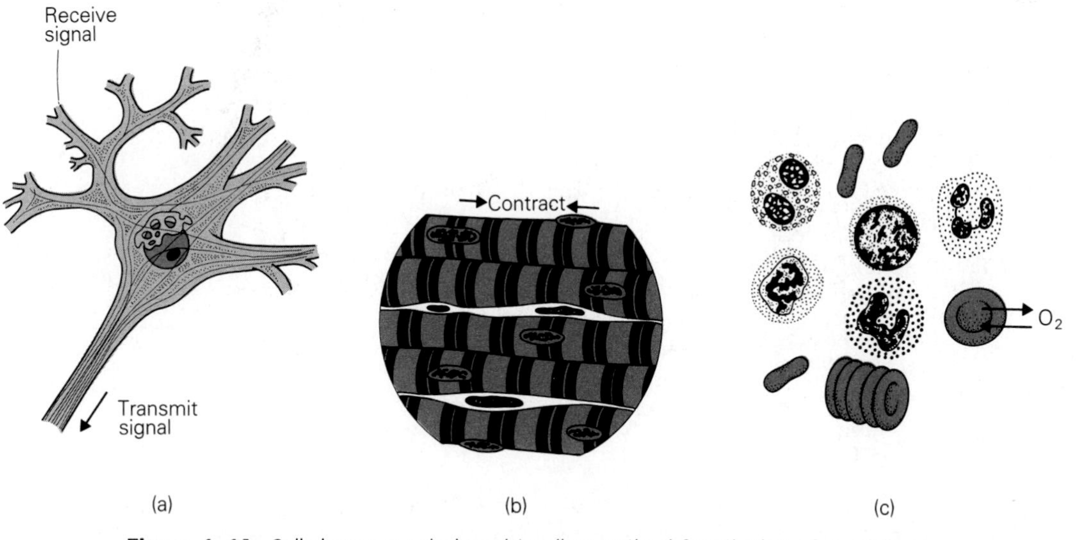

Figure 1–14 Cell shapes are designed to allow optimal functioning of specialized cells. The long extensions of the neuron enable it to receive and transmit nervous signals. The fibers of a muscle cell allow it to change shape during contraction and expansion. The bi-concave disc shape of a red blood cell assists it in carrying oxygen to the tissues and cells of the body.

bears numerous cellular extensions that act to receive and transmit the chemical and electrical signals that form the basis for nervous function. A skeletal muscle cell has a long cylindrical shape, which helps the cell move other parts of the body when it contracts. The "bi-concave disc" shape of a red blood cell helps it in its work of transporting oxygen through the body.

When one or more types of specialized cells become closely associated, they form a **tissue.** There are four classes of tissues in animals: epithelial tissue, connective tissue, muscle tissue, and nervous tissue (Table 1–2).

Epithelial Tissue

Epithelial tissue consists of closely packed cells arranged in flat sheets ranging from one to several layers in thickness. This type of tissue covers exposed body surfaces and lines body cavities, tubes, and organs. Epithelial tissue protects the body parts and aids the transport of materials to and from the structures they protect by absorbing, filtering, and secreting substances. The skin is made of epithelial tissue.

Table 1–2

Examples of Cell Types	Examples of Cell Functions	Examples of Tissue Functions	Tissue	Subtypes	Major Tissue Type
squamous cells	formation of protective layers	lining of body surfaces	simple epithelia	EPITHELIAL SHEETS	EPITHELIAL TISSUE
cuboidal cells			stratified epithelia		
columnar cells					
lacrimal gland cells	secretion of tears	secretion of substances through ducts	exocrine glands	GLANDS	
sweat gland cells	secretion of sweat				
intestinal gland cells	secretion of digestive juices				
liver cells	secretion of bile				
sebaceous gland cells	secretion of sebum				
parathyroid cells	secretion of parathyroid hormone	secretion of hormones directly into blood	endocrine glands		
thyroid cells	secretion of thyroxine				
hypothalamic cells	secretion of oxytocin				
adrenal cells	secretion of epinephrine				
pituitary cells	secretion of human growth hormone				
ovarian cells	secretion of estrogen				
testicular cells	secretion of testosterone				

Table 1–2 (continued)

Examples of Cell Types	Examples of Cell Functions	Examples of Tissue Functions	Tissue	Subtypes	Major Tissue Type
fibroblasts	production of polymers that form connective tissue fibers	firm support and transmission of mechanical forces	dense regular connective tissue dense irregular connective tissue	DENSE CONNEC-TIVE TISSUE	
macrophages (histiocytes)	protection against infection	soft support and protection	areolar connective tissue adipose connective tissue reticular connective tissue	LOOSE CONNEC-TIVE TISSUE	
lymphocytes	immunization				
mast cells	secretion of heparin				
osteoblasts	secretion of bone matrix	rigid support of body structures		BONE	
osteocytes	bone metabolism				
osteocytes	removal of old bone				CONNECTIVE TISSUE
chondrocytes	secretion of intercellular substance	flexible support of body structures		CARTILAGE	
chondroblasts	production of chondrocytes				
bone marrow cells	production of all red and most white cells	production of blood cells	myeloid tissue	BLOOD-FORMING TISSUE	
lymph node cells	production of some white cells		lymphoid tissue		
red blood cells	transport of oxygen	circulation of oxygen and immune factors		BLOOD	
white blood cells	protection against foreign substances				

Connective Tissue

The cells that make up **connective tissue** are somewhat loosely arranged and separated by an intercellular **matrix** (ground substance). The matrix material, made by the connective tissue cells themselves, frequently contains cellular products such as fibers, soluble proteins, and crystalline complexes. Connective tissue is the most widely distributed tissue in the body. It supports, protects, binds, and partitions nearly all bodily components. Specific examples are cartilage, bone, ligaments, blood, and adipose (fat) tissue.

Muscle Tissue

The cells of **muscle tissue** are elongated and contain many parallel contractile fibers called **myofibrils.**

Table 1–2 (continued)

Examples of Cell Types	Examples of Cell Functions	Examples of Tissue Functions	Tissue	Subtypes	Major Tissue Type
skeletal muscle fiber	voluntary movement of bones			SKELETAL MUSCLE	
smooth muscle fiber	involuntary movement of blood vessels, ducts, and other structures			SMOOTH MUSCLE	MUSCLE TISSUE
cardiac muscle fiber	rhythmic movement of heart			CARDIAC MUSCLE	
unipolar neurons	transmission of sensory impulses				
bipolar neurons	transmission of sensory impulses	transmission of nervous impulses		NEURONS	
multipolar neurons	transmission of motor impulses				
astrocytes	support of neurons		neuroglia of the central nervous system		NERVOUS TISSUE
oligodendrocytes	formation of myelin sheaths	support and protection of neurons		NEUROGLIA	
microglia	response to injury				
Schwann cells	formation of myelin sheaths		neuroglia of the peripheral nervous system		
satellite cells					

Muscular tissue in general is specialized for contraction and functions to accomplish movement of the organism. There are three types of muscle tissue: skeletal muscle, cardiac muscle, and smooth muscle. **Skeletal muscle** is attached to the bones and is used for conscious, voluntary movements of the body such as walking. **Cardiac muscle** is present only in the walls of the heart and produces the autonomic pumping contractions of the organ. **Smooth muscle** lines the hollow internal organs, such as the blood vessels and allows these structures to undergo contraction and expansion, which are also autonomic, ''involuntary'' movements.

Nervous Tissue

The chief components of **nervous tissue** are **neurons,** cells that are specialized for receiving and conducting electrical impulses. The characteristic shape of a neuron is an enlarged main cell body with hairlike extensions that receive and transmit the nerve impulses. **Glia**, sometimes called **glial cells** or **neuroglia**, provide neurons with structural and metabolic support and protection and also form part of nervous tissue. This type of tissue is the primary component of the brain and spinal cord. In addition, bundles of nervous tissue called simply **nerves** are found in all parts of the body.

Organs and Systems

Combinations of the primary tissues make up the organs of the body. An **organ** is defined as a group of tissues that have been combined for the performance of a specific function or series of related functions. The stomach, for example, contains all four types of primary tissues, organized for the purpose of receiving, storing, and digesting food and for moving partially digested food into the small intestine for further processing. No single tissue can perform as well as can several tissues working together.

In a similar manner, organs and other structures that share in the performance of related tasks are grouped into **systems.** The urinary system, for example, consists of the kidneys, ureters, urinary bladder, and urethra. These organs work toward the common purpose of removing waste products from the blood and eliminating them from the body.

The grouping of organs and related structures into systems provides a logical approach for a textbook such as this. Keep in mind as you proceed through the various chapters, however, that although many levels of organization and function exist in the human body, all of the body's components function collectively to maintain an individual's integrity and life.

Bodily Fluid Compartments

The body can be thought of as mostly a watery solution. Approximately 60% of the body's weight is water. The body fluids contain various mineral ions (e.g., sodium, potassium, chloride) and organic substances (e.g., proteins, glucose) dissolved in the body water. Bodily fluids usually are grouped into two major divisions or compartments: **intracellular fluid** and **extracellular fluid.**

The two compartments are separated by the plasma membranes of each of the body's cells, and the chemical compositions of the two compartments differ strikingly. The differences in chemical composition are maintained by the cells themselves and have important effects on the cell membrane potential, cell excitability, cellular metabolism, and the ability of the cell membrane to transport substances in and out of the cell. Much of cell physiology is concerned with the mechanisms for maintaining differences in the ionic composition of intra- and extracellular fluids and the consequences of these differences for cell function.

Intracellular Fluid

The intracellular fluid is the fluid within cells. The intracellular fluid, therefore, is not one single continuous compartment but a conglomeration of trillions of cellular subcompartments. Intracellular fluid has a relatively low concentration of sodium ions and a high concentration of potassium ions.

Extracellular Fluid

The extracellular fluid is the fluid outside of the cells. Extracellular fluid, in contrast to intracellular fluid, has a high concentration of sodium ions and a low concentration of potassium ions. The extracellular fluid that bathes our cells is a medium of exchange between one cell and another and between cells and the external environment.

Extracellular fluid consists of three types: interstitial fluid, plasma, and lymph. **Interstitial fluid** is the fluid found between the spaces of body cells. **Plasma** is the fluid portion of the blood, and **lymph** is the fluid in lymphatic vessels. Together these extracellular fluids make up the **internal environment** of the body, whose constant regulation is the purpose of the complex physiological processes, as we are about to discover.

Homeostasis: Control of the Internal Environment

In a complex, multicellular organism such as a person, most of the living cells of the body are not directly exposed to the gaseous external environment (the atmosphere) but exist in the liquid internal environment, the extracellular fluid consisting of lymph, plasma, and interstitial fluid. Conditions in this extracellular fluid must be maintained within certain narrow limits in order to permit our cells to live and function. Conditions such as oxygen tension, pH, osmotic pressure, temperature, and the concentrations of various metabolic substrates, hormones, and waste products are all closely regulated by the cooperative workings of tissues, organs, and systems. At a certain temperature, the human body will function well; at another temperature, its processes will fail. When certain concentrations of ions exist in its internal environment, the body will thrive; at lower or higher concentrations, it will be unable to live or function well.

In many diseases, the composition and/or volume of the internal environment becomes abnormal. Consider the person with pulmonary disease, for example, in whom arterial blood is not oxygenated adequately. The oxygen tension in such a person's internal environment will be reduced. Consider the person with renal failure, who cannot get rid of metabolic waste products at an adequate rate. In such a person, toxic waste materials and hydrogen ions will accumulate in the extracellular fluid;

essentially, the body will be poisoned by its own internal environment. The functioning of body cells will be disturbed under such conditions as these. The organism will be forced to lead a more restricted existence.

The famous nineteenth-century French physiologist Claude Bernard was the first to point out that **constancy,** or stability, in the internal environment—the extracellular fluid—of an organism is the requisite condition for a free and independent existence. Coining the phrase *milieu intérieur,* "internal environment," Bernard paved the way for our current understanding of **homeostasis,** the stable state of the internal environment that is maintained by physiologic processes. Despite the many changes that may occur in the external environment, the temperature and composition of the extracellular fluid remain constant. Factors such as ion concentrations, oxygen content, nutrient and waste product concentrations, and temperature change appreciably neither minute by minute nor day by day.

Indeed, much of the study of physiology is taken up with the analysis of the regulatory processes by which the body maintains the constancy of the internal environment. Some of these regulatory processes are simple, whereas others are highly complex. As you read about each of the various mechanisms in later chapters of this book, bear in mind that each process serves to accomplish, at least in part, one common goal, which is to maintain homeostasis of the internal environment of the body. Thus, for example, although the regulation of breathing in the lungs and the secretion of digestive enzymes into the intestine would seem to have little in common, each of these processes serves in the end to maintain homeostasis.

The Advantages of an Internal Environment

In a single-celled, primitive organism (e.g., an amoeba living in a pond), there is no internal environment, simply because there is no *extra*cellular fluid that is still part of the organism. The entire environment of the unicellular creature is outside it, beyond its borders and therefore beyond its control. Such an organism is subject to the vagaries of the external world. Changes in temperature, light, and concentrations of various chemicals in the external environment dictate its existence. It has little freedom since it has no internal environment that it can manipulate and use as a buffer or mediator between itself and the external world. If the water (its external environment) freezes, the amoeba freezes also. If it does not die as a result, it is at least immobilized.

A more highly developed organism, such as a frog, has a greater degree of independence from external conditions. By virtue of its multicellularity, it has both an external *and* an internal environment. Its internal environment is the extracellular fluid that surrounds its cells; this fluid is *outside* the cells but still *inside* (and therefore part of) the frog. The frog, however, is unable to regulate its own body temperature. When the weather turns cold, it may not freeze, but it will be forced into inactivity. Reptiles such as snakes and alligators must spend a great deal of time simply sitting in the sun after periods of exposure to cold air or water, so that the temperature of their internal environment can be raised to a level that permits life processes to continue at full speed.

In contrast to reptiles, mammals have developed a great degree of freedom from external conditions by evolving sophisticated mechanisms for maintaining a stable internal environment. Most noticeably, their internal body temperatures remain virtually constant under a wide range of external temperature conditions, by virtue of physiological and behavioral mechanisms that counteract the effects of cold or heat in the external environment. Mammals are able to keep their body temperature close to 37°C whether the air around them is a 0°C or 38°C. Furthermore, unlike frogs, many mammals are fully active under all climatic conditions; they are not required to undergo periods of physiological inactivity while waiting for their bodies to warm.

Homeostasis and Protein Conformation

Intuitively, one can appreciate that cells, as living things, can flourish and prosper in certain sets of conditions and not in others. By providing a constant environment for the cells of the body, the various homeostatic mechanisms cooperate to create optimal conditions for cell function.

Beyond this, however, one might ask, "What is the molecular basis of this need for a constant internal environment?" The answer to this question lies in the fact that the primary components or building blocks of cells are macromolecules called **proteins.** Proteins serve a variety of functions in cells. They form enzymes, which are responsible for the various metabolic processes by which cells obtain energy and synthesize other macromolecules in the cell. They also serve as structural components in the interior of cells and as components of the cell membrane.

Every protein in the cell has its own distinctive **conformation,** or shape, that allows it to best carry out its own particular function. The shape, and therefore the function, of proteins are affected by such factors as ion concentrations, temperature, and pH. Changes in these factors can change protein conformation. With the right conditions of ion composition, temperature, and pH, proteins assume their proper conformations and carry out their tasks perfectly. Under abnormal chemical, temperature, and pH conditions, however, they lose their shape and their ability to function. Protein malfunction in turn prevents the cells as a whole from carrying out their own specialized tasks, and the malfunctions accumulate until they become manifest in a disease condition in the organism.

But proteins function *inside* cells. The internal environment is *extra*cellular fluid. What is the connection? As we shall see in Chapter 4, the connection is everywhere, anywhere cells and their membranes are found. The plasma membranes of cells serve as the boundary between the two body fluid compartments—the extracellular fluid and the trillions of subcompartments of the intracellular fluid. At the same time, they serve as a transport system between the two compartments, constantly permitting or encouraging movement of substances between cells and the extracellular fluid. The composition of that fluid determines whether or not the cells in turn can obtain from it what they need to make proteins and other molecules essential for their existence.

Thus, the molecular basis for the homeostatic process is, in essence, the maintenance of protein conformation. By constantly ensuring that proper conditions exist in the extracellular fluid, homeostatic mechanisms allow proper conditions inside cells, thereby permitting cell proteins to assume their proper shapes and carry out their special functions. The optimal functioning of specialized cells, in turn, ensures the coherent functioning of tissues, organs, and organ systems and the health of the body as a whole.

Control Systems

The mechanisms by which the body maintains homeostasis are known collectively as **control systems.** Control systems share several features. These systems use mechanisms called sensors to *detect* conditions in the body and effectors to *change* conditions in the body.

Sensors and Effectors

In order to monitor and control a variable of the extracellular fluid such as temperature, the body must first detect the variable. This is accomplished most often by **sensors,** which determine the level of the variable with respect to a **reference point.** For example, in monitoring a physiological parameter such as concentration of sodium in the extracellular fluid, the sensor obtains information about the quantity of sodium present at a particular time and compares it to the level of sodium that *should* be present. If too much or too little sodium is present, this information is sent from the sensors to effectors, which then attempt to change the level of the parameter so that it is at the desired level. Thus, control systems detect discrepancies between *desired* levels or quantities and *actual* levels or quantities.

Two general classes of control systems are active in the body: negative feedback systems and positive feedback systems. **Negative feedback systems** work to restore normal values of a variable and thus exert a stabilizing influence. **Positive feedback systems** destabilize and thus have limited value in maintaining homeostasis.

The Nervous System in Homeostatic Control

The nervous system is one of the major homeostatic control systems in the body. For example, it monitors the temperature of the body and attempts to maintain it at 37°C. It monitors the mean arterial blood pressure and tries to keep it at approximately 90 mm Hg. It monitors the concentration of hydrogen ions in the extracellular fluid and attempts to maintain it at a pH of 7.4.

The sensors that are part of the homeostatic function of the nervous system are associated with the autonomic nervous system (see Chapter 10). Sensors called **baroreceptors** are nerve endings that wrap themselves around blood vessels to monitor blood pressure. The information they transmit to the brain causes the blood pressure either to rise or to fall. This adjustment is accomplished by the activation or inactivation of effector systems such as the heart, causing it to change its rate and force of contraction, and the smooth muscles, causing them to expand or contract blood vessels.

Chemical sensors, also called **chemoreceptors,** of the autonomic nervous system monitor the concentrations of hydrogen ions and carbon dioxide in the extracellular fluid. These chemicals affect the pH of the body, which in turn affects the ability of cells to carry out essential enzymatic reactions. Alterations in the levels of hydrogen ions and carbon dioxide are detected by chemoreceptors in the brain and peripheral autonomic nervous system. The information these chemoreceptors transmit to the brain eventually affects the ways in which the lungs and kidneys work to maintain a constant pH in the body.

Sensors within the brain are also capable of monitoring body temperature. These **thermorecep-**

tors, located in the hypothalamus, detect the temperature of the blood flowing in the brain. When the temperature falls below 37°C, effector mechanisms are activated. These mechanisms reduce the amount of blood that flows to the skin, thus minimizing heat loss. Heat is retained in the body, thereby raising the temperature. In addition, hormones are released to increase the rate of cellular metabolism, which releases energy in the form of heat. Moreover, neural mechanisms go into action to produce shivering, which generates even more heat. Together these mechanisms generate and conserve heat in the body until the temperature is raised to 37°C. If too much heat is retained, different effector mechanisms take action to dissipate heat. Blood flow to the skin increases, and sweat glands are activated. Heat is lost from the body, and the temperature falls to 37°C.

Summary

I. A. Aristotle and the investigators of Alexandria searched for explanations of blood flow, temperament, and other aspects of physiology. Galen's "pneumatology," based on these early studies, dominated the Western view of body processes until well into the Renaissance.

B. Renaissance scientists corrected many errors in Galen's teachings. William Harvey's discovery of the closed circulation marked the end of Galen's dominance and the beginning of modern physiology.

C. Scientists of the eighteenth and nineteenth centuries studied the breathing process, the nature of the cell, and the phenomenon of homeostasis. Twentieth-century frontiers in physiology include genetics, bioenergetics, and control systems.

II. A. The scientific method, which emphasizes experiment, quantitative results, retesting, and information exchange, forms the basis of all modern scientific research. In conducting investigations, scientists use both inductive and deductive reasoning.

B. Through measurement, computation, and the use of graphs, scientists apply quantitative methods to the study of physiology.

C. Physiologists use standard terms of reference to describe planes and positions of the body.

III. A. The body consists of protoplasm organized into cells of different types and functions.

B. Cells that are similar in shape and function form groups called tissues. Muscle, nerve, epithelial, and connective tissues are the four major tissue types in the body.

C. Groups of tissues that work together form organs; organs in turn combine their functions to maintain systems.

D. The body comprises two fluid compartments: the fluid within each plasma membrane, the intracellular fluid; and the extracellular fluid (plasma, lymph, and interstitial fluid) that surrounds the cells.

IV. A. The extracellular fluid forms the "internal environment" of the body, allowing it to maintain constant conditions such as temperature.

B. Through homeostasis, or self-regulation, the body protects the shapes of proteins, allowing them to function optimally in cells.

C. Many control systems work to maintain homeostasis. They employ sensors to detect changes in bodily conditions and effectors to adjust those conditions.

Review Questions

1. What physiological phenomena, according to the early Greeks, produced characteristic temperaments in people?
2. What was Galen's view of blood and air flow?
3. How did William Harvey revise Galen's concepts of the circulation?
4. How did the thermometer contribute to the modern understanding of homeostasis?

5. What are some examples of deductive and inductive reasoning?
6. Why are retesting and publication important steps in the scientific research process?
7. What is the purpose of scientific notation?
8. How do graphs aid the study of physiology?
9. What are the major planes of the body?
10. What are the two major fluid compartments of the body, and how do they differ?
11. What are the advantages of an internal environment?
12. What is the purpose, at the molecular level, of maintaining a constant internal environment?
13. What are the major types of control systems in the body?

Suggested Readings

Boas, M. *The Scientific Renaissance 1450–1630.* New York: Harper & Row, 1962.

Broad, W., and Wade, N. *Betrayers of the Truth.* New York: Simon & Schuster, 1982.

Comroe, J.H., Jr. *Retrospectroscope—Insights into Medical Discovery.* Menlo Park, Calif.: Von Gehr Press, 1977.

Davis, W. *The Serpent and the Rainbow.* New York: Simon & Schuster, 1985.

Dyson, F. *Disturbing the Universe.* New York: Harper & Row, 1981.

Feynman, R.P. *Surely You're Joking, Mr. Feynman!* New York: Bantam Books, 1986.

Fishman, A.P., and Richards, D.W., eds. *Circulation of the Blood—Men and Ideas.* New York: Oxford University Press, 1964.

Knight, B., M.D. *Discovering the Human Body.* New York: Lippincott and Crowell, 1980.

Miller, J. *The Body in Question.* New York: Random House, 1978.

Poncins, G. *Kabloona.* New York: Reynal and Hitchcock, Inc., 1941.

Rothschuh, K.E., M.D. *History of Physiology* (Guenter, B.R., M.D., ed.). New York: Robert E. Krieger Publishing, 1973.

Sagan, C. *Cosmos.* New York: Random House, 1980.

Saunders, J.B. deC.M., and O'Malley, C.D. *The Anatomical Drawings of Andreas Vesalius.* New York: Bonanza Books, 1982.

Scott, L.M., and Waterhouse, J.M. *Physiology and the Scientific Method.* New Hampshire: Manchester Press, 1987.

Simon, T. *The Heart Explorers.* New York: Basic Books, Inc., 1966.

Thomas, L. *The Youngest Science.* New York: Bantam Books, 1984.

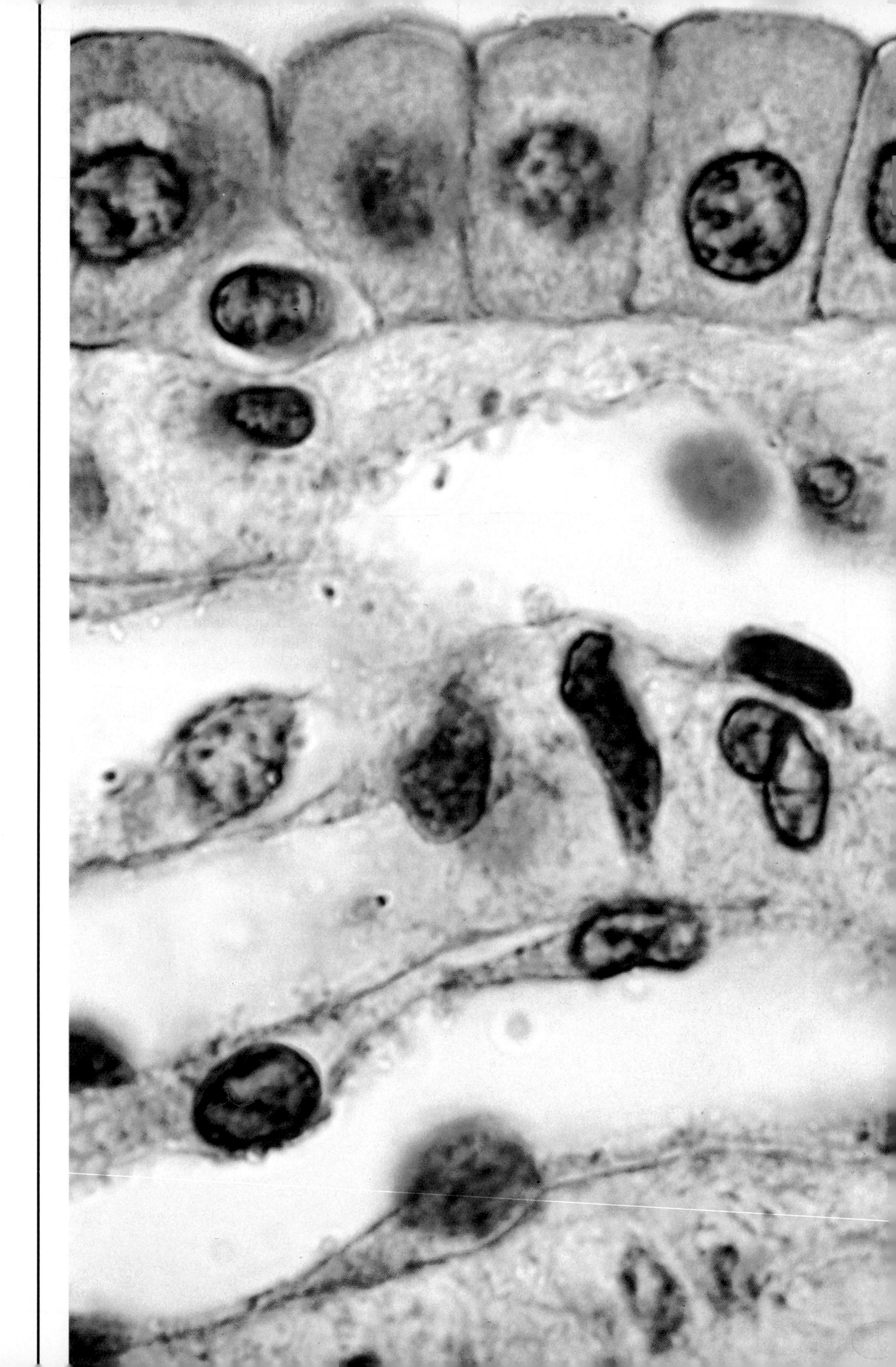

To determine why certain chemical structures dominate all forms of life so regularly, the current chapter provides a survey of general chemical and physical principles. Then the chapter focuses more closely on the specific principles that influence the life processes. The chapter begins with the simplest components of matter—atoms and small molecules—and the chemical and physical laws that govern their behavior. It then proceeds to the major classes of giant molecules such as sugars, proteins, lipids, and nucleic acids. A look at their components and properties provides a sound basis for the description, in later chapters, of their unique roles in the life of cells.

Atoms and Elements

An **atom** is the smallest unit of an element that retains all the chemical properties of the element. In solid matter, atoms are closely packed. In liquid and gaseous matter, atoms are more loosely arranged. The same element can appear in different forms depending on the temperature and pressure of its surroundings. For example, at room temperature and pressure, the element carbon is a solid because its atoms are packed closely together. At higher temperatures, carbon becomes liquid; at still higher temperatures, it takes the form of a gas as its atoms become separated from each other.

Subatomic Particles

Each atom, regardless of its element, is composed of certain subatomic particles. Three types of particles are important to an understanding of the behavior of atoms in life processes: neutrons, protons, and electrons. **Neutrons** are electrically neutral, whereas **protons** are positively charged and **electrons** negatively charged. Protons and neutrons stay in the **nucleus,** or innermost area, of the atom, whereas electrons constantly orbit the nucleus (Fig. 2–1). The electrical attraction between the positively charged protons in the nucleus and the negatively charged electrons in orbit around the nucleus stabilizes the atom's structure.

Because atoms are so small, they most often occur together in large groups. In order to be able to count large numbers of atoms conveniently, scientists have come up with a counting method called the **mole.** Just as in everyday speech we talk of a dozen eggs, a dozen pencils, or a dozen pages—all the time meaning 12 eggs, 12 pencils, or 12 pages—scientists speak of a mole of atoms to mean

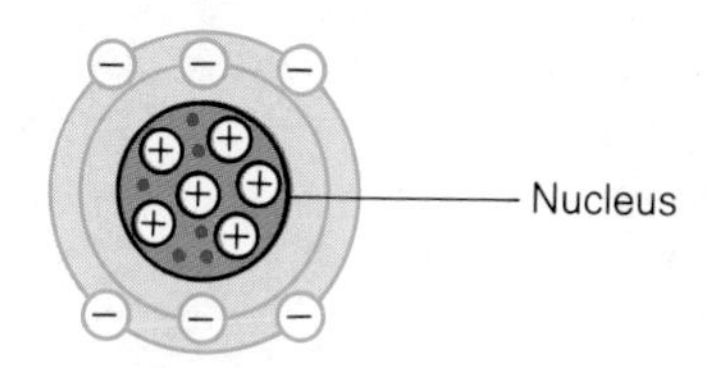

Figure 2–1 The basic structure of an atom. Protons (positively charged particles) and neutrons (neutral particles) are found in the nucleus. The number of protons is the atomic number. Electrons (negatively charged particles) orbit the nucleus. A neutral atom contains equal numbers of protons and electrons.

6.022045×10^{23} atoms. The number 6.022045×10^{23}, called **Avogadro's number,** has been found to be a convenient and accurate way to express the same number of atoms of any substance. Thus a mole of sodium atoms would be 6.022045×10^{23} sodium atoms; a mole of chlorine atoms would be 6.022045×10^{23} chlorine atoms.

Uniqueness of Elements: Protons and Atomic Number

Atoms of different elements are distinct from one another because of the different numbers of each subatomic particle they contain. In fact, the essential identity of an atom is found in the number of protons in its nucleus, called the **atomic number.** In the uncharged or electrically neutral atom, the number of electrons is the same as the number of protons. An atom of the element hydrogen has one proton in its nucleus and one electron in orbit around the nucleus. An oxygen atom typically has 8 protons and 8 electrons. A carbon atom has 6 protons and 6 electrons, a nitrogen atom has 7 of each particle, and a uranium atom has 92 of each particle.

Variation Within Elements: Neutrons, Mass Number, and Isotopes

The actual mass of an atom is too small to be measured or expressed conveniently. Instead, scientists refer to the **mass number** of an element. The mass number is the total number of protons and neutrons in the nucleus. Since the number of protons is different for every element, the atomic mass is also different for every element. For example, the atomic mass of the most common form of hydrogen is 1 because the hydrogen atom has one proton and no neutrons in its nucleus (Fig. 2–2). Oxygen has 8 protons and 8 neutrons, so its mass number is 16. Nitrogen has 7 protons and 7 neutrons, so its mass number is 14.

Moles and grams (the metric unit of mass) can be related in the following way: one mole of an ele-

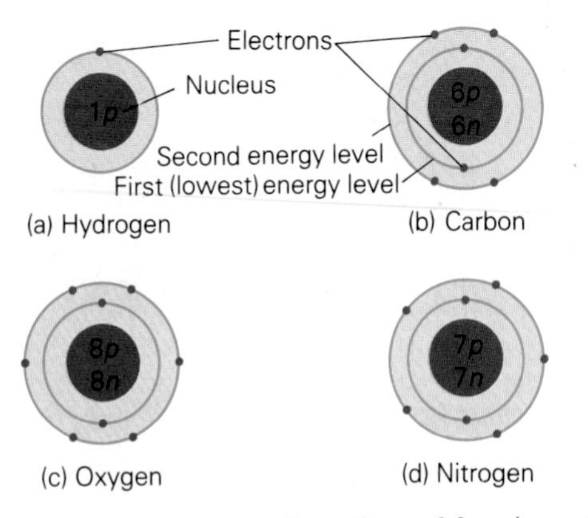

Figure 2–2 The electron configurations of four important elements. Hydrogen needs one more electron to complete its only shell. Carbon needs four electrons, oxygen two, and nitrogen three electrons, to complete the outer shell. The electron orbits are depicted as concentric circles for the sake of simplicity, but their true movements are more complex.

ment, or 6.022045×10^{23} atoms, contains x grams of the element, x being equal to the mass number of the element. For example, one mole of sodium atoms (6.022045×10^{23} atoms) weighs 23 grams, 23 being the mass number of sodium. One mole of chlorine atoms weighs 35 grams, since the mass number of chlorine is 35.

While all atoms of the same element have the same atomic number (number of protons), atoms of the same element may have differing mass numbers because they may have differing numbers of neutrons. For example, hydrogen consists in three forms: one form contains no neutrons in its nucleus, another contains one neutron, and another two neutrons. These three forms of hydrogen are called **isotopes** of hydrogen. Since the mass number of an atom takes into account the number of neutrons, different isotopes of an element have different mass numbers.

An isotope is designated with both the atomic number (number of protons) and the mass number (number of protons plus neutrons) written to the left of the chemical symbol. For example, the isotope of hydrogen that has no neutrons has an atomic number of 1 (because all hydrogen atoms have 1 proton) and a mass number of 1, written as $^{1}_{1}H$. The isotope of hydrogen that has one neutron (deuterium) would be written as $^{2}_{1}H$; the isotope with two neutrons (tritium) as $^{3}_{1}H$. Isotopes are important in biomedical research, as will be discussed in Chapter 3.

Ions: Atoms with Electrical Charges

A neutral, or uncharged, atom has the same number of protons (positive charges) and electrons (negative charges). When a neutral atom gains or loses an electron, it acquires an electric charge and becomes an *ion*. An atom that loses one or more electrons acquires a positive charge because it now has more protons (positively charged particles) than electrons (negatively charged particles). Such an atom is called a **cation** (CAT-eye-on). An atom that gains one or more electrons acquires a negative charge and is called an **anion.**

For example, a potassium cation is formed when neutral potassium loses one electron. The single charge is denoted by a plus sign written to the upper right of the chemical symbol for potassium: K^{+}. The chloride anion, formed when chlorine gains an electron, is written as Cl^{-}. A calcium ion, formed by the loss of two electrons, would be denoted by a number 2 and a plus sign: Ca^{2+}. An oxide ion, formed by the gain of two electrons, is written as O^{2-}.

Ions have a strong impact on the ability of certain substances to dissolve in water. The potassium ion is a very important ion in the study of physiology because it is the major cation found inside most cells, while the sodium ion (Na^{+}) is the major cation present outside cells in fluids such as plasma. Chloride ions are the most abundant anions in plasma.

Electron Configuration and the Outer Shell

The electrons, in their orbit of the nucleus, move in a specified **electron configuration,** consisting of a series of **shells** that can be thought of as concentric circles (although this is a very simplified view of the way electrons actually move). The shells represent different energy levels and can hold different numbers of electrons. The shell with the lowest energy level, located closest to the nucleus, can accommodate 2 electrons. The shell with the next highest energy level can hold 8 electrons. The electron configurations of hydrogen, oxygen, carbon, and nitrogen atoms are illustrated in Figure 2–2. Larger atoms have more electrons occupying several different energy levels. For example, the 92 electrons in the uranium atom are arranged in seven different shells, each with a different energy level.

The number of electrons in the outermost shell of an atom determines many of the chemical properties of the element. Although an atom may be stable and neutral even if its outermost shell is not filled to capacity, it will try to fill this outermost shell by interacting in certain ways with other atoms both of the same element and of different elements. An atom of carbon, for example, has four electrons in its outermost shell. Since this is the second shell from the nucleus, it can hold up to eight electrons. The carbon atom will try to fill that shell with four more electrons.

An atom can fill its outermost shell in one of three ways. It can share electrons with one or more atoms that do not have complete outer shells. It can obtain electrons outright from other atoms. Or it can donate all of the electrons in its currently incomplete outer shell to other atoms so that its next-lower shell, filled to capacity, becomes its outermost shell.

Molecules, Groups, and Compounds

The process of sharing, gaining, or losing electrons leads to the creation of a **chemical bond** between atoms. A chemical bond represents potential energy, energy being stored in the molecule. The release of this potential energy in chemical reactions for use by the cell is an important aspect of cellular physiology.

Atoms combine in definite proportions to form variations of elemental substances or new substances called compounds. Two atoms from the same element may join to form a "diatomic" molecule (see Fig. 2–3). Oxygen and hydrogen, for example, rarely exist as single atoms; instead, they are found as O_2 and H_2, pairs of atoms bonded together.

Two or more atoms from different element may come together to form a new compound. Depending on the type of bond they form, the result may be a molecule (a discrete group of atoms) or a larger structure known as a **crystal lattice,** which is made up of atoms arranged in precise patterns. Whether or not separate molecules are formed, atoms bond in definite proportions, which are indicated by a molecular formula for the compound. For example, even though sodium chloride exists as a crystal lattice made up of many alternating sodium and chloride ions, rather than discrete molecules, the compound has the formula NaCl, indicating that sodium and chloride are always present in a ratio of 1:1.

The molecular formula of a compound shows the relative quantities of atoms; the **structural formula** shows their arrangement in space. For example:

	Water	Methane
Molecular formula	H_2O	CH_4
Structural formula	H—O—H	H above, H—C—H, H below

A **chemical group** is a small cluster of bonded atoms that functions almost as a single atom. Chemical groups behave in distinctive ways and have special names. Examples of chemical groups are —OH (the "hydroxide group"), —COOH (the "carboxyl group"), and —PO_4 (the "phosphate group"). The dash written before the chemical symbols indicates that the group is not a free molecule but is attached to a larger molecule.

Just as atoms are counted in groups called moles, so are molecules. A mole of molecules, regardless of the substance, is 6.022045×10^{23} molecules. A mole of oxygen is 6.022045×10^{23} O_2 molecules. A mole of water is 6.022045×10^{23} H_2O molecules. The **molecular mass** of a compound is the sum of the atomic masses, or mass numbers, of its elements.

Regardless of whether a molecule is made of atoms from the same element of different elements, the type of bond formed depends on the manner in which the atoms interact. If they complete their outermost shells by sharing electrons, the bond formed is a covalent bond. If the atoms gain or lose electrons outright, the bond formed is an ionic bond.

Covalent Bonds: Polar and Nonpolar Molecules

Chemical bonds formed when two or more atoms share electrons are known as **covalent bonds.** These bonds are very stable and strong and are the type of bond most frequently found in biological compounds. Hydrogen, oxygen, carbon, and nitrogen, the four elements most prevalent in living cells, readily form covalent bonds with each other (Fig. 2–3). Atoms of these four elements all have an incomplete outer shell: hydrogen lacks one electron, oxygen two, carbon four, and nitrogen three. Not only can they form single bonds by sharing one pair of electrons; they can form double and (in the case of carbon) triple bonds. This ability increases their versatility in chemical reactions.

In addition, carbon atoms can form covalent bonds with each other. If a carbon obtains each of the four electrons it needs by sharing a pair of electrons with four other carbon atoms, it will form four distinct covalent bonds. The ability to bond with as many as four other carbon atoms allows carbon atoms to form the enormous variety of linear, branched, and cyclic structures that serve as the backbones of large molecules (Fig. 2–4). More than a million compounds are known to contain carbon, and many thousands of them are vital for the maintenance of life. A very large proportion of the molecules found in cells contain carbon.

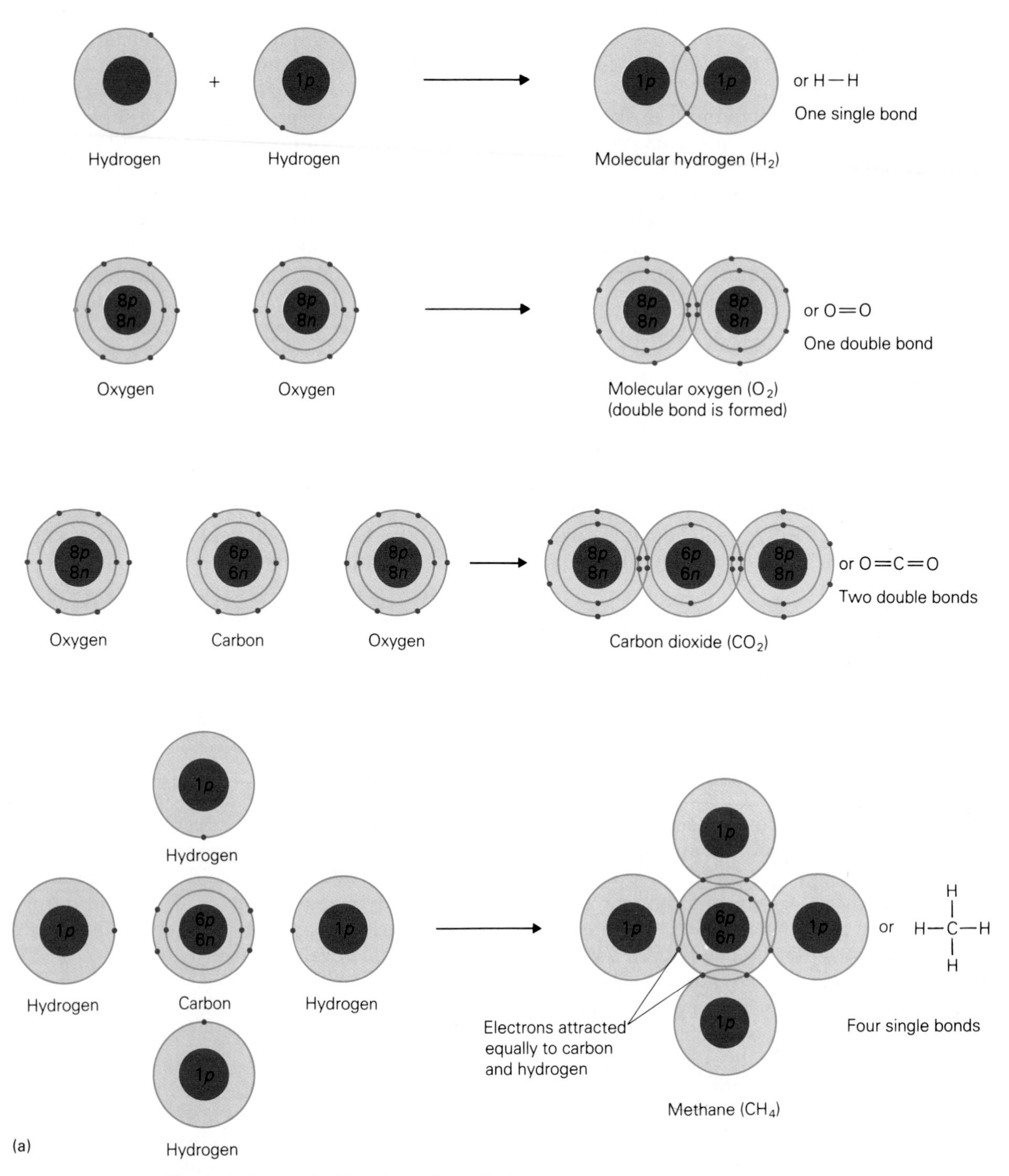

Figure 2–3 Covalent bonds are formed when two or more atoms share electrons. (*a*) Nonpolar covalent bonds are formed when the shared electrons are attracted equally to the nuclei of the original atoms. Hydrogen (H_2) and oxygen both exist as diatomic molecules formed by single nonpolar bonds. Carbon dioxide (CO_2) and methane (CH_4) are compounds by double and single nonpolar covalent bonds, respectively. (*b*) Polar bonds are formed when the shared electrons are attracted more strongly to one nucleus than to another. In ammonia (NH_3), the three shared pairs are more strongly attracted to the nitrogen nucleus, creating a partial negative charge there (symbolized by δ^-) and a slight positive charge (symbolized by δ^+) in the region of the molecule near the three hydrogen nuclei. A similar polarity, formed by two polar bonds, exists in the water molecule.

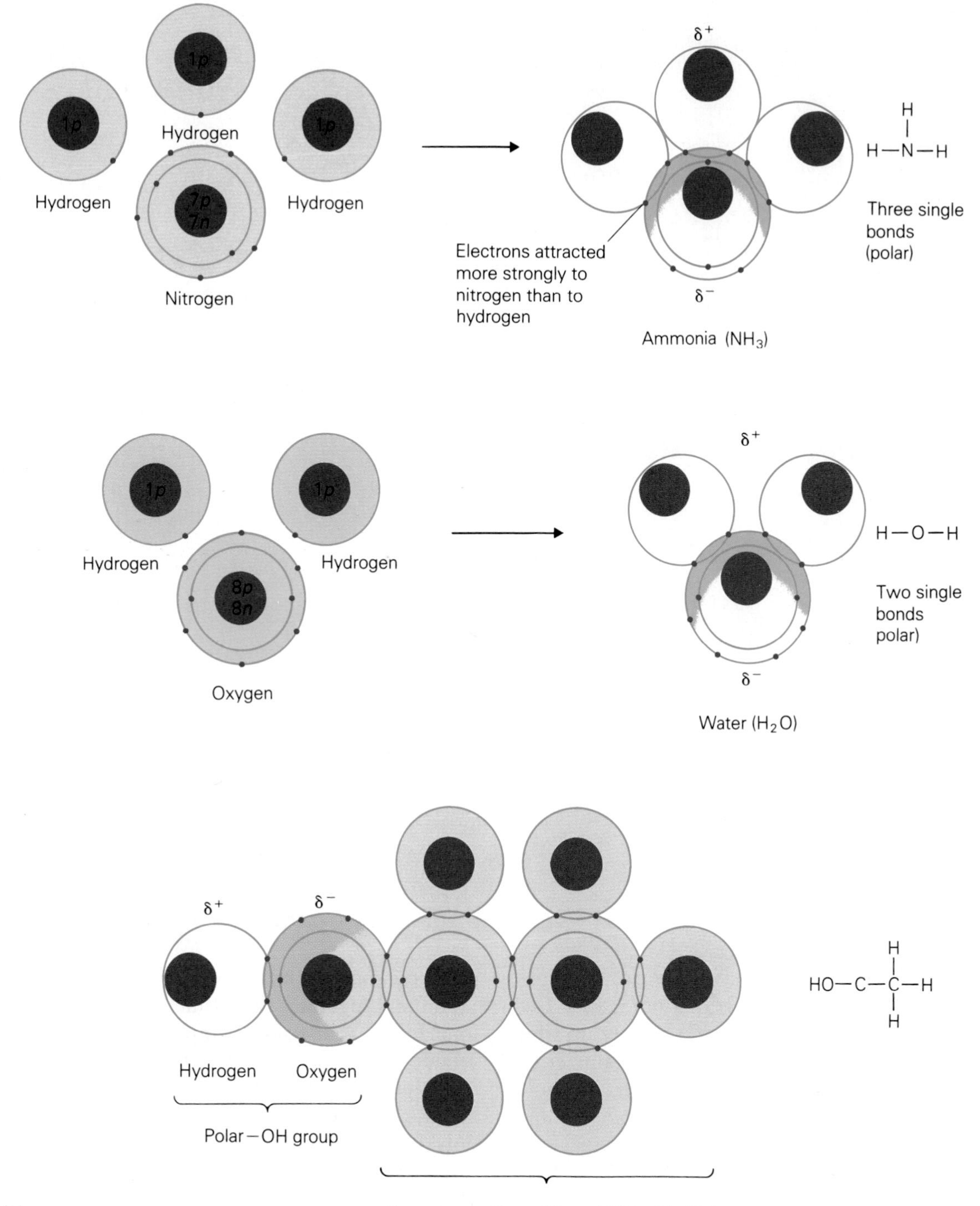

Covalent bonds and the molecules they form may be either polar or nonpolar (Fig. 2–4). In a **nonpolar covalent bond,** the electrons in the pair are shared equally by each atom in a pair, so the molecule overall is completely neutral. In general, compounds formed with carbon and hydrogen are nonpolar covalent compounds. Their overall shapes are square or rectangular. In a **polar covalent bond,** a pair of electrons is shared unequally; that is, the electrons in the pair are attracted more strongly to the nucleus of one bonded atom than they are to the nucleus of the other bonded atom. The electrons can be thought of as spending more time near the nucleus of one atom, creating a partial negative charge, or pole, at one end of the molecule and a partial positive charge, or pole, at another end. The small charges are symbolized as δ^- and δ^+. Water is an example of a **polar covalent compound.** The two

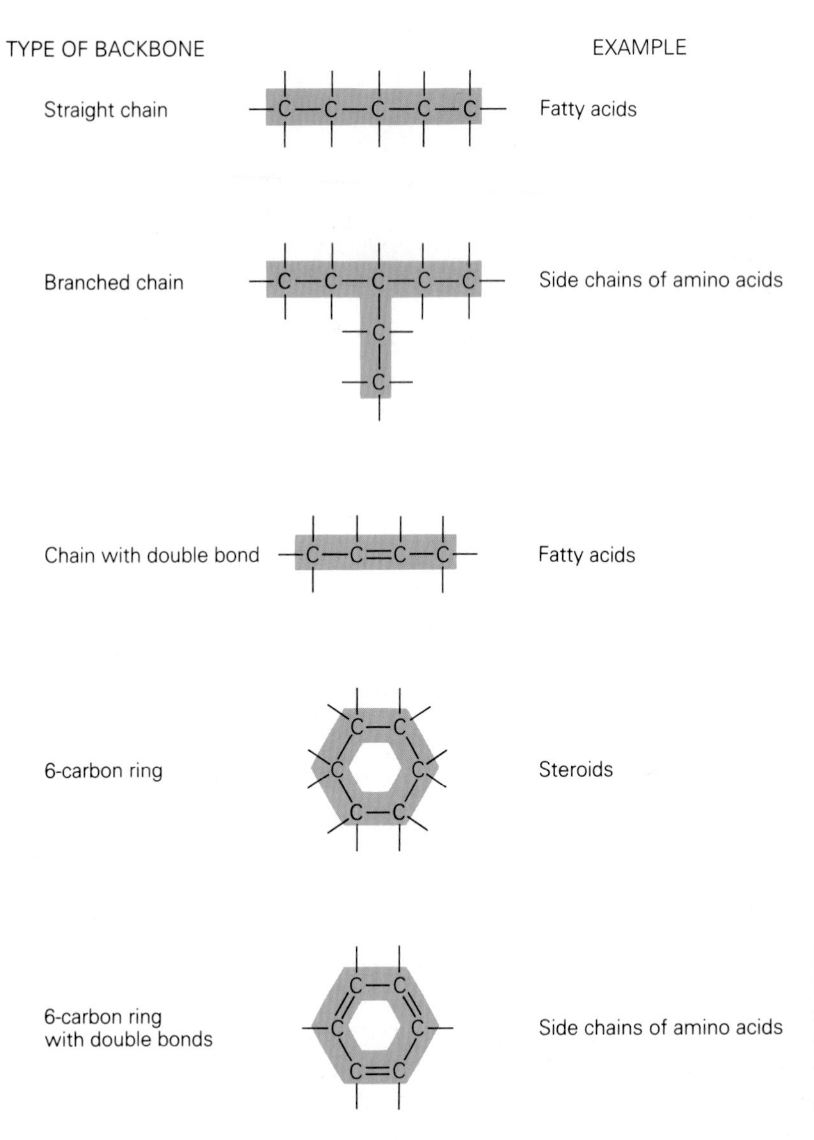

Figure 2–4 The ability of carbon to form bonds with four other carbon atoms at once allows it to form a large variety of different linear, branched, and ring structures that serve as the backbones for organic macromolecules.

hydrogen atoms form an assymmetrical, triangular shape with the one oxygen atom, with the shared electrons attracted more strongly to the nucleus of the oxygen atom than to the nuclei of the hydrogen atoms. This polarity of certain molecules, and the lack of it in others, has important consequences for chemical reactions, as the section on solutions will show.

Ionic Bonds and Ionic Compounds

When atoms complete their outer shells by either gaining or losing electrons instead of sharing them, they form **ionic bonds** to create **ionic compounds.** When two atoms form an ionic bond, one atom acquires one or more electrons and therefore one or more negative charges. The other atom, by losing the electrons to the first atom, acquires an equal number of positive charges. The equal and opposite electrical charges of the two atoms attract them and hold them together; this attraction is the basis of the ionic bond. A specific example (Fig. 2–5) is the formation of sodium chloride. The neutral sodium (Na) atom has 11 protons and 11 electrons. After donating the lone electron in its outermost shell, it has only 10 electrons left—one more proton than electrons—and a net electrical charge of +1. The neutral chlorine (Cl) atom, with 17 protons and 17 electrons, is able to fill its outermost electron shell by accepting the single electron the sodium atom has given up. The chlorine atom then has 18 electrons—one more electron than protons—and a net electrical charge of −1. An ionic bond has been formed between the oppositely charged sodium and chlorine atoms, and the chemical compound sodium chloride (NaCl) results.

Ionic bonds between atoms do not produce molecules per se; rather, they make crystal lattices, or orderly patterns of cations and anions. As long as

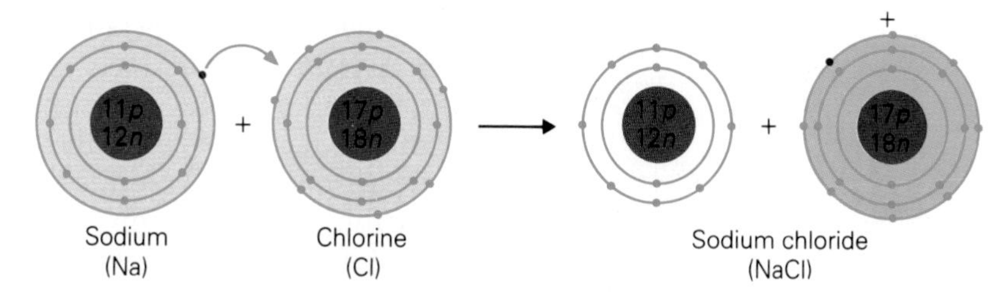

Figure 2–5 Sodium (Na) forms an ionic bond with chlorine (Cl) by donating the single electron in its outer shell to the chlorine atom. The sodium atom then has a complete inner shell, but it has become a cation because it has one more proton than electron. The chlorine atom also has a complete outer shell after gaining the electron from sodium, but it has become an anion because it has one more electron than proton.

the cations and anions are bound together in the lattice, the compound overall is neutral. If the ions become dissociated, as in a solution (see Fig. 2–9*a*), they exist as charged particles.

Van der Waals Bonds

Another type of attractive force develops when any two atoms are close together, about 0.3 to 0.4 nm apart. The **van der Waals bond** forms when the negatively charged electron "cloud" formed by an asymmetrical orbit of electrons in one atom is attracted to the positively charged nucleus of a neighboring atom (Fig. 2–6). The electrons do not leave one atom for another, however. Van der Waals bonds are a much weaker attractive force than covalent and ionic bonds. The amount of energy needed to break a van der Waals bond is less than 1% of the energy necessary to disrupt a covalent bond between hydrogen and oxygen.

Although individual van der Waals bonds are relatively weak, they become important when large numbers of such bonds form at the same time. This situation often occurs in large molecules when the specific structure of the molecule permits the close proximity of many of the atoms. The van der Waals bonds helps to stabilize the three-dimensional shape, which is often important for the function of the molecule.

Hydrogen Bonds

A **hydrogen bond** is similar to a van der Waals bond but slightly stronger. Hydrogen bonds occur among polar covalent molecules, such as water, made up of hydrogen and certain elements such as nitrogen, oxygen, and fluorine. In water, for example, the partial positive charge of the hydrogen atom of one water molecule is attracted to the partial negative charge of the oxygen atom of another water molecule (Fig. 2–7). Hydrogen bonds can occur between molecules or within the same molecule.

Hydrogen bonds are still relatively weak compared to covalent bonds. The input of energy needed to disrupt the hydrogen bonds between water molecules is only 4% of that needed to break the covalent bond between the hydrogen and oxygen in the same molecule. Hydrogen bonds form from the hydrogen atoms in many biological compounds. They are important for maintaining the shapes of large molecules such as proteins and nucleic acids.

Chemical Reactions: Basic Principles

All atoms and molecules are moving continuously and therefore frequently collide with one another. These frequent collisions sometimes disrupt the

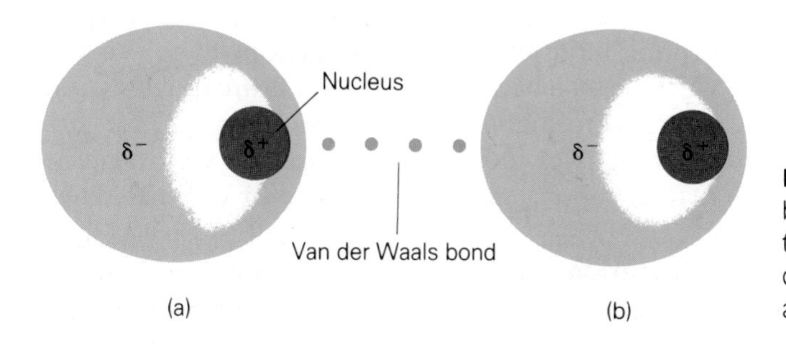

Figure 2–6 Van der Waals bonds form between two atoms because, if the two atoms are close enough, asymmetry in the distribution of electrons in atom A allow the positively charged nucleus to attract negatively charged electrons in atom B.

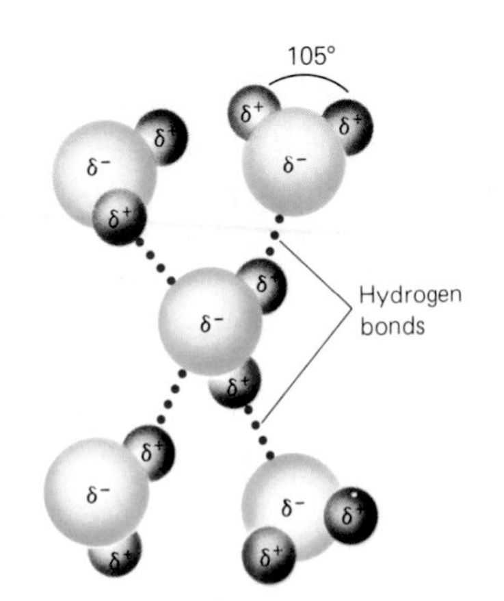

Figure 2–7 The polarity of the water molecule allows it to form hydrogen bonds with other water molecules.

electron configurations of the atoms involved, thereby breaking the covalent or ionic bonds the atoms have formed with other atoms. When bonds are broken, compounds break down into their elements; when these elements recombine in new ways, new compounds are formed. This process is called a **chemical reaction.** According to the Law of Conservation of Mass, the total number of atoms is the same before and after a chemical reaction; the atoms are simply rearranged into different groups, which represent new compounds with different chemical properties than the original compounds.

A chemical reaction is represented by a **chemical equation,** which shows the relative quantities of each compound, in terms of their molecular formula. A simple example is the reaction of perchloric acid ($HClO_3$) and potassium carbonate (K_2CO_3) to form potassium perchlorate ($KClO_3$), carbon dioxide (CO_2), and water (H_2O). The two original compounds are called **reactants;** the resulting compounds are called **products.** Note that the same number of atoms of each element are present both before and after the reaction. This is ensured by providing two molecules (indicated by the number 2 to the left of its molecular formula) of the reactant perchloric acid for each molecule of potassium carbonate.

Reactants | **Products**

$$2HClO_3 + K_2CO_3 \longrightarrow 2KClO_3 + CO_2 + H_2O$$

Perchloric acid, Potassium carbonate → Perchlorate, Carbon dioxide, Water

Total Atoms (Reactants)	Total Atoms (Products)
2 H	2 H
2 Cl	2 Cl
9 O	9 O
2 K	2 K
1 C	1 C

Sometimes, the reaction proceeds in one direction, giving stable products. Some reactions, however, form unstable products. Either they break down and recombine into the original reactants or break down into new products altogether. An example of this is the reaction of carbon dioxide (CO_2) and water to form carbonic acid (H_2CO_3). The product, H_2CO_3, is unstable, breaking down into the original reactants. Hence, the reaction is called *reversible* and is indicated by a set of opposite arrows. Many metabolic reactions that occur in the cell are reversible.

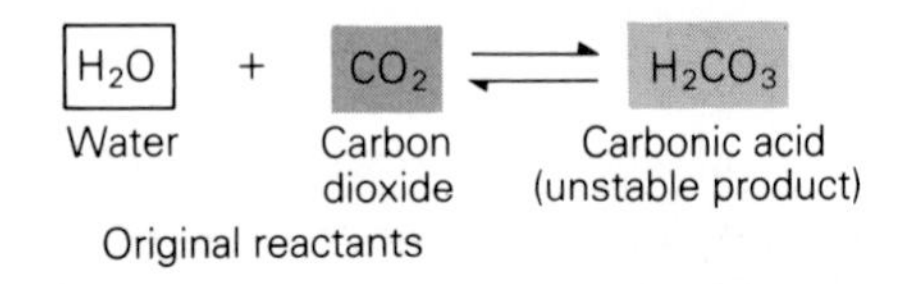

When a reaction reaches a point at which the formation of products is exactly balanced by the reformation of the original reactants, the reaction is said to be in **equilibrium.**

The rate or speed at which a reaction occurs depends on a number of factors. Some factors that affect rates of reaction are temperature, concentration, and physical orientation of molecules.

Temperature. An increase in temperature generally increases the rate of a chemical reaction by increasing the velocity at which the reactant molecules and atoms are moving. The particles collide more frequently, increasing the possibility that electron configurations will be disrupted. The rate of many reactions doubles when the temperature increases by only 10°C. The metabolic reactions that occur in cells occur at 37°C, a temperature higher than most environments. A further increase in temperature would be harmful to cells because reaction products such as proteins would become unstable, breaking down and losing the ability to perform their functions.

Concentration. At least initially, an increase in the concentration of reactants usually increases the rate of reaction. The presence of more reactant molecules in the same space will increase the frequency with which moving particles collide, thereby disrupting more bonds. Eventually, however, the rate will decrease because the reactants are quickly used up in the formation of products. In the cell, the concentrations of molecules are carefully controlled so that metabolic reactions proceed at optimal rates.

Orientation. The way in which molecules are oriented, or arranged in space, often affects the rate at which their atoms will break off and form new bonds. The atoms also must be near enough for the reaction to occur. Certain important bonds occur only between specific chemical groups (e.g., peptide bonds that form between amino acids in protein molecules).

Some chemical reactions fall into a special category known as **oxidation-reduction reactions,** or **redox reactions.** An example is the reaction in which iron changes to ferric oxide (rust):

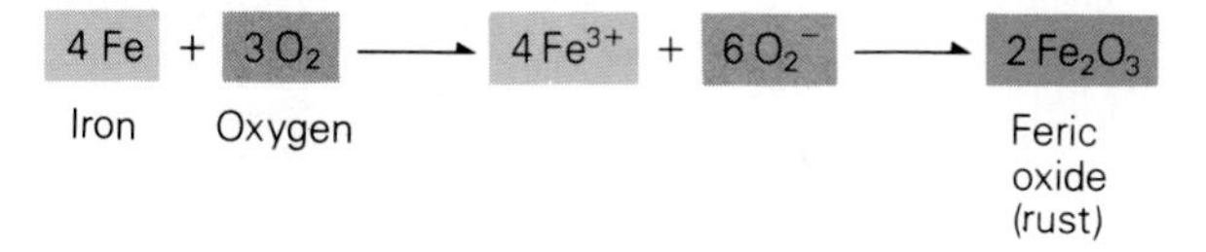

In this reaction, the iron loses electrons, a process known as **oxidation.** Each iron (Fe) atom loses 3 electrons, leaving an excess of three positively charged protons and creating iron cations (Fe^{3+}). Each oxygen accepts 2 of the electrons from the iron atoms; the extra electrons confer two negative charges on each oxygen atom, creating oxide anions (O^{2-}). The process of gaining electrons is known as **reduction.** The charged iron and oxygen atoms then form ionic bonds and produce two molecules of the compound ferric oxide. Three iron atoms lose a total of 12 electrons, which are gained by a total of six oxygen atoms. The iron is being **oxidized** at the same time that the oxygen is being **reduced.** Oxidation and reduction always occur simultaneously, one reactant accepting the electrons removed from another.

Water and Aqueous Solutions

Water is essential for living organisms and is the major chemical compound in most cells, accounting for up to 70% of cell weight. Many biologically important compounds are soluble in water; in fact, the chemical reactions they undergo require an **aqueous** (watery) medium or environment. Here we will consider some special properties of water that make it an essential substance for living cells.

Polarity of the Water Molecule

The water molecule is formed by the covalent bonding of two hydrogen atoms to one oxygen atom. To complete its outer shell, oxygen needs two electrons. It obtains these electrons by sharing an electron with each of two hydrogen atoms. These bond to the oxygen atom to form a triangular structure. The covalent bonds formed between the atoms are polar bonds; that is, the shared electrons are attracted more strongly to the oxygen nucleus than to the hydrogen nuclei. This creates a small negative charge in the area of the water molecule near the oxygen nucleus and a small positive charge near the two hydrogen nuclei. Because it has these oppositely charged areas at opposite ends, the water molecule is said to be **polar** (see Fig. 2–3).

Arrangements of Water Molecules

The polarity of water molecules leads them to form hydrogen bonds with each other when the partial positive charge at one end of one molecule attracts the partial negative charge at the opposite end of another molecule (Fig. 2–7). Hydrogen bonds are responsible for the higher melting and boiling points of water compared to other substances with similar structures that do not form hydrogen bonds (e.g., ammonia [NH_3]). In solid water (ice), for example, the molecules are arranged in a regular three-dimensional pattern so that each molecule is hydrogen-bonded to four other water molecules. When ice melts by heating, and liquid water forms, only about 15% of the hydrogen bonds are broken. Thus, liquid water contains clusters of molecules that have retained a highly organized structure. Energy in the form of heat is required to change ice to liquid water and liquid water to a gas because these changes require the disruption of the hydrogen bonds between water molecules. Similar changes in the physical state of ammonia require less energy because there are no hydrogen bonds to be broken.

Water as a Solvent

In chemistry, a **solution** is a type of mixture in which one substance, called the **solute,** is uniformly distributed throughout another substance, usually a liquid, called the **solvent.** A solvent dissolves, or breaks up, the solute without reacting chemically with it: that is, electrons are not shared or exchanged between the two substances. The polarity and hydrogen-bonding tendencies of water make it an excellent solvent of both ionic compounds and polar covalent compounds.

Any substance that dissolves in water is said to be **hydrophilic** ("water-loving"). Ionic compounds

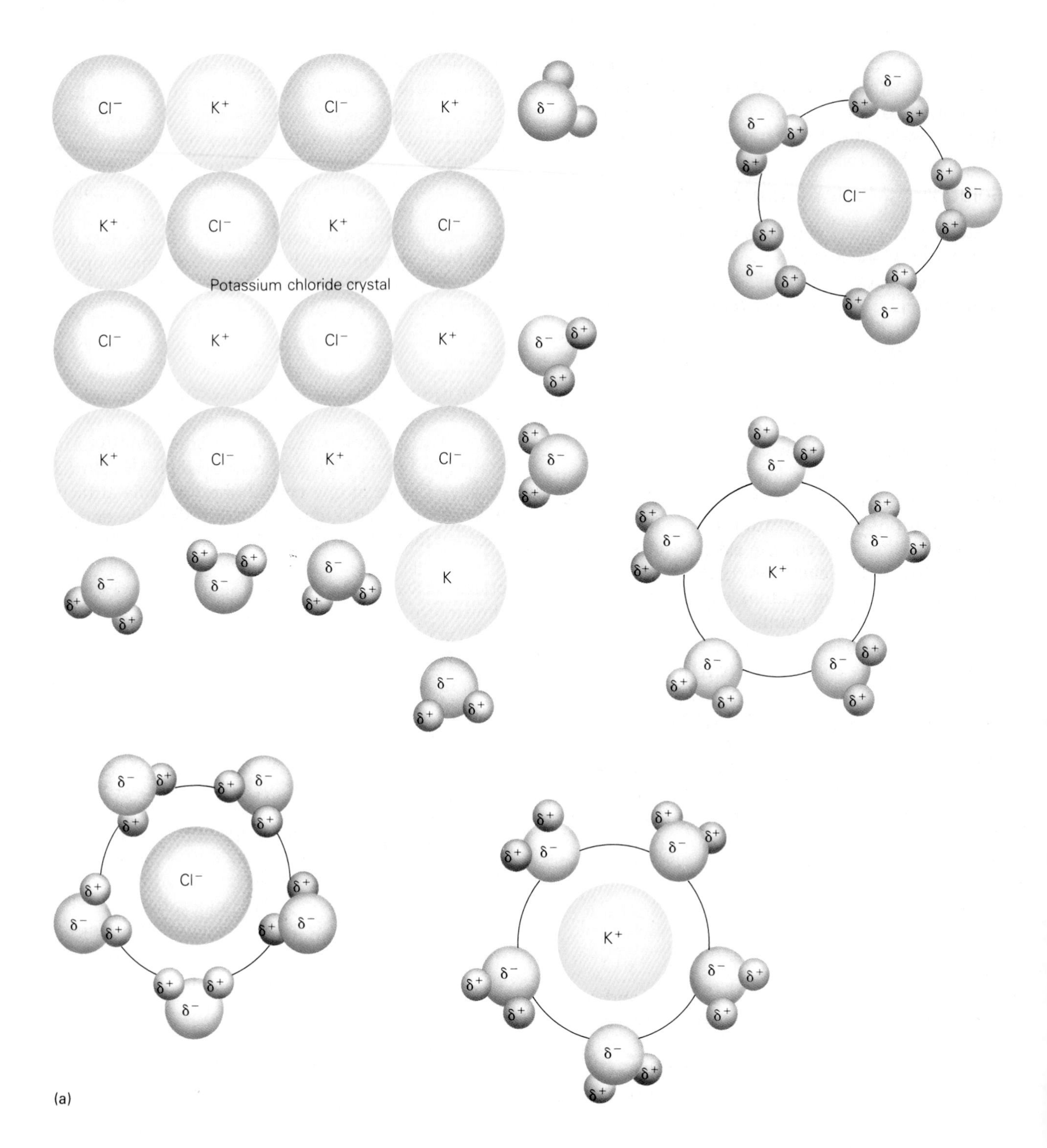

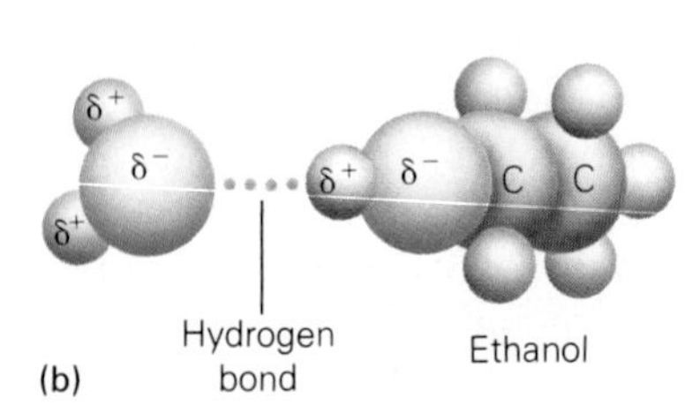

Figure 2–8 (*a*) Compounds such as potassium chloride (KCl) that form crystal lattices stabilized by ionic bonds dissolve readily in water because the small polar water molecules can slip in between and surround the K^+ and Cl^- ions. Some water molecules point their positively charged poles toward the Cl^- anions, forming shells around them. Other water molecules point their negatively charged poles toward the K^+ cations, forming shells around them. (*b*) The ethanol molecule consists mostly of nonpolar covalent bonds, but the —OH (hydroxide) group is polar, so it forms hydrogen bonds with water. These bonds allow ethanol and other polar covalent compounds to dissolve in water.

and polar covalent compounds are generally hydrophilic. Nonpolar covalent compounds and groups are not easily soluble in water; they are said to be **hydrophobic** ("water-hating").

Solutions of Ionic Compounds

Water weakens the ionic bonds that occur in ionic compounds. The ionic bonds in potassium chloride, for example, are readily broken when solid potassium chloride is added to water because the polar water molecules are able to form shells around the potassium and chlorine ions (Fig. 2–8*a*). Some water molecules orient their positively charged hydrogen atoms toward the negatively charged chlorine atoms; others point their negatively charged oxygen atom toward the positively charged potassium atoms. The net force of all the water molecules attracting each ion overcomes the bonds between the ions. As a result, the potassium and chloride ions break away, or **dissociate,** from one another. However, in the process they do not take or give back the electrons they lost or donated in the first place, so they still have unequal numbers of protons and electrons. Therefore, they remain ions in solution as they were in the crystal lattice.

Solutions of Polar Covalent Compounds

Uncharged polar molecules also dissolve readily in water because they can form hydrogen bonds with water molecules. Ethanol is an example (Fig. 2–8*b*). The hydrogen atom in the —OH group of ethanol has a small positive charge because the shared electrons in the covalent bond are more strongly attracted to the oxygen atom. The polar —OH group provides the hydrogen atom for the hydrogen bond, while the water provides the oxygen atom.

Solutions of Nonpolar Compounds: Hydrophobic Interactions

Uncharged molecules that are not polar are poorly soluble in water because they have no positive or negative poles to attract the polar water molecules. Thus, water molecules have no effect on the covalent bonds that hold the solute molecules together. The polar water molecules can form neither shells around the atoms nor hydrogen bonds with them (Fig. 2–9*a*).

Hydrophobic substances are common in biological processes. Important examples are the chains of carbon atoms with only hydrogen atoms attached (called *hydrocarbon chains*). Many important macromolecules are partially composed of these hydrocarbon chains. When the macromolecules are placed in water, the hydrocarbon chains—the hydrophobic portions—form clusters or "pockets" to minimize their contact with the water molecules (Fig. 2–9*b*). This arrangement causes the least disruption of

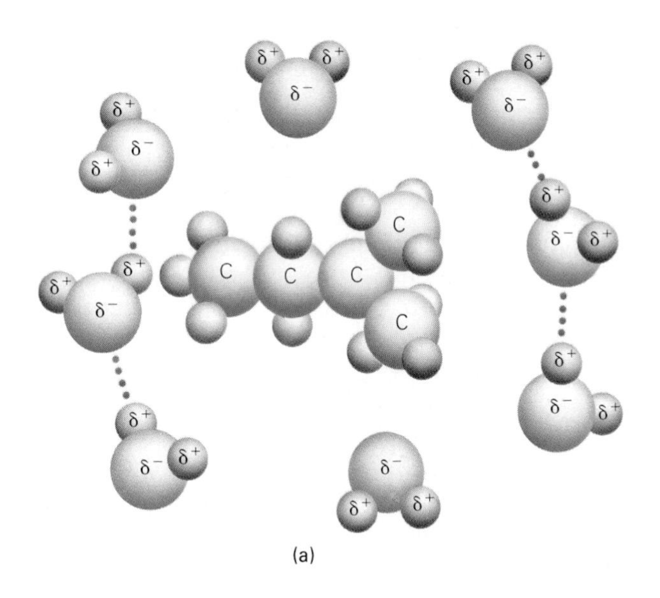

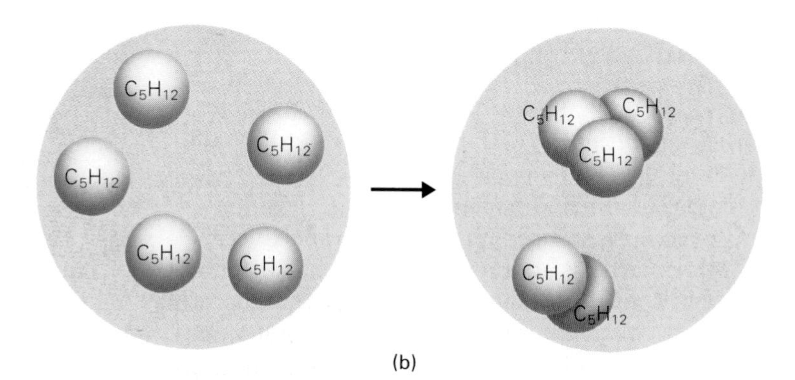

Figure 2–9 Nonpolar covalent molecules and groups, such as those formed from only carbon and hydrogen (hydrocarbons), are not able to dissolve in water because they lack positive and negative poles. The polar water molecules are unable to attract their atoms away from each other since they cannot form hydrogen bonds with them or shells around them. Instead, nonpolar molecules cluster together in hydrophobic "pockets"; these reactions are called hydrophobic interactions. Such pockets allow water to retain its hydrogen-bonded structure.

hydrogen bonds in the water. By creating such collections of molecules, hydrophobic interactions play an important role in maintaining the structures of certain cell organelles and of the plasma membrane that surrounds the cell. The strong tendency to cluster means that these hydrophobic groups can create a sturdy mass.

Concentrations of Solutions

We have said that some chemical reactions require an aqueous medium in order to occur and that the concentration of reactants can be an important factor in determining the rate of reaction. A solution may contain several different solutes, which may then react with one another; indeed, the purpose of their being in solution is to take part in important metabolic reactions. The concentration of reactants in these metabolic processes is determined by their concentration in solutions. Scientists have devised convenient ways of expressing these concentrations in order to understand the rates at which certain reactions occur both in the laboratory and in the living cell.

The concentration of a solution can be expressed in several ways. One way is to compare the number of moles of solute to the kilograms of solvent: this is known as **molality.** In aqueous solutions, this is the same as comparing moles of solute to liters of solution, since one kilogram of water is one liter. For example, if one mole (58.5 g) of sodium chloride (the solute) is dissolved in one liter of water, concentration is expressed as 1 mole per kilogram (abbreviated to 1 mol/kg). This is a "1 molal" solution, abbreviated as "1 m" solution.

An alternative way to express concentration is **molarity.** Molarity is a unit of concentration, based on the number of moles of solute per liter of solution. A solution with a concentration of 1 mol/L is called a "1 molar" solution, abbreviated as "1 M" solution. The solution is said to have a molarity of 1. A solution with a concentration of 2 mol/L is a 2 M solution.

Very low concentrations can be expressed with the same prefixes used in metric measures. A 1 mM solution of sodium chloride contains 1 millimole (58.5 mg) of sodium chloride per liter of solution. The solutions that occur in living organisms have a variety of concentrations, but most are in the millimolar range. Human blood plasma, for example, contains sodium and potassium at concentrations of approximately 140 mM and 4 mM, respectively.

Osmolarity is another expression of concentration that considers the total number of solute **particles** rather than the relative weights of solute and solvent. Osmolarity is discussed in the section on osmosis.

The pH Scale

Water (H_2O or HOH) has a slight tendency to dissociate into hydrogen ions (H^+) and **hydroxide** (OH^-) **ions.** This dissociation process can be represented as:

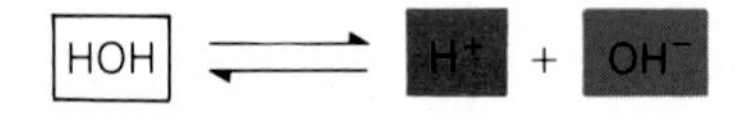

At the same time, the ions tend to reassociate, forming new water molecules. When the rate of dissociation equals the rate of reassociation, the reaction is said to be in **equilibrium.** At this point, the net concentrations of HOH molecules and of H^+ and OH^- ions remain constant, even while individual molecules and ions may be breaking down and reforming. The dissociation of water reaches equilibrium very rapidly, so that only very few of the water molecules are dissociated at any instant.

Ideally, the concentrations of H^+ and OH^- ions in water are exactly equal and very small, making the water a neutral compound. In reality, however, the concentrations of these ions are not equal because of the presence of solutes in the water that, by dissociating in it, increase or decrease the amount of H^+ ions relative to the number of OH^- ions. A solution in which the H^+ concentration is greater than the OH^- concentration is said to be **acidic.** A solution in which the number of OH^- ions is greater than the number of H^+ ions is a **basic** solution.

We can express the acidity or basicity of a solution by specifying the concentration of H^+ ions. Because this concentration is very small, scientists have devised a logarithmic method of expressing it, called **pH.** This expression is the negative of the base-10 logarithm (log) of the H^+ concentration:

$$pH = -\log [H^+]$$

where [ ] indicates the H^+ concentration in moles per liter.

In one liter of neutral water, there are 1×10^{-7} moles of H^+ ions (and 10^{-7} moles of OH^- ions). The pH of neutral water then is:

$$\begin{aligned} pH &= -\log (1 \times 10^{-7}) \\ &= -[\log (1.0) + \log (10^{-7})] \\ &= -[(0.00) + (-7)] \\ &= 7 \end{aligned}$$

Suppose the concentration of H^+ ions were to be increased 10-fold, to 1×10^{-6} mol/L. The pH of the solution would then be:

$$\begin{aligned} pH &= -\log (1 \times 10^{-6}) \\ &= -[\log (1.0) + \log (10^{-6})] \\ &= -[(0.00) + (-6)] \\ &= 6 \end{aligned}$$

The solution is more acidic, and the pH has decreased. Conversely, if the H^+ concentration were to decrease, the solution would become more basic, and the pH would increase.

The pH scale goes from 0 to 14. A solution with a pH of 0 is highly acidic; one with a pH of 14 is highly basic. In terms of pure numbers, the scale is small, so that a 1000-fold increase in H^+ concentration produces a pH decrease of only 3 units. It is important to remember, therefore, that a *small decrease* in pH represents a *large increase* in the H^+ concentration, and vice versa.

Acids and Bases in Solution

An **acid** is any substance whose dissociation in water releases H^+ ions. The hydrogen ion does not contain any neutrons and can be thought of as essentially equivalent to a proton. Therefore, an acid is often referred to as a **proton donor** because it gives off protons in the form of H^+ ions. A **base** is a **proton acceptor,** a substance that accepts H^+ ions and thereby decreases the H^+ concentration in a solution.

Acids: Proton Donors

The addition of an acid to a solution greatly increases the concentration of free H^+ in the solution, producing a more acidic solution. The dissociation of an acid, HA, can be represented symbolically by the following equation:

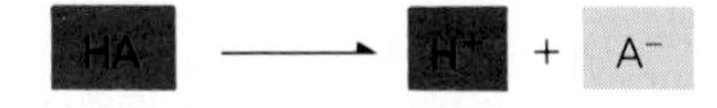

where A^- represents the anion that remains when the hydrogen cation is formed. The strength of an acid is a function of its tendency to give up H^+ ions. In the case of a strong acid, such as HCl, all the molecules will dissociate in water and the dissociation will not be reversible:

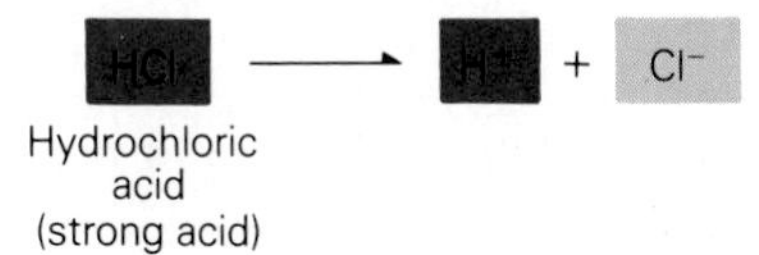

Weak acids, in contrast, give up H^+ ions less readily and do not dissociate completely in water. Thus, a solution of a weak acid contains both dissociated ions and complete molecules. Most of the acids produced in living organisms are weak acids.

Bases: Proton Acceptors

Usually a base gives off hydroxide ions (OH^-) in the process of dissolving in water and these decrease the H^+ concentration by accepting free H^+ ions. Sodium hydroxide (NaOH) and ammonium hydroxide (NH_4OH) are examples of strong bases that dissociate completely in water to yield free OH^- ions. The hydroxide anions then accept H^+ ions, and the result is water:

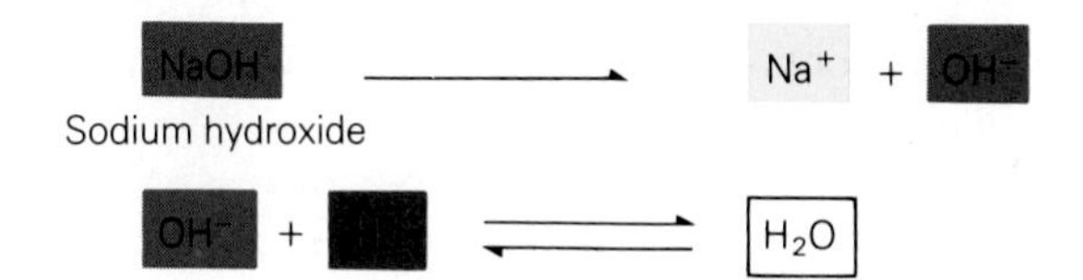

This reaction occurs when NaOH is added to water. The OH^- ions combine with the free H^+ ions already present in the water, decreasing the concentration of H^+ in the solution. In the case of a weak base, such as the bicarbonate ion (HCO_3^-), H^+ also is used up as follows:

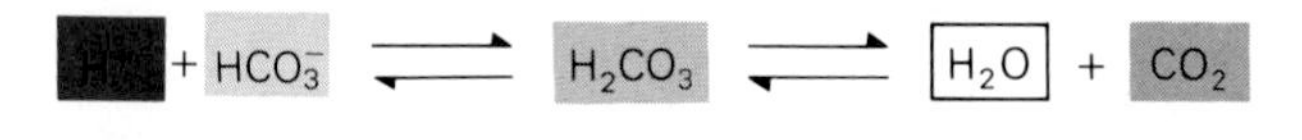

Bases use up H^+ ions and decrease the H^+ concentration, thereby producing basic solutions.

If a weak base such as $NaHCO_3$ is present in solution with a strong acid such as HCl, the base will absorb the H+ ions released by the acid. The following four processes will be occurring:

Dissociation of the strong acid

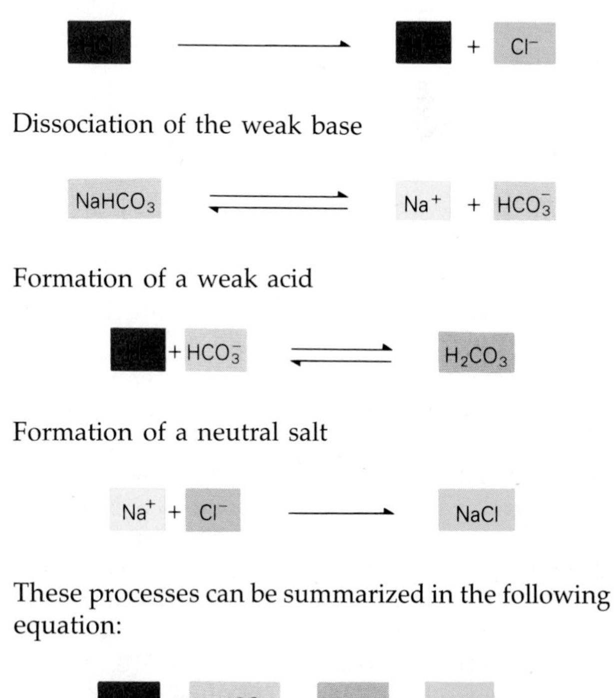

These processes can be summarized in the following equation:

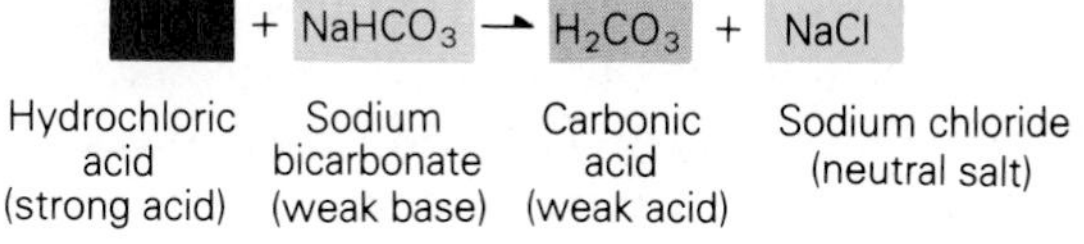

Buffers

The pH of blood is controlled very closely and does not vary outside a small range, usually 7.38 to 7.42. This is a remarkable achievement because the chemical reactions in the cells of the body produce more than 10,000 nmol of H^+ ions each day. An average blood pH of 7.4 represents an H^+ concentration of only 40 nmol/liter, so that highly efficient mechanisms must exist to absorb the H^+ ions added to the blood. These mechanisms will be addressed in detail in later chapters. A general discussion of buffers will serve as an introduction to the "buffer" systems of the blood.

The purpose of a **buffer system** is to minimize the change in pH that occurs when an acid or base is added to any solution. For example, the dissociation of carbonic acid, a weak acid, occurs as follows:

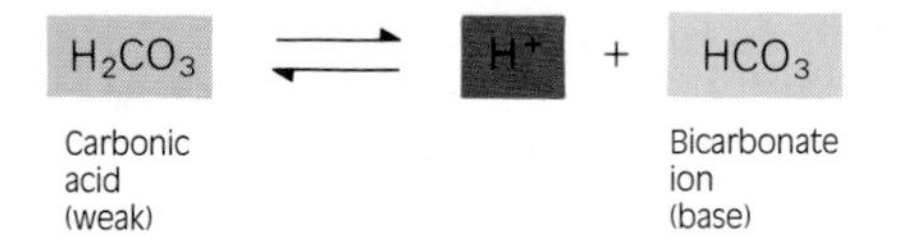

If the H^+ concentration of this solution were to be decreased by the addition of a strong base such as sodium hydroxide (NaOH), the equilibrium of the solution will be disturbed because the base would use up H^+ ions. The equilibrium would be re-established by dissociation of some more carbonic acid molecules (H_2CO_3), yielding additional H^+ ions to replace those consumed by the base. The overall reaction can be summarized as follows, using NaOH as an example of a strong base:

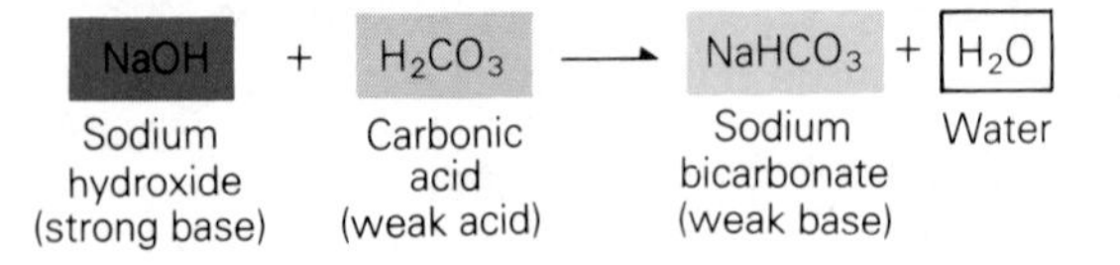

This is actually a combination of two reactions:

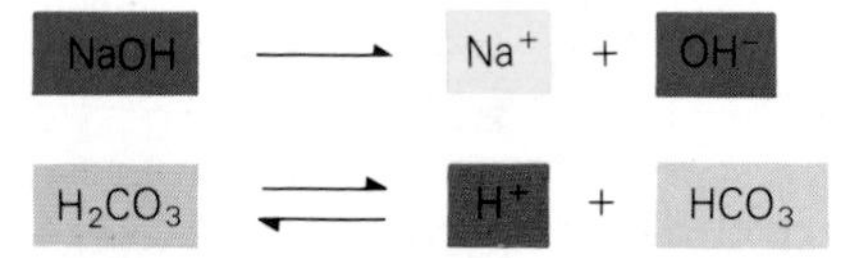

The sodium cations, Na^+, combine with the bicarbonate anions, HCO_3^-, and the H^+ ions combine with the OH^- ions, producing sodium bicarbonate and water. Thus, the strong base NaOH is converted to a weak base, $NaHCO_3$, and the reaction prevents a marked change in pH. The carbonic acid in this reaction is the buffer because its presence produces new H^+ to replace used H^+, thus maintaining the H^+ concentration.

Addition of an acid also disturbs the equilibrium of a solution because the H^+ concentration is increased. Bicarbonate ions will combine with the added H^+ to form carbonic acid, and in this way the total H^+ concentration is reduced to the level present before the acid was added. The overall reaction with HCl, a strong acid, can be summarized as:

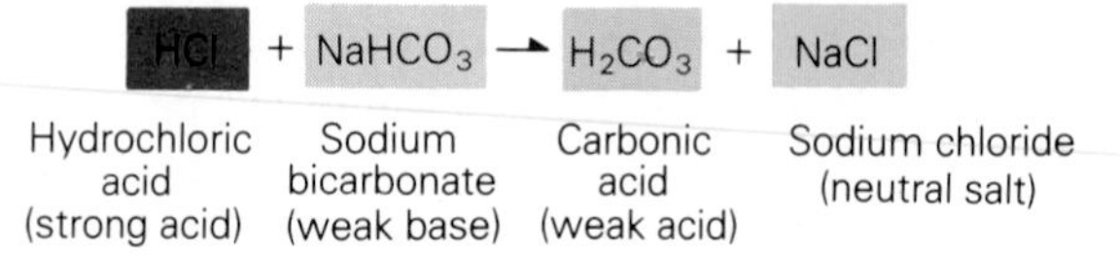

The bicarbonate ion produced by the dissociation of the weak base $NaHCO_3$ uses up excess H^+ ions released by the dissociation of the strong acid HCl. The strong acid HCl is converted to the weak acid H_2CO_3 and excess H^+ ions are used up to minimize the pH change. As $NaHCO_3$ is consumed, carbonic acid is generated.

It is through these reactions that a mixture of $NaHCO_3$ and H_2CO_3 functions as the bicarbonate buffer system that prevents harmful changes in blood pH when acids or bases are added to blood. Destruction of one component of the buffer system always regenerates the other component, so that the buffer capability is maintained. This aspect of a buffer system distinguishes it from a simple neutralization reaction. A neutralization reaction also prevents a pH change, but the product of the reaction is a neutral **salt,** a compound in which the hydrogen atom of an acid (e.g., HCl) has been replaced by a different cation (e.g., NaCl). The salt NaCl is produced by the following neutralization reaction:

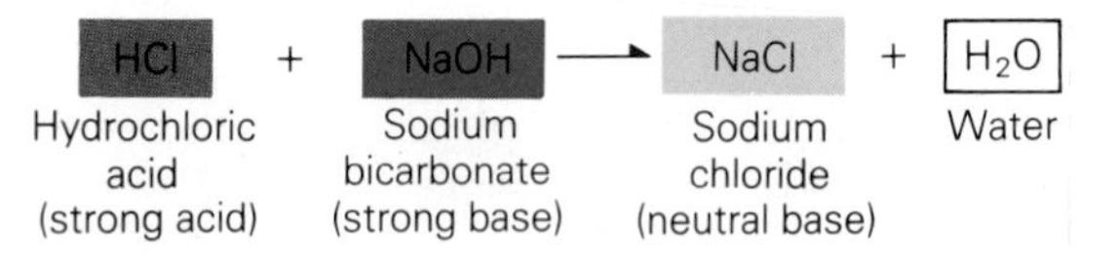

NaCl cannot serve as a buffer. The resulting solution, although at neutral pH, has no capacity for dealing with further threats to the pH.

The effect of a buffer system is illustrated in Figure 2–10. Addition of 1 mmol (1000 nmol) of a strong acid, which dissociates completely in water, to 1 liter of pure water increases the concentration of H^+ from 100 nmol/liter to 1100 nmol/liter. The pH will fall from 7.00 to 2.96. If an appropriate buffer is present at 10 mmol/liter, addition of 1 mmol strong acid will decrease the pH of the solution only to 6.90, a change of only 0.10 unit compared to the change of 4.04 units when no buffer was present. The buffer minimizes the decrease in pH until about 8 mmol of acid have been added. At this point, all the free buffer anions have been consumed by combining with the added H^+ ions. The buffer system no longer has an effect and the pH declines quickly. The buffer in this example was most effective in the range from 6.00 to 7.00. Most chemical buffers work best over a limited pH range, and each buffer has its own range.

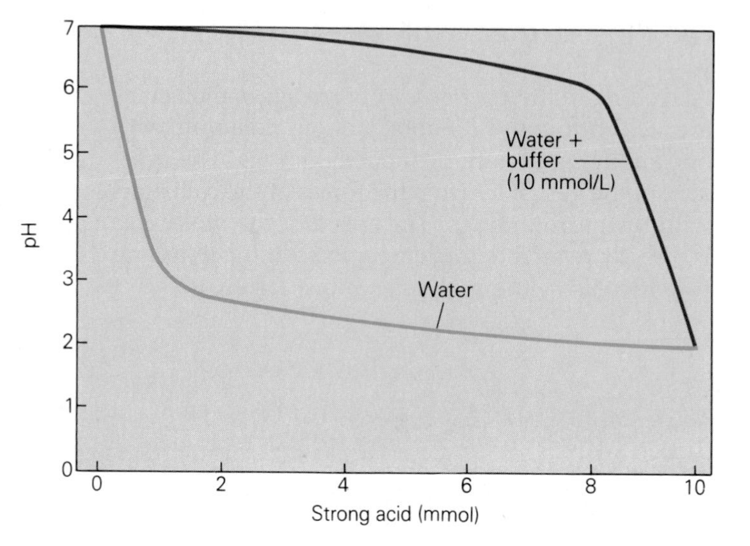

Figure 2–10 The presence of a buffer in a solution maintains a stable pH in spite of the sudden increase in H^+ concentration caused by the dissociation of a strong acid.

The Behavior of Gases

When a liquid is heated sufficiently, it begins to evaporate; its molecules break apart and begin moving about rapidly in space. The substance has passed into the gas phase. Scientists describe four physical properties of any gas: (1) the **pressure** (P) or the force the gas molecules exert on their surroundings; (2) the volume (V) or the amount of space the gas molecules occupy; (3) the **amount,** or number (n) of moles of gas; and (4) the **temperature** (T) expressed in Kelvins (273 plus the Celsius temperature).

The Ideal Gas Laws

Three factors can affect the volume of a gas: pressure, temperature, and amount. At a constant temperature, the same amount of gas will decrease in volume with an increase in pressure. At constant pressure, the same amount of gas will increase in volume with an increase in temperature. At a constant temperature and pressure, the same amount of gas will remain at constant volume; the only factor that would change the volume would be an increase in the number of moles of gas present.

These relationships can be summarized as follows: (1) volume is inversely proportional to pressure; (2) volume is directly proportional to temperature; and (3) volume is directly proportional to amount. They are further summarized in an equation known as the **Ideal Gas Law:**

$$PV = nRT$$

in which R is a constant, known as the **ideal gas constant,** equal to 0.082 liter · atm/deg · mol for any gas.

Partial Pressures in Gaseous Solutions

Unlike solids and liquids, which do not all readily dissolve in one another, all gases are completely soluble in all other gases. Air is a solution of several gases, including oxygen, nitrogen, carbon dioxide, and water vapor. According to the **law of partial pressures,** the total pressure in a container holding a solution of gases is the sum of the partial pressures of the component gases. **Atmospheric** (or **barometric) pressure,** which is the pressure exerted by air on the surface of the earth, is the sum of the partial pressures of all the gases that are present in air.

The partial pressure of a gas is proportional to the amount, in moles, of the gas solution it makes up. Oxygen, for example, makes up 21% of air. Atmospheric pressure is 760 mm Hg (so determined because, at sea level, air causes liquid mercury, Hg, to rise 760 mm in a closed tube). The partial pressure of oxygen, or P_{O_2}, is 21% of 760 mm Hg, or 160 mm Hg. If all other gases were removed from a container of air, with only oxygen remaining, it would still exert a pressure of 160 mm Hg. (Atmospheric pressure is used as a standard unit of pressure called the "atmosphere." One atmosphere—1 atm—is equal to 760 mm Hg.)

Partial pressures are designated as P_{O_2}, P_{CO_2}, P_{N_2}, P_{H_2O}, and so on. The idea of partial pressure is very useful for dealing with concentration gradients and other aspects of physiology. Because partial pressures are directly proportional to the amount of gas present, they can be used to determine the number of moles of any constituent gas in a solution of gases.

The Kinetic Theory of Gases

According to the *kinetic theory of gases,* gas molecules are in continuous, random motion, colliding with one another and with their surroundings. The pressure of the gas arises from the forces of the collisions with the surroundings. The speed of the molecules is directly related to the temperature; if the temperature increases, the molecules move faster.

Gradients and the Movement of Substances

A **gradient** is a difference between two quantities divided by the physical distance (measured in units of length) between them. Gradients are responsible for the movement of various substances throughout the body, including the movement of blood through the circulatory system, air into lungs, and ions through membranes of cells. These gradients take a variety of forms and manifest themselves as pressure gradients, electrical gradients, or concentration gradients. Concentration gradients are part of a larger topic, diffusion and osmosis.

Pressure Gradients

The movement of blood through arteries and veins is a result of a **pressure gradient.** Just as water will flow through a pipe only if the pressure at one end is greater than the pressure at another end, pressure in one region of the circulatory system must be greater than that at another region in order for blood to flow between them.

The larger the difference in pressure between two points, the greater the flow. In other words, blood flow is directly proportional to the difference in pressure between the two points. The difference in pressure between any two points can be symbolized as $(P_1 - P_2)$, where P_1 is the pressure at the starting point and P_2 the pressure at the final point. (This difference, or pressure change, can also be expressed as ΔP, where the Greek letter Δ or "delta" symbolizes a change.) The relationship between flow, F, and pressure can be represented as:

$$F \propto (P_1 - P_2)$$

where $\propto$ means "is proportional to."

Flow is met with **resistance** in the form of veins and arteries, or, more precisely, their diameters. Vessels with small diameters impede or resist the flow of blood to a greater degree than vessels with large diameters. Other sources of resistance are the length of the vessel and viscosity of the fluid. Flow is inversely related to resistance, that is, the greater the resistance, the smaller the flow.

$$F \propto 1/R$$

We can express the combined effects of pressure and resistance on flow as:

$$F = (P_1 - P_2)/R$$

The movement of air into the lungs is subject to the same influences. Gradients established by pressure differences between the external environment and the interior of the lung cause air to rush in or out of lungs. The rate of air flow is greatly affected by the lung airway resistance. In people with asthma, for example, the constriction of the airways, of decrease in their diameters, creates resistance to the flow of air, and breathing becomes difficult.

Voltage Differences and Ion Flow

Gradients of an electrical nature are responsible for the movement of charged ions through cell membranes. **Electrical gradients** are also known as **potentials** or **potential differences.** These potential differences result from differences in the number of positive and negative charges on either side of a membrane. Typically, the interior of a cell is more negatively charged than the outside. The resulting electrical gradient causes positively charged ions in the extracellular fluid to be attracted to the negatively charged intracellular space. Likewise, negatively charged ions in the intracellular fluid are attracted to the positively charged extracellular space.

Ions can move across cell membranes by way of specific channels. The diameter of a channel in relation to the size of the ion passing through it helps determine the amount of resistance the ion encounters. The flow of ions is referred to as ionic current, I. The relationship of ionic current flow to potential differences and resistance can be expressed as:

$$I = (V_1 - V_2)/R$$

in which $(V_1 - V_2)$ is the potential difference or electrical gradient and R is the resistance of the channel. This expression (like that for pressure) is a variation of **Ohm's Law,** which relates flow to gradients and resistances in a variety of physical and biological systems.

Concentration Gradients: Diffusion and Osmosis

Diffusion is a process that involves the movement of particles such as ions and molecules in solution. An understanding of these processes in general is essential for an understanding of the movement of particles in and out of cells. In diffusion, a substance such as a solute moves through and gradually becomes evenly mixed with another substance such as a solvent. Osmosis is a special kind of diffusion in which water moves through a membrane.

Diffusion of Solutes in Liquid Solvents

In solutions as well as elsewhere in nature, molecules, ions, and other particles move from one location to another by random thermal motion sometimes called **Brownian motion.** The magnitude and direction of the net movement of particles is determined by **concentration gradients.** A concentration gradient exists when two adjacent regions have different concentrations of particles. The larger the difference in concentrations, the greater the net movement between them. Particles will tend to move from regions of higher concentration to regions of lower concentration, that is, "down" the concentration gradient. If there is no concentration gradient (a situation that will occur when particles have dispersed throughout the solution), then the net movement of particles will be zero, even though individual particles may move back and forth between the two regions. Such a solution is said to be in **dynamic equilibrium.**

The movement of sugar molecules in water illustrates the effects of a concentration gradient (Fig. 2–11). In Figure 2–11*a,* the sugar concentration in the lower region of the beaker is higher than elsewhere in the beaker. Although the sugar molecules are moving constantly, and some are moving toward the bottom of the beaker, most will move up and away from the region of concentration until they are evenly dispersed throughout the water (Fig. 2–11*d*). Even at this point, individual particles will still be moving in all directions, but there will be no net movement and no net change in concentration. The concentration gradient that existed when the particles were more concentrated in one region has been abolished by diffusion.

The driving force for solute diffusion is simply the difference in solute concentration, that is, the concentration gradient, between two regions or compartments. Diffusion of solutes abolishes the concentration differences. At this point, there is no net movement because there is no driving force.

Flux is a general term used to describe the rate of movement of solute molecules. It is expressed in terms of amount of solute moving per unit time, for example, mmol/minute. **Fick's Law** of diffusion states that the flux (J) of an uncharged solute is directly proportional to the driving force (dc/dx) and the area (A) through which diffusion can occur. Thus:

$$J = -DA \cdot dc/dx$$

D is the **diffusion coefficient,** which considers the mobility of the solute molecules in solution. This mobility will vary with both the specific solute and the specific solvent. The driving force (dc/dx) is the concentration gradient—literally, the change in concentration relative to the change in distance. The minus sign is a standard way of showing that the direction of net solute movement is always from the region of high concentration to the region of low concentration, that is, "down" the concentration gradient.

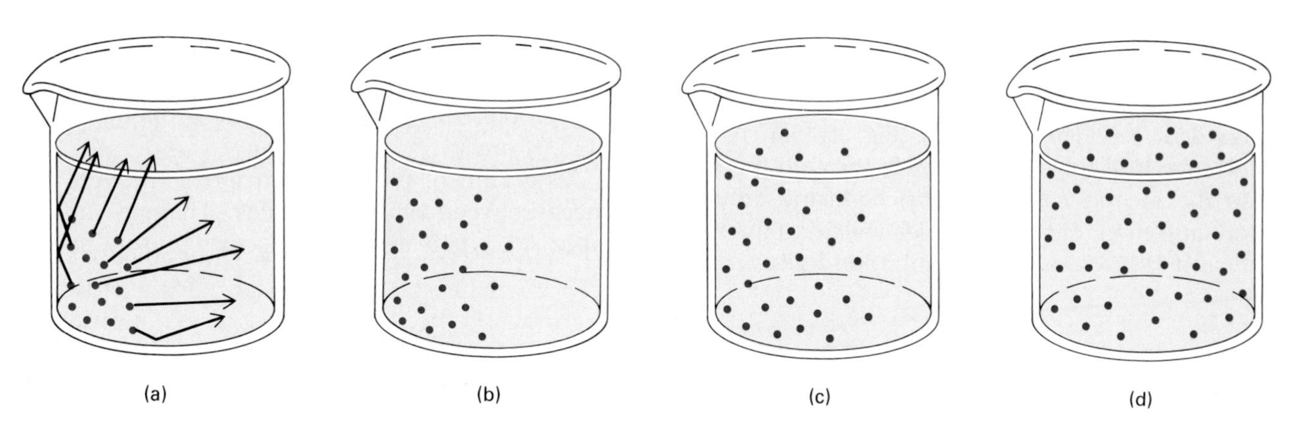

Figure 2–11 Diffusion occurs when one substance moves randomly throughout another and becomes evenly mixed with another. Diffusion occurs only when a concentration gradient exists, such that the substance to be mixed is more concentrated in one region than another. Here sugar molecules move from a region of high concentration to regions of low concentration in a beaker of water. When they are fully dispersed, a gradient no longer exists and diffusion ceases. Although the individual sugar particles are still moving in all directions, there is no net movement in any one direction because there is no gradient.

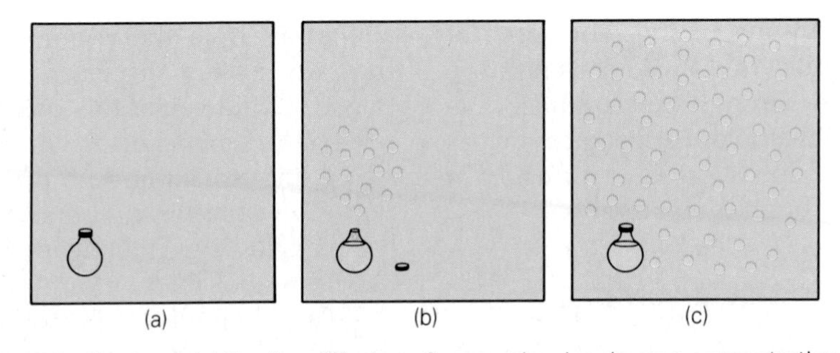

Figure 2–12 A model for the diffusion of gas molecules down a concentration gradient. When a perfume bottle is opened in a closed space, some of the liquid evaporates, and the gaseous perfume molecules pass into the surrounding space. A concentration gradient then exists because the molecules are more concentrated near the bottle than in the rest of the room. The bottle is recapped, and the perfume molecules slowly diffuse throughout the room, until they are evenly distributed, and the concentration gradient no longer exists.

Diffusion of Gases in Gaseous Solutions

Gas molecules diffuse in the same way as do solid or liquid solute particles. Imagine a room in which there is no movement of air molecules (Fig. 2–12). If a bottle of perfume is opened, some of the liquid perfume evaporates. The gas molecules come out of the bottle into the room. Slowly, the perfume molecules spread and eventually reach an equilibrium in which the distribution of the scent is even throughout the room. The perfume has spread by diffusion, owing to the concentration gradient that exists because initially the molecules are more concentrated near the bottle than in the rest of the room. The molecules move from this region of high concentration to regions of low concentration.

The movement of gas along a concentration gradient can be pictured as the movement of molecules in two boxes with a common wall (Fig. 2–13*a*). Suppose one box is filled with oxygen molecules. The molecules move randomly about in the box, bouncing off the walls and ricocheting off of each other. Now, suppose a hole is opened in the common wall between the boxes (Fig. 2–13*b*). A few molecules will travel immediately through the hole into the second box and begin bouncing around. More and more of the molecules will accumulate in the second box. As the concentration builds in the second box, some of the molecules will ricochet back through the hole into the first box. Eventually, the same number of molecules will be traveling in both directions through the hole, and the number of molecules will be the same in both boxes. At that time, the gas in the two boxes will be in dynamic equilibrium (Fig. 2–13*c*).

This diffusion process can take place even if the overall pressure is the same in each box. It can also occur if there are other kinds of molecules in each box. For example, nitrogen, carbon dioxide, and water vapor might be present in each box in equilibrium. If oxygen is introduced, it will move along its own concentration gradient as if it were the only type of molecule present (Fig. 2–14*a*, *b*, and *c*).

Likewise, different gases can be present in different concentrations in each box. For example, there can be a lot of carbon dioxide in one box and a lot of oxygen in the other. Even though the total pressure is the same in each box, the gases will diffuse simultaneously along their respective concentration gradients and move in opposite directions between boxes as soon as the hole is opened (Fig. 2–14*d*, *e*, and *f*).

One kind of molecule can move through the spaces between other molecules. There is mostly empty space between the molecules present in air.

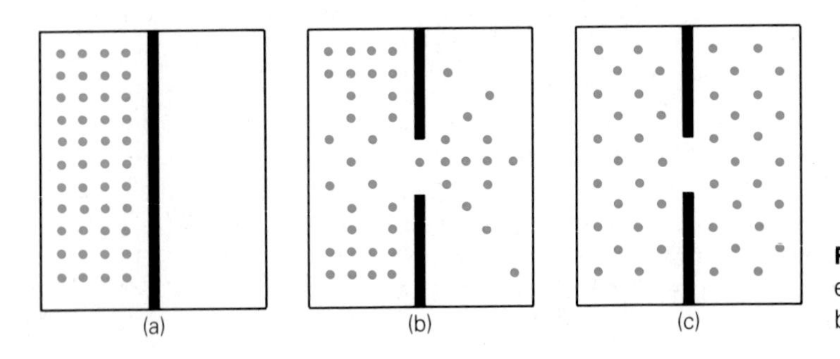

Figure 2–13 In response to a concentration gradient, gas molecules move through a hole in a wall between two boxes until they are evenly distributed.

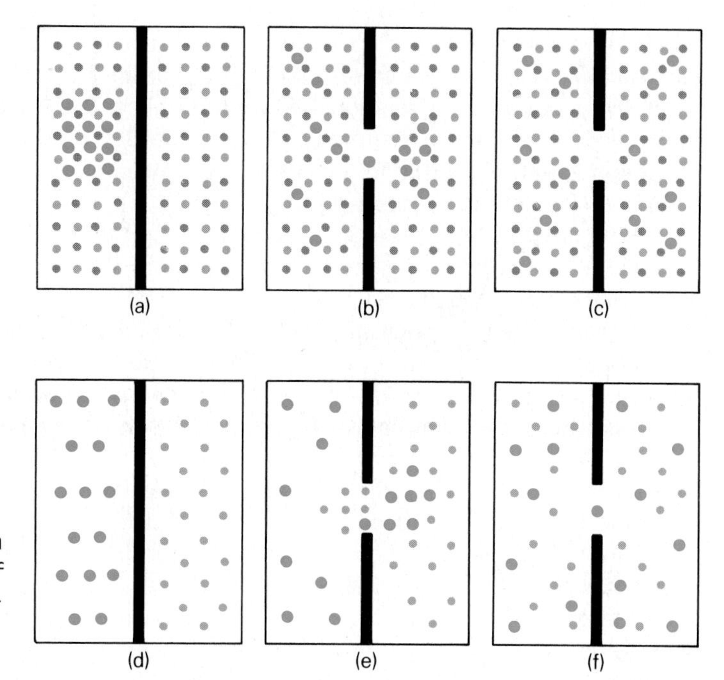

Figure 2–14 (*a, b,* and *c*) In a solution of gases, one constituent gas will diffuse independently along its own concentration gradient to become evenly distributed throughout the solution. (*d, e,* and *f*) Two concentration gradients can exist in two adjoining spaces. Molecules of each gas will move along their own concentration gradients at the same time until both gases are evenly spread throughout the entire space.

The movement of perfume molecules between air molecules is analogous to the movement of oxygen molecules through the hole in the wall. In air, there are a very large number of spaces through which molecules can move, and this speeds up any diffusion processes that occur. The wall in the box is analogous also to various membranes in the body through which gas molecules diffuse in response to concentration gradients.

Osmosis: Diffusion of Water Through a Membrane

When the molecules that are moving in diffusion are water molecules, the movement is known as **osmosis.** Osmosis is the flow of water across a barrier in response to a difference in solute concentration on either side of the barrier. Water flows from the side of the barrier with the lower concentration of solutes (the more dilute region) to the side of the barrier with the higher concentration (the less dilute region). The water molecules simply move down their own concentration gradient, like any other type of molecule. More water molecules are present in the dilute solution than in the concentrated solution, which is why there is net movement of water toward the region of high solute concentration.

Osmosis is involved in the preservation of meat by salting. When bacteria contact salt spread on the surface of meat, a concentration gradient exists because the salt is more highly concentrated outside the bacterial cell than inside. The plasma membrane of the cell is impermeable to the salt; salt cannot diffuse into the cell to abolish the gradient. The membrane is permeable to water, however, and water flows out of the cell toward the region in which the salt is more concentrated. The bacterial cell dies before it can infect the meat.

Osmotic Concentration. When discussing osmosis, we use a special measure of concentration in order to better understand the relationship of water and solute particles. The number of solute particles a solute releases in solution is not related to the mass of the solute. Chemical concentration, e.g., molality or molarity, is based essentially on weight and does not always reveal the extent to which solute particles are present in the solution. For example, in a solution of glucose in water, the glucose molecules ($C_6H_{12}O_6$) stay intact rather than dissociating into atoms. In a solution of sodium chloride in water, by contrast, the Na^+ and Cl^- ions dissociate completely and this doubles the total number of particles that are present in the solution.

Osmotic concentration is expressed in units called **osmoles (Osm).** For physiological solutions, we use **milliosmoles (mOsm)** since the concentrations are generally low.

Just as molarity expresses moles of solute per liter of solution, osmolarity expresses osmoles of solute per liter of solution. The relationship between molarity and osmolarity can be summarized as:

$$\text{osmoles} = n \times \text{moles}$$
$$(\text{milliosmoles} = n \times \text{millimoles})$$

where n is the number of particles derived from each molecule of solute when in solution. It is important to note that this relationship applies only to dilute solutions that are assumed to behave ideally. Deviation from ideal behavior occurs as the concentration of the solution increases.

For example, the osmotic concentration of a glucose solution with a chemical concentration of 60 mmol/liter is 60 mOsm/liter because n equals 1 for glucose. Glucose exists as single, intact molecules in water, producing only one solute particle per molecule.

In contrast, the osmotic concentration of a solution of NaCl whose chemical concentration is 60 mmol/liter is 120 mOsm/liter because $n = 2$. Each molecule of NaCl produces two solute particles (Na^+ and Cl^- ions) and both become osmotically active solute particles. A $CaCl_2$ solution with a chemical concentration of 20 mmol/liter will have an osmotic concentration of 60 mOsm/liter because each $CaCl_2$ group dissociates completely in solution into three ions: one Ca^{2+} ion and two Cl^- ions. A 20 mM solution of $CaCl_2$ is said to have an osmolarity of 60 mOsm.

Osmotic Pressure. **Osmotic pressure** is a property of a solution that is proportional to the solute concentration. A dilute solution will have a lower osmotic pressure than a concentrated solution. Thus, water will move by diffusion from a region of low osmotic pressure to a region of high osmotic pressure.

The osmotic pressure difference between two solutions of different solute concentrations is illustrated in Figure 2–15. A solution of water containing 1.0 M sucrose is placed inside an inverted thistle funnel. The thistle funnel is placed in a bath of pure water. The water and the sucrose solution are separated by a membrane placed across the mouth of the funnel. The membrane allows only pure water, not sugar molecules, to pass through. The apparatus is arranged so that, initially, the level of the sucrose solutions is the same as the level of the water in the bath.

The sucrose solution has a higher osmotic pressure than the water because it has a higher solute concentration. Consequently, water will begin to move from the pure water bath across the membrane into the thistle funnel. The movement of water into the sucrose solution gradually dilutes the solution and raises the level of the solution inside the thistle tube.

Eventually, the increased fluid level in the funnel generates a new pressure, hydrostatic pressure, which now opposes the movement of more water into the tube. Soon there is enough hydrostatic pressure to completely stop the osmosis of water into the thistle tube. At this point, the hydrostatic pressure reflects the difference in osmotic pressure between water and the original sucrose solution.

The increase in the fluid level in the funnel can be prevented by applying pressure down the funnel at the start of the experiment. This pressure blocks the osmotic movement of water into the sucrose solution. The pressure applied is the osmotic pressure of 1.0 M sucrose.

If we know the solute concentration of a solution, it is relatively simple to calculate the osmotic pressure. Osmotic pressure is related to the solute concentration by the **van't Hoff law,** which is described by the equation:

$$rr = RTC$$

in which rr is the osmotic pressure, R is the ideal gas constant (0.082 liter · atm/degree · mole), T is the absolute temperature in Kelvins (°C + 273), and C is

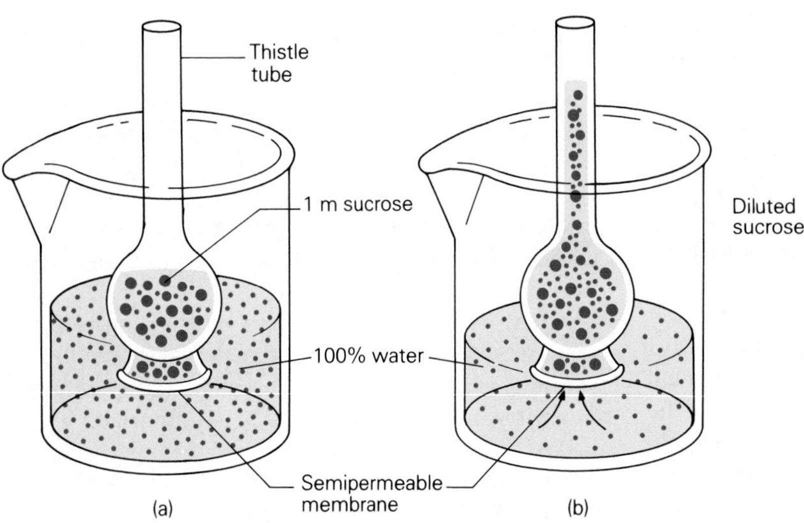

Figure 2–15 When a thistle tube containing a sucrose solution is placed in a bath of pure water, the water moves across a membrane up into the tube, diluting the sucrose solution and raising the level of the solution in the tube. No sucrose flows from the tube into the bath because the membrane is impermeable to sucrose.

the total solute concentration expressed as osmoles of solute per liter of solution. For example, the osmotic pressure of a solution of 1.0 Osm/liter at 0°C is calculated as follows:

$$\begin{aligned}\pi &= RTC \\ &= 0.082\ \frac{\text{liter} \cdot \text{atm}}{\text{degree} \cdot \text{mole}} \times 273° \times 1.0\ \frac{\text{Osm}}{\text{liter}} \\ &= 22.4\ \text{atm}\end{aligned}$$

Thus, a 1.0 Osm solution exerts an osmotic pressure of 22.4 atm at 0°C. A similar calculation reveals that human blood plasma, which is normally a 0.31 Osm solution, will have an osmotic pressure of 7.9 atm at 37°C. The osmotic pressure of a solution depends on the total osmolarity of the solution and not on the nature of the specific solute or solutes. The value of C for pure water containing no dissolved solutes is zero; hence, the osmotic pressure of pure water is zero.

Since R and T are constants, if there is no temperature change, the osmotic pressure will be directly related to osmolarity. In this case, there is no need to calculate osmotic pressure in order to determine which of several solutions has the highest osmotic pressure. A comparison of the osmolarity of the solutions will suffice; the solution with the highest osmolarity will have the highest osmotic pressure.

The osmolarity of a solution can be directly determined by use of a freezing point **osmometer,** even if the identity of the specific solute is unknown. The freezing point of pure water is depressed when solutes are present, and the extent of the depression is proportional to the total amount of all the solutes present. Most osmometers are calibrated to give a readout directly in osmoles or milliosmoles, and they provide a rapid and accurate method for determining the osmolarity of solutions.

Two solutions that have the same osmolarity are said to be **iso-osmotic.** Thus, 300 mOsm NaCl is iso-osmotic with 300 mOsm sucrose. A solution with a higher osmolarity than another is said to be **hyperosmotic;** a solution with a lower osmolarity is said to be **hypo-osmotic** relative to one with a higher osmolarity.

Osmotic pressure is only one property of a solution that is affected by the presence of solute particles. Dissolved solutes lower the vapor pressure and increase the boiling point of solutions. They also lower the freezing point. When salt is spread on a wet road in winter, it dissolves in the water, preventing the water from freezing at 0°C where pure water would normally freeze. $CaCl_2$ is used instead of NaCl because it releases more solute particles (three rather than two) per molecule and has a greater effect on the freezing point.

Macromolecules

The unique properties of the carbon atom allow it to form four bonds at once and lead to a huge variety of molecules of different shapes and enormous sizes. These molecules are called **macromolecules.** These substances play important roles in the life processes of all cells and organisms. The most important macromolecules in cell physiology are carbohydrates, lipids, proteins, and nucleic acids. The following brief survey provides a glimpse of these compounds and a review of their chief functions in cells.

Carbohydrates

Carbohydrates are often referred to as **sugars.** All **carbohydrates** are composed of carbon, hydrogen, and oxygen atoms. Carbohydrates are classed in three broad groups: monosaccharides, disaccharides, and polysaccharides. In addition, they form compounds with proteins and lipids to form glycoproteins and glycolipids.

Monosaccharides

The simplest type of carbohydrates are called **monosaccharides.** The general chemical formula for a monosaccharide is $(CH_2O)_n$, where n is a whole number indicating the number of carbons. A monosaccharide with three carbons, such as glyceraldehyde, is a **triose** (Fig. 2–16). A monosaccharide with four carbons is a **tetrose;** one with five carbons a **pentose,** and one with six carbons a **hexose.**

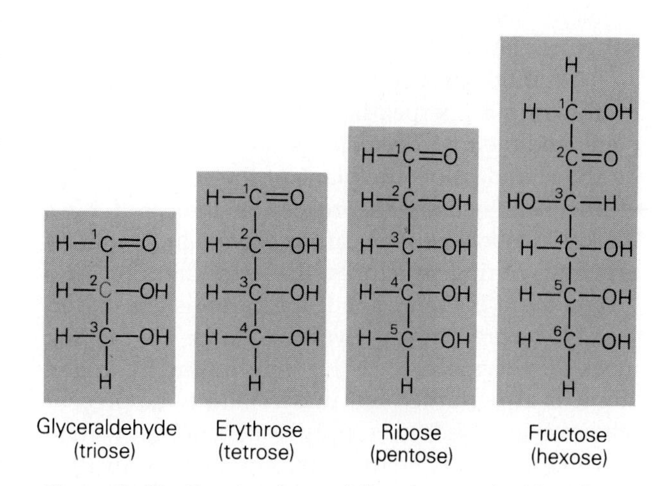

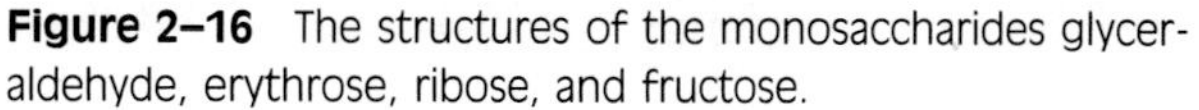

Figure 2–16 The structures of the monosaccharides glyceraldehyde, erythrose, ribose, and fructose.

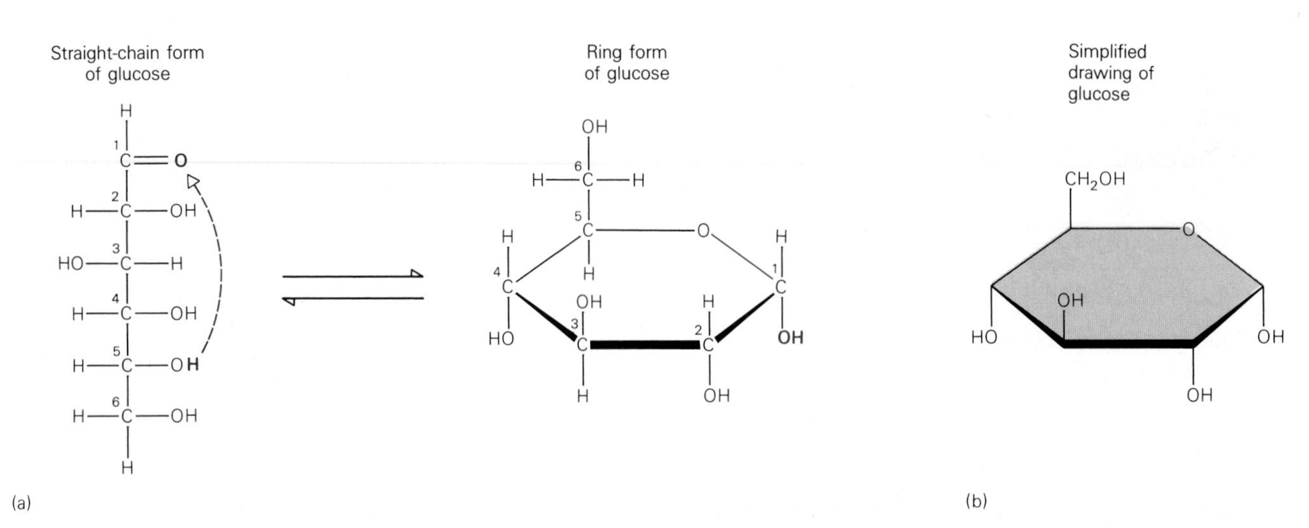

Figure 2–17 The linear and ring structures of the sugar glucose, a common hexose. When the —OH group on carbon 5 comes near the =O group on carbon 1, the two carbons are joined by the —O— atom, and a ring is formed.

In structural formulas of sugars, and other "organic" molecules, the carbon atoms are often numbered for identification. Specific carbons often have specific roles based on their position in the molecule and the other atoms that are attached to them. A carbon atom, such as the second carbon atom ("carbon 2") in the glyceraldehyde molecule, is described as **asymmetrical** when it is linked covalently to four different chemical groups. Carbon 2 in glyceraldehyde (Fig. 2–16) is linked to a hydrogen (H), an aldehyde (—CHO) group, a hydroxide (—OH) group, and —CH_2OH.

Ring Structures. Monosaccharides can have different forms, either linear chains or rings. In solutions, monosaccharides most often form rings. One common hexose is **glucose,** $C_6H_{12}O_6$, or $(CH_2O)_6$. In glucose, ring formation occurs when the aldehyde (—CHO) group on carbon 1 reacts with the hydroxyl (—OH) group on carbon 5 to form a stable covalent bond, giving rise to a six-member ring structure (Fig. 2–17).

Although drawn as a flat structure, the plane of the ring is perpendicular to the plane of the page, and the hydrogen atoms and hydroxyl groups lie above and below the plane of the ring. The edge of the ring nearest the reader is indicated by a thicker line between the carbon atoms. In simplified drawings of ring structures, the letter *C* indicating a carbon atom is omitted, but the carbons are understood to be present wherever two "sides" of the ring meet. Furthermore, the letter *H* indicating hydrogen atoms is also omitted but understood to be present wherever needed to supply each carbon with a total of four bonds. In the most simplified drawings, only the —OH groups are identified by letters (Fig. 2–17).

Pentoses form five-member ring structures, as do some hexoses such as fructose (Fig. 2–18). Two important pentoses are ribose and deoxyribose, which are present in nucleic acids.

When a hexose forms a five-member ring, carbon 1 is left outside the ring because the ring forms when the carbonyl (C=O) group on carbon 2 reacts with the hydroxyl (—OH) group on carbon 5 (Fig. 2–18).

Monosaccharides have a unique ability to form chemical bonds at more than one carbon atom. One

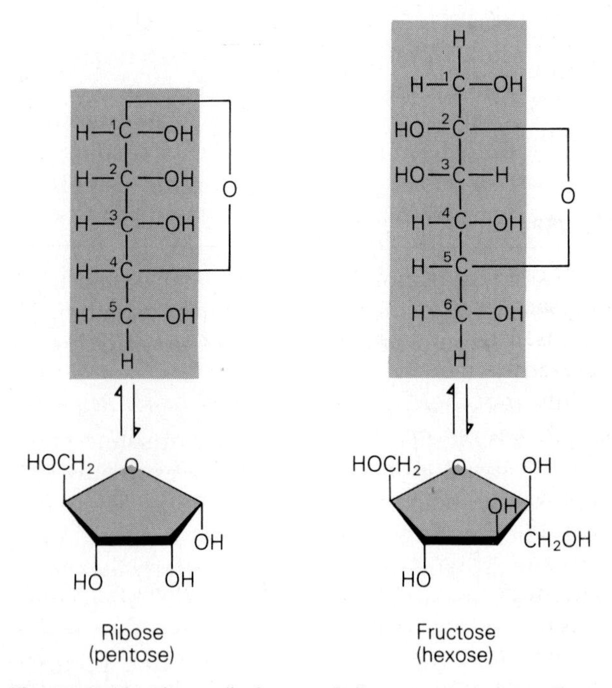

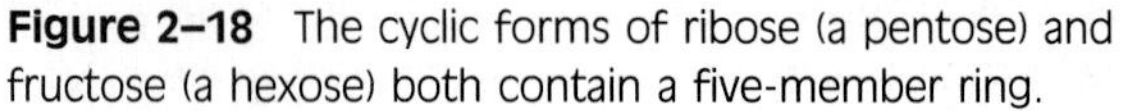
Figure 2–18 The cyclic forms of ribose (a pentose) and fructose (a hexose) both contain a five-member ring.

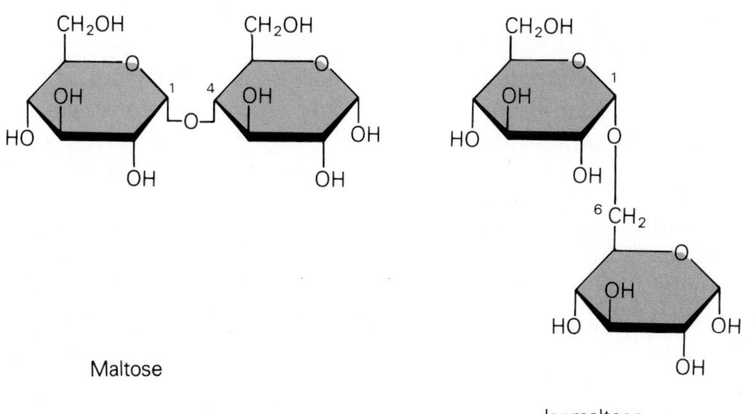

Figure 2–19 Maltose and iosomaltose are isomers. Both are formed from two molecules of glucose; both have the same molecular formula. They have different structural formulas, however, and therefore constitute two different disaccharides.

monosaccharide can combine with several others to create branched chains of monosaccharide units. In contrast, other macromolecules such as proteins and nucleic acids can assemble only in linear, unbranched chains.

Disaccharides

Two monosaccharides can combine to form a **disaccharide.** The most common disaccharides are **maltose** (two glucose molecules), **lactose** (one glucose molecule plus one galactose molecule), and **sucrose** (glucose plus fructose).

Isomaltose, like maltose, is formed from two glucose molecules, and both isomaltose and maltose have the same chemical formula. However, they have different structures because their monosaccharide units join at different carbon atoms. In maltose, the glucose molecules join at carbons 1 and 4; in isomaltose, they join at carbons 1 and 6 (Fig. 2–19). Maltose and isomaltose are called **isomers,** molecules with the same chemical formula but different structural arrangements of the atoms.

Polysaccharides

Most of the carbohydrates found in nature occur as **polysaccharides,** molecules composed of many monosaccharide units arranged in branched chains. Polysaccharides are distinguished from one another mainly by their monosaccharide units and the length and amount of branching of their chains. Glucose is the most common monosaccharide unit of polysaccharides. Glucose is the fuel that most animal cells require, and it can be stored in certain tissues in the form of the polysaccharide **glycogen.** This important polysaccharide is especially abundant in the liver, where it can account for up to 10% of the wet weight of the tissue. It consists of a chain of glucose units that are linked by chemical bonds between carbons 1 and 4 of adjacent glucose units. Branches occur about every 10 glucose units, arising from a bond between carbons 1 and 6 (Fig. 2–20). The molecular mass of glycogen has been estimated to be in the range of from 3×10^5 to 3×10^6. This means that the number of glucose units present in a glycogen molecule could be as high as 16,667.

Sugar Combinations

Many of the polysaccharides present in living organisms occur in covalent combination with proteins or lipids. **Glycoproteins** and **glycolipids** are present in cell membranes. The polysaccharide chains in these compounds tend to be short but highly branched, containing several different monosaccharide units arranged in complex sequences. They are thought to be important in recognition processes such as those involved in the action of hormones. A hormone released into the blood will be distributed to all the cells in the body, but will affect only those cells that have an external site that the hormone "recognizes" and binds. The carbohydrate portions of membrane glycoproteins and glycolipids are thought to be important parts of these binding sites for hormones.

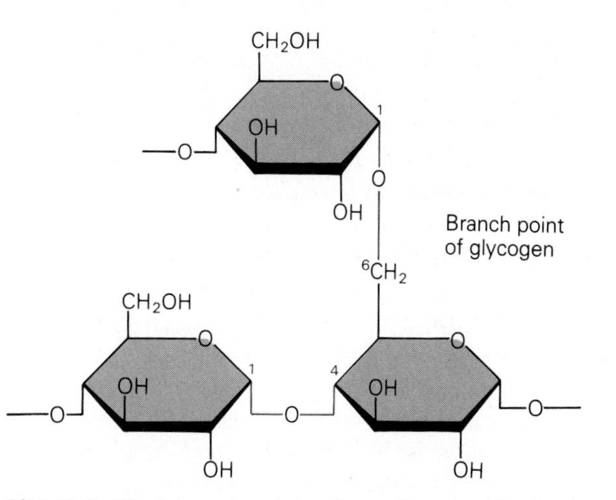

Figure 2–20 A branch point in the polysaccharide glycogen.

Lipids

Lipids are sometimes called **fats** and are distinguished from the other cellular macromolecules because they are poorly soluble in water and highly soluble in organic liquids such as ether and chloroform. They are composed of carbon, hydrogen, and oxygen atoms, the same elements that make up carbohydrates; however, the proportions of these elements differ in lipids. The lipids contain only small amounts of oxygen relative to carbon and hydrogen.

Three major classes of lipids will be considered here: fatty acids, lipids containing glycerol, and lipids that do not contain glycerol.

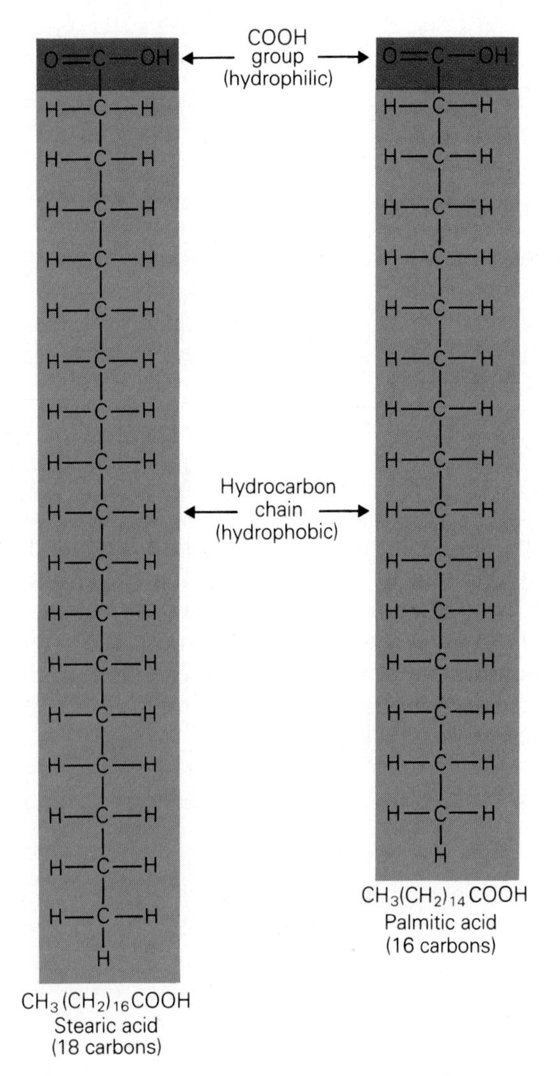

Figure 2–21 The structures of stearic acid and palmitic acid, two common saturated fatty acids. The molecules consist mostly of hydrophobic hydrocarbon chains. The carboxylic acid (—COOH) group at one end, however, is water-soluble or hydrophilic. Fatty acids are sometimes given the general symbol RCOOH, *R* representing the hydrocarbon chain.

Fatty Acids

A **fatty acid** is a long chain of covalently linked carbon atoms to which only hydrogen atoms are attached, except for a carboxylic acid (—COOH) group at one end. The long hydrocarbon chain is hydrophobic, or poorly soluble in water, whereas the —COOH group is highly water soluble and reacts readily with other chemical groups, forming covalent bonds. Most of the fatty acids in animal cells are covalently linked to other molecules via the carboxylic acid group. Two of the most abundant fatty acid molecules in animal cells are **palmitic acid** and **stearic acid** (Fig. 2–21).

Specific fatty acids are distinguished from one another by the number of carbon atoms in their hydrocarbon chains and the number of carbon-carbon double bonds. When only single bonds exist between all the carbons in a fatty acid chain, the fatty acid is said to be **saturated,** i.e., saturated with hydrogen, because as many hydrogen atoms as possible are bonded to each carbon. When one or more double bonds are present between carbon atoms, the fatty acid is **unsaturated.** The most abundant unsaturated fatty acids in animal cells are oleic acid, linoleic acid, and arachidonic acid, which contain one, two, and four double bonds, respectively:

Oleic acid	$CH_3(CH_2)_7CH{=}CH(CH_2)_7COOH$
Linoleic acid	$CH_3(CH_2)_4CH{=}CHCH_2CH{=}CH(CH_2)_7COOH$
Arachidonic acid	$CH_3(CH_2)_4(CH{=}CHCH_2)_4(CH_2)_2COOH$

Fatty acids are a valuable source of fuel for cells. Considerably more energy can be obtained from the breakdown of 1 g of a fatty acid than from 1 g of carbohydrate. Arachidonic acid is an intermediate structure produced during the synthesis of an important group of molecules called **prostaglandins,** which affect a wide variety of physiological processes.

Glycerol Compounds and Phospholipids

Glycerol is a molecule with three hydroxyl groups (Fig. 2–22). These hydroxyl groups can form bonds with the carboxylic acid groups of fatty acids. The attachment of one, two, or three fatty acids to glycerol produces a **monoacylglycerol,** a **diacylglycerol,** or a **triacylglycerol,** respectively. Both mono- and diacylglycerols are important natural compounds, but the triacylglycerols (Fig. 2–22) are the most abundant and represent an important means by

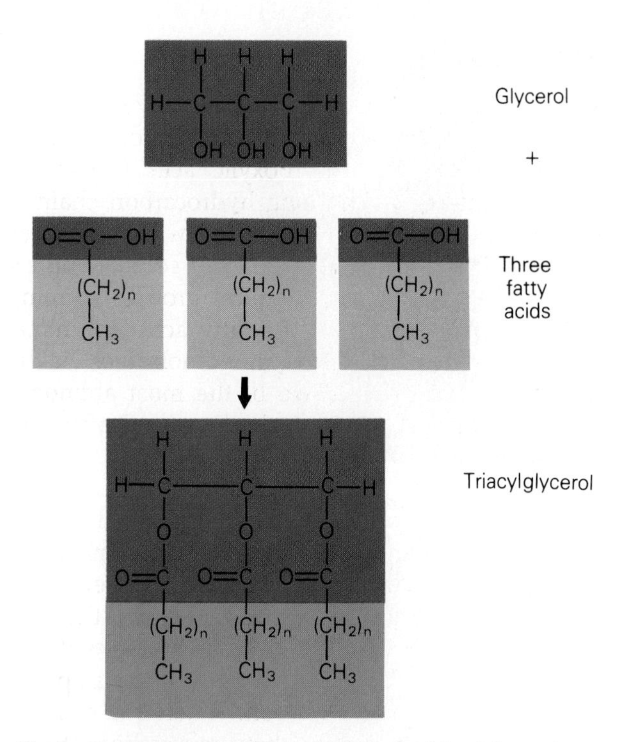

Figure 2–22 Steps in the formation of a triacylglycerol from glycerol and three fatty acids. The fatty acids may be identical or different. *n* stands for a whole number, the number of carbon atoms in the parentheses representing all but the last carbon in the fatty acid hydrocarbon chain. For stearic acid, *n* is 16 [see Fig. 2–21].

which cells are able to store fatty acids until they are needed. Cells can convert excess carbohydrate and protein to fatty acids and then store the fatty acids in triacylglycerols. The continued accumulation of triacylglycerol deposits in the body, however, can lead to obesity.

Another group of glycerol-containing lipids are the **phospholipids.** The parent compound of this group is **glycerol-3-phosphate,** a glycerol molecule to which a phosphate (—PO_4) group has been attached (Fig. 2–23*a*). A phospholipid is formed from this basic structure by (1) the addition of fatty acids to the free hydroxyl groups on carbons 1 and 2 to give **phosphatidic acid** and (2) the addition of another chemical group to the phosphate on carbon 3. In Step 1, the fatty acid added to carbon 1 of the glycerol unit is usually saturated and contains 16 to 18 carbon atoms, whereas the fatty acid added to carbon 2 is usually unsaturated and contains 16 to 20 carbon atoms. The most common groups added in Step 2 are choline, ethanolamine, serine, and inositol. The phospholipids that result are called phosphatidylcholine, phosphatidylethanolamine, phosphatidylserine, and phosphatidylinositol, respectively (Fig. 2–23*b*).

At physiological pH (7.40), the phospholipids are in the charged forms shown, and one end of the molecule is a polar (hydrophilic) region. In contrast, the long hydrocarbon chains of the two fatty acid groups constitute a highly hydrophobic region of the molecule (Fig. 2–23*c*). The dual nature of these phospholipids—hydrophilic at one end and hydrophobic at the opposite end—plays an important structural role in the plasma membrane surrounding all cells. The membrane is made up of a "lipid bilayer": two layers of phospholipid molecules oriented so that their hydrophobic hydrocarbon chains point toward each other, away from the aqueous surroundings both inside and outside the cell (Fig. 2–23*d*). The hydrophilic parts of the phospholipids in each layer are oriented outward. This arrangement represents a very stable physical structure with unique properties, as we will learn in Chapter 4.

Sphingomyelin is a phospholipid, the only phospholipid not derived from glycerol. The basic unit of sphingomyelin is **sphingosine:**

$$CH_3(CH_2)_{12}\text{-}CH{=}CH\text{-}CHOH\text{-}CHNH_3^+\text{-}CH_2OH$$

Sphingosine contains an unsaturated hydrocarbon chain and is converted to sphingomyelin by the addition of a long-chain fatty acid, a phosphate group, and a choline group. Overall, the structure of sphingomyelin is very similar to that of phosphatidylcholine (Fig. 2–23*c*), since it has two hydrophobic hydrocarbon chains and a hydrophilic phosphorylcholine group.

Steroids

Some of the most important lipids do not contain glycerol; these are the **steroids.** The basic structure of a steroid is four interlocking rings of carbon atoms (Fig. 2–24) and is quite different from that of other lipids. The male and female sex hormones and some hormones secreted by the adrenal glands, such as aldosterone, are steroids. The term **sterol** is often used when a hydroxyl group is present. **Cholesterol** is an example of a sterol that plays an important role in the structure of plasma membranes. Like the phospholipids, cholesterol has a polar hydrophilic region (the hydroxyl group) and a nonpolar hydrophobic region (the rest of the molecule).

Glycolipids

Glycolipids are lipids, also derived from sphingosine, that contain monosaccharide units. They have two hydrocarbon chains, as sphingomyelin does, but between one and seven monosaccharide units

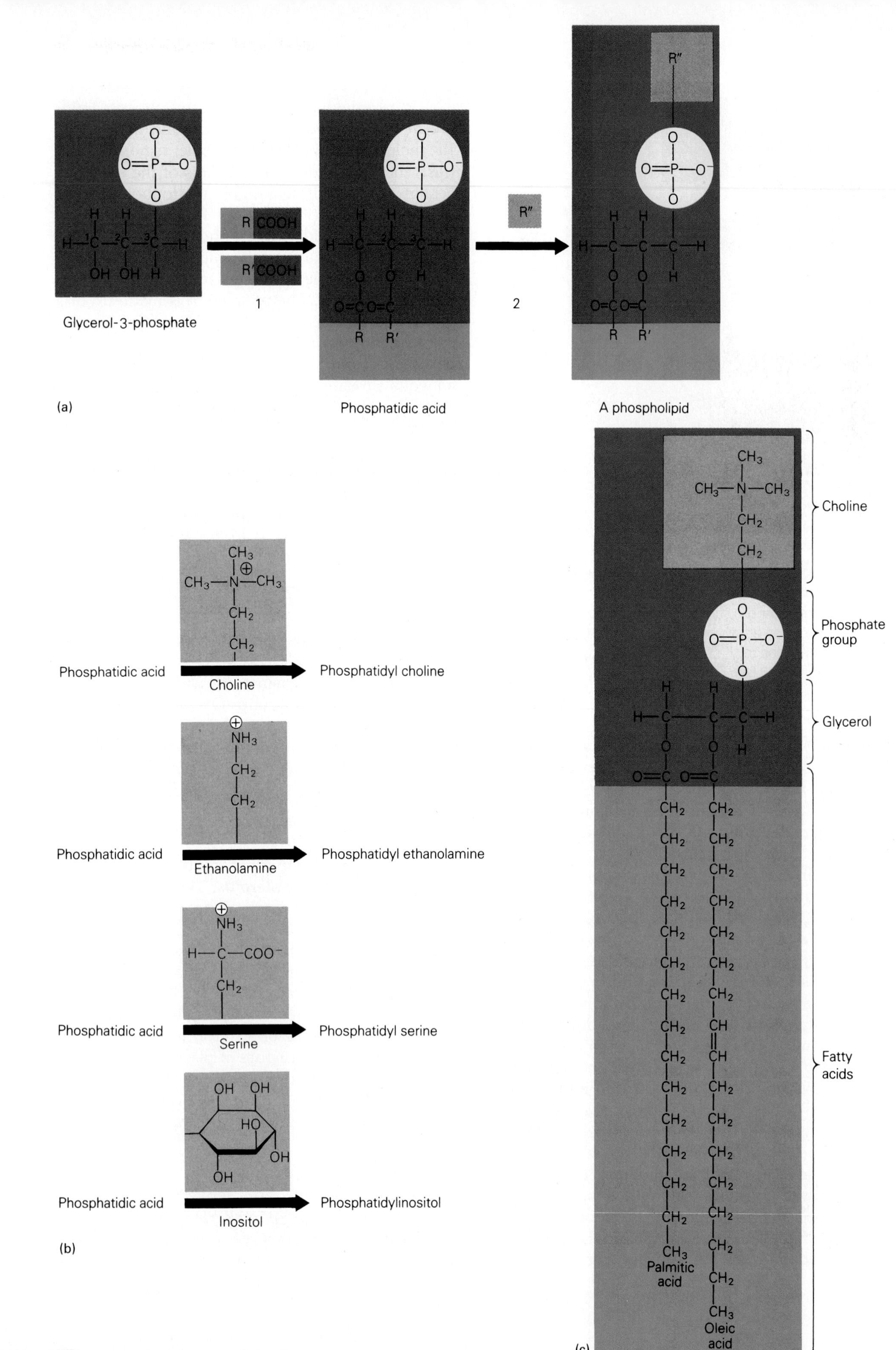

Glycerol-3-phosphate
R COOH
R′ COOH
1
Phosphatidic acid
R″
2
A phospholipid
(a)
Phosphatidic acid
Choline
Phosphatidyl choline
Phosphatidic acid
Ethanolamine
Phosphatidyl ethanolamine
Phosphatidic acid
Serine
Phosphatidyl serine
Phosphatidic acid
Inositol
Phosphatidylinositol
(b)
Choline
Phosphate group
Glycerol
Fatty acids
Palmitic acid
Oleic acid
(c)

Summary

I. Matter is composed of elements, the smallest units of which are atoms. Atoms combine to form molecules, and elements combine to form groups and compounds. Carbon, hydrogen, nitrogen, and oxygen are the four predominant elements in the human body. Living things require a constant supply of energy to maintain order in the matter that comprises them.

II. A. All atoms are composed of the same subatomic particles. Atoms of the same element have the same number of protons in their nuclei. Uncharged atoms have equal numbers of protons and electrons. Isotopes are atoms of the same element with different numbers of neutrons.

B. Ions are atoms with electrical charges due to unequal numbers of electrons and protons.

C. The electron configuration of an atom, especially the arrangement of electrons in its outer shell, determines its reactivity. An atom will try to fill its outer shell with electrons.

III. A. Atoms share, gain or lose electrons to form bonds with other atoms, producing molecules, groups, and compounds. Atoms that share electrons form polar or nonpolar covalent molecules.

B. Atoms that form bonds by gaining or losing electrons produce ionic compounds, often arranged as crystal lattices.

C. Van der Waals bonds develop between atoms that are close together, owing to the attraction of the nucleus of one atom to the electrons of another.

D. Hydrogen bonds form between hydrogen atoms of polar molecules and groups and small atoms such as oxygen and nitrogen of other polar molecules.

IV. Chemical reactions occur when compounds react with other compounds to form new products. The rate of a chemical reaction depends on such factors as temperature, concentration, and orientation.

V. A. The polarity of the water molecule makes water a strong solvent of ionic and polar covalent compounds. These compounds are therefore called "hydrophilic," in contrast to nonpolar compounds, which are "hydrophobic," or insoluble in water.

B. Because of hydrogen bonds, water molecules form three-dimensional patterns in the solid and liquid phases.

C. Water molecules dissolve ionic compounds by forming shells around the ions, causing them to dissociate from the crystal lattice. Water dissolves polar covalent compounds by forming hydrogen bonds with them. Nonpolar covalent compounds do not dissolve in water; instead, the nonpolar molecules undergo hydrophobic reactions, clustering together in pockets in the water.

D. Concentration of a solution is measured in terms of molality (moles of solute per kilograms of solvent) or molarity (moles of solute per liter of solution).

E. The pH scale is a means of measuring the relative acidity or basicity of a solution by expressing logarithmically the H^+ concentration.

F. An acid is a substance that dissociates in solution to release H^+ ions, or protons. A base is a substance that accepts H^+ ions in solution, often by releasing OH^- ions. An acidic solution is one with a high H^+ concentration; a basic solution is one with a low H^+ concentration. Buffers are weak acids or bases added to solutions to minimize changes in pH due to the presence of a strong acid or base.

VI. A. Gases can be described in terms of pressure, temperature, volume, and amount (in moles).

B. These four properties are related in a constant way for all gases, as described in the Ideal Gas Equation, $PV = nRt$.

C. In a solution of gases, the total pressure of the solution is equal to the sum of the partial pressures of the constituent gases.

D. According to the kinetic theory of gases, gas molecules are moving in continuous random motion, colliding with each other and with their surroundings, and increasing in speed with increasing temperatures.

VII. A. A gradient is a difference in two quantities divided by the distance between them. Some substances move through the body in response to pressure gradients.

B. Other substances move in response to voltage gradients, or differences in electrical charges between regions.

C. Concentration gradients are responsible for the movement of substances by diffusion and osmosis. Solute particles in both liquid and gaseous solutions spread evenly throughout solvents by means of diffusion. Water molecules pass through selectively permeable membranes by means of osmosis in order to dilute regions that have high concentrations of solute particles.

VIII. A. The unique covalent bonding capabilities of the carbon atom allow it to form molecules of huge proportions and intricately varied structures. These macromolecules are important in all cell processes. One class of macromolecules, called carbohydrates, are made up of smaller units called monosaccharides. Ring structures are common among carbohydrates. Monosaccharides combine in linear and branched chains to form di- and polysaccharides. Sugars are important sources of chemical energy in cells.

B. Lipids are macromolecules that often consist of long hydrocarbon tails that are not soluble in water. The hydrophobic behavior of phospholipids in water makes them ideal structural components of cell membranes. Lipids are also important sources of chemical energy in cells. The major classes of lipids are fatty acids, glycerol compounds, and nonglycerol compounds.

C. Proteins are complex macromolecules that carry out diverse and important functions in cells. Proteins consist of amino acids arranged in four levels of structure. The sequence of amino acids and the shape of protein macromolecules are highly specific features of proteins that determine their functions in the cell.

D. Nucleotides are composed of a sugar, a phosphate group, and a special group called a base. Nucleic acids (DNA and RNA) are composed of nucleotides linked together in long strands. They store genetic information.

IX. A. Thermodynamic principles play an important role in cell physiology. All chemical reactions involve a free energy change between the reactants and the products.

B. Enzymes increase the rates of specific reactions by bringing reactant molecules together and straining existing bonds, thus lowering the initial activation energy barrier for the reaction.

C. Endergonic reactions in the cell absorb the free energy they need from exergonic reactions.

Review Questions

1. How many nanometers (nm) are there in one kilometer (km)?
2. How is a covalent bond formed between two atoms? How is a polar covalent bond different from a nonpolar bond?
3. How is an ionic bond formed? How is it different from a covalent bond?
4. What are some factors that affect the rate of a chemical reaction?
5. How does water dissolve hydrophilic compounds?
6. How many grams of potassium chloride must be dissolved in 100 ml of water to produce a concentration of 300 mM?
7. What is the relationship between pH units and H^+ concentration? Calculate the pH of a solution in which the H^+ concentration is 1 nmol/l.
8. Why is an acid called a "proton donor"?
9. At constant pressure, how will an increase in temperature affect the volume of a gas?
10. What is a gradient?
11. Why is osmosis a special type of diffusion?
12. How is a molecule of glucose linked to other glucose molecules in glycogen?
13. What is the difference between a saturated and an unsaturated fatty acid?

14. What are the subunits of phosphatidylinositol?
15. How do phospholipid molecules behave together in water?
16. How does a peptide bond form between two amino acids?
17. How would an amino acid such as leucine or phenylalanine behave differently than an amino acid such as aspartic acid or serine in aqueous solution?
18. What are the bonds that help to stabilize the secondary structures of proteins?
19. What are the differences in structure and chemical composition between DNA and RNA?
20. What does it mean to say that the nucleotide strands in DNA are complementary?
21. Why is some energy always lost in a chemical reaction?
22. What is the difference between an endergonic and an exergonic reaction?
23. How does an enzyme increase the rate of a reaction?

Suggested Readings

Smith, E.L., Hill, R. L., Lehman, I.R., Lefkowitz, R.J., Handler, P., and White, A. *Principles of Biochemistry: General Aspects*, 7th ed. New York: McGraw-Hill, 1983.

Stryer, L. *Biochemistry*, 3rd ed. San Francisco: Freeman, 1988.

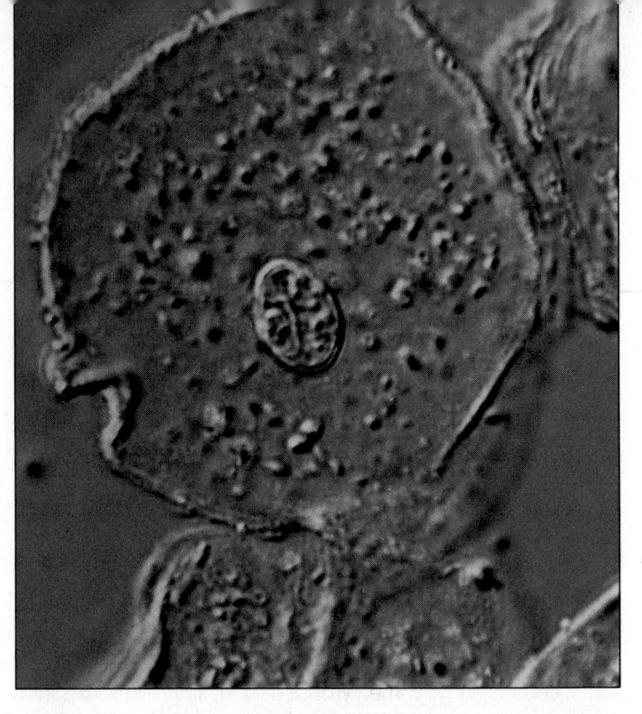

The Structure and Functions of Cells

According to the **cell theory,** which forms the basis of all modern biology and physiology, (1) all organisms are made up of cells and their products; (2) new cells arise only from pre-existing cells; (3) all cells have the same fundamental chemical makeup and metabolic processes; and (4) the activities and processes of the organism as a whole result from the interdependent and cooperative workings of groups of cells.

In all of nature there are two broad classes of cells: eukaryotic cells and prokaryotic cells. The primary distinction between these cell types is the presence or absence of a membrane-bounded compartment surrounding the genetic material, the DNA. In **prokaryotic cells,** the DNA is not in a separate compartment from the rest of the cell, whereas all **eukaryotic cells** have a well-defined **nucleus,** or central area, contained within its own nuclear membrane.

Organisms in turn are broadly classified according to whether they have prokaryotic or eukaryotic cells. Many bacteria are prokaryotic organisms, for example; all mammals are eukaryotic organisms. Eukaryotic cells are generally larger (10 to 100 μm in diameter) than prokaryotic cells (1 to 10 μm).

In addition to a membrane-enclosed nucleus, all eukaryotic cells contain many small subcellular structures, most of which are also enclosed in a membrane. These structures are grouped under the

(see Fig. 3–2). The membrane of the endoplasmic reticulum of a liver cell makes up more than 50% of the total membrane of the cell. There are seven different subcompartments, or organelles, formed by the intracellular membranes of most eukaryotic cells. Two of these organelles, the nucleus and the mitochondrion, are surrounded by a double membrane. These membranes are composed, as is the plasma membrane, primarily of two layers of phospholipid molecules.

Continuity exists to some extent between the subcompartments in that some of the materials present in one compartment may be passed on to another, but direct physical links between the compartments are probably very limited. Each of the compartments within the cell has a unique role carried out at the molecular level by the enzymes that are either packed inside the organelle or are components of the membrane of the organelle. Since each organelle has a different function, the set of enzymes within a specific organelle will not be found in any of the other organelles. **Catalase,** for example, is found only in organelles called peroxisomes. The enzyme that breaks down glycogen into glucose molecules occurs only in the cytosol. The enzymes that catalyze the sequence of reactions in the citric acid cycle are present only within mitochondria.

Enzymes found in lysosomes, called **lysosomal enzymes,** are essential to normal cell function. The absence of just one of these enzymes, as occurs in some genetic diseases, has severe consequences for the individual. However, this group of **degradative enzymes** can break down all of the macromolecules responsible for the structure and function of the cell. Clearly, these enzymes cannot be allowed direct contact with other organelles or with the components of the cytosol because they would cause irreversible, lethal damage. The cell solves the problem very neatly by keeping the enzymes within lysosomes so that they are always available within the cell. They are allowed access to structures and molecules targeted for degradation (breakdown) only when the lysosomes fuse with endocytotic vesicles containing material to be broken down. This process ensures that the degradative process is confined locally within a specific compartment, the secondary lysosome (see Fig. 3–11).

Similarly, the presence of catalase within peroxisomes ensures that the toxic hydrogen peroxide produced within these organelles is destroyed on the spot before it has a chance to leak into the cytosol, where it would cause serious damage.

The presence of intracellular compartments also permits separation of the processes of synthesis and degradation of molecules. The advantage derived from physically separating these processes is that they can occur simultaneously within the cell and can be controlled independently. The metabolism of fatty acids provides a good example. These molecules are continually synthesized by the cell because they are components of the phospholipid molecules needed for the structure of cell membranes. At the same time, the cell also is degrading other fatty acids because they are an excellent source of chemical energy; in fact, they yield more energy than equal weights of carbohydrates such as glucose. The goal of synthesis is the opposite of the goal of degradation, and the two processes would not be compatible if both occurred openly in the cytosol. Eukaryotic cells solve the problem by locating the reactions that degrade fatty acids in the mitochondria and the synthesis of fatty acids in the cytosol.

Finally, compartmentalization of cellular functions allows for intracellular processing and "sorting" of some of the synthesized molecules. We will discuss shortly some of the sorting and processing that occurs in the endoplasmic reticulum (ER) and the Golgi complex. For example, proteins destined to become components of membranes are trapped within the ER membrane during synthesis, whereas proteins destined for export from the cell are discharged into the lumen of the ER, where they are attached to carbohydrates to become glycoproteins. From there, they are transported to the Golgi complex, where the carbohydrate portion of the molecules is chemically modified. The Golgi complex also carries out another sorting step based on the specific structures present in the glycoprotein molecules. It distinguishes between the molecules that must be packaged inside secretory vesicles for export from the cell, and the molecules that are destined to become lysosomal enzymes and must be directed into primary lysosomes for use inside the cell.

Cell Compartments and Their Functions

The Nucleus: The Library of the Cell

The nucleus (Fig. 3–4) is the site where DNA and RNA are made. The information stored in DNA and RNA molecules directs the synthesis of proteins that determine the shape and function of a cell. (The red blood cell is the only common mammalian cell that lacks a nucleus, and it survives for only a few months.) The nucleus also contains proteins. The approximate composition of the nucleus in

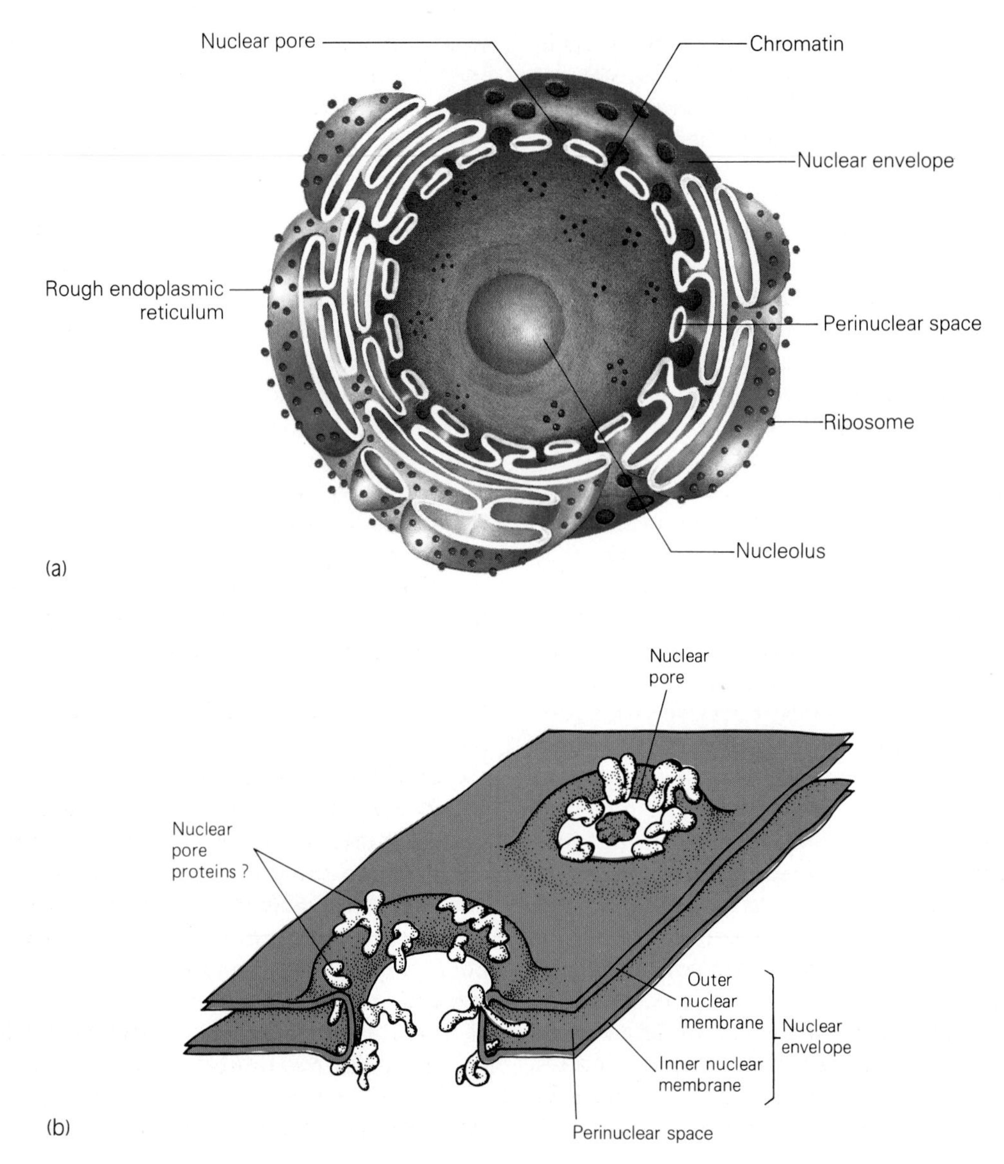

Figure 3–4 (*a*) The cell nucleus is enclosed in a double membrane called the nuclear envelope. Pores in the envelope permit the passage of molecules in and out of the nucleus. The outer layer of the nuclear envelope is continuous with the endoplasmic reticulum, so that the lumen of the ER is continuous with the perinuclear space. In the nondividing nucleus, DNA is visible as chromatin. The nucleolus plays a role in the synthesis of ribosomes from RNA. (*b*) A nuclear pore is formed from the fusion of the two layers of the nuclear envelope. Proteins are thought to be located in the pores.

a liver cell, expressed as a percentage of the dry weight of the nucleus, is 80% protein, 15% DNA, and 5% RNA.

DNA: The Permanent Files

Recall from Chapter 2 that DNA (deoxyribonucleic acid) is an enormously long, unbranched, double strand of nucleotides arranged in an alpha-helix. A typical animal cell contains more than 1 meter of DNA, all of which must fit inside a nucleus with a diameter of only 0.000006 meter (6 μm). Each section of the chromosomal DNA that carries the information necessary for synthesis of a single polypeptide is known as a **gene.** The sum of all the chromosomal genes is termed the **genome** of the cell. All cells in the body, except the reproductive cells, normally contain the same DNA. The DNA molecules remain unchanged throughout the life of the organism unless they are damaged in some way.

The tight packing of DNA (Fig. 3–5) is achieved by specialized proteins called **histones,** which bind to the DNA and organize its structure in the nucleus. Histones are by far the most common protein found in the nucleus. They are rich in amino acids with basic (positively charged) side chains, such as lysine and arginine (see Fig. 2–26). These positively charged side chain groups may be able to interact with and bind to the negatively charged phosphate groups in DNA (see Fig. 2–33), regardless of the specific nucleotide sequence.

Chromatin. Most of the DNA is associated with histones and other nuclear proteins, giving rise to a complex known as **chromatin.** The basic packing unit of chromatin is the **nucleosome** (Fig. 3–5*b*), in which parts of the DNA double helix are wrapped around histone molecules and linked by stretches of "free," or **linker,** DNA. The nucleosome is still a relatively extended form of chromatin, and it is unlikely that much of the chromatin in the nucleus is present in this form.

A higher level of organization is produced by the packing together of nucleosomes in a regular array known as a **chromatin fiber.** Loops in these fibers give rise to **looped domains,** which reduce the initial length of the DNA double helix about 500-fold (Fig. 3–5*d*).

Chromosomes. Further levels of folding must occur, perhaps involving close packing of the loops, in order to condense the DNA into a single **chromosome** (Fig. 3–5*f*). Thus, each chromosome of a eukaryotic cell consists of a single very large molecule of DNA arranged so that the double helix is folded in a compact and highly organized way. Each species has a characteristic number of chromosomes: For example, human cells have 23 pairs. When cells are not dividing, the chromosomes are in a relatively uncoiled state and are not readily identifiable as individual units. Only when cell division is about to occur do the chromosomes become apparent.

RNA: Selective Retrieval of Information

The RNA (ribonucleic acid) present in the nucleus is destined for export into the cytoplasm, where the machinery for protein synthesis is located. In cells such as rat liver cells, less than 10% of the total RNA in the cell is located at any one time in the nucleus.

An RNA molecule consists of a single strand of nucleotides that represent a copy of a limited region of the nucleotide sequence of one of the strands of DNA. Three major functional types of RNA interact in the cytoplasm to synthesize proteins according to the information stored in DNA: **messenger RNA (mRNA); ribosomal RNA (rRNA);** and **transfer RNA (tRNA).** For now, the related roles of these three types of RNA can be summarized as follows (more detailed discussions will follow in Chapter 5): Ribosomal RNA is RNA that joins with certain proteins in the cytosol to form ribosomes, the protein-making structures that occur in the cytosol or are attached to the endoplasmic reticulum. Messenger RNA is RNA that contains the information about which amino acids should be joined together, and in which specific sequence, on the ribosomes. Finally, transfer RNA seeks out the required amino acids in the cytosol and brings them to the ribosome to be assembled into polypeptide chains.

One of the most obvious structures within the nucleus is the **nucleolus,** which is particularly prominent in cells that synthesize a lot of protein. The nucleolus is not limited by a membrane; it con-

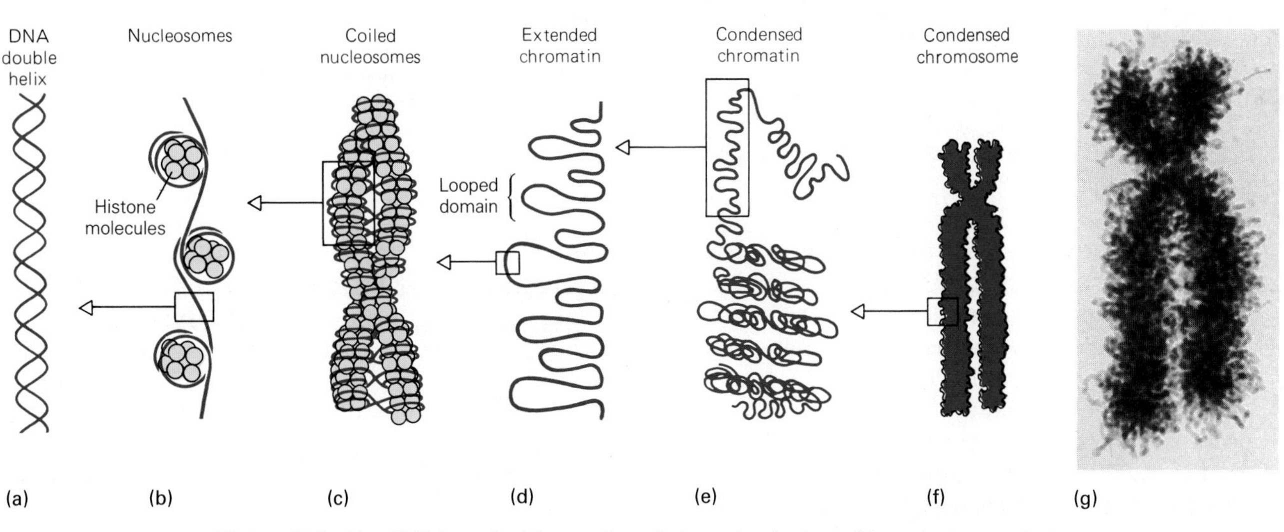

Figure 3–5 The DNA is packed in a series of stages beginning with nucleosomes. In a nondividing cell, DNA appears as chromatin; in a dividing cell, it appears in the more tightly coiled form of chromosomes.

tains a lot of protein and large loops of DNA from which rRNA is rapidly synthesized. Cells begin assembling rRNA and proteins into ribosomes in the nucleolus. However, ribosomes are not mature, or completely formed, until they reach the cytoplasm. This aspect of ribosome formation will be discussed in Chapter 5.

The Nuclear Envelope: A Porous Double Membrane

The interior of the nucleus is separated from the cytoplasm by the **nuclear envelope,** a set of two separate membranes. These membranes are separated by a space about 10 to 50 nm wide termed the **perinuclear space.** The outer membrane of the nuclear envelope often has attached ribosomes that are engaged in protein synthesis. This outer membrane appears to be continuous with the endoplasmic reticulum (Fig. 3–4*a*).

Much exchange occurs between the interior of the nucleus and the cytoplasm. Since DNA and RNA are synthesized only in the nucleus, the individual nucleotides and other parts required for their synthesis must be obtained from the cytoplasm. Conversely, RNA molecules and ribosomes made in the nucleus must move out to the cytoplasm in order to carry out their roles in protein synthesis. The rapid exchange of materials between the nuclear and cytoplasmic compartments probably occurs through the **nuclear pores.** These are open channels in the nuclear envelope that form in areas where the inner and outer membranes are pinched together (Fig. 3–4*b*). A mammalian cell has about 3000 to 4000 pores per nucleus. The diameter of each pore is 9 nm. This size permits the passage of small water-soluble molecules but excludes large macromolecular structures.

The Cytoplasm: Synthesis and Energy Release

The normal structure, growth, and function of a cell requires a constant supply of macromolecules, which are made from simpler molecules, or **precursors.** Synthesis of precursors such as amino acids or nucleotides, for example, is achieved by a set of chemical reactions that are part of the cell metabolism. Also important to intermediary metabolism are the reactions that break down small molecules and trap their chemical energy as ATP, as discussed in the previous section. All the reactions of metabolism take place outside the nucleus (Fig. 3–1).

The Cytosol: A Wealth of Enzymes and Precursors

The cytosol of a mammalian liver cell is a large compartment, representing about 50% of the total volume of the cell. It contains the small molecules that are precursors for macromolecules. Also, it contains thousands of enzymes, which account for its high protein content (about 20% by weight). Many of the enzymes are catalysts for the reactions of cell metabolism. Thus, the cytosol resembles a highly viscous fluid rather than a watery solution.

The cytosol contains some structures that are not organelles but are worthy of mention, including glycogen granules and lipid droplets. **Glycogen,** a polysaccharide, is the storage form of carbohydrate. The lipid droplets contain **triglycerides,** which are the storage form of fatty acids (see Chapter 2).

Ribosomes and the Endoplasmic Reticulum: Synthesis of Proteins and Lipids

Only one endoplasmic reticulum (ER) is present in each cell. It is a membrane sheet, folded repeatedly to form a large interconnected system (see Fig. 3–2). Two regions of the ER are structurally and functionally different, however: the **rough,** or **granular, ER** and the **smooth,** or **agranular, ER** (Table 3–1). The rough ER is so called because ribosomes are bound to its outer surface; smooth ER has no ribosomes attached.

The ER makes new proteins and lipids, including molecules destined to be released outside the cell by a process known as **secretion.** Rough ER is particularly extensive in cells that synthesize and secrete a lot of proteins, such as certain hormones. Smooth ER is well developed in cells specializing in lipid synthesis or the synthesis and secretion of steroid hormones. Muscle cells have a highly specialized smooth ER known as the **sarcoplasmic reticulum.** Its principal function is to remove calcium ions from the cytosol.

Table 3–1
Structural and Functional Differences Between Rough and Smooth Endoplasmic Reticulum (ER)

	Rough ER	Smooth ER
Ribosomes present	Yes	No
Protein synthesis	Yes	No
Lipid synthesis	Yes	Yes
Steroid synthesis	No	Yes

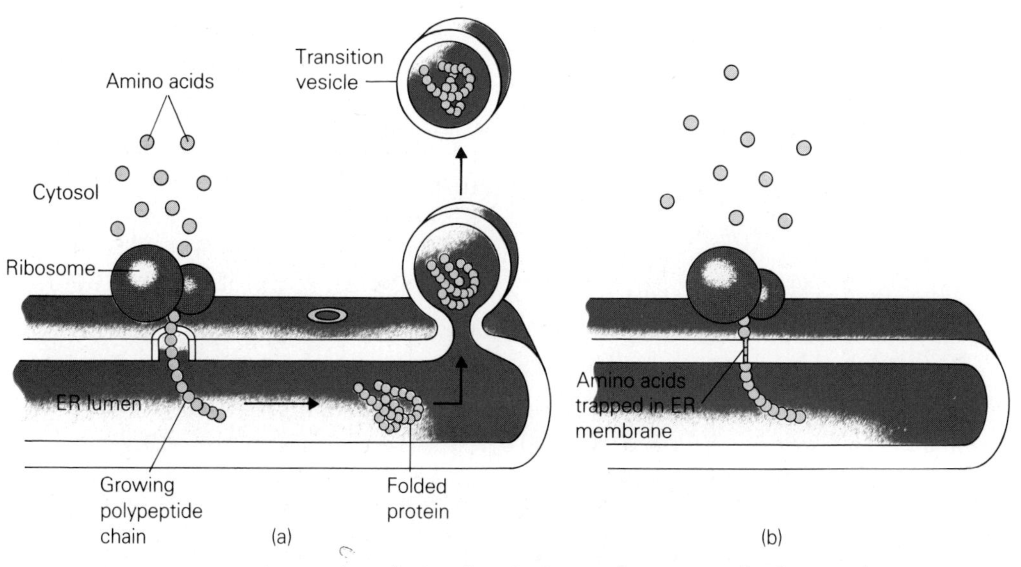

Figure 3–6 In the rough endoplasmic reticulum, a ribosome synthesizes a polypeptide from amino acids found in the cytosol. (*a*) A protein destined to be a secretory product or an enzyme enters the ER lumen as it is assembled, folding into its various levels of structure. It is enclosed in a transition vesicle for transport to the Golgi complex. (*b*) A protein destined to be a membrane protein is partially trapped in the ER membrane as it is synthesized.

Protein Synthesis on the Rough ER. Individual amino acids are assembled into polypeptides on ribosomes. Some of the ribosomes occur freely in the cytosol, but many are attached to the ER. The sequence of the amino acids in each polypeptide is dictated by mRNA. Polypeptide chains formed in the ribosomes are directed into the lumen of the ER and separated from the cytosol by the ER membrane (Fig. 3–6). Most of the proteins synthesized in the rough ER become modified by the addition of monosaccharide units to form glycoproteins. In contrast, the proteins synthesized on free ribosomes in the cytosol are excluded from the ER lumen and cannot be modified by addition of carbohydrate.

The proteins that are destined to become structural components of membranes (see Chapter 4) are trapped within the rough ER membrane during synthesis. The mechanism for trapping the polypeptide chain may be a specific sequence of amino acids that interacts strongly with the interior of the ER membrane and prevents further movement of the polypeptide into the ER lumen (see Fig. 3–6). The part of the polypeptide that is synthesized after the trapping sequence is left on the cytosolic side of the ER membrane. Movement of ER membrane to other membrane-bounded organelles and to the plasma membrane will be discussed shortly.

Lipid Synthesis in the Smooth ER. The endoplasmic reticulum in many cells is the major site of the synthesis of lipids, including triacylglycerols, phospholipids, and steroids. Lipids from each of these three general classes form complexes with proteins in the ER lumen. The resulting structures are the compounds known as **lipoproteins.** The lipoproteins that are secreted from the cell and that enter the blood have a critical role in the transport and distribution of lipids, such as cholesterol, that are absorbed by the intestine.

Lipids such as phospholipids and cholesterol are also important structural components of membranes. The successive steps leading to the formation of a phospholipid occur with the ER membrane because the enzyme proteins that catalyze the synthesis are part of the membrane. Thus, synthesis of phosphatidylcholine (see Fig. 2–23*c*) occurs on the cytosolic surface of the ER membrane; the molecules of phosphatidylcholine that are produced will remain within the ER membrane (Fig. 3–7). Clearly, synthesis of the proteins and lipids that are required for production of new cellular membrane structures is another important function of the ER.

A macromolecule formed in the ER to be secreted by the cell first moves to the Golgi complex, possibly via a transition vesicle formed from a portion of the ER membrane. In the Golgi complex, the macromolecule is processed and packaged for export out of the cell or for storage or use within the cell.

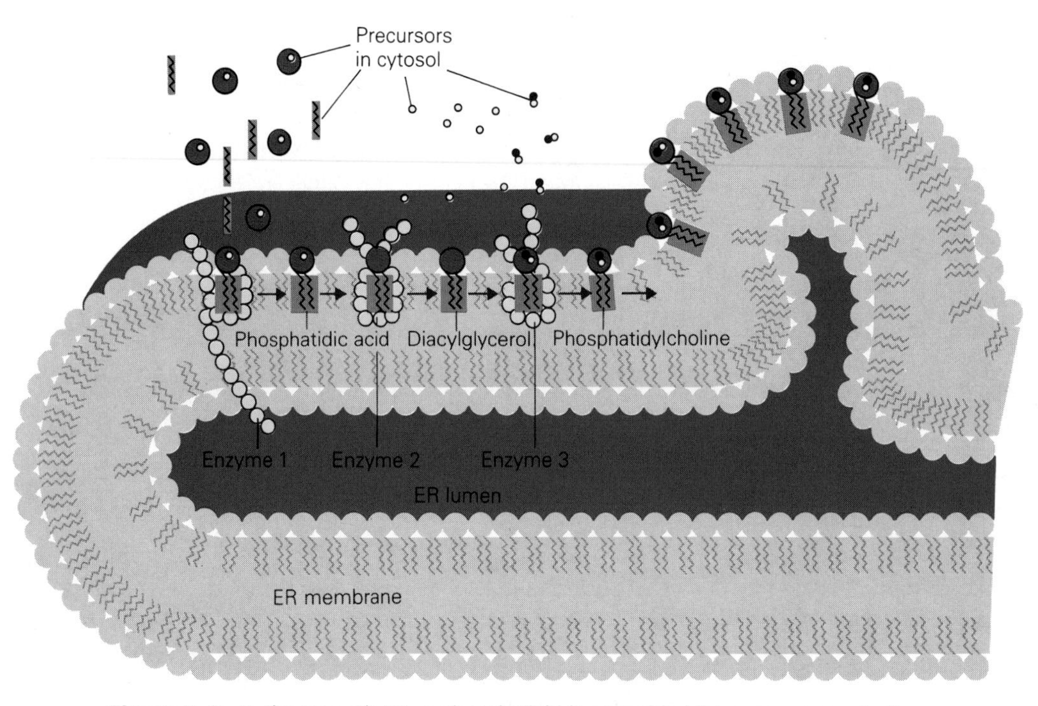

Figure 3–7 In the smooth ER, a phospholipid is assembled from precursors in the cytosol through a series of reactions catalyzed by enzymes located in the ER membrane. The developing phospholipid molecule moves through the layer of existing phospholipid molecules that make up one half of the ER membrane.

The Golgi Complex: Processing and Packaging

Generally, the function of the Golgi complex is to modify the macromolecules synthesized in the ER, especially in secretory cells. The Golgi complex processes and packages the macromolecules inside small membrane bags, or **vesicles,** ready for transport to the cell surface or the cytosol.

The Golgi complex is a collection of smooth membranes that look like flattened sacs. Each sac is called a **cisterna.** Commonly, about six cisternae will be stacked together to form a structure known as a **dictyosome** (Fig. 3–8). Small vesicles surround and radiate away from the dictyosome.

The **outer,** or **forming, face** of the dictyosome is oriented toward and closely associated with rough ER. The two structures are thought to be connected by the small membrane-bounded **transition vesicles** that shuttle newly synthesized macromolecules from the rough ER to the Golgi complex (see Fig. 3–8).

The **inner,** or **mature, face** of the Golgi complex, on the other hand, is oriented toward the plasma membrane and away from the nucleus, especially in secretory cells. Membrane-bounded **secretory vesicles** appear to arise from the inner face (see Fig. 3–8) to transport newly processed and packaged macromolecules from the Golgi complex to the plasma membrane for export.

There is good evidence that the Golgi complex modifies the macromolecules destined for secretion. One of the important activities involves modification of the polysaccharide chains that were added covalently to the proteins in the rough ER. The processing of these glycoproteins within the Golgi complex involves a considerable amount of "trimming" as well as the addition of new monosaccharide units.

The form of the Golgi complex can vary considerably. In some cell types it may be compact and limited, whereas in others it may be more spread out, almost like a net. The position of the Golgi complex within the cytosol may also vary. In nerve cells it often surrounds the nucleus, whereas in secretory cells it is usually found between the nucleus and the side of the cell that is engaged in the secretory process. The number of dictyosomes in a cell varies considerably with the type of cell. The Golgi complex in a liver cell accounts for about 7% of the total cell membrane, and in this cell type it is a much less extensive structure than the ER.

Secretory Vesicles: Exocytosis and Membrane Recycling

A secretory vesicle containing a packaged macromolecule moves from the inner face of the Golgi complex to the plasma membrane. The vesicle mem-

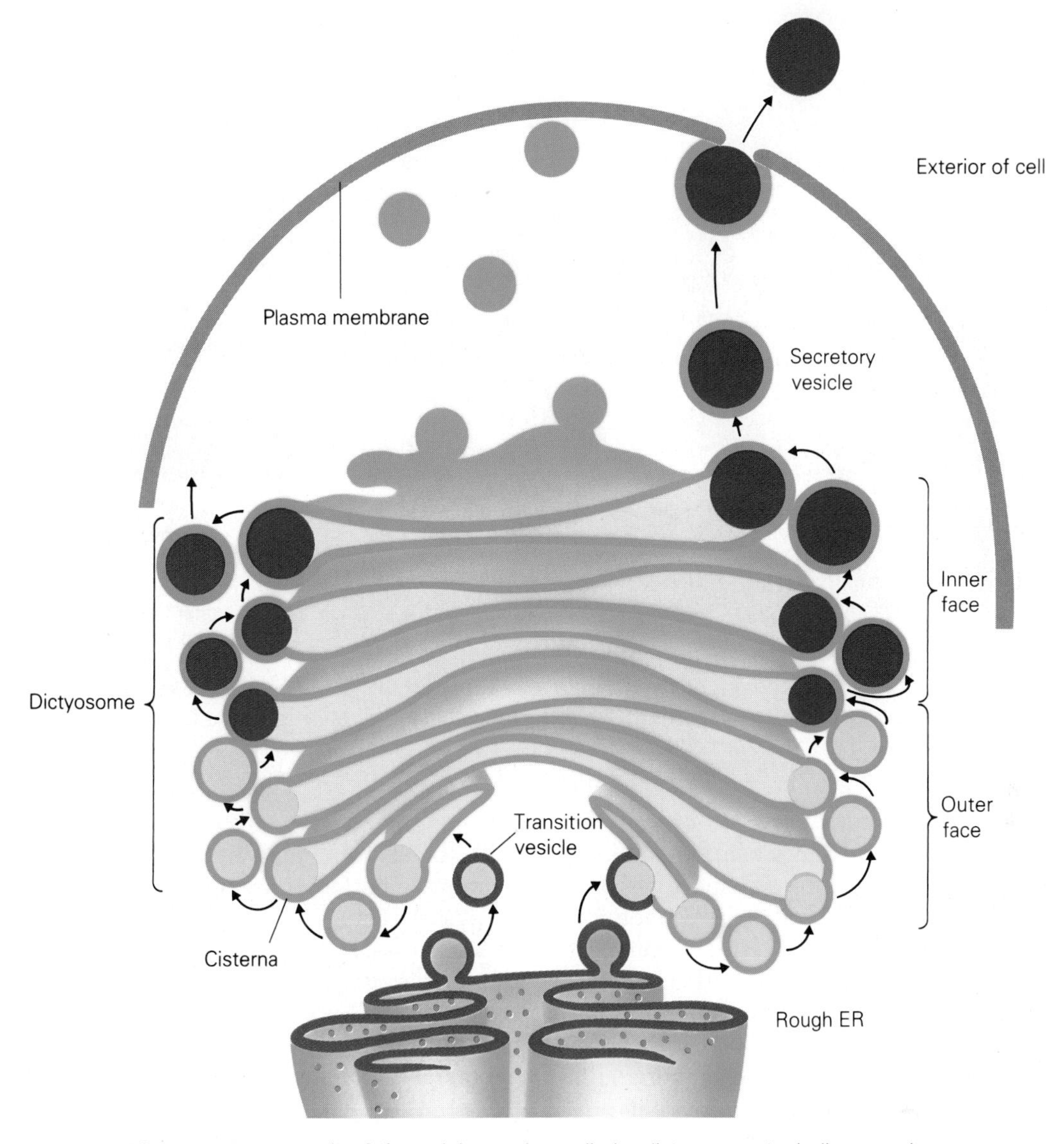

Figure 3–8 One unit of the Golgi complex, called a dictyosome, typically comprises about six folded sacs called cisternae. Macromolecules formed in the ER enter the outer face of the Golgi complex and pass through to the inner face via a series of vesicles. In the process, the lipids or proteins are modified. They leave the Golgi complex in secretory vesicles.

brane fuses with the plasma membrane in a process known as **exocytosis,** in which the secretory vesicle is allowed to open and expel its contents into the extracellular fluid. The vesicles are often designed so that the exocytotic step occurs very rapidly whenever the plasma membrane is stimulated from the outside by a signal such as a specific hormone.

At the same time that the products of ER synthesis and Golgi complex packaging are conveyed to the exterior of the cell, the membrane of the transition vesicle becomes part of the plasma membrane. Recall that some proteins have been synthesized in the ER for use as membrane proteins. They have also been packaged in the Golgi complex and form the membranes of the same transition vesicles that carry the proteins destined for export. In this way, the new membrane proteins and lipids synthesized and trapped within the ER membrane can be transported to the cell surface and inserted into the plasma membrane by the fusion step to replace damaged or "worn out" proteins and lipids.

Not all of the membrane lost from the inner face of the Golgi as secretory vesicles is replaced by new membrane arriving at the outer face in the form of transition vesicles from the ER. Some of the replacement membrane is derived from pieces of the plasma membrane that are internalized and delivered to the inner face of the Golgi by processes such as **endocytosis,** which are the reverse of exocytosis (Fig. 3–9). The internalized membrane may be returned to the plasma membrane by a subsequent exocytotic step. Exchange of membrane in this man-

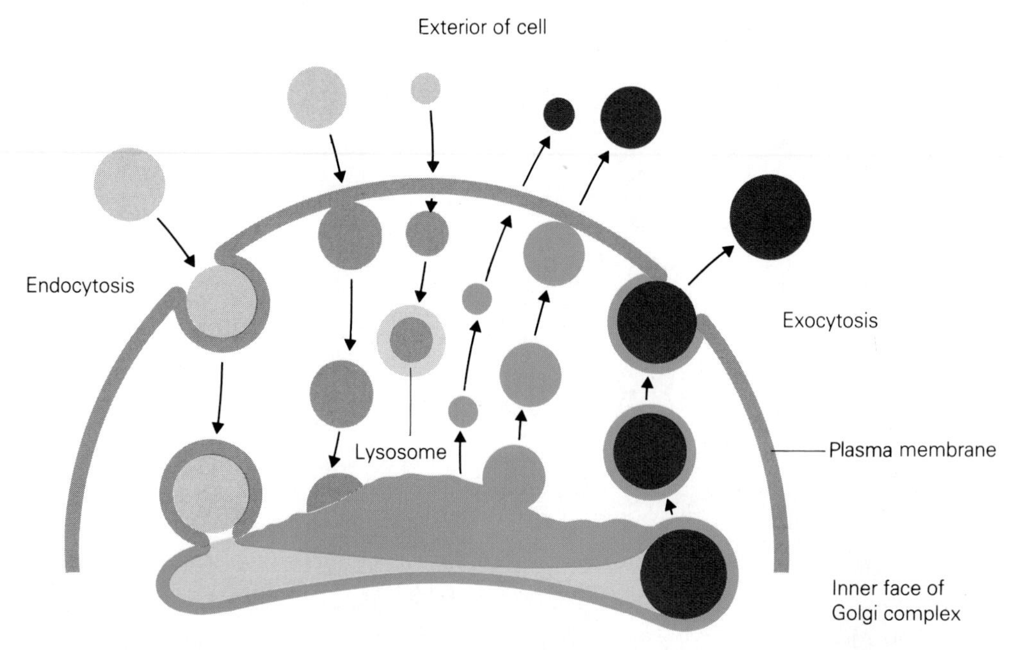

Figure 3–9 As portions of the Golgi complex fuse with the plasma membrane in the secretion process, portions of the plasma membrane are returned to the Golgi complex through various endocytotic events. Thus, the plasma membrane maintains the same size despite the continual addition of membrane from secretory vesicles.

ner between the Golgi and the plasma membrane conserves the membranes of the transport vesicles, which may be highly specialized. It is economical for the cell to use them for several series of transport events rather than to synthesize new vesicle membranes each time. This **membrane recycling** mechanism can provide the cell with reusable raw materials because some of the internalized membrane is directed to **lysosomes.** There the component proteins and lipids are taken apart and enter the cytosolic pool of precursor molecules to await a new round of synthesis. Membrane recycling also helps to avoid an increase in the surface area of secretory cells when large numbers of secretory vesicles fuse with the plasma membrane.

Lysosomes and Peroxisomes: Digestion of Food and Waste

Lysosomes and peroxisomes are small membrane-bounded bags, about 0.5 μm in diameter. Together they account for less than 1% of the total membrane in a cell, but they are found in almost all cells in large numbers. There are about 300 to 400 of each type of organelle in one mammalian liver cell.

Lysosomes. The **lysosome** is an organelle that contains about 50 different digestive enzymes. As a group, these enzymes allow cells to break down completely any of the macromolecules in the body (Fig. 3–10). Since these enzymes would destroy the cells themselves, they are kept within a specific membrane-bounded organelle. Two general classes of lysosomes have been distinguished: newly formed **primary lysosomes** contain only the digestive enzymes; **secondary lysosomes** contain both enzymes and molecules (substrates) that the cell needs to break down. (Recall that "substrate" is a general term for any material on which an enzyme is designed to work.)

Primary lysosomes are thought to arise by budding from the inner face of the Golgi complex (Fig. 3–11). Most of the lysosomal enzymes they contain are glycoproteins that originate in the rough ER. When they enter the Golgi complex, the polysaccharide parts of the enzyme molecules present a unique signal to the Golgi complex, which responds by packaging these enzymes into the vesicles that become primary lysosomes.

There are various types of secondary lysosomes. Among the largest are the **digestive vacuoles** that arise after a primary lysosome fuses with an endocytotic vesicle containing particulate material such as a bacterium. Internalization of particulate material is a special type of endocytosis known as **phagocytosis.** The resultant endocytotic vesicle is described sometimes as a **phagosome.** The process of fusion allows the digestive enzymes of the primary lysosome to gain access to the material to be broken down, whether it is bacteria or food.

After the lysosomal enzymes have performed their digestive task, the products of digestion can

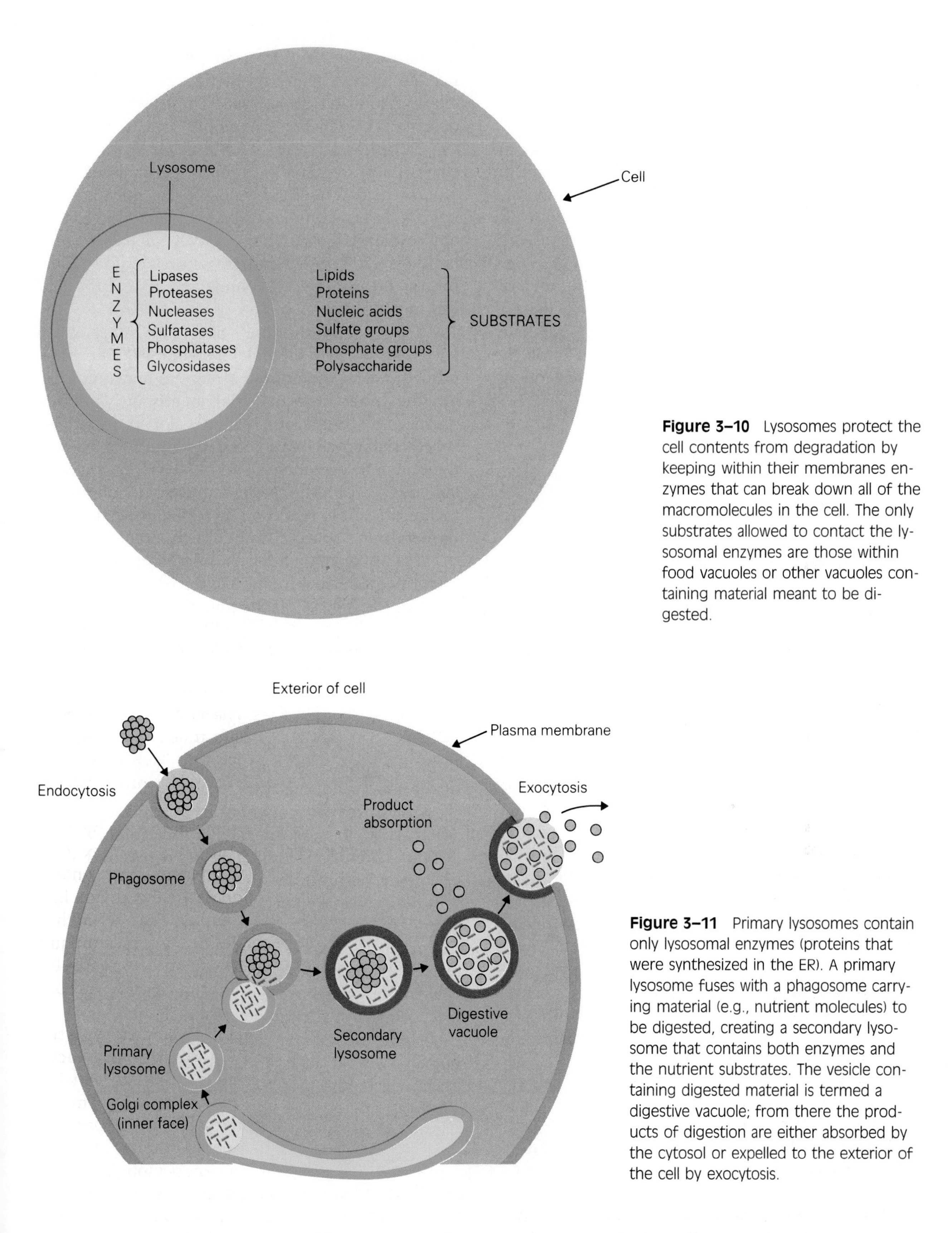

Figure 3–10 Lysosomes protect the cell contents from degradation by keeping within their membranes enzymes that can break down all of the macromolecules in the cell. The only substrates allowed to contact the lysosomal enzymes are those within food vacuoles or other vacuoles containing material meant to be digested.

Figure 3–11 Primary lysosomes contain only lysosomal enzymes (proteins that were synthesized in the ER). A primary lysosome fuses with a phagosome carrying material (e.g., nutrient molecules) to be digested, creating a secondary lysosome that contains both enzymes and the nutrient substrates. The vesicle containing digested material is termed a digestive vacuole; from there the products of digestion are either absorbed by the cytosol or expelled to the exterior of the cell by exocytosis.

either be released outside the cell by exocytosis for re-use by other cells or absorbed from the digestive vacuole and used again within the same cell. Lysosomes are used not only to digest material originating from outside the cell, but also to break down the cell's own organelles and other matter in a process known as **autophagy.** This process can be a response to cell injury, or it can occur during "remodeling" of the cell or during adverse nutritional conditions. The secondary lysosomes formed during autophagy are called **autophagic vacuoles.**

Peroxisomes. **Peroxisomes** also contain specialized enzymes. These enzymes are synthesized in the cytosol and transported into peroxisomes as they form, probably as buds from the smooth ER. The enzyme **catalase** is contained almost exclusively within the peroxisomes. Other enzymes within peroxisomes use oxygen (O_2) to carry out oxidative reactions. One of the products of these reactions is hydrogen peroxide (H_2O_2), which is potentially harmful to the cell. Catalase helps to degrade the hydrogen peroxide to harmless water by the following reaction:

$$2\ H_2O_2 \xrightarrow{\text{Catalase}} 2\ H_2O + O_2$$

The peroxisomes in many body cells are important in detoxifying a number of molecules.

Mitochondria: Energy Transformation

The mitochondrion is thought of as a "powerhouse" for the cell because its metabolic activities provide the energy needed to sustain the life of mammals and other **aerobic** (oxygen-breathing) organisms.

The principal fuels for most life forms are carbohydrates. After digestion, carbohydrates are stored in cells in the form of glycogen. In the cytoplasm, glycogen is broken down into glucose. The primary mechanism for processing these molecules and extracting the energy stored in them is **aerobic respiration.** This metabolic pathway is completed within the mitochondrion by a process that uses most of the oxygen that mammals take in with each breath. During aerobic respiration, about 36 molecules of adenosine triphosphate (ATP) are produced for each molecule of glucose that is broken down. Without the participation of the mitochondrion, only 2 molecules of ATP are produced. Clearly, the mitochondrion is the major site of ATP production within the cell. ATP is the form in which the cell is able to trap chemical energy and transfer it to sites where it is needed to drive many synthetic reactions and other energy-requiring processes. These processes are described in detail in Chapter 6.

Mitochondria are much smaller than the nucleus. They are usually cylindrical and about 0.5 to 1.0 μm in diameter. The number of mitochondria per cell varies with the cell type but animal cells usually contain large numbers. A liver cell may have as many as 1700 mitochondria, which would account for one fifth of the cell volume and 40% of the total cell membrane.

The mitochondrion, like the nucleus, is bounded by two separate membranes. The **outer mitochondrial membrane** is smooth and unfolded, but the **inner mitochondrial membrane** is distinguished by numerous well-developed infoldings called **cristae,** which project deep into the interior of the organelle and increase its surface area considerably (Fig. 3–12). The two membranes divide the mitochondrion into two compartments, the narrow **intermembrane space** and the much larger **matrix.**

The inner mitochondrial membrane is a highly impermeable structure, unlike the outer membrane. Most of the "work" of the mitochondrion is carried out by enzymes and other molecules located in the inner membrane and the matrix. The matrix compartment also contains ribosomes and 5 to 10 small double-helical DNA molecules that the mitochondria use to synthesize some of the mitochondrial proteins. Mitochondrial DNA has a circular structure. In mammalian cells, it makes up less than 1% of the total cellular DNA.

The major component of mitochondria, besides water, is protein. Most (67%) of the protein of liver cell mitochondria is located in the matrix, and about 21% is in the inner membrane. Many of these protein molecules are specific enzymes; about 120 different enzymes are present. Some of the enzymes in the matrix participate in a series of reactions known as the **citric acid cycle,** also referred to as the **Krebs cycle** or the **tricarboxylic acid cycle.** This cycle is the final common pathway for the breakdown of all the fuel molecules in the cell, not only carbohydrates but also amino acids and fatty acids. These processes will be discussed in detail in Chapter 6.

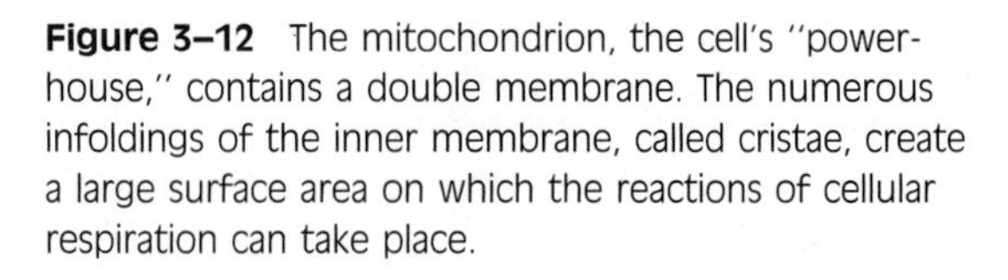

Figure 3–12 The mitochondrion, the cell's "powerhouse," contains a double membrane. The numerous infoldings of the inner membrane, called cristae, create a large surface area on which the reactions of cellular respiration can take place.

The Cytoskeleton: Shape and Structure

Specialized proteins in the cytosol are organized into the **cytoskeleton** of the cell, which is responsible for maintaining the cell shape and the positions of the internal organelles and for mediating movement of the cell itself or of the organelles within the cell. Specific examples of organelle movement that have been discussed already in this chapter are (1) the movement of secretory vesicles from the Golgi complex to the plasma membrane and (2) the movement of endocytotic vesicles or phagosomes from the plasma membrane to the cell interior.

The cytoskeleton is made up of a dense network of different types of protein fibers, the **microtubules, intermediate filaments,** and **microfilaments.** Their structures and sizes are compared in Table 3–2. Microtubules and intermediate filaments are found in deeper regions of the cell, whereas microfilaments generally are located just beneath the plasma membrane.

Table 3–2
Cytoskeletal Fibers

	Diameter (nm)	Structure	Protein Subunit
Microtubules	24	Hollow	Tubulin
Intermediate filaments	10	Hollow	Several types
Microfilaments	6	Solid	Actin
Myosin filaments	15	Solid	Myosin

Microfilaments also extend into cell processes. Good examples are the fingerlike projections called **microvilli,** which cover the inner surfaces of the intestine and the kidney proximal tubule. Figure 3–13 (drawn to scale) gives an indication of the huge size of the cytoskeletal protein fibers relative to the protein molecules present in the membrane of the microvillus.

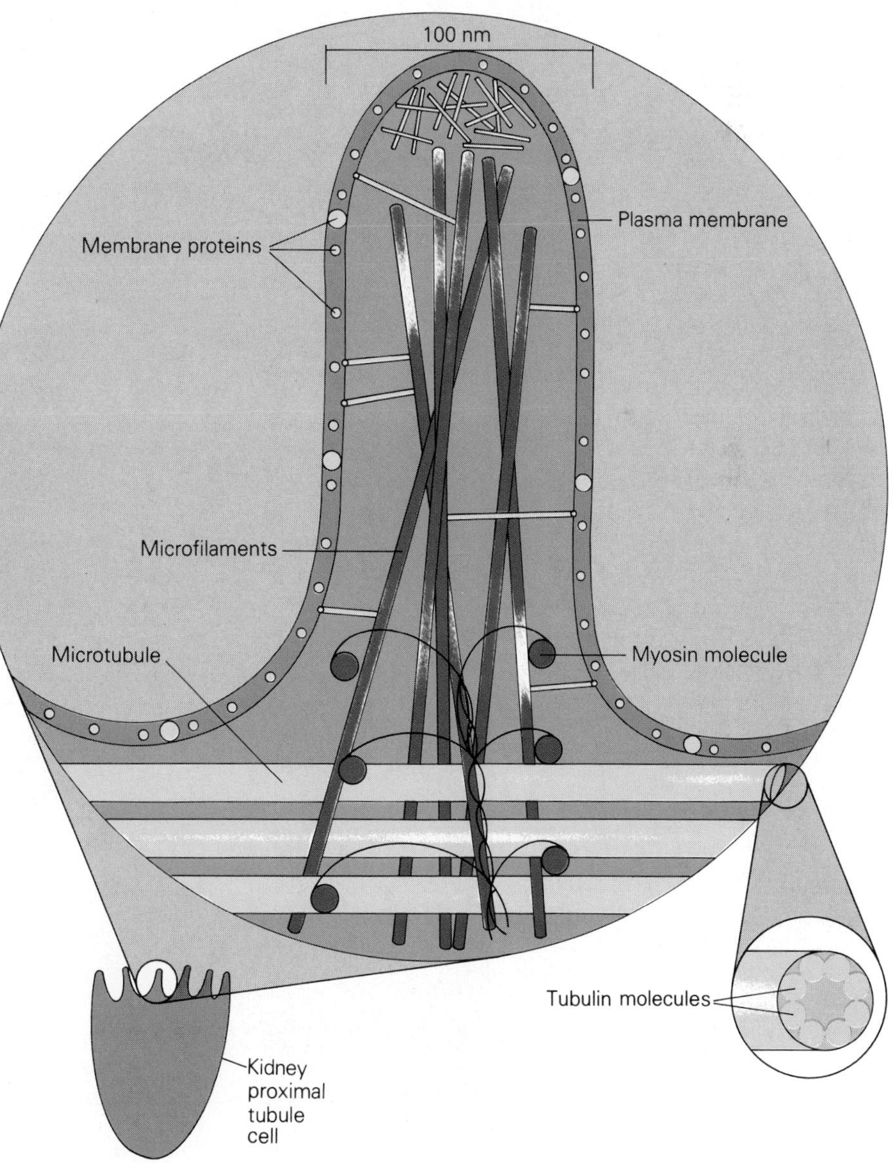

Figure 3–13 Microvilli are found in cells of the digestive system and kidney. The internal structure of a microvillus, drawn to scale, shows that the microtubules and microfilaments that create its fingerlike shape are relatively large structures in the cell. They are composed of and connected by protein molecules (e.g., actin, myosin, tubulin) specialized for contraction and support. For example, a microtubule is formed by a circular collection of tubulin molecules. Tubulin is a protein that can be detected through the use of fluorescent antibodies (see Fig. 3–17).

Microtubules and microfilaments can be described as dynamic structures because their formation is readily reversed (and often is), and because the distribution of these structures within a cell is subject to change as the physiological conditions change. In contrast, the intermediate filaments are much more stable structures, and their formation may be irreversible. Most of the cytoskeletal fibers are made up of large numbers of a single type of protein molecule or subunit (see Table 3–2).

Myosin filaments (also known as **thick filaments**) are a fourth type of fiber found in almost all cell types. They are most abundant in muscle cells where, together with the thin actin filaments, they form part of the contractile machinery of these cells. The presence of **actin** and **myosin** molecules in nonmuscle cells suggests that nonmuscle cells may possess a contractile mechanism such as that of muscle cells and use it for some internal movements.

The organization of the cytoskeleton is complex because the major fibers are linked by another set of thin filaments, the **microtrabecular lattice.** These filaments appear to hold the fibers and the internal organelles in their places (Fig. 3–14). They even

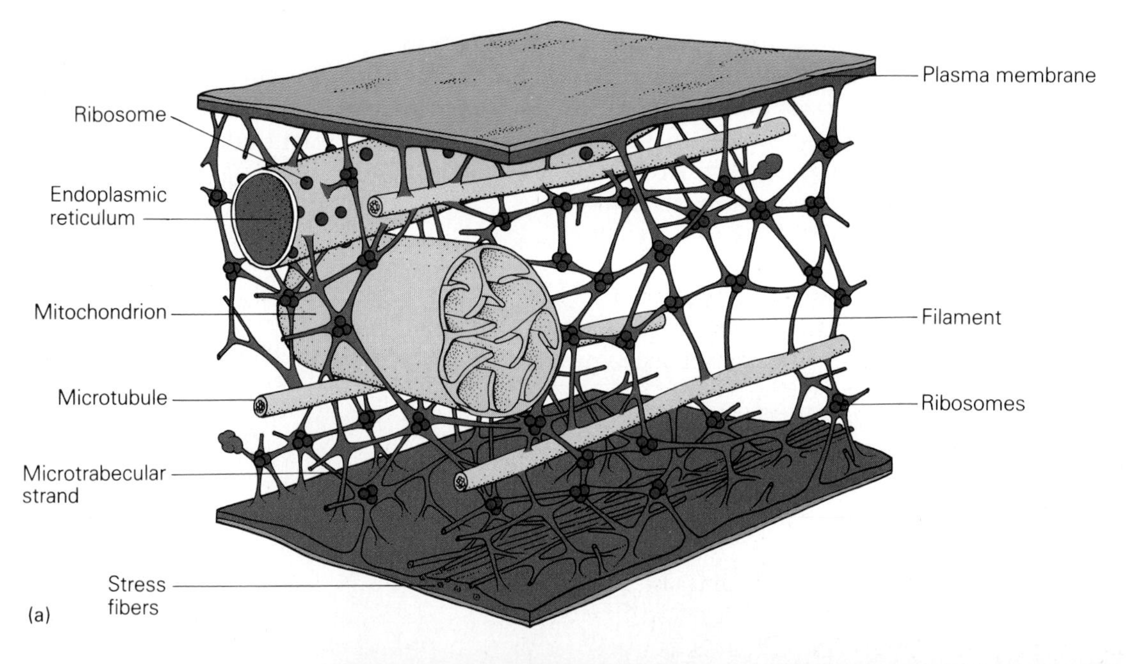

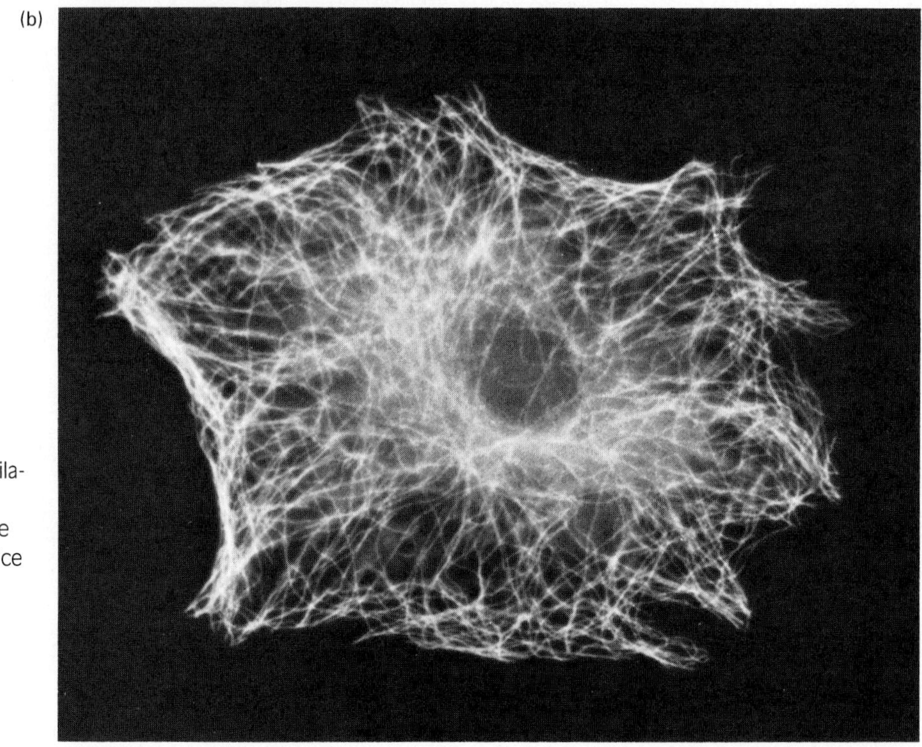

Figure 3–14 Microtubules and filaments are found throughout the cell, not only in microvilli. They are linked by the microtrabecular lattice (*a*) that shapes the cell and holds the internal organelles in place in the cytosol. (*b*) Fluorescent antibodies show the position of microtubules in the cell (see also Fig. 3–17).

hold ribosomes in suspension. The microtubules may have a major influence on the overall design of the cytoskeleton because they appear to influence the distribution of microfilaments and intermediate filaments. They may provide the framework on which the rest of the cytoskeleton is erected. The molecular mechanisms by which the cytoskeleton can mediate the movement of cells and the movement of organelles within cells have not been determined.

The Life Cycle of the Cell

The many different cell types present in the human body have different patterns and rates of division. Cells such as red blood cells and nerve cells stop dividing when they reach maturity, whereas certain epithelial cells in the intestine and skin divide rapidly and continuously.

The **cell cycle** is the period of time from the beginning of one cell division to the beginning of the next division (Fig. 3–15). Cell cycle times vary from just 8 hours for rapidly dividing cells to more than 100 days for rarely dividing cells.

The cell cycle consists of two main phases: interphase and cell division. **Interphase** is the period between the end of one cell division and the beginning of the next, during which the cell carries out all the normal cell processes of growth and metabolism, including the replication of DNA. **Cell division** is the period during which the nucleus and then the cytoplasm divides to form two new cells, each containing one of the copies of the original DNA produced during the preceding interphase.

The process of cell division, sometimes called the **"M"-phase,** involves two steps, mitosis and cytokinesis. In **mitosis,** the nucleus divides into two new nuclei. In **cytokinesis,** the cytoplasm divides, each half taking with it one of the two new nuclei to form a new cell.

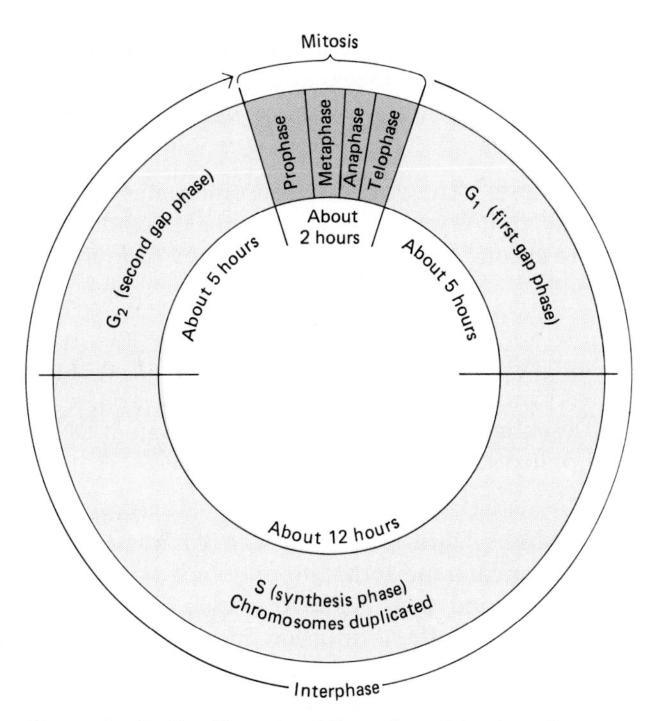

Figure 3–15 The life cycle of the cell contains two phases: interphase and mitosis. During interphase, the cell carries out normal metabolic activities and duplicates DNA. During mitosis, the cell divides in four steps called prophase, metaphase, anaphase, and telophase.

Interphase: Growth and DNA Replication

Much of the preparation for cell division occurs during the growth phase of the cell cycle, a phase that is known as **interphase** (see Fig. 3–15). It occupies 90% or more of the cell cycle. Interphase itself is divided into three phases: the synthesis, or S, phase, and two gap, or G, phases.

The period of interphase that directly follows cell division is the **first gap,** or **G_1, phase.** During this period the new cell begins its growth as the organelles it inherited from the parent cell resume their biosynthetic activities (which slowed during the M phase).

The **S,** or **synthesis, phase** follows the first gap phase. This is the part of interphase during which the cell synthesizes DNA in the nucleus in order to duplicate the DNA in preparation for cell division.

Directly following the S phase is the **second gap phase (G_2),** during which the cell increases protein synthesis in final preparations for division.

The time taken to progress from the beginning of the S phase to the end of the M phase is markedly similar in all cell types. The principal difference between cells that have a short cell cycle and those that have a long cell cycle is the amount of time spent in the G_1 phase; slowly dividing cells stay in G_1 for days or years.

Cell Division: Mitosis and Cytokinesis

The process of mitosis occurs in four stages: prophase, anaphase, metaphase, and telophase. Each of the stages merges smoothly with those coming before and after. The stages are distinguished mainly by the different movement of the DNA.

Prophase. In the first state of mitosis, called **prophase,** the DNA duplicated during the preceding S phase condenses into chromosomes. The nuclear membrane breaks down, and the contents of the nucleus become mixed with the cytoplasm.

Metaphase. The duplicated chromosomes, not yet separated, line up at the center of the cell and attach to the **mitotic spindle,** which is formed from microtubules.

Anaphase. The third stage begins when the duplicated chromosomes separate. The two sets, each representing a complete copy of the original DNA, are pulled to opposite ends of the cell.

Telophase. During the final stage of mitosis, a new nuclear membrane forms around each set of chromosomes, and they begin to uncoil. **Cytokinesis,** or division of the cytoplasm into two daughter cells, also occurs during this stage. Following cytokinesis, the new cells begin interphase.

Abnormal Cell Cycles

The rate of cell division in normal tissues is controlled so that the cells will divide only when new cells are needed. Removal of part of the liver, for example, will stimulate rapid division and growth of the remaining cells. This new growth ceases when the normal liver mass has been restored. The normal functions of tissues and organ systems would be destroyed rapidly if cell division were uncontrolled.

Cancer cells are the cells of any malignant **neoplasm.** "Neoplasm" means "new growth." The term *malignant neoplasm* is used because the cells divide and multiply rapidly and behave abnormally. In contrast to the highly organized growth of normal cells, cancer cells grow upon one another, invading normal tissues and threatening their functions. Cancer cells do not respond to the normal controls of cell division and growth, and the body is simply a very favorable environment to support their growth.

A cancer cell probably develops when the DNA of a cell experiences **mutation,** or alteration, after exposure to radiation, chemicals, or viruses. When the cell divides, all the cells derived from it contain copies of the altered DNA and inherit the defect. An accumulation of several DNA mutations arising independently from various insults may complete the transformation of a normal cell into a cancer cell. The development of a means to arrest the growth of cancer cells requires a greater understanding of the communication systems of cells and their intracellular controls.

Research Techniques in Cell Physiology

The information about cell structure and function described in the preceding paragraphs has been obtained through a variety of research methods, some as simple as observation through a microscope, others involving the sophisticated analysis of complex chemical reactions. Following is a brief survey of the research methods still being used to elucidate the details of cell physiology.

Cell Cultures

Cells taken out of mammalian tissues will grow and multiply in the laboratory if provided with appropriate nutrients and other factors. Human cells have been used almost routinely for **cell culture** since 1952. Many types of cells can be grown as a single layer attached to glass or plastic surfaces and covered with a liquid **growth medium,** or mixture of nutrients. Under these conditions, the cells are ideally suited for direct study by various techniques of microscopy.

Alternatively, cells can be grown in suspensions in liquid media for the purpose of producing large numbers of cells. Special conditions are used to avoid bacterial growth in the liquid medium, and antibiotics are frequently added as an extra precaution. The growth medium contains glucose, various salts, amino acids, and vitamins and is often supplemented by serum derived from the blood of a horse or calf. (Serum is the fluid that remains in the blood after the blood cells have clotted together.) The precise composition of serum is always uncertain, but it is an adequate supplement to the growth medium for routine cell culture.

Cells that are transferred directly from the tissues of an animal to a plastic dish containing artificial growth medium are known as *primary* cultures. As they grow and multiply, by dividing to produce daughter cells, they can be **subcultured**—that is, the cells can be removed and dispersed among many new culture dishes to produce a large number of cells.

One of the purposes of growing cells in culture is to obtain cells that still exhibit the specific functions of the tissue from which they were removed. Ideally, a culture contains only one specific cell type. Thus, cell cultures have an advantage over studies of intact tissues, in which several different cell types, products, and processes are usually at work complicating the picture. When they contain only the cells under study, cell cultures are extremely useful for studies of unique cellular functions. The unique studies are possible because the cells are now growing under carefully controlled conditions and are far more accessible to study than when they were part of an intact tissue inside a living animal.

Most normal cells will divide only a limited number of times in culture before dying. Some researchers believe that studies of this built-in "senescence" will provide important clues about the aging process in humans and other animals. Occasionally, however, a few apparently normal cells will grow indefinitely in culture: they are known as a **cell line.** A cell line, and also a primary culture, can be stored frozen in liquid nitrogen (−200°C) for long periods of time, even years. When warmed to 37°C, some of the cells will resume growth and division. This ability is very useful because once a suitable group of cultured cells has been obtained, some of them can be stored frozen for studies in the future, creating an almost inexhaustible supply of cells.

Among their many research purposes, cell cultures are used in investigations of the interactions that allow similar cell types to "recognize" one another and to form the close associations that lead to the development of a specific tissue. When cells are cultured on a plastic surface, they grow and multiply only to the point at which most of the surface is covered by a monolayer of cells. Further growth is inhibited by close contact with other cells (so-called **contact inhibition**), and the monolayer is said to be **confluent.** Interestingly, cancer cells that grow in an uncontrolled way in the body exhibit the same growth pattern in culture. They are much less sensitive to contact inhibition of growth.

Finally, the study in culture of **mutant,** or abnormal, cells that cannot synthesize a specific protein often reveals a great deal of information about the role of that protein in normal cells because the specific functions that protein would normally direct are absent from the cultured mutant cell. The altered behavior of the mutant cell is a strong clue to the function of the protein in normal cells of the same type.

A word of caution should be added to this discussion. Cells in culture—especially cell lines that have been cultured for a number of years—are not growing in their normal physiological environment. Therefore, despite the usefulness of the technique, the behavior of cells in culture may be different from that in an intact tissue. The observations made on cultured cells must be checked eventually against cells in their normal situation.

Microscopy and Cytochemistry

First used to study cells and tissues in the seventeenth century, the microscope in its modern form is essential in physiological research. A significant advantage of microscopic studies is that the researcher can see the complete cell and its organelles in a relatively undisrupted state.

The Light Microscope

The **light microscope** (Fig. 3–16) is familiar to all students of science. It is extremely useful for examining basic sizes and shapes of cells, and it can be used to detect the presence of nuclei and mitochondria when the cells are stained with dyes that change the color of only these organelles. The tissue of interest is usually **fixed** with a solution that preserves the cell structure and **embedded** in a hard wax to provide support so that thin slices can be prepared for viewing.

Students and researchers are often concerned that these treatments may distort the appearance of the cells. This problem can be avoided by using a different optical system, such as that used in a **phase contrast microscope.** This instrument enhances the contrast between the internal organelles so that mitochondria, lysosomes, chromosomes, nucleoli, and other structures can be seen quite clearly in living, unstained, and non-fixed cells. Light microscopes are rarely used for magnifications greater than 1000 to 1500 times normal size because beyond this point the **resolution,** or the clarity of the image, is poor.

The Electron Microscope

The development of the **transmission electron microscope** (see Fig. 3–16) in the 1940s and 1950s made greater magnifications (more than 100,000 times normal size) possible. The use of this type of microscope has provided a wealth of information about the organization of cells and the structure of their organelles.

In an electron microscope, magnetic lenses focus a beam of electrons in much the same way that glass lenses in ordinary microscopes focus a beam of light (see Fig. 3–16). To be seen by an electron microscope, specimens must be fixed and then stained with solutions that introduce atoms such as lead or uranium into the various structures of the cell. The presence of these large atoms permits fewer electrons to pass through the specimen and gives rise to a dark image of the specimen on a bright fluorescent screen like a television screen. This image can then

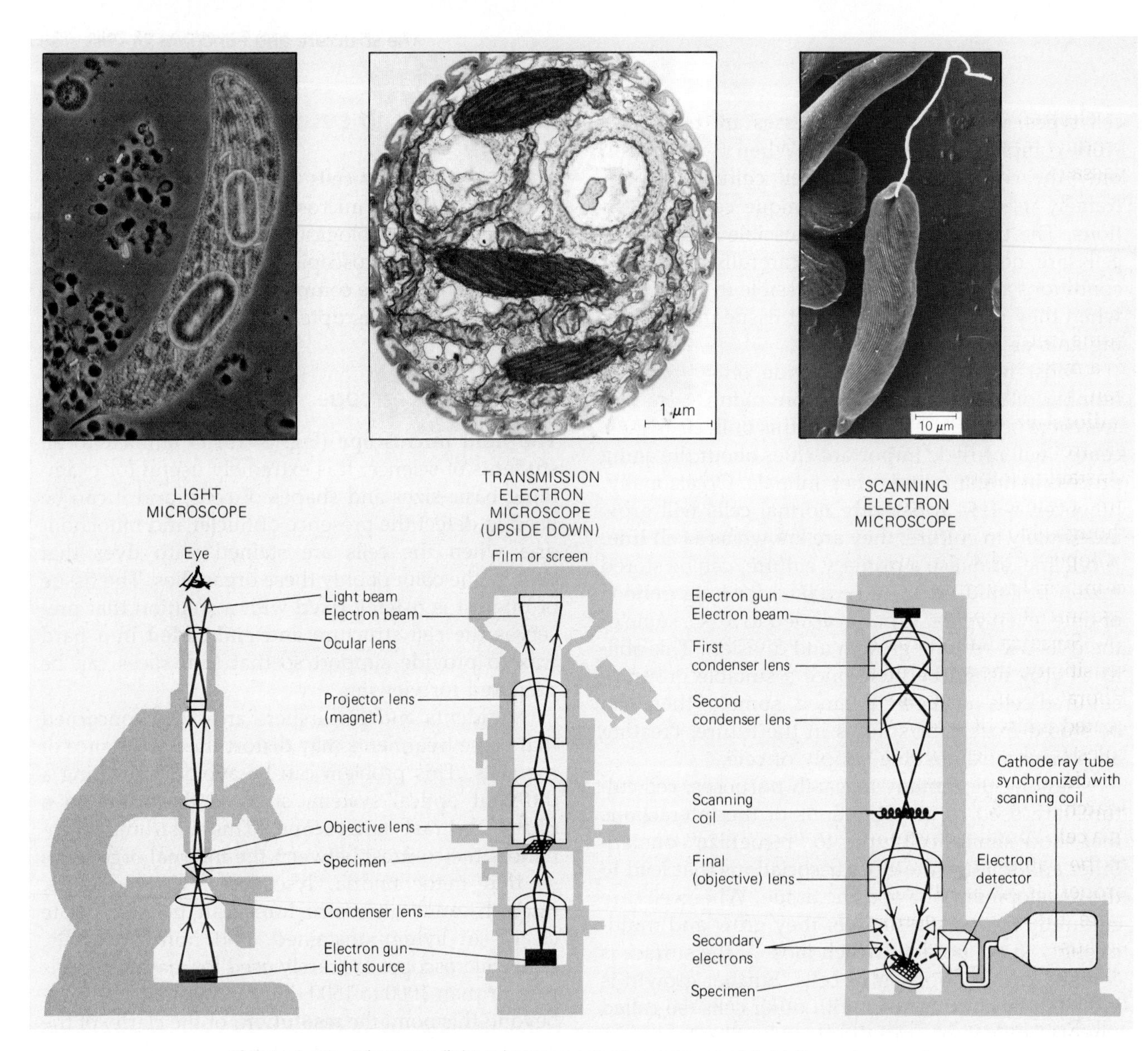

Figure 3–16 Whereas a light microscope focuses a beam of light through curved glass to produce an image, a transmission electron microscope focuses a beam of electrons using an electromagnet. The electrons pass through the specimen to produce an image that can be made visible to the human eye on a screen or photographic film. A scanning electron microscope images the deflection of electrons by the specimen. The photographs show views of the unicellular organism **Euglena.** The light microscope shows the whole cell; the transmission electron microscope shows a thin cross section; the scanning electron microscope shows the cell's outer surface with its pattern of fine ridges.

be recorded on a photographic plate to provide a permanent record.

The **scanning electron microscope** (see Fig. 3–16) is especially useful for examining the surface of a specimen. These microscopes image the deflection of electrons by the specimen. The **high-voltage electron microscope** uses an energy source of one million volts and stands about 32 feet high. It allows much thicker tissue sections to be viewed with high resolution and in considerable depth, producing an almost three-dimensional view of the cell interior. The use of this technique led to the concept of the microtrabecular lattice in the cytosol.

Cytochemical Studies

Cytochemistry is used in combination with microscopy to produce a color or enhanced contrast at the sites of specific molecules. Specific stains for DNA and RNA have been used to demonstrate the location of these molecules within the cell. If the molecule being studied is an enzyme, chemicals can be added to sections of a tissue to cause the enzyme to generate an insoluble reaction product. This material will be deposited wherever the enzyme is located. Researchers can identify these sites precisely by looking at the section through a light microscope or electron microscope. They can see the reaction product readily because it is too dense to allow the passage of light or electrons.

Radioactivity in Cell Research

Some isotopes are unstable and tend to break down to a more stable form. Tritium, an isotope of hydrogen, is a specific example. This process is termed **radioactive decay** because it releases high-energy electrons or other particles known as **radiation.** Unstable isotopes that undergo radioactive decay are **radioactive isotopes,** or **radioisotopes.** Although radioisotopes are rare in nature, radioisotopes of phosphorus, iodine, sulfur, carbon, calcium, and hydrogen are readily manufactured for use in cell research.

Since radiation can be detected with extreme sensitivity, the amounts that need to be used in the laboratory are very small and well within the accepted limits of safety. The use of radioactive molecules (molecules to which a radioisotope has been attached) is a powerful technique, which allows researchers to follow the course of almost any process in a cell. A radioactive molecule performs identically to the naturally occurring molecule. Its biochemical properties are unaltered, and the cell reacts as if it were the natural nonradioactive form. However, because it is "labeled" with a radioisotope, any changes in its chemical form or its location can be followed as the molecule is processed within the cell. Researchers worked out the complex metabolic pathways of cells largely with the aid of this technique.

The location of radioactive molecules within cells or tissues is best determined by the technique of **autoradiography.** After incubating the cells with an appropriate radioactive compound to allow its incorporation into the macromolecules of the cells, the researcher washes the cells or slices of tissue to remove unincorporated radioactivity, fixes and embeds them, and cuts them into thin sections for either light or electron microscopy. Each preparation is then overlaid with a film of photographic emulsion and left in the dark. The radioactivity in the preparations reacts with the emulsion in much the same way as photographic film reacts to light. After developing the film, the researcher can determine the position of the radioactivity in the cells from the position of the dark spots on the emulsion when viewed in the microscope. Autoradiography has been very useful in discovering the secretory pathways in cells and in determining the role of the Golgi complex in this process.

The Immune System and Cell Studies

The mammalian **immune system** is an important defense mechanism that recognizes and destroys invading microorganisms and cells not normally present in blood. An important phase of this reaction is the synthesis and release of special proteins, known as **antibodies,** into the bloodstream. A molecule on the surface of the foreign cell triggers their production. A foreign molecule, such as a protein, polysaccharide, or nucleic acid can elicit a similar response. Any molecule that stimulates the production of antibodies is termed an **antigen.** The antibodies that are released are highly specific molecules that recognize and bind only to the antigen. This reaction may inactivate the cell bearing the antigen, or, alternatively, the antibody-antigen complex on the surface of the cell may stimulate removal of the cell from the bloodstream by phagocytic cells specifically designed for this purpose.

Immunocytochemistry

Mammals can produce an enormous variety of antibodies. The specificity of these molecules makes them powerful research tools because they can be used to pick out specific molecules within a cell. For example, antibodies to tubulin can be used to study the organization of microtubules in a cell. A researcher labels the antitubulin antibodies by attaching a fluorescent molecule. The antibodies then are allowed to react with thin sections of cells or with intact cells that have been treated to allow them to enter. The antibodies will bind only to microtubules, since only the microtubules contain tubulin. Because the antibodies are fluorescent, they can be traced in a microscope fitted with an ultraviolet light source. Only the fluorescent label will be seen, but it suggests the presence of the antitubulin antibodies, which in turn indicate the microtubules to which they are bound (Fig. 3–17*a* and *b*). Thus, the fluorescent sites within the cell indicate the location of microtubules. If the antibodies to an antigenic molecule are labeled with **ferritin,** an iron-protein complex that does not permit the passage of electrons, the location of the antigen within cells can be seen in great detail in the electron microscope.

Radiolabeled Antibodies

In a method similar to using fluorescent markers, a radioactive isotope can be attached to an antibody to give a **radiolabeled** antibody. The radiolabeled antibody can be used in combination with autoradiography to pinpoint the location of specific molecules in

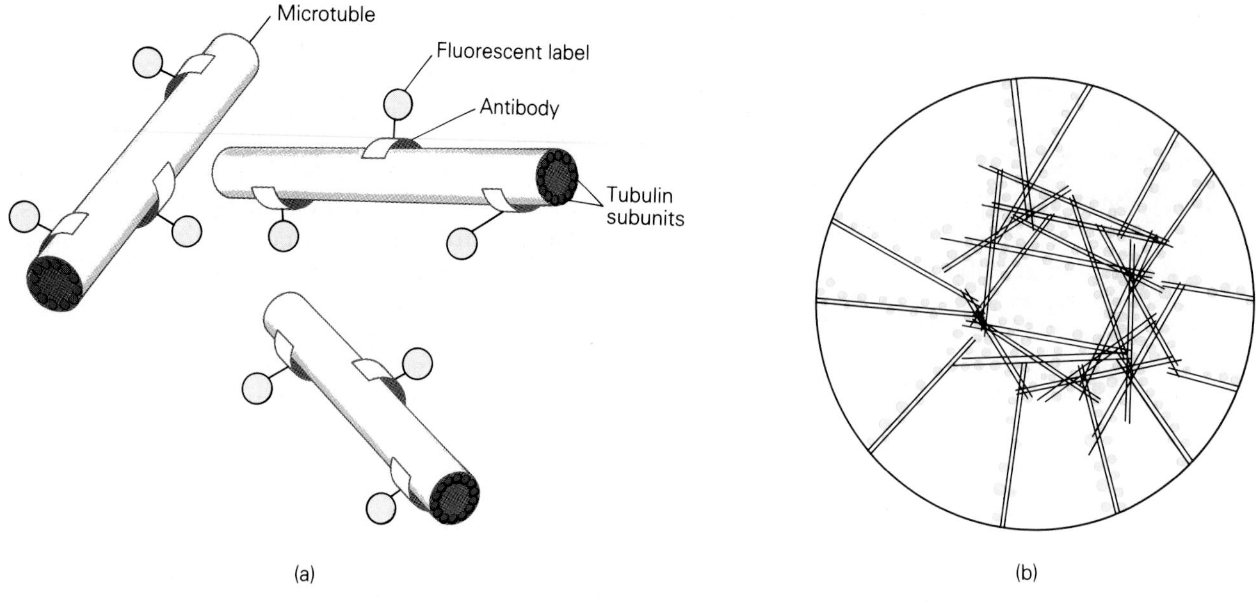

Figure 3–17 In the fluorescent-labeled antibody technique, an antibody to tubulin, which is the protein that makes up microtubules, is injected into a cell. The antibody has been labeled with a fluorescent marker. *(a)* The antibody attaches to the tubulin molecules as they form microtubules throughout the cell. A special microscope allows a view of the fluorescent markers, which indicate the presence of the antibodies, in turn showing the presence of microtubules *(b)*.

the cell. Researchers have used this approach to follow the movement and fate of the plasma membrane that is internalized by cells during endocytosis. In this case, the antibody used is one that binds to molecules that are components of the plasma membrane.

Radioimmunoassays

Another use of radioisotopes and immunochemistry provides the means to measure very small amounts (1×10^{-12} mole) of certain molecules with great accuracy. The measuring procedure is known as a **radioimmunoassay.** A specific example of its use is the measurement of cyclic AMP in various tissues, in blood, and in urine.

The basic principal of the method is to extract the cyclic AMP and to allow it to bind to an anti-cyclic AMP antibody. A trace amount of radiolabeled cyclic AMP is added to the mixture, which is then set aside for several hours. The unlabeled and the labeled cyclic AMP are identical molecules and they compete for binding to the antibody. The amount of radiolabeled cyclic AMP that is present is always the same. The only variable is the amount of unlabeled cyclic AMP in the sample; this determines how much labeled cyclic AMP can be bound to the antibody. When there is very little cyclic AMP in the sample, a lot of **labeled** cyclic AMP can bind to the antibody because only a few unlabeled cyclic AMP molecules will compete for binding. If there is a lot of cyclic AMP in the sample, however, much more will bind to the antibody and only a small amount of labeled cyclic AMP will be able to bind (Fig. 3–18). A researcher can determine the amount of radiolabeled cyclic AMP bound by the antibody by isolating the antibody-antigen complex and measuring the radioactivity present. The amount of radiolabeled cyclic AMP that is bound is inversely related to the amount of cyclic AMP present in the sample. The latter is easily calculated.

Affinity Chromatography

Another technique that uses the specificity of antigen-antibody reactions is **affinity chromatography.** Following is an example of how this technique might be used to study a particular antibody. A researcher produces the antibody to be studied by injecting a purified sample of an antigen several times into an animal (such as a goat or a rabbit) in which the antigen is a foreign molecule. The researcher draws blood from the animal: the blood now contains the antibodies the researcher wants to study as well as proteins and other antibodies normally present in the blood. The antibody in question must be separated out. The researcher allows the blood sample to clot in order to remove red blood cells. The remaining serum is passed through a column of an insoluble carbohydrate matrix to which an antigen

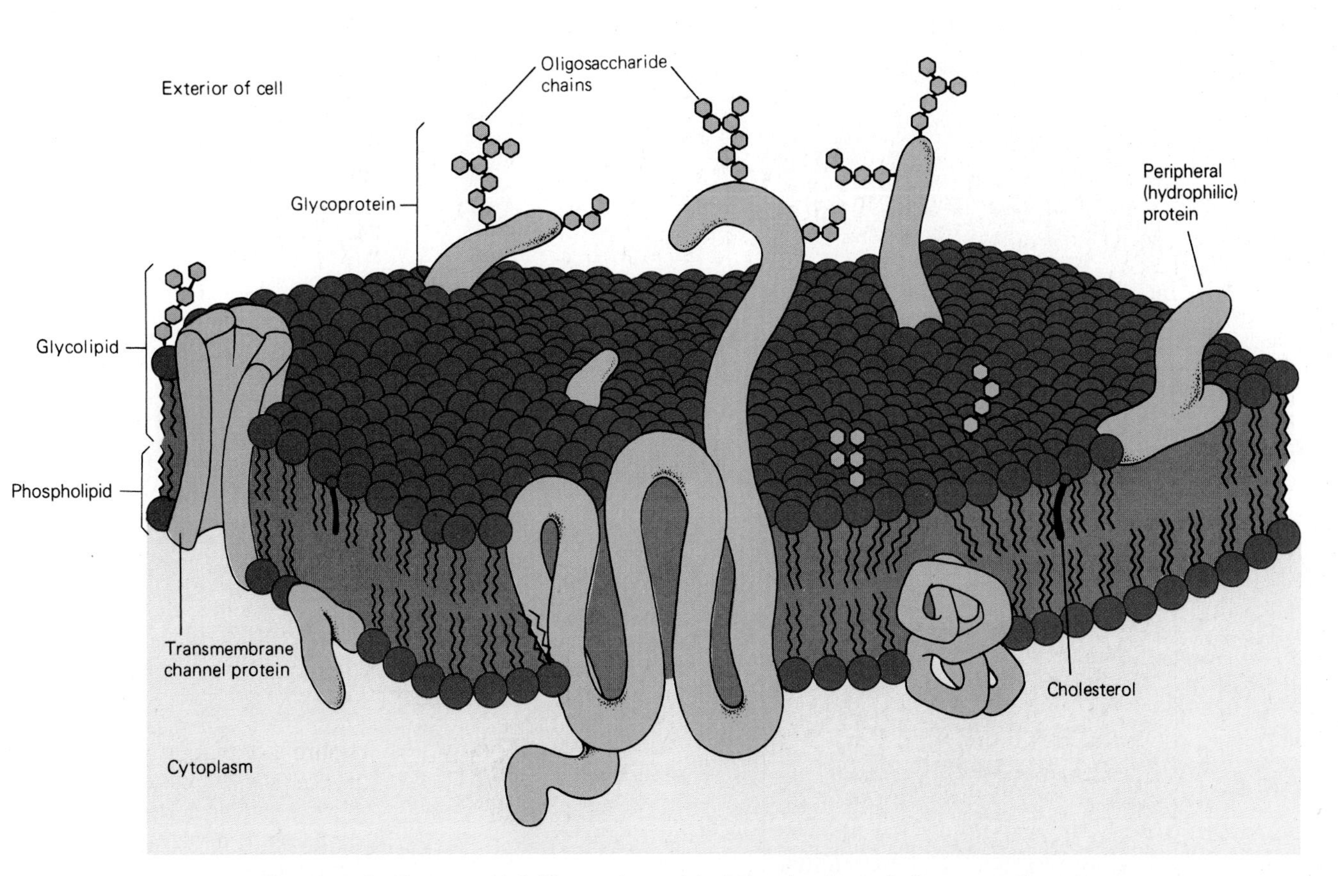

Figure 4–2 The current fluid-mosaic model of the structure of plasma membranes.

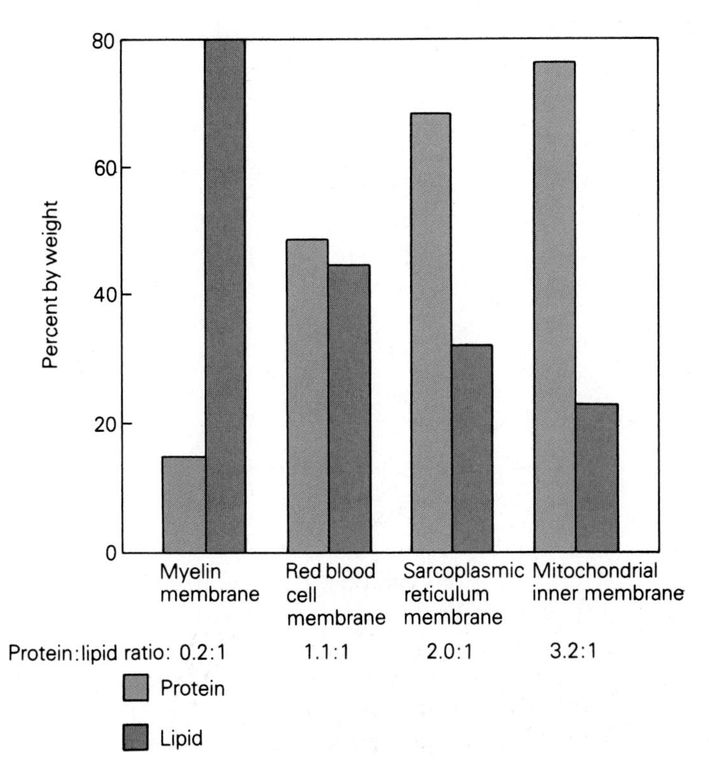

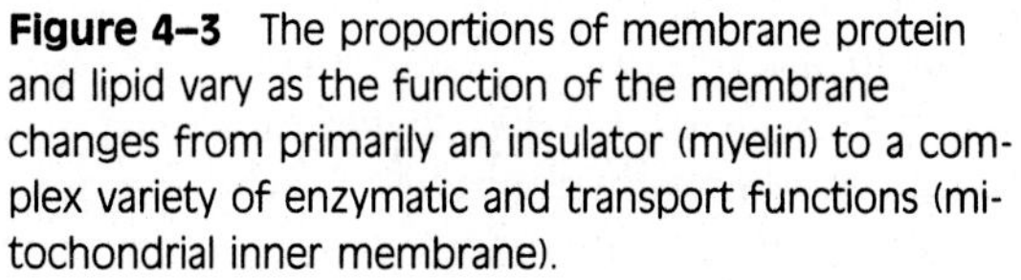

Figure 4–3 The proportions of membrane protein and lipid vary as the function of the membrane changes from primarily an insulator (myelin) to a complex variety of enzymatic and transport functions (mitochondrial inner membrane).

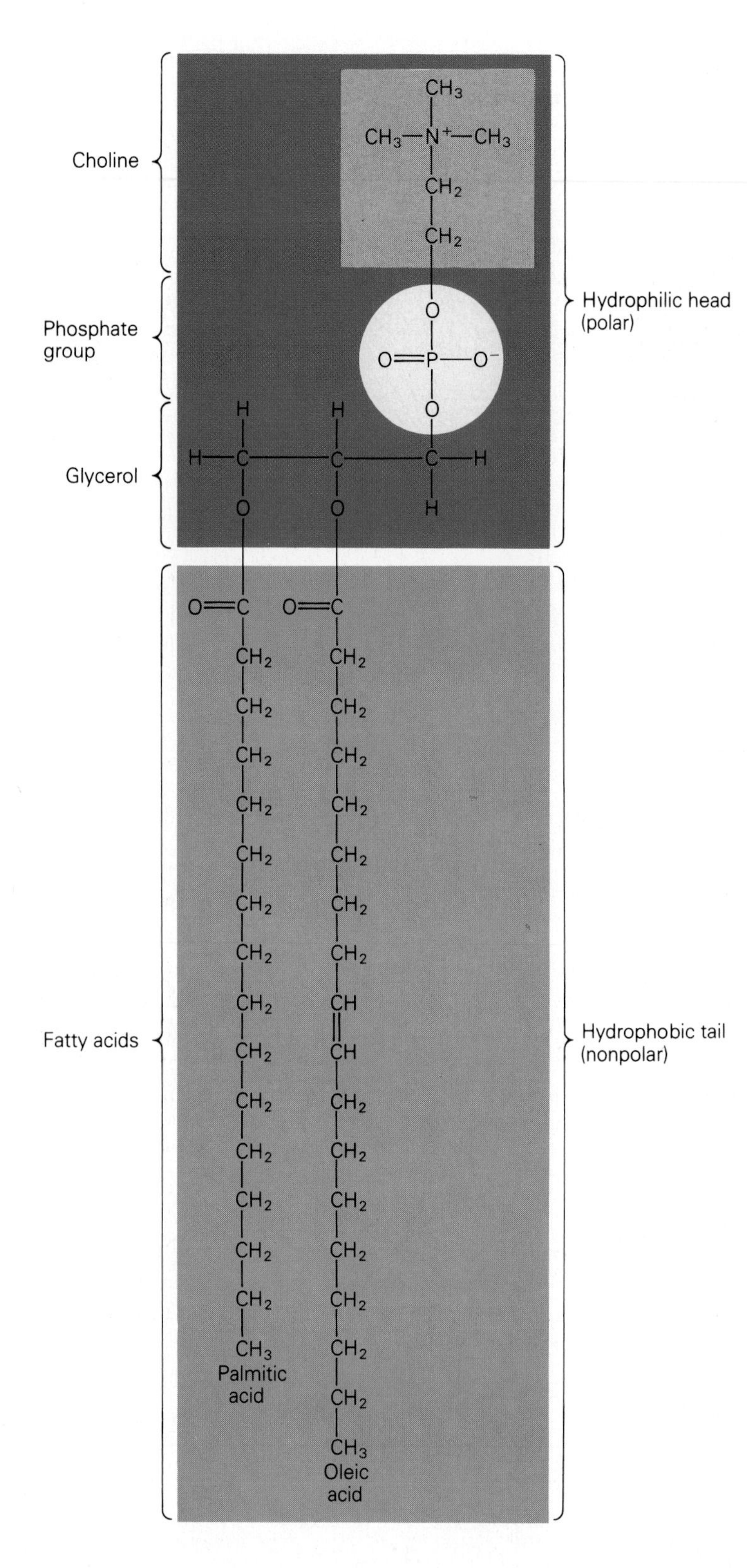

Figure 4–4 The structure of a typical phosphatidylcholine molecule.

philic head group and two long hydrophobic fatty acid tails, each of which may contain 14 to 24 carbon atoms in an unbranched chain. (The 16- and 18-carbon fatty acids are the most common fatty acid chains found in phospholipids; typical examples are **palmitic acid** and **oleic acid.**)

Recall from Chapter 2 that nonpolar molecules and groups, when placed in aqueous solution, cluster together in hydrophobic pockets. Since they consist mainly of hydrophobic fatty acid chains, phospholipids form clusters in order to "hide" their hydrophobic tails from water molecules. They form either spheres, called **spherical micelles,** or **bimolecular sheets (bilayers)** (Fig. 4–5). In both types of clusters, the hydrophobic tails of the phospholipids are buried in the interior of the structure, and only the hydrophilic head groups are exposed to the water.

The micelles are rather small structures, less than 20 nm in diameter, whereas a bimolecular sheet or bilayer is a much more extensive structure. When enough phospholipid molecules occur together, they will be more likely to form a bilayer than a micelle. A bilayer is a more efficient way for the phospholipids to protect their hydrophobic portions because it tends to close in on itself (forming a circle) to eliminate the free edges where the hydrophobic tails would be exposed to water. For the same reason, it will reseal itself if punctured.

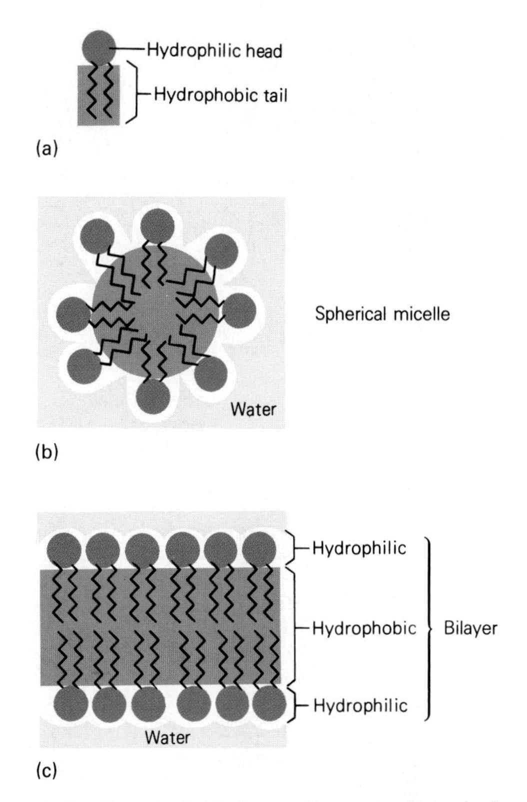

Figure 4–5 Phospholipids form either spherical micelles or bilayers in aqueous solution.

The Movement of Phospholipids Within the Bilayer

Studies on the movement of individual phospholipid molecules within a lipid bilayer have revealed that the bilayer is a dynamic structure. Phospholipid molecules move laterally along one layer of the membrane as well as crossing from one layer to the other.

Movement Within Molecules: Flexing and Rotation. Within each phospholipid molecule there is a certain amount of movement. The phospholipid molecules are able to rotate about their axes (Fig. 4–6*a*). In addition, at physiological temperatures (37°C), the hydrocarbon tails are mobile and undergo rapid flexing movements. The rate of this movement is higher if the hydrocarbon tail is short and contains double bonds.

Stabilization by Cholesterol. In eukaryotic cell membranes, as much as 20% of the total lipid may be cholesterol. Cholesterol stabilizes the lipid bilayer. Like the phospholipids, cholesterol is an amphipathic molecule; however, its steroid ring structure is more rigid than the long fatty acid tails of the phospholipids. The situation of cholesterol in the lipid bilayer allows its ring structure to interact with and reduce the motion of the neighboring phospholipid hydrocarbon chains (Fig. 4–6*b*).

Movement Within Layers. Phospholipid molecules exchange positions with their neighbors in the same half of the bilayer by very rapid lateral diffusion (Fig. 4–7*a*). By this process a lipid molecule takes only 1 sec to move about 2 μm, the entire length of a large bacterial cell.

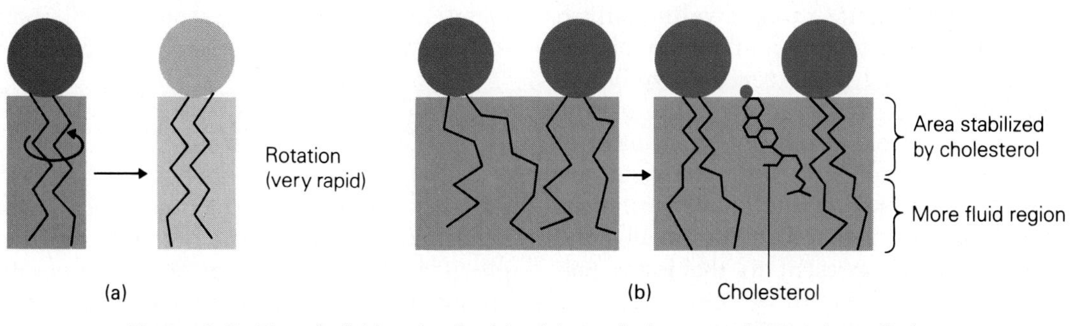

Figure 4–6 Phospholipid molecules (*a*) rotate on their axes and (*b*) undergo flexing motions. The rigid steroid ring structures of cholesterol stabilizes certain portions of the bilayer.

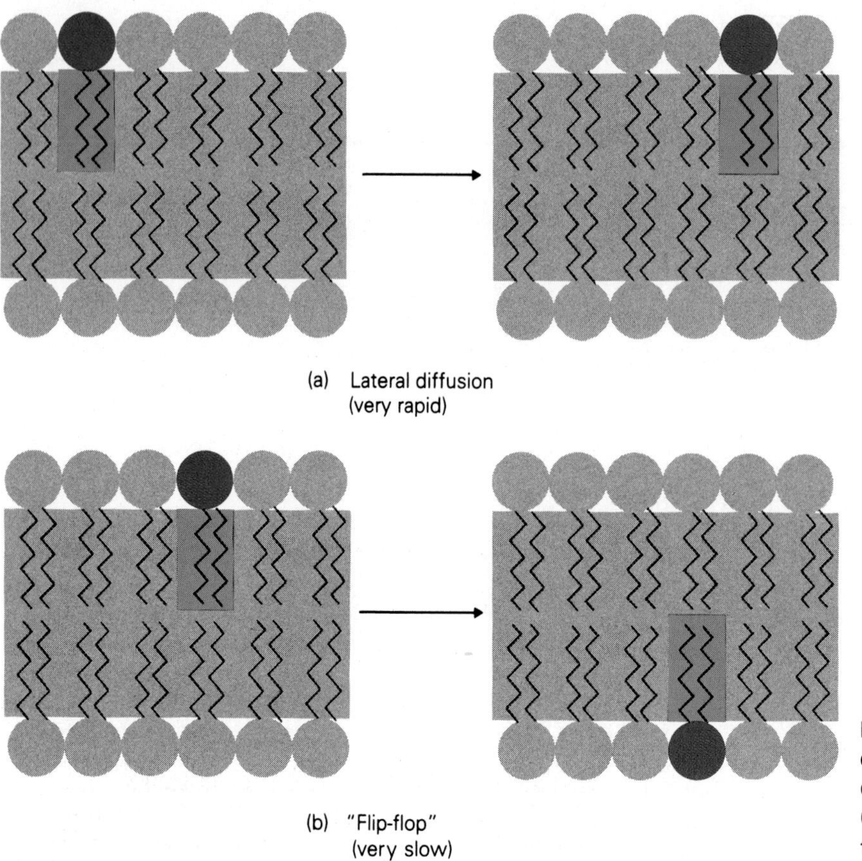

Figure 4–7 (*a*) Phospholipid molecules diffuse laterally very rapidly within one layer of the plasma membrane. (*b*) Phospholipids move very slowly from one layer to the other.

Movement Between Layers. Lipids also move from one half of the lipid bilayer to the other in what is called "**flip-flop.**" This motion requires that the polar head groups of the phospholipids in one half of the bilayer must pass through the hydrophobic interior of the bilayer to reach the other half of the bilayer (Fig. 4–7*b*). This is a difficult task; hence, flip-flop movement is extremely slow.

Lipid Asymmetry in the Bilayer

Phosphatidylcholine is not the only phospholipid in biological membranes. Other important membrane phospholipids are **phosphatidylethanolamine, phosphatidylserine,** and **phosphatidylinositol** (all derived from glycerol), and **sphingomyelin** (derived from sphingosine). These different phospholipids are not distributed equally between the two halves of the lipid bilayer. A good example is the plasma membrane of the red blood cell, in which most of the phosphatidylcholine and sphingomyelin are present in the outer half of the bilayer, whereas most of the phosphatidylethanolamine and all of the phosphatidylserine are present in the inner half (Fig. 4–8).

The persistence of this **lipid asymmetry** in membranes is good evidence that the rate of phospholipid flip-flop is very slow. If flip-flop were rapid, the various phospholipids would probably mix randomly and thus be distributed equally between the bilayers. The phospholipid asymmetry must originate during the initial assembly of the bilayer, but scientists do not understand the mechanism and possible advantages of this arrangement.

The Texture of the Plasma Membrane

The net result of the movements of the phospholipids and their interactions with cholesterol is that, at 37°C, the interior of the lipid bilayer of biological membranes is more like a viscous fluid such as olive oil than a rigid crystalline matrix. This fluidity of the membrane structure is probably very important for the normal function of many membrane proteins.

Membrane Proteins

Proteins are an important component of most biological membranes. Some proteins, called extrinsic proteins, are found only on the surface of the membranes, whereas others, called intrinsic or integral proteins, partially or completely penetrate the lipid bilayer.

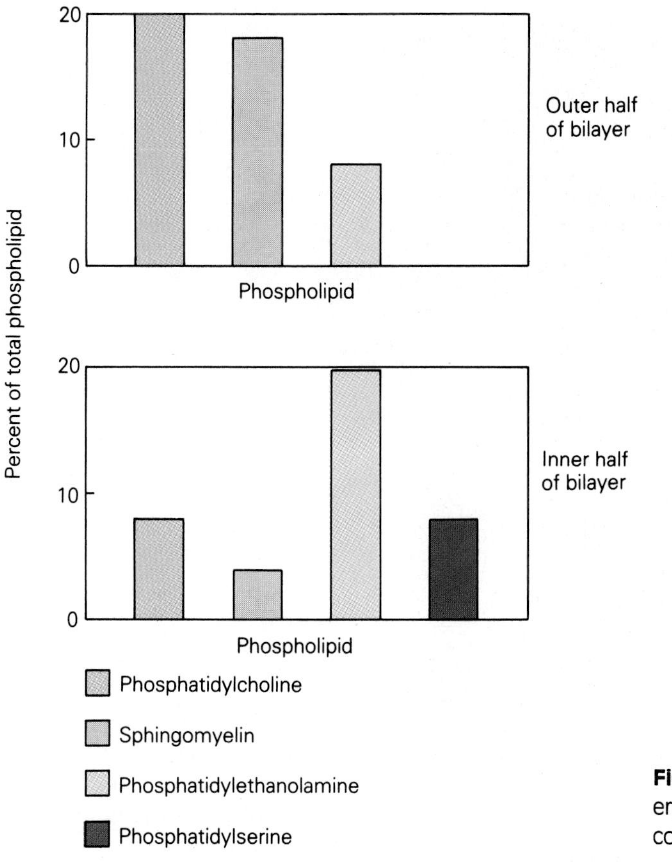

Figure 4–8 Asymmetrical distribution of lipids in the erythrocyte membrane. Note that each half of the bilayer contains 50% of the total lipids present in the membrane.

Intrinsic Proteins: Amphipathic Molecules

The proteins that penetrate the bilayer, called **intrinsic** (or **integral**) **proteins,** account for about 70% of total membrane protein. Most of them probably span the bilayer completely, so that parts of the protein are exposed on both sides of the bilayer.

Intrinsic proteins are usually amphipathic molecules. Their hydrophobic regions cluster in the interior of the bilayer with the hydrophobic portions of the lipid bilayer, whereas their hydrophilic regions remain exposed to the aqueous medium outside the bilayer (Fig. 4–9).

The three-dimensional structure or conformation (see Chapter 2) of much of the hydrophobic part of the protein is usually an alpha-helix. Any polar (hydrophilic) amino acid side chains in this portion of the molecule are sequestered within the alpha-helix. Only the nonpolar (hydrophobic) side chains are exposed to the hydrophobic interior of the bilayer. In contrast, in regions of the protein that project outside the lipid bilayer, the nonpolar side chains are buried in the interior of the protein conformation, whereas the polar and ionic side chains are exposed to the hydrophilic environment.

Many of the intrinsic membrane proteins display the same dynamic properties as the membrane lipids, such as rapid lateral diffusion and rotation about an axis perpendicular to the plane of the membrane. The fluidity of the lipid bilayer makes these movements possible. However, flip-flop of intrinsic membrane proteins is even more difficult than it is for phospholipids; in fact, it probably does not occur.

The lateral mobility of these membrane proteins can be prevented sometimes by interactions with peripheral components of the membrane (such as other proteins) or with components of the cell cytoskeleton (microtubules and microfilaments).

Extrinsic Proteins: Hydrophilic Molecules

Membrane proteins that do not penetrate the lipid bilayer are termed **extrinsic** (or **peripheral**) **proteins** (Fig. 4–9). They are usually not amphipathic molecules. They are surrounded by the aqueous environment of the cell and are associated with the membrane primarily through interactions (electrostatic and hydrogen-bonding) with the polar portions of intrinsic membrane proteins and with the polar head groups of the phospholipids. These are relatively weak interactions. Consequently, extrinsic proteins can be easily extracted from membranes by relatively simple procedures.

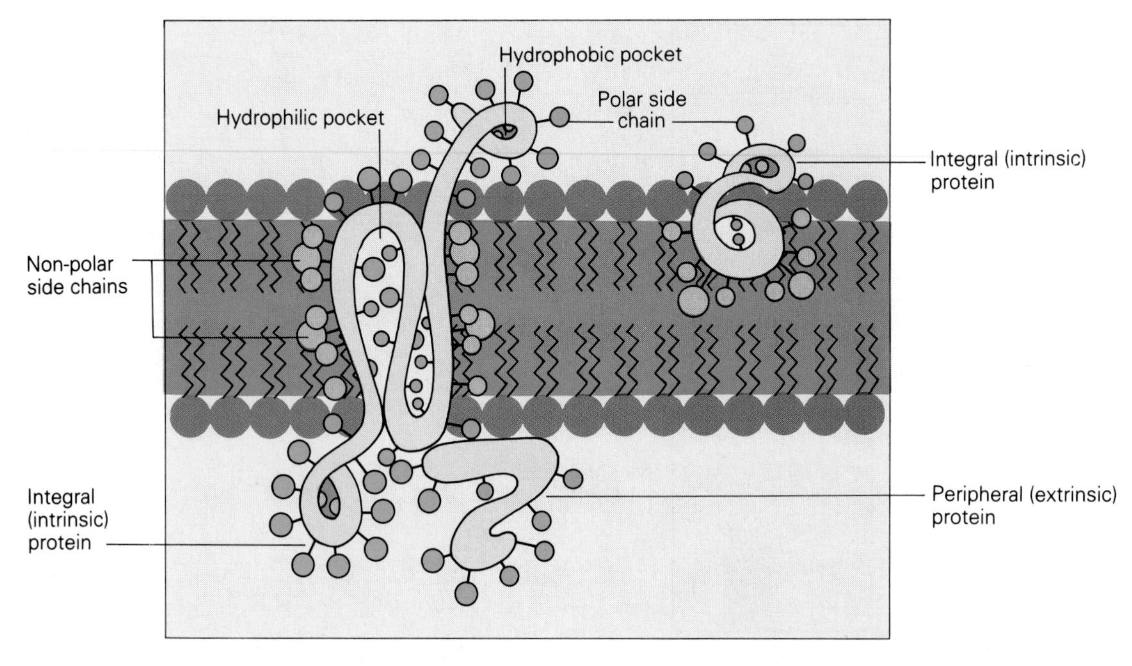

Figure 4–9 The arrangement of integral and peripheral proteins in a biological membrane.

Protein Asymmetry in the Membrane

Both extrinsic and intrinsic membrane proteins, like the phospholipids, are often distributed unequally between the two halves of the bilayer. That is, a specific protein may be associated with only one side of the bilayer. Specific advantages of protein asymmetry are better understood than are those of phospholipid asymmetry.

This asymmetry is related to cell function. In the membrane of the red blood cell, for example, the extrinsic protein called **spectrin** is found only on the cytoplasmic (inner) side of the membrane. It is part of a filamentous network inside the cell that may be important in maintaining the biconcave shape but is flexible enough to allow the cell to change shape as necessary during its passage through narrow capillaries.

In contrast, many receptors for specific chemical signals such as hormones are located only on the external surface of the cell membrane and often represent part of an intrinsic membrane protein. If such a protein spans the lipid bilayer, a potential mechanism exists for converting an extracellular signal into an intracellular response. For example, binding of a hormone to the receptor could trigger a conformational change in the protein, especially if the protein is composed of several separate subunits. This conformational change could open a hydrophilic channel between the protein subunits and allow a specific ion to cross the plasma membrane.

Carbohydrate on the Surface of the Membrane

Carbohydrate is present in the plasma membrane of all eukaryotic cells and accounts for 2% to 10% of the mass of the membrane. In terms of quantity, carbohydrate is a minor component of the plasma membrane, yet it has important functions in membrane physiology.

Carbohydrates, like the membrane phospholipids and proteins, are distributed unequally between the two layers of the membrane. The carbohydrate molecule is a hydrophilic structure that does not penetrate the lipid bilayer but is found only on the surface of the membrane (see Fig. 4–2). Most of the membrane carbohydrate is bound either to intrinsic membrane proteins, forming glycoproteins, or to lipids in the bilayer, forming glycolipids.

Glycolipids are present only in the external half of the bilayer. The glycoproteins are more abundant than the glycolipids. Most of the proteins that are exposed on the external surface of the membrane have carbohydrate attached, but fewer than 10% of the lipid molecules in the external half of the bilayer have carbohydrate attached.

The amount of carbohydrate present in an individual glycoprotein may be large relative to the amount of protein. An extreme example is **glycophorin,** an intrinsic glycoprotein in the red blood cell membrane. This protein is composed of 131 amino acids and about 100 monosaccharide units. The monosaccharides together make up approximately 60% of the mass of the glycophorin molecule.

The location of membrane carbohydrates on the surface of the plasma membrane is probably an important factor in cell-to-cell recognition processes. These carbohydrates also probably serve as specific membrane receptors for extracellular messengers such as hormones. Unlike amino acids, which form only linear chains, monosaccharide units can be assembled in branching chain structures. This property provides almost limitless possibilities for construction of membrane receptor sites to which only one specific extracellular molecule can bind.

The Types of Movement Through the Membrane

Two types of movement occur as substances pass through the plasma membrane of a cell: (1) **passive,** or **spontaneous,** processes that occur in response to gradients and produce a *less* concentrated state overall, and (2) **active,** or **energy-requiring,** processes that occur in opposition to gradients and produce a *more* concentrated state where a high degree of concentration already existed. **Passive transport** processes include both **simple diffusion,** such as **osmosis,** and **facilitated diffusion. Active transport** processes include **primary** and **secondary active transport** (Fig. 4–10). Note that the same substance—for example, sodium ions—can move one way through the membrane by a passive process and the other way by an active process.

Passive Transport: Spontaneous Movement down a Gradient

Spontaneous, random movements of substances into or out of the cell include simple diffusion and facilitated diffusion. **Simple diffusion** takes place when a substance, usually one that can dissolve in lipids, passes directly through the lipid interior of the plasma membrane. **Facilitated diffusion** occurs in the presence of a protein that enables the passage of the substance. However, both processes are passive and spontaneous movements that occur in response to a gradient.

Simple Diffusion of Lipid-Soluble Substances

Diffusion in free solution or any other type of mixture occurs because of the random movement that all particles exhibit. This motion is an expression of the thermal energy of the molecules. As a result of diffusion, a substance such as a solute will tend to become distributed in an equal concentration throughout the entire space available.

A Selectively Permeable Membrane. Chapter 2 discussed the diffusion of solute particles in free solution, from a region of high solute concentration to an adjacent region of low concentration. There was no physical division between these two regions; they were more or less continuous.

Diffusion can also occur in the presence of a **selectively permeable membrane.** If a solute is highly concentrated on one side of a membrane but is able to pass through the membrane, it will do so, diffusing into the adjoining region of low concentra-

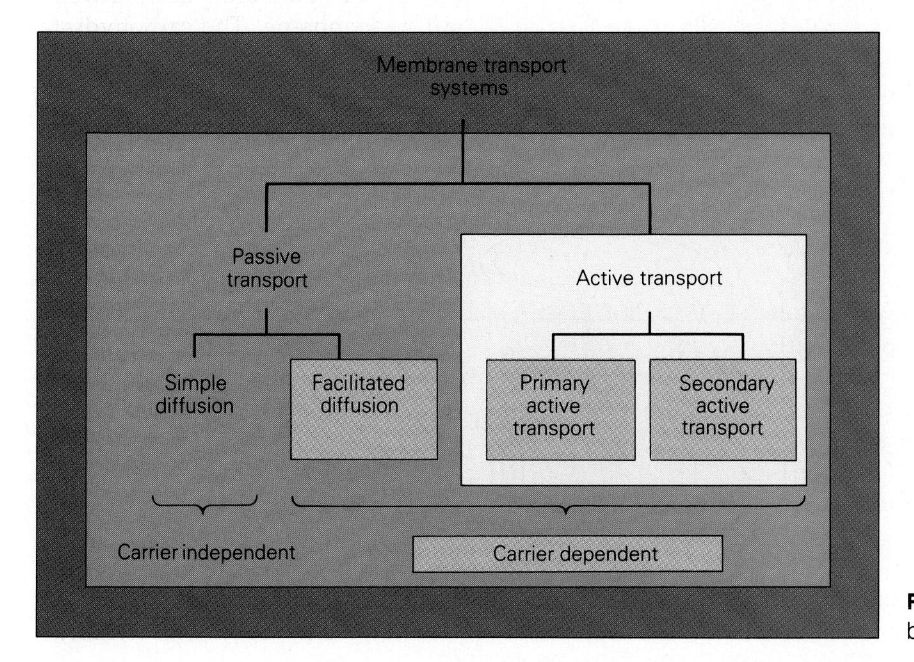

Figure 4–10 An overview of membrane transport systems.

tion by the same process that would occur if the membrane were not there (Fig. 4–11). If another solute were present to which the membrane were not permeable, that solute would remain highly concentrated on one side of the membrane.

The plasma membrane, as a mixture of lipids and proteins, represents a barrier to the diffusion of some solutes into or out of the cell. An inherent, important property of any lipid bilayer is that it is nearly impermeable to ions and most polar molecules. (Water is a notable exception to this general rule because it readily crosses lipid bilayers even though it is a polar molecule.) In contrast, a lipid bilayer is permeable to nonpolar molecules—that is, molecules that can dissolve in its nonpolar interior. The lipid bilayer of the plasma membrane is selectively permeable: it readily permits some substances to cross it but prevents or retards the free passage of others.

Consider the diffusion of glycerol, a polar molecule, from the plasma to the interior of a red blood cell. The diffusion coefficient D for glycerol in water is 1.0×10^{-5} cm^2/sec. This is several orders of magnitude larger than the diffusion coefficient for glycerol inside the plasma membrane: 1.7×10^{-10} cm^2/sec. This large difference reflects the fact that the mobility of glycerol is greatly diminished once it enters the membrane.

The Permeability Coefficient. The rate of diffusion of solute through a cell membrane is governed by the thickness of the membrane and by the lipid solubility of the solute. The combined effect of these two factors and the diffusion coefficient can be expressed as the **permeability coefficient *P*.** The value of P varies with the specific solute and the specific cell membrane.

In spite of the barrier represented by the cell membrane, an uncharged solute will still tend to move across it if the solute concentration inside the cell is different from the solute concentration outside—that is, if there is a concentration gradient. In simple diffusion, the solute will always move down the concentration gradient, from the region of high solute concentration to the region of low concentration, just as in free solution. Fick's Law can be adapted to represent the rate of solute diffusion across a membrane as follows:

$$J = -PA[C_1 - C_2]$$

C_1 and C_2 are the solute concentrations on either side of the membrane. Thus, when C_1 is greater than C_2, there will be net movement of solute from Side 1 to Side 2 (Fig. 4–12). As the solute concentration builds up on Side 2, the rate of movement back to Side 1 will increase. Eventually the solute concentration gradient will be abolished, and these opposing fluxes will become equal. At this point, the system will be in a dynamic equilibrium—that is, individual solute particles will continue to move from one side of the membrane to the other, but the net movement of solute will cease. This can be seen also from the equation: When $C_1 = C_2$, the value of J is zero.

The Electrochemical Gradient. When the solute is electrically charged, as are Na^+ and Cl^- ions, for example, the flux of solute will be influenced not only by a concentration gradient but also by an electrical gradient. A typical animal cell has a small negative potential (about −70 millivolts) on the inside relative to the outside. This difference tends to cause positively charged solutes to move across the plasma membrane and into the cell while it opposes the entry of negatively charged solutes. This electrical gradient represents an additional factor in diffusion that must be considered when the solute bears a charge. The total gradient is now the sum of the chemical (concentration) and electrical components and is termed the **electrochemical gradient.**

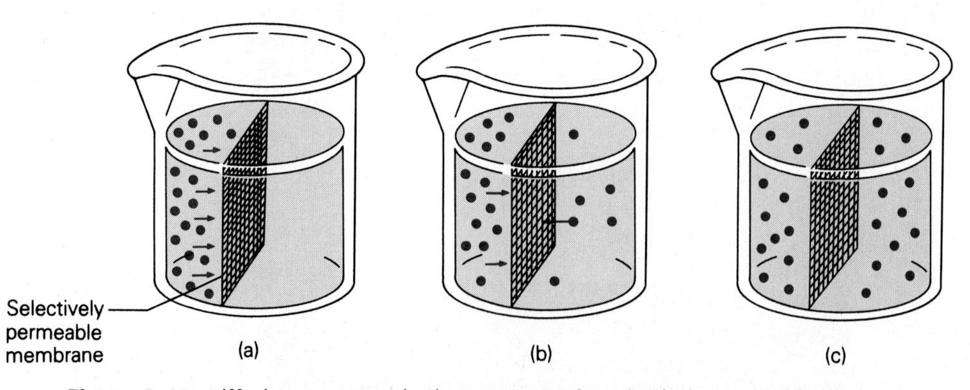

Figure 4–11 Diffusion can occur in the presence of a selectively permeable membrane. As long as solute particles can pass through the membrane, they will diffuse from a region of high concentration to one of low concentration.

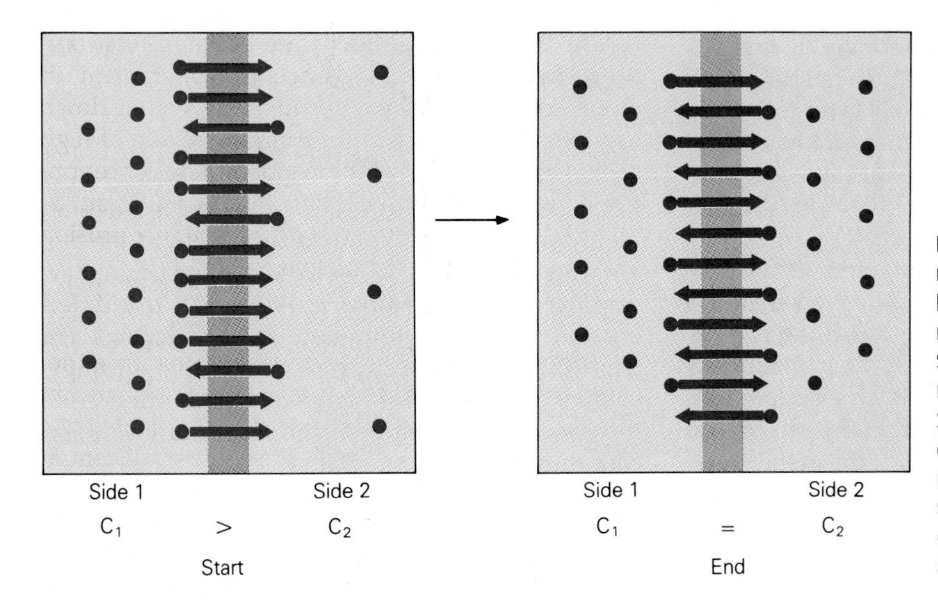

Figure 4–12 Diffusion of solute molecules across a biological membrane. At the start there are 25 molecules on Side 1 and only 7 on Side 2, so there will be net movement of solute from Side 1 to Side 2. The end point is reached when each side has 16 molecules. At this point, there is no net movement of solute across the membrane because equal numbers of molecules are moving in each direction.

Considered separately, the chemical and electrical gradients often favor solute movement in the same direction. For example, the passive entry of Na^+ into cells is favored by both the concentration gradient (low Na^+ concentration inside the cell) and the electrical gradient (negative potential inside the cell).

On the other hand, electrical and chemical gradients can oppose each other. For example, the passive leak of K^+ ions out of cells goes against, or "up," the electrical gradient (there is a positive potential outside the cell). However, the concentration gradient is so large—because of the high K^+ concentration inside the cell—that the net electrochemical gradient force favors the exit of K^+ from the cell.

Overton's Rules. The principal barrier to solutes that cross cell membranes by simple diffusion is the lipid bilayer. The rate at which a solute diffuses through the membrane depends in large part on two factors: its ability to dissolve in lipids and its molecular size. In general, a nonpolar, and therefore lipid-soluble, solute will diffuse across a cell membrane much more rapidly than a polar (water-soluble) solute of the same size. The effect of the size of the solute molecule is particularly important for polar solutes. Membrane permeability for these molecules, already limited by their polarity, decreases further as their molecular size increases.

These relationships are summarized in **Overton's Rules,** as follows: (1) the permeability of cell membranes to small nonpolar solutes is directly proportional to the lipid solubility of the solute; and (2) the permeability of cell membranes to polar solutes is inversely proportional to the molecular size of the solute. Thus, a small, highly lipid-soluble molecule has the best chance of rapidly penetrating cell membranes by simple diffusion.

An exception to these general statements is water itself. Despite its small size, a water molecule is highly polar (lipid-insoluble) and yet it crosses cell membranes very rapidly. The precise mechanism is not fully understood, but it has been suggested that membranes contain small nonspecific pores that allow water and small water-soluble solutes to cross the bilayer without entering the lipid phase. The diffusion of water will be considered in the section on osmosis.

An electrical charge on the solute is another factor that will decrease its lipid solubility. Thus, membranes are almost impermeable to ions, even though an ion represents a solute of small size. Any ionic diffusion that occurs probably uses specific channels in proteins that span the lipid bilayer (see the section on facilitated diffusion later in this chapter).

Diffusion of Acids and Bases. The very low permeability of cell membranes to solutes bearing an electrical charge has important consequences for the diffusion of **weak acids** and **weak bases.** These compounds, unlike strong acids and bases, do not dissociate completely when dissolved in water (see Chapter 2). Recall that an acid is a proton donor and a base a proton acceptor. We can write the following general equations to represent the behavior in solution of a weak acid and a weak base:

$$HA \rightleftharpoons H^+ + A^- \quad \text{(weak acid)}$$
$$B^- + H^+ \rightleftharpoons BH \quad \text{(weak base)}$$

A solution of a weak acid or weak base will contain a mixture of the undissociated form and the dissociated form, the proportions of these two forms varying with the pH of the solution. In the case of the weak base, for example, a low pH (high H^+ concentration) will shift the equilibrium to the right, resulting in more BH. Conversely, a high pH (low H^+

concentration) will shift the equilibrium to the left, resulting in more B^- and more H^+. The pH at which the concentration of the associated form is exactly equal to that of the dissociated form is defined as the *pK* and is a characteristic property of the weak base or the weak acid.

Diffusion Trapping. The neutral form of a weak base will diffuse across a cell membrane whereas the charged form will not, which gives rise to the phenomenon of **diffusion trapping.** This is illustrated in Figure 4–13 using ammonia as an example of a weak base. Ammonia is produced in the cells of the kidney tubules as an end product of the metabolism of amino acids, especially glutamine. Ammonia is neutral and lipid-soluble and easily diffuses out of the cell, moving down its concentration gradient, and into the fluid present in the lumen of the tubule. Typically, this fluid has a low pH; it is slightly acidic relative to the cell cytosol. When the ammonia contacts the excess H^+ in this acidic solution, it gains a proton to form the ammonium ion. The ammonium ion is positively charged and lipid-insoluble and cannot diffuse back into the tubule cell. It is effectively "trapped" in the lumen and eventually excreted in the urine.

Protonation of available ammonia in the tubular fluid maintains a low concentration of ammonia outside the tubule cells, and continued production of ammonia by cellular metabolism maintains a relatively high concentration of ammonia inside the tubule cells. The combination of these processes preserves the concentration gradient that favors the passive diffusion of ammonia out of the cells.

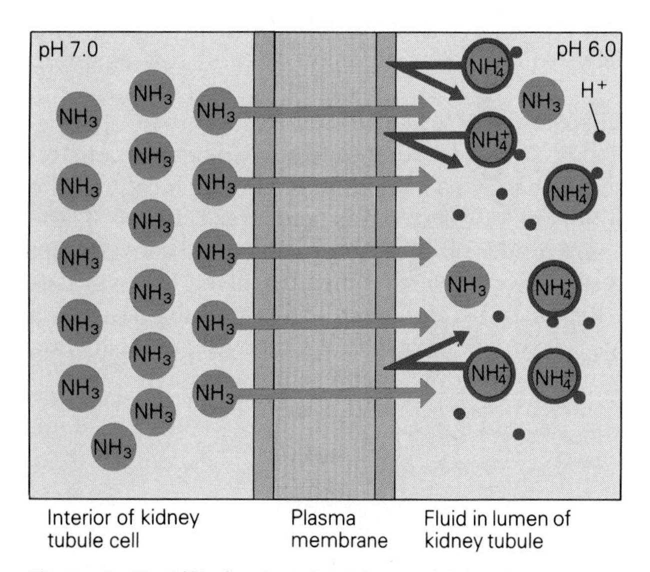

Figure 4–13 Diffusion trapping of a weak base. NH_3 can cross the plasma membrane because it is a neutral molecule. When it gains a proton in acidic solution, it becomes the charged ammonium ion NH_4^+, which cannot cross the membrane.

Simple Diffusion of Water: Osmosis

Recall that **osmosis** is the term used to describe simple diffusion when the molecules that are moving are water molecules. Osmosis is defined as the flow of water across a membrane in response to a difference in solute concentration on each side of the membrane. The water always moves from the dilute solution (the region of high water concentration) to the concentrated solution (the region of low water concentration).

Changes in Cell Volume: The Erythrocyte. Nearly all cell membranes are permeable to water and must regulate the amount of water that enters the cell. Plant cells and many bacteria have a rigid cell wall that surrounds the plasma membrane and prevents an increase in the volume of the cell due to an influx of water. Animal cells lack an exterior cell wall and have developed other mechanisms to deal with osmotic stress.

We will use the red blood cell, or **erythrocyte,** to illustrate the changes that occur in animal cells when the osmolarity of the external medium changes. The erythrocyte is a favorite experimental tool of cell physiologists because of its simple structure. It has no intracellular membranes and is available in large numbers.

Previously in this chapter, we stated that many biological membranes are permeable to solutes as well as to water. This is certainly true of the erythrocyte plasma membrane, but the net movement of the major solutes, such as Na^+, K^+, and Cl^- ions, will be relatively small during osmotic experiments of short duration.

The osmolarity of the erythrocyte cytosol is close to 300 mOsm. Suppose an erythrocyte is placed in an iso-osmotic solution of 300 mOsm NaCl. The total osmolarity (and hence the osmotic pressure) of the solution inside the cell will be the same as that of the solution outside. There will be no concentration gradient favoring osmotic water flow and no net movement of water into or out of the cell; the intracellular volume will not change.

Next, suppose the erythrocyte is placed in hyperosmotic 400 mOsm NaCl. The erythrocyte will no longer be in osmotic equilibrium because the external solution will have a higher osmolarity than the cell cytosol. In response to the higher solute concentration outside the cell, water will flow out of the cell (Fig. 4–14*b*). However, solutes will remain in the cell, and therefore the osmolarity of the cell cytosol will increase. Eventually the osmolarity of the cytosol will increase to the point where it is equivalent to that of the external solution and osmotic flow of water will cease. The cell will return to an osmotic equilibrium with the external medium, but this will occur at the expense of a decrease in intracellular volume. The decrease in volume is not accompanied by a decrease in surface area, and the cell assumes a spiky, or **crenated,** shape (Fig. 4–14*a*).

When a normal erythrocyte is placed in a hypo-osmotic solution of 200 mOsm NaCl, the osmolarity of the cytosol is higher than that of the external solution. Water will enter the cell, leading to an increase in intracellular volume (Fig. 4–14*d*). Osmotic flow of water into the cell will continue until the osmolarity of the cytosol has been diluted to 200 mOsm. At this point net movement of water into the cell will cease, and osmotic equilibrium will be re-established, but with an increase in cell volume (Fig. 4–14*e*).

The erythrocyte has a unique characteristic that enables it to accommodate a considerable increase in cell volume. The normal cell is bi-concave, not spherical. When water enters from a hypo-osmotic solution, the cell simply fills out to become more like a sphere (Fig. 4–14*e*). When the cell has become a perfect sphere, the intracellular volume will have increased by 67% over the volume of the original bi-concave cell. In this way the erythrocyte can undergo a large increase in volume without experiencing a change in the surface area of the cell. Most other types of animal cells are not bi-concave and cannot respond in this way to osmotic entry of water. The plasma membrane is not elastic—it cannot stretch—and so the cells can accommodate only a very small increase in intracellular volume. If osmotic entry of water continues, the cells will undergo osmotic **lysis**—that is, holes develop in the plasma membrane and the intracellular contents are lost. The cells literally burst open.

Osmotic Fragility. An erythrocyte will burst open, or **lyse,** if it is placed in a medium that is very dilute (e.g., 100 mOsm NaCl). The difference in osmolarity between the inside and the outside of the erythrocyte is now so large that the cell cannot achieve osmotic equilibrium. Water will continue to enter even when the cell has become a sphere, and so the cell will lyse.

Osmotic fragility is a term used to characterize how a sample of erythrocytes responds to osmotic stress. It represents the osmolarity of NaCl that causes 50% of the erythrocytes to lyse. In the case of normal erythrocytes, 50% of the cells lyse at an osmolarity of 150 mOsm (Fig. 4–15). Some cells, pre-

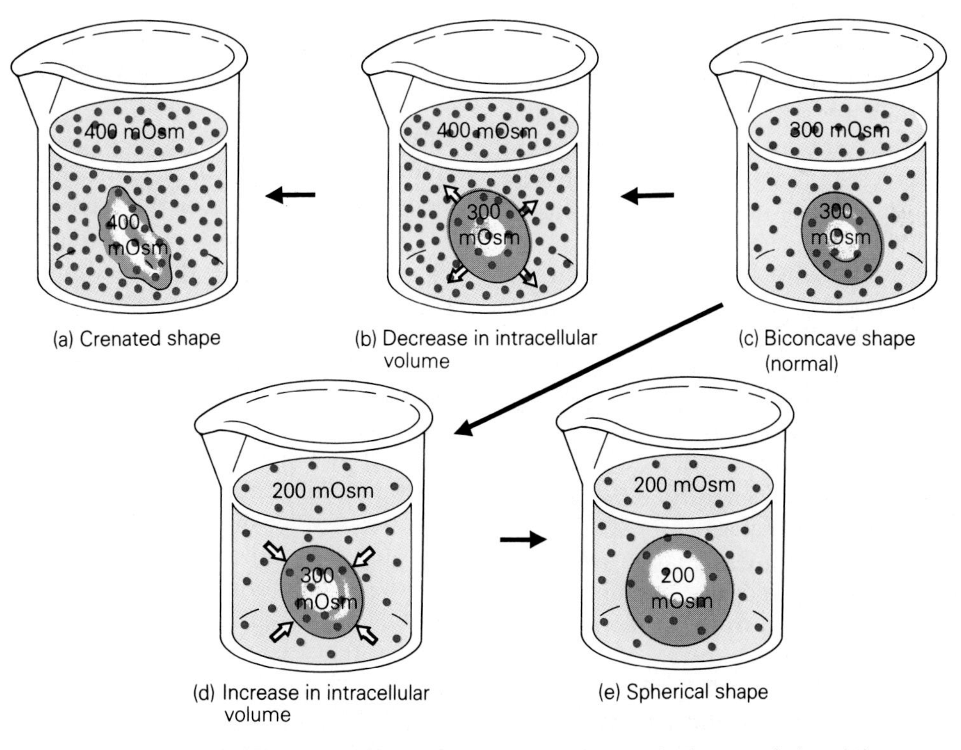

Figure 4–14 Response of a human erythrocyte to changes in the osmolarity of the extracellular medium. The normally bi-concave disc (*c*) becomes crenated (*a*) when water flows out of the cell because of a greater osmolarity in the surrounding fluid (*b*). The cell becomes a sphere (*e*) when water flows into the cell from a solution with a lower osmolarity (*d*).

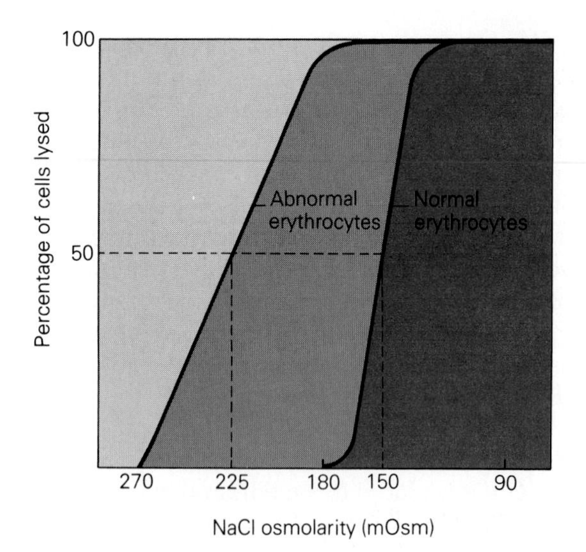

Figure 4–15 Abnormal erythrocytes are more fragile and lyse in solutions of higher osmolarity compared to normal erythrocytes.

sumably the older ones, are more fragile and lyse at a slightly higher osmolarity than 150 mOsm. Other cells, presumably the younger ones, are less fragile and lyse at a slightly lower osmolarity. However, almost all of the population of cells lyse over a fairly narrow range of osmolarities as shown by the steep slope of the curve in Figure 4–15.

Osmotic fragility can be an important test for diseases that involve a change in erythrocyte behavior. When the erythrocytes become more resistant to osmotic stress, they are said to be less fragile, and the curve (Fig. 4–15) shifts to the right. When the population is less resistant to osmotic lysis, the cells are said to be more fragile, and the curve shifts to the left.

A specific example of a fragile population are the erythrocytes from mice with **hereditary spherocytosis.** These cells tend to be spherical rather than bi-concave. As a result, the cells can no longer accommodate the volume increase caused by osmotic flow of water into the cells. As seen in Figure 4–15, some of the cells lyse when the osmolarity of the NaCl medium decreases only slightly. Fifty percent of the cells lyse at an osmolarity of 225 mOsm, which represents a considerable change from the behavior of the normal cell population. The erythrocytes in outdated blood from a blood bank provide another example of increased osmotic fragility. These cells are less resistant to osmotic stress than erythrocytes in freshly drawn human blood.

After a discussion of some specialized transport mechanisms, we will return to the concept of osmosis and the way in which the cell transports materials in such a way as to maintain appropriate cell volume.

Facilitated Diffusion of Lipid-Insoluble Substances

A lipid bilayer is essentially impermeable to ions and large water-soluble molecules such as D-glucose, amino acids, and other metabolites. Still, these substances diffuse across the membrane in response to electrochemical gradients. They do so by a process of **facilitated diffusion,** so called because their passage is facilitated by intrinsic membrane proteins that act as **carriers** or **channels.** Ions and other lipid-insoluble substances normally would not be able to pass through the nonpolar interior of the membrane. The proteins enable them to get from one side of the lipid bilayer to the other in part by shielding them from the lipid molecules. The molecular mechanisms of action of membrane carriers and channels are the subjects of intense investigation.

Simple diffusion and facilitated diffusion both occur spontaneously and passively in response to an electrochemical gradient. Similarly, both processes cease once the gradient has been abolished. Since facilitated diffusion is mediated by a carrier or channel protein, however, a number of characteristics distinguish it from simple diffusion: (1) facilitated diffusion allows a very high rate of solute transport; (2) it is a saturable process; there is a limit to the transport rate that cannot be exceeded; (3) it is a highly specific process; and (4) solute transport can be blocked by competitive inhibitors.

Carrier Proteins. A hypothesis to explain the operation of a carrier is depicted in Figure 4–16. This model proposes (1) that the carrier protein is composed of at least two subunits and (2) that binding of solute to a site exposed to the cell exterior triggers an intramolecular rearrangement in the structure of the protein such that the solute binding site is now exposed to the cell interior. The solute is always surrounded by hydrophilic regions of the protein subunits and is never exposed to the hydrophobic lipid phase of the membrane. Once the solute is exposed to the interior of the cell, where the solute concentration is low, it dissociates from the binding site. This step allows the carrier protein to revert back to its original conformation with the empty binding site again exposed to the cell exterior. The cycle can then be repeated to move another solute molecule into the cell.

Channel Proteins and Gating. A channel mechanism is particularly important for the rapid transmembrane movement of ions such as Na^+ and K^+. One hypothesis is that a hydrophilic channel across the membrane is formed between the subunits of a protein. The open channel permits a much higher

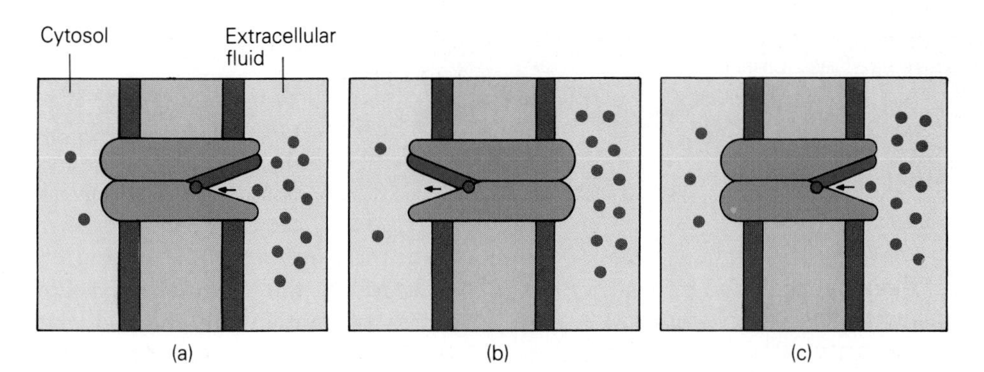

Figure 4–16 A possible mechanism for carrier-dependent transport. (*a*) Solute concentration is high outside and low inside the cell. A solute molecule binds to the carrier protein at a specific site that is exposed on the exterior surface of the cell membrane. (*b*) Binding of the solute to the carrier triggers a change in conformation, and the solute binding site becomes exposed to the cell interior. (*c*) The solute easily dissociates from the carrier because of the low concentration of solute inside the cell. This dissociation allows the carrier protein to revert to its original conformation. The solute binding site is again exposed to the exterior of the cell and the cycle can be repeated. Probably the changes in protein conformation are much more subtle than depicted here.

Focus on Cystic Fibrosis

A disorder that is specifically linked to cell transport and secretion is **cystic fibrosis (CF).** CF is an inherited genetic disorder occurring in children in which the secretions of the body glands are abnormally thick and sticky. They clog up the glands and prevent certain organs from working properly. The pancreas cannot function, and without pancreatic enzymes it is difficult to absorb certain types of food. The liver and lungs also become clogged with sticky mucus. The obstruction in turn leads to cystic and fibrotic changes in these organs. In addition, children with cystic fibrosis have abnormally high concentrations of electrolytes (salt) in their sweat and saliva.

The most severe effect of CF is upon the lung. Sticky mucus not only blocks the air passages but also serves as a culture (growth) medium for harmful bacteria. Consequently, the small airways become obstructed, and the bacteria or their toxins cause respiratory infections that often lead to pneumonia. The repeated infection causes scarring and promotes fibrosis of the lung.

Another complication in CF patients, as a result of the clogged airways, is the low concentration of oxygen in the lungs. This decreases the amount of oxygen that gets into the blood, and it also triggers a response that causes the pulmonary arteries to constrict. The constriction causes pulmonary hypertension (an abnormal high blood pressure in the lungs) and puts an increased strain on the heart. In fact, 98% of all CF patients die from cardiopulmonary complications.

Prior to the 1940s, babies with CF died early from respiratory failure. In the 1950s, with the advances in antibiotic therapy, half of all CF infants lived to be 3 to 5 years old. Today, with strong antibiotics and proper therapy, half of all children with CF live into their twenties, which for everyone else is the prime of life. The disease is an autosomal recessive disorder in Caucasian children and occurs in about one in 1,500 live births. CF is rare in black children and essentially absent among Asiatics. Scientists are rapidly closing in on the disease process of cystic fibrosis. They have discovered that an abnormal gene leads to a defective transport of chloride ion in the mucous gland; they are now trying to find a way to correct the defective gene.

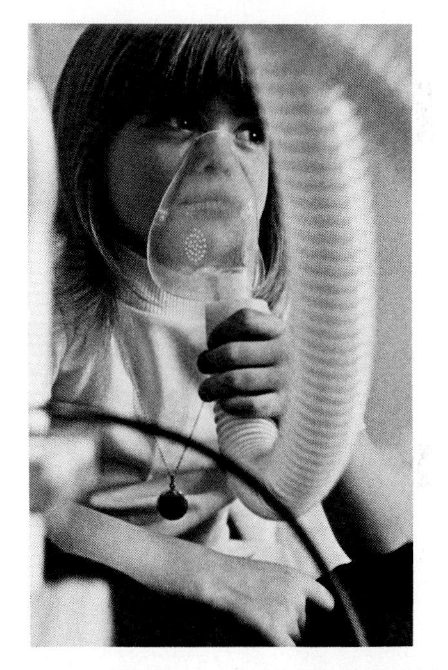

A child with cystic fibrosis using a nebulizer which disperses medications into a fine mist that can be inhaled. (Courtesy of the Cystic Fibrosis Foundation.)

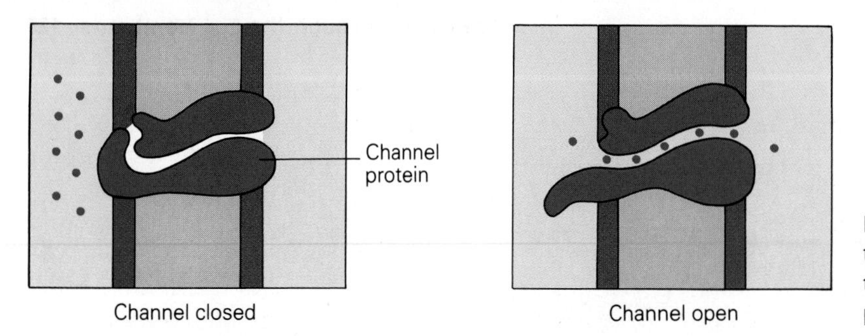

Figure 4–17 A highly simplified view of the gating mechanism for regulating ion transport through a membrane channel protein.

rate of transport than that provided by a carrier. For example, some channels enable ions to pass through them at the rate of 10^8 ions/sec, whereas the fastest carriers move solutes across membranes at a rate no greater than 10^5 molecules/sec.

It is unlikely that an ion moves through a channel by simple diffusion down the electrochemical gradient. Most likely the ion undergoes specific interactions with charged groups along the sides of the channel. These reactions help to move the ion through the channel. The types of charged groups present in the channel may be important also in determining the specificity of the channel for one ion type.

Another difference between a channel and a carrier, in addition to the overall rate of transport, is that a channel can be kept closed by a "**gate**" (Fig. 4–17). A closed channel will not permit movement of ions even though the electrochemical gradient may favor it. The mechanisms that regulate the

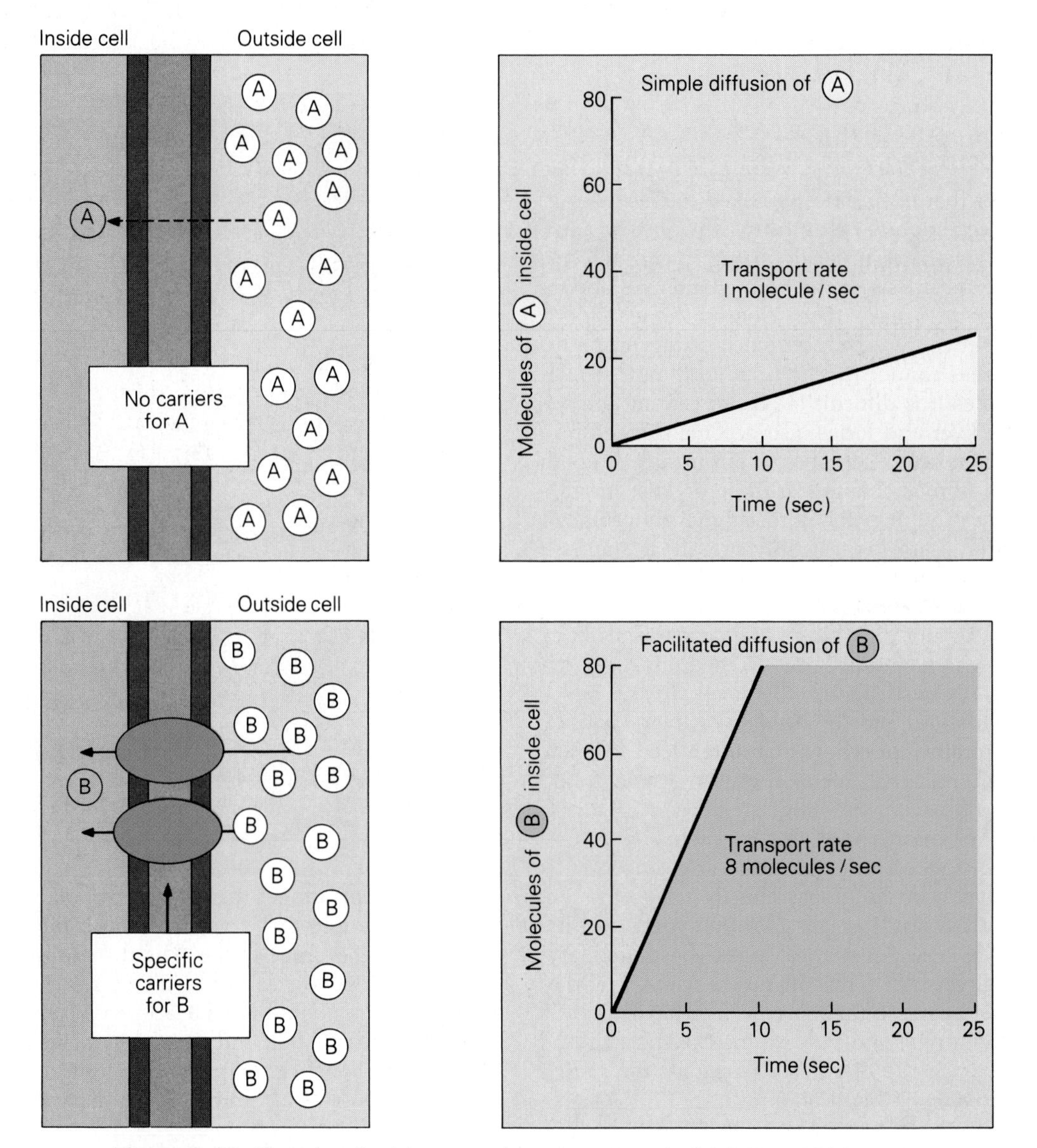

Figure 4–18 The rate of net transmembrane movement of Solute B, which occurs by facilitated diffusion, is much higher than the rate of simple diffusion of Solute A, even though Solutes A and B have similar molecular sizes and lipid solubilities.

opening and closing ("**gating**") of ion channels will be discussed in a later chapter.

The High Rate of Facilitated Diffusion. A solute that crosses a cell membrane by facilitated diffusion does so at a rate that is much more rapid than that of a solute crossing by simple diffusion, even though both solutes are of the same molecular size and have the same lipid solubility. Consider a hypothetical example in which there are 17 solute molecules outside a cell and only 1 inside; assume that the solute is electrically neutral (Fig. 4–18). Net movement of solute inside the cell will cease when 8 more molecules have entered—that is, when there are 9 molecules on each side of the membrane. It will take 8 sec to achieve this if the solute moves by simple diffusion at a rate of 1 molecule entering the cell per second. In contrast, it will take only 1 sec to do this if the solute moves by facilitated diffusion, which allows solute entry at a rate of 8 molecules/sec.

This example illustrates that the end point of both processes is the same but is achieved faster by facilitated diffusion. In this way the carrier protein serves a function analogous to that of an enzyme in a biochemical reaction. The rates of transport used in this example are arbitrary and are not meant to represent experimentally determined values.

Saturation in Facilitated Diffusion. A facilitated diffusion system can become saturated because the number of binding sites on the carrier for solute molecules is limited. The rate of transport will be at a maximum when all the sites are occupied by solute (Fig. 4–19). A graph drawn to relate transport rate to solute concentration (see Fig. 4–20) shows that initially the rate of solute transport increases in proportion to the solute concentration. Once the solute concentration has reached a level at which the carrier binding sites are always full (saturation), however, any further increases in solute concentration will produce no change in the solute transport rate. The rate of transport is now at its maximum; this is termed the $\boldsymbol{V_{max}}$. The solute concentration that gives rise to one half of the V_{max} is known as the $\boldsymbol{K_m}$. These kinetic parameters are a very useful way of comparing the characteristics of different transport systems. It is worth adding that transport by simple diffusion is not saturable. The rate of transport increases in proportion to the solute concentration (see Fig. 4–20). The limitation posed by carrier saturation is rarely a problem because the overall rate of transport by this mechanism is so much faster than that by simple diffusion. The V_{max} for transport by facilitated diffusion is analogous to an enzyme-catalyzed reaction, which proceeds at a maximum rate once the enzyme is fully saturated with substrate (see Chapter 2). Any further increase in substrate concentration will have no effect on the reaction rate. The analogy with enzymes is limited, however, because the solute is not permanently altered as a consequence of binding to the carrier protein.

The Specificity of Facilitated Diffusion. Unlike simple diffusion, facilitated diffusion is a highly specific process. Since transport of the solute involves binding to a carrier protein, it is likely that there are

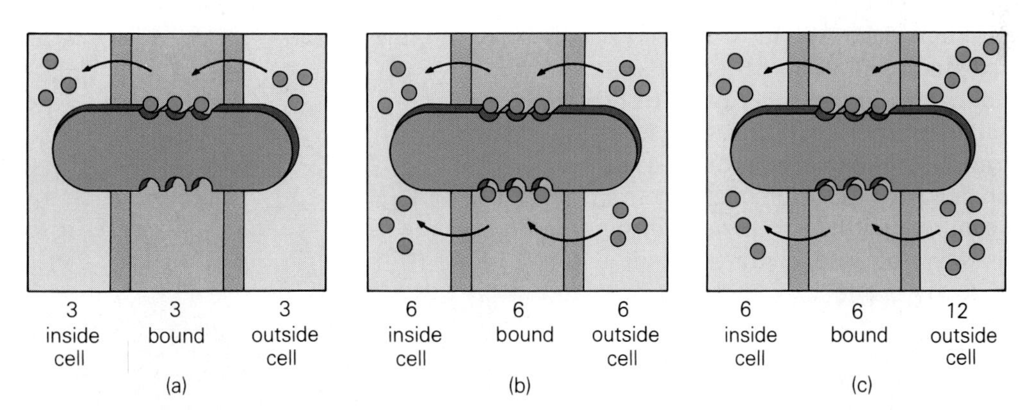

Figure 4–19 The saturation of a membrane carrier with solute. (*a*) Assume that a carrier protein in a membrane has specific binding sites for 6 solute molecules. If 3 solute molecules inside a cell arrive at the carrier, all 3 molecules will be transported outside the cell in one cycle of the carrier. (*b*) If 6 molecules inside the cell arrive at the carrier, all 6 molecules will be moved outside the cell in one cycle. That is, the rate of net transmembrane transport of solute doubles when the amount of solute doubles. (*c*) When 12 solute molecules arrive at the carrier, the rate of solute movement will not increase because the carrier can bind and transport only 6 molecules per cycle. Once the carrier is saturated with solute, any further increase in solute concentration inside the cell will have no effect on the overall rate of solute transport. At this point the rate of solute transport has reached a maximum.

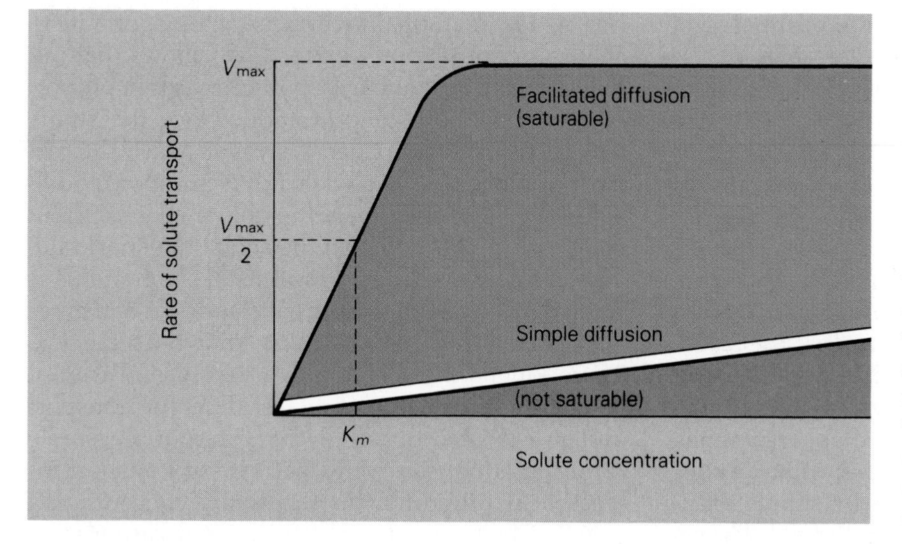

Figure 4–20 Graphical representation of the saturability of facilitated diffusion. The maximum rate of transport possible is termed the V_{max}, and the K_m is the concentration of solute necessary to reach one half of the V_{max}. Although simple diffusion is not saturable, the rate of transport is much slower at the physiological range of solute concentration.

specific carriers in the cell membrane for each of the solutes that enter or leave the cell by facilitated diffusion. The specificity of the carrier is not absolute. For example, structurally related amino acids often are able to share the same carrier. The transport of sugars across the erythrocyte plasma membrane provides another example. The transport system works best when glucose is the solute being transported, but related sugars such as galactose can utilize the same system to cross the membrane.

Competitive Inhibition in Facilitated Diffusion. Structurally related compounds will compete for the same binding site on a membrane carrier, giving rise to the phenomenon of **competitive inhibition** of solute transport, as characterized by an increase in the K_m without a change in the V_{max} of the transport system. Thus, amino acid A, which shares the same transport system as amino acid B, will act as a competitive inhibitor of the transport of B if both are present simultaneously (Fig. 4–21). The inhibitory effect of A can be overcome by an increase in the concentration of B such that more molecules of B are available to compete with A for the carrier binding sites. In this way the rate of transport of B can be increased to the same V_{max} that was achieved in the complete absence of A, but a higher concentration of B (i.e., an increased K_m) will be required (see Fig. 4–21).

Active Transport: Energized Movement Against a Gradient

Thus far we have discussed passive transport processes, in which molecules and ions move in or out of the cell from regions of high to regions of low concentration. These movements, while sometimes requiring the presence of a special protein molecule, do not require energy. They occur in response to electrochemical gradients. The result is always a *decrease* in concentration of a once highly concentrated region.

Frequently, however, the cell requires an increase in concentration of an already highly concentrated solute. The solute must move from a less concentrated region outside the cell to a more concentrated region inside—that is, *against* the prevailing gradient. This process requires input of energy. The transport system will continue to operate for as long as the energy supply and the solute are available. Active transport, like facilitated diffusion,

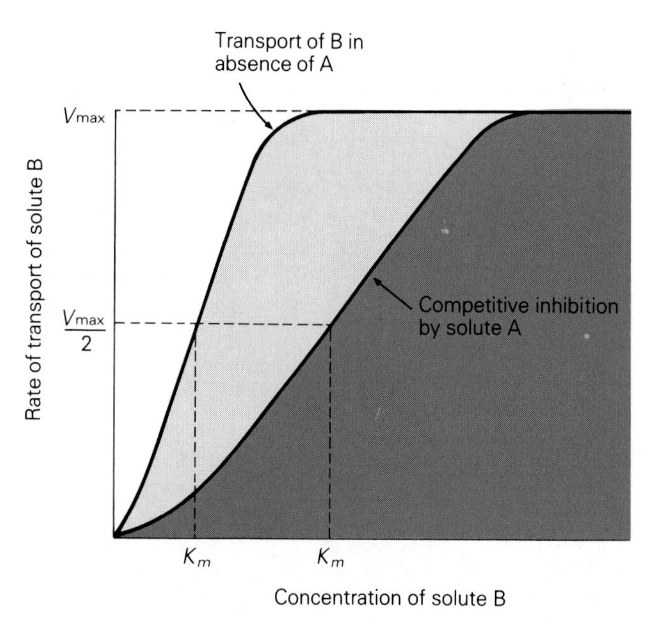

Figure 4–21 Competitive inhibition of facilitated diffusion of Solute B by the structurally similar Solute A. In the presence of A, the concentration of B must increase to reach the same V_{max} as that attained in the absence of A.

requires carrier proteins. The energy supply for active transport is derived from cellular metabolism; hence, active transport systems are susceptible to inhibition by compounds that act primarily as metabolic inhibitors and have no direct effect on the carrier proteins.

The mechanisms of active transport are of two general types depending on whether the movement of solute is linked directly or indirectly to energy-yielding reactions (see Fig. 4–10). When the movement of solute is coupled directly to an energy-yielding reaction, the transport process is termed **primary active transport.** When active transport of a solute is not coupled directly to the energy-yielding reactions, the transport mechanism is described as **secondary active transport.**

Primary Active Transport

The best known example of a primary active transport system is probably the **sodium pump.** This pump is present in the plasma membrane of most animal cells and is known also as the **Na^+/K^+-ATPase pump** because Na^+ is pumped out of the cell at the same time as K^+ is pumped in and because the ion movements are coupled directly to breakdown of ATP. Removal of one of the phosphate groups from ATP releases free energy that is used to drive the movement of both Na^+ and K^+ against their respective electrochemical gradients. It is chiefly through this active transport system that the cell is able to maintain a high K^+ concentration and a low Na^+ concentration in the cytosol.

The breakdown of ATP is tightly coupled to the transport of Na^+ and K^+; the one cannot occur without the other. For every molecule of ATP that is split into ADP and phosphate, 3 Na^+ are pumped out of the cell and 2 K^+ are pumped in (Fig. 4–22). The molecular events that link ATP breakdown to ion transport are not fully understood, but it seems clear that the enzyme protein is temporarily phosphorylated by attachment of the phosphate group released from the ATP. The phosphorylation step is Na^+-dependent—that is, it occurs only when Na^+ is present, and presumably leads to a change in protein conformation that moves bound Na^+ out of the cell. The subsequent dephosphorylation (removal of phosphate) is K^+-dependent and probably allows the protein to return to its original conformation moving bound K^+ into the cell. These steps are depicted in the scheme in Figure 4–23. The Na^+/K^+-ATPase is represented as a protein with four subunits. This active transport system is so important to the cell that about one third of the cell's energy supply is used to maintain its activity.

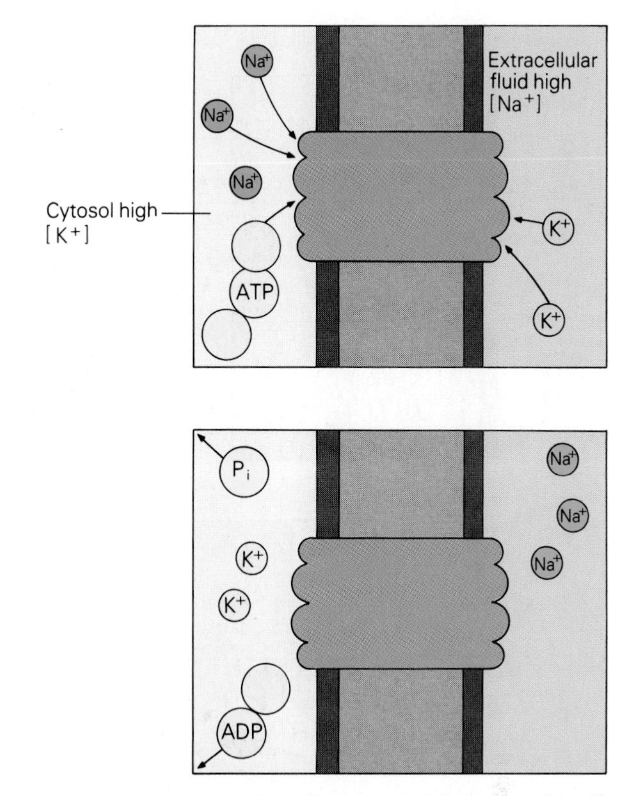

Figure 4–22 An overview of primary active transport by the Na^+/K^+-ATPase. The free energy derived from hydrolysis of ATP is used to drive transport of Na^+ out of the cell and K^+ into the cell. Both ions are moved against their electrochemical gradients. Three Na^+ exit for every two K^+ that enter the cell.

Secondary Active Transport

When transport of a solute is not coupled directly to energy-yielding reactions it is described as **secondary active transport.** A common example of this kind of mechanism is a transport system that is driven by the energy stored in the electrochemical gradient for another solute. In animal cells, this solute is usually Na^+, which enters cells passively by moving down a very favorable electrochemical gradient.

Symport and Antiport. Recall that there is a low sodium concentration and a negative potential inside the cell relative to the outside. Many different solutes are transported across cell membranes against their electrochemical gradients by coupling to the movement of Na^+. Movement of the solutes in the same direction as Na^+ is known as **symport;** movement in the opposite direction is known as **antiport** (Fig. 4–24). Solute transport systems that are coupled to Na^+ movement have a very specific requirement for Na^+; other ions of similar size such as K^+ and Li^+ cannot be used in place of Na^+.

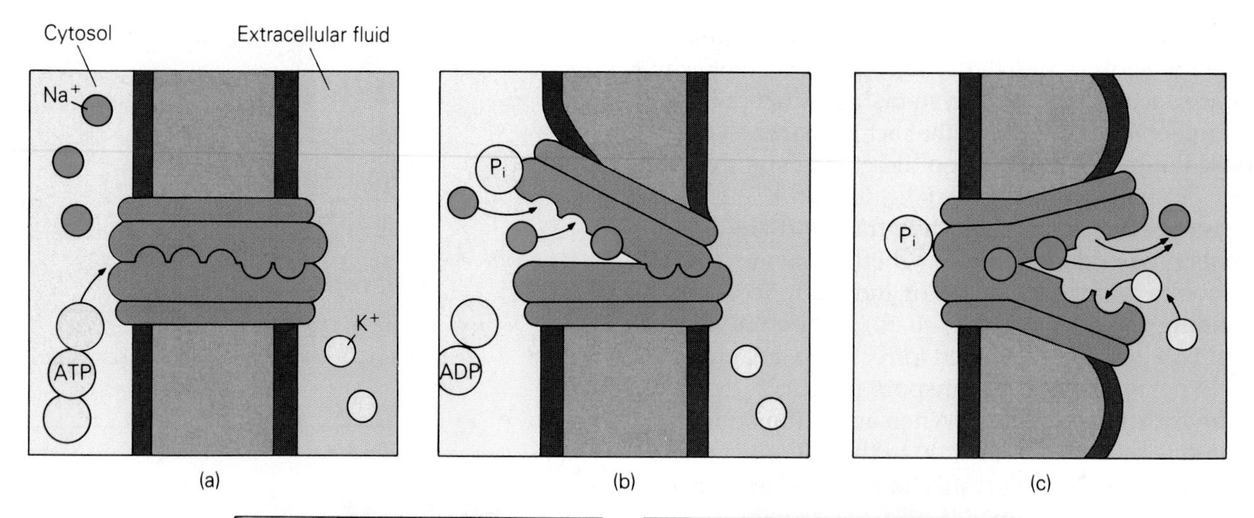

Figure 4–23 Hypothetical scheme of action of Na^+/K^+-ATPase. (*a, b*) In the presence of Na^+ inside the cell, ATP binds to and phosphorylates one of the protein subunits. Na^+ then binds to specific sites on the phosphorylated protein. (*c, d*) These events trigger a conformational change that exposes the bound Na^+ to the cell exterior, and Na^+ dissociates from the protein. The binding site for K^+ is now exposed, and external K^+ binds. (*e*) Binding of K^+ promotes dephosphorylation (removal of the phosphate) of the protein, which allows it to revert to the original conformation. The conformational change moves the bound K^+ inside the cell, where it dissociates from the protein. The cycle now can be repeated. The conformational changes are greatly simplified in the drawing.

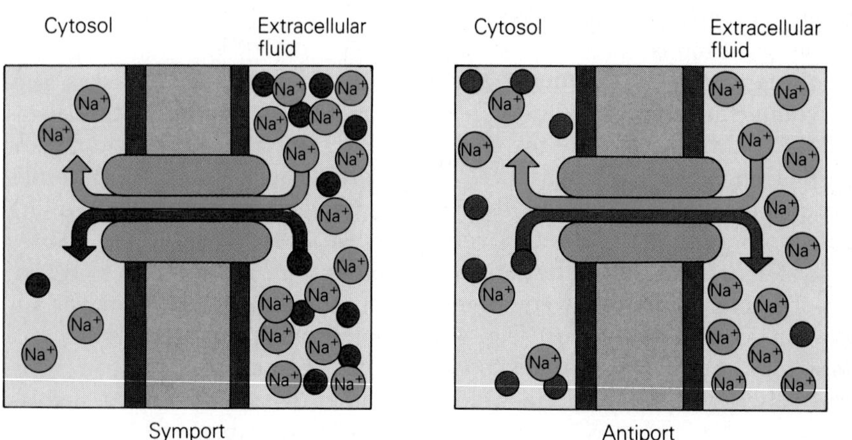

Figure 4–24 Secondary active transport occurs when solute transport is driven by the electrochemical gradient for Na^+. Both solute and Na^+ bind to the membrane carrier. The Na^+ gradient is maintained by primary active transport of Na^+.

The Na^+ gradient across the cell membrane is maintained by the Na^+/K^+-ATPase, which actively pumps Na^+ out of the cell against its electrochemical gradient. As we have just discussed, this process is driven directly by a metabolic energy supply in the form of ATP. Thus, secondary active solute transport that is driven by the Na^+ electrochemical gradient depends ultimately on a supply of ATP but is not linked directly to the reactions that break down ATP.

The mechanism of Na^+-coupled transport systems is believed to involve a two-site carrier that binds both Na^+ and the solute. A possible mechanism is illustrated in Figure 4–25. Na^+ outside the cell binds readily to the carrier and encourages binding of solute to the carrier even though the solute concentration is low outside the cell. A conformational change in the carrier protein now exposes the bound Na^+ and solute to the cell interior. Na^+ readily dissociates from its site because the intracellular Na^+ concentration is very low. Loss of Na^+ from the carrier may alter the binding of solute to the carrier and cause the solute to dissociate from the carrier even though there is already a high intracellular concentration of the same solute.

Transport Processes and Cell Volume

The sodium pump, which uses ATP to pump Na^+ ions out of the cell in exchange for K^+, is a key factor in the mechanism for controlling the osmolarity of the cytosol, and hence the intracellular volume, of animal cells. By effectively adding this solute to the extracellular fluid and simultaneously keeping the intracellular Na^+ concentration low, the cell is able to prevent osmotic entry of water and maintain the intracellular volume. The other principal solutes that contribute to the osmotic pressure of cell cytosol include proteins, organic phosphates such as ATP, metabolic intermediates, and K^+ ions.

The importance of the sodium pump in the control of cell volume is evident in the fact that animal cells swell and sometimes burst when the pump is inhibited. Decreased activity of the Na^+ pump probably contributes to the increased osmotic fragility of erythrocytes in stored blood. During storage, the ATP supply tends to fall. Since the Na^+ pump cannot operate without ATP, any Na^+ ions that "leak" (diffuse) back into the cell will not be pumped out. This abnormal buildup in intracellular Na^+ leads to a gradual rise in the osmolarity of the cell cytosol,

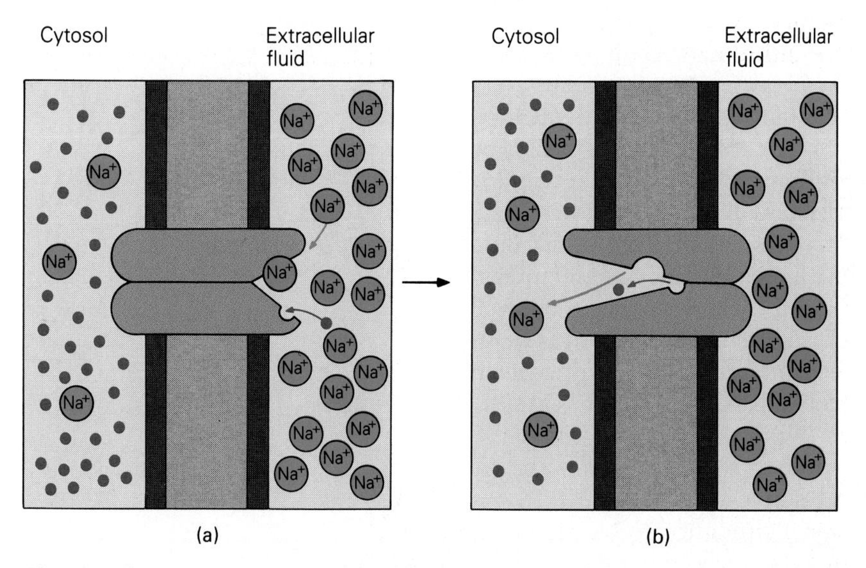

Figure 4–25 Hypothetical scheme of action of a two-site carrier for Na^+-coupled solute symport. (*a*) Binding of Na^+ outside the cell to the carrier increases the affinity for the solute, which also binds even though it is present at a low concentration. (*b*) A conformational change in the carrier protein exposes the binding sites to the cell interior. Na^+ dissociates from the carrier because the Na^+ concentration is low inside the cell. This dissociation step leads to a decrease in carrier affinity for solute, so the latter also dissociates from the carrier even though the solute concentration is high inside the cell. The conformational changes are probably much more subtle than those shown in these diagrams.

with osmotic entry of water causing the cell to swell. In this swollen state, the cells are less resistant to osmotic stress than normal biconcave erythrocytes because they cannot accommodate the same increase in intracellular volume.

Typically, animal cells are exposed to a high concentration of external Na^+. There is no net increase in intracellular Na^+ because the rate at which Na^+ leaks into the cell is closely matched by the rate at which Na^+ leaves the cell via the Na^+ pump. This mechanism prevents the external and internal Na^+ concentrations from equilibrating even though Na^+ constantly enters and leaves the cells.

Thus, Na^+ outside the cell can be considered to behave as if it were a **nonpermeant solute.** Similarly, the rate at which K^+ is pumped into the cell matches the rate at which K^+ leaks out, so the high internal K^+ concentration and the low external K^+ concentration are maintained. Thus, K^+ inside the cell behaves effectively as a nonpermeant solute because, as with Na^+, there is no net movement of K^+ into or out of the cell even though the internal and external concentrations are markedly different.

Some solutes, such as urea and glycerol, penetrate the plasma membranes of an animal cell very rapidly. The cell does not have a mechanism for pumping these solutes back out again. Such **permeant solutes** exert only a transient effect on cell volume, however. Unlike Na^+ and K^+, their external and internal concentrations rapidly equilibrate.

Consider an erythrocyte placed in a large volume of a solution containing 300 mOsm NaCl and 60 mOsm urea (Fig. 4–26). Initially the cell will experience a hyperosmotic environment. Water will flow out, and the cell volume will decrease. However, as the urea rapidly enters the cell and the intracellular urea concentration increases toward 60 mOsm, water also will enter and the cell will begin to swell. The final volume of the cell at equilibrium is determined only by the osmolarity of the impermeant solutes in the extracellular medium. In this case, the impermeant solutes (NaCl) have a total osmolarity of 300 mOsm, the same as the cell cytosol, and the cell will assume its original volume.

Na^+ is the most abundant solute present in blood plasma. As an effectively impermeant solute, it is the major determinant of intracellular volume in the animal, even though plasma contains many membrane permeable solutes. A drop in plasma Na^+ concentration will reduce plasma osmolarity, and osmotic flow of water into cells will cause them to swell. Conversely, a rise in plasma Na^+ concentration will lead to cell shrinkage owing to loss of intracellular water. This is one of the reasons why plasma Na^+ concentration is closely

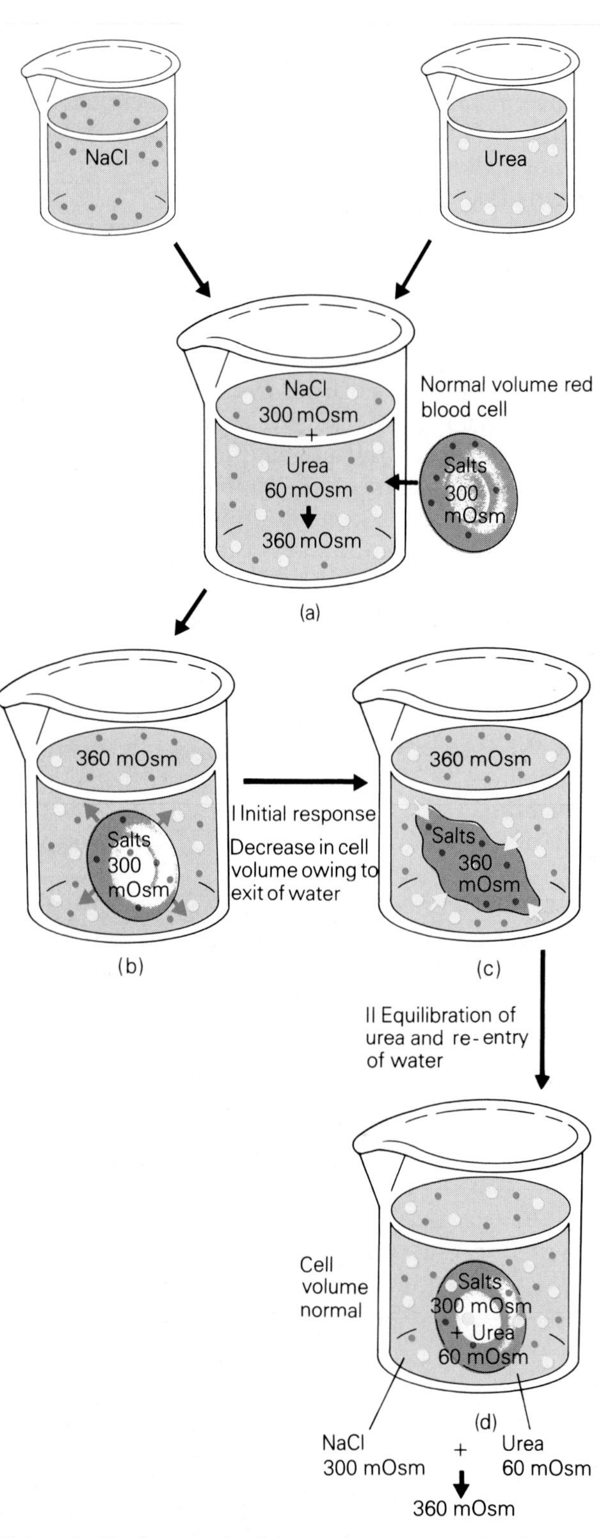

Figure 4–26 Permeant solutes such as urea exert only a transient effect on the volume of an erythrocyte because their intracellular and extracellular concentrations rapidly equilibrate. In reality, Steps I and II probably overlap considerably.

regulated. A change in Na^+ concentration induces a series of rapid responses designed to return it to the normal range.

Epithelial Transport

The transport systems in the plasma membrane of the cell are important for supplying nutrients and removing waste products. In addition to these roles, the specific arrangement of these transport systems in epithelial cells permits the net movement of solutes and water from one side of the epithelium to the other. This organization is crucial for the function of epithelial cells such as those lining the lumen of the small intestine and the renal proximal tubule, tissues that are specialized for reabsorption. These cells allow **transcellular transport** of water and solutes because the entry and exit pathways are on opposite sides of the cell (Fig. 4–27).

The entry step is at the apical membrane. This is the membrane where the Na^+-coupled symport systems for amino acids, glucose, and phosphate are located. The exit step is across the basolateral membrane where the Na^+/K^+-ATPase is located. The Na^+ gradient across the apical membrane is maintained because the ATPase in the basolateral membrane constantly pumps Na^+ out of the other side of the cell to keep intracellular Na^+ low.

Junctional Complexes. The asymmetrical distribution of transport systems between the apical and basolateral plasma membranes persists because of the **junctional complexes** between the cells. The junctional complexes occur at the point where the plasma membranes of neighboring cells come into close contact, and they prevent lateral diffusion of the proteins in the cell membrane. Thus, the carrier proteins in the apical membrane are physically prevented from leaving the apical membrane and mixing with the carrier proteins in the basolateral membrane.

In some types of epithelia, these junctions are relatively permeable and allow water and ions to move between the epithelial cells, providing an additional or **paracellular** pathway across the epithelium (see Fig. 4–27). This type of epithelium is termed a "**leaky**" epithelium because water and some solutes can leak between the cells in addition to moving through the transcellular pathway. A good example of this type of epithelium is the proximal tubule of the kidney. In other epithelia, such as the small intestine and the collecting tubule of the kidney, the cell junctions are impermeable; the only route by which water and solutes can cross the epithelia is the transcellular pathway. These types of epithelia are termed "**tight**" epithelia because they do not permit leakage between the cells.

Water movement across epithelia is always passive and is driven by the osmotic gradient. This gradient is set up by the active transport of solutes from the lumen and their addition to the fluid in the intercellular space.

A problem that leaky epithelia face is **backflux** of solute through the paracellular pathway. Thus, in the proximal tubule, the Na^+ that is reabsorbed from the lumen by the transcellular route is added to the fluid in the intercellular space (see Fig. 4–27), tending to increase the Na^+ concentration of the fluid. If the Na^+ concentration increases to the point at which it exceeds the concentration in the fluid in the tubular lumen, a Na^+ concentration gradient will exist. In addition to favoring water reabsorption from the lumen, this Na^+ gradient will drive movement of some of the Na^+ back into the lumen through the paracellular pathway. This backflux will limit the net reabsorption of Na^+ from the lumen. Movement of ions through the paracellular pathway is entirely passive and is not under any direct con-

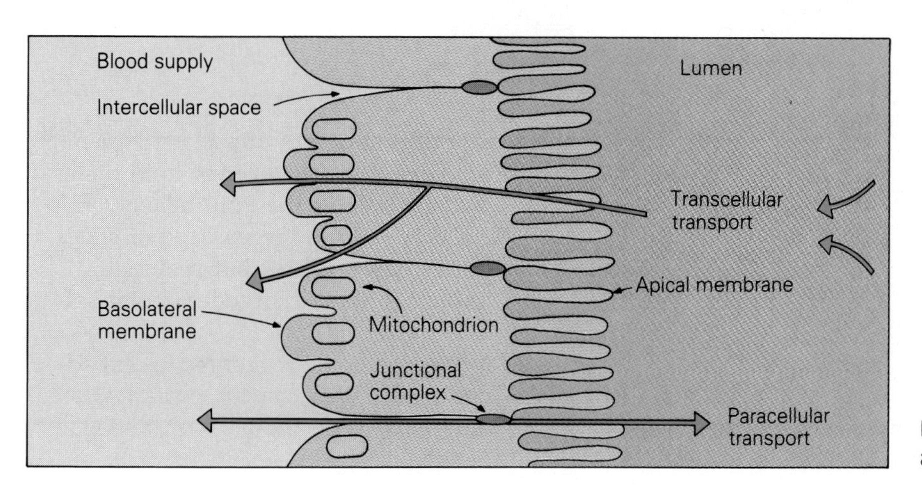

Figure 4–27 Transport pathways across a layer of epithelial cells.

trol. Ions will move in the direction favored by their electrochemical gradient.

The transcellular route achieves net movement from the lumen at one side of the cell to the blood supply at the other side. It is the only route across tight epithelia, and it can be regulated by hormones. This regulation allows almost complete control of the reabsorptive process by specific hormones that are able to regulate either the entry step at the apical membrane or the exit step at the basolateral membrane. For example, the reabsorption of water in the collecting tubule of the kidney is regulated by antidiuretic hormone, which increases the water permeability of the apical membrane of the tubular cells.

In addition to functional specialization of reabsorptive epithelia, there is also marked structural specialization. The surface area of the apical membrane often is increased considerably by numerous **microvilli.** Similarly, the surface area of the basolateral membrane is increased by extensive **infoldings.** Large numbers of mitochondria are present. They are found frequently between the basolateral infoldings (see Fig. 4–27), where presumably they provide a plentiful supply of ATP to maintain the activity of the Na^+/K^+-ATPase in the basolateral membrane. In other words, the cells are adapted for the rapid transport of large volumes of fluid containing a mixture of solutes and water.

We can understand how the different transport systems interact to achieve net movement of a solute across an epithelial cell by considering how inorganic phosphate crosses the proximal tubule in the kidney (Fig. 4–28). The initial step in this process is entry of phosphate into the cell across the apical membrane. This step requires active transport because the negatively charged phosphate ion must move against both an electrical gradient (a negative potential inside the cell relative to the lumen) and a chemical gradient (a higher phosphate concentration inside the cell relative to the lumen). Phosphate enters the cell via a symport mechanism in the apical membrane that couples the movement of phosphate to the movement of Na^+. There is a very favorable electrochemical gradient driving Na^+ entry. Thus, phosphate entry is an example of secondary active transport. Once inside the cell, the phosphate is able to leave by a passive mechanism, possibly facilitated diffusion, located in the basolateral membrane. The electrochemical gradient that opposed phosphate entry at the apical membrane is used now to drive this exit step at the basolateral membrane. After diffusing across the intercellular space, the reabsorbed phosphate enters the blood in the capillary system and is returned to the general circulation. The Na^+ that enters the cell across the apical membrane is pumped out across the basolateral membrane by the Na^+/K^+-ATPase. The Na^+ leaving the cell must be moved against the gradient that enabled it to enter the cell at the apical membrane (see Fig. 4–28). The exit of Na^+ is an example of primary active transport because it is coupled directly to ATP hydrolysis. Na^+ must be removed from the cell in order to maintain a low intracellular Na^+ concentration and a favorable gradient for continued Na^+ entry across the apical membrane.

An analogous arrangement of the appropriate transport systems is used to achieve the reabsorption of glucose and amino acids across the cells of the renal proximal tubule. The solutes enter the cell from the tubule lumen via secondary active Na^+-coupled symport across the apical membrane. Then they exit from the cell via passive facilitated diffusion across the basolateral membrane.

Figure 4–28 Transepithelial movement of inorganic phosphate (Pi) in the renal proximal tubule.

Summary

I. A. The core structure of a plasma membrane is a double layer of lipid molecules known as the lipid bilayer. This structure allows the passage of small lipid-soluble solutes but it is highly impermeable to charged or polar solutes.

B. Most membrane proteins are embedded in the lipid bilayer, but a few are only loosely associated with one side of the bilayer. The proteins carry out most membrane functions by acting as specific transporters, receptors, or enzymes.

C. Carbohydrates are a relatively minor component of the membrane and occur as both glycoproteins and glycolipids. The carbohydrate portion of the molecule is located exclusively on the exterior surface of the membrane.

II. A. A solute crosses a membrane by simple diffusion in response to an electrochemical gradient for the solute. Lipid-soluble, nonpolar molecules cross the plasma membrane this way. Specific membrane proteins, most likely spanning the lipid bilayer, bind charged or polar solutes and "facilitate" their diffusion across the membrane. Cells respond to changes in the osmolarity of their surroundings by gaining or losing water, sometimes lysing in the process.

B. Active transport is the movement of solute against its electrochemical gradient, achieved by input of energy (usually in the form of ATP) derived from cellular metabolism. The solute is moved across the membrane by a specific membrane protein carrier. This step can be linked either directly or indirectly to the energy-yielding reactions.

C. Animal cells resist osmotic entry of water and, hence, an increase in intracellular volume by pumping Na^+ ions out so that the intracellular Na^+ concentration remains low relative to the extracellular concentration.

III. The different types of transport mechanisms act in concert in epithelial cells to achieve directional movement of solutes and water from one side of the cell to the other. This process is aided by the structural and functional organization of the cells.

Review Questions

1. How do the terms *asymmetry* and *mobility* apply to the behavior of the components of cell membranes?

2. What is the arrangement of lipids, proteins, and carbohydrates in a typical plasma membrane?

3. What will happen to the volume of an erythrocyte when it is transferred from an iso-osmotic solution to a hyperosmotic solution?

4. What is the physiological importance of membrane transport processes?

5. What are Overton's Rules? Why is water an exception to these rules?

6. What characteristics of facilitated diffusion distinguish it from simple diffusion?

7. What are microvilli? How do they aid epithelial transport?

8. What is the principal difference between a tight epithelium and a leaky epithelium with respect to transcellular and paracellular routes of transport?

9. How are the transport systems arranged for the movement of solutes, such as glucose, phosphate, and amino acids, from one side of an epithelial cell to the other?

Suggested Readings

Armstrong, W.McD. *The cell membrane and biological transport. Physiology,* 5th ed. Selkurt, E.E., ed. Boston, Mass.: Little, Brown, 1984. pp. 1–26.

Finean, J.B., Coleman, R., and Michell, R.H. *Membranes and Their Cellular Functions,* 2nd ed. Oxford: Blackwell Scientific Publications, 1978.

Hille, B. *Ionic Channels of Excitable Membranes.* Sunderland, Mass.: Sinauer Associates Inc., 1984.

Rothman, J.E., and Lenard, J. "Membrane asymmetry." *Science* (1977) 195: 743–753.

Singer, S.J., and Nicolson, G.L. "The fluid mosaic model of the structure of cell membranes." *Science* (1972) 175: 720–731.

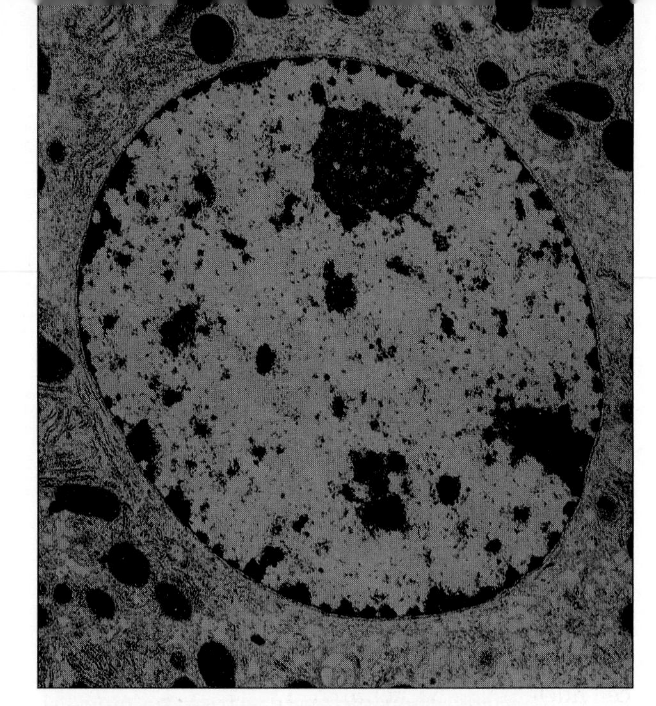

5 Cellular Control Mechanisms

Much of the success of multicellular organisms is enabled by the organization of groups of differentiated cells into specialized tissues and organs. The tissues coordinate to carry out all the functions necessary for the organism's survival.

The specialization and coordination of cell functions require very effective communication both within each cell and among the different tissues. A good example of the importance of efficient communication is the process involved in avoiding a predator. When the tissues responsible for sight, sound, or smell detect the presence of a predator, the survival of the organism often depends on the almost-instantaneous response of other tissues that can carry out rapid evasive movement. Mammals have systems that provide communication and coordination between the different organs. These are the **nervous system** and the **endocrine system.** A large section of the book is devoted to descriptions of these very sophisticated systems.

This chapter emphasizes the ways in which some of the intracellular structures and macromolecules described in the preceding chapters participate in specific control systems that reside and operate within a cell, as well as ways in which these controls are influenced by signals from other cells. We begin by discussing the use of the genetic information stored in DNA, perhaps the fundamental **intracellular control system.** We will learn how this system is subject to regulation primarily by intracellular influences. The latter part of the chapter introduces the concept of **extracellular control systems,** or **cell-to-cell signaling,** the ways in which one cell communicates with and influences the function of another.

An Intracellular Control System: DNA, RNA, and Protein Synthesis

The cells of a complex eukaryotic organism such as a human being have developed a variety of structural and functional specializations, giving rise to different types of tissues, such as epithelial, connective, muscle, and nervous tissue. As previously mentioned in the brief discussion of the functions of the cell nucleus in Chapter 3, the molecular basis for the differentiation of functions and structures of cells is the structure of their proteins. In Chapter 4 we learned, for example, that different cells have different types and amounts of proteins in their membranes. These proteins are distributed in specific ways in the membrane so as to effectively carry out their roles as carriers, channels, receptors, or shaping agents. The red blood cell produces hemoglobin, whereas a nerve cell makes none of this protein. All cells in the body make and use Na^+/K^+-ATPase, but different cells make different amounts of it. Most cells in the body produce myosin (a contractile protein), but only muscle cells make it in the abundance necessary to produce the type of motion we associate with visible body movements.

Overview: Protein Synthesis as Genetic Expression

With few exceptions, the markedly different types of cells contain the same **genome** in their nuclei: that is, each cell, regardless of tissue type, has the same DNA molecules, the same "permanent files" or master set of protein-synthesis instructions. These DNA molecules are bound in groups called **chromosomes.** Every cell in the human body, except the reproductive cells, has the same number (46) of chromosomes containing the same DNA molecules (Fig. 5–1*a*). These DNA molecules are made up of **genes** (Fig. 5–1*b*). Each gene consists of a certain sequence of nucleotides that dictates the sequence of amino acids in a specific protein produced by the cell. Every cell in the body, except the sex cells, can potentially make every protein found in the body.

It would seem that if all cells in the body contain the same genetic information, they all would make the same proteins in the same amounts. How then do cells become **differentiated,** producing some proteins and not others? How does one cell type, such as a pancreatic cell, know to produce an abundance of insulin, whereas another cell, such as a red blood cell, knows *not* to produce insulin (even though it possesses the gene for insulin), but to produce hemoglobin instead? How does a cell in the kidney know to produce myosin in the controlled amount required for its own shape and internal movements, but not in the abundant amounts that would produce movement of the entire organ? How does a skeletal muscle cell know to produce myosin in abundance, enough to produce movement of other tissues?

Genetic Expression: Selective Use of the Permanent Files

The answer to the previous questions lies in the phenomenon of **genetic expression** or, more precisely, the **control of genetic expression.** An individual cell has much more genetic information than it ever needs. Most of the genetic information in the cell is unused, or **suppressed.** At any given time in a particular cell, only a small part of the DNA is being used as a source for information about protein synthesis. In other words, only certain genes are being tapped for information about how to make the specific protein they describe; only certain genes in each cell are **expressed** in the form of **protein products.**

Thus, gene expression is the principal factor that determines the structural, functional, and behavioral characteristics of a cell. In complex eukaryotes, different genes are being expressed as the characteristic protein products of each different cell type. Furthermore, in a given cell, certain genes may be expressed only at certain times (e.g., when the cell receives the appropriate stimulus, possibly a hormone). Some of the mechanisms that control the expression of genes in a cell are discussed in detail in a later section.

The Genetic Code: From Nucleic Acids to Amino Acids

The term **genetic code** refers to the way in which information about protein synthesis is stored in DNA, "translated" into RNA, and then used to make proteins. The information in a DNA molecule specifies the sequence in which amino acids must be assembled to form a functional polypeptide molecule. A relationship exists between the nucleotides of RNA and DNA and the amino acids of a polypeptide chain, such that the order of nucleotides dictates the order of amino acids.

DNA and RNA are each composed of only four different nucleotides, whereas polypeptides can contain up to 20 different amino acids. Since there are only four nucleotides in each type of nucleic acid to specify the 20 different amino acids, it must be that **combinations** of nucleotides, rather than single nucleotides, are used to describe the amino acids. The genetic code must specify, in the sequence in which the four nucleotides are arranged, the sequence in which the 20 common amino acids should be arranged. Think of DNA, RNA, and proteins as three different languages, each with a different alphabet and vocabulary. Protein synthesis then is a process of translation, from the language of nucleic acids to the language of proteins.

The Four Letters of the DNA and RNA Alphabets. DNA molecules are formed from **deoxyribonucleic acids,** four different molecules distinguished by four different **bases,** adenine, cytosine, guanine, and thymine (Fig. 5–2*a*). The DNA "alphabet" consists of four letters—A, C, G, and T—representing the four DNA bases (Table 5–1). These four letters can be combined in a total of 64 different "words," or **triplets,** such as CGA, CGT, CGG, and so on.

RNA molecules are formed from **ribonucleic acids**—again, four different molecules distinguished by four different bases, adenine, cytosine,

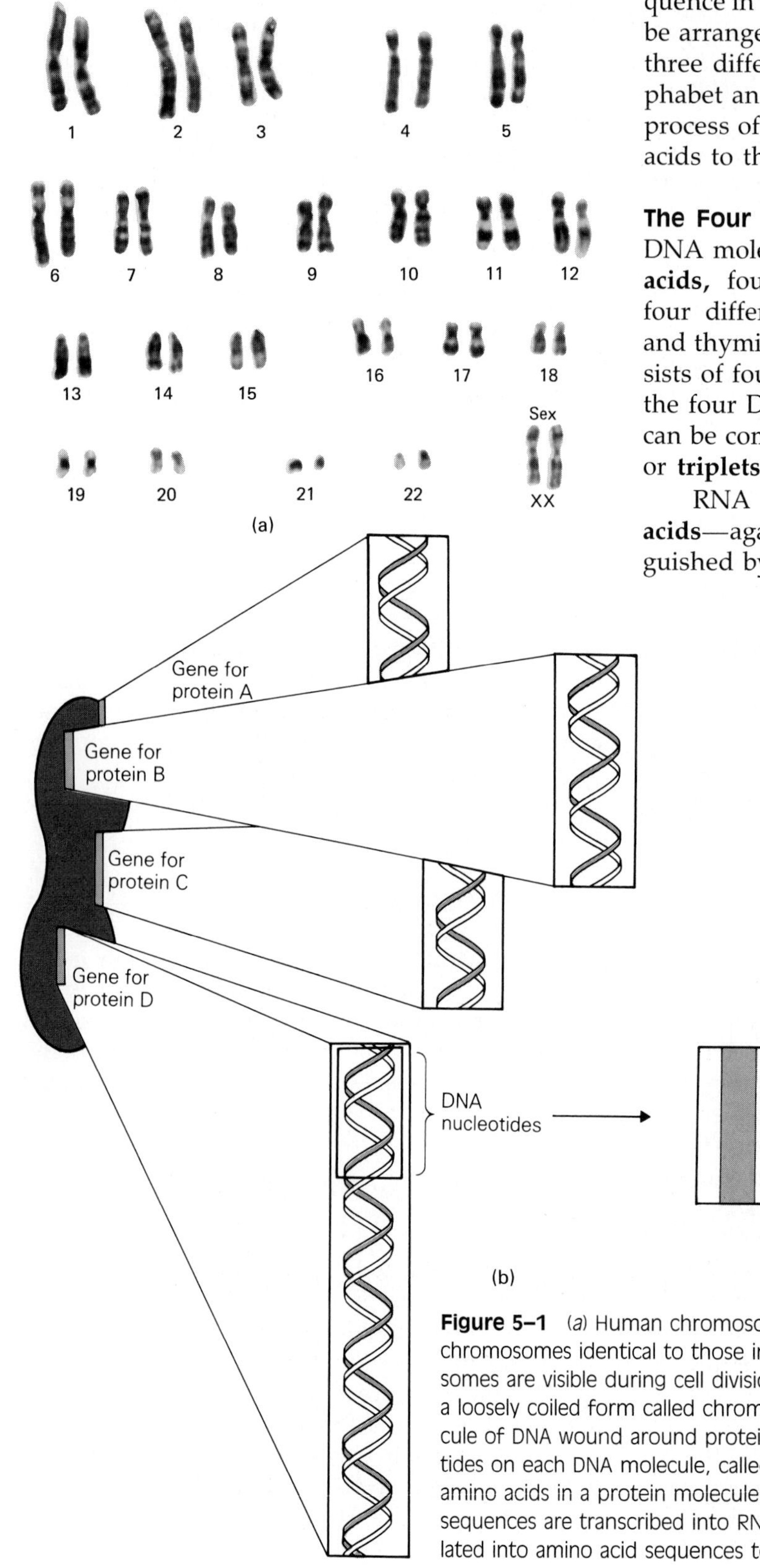

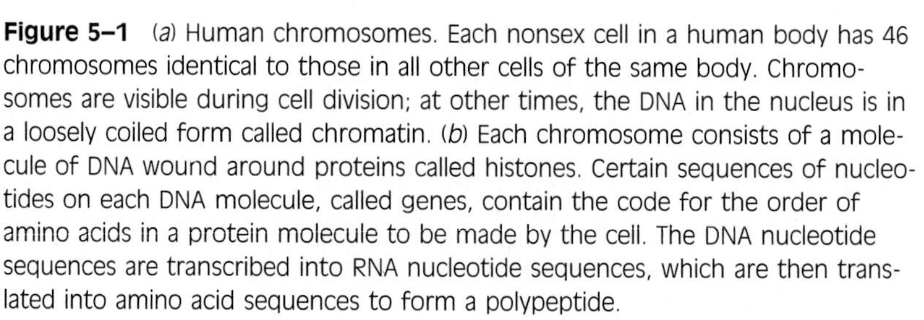

Figure 5–1 (*a*) Human chromosomes. Each nonsex cell in a human body has 46 chromosomes identical to those in all other cells of the same body. Chromosomes are visible during cell division; at other times, the DNA in the nucleus is in a loosely coiled form called chromatin. (*b*) Each chromosome consists of a molecule of DNA wound around proteins called histones. Certain sequences of nucleotides on each DNA molecule, called genes, contain the code for the order of amino acids in a protein molecule to be made by the cell. The DNA nucleotide sequences are transcribed into RNA nucleotide sequences, which are then translated into amino acid sequences to form a polypeptide.

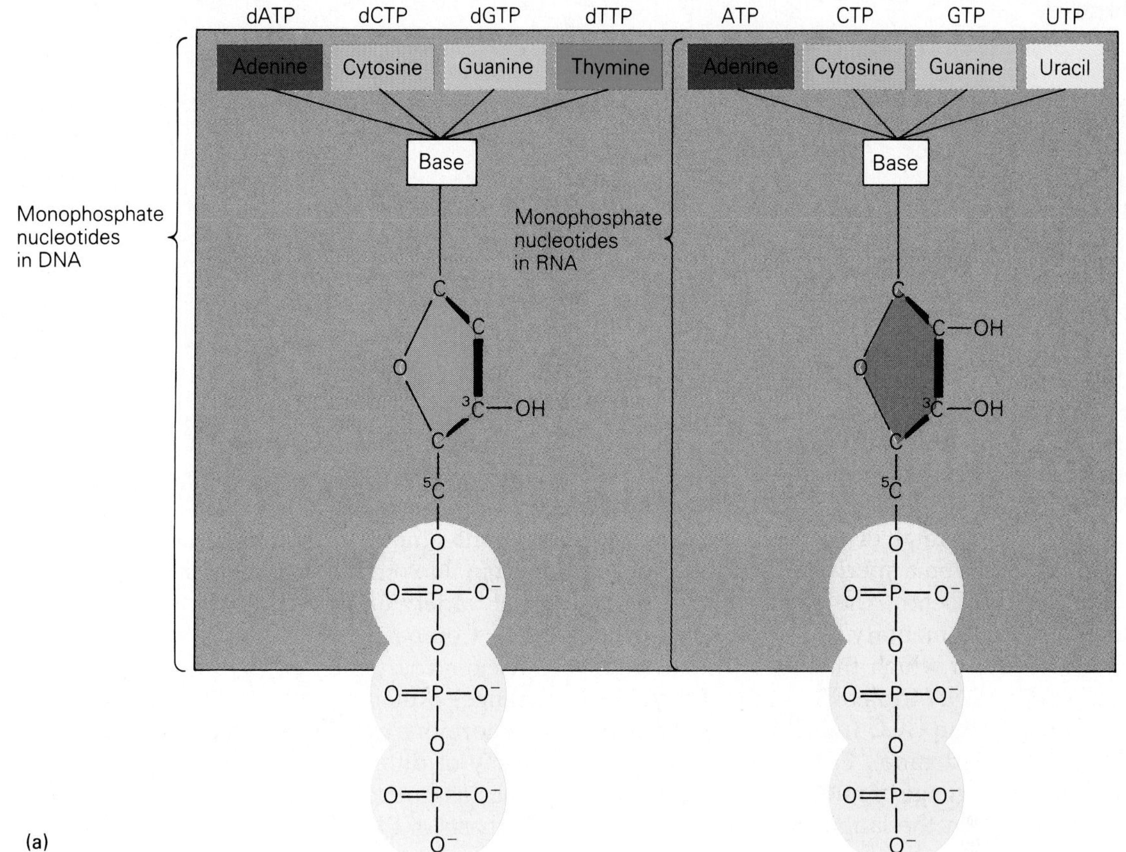

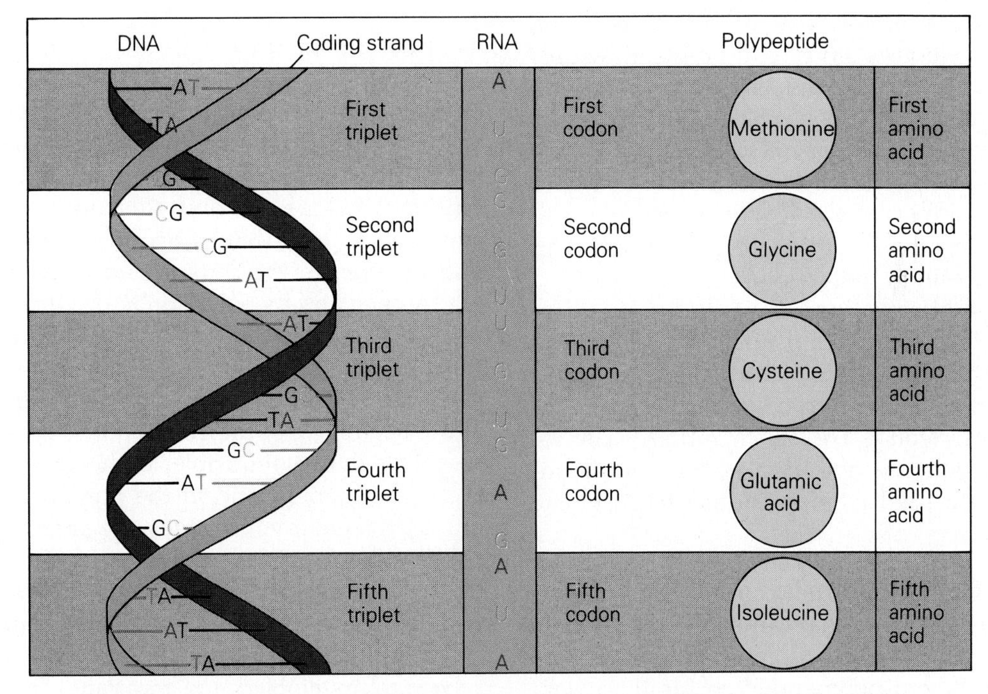

Figure 5–2 (*a*) The DNA "alphabet" consists of four nucleotides, derived from triphosphate nucleotides and distinguished by their bases: adenine, cytosine, guanine, and thymine. The RNA "alphabet" contains nucleotides formed from a different sugar molecule than that in DNA and distinguished by a slightly different set of bases: adenine, cytosine, guanine, and uracil. (*b*) The genetic code on DNA is copied in a complementary strand of RNA, and then a matching chain of amino acids is formed. Each set of three bases on a DNA strand, called a triplet, is transcribed into a complementary set of RNA bases, called a codon. Each codon refers to a specific amino acid.

Table 5–1
DNA, RNA, and Protein "Alphabets"

DNA Bases	RNA Bases	Amino Acids				
A	A	Alanine	Cysteine	Histidine	Methionine	Threonine
C	C	Arginine	Glutamic acid	Isoleucine	Phenylalanine	Tryptophan
G	G	Asparagine	Glutamine	Leucine	Proline	Tyrosine
T	U	Aspartic acid	Glycine	Lysine	Serine	Valine

guanine, and uracil (see Fig. 5–2*a*). Thus, the RNA alphabet also contains four letters—A, C, G, and U (Table 5–1). These also can be combined in groups of three to make a total of 64 different "words" called **codons** found on a special type of RNA called mRNA. As we will learn, the codons on an RNA molecule are **complementary** to the triplets on the DNA molecule from which they were transcribed. For example, the DNA triplet CGA would have the complementary codon GCU (uracil is the RNA base complementary to adenine). Each codon represents a "word" in the language of mRNA, translated from the triplet "word" in the language of DNA.

The Twenty Letters of the Protein Alphabet. Each mRNA codon in turn refers to one of the 20 "letters" in the protein alphabet, one of the 20 common amino acids. Scientists have deduced the individual codons that represent the links between the nucleic acid strands and the polypeptide chains. The DNA triplets and complementary mRNA codons for each of the 20 amino acids are listed in Table 5–2. Figure 5–2*b* gives examples of the way in which the relationship among the three languages results in an RNA molecule and a polypeptide.

Of the 64 different codons, 61 specify a particular amino acid (Table 5–2). The codon for **methionine (AUG)** has a dual role because it also serves as a **start codon,** which is involved in initiating **translation** of the nucleotide triplets in mRNA. The remaining 3 codons do not specify amino acids; instead, they are used as **stop codons** to terminate translation.

The Vast Language of Proteins. Consider the enormous number of words that can be generated from our alphabet of 26 letters. Then consider the huge number of different polypeptide chains that could be constructed from nearly that many amino acids, without the constraints on "word" length found in verbal language. This will give you some sense of the variety of proteins that can be synthesized in the cell and the great possibilities for specialization of function that this variety allows. Approximately 2000 different proteins are thought to exist in the body.

Think of the difference in meaning one letter change makes in a word; for example, the change from *deride* to *decide*. This is similar to the difference that one amino acid could make in the nature of a protein. No other type of macromolecule allows this great variety and specificity of structure and function. Compared with lipids, carbohydrates, and nucleic acids (which have only a few subunits to arrange) proteins (with 20 different subunits) have an enormous evolutionary advantage because of the variety of different functional macromolecules that can be assembled. Indeed, it is likely that the molecular basis of the evolution of species lies in the ability of cells to make new proteins in response to changes in their environment.

A Consistent and Universal Code. The genetic code is said to be **consistent** because each codon specifies only a single amino acid. It is essentially **universal** because it is the same in diverse types of organisms including plants and humans. Because there are 64 different codons for only 20 amino acids, some amino acids are specified by more than one codon. For this reason, the code is said to be **degenerate.**

A gene and its polypeptide product are usually regarded as **colinear molecules.** This term means that the codon representing the beginning of the gene determines the amino acid that will be present at the beginning of the polypeptide, the codon representing the second triplet in the gene determines the second amino acid in the polypeptide, and so on. Many eukaryotic genes are interrupted with nucleotide sequences known as **introns,** the functions of which have yet to be determined. Introns will be discussed again later in this chapter.

Let us now look at how the nucleic acids themselves are assembled and translated into proteins. Figure 5–3 gives a simplified overview of the flow of information during these processes. We will discuss **DNA synthesis** (also known as **DNA replication**) and **RNA synthesis** (also known as **DNA transcription**). We also will look at methods of altering newly synthesized RNA, known as **RNA processing,** before it leaves the nucleus, which result in the production of functional RNA molecules. Finally, we

Table 5–2
The Genetic Code

DNA Triplets	mRNA Codons		Amino Acids
AAA	UUU	1	Phenylalanine
AAG	UUC	2	
AAT	UUA	3	Leucine
AAC	UUG	4	
GAA	CUU	5	
GAG	CUC	6	
GAT	CUA	7	
GAC	CUG	8	
TAA	AUU	9	Isoleucine
TAG	AUC	10	
TAT	AUA	11	
CAA	GUU	12	Valine
CAG	GUC	13	
CAT	GUA	14	
CAC	GUG	15	
AGA	UCU	16	Serine
AGG	UCC	17	
AGT	UCA	18	
AGC	UCG	19	
GGA	CCU	20	Proline
GGG	CCC	21	
GGT	CCA	22	
GGC	CCG	23	
TGA	ACU	24	Threonine
TGG	ACC	25	
TGT	ACA	26	
TGC	ACG	27	
CGA	GCU	28	Alanine
CGG	GCC	29	
CGT	GCA	30	
CGC	GCG	31	
ATA	UAU	32	Tyrosine
ATG	UAC	33	

DNA Triplets	mRNA Codons		Amino Acids
GTA	CAU	34	Histidine
GTG	CAC	35	
GTT	CAA	36	Glutamine
GTC	CAG	37	
TTA	AAU	38	Asparagine
TTG	AAC	39	
TTT	AAA	40	Lysine
TTC	AAG	41	
CTA	GAU	42	Aspartic acid
CTG	GAC	43	
CTT	GAA	44	Glutamic acid
CTC	GAG	45	
ACA	UGU	46	Cysteine
ACG	UGC	47	
ACC	UGG	48	Tryptophan
GCA	CGU	49	Arginine
GCG	CGC	50	
GCT	CGA	51	
GCC	CGG	52	
TCT	AGA	53	
TCC	AGG	54	
TCA	AGU	55	Serine
TCG	AGC	56	
CCA	GGU	57	Glycine
CCG	GGC	58	
CCT	GGA	59	
CCC	GGG	60	
TAC	AUG	61	Methionine START
ATT	UAA	62	STOP
ATC	UAG	63	
ACT	UGA	64	

will examine the process of **protein synthesis** (also known as **RNA translation**), in which the information in RNA is translated into a polypeptide chain.

Synthesis of Nucleic Acids

As discussed in Chapter 2, the macromolecules deoxyribonucleic acid (DNA) and ribonucleic acid (RNA) are each composed of molecules called nucleotides. Nucleotides contain a ribose sugar and one to three phosphate groups, in addition to one of five bases. DNA and RNA are distinguished by the type of ribose sugar and bases their nucleotides contain. All nucleotides forming part of a nucleic acid strand have one phosphate group.

Synthesis of DNA: Replication of the Permanent Files

When a cell divides into two, the contents of both the cytoplasm and the nucleus divide (Fig. 5–4). Division of the nucleus, a process known as **mitosis** (Chapter 3), ensures that both of the new cells have an identical copy of the DNA of the parent cell. Thus, the new cells will be identical in form and function to the parent.

Before the nucleus can divide, the DNA molecules within the nucleus (in the form of **chromatin**) must be copied, or **replicated.** The replication process must be accurate so that the genetic information is not altered, and it must be fast so that it can occur in time for a cell to divide. DNA replication, as mentioned in Chapter 3, occurs during the interphase

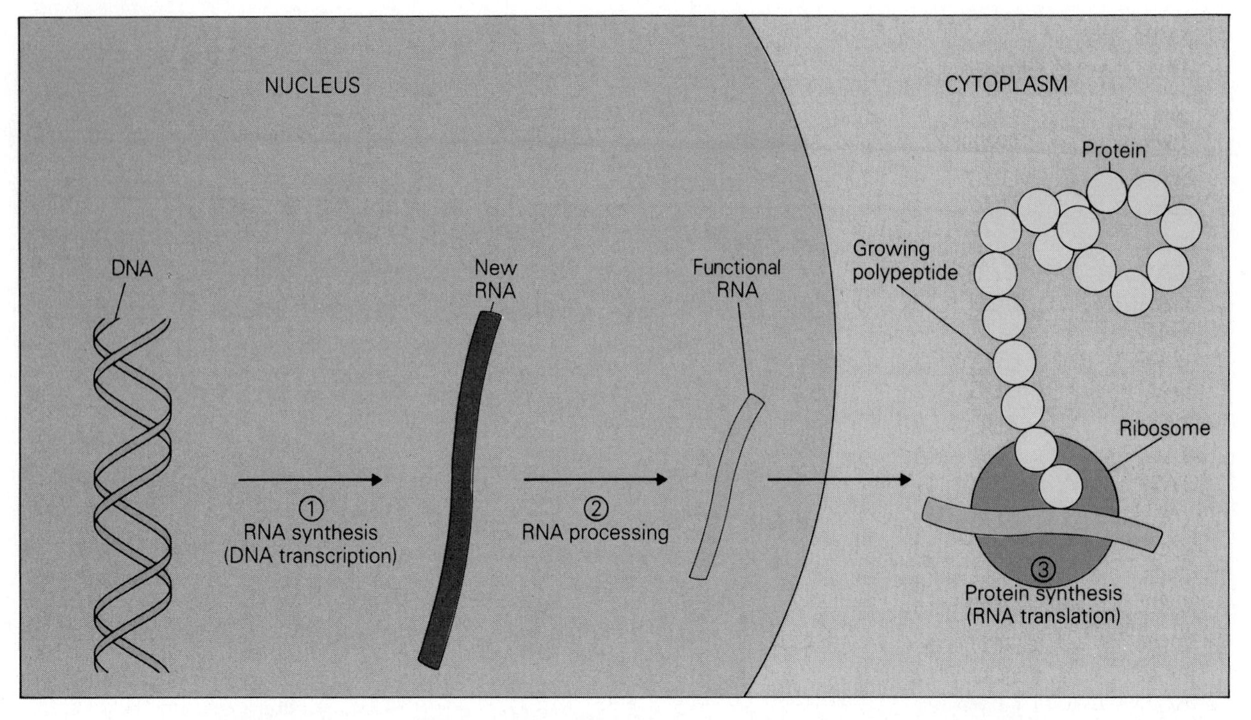

Figure 5–3 A simplified overview of the flow of information in the expression of the genetic code. The information on a DNA molecule is transcribed, or reproduced in coded form, on an RNA molecule. After the RNA molecule is processed, it leaves the nucleus and travels to a ribosome, where the coded information is translated into a sequence of amino acids that forms a polypeptide chain.

preceding mitosis. Each chromosome, a long DNA molecule, is copied to produce two identical DNA molecules called **chromatids** (see Fig. 5–4).

A DNA molecule consists of the two long strands of nucleotides arranged in a double helix (see Chapter 2). The two strands are held together by hydrogen bonds between the bases of the nucleotides. Adenine always forms bonds with thymine and guanine with cytosine. The specific base pairing produces two strands that complement each other.

Semiconservative Replication. During the process of DNA replication, the two strands of the DNA molecule are first separated. Each one then acts as a mold, or **template,** upon which a new complementary strand of nucleotides is assembled. In this way, two new molecules of DNA are produced. Each molecule contains one strand that was the template derived from the parent molecule and one newly synthesized strand (Fig. 5–5). The process is described as **semiconservative** because the structure of the parent DNA molecule is disrupted but the structure of the subunits—that is, the nucleotide sequence of each strand—is conserved for many generations. Maintenance of the nucleotide sequence of DNA is a critical step that ensures that the new cells receive an exact copy of the genetic information in the parent cell.

Separating the Strands of the Double Helix. Before replication of DNA can take place, the double helix must be unwound and "unzipped" to separate the two strands. Unwinding is aided by an enzyme known as a **DNA topoisomerase** and unzipping (breaking the hydrogen bonds between bases) by a **DNA helicase.** The separated strands are stabilized by proteins that bind to the single DNA strands and therefore assist the opening of the helix. They are known as **helix-destabilizing proteins.** Each strand then acts as a template for the assembly of a complementary chain. Separation of the two original DNA strands produces a structure called a **replication fork** at the site of active synthesis of the new strands (Fig. 5–6*a*). Many replication forks can exist on a DNA molecule at once (Fig. 5–6*b*).

Replicating the Strands: DNA Polymerase. DNA replication is catalyzed by an enzyme called **DNA polymerase,** which takes instructions from the DNA strand that is acting as the template for the new DNA strand. The specific reaction catalyzed by DNA polymerase is the joining together of the various nucleotides that are components of DNA. The individual nucleotides must contain deoxyribose and three phosphate groups; otherwise, the enzyme cannot use them. Thus, DNA polymerase takes the nucleotides

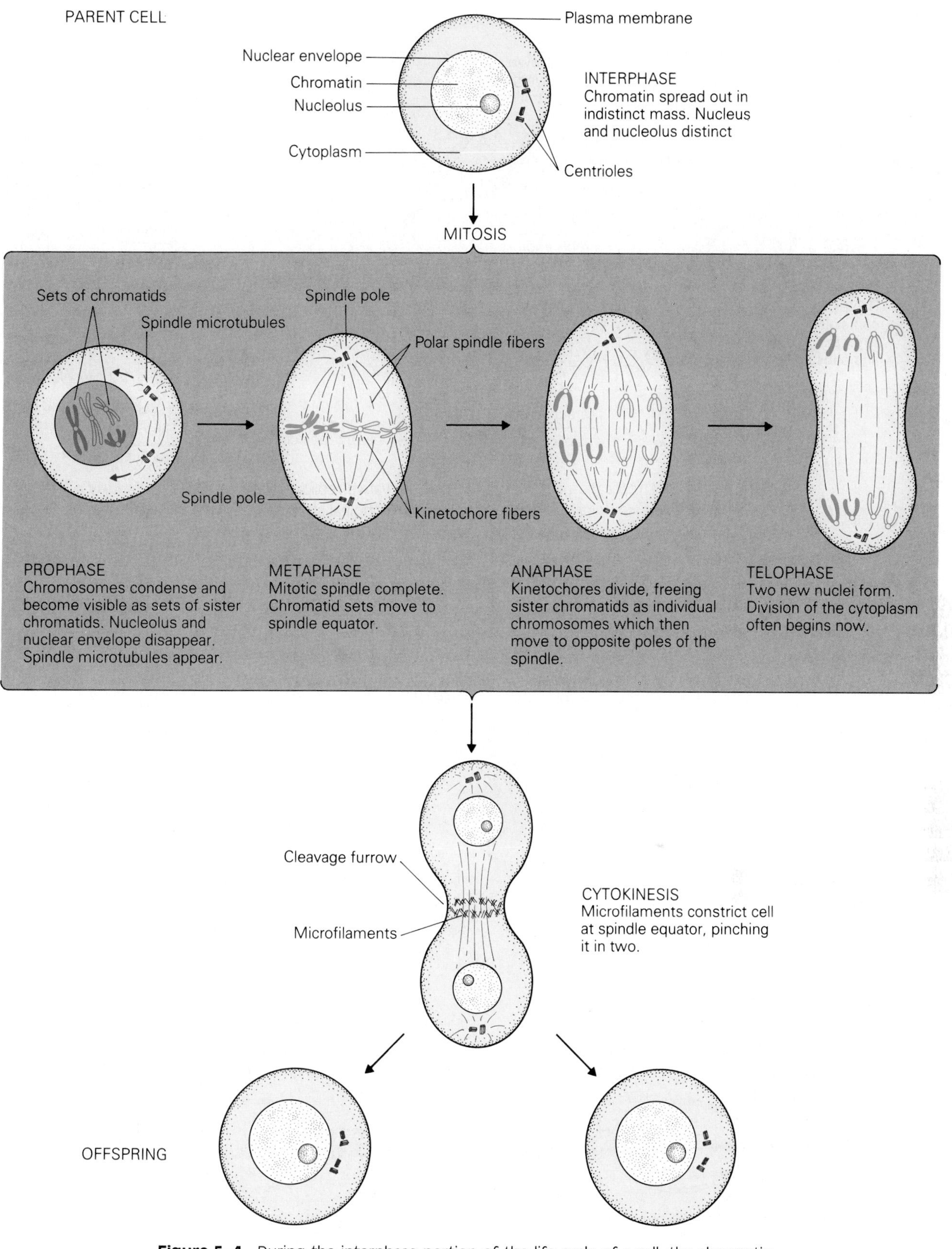

Figure 5–4 During the interphase portion of the life cycle of a cell, the chromatin form of the DNA in the nucleus replicates, or copies itself. In mitosis, the chromatin forms the tightly coiled bundles known as chromosomes, with the pairs of identical chromosomes linked together to form chromatids. The chromatids separate such that one copy of each chromosome becomes part of one of the nuclei of the two new cells. Thus, each new cell has the same genetic information contained in the parent cell.

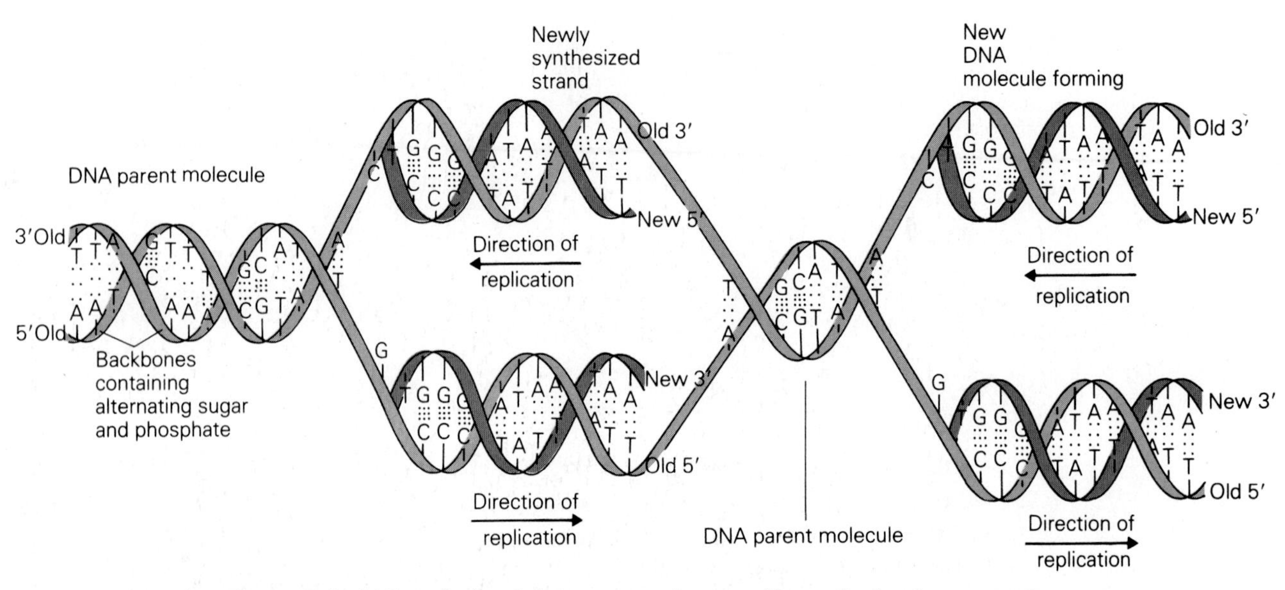

Figure 5–5 DNA replication is known as semiconservative replication because each new DNA molecule contains one new strand and one original strand. Thus, single strands of DNA are preserved and passed on for many generations.

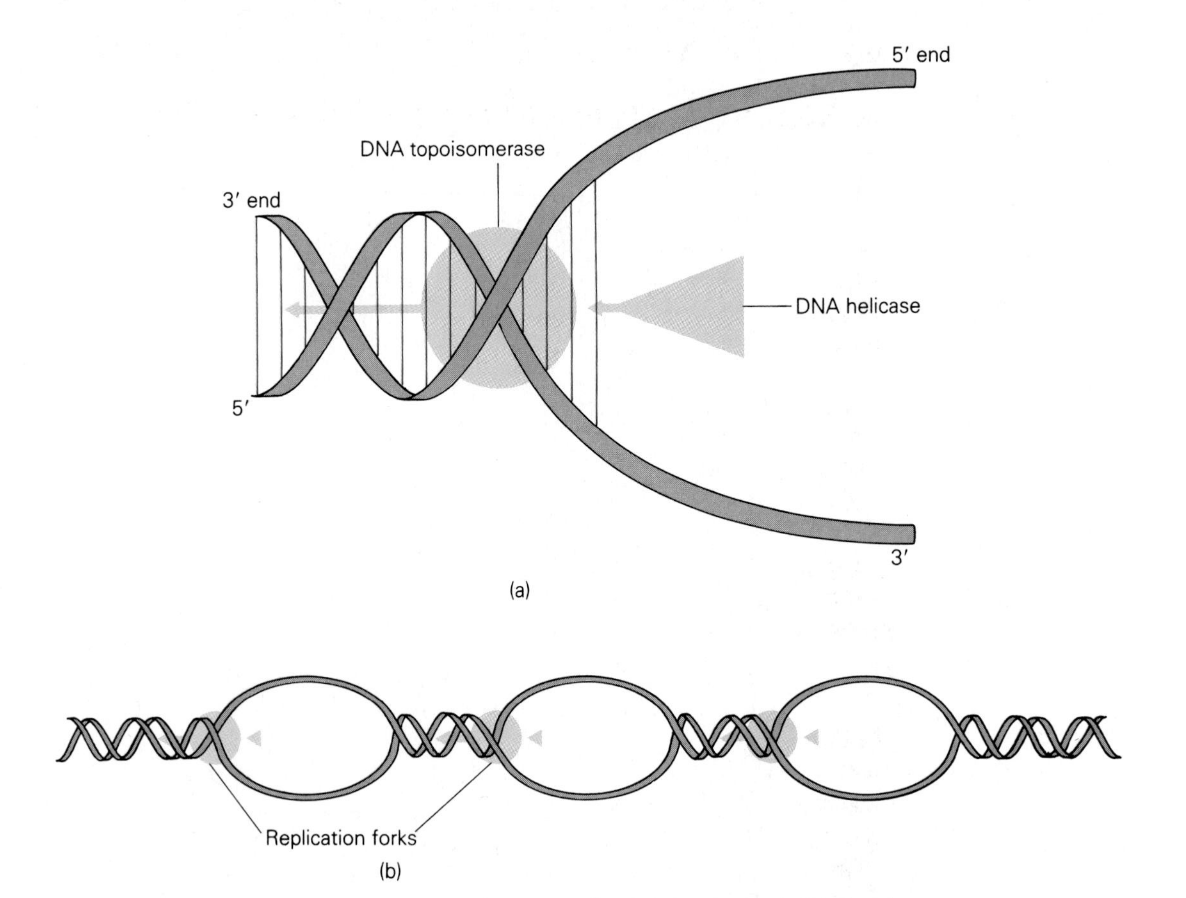

Figure 5–6 *(a)* Before DNA replication can begin, the double helix must be unwound and the two strands separated. Unwinding is accomplished by an enzyme called DNA topoisomerase, and another enzyme, DNA helicase, breaks the hydrogen bonds holding the base pairs—and hence the two strands of the molecule—together. *(b)* DNA topoisomerase and DNA helicase may be active at several sites on the same DNA molecule, thereby forming many replication forks.

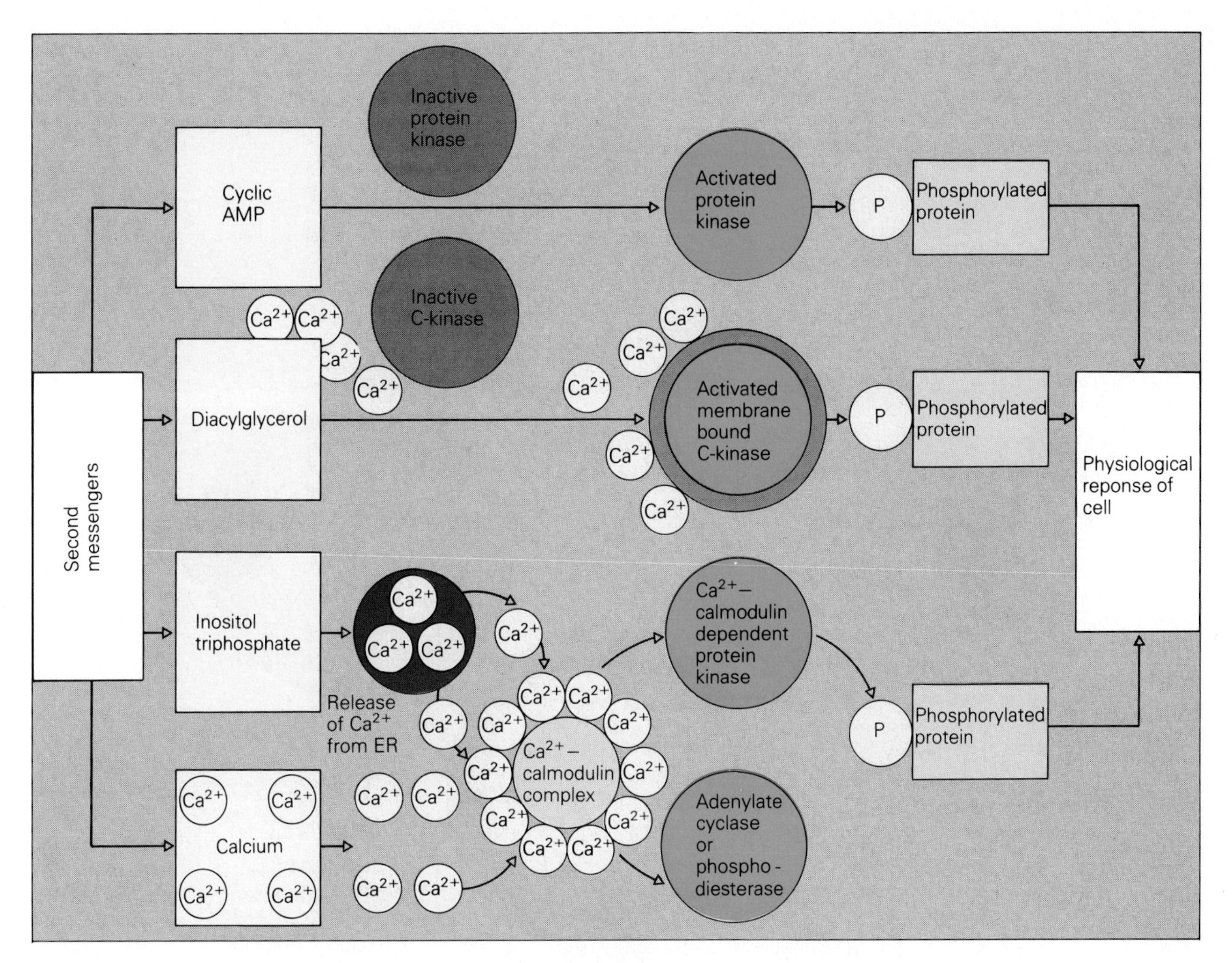

Figure 5–30 Second messengers may activate certain enzymes that catalyze the phosphorylation of certain proteins, which in turn produce the physiological response of the cell to the extracellular signal.

molecules are continuously and rapidly destroyed or removed from the cytosol (Fig. 5–31). The cytosolic concentration of a second messenger is altered principally by changes in its rate of synthesis or influx into a cell. This ensures that an increased rate of synthesis or influx will rapidly increase the cytosolic concentration of the messenger, allowing the target cell to respond quickly to the extracellular signal (see Fig. 5–31). The concentration of cyclic AMP is usually only 1 μM, but this can be increased fivefold within seconds after a hormone binds to the plasma membrane receptor and stimulates adenylate cyclase. Synthesis or influx of the second messenger will fall to zero when the extracellular signal is "turned off." Since the rapid rate of destruction or removal of the messenger will continue, the cytosolic concentration of the messenger molecules will decline rapidly to the point where the cell will cease to respond. In other words, the response of the cell to the second messenger is reversed rapidly once the second messenger is no longer present (see Fig. 5–31).

Cyclic AMP, for example, is degraded very rapidly to adenosine 5′-monophosphate (AMP) by a reaction catalyzed by an enzyme known as **cyclic AMP phosphodiesterase** (see Chapter 12). Likewise, inositol triphosphate and diacylglycerol are rapidly degraded by specific enzymes.

To give another example, Ca^{2+} is removed from the cytosol by several mechanisms acting together (Fig. 5–32). One group actively pumps Ca^{2+} out of the cell, moving the ion against its electrochemical gradient. This is achieved by a specific Ca^{2+}-ATPase that utilizes chemical energy (as ATP) to pump Ca^{2+} out of the cell. In antiport, Ca^{2+} leaves the cell in exchange for Na^+ that enters the cell by diffusing down a favorable electrochemical gradient.

Free Ca^{2+} is removed also by binding to cytosolic calcium-binding proteins and other molecules, and by sequestration inside intracellular organelles

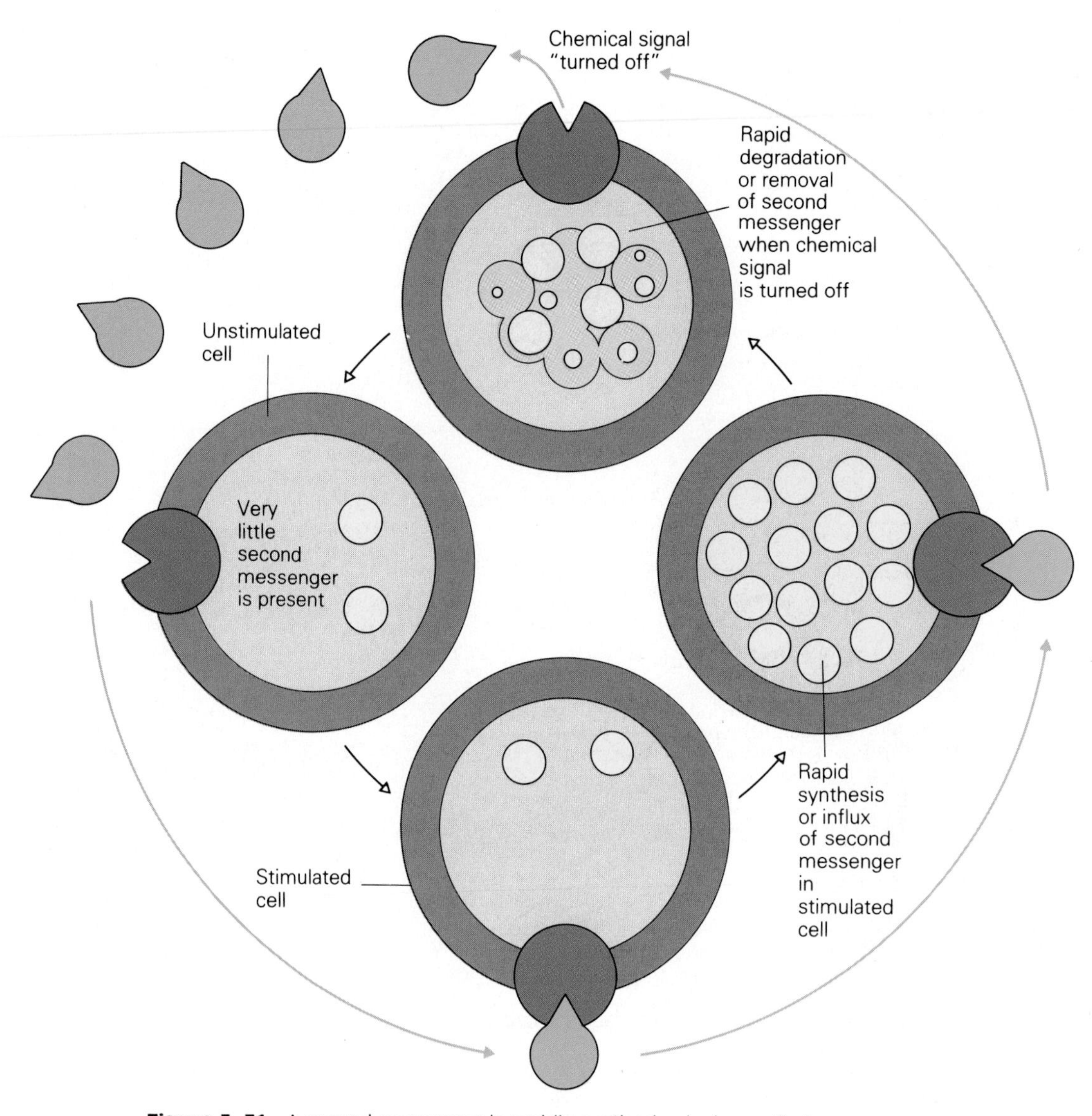

Figure 5–31 A second messenger is rapidly synthesized when a first messenger molecule (an extracellular signal) binds to a receptor. When the extracellular signal is removed, however, the second messenger is rapidly destroyed or removed from the cell. Thus, the next time a signal binds to the receptor, the second messenger must be synthesized anew.

such as mitochondria and the endoplasmic reticulum. The large number of different mechanisms that can regulate cytosolic Ca^{2+} levels may be, at least in part, a defense system to guard against possible toxic effects of uncontrolled concentrations of this ion.

Case Study: Glycogen Metabolism and Protein Phosphorylation. Protein phosphorylation is one example of a way in which the production of second messengers can mediate a change in cell function that represents the cellular response to a chemical signal. The metabolism of **glycogen** in skeletal muscle cells provides clues to a possible sequence of events. Recall from Chapter 2 that glycogen is a polysaccharide, the form in which glucose is stored until its chemical energy is needed by the muscle cell. The hormone **epinephrine** stimulates glycogen breakdown and shuts off glycogen synthesis—the net effect being to provide the cell with an adequate supply of glucose for strenuous activity.

Epinephrine works through stimulation of adenylate cyclase in the plasma membrane and elevation of intracellular cyclic AMP. The cyclic AMP activates a protein kinase, and a sequence of phosphorylation reactions is initiated (Fig. 5–33*a*). Phosphorylation by the protein kinase of Enzyme 1 activates this enzyme and, once activated, Enzyme 1 phosphorylates Enzyme 2. This reaction also activates Enzyme 2, which begins to remove individual glucose molecules from glycogen. The same protein kinase phosphorylates a third enzyme (Enzyme 3)

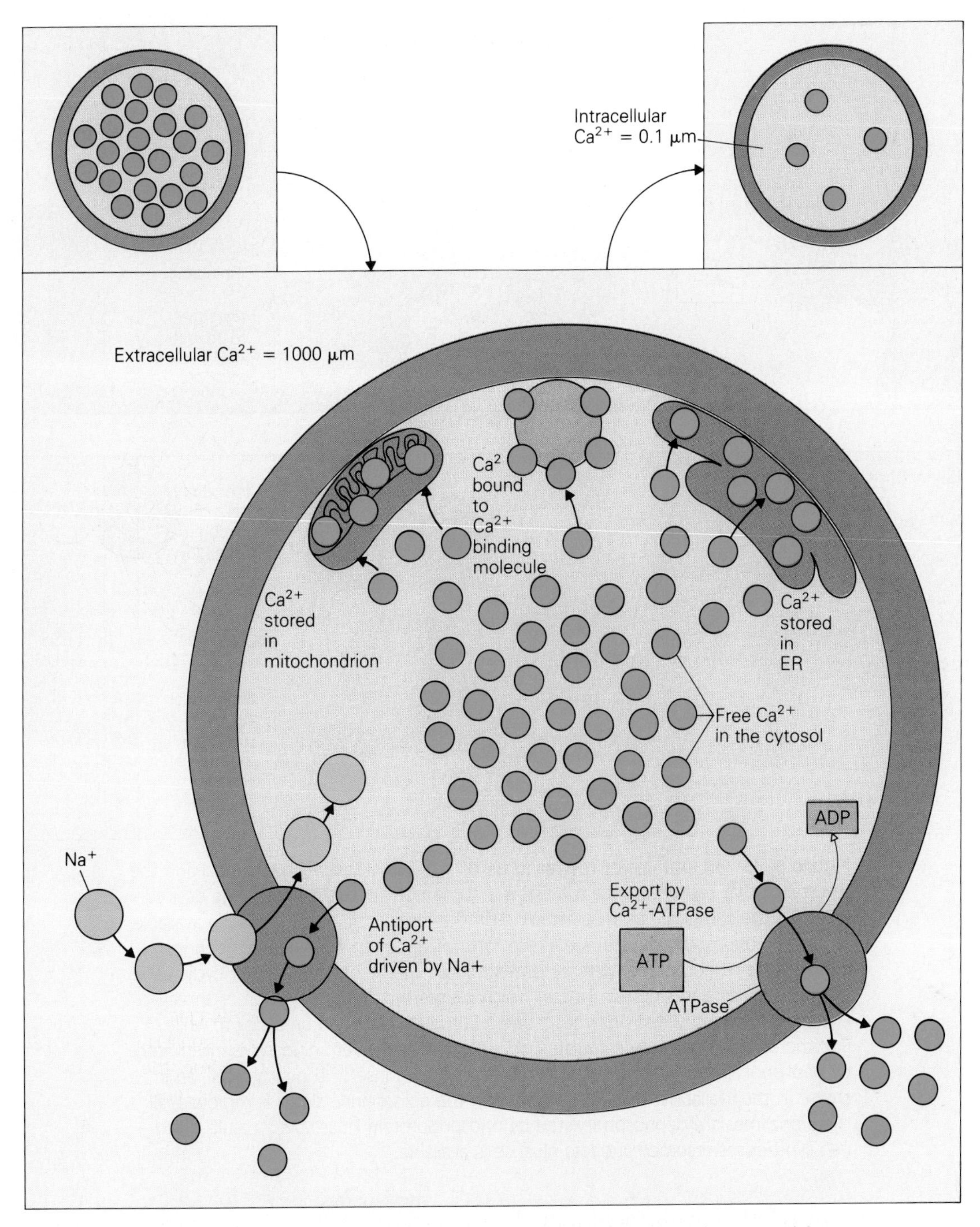

Figure 5–32 Examples of the ways in which a second messenger is removed from a cell. The second messenger calcium ions in the cytosol may be removed by active transport, by antiport, by sequestration inside cytoplasmic organelles, or by binding to special molecules.

that is involved in synthesis of glycogen from glucose. In contrast to Enzymes 1 and 2, phosphorylation of Enzyme 3 causes inactivation.

The overall effect of these protein phosphorylations is to make available a maximum supply of glucose by simultaneously stimulating breakdown of glycogen into glucose molecules while inhibiting the use of glucose for synthesis of glycogen.

When the epinephrine signal is turned off, the phosphorylation steps just described are rapidly reversed by reactions catalyzed by the enzyme **phosphoprotein phosphatase** (see Fig. 5–33*b*). In other words, the phosphate groups on each of the three key enzymes are removed by the phosphatase, with the net effect of inhibiting glycogen breakdown (Enzymes 1 and 2 are inactivated) and stimulating glycogen synthesis (Enzyme 3 is activated). The opposing action of the phosphoprotein

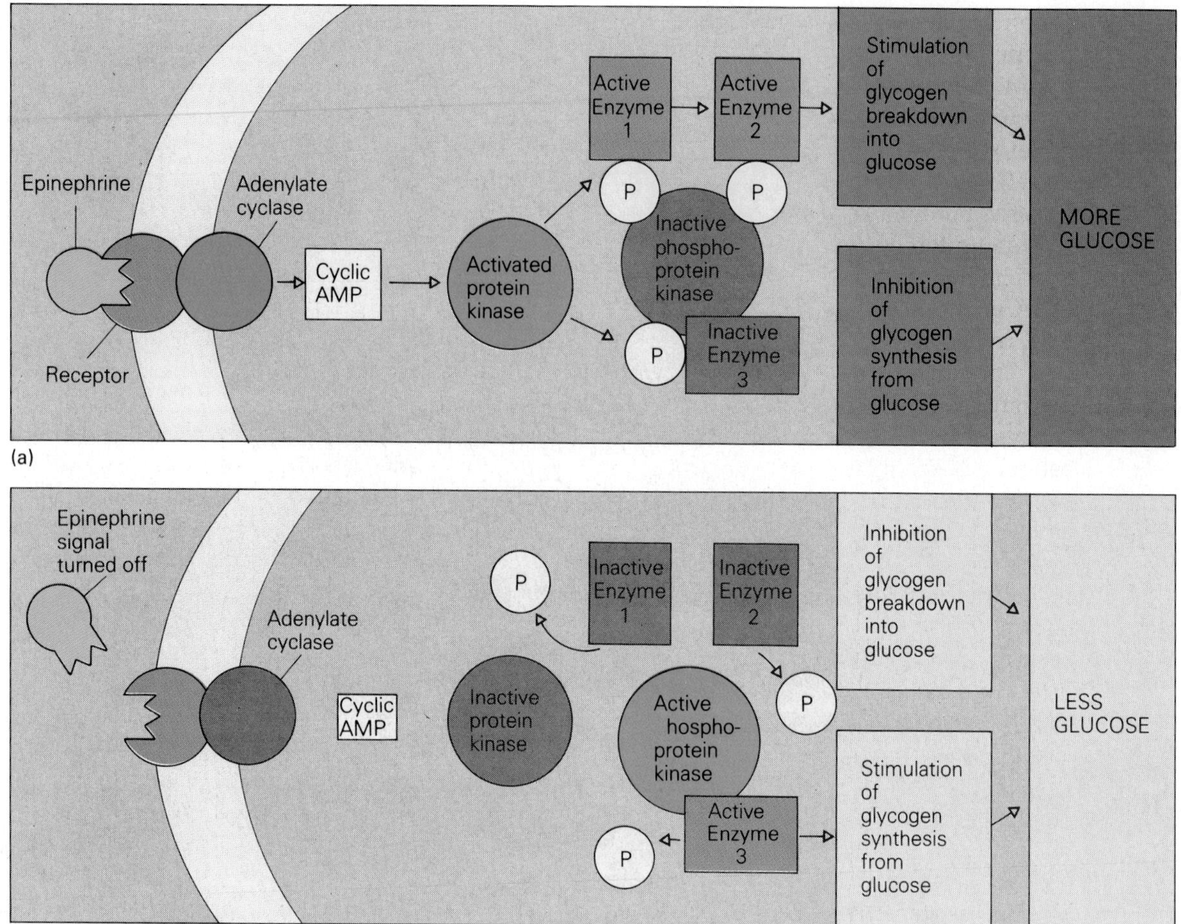

Figure 5–33 An example of the response of a cell to an extracellular signal and the reversal of that response once the signal is turned off. In this example, the desired effect of the chemical signal epinephrine on the target cell is an increase in available glucose as opposed to its storage in the form of glycogen. (*a*) When epinephrine binds to the receptor, adenylate cyclase produces the second messenger cyclic AMP. This, in turn, activates protein kinase, which causes the phosphorylation of three enzymes. Phosphorylated Enzymes 1 and 2 stimulate glycogen breakdown, while Enzyme 3, which normally stimulates glycogen synthesis from glucose, is inactivated by phosphorylation. The net result is the desired response of the target cell, an increase in the availability of glucose. (*b*) When the epinephrine signal is removed, all three enzymes are dephosphorylated by phosphoprotein kinase. As a result, glycogen synthesis is resumed and less glucose is available.

phosphatase does not interfere with the action of cyclic AMP because the phosphatase is inhibited while the level of cyclic AMP is elevated.

Regulation of both breakdown and synthesis of glycogen is an important feature of this cyclic AMP control system. It increases the response to an elevated level of cyclic AMP in the cell, the specific response being a readily available supply of glucose molecules. There are now more than 25 different enzymes known to be regulated by a reversible phosphorylation process. These and other types of covalent modifications of proteins are so abundant that most of the proteins that mediate physiological processes are probably regulated in this way.

Summary

I. A. Genetic information, the instructions a cell uses to synthesize proteins, is stored in both DNA and mRNA by a code formed from the specific linear sequence in which the four different nucleotides are arranged. Each triplet of DNA nucleotides is transcribed into a codon of RNA nucleotides; each

4. How is the rate of glycolysis regulated by ATP? Beginning with one molecule of glucose, how many molecules of ATP are used to drive glycolysis and how many are generated under both aerobic and anaerobic conditions?
5. In what form is pyruvic acid fed into the citric acid cycle? How does ATP slow down the cycle?
6. Where in the cell is the electron transport chain located? Explain the chemiosmotic mechanism for phosphorylation of ADP to produce ATP.
7. What is the primary purpose of the pentose phosphate pathway? What is the fundamental difference in the role of the reduced coenzymes NADH and NADPH?
8. In what metabolic pathways does acetyl coenzyme A participate?

Suggested Readings

Alberts, B., Bray, D., Lewis, J. Raff, M., Roberts, K., and Watson, J.D. *Molecular Biology of the Cell.* New York and London: Garland, 1983.

Karp, G. *Cell Biology,* 2nd ed. New York: McGraw-Hill, 1984.

Smith, E.L., Hill, R.L., Lehman, I.R., Lefkowitz, R.J., Handler, P., and White, A. *Principles of Biochemistry: General Aspects,* 7th ed. New York: McGraw-Hill, 1983.

Stryer, L. *Biochemistry,* 3rd ed. San Francisco: Freeman, 1988.

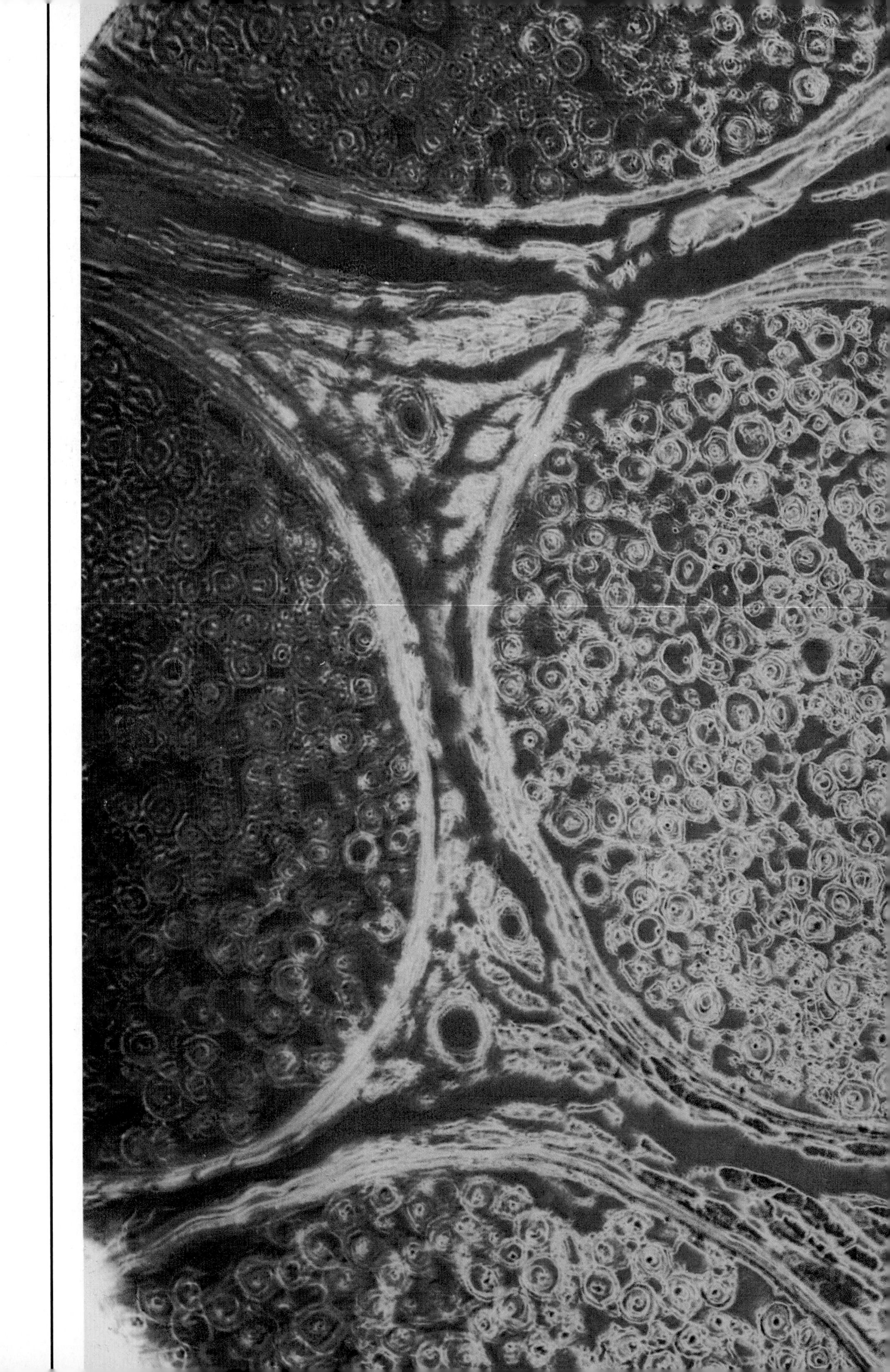

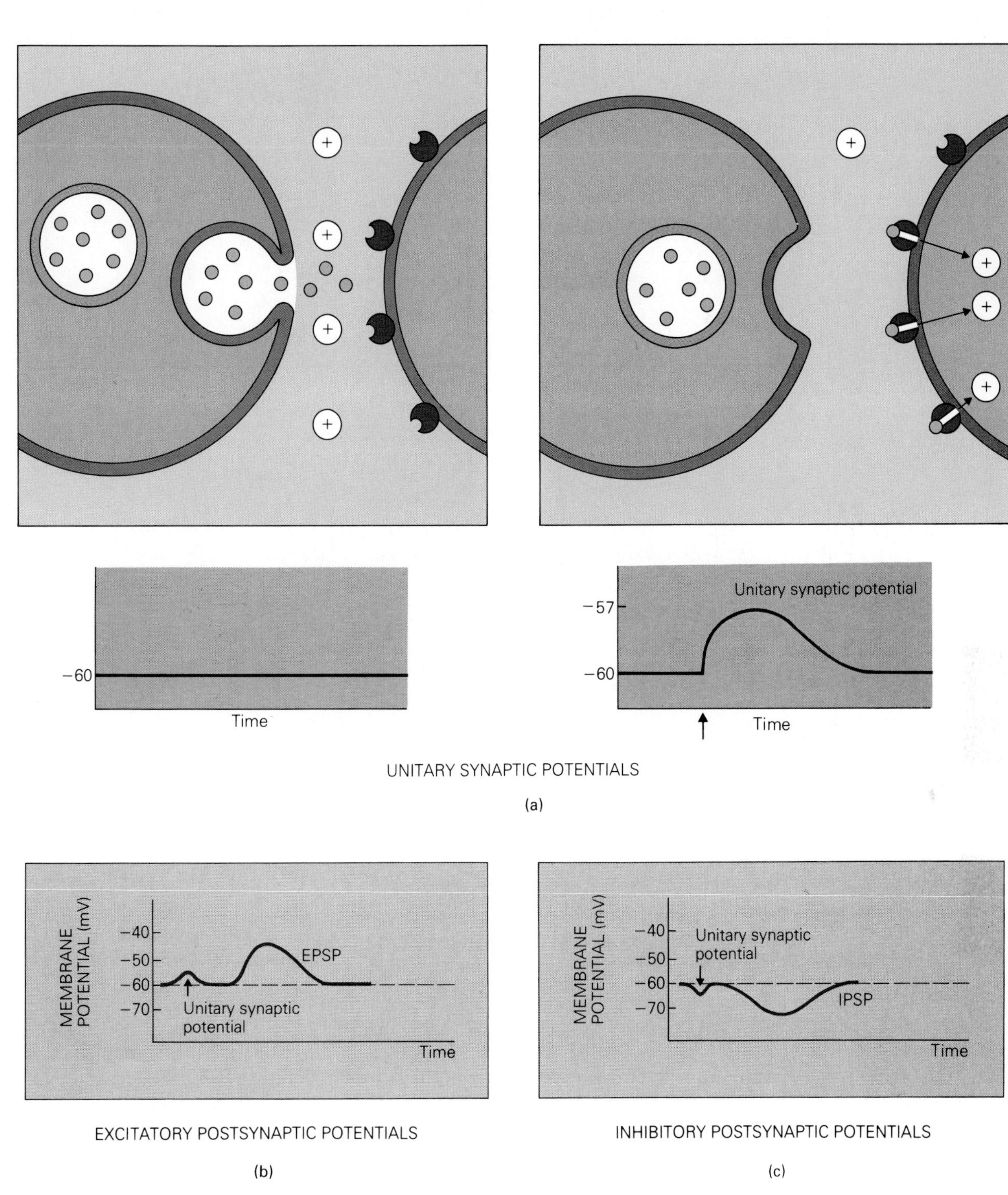

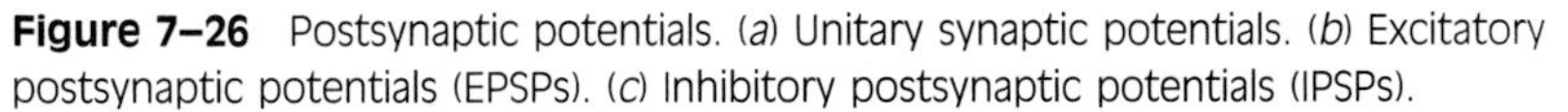

Figure 7–26 Postsynaptic potentials. (*a*) Unitary synaptic potentials. (*b*) Excitatory postsynaptic potentials (EPSPs). (*c*) Inhibitory postsynaptic potentials (IPSPs).

the K^+ efflux is much smaller than the magnitude of the Na^+ influx (Fig. 7–27). Thus, when this type of acetylcholine receptor is activated, the total influx of Na^+ ions will be substantially greater than the efflux of K^+ and will therefore result in the depolarization of the membrane.

Inhibitory Postsynaptic Potentials. As with the EPSPs, there are a number of ways that an IPSP can be produced. One mechanism is the synaptic activation of a receptor that will open its channel to K^+ ions. Because the concentration gradient and electrical gradient for K^+ cause it to move outward, the

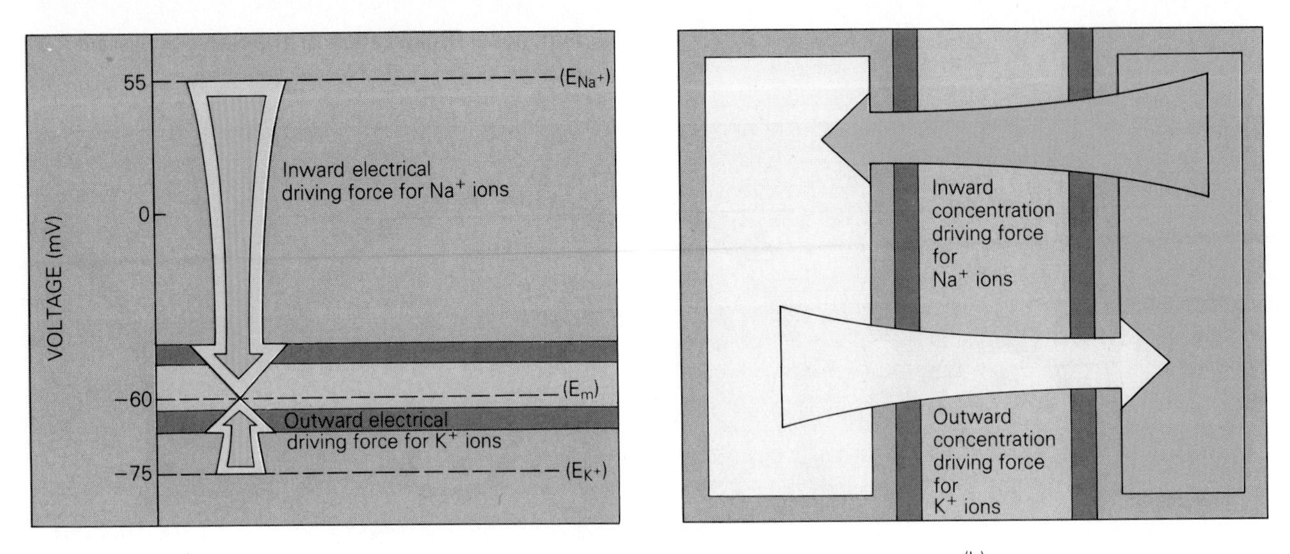

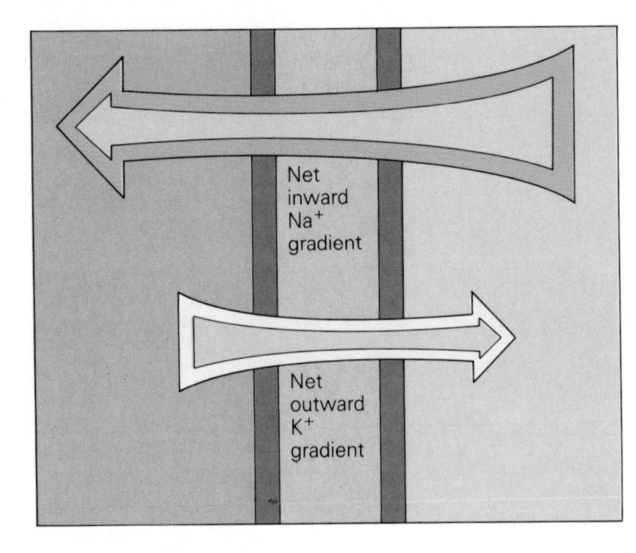

Figure 7–27 The net gradient for ion flow comprises the electrical gradient (*a*) and the chemical (concentration) gradient (*b*). The net gradient for Na^+ flow into the cell is greater than the net gradient for K^+ flow out of the cell (*c*).

interior of the cell will lose positive charges and become more negative. Consequently, the membrane potential will hyperpolarize, and an inhibitory postsynaptic potential will develop. Other receptors that produce IPSPs are those that permit Cl^- to flow across the membrane. The increase in the flow of negative charges into the cell also acts to hyperpolarize the membrane potential.

Closing of Ion Channels by Receptors. Our discussion of transmitter-receptor interactions has thus far focused only on the opening of ion channels. There is also a class of receptors, however, that close their ion channels when coupled to a transmitter. The closing of these chemically activated ion channels can also produce depolarizing or hyperpolarizing changes of the membrane potential.

For example, the closing of Na^+ channels that are normally open would prevent the influx of Na^+ ions. As a consequence, the efflux of K^+ would be the predominant effect, and the membrane would hyperpolarize. In a similar fashion, transmitter-receptor interactions that close K^+ channels would prevent the efflux of K^+ ions. In this situation, the influx of Na^+ would predominate, and the membrane would depolarize.

The mechanisms of transmitter-activated channel closings are typically carried out by second messenger systems from within the postsynaptic cell and last longer than transmitter mechanisms that involve the opening of ion channels. The details of the ways in which channels close vary with different types of receptors and will be discussed in relation to their associated neurotransmitters.

Termination of Synaptic Transmission

The termination of synaptic transmission occurs when the transmitter is removed from the synaptic cleft. This process is accomplished in most neurotransmitter systems by the transport of the transmitter back into the presynaptic terminal. A **re-uptake** mechanism pumps most transmitters back into the presynaptic terminal. Other transmitters are removed from the synaptic cleft by degrading enzymes, and the metabolic products are then transported back into the presynaptic terminal. The pumping mechanism is specific for each type of transmitter substance or metabolite and can be affected by selective drugs.

Recycling of Neurotransmitters

After a transmitter has been transported back into the presynaptic terminal, it is again packaged into vesicles for storage in preparation for release. In

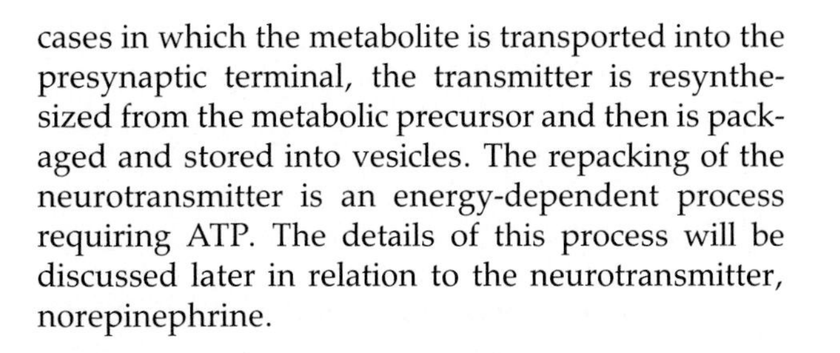
cases in which the metabolite is transported into the presynaptic terminal, the transmitter is resynthesized from the metabolic precursor and then is packaged and stored into vesicles. The repacking of the neurotransmitter is an energy-dependent process requiring ATP. The details of this process will be discussed later in relation to the neurotransmitter, norepinephrine.

Diffuse and Discrete Chemical Synapses

Chemical synapses have been shown to have either discrete or diffuse actions. In **discrete synapses,** the chemical neurotransmitter is released from restricted areas of the presynaptic terminal (called **active zones**) into a small synaptic cleft; only 30 nm separate the pre- and postsynaptic membranes (Fig. 7–28). An example of a discrete synapse is the junction between nerve and striated muscle cells, which tends to activate a small region on the muscle fiber.

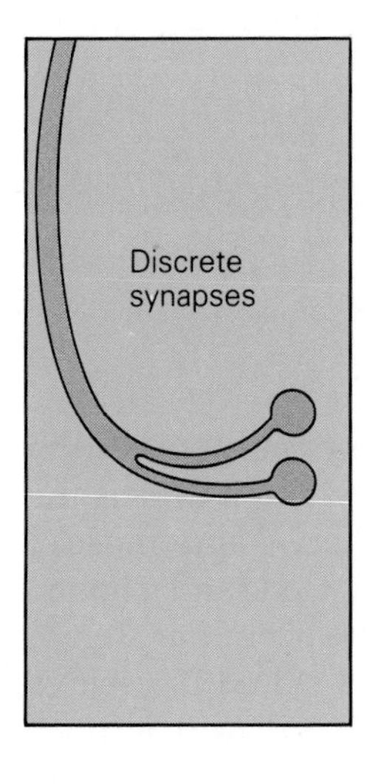

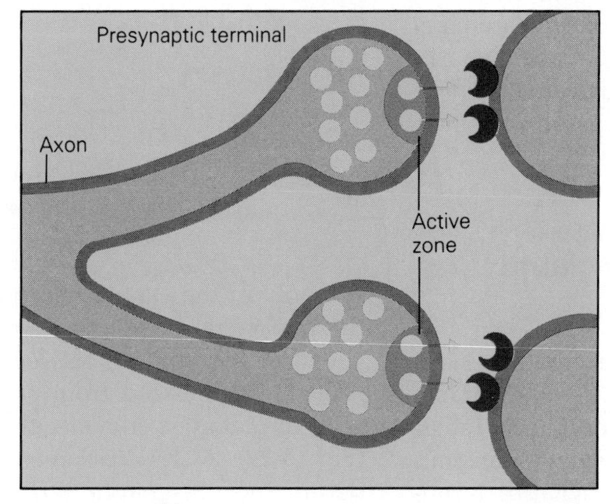

(a)

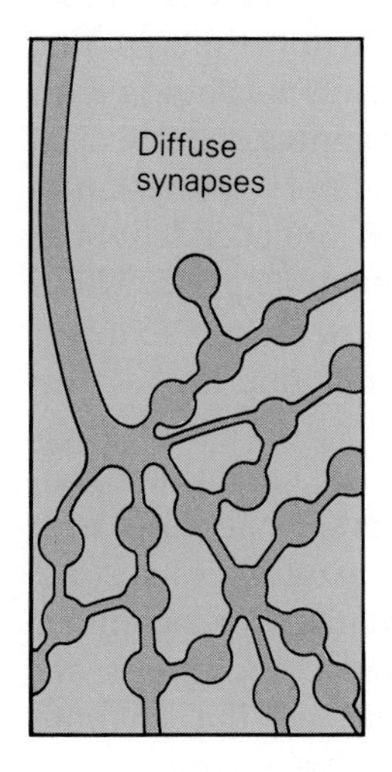

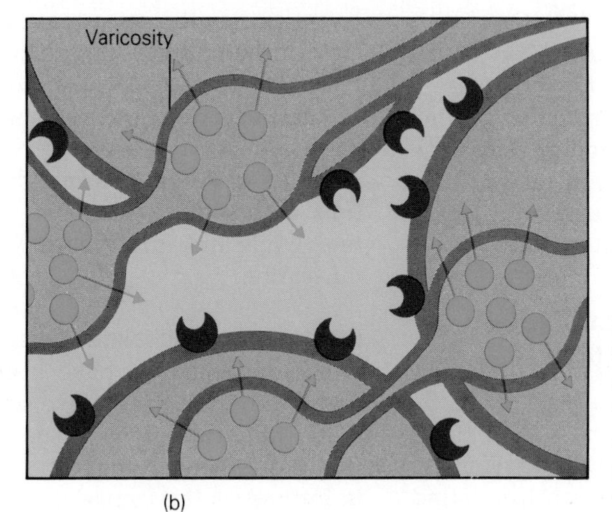

(b)

Figure 7–28 In discrete synapses, transmitters are released only through active zones of the axon terminal and travel across very small clefts to target cells (*a*). In diffuse synapses, axon terminals have varicosities that release transmitters over a wider range (*b*).

In **diffuse synapses,** on the other hand, transmitter release is not limited to specific active zones, and the distance between the presynaptic and postsynaptic membrane can be as much as 150 nm. These synapses have the form of beads on a chain called **varicosities.** The varicosities are an extension of the axon and form an overlapping network of synapses **en passant** or synapses in passage. As the action potential invades each varicosity, vesicles fuse with the presynaptic membrane, and the action potential continues on to the next varicosity. The effect of this type of synapse is to activate a large surface area of one cell or a large number of cells in a diffuse manner. The diffuse synapse is typical of nerve terminals of the sympathetic autonomic nervous system and of nerve cells containing catecholamines within the central nervous system.

Metabotrophic Receptors

As we have seen, the activation of the receptor leads to the opening and closing of ion channels. This activation can also affect the metabolic activity of the cell by forming secondary intracellular messengers such as cAMP, cGMP, diacylglycerol, Ca^{2+}, and calmodulin (see Chapter 5). The mechanisms of these metabotrophic receptors will be discussed with their associated neurotransmitters.

Table 7–1
Low-Molecular-Weight Transmitters

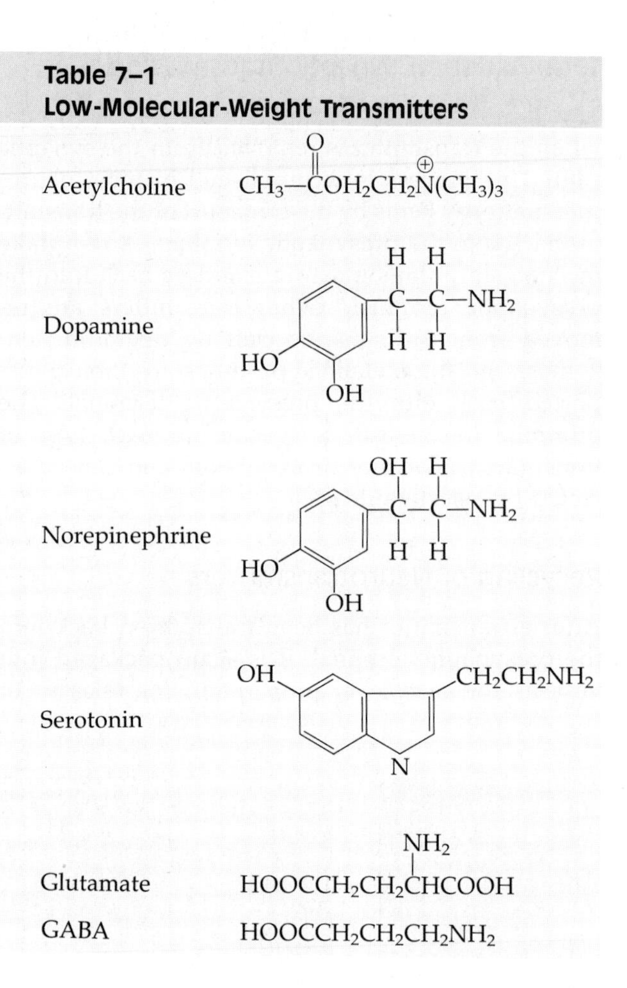

Transmitter	Structure
Acetylcholine	$CH_3{-}\overset{O}{\overset{\|}{C}}OH_2CH_2\overset{\oplus}{N}(CH_3)_3$
Dopamine	
Norepinephrine	
Serotonin	
Glutamate	$HOOCCH_2CH_2\overset{NH_2}{\overset{\|}{C}}HCOOH$
GABA	$HOOCCH_2CH_2CH_2NH_2$

Neurotransmitters and Associated Receptors

Chemical neurotransmitters can be broadly classified into two groups: **low-molecular-weight transmitters** and **neuropeptide transmitters** (see Tables 7–1 and 7–2). The low-molecular-weight transmitters are synthesized within the presynaptic terminal. The necessary synthesizing enzymes are produced within the soma and transported to the terminal region. Neuropeptides, on the other hand, are fabricated in the soma and carried via axonal transport mechanisms to the synaptic terminal.

Although the processes of transmitter synthesis, release, degradation, uptake, and physiological effect are not completely understood for the peptides, they are fairly well characterized for many of the low-molecular-weight transmitters.

Low-Molecular-Weight Transmitters: Products of the Axon Terminal

Acetylcholine

One of the first low-molecular-weight transmitters to be studied was acetylcholine (ACh). It is found not only in the central nervous system, but also in the peripheral nervous system, where it acts as the chemical transmitter between nerve and muscle. ACh is synthesized from acetyl CoA and choline with the catalytic enzyme, choline acetyltransferase (Fig. 7–29). ACh is packaged and stored in vesicles within the presynaptic terminal, where it is positioned for release near the active zones. The fusion of the synaptic vesicle with the presynaptic membrane causes the release of ACh into the synaptic cleft. It then diffuses across the cleft and binds with receptors on the postsynaptic membrane.

In the central and peripheral nervous systems, there are two types of receptors for acetylcholine: nicotinic receptors and muscarinic receptors. Nicotinic acetylcholine receptors are sensitive to nicotine, while muscarinic receptors respond to the drug muscarine.

Nicotinic Acetylcholine Receptors. Nicotinic acetylcholine receptors have been well characterized and found to consist of five subunits: beta, gamma, delta, and two alphas. The five components are thought to be arranged so that an ion channel is formed at the central core of the receptor (Fig.

Table 7–2
Neuropeptides

Neuroactive Peptides: Mammalian Brain Peptides Categorized According to Principal Tissue Localization

Hypothalamic-releasing hormones	Gastrointestinal peptides
Thyrotropin-releasing hormone	Vasoactive intestinal polypeptide
Gonadotropin-releasing hormone	Cholecystokinin
Somatostatin	Gastrin
Corticotropin-releasing hormone	Substance P
Growth hormone–releasing hormone	Neurotensin
Neurohypophyseal hormones	Methionine-enkephalin
Vasopressin	Leucine-enkephalin
Oxytocin	Insulin
Neurophysin(s)	Glucagon
Pituitary peptides	Bombesin
Adrenocorticotropin	Secretin
β-Endorphin	Somatostatin
α-Melanocyte-stimulating hormone	Thyrotropin-releasing hormone
Prolactin	Motilin
Luteinizing hormone	Others
Growth hormone	Angiotensin II
Thyrotropin	Bradykinin
Invertebrate peptides	Sleep peptide(s)
FMRF amide*	Calcitonin
Hydra head activator	CGRP (calcitonin gene-related peptide)
	Neuropeptide Yy

*Phe-Met-Arg-Phe-NH_2.
Source: From Krieger, 1983.

7–29*b*). The active binding sites for the nicotinic receptor are on the two alpha subunits. Both binding sites must be occupied by an acetylcholine molecule in order for the receptor to become activated. Once activated, the receptor opens its gate to permit the simultaneous influx of Na^+ ions and efflux of K^+ ions.

As we have seen, the driving forces for Na^+ influx are much greater than for K^+ efflux. Consequently, the excess positive charge movement into the cell causes the membrane potential to depolarize. The ion channel of the nicotinic receptor remains in the open state until ACh uncouples from its receptor. After ACh dissociates from its receptor, the receptor channel closes and Na^+ and K^+ are no longer able to pass through the channel. ACh is then free to diffuse within the synaptic cleft, where it binds with the membrane-bound enzyme acetylcholinesterase, AChE. The acetylcholinesterase enzyme degrades ACh by hydrolysis to produce choline and acetate. Choline is then taken up into the presynaptic terminal by a high-affinity uptake mechanism and recycled to be used again to synthesize ACh. This process of cholinergic nicotinic transmission is found at the neuromuscular junction in addition to various locations in the central nervous system.

Muscarinic Acetylcholine Receptors. The structure of the second type of acetylcholine receptor, the muscarinic receptor, is less well characterized. Information is available, however, about its functional properties. The binding of acetylcholine to the muscarinic receptor causes the associated ion channel to close, thus preventing the efflux of K^+ ions. The blocking of K^+ efflux results in the depolarization of the membrane potential, which leads to the excitation of the nerve cell. It can be seen, therefore, that the effect a neurotransmitter has on a nerve cell depends upon the type of receptor it activates.

Biogenic Amines

Biogenic amines are a class of low-molecular-weight transmitters characterized by the presence of an amine group. A subgroup of the biogenic amines, the catecholamines, have a catechol ring. This subgroup includes the transmitters **dopamine, norepinephrine,** and **epinephrine.** The synthesis of each catecholamine follows a similar biochemical pathway that starts with the synthesis of dopamine.

Dopamine. Dopamine is synthesized from the amino acid tyrosine, which is converted to DOPA by the enzyme tyrosine hydroxylase (TH). Tyrosine

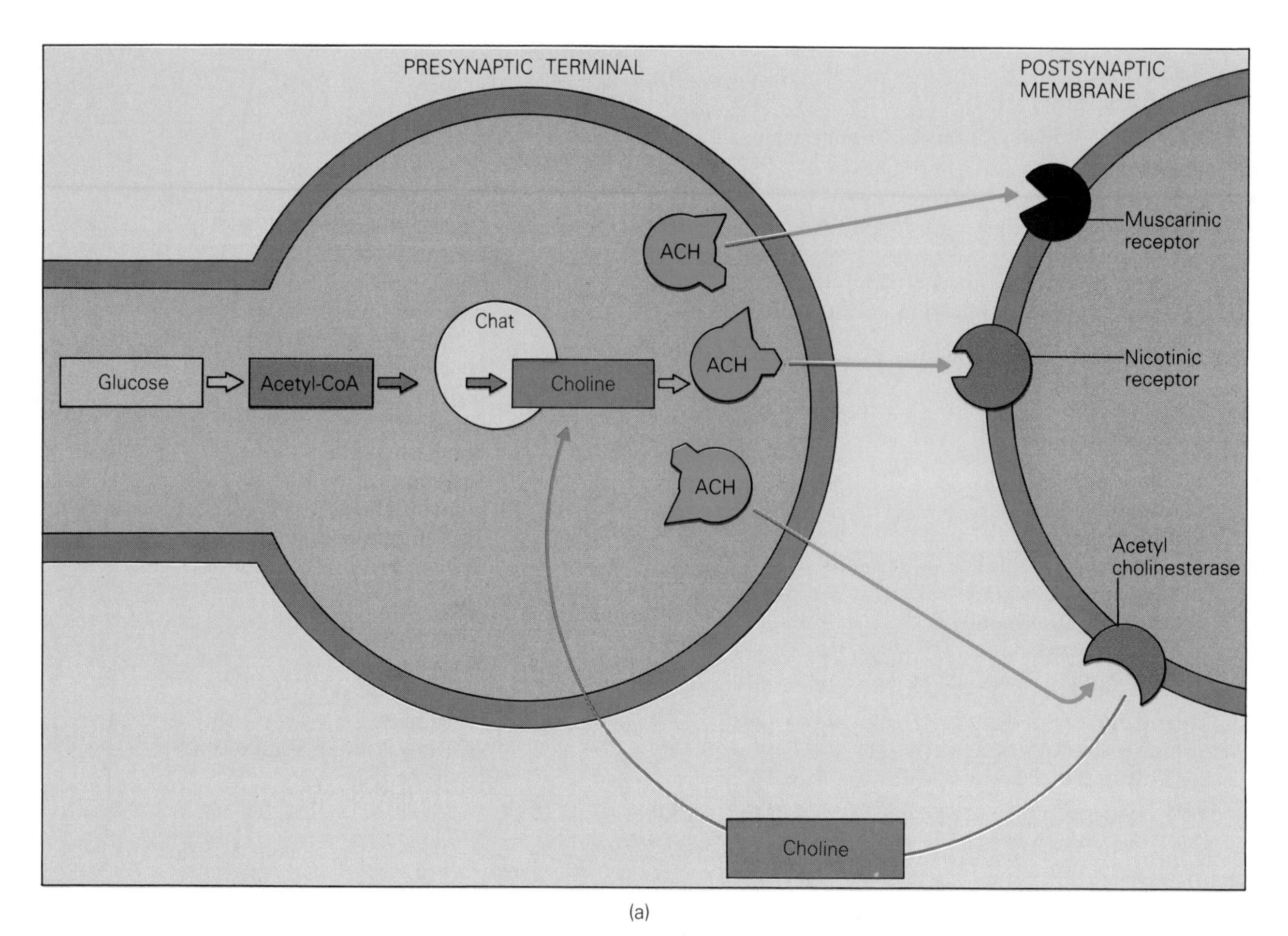

Figure 7–29 Acetylcholine is synthesized from acetyl CoA in the axon terminal and binds to two types of receptors: muscarinic and nicotinic receptors. After binding to acetyl cholinesterase, it is degraded to choline and taken up into the presynaptic terminal for resynthesis of ACh.

Focus on Alzheimer's Disease: The Loss of Acetylcholine Neurons and Memory

Alzheimer's disease is characterized by the loss of recent memories. Initially, a person's more distant memories are preserved, but as the disease progresses even the earliest of memories are lost. Eventually, the victim is unable to recognize close family members. As the disease progresses further, other deficits begin to appear. Along with loss of memory, speech is often impaired; gradually, difficulties in reading, writing, and performing complex movements occurs. In later stages, Alzheimer victims are unable to perform even the simplest of tasks such as eating, dressing, and caring for themselves. Most patients die within 5 to 15 years after the onset of the disease.

Postmortem studies of brain tissue from Alzheimer's patients have revealed a selective loss of neurons that synthesize acetylcholine. These acetylcholine neurons are located at the base of the brain (the nucleus basalis of Meynert, the diagonal band of Broca, and the medial septal nucleus) and send axons to all areas of the neocortex and hippocampal formation. These areas of brain are responsible for sensory perception, movement, speech, and language, as well as learning and memory (see Chapters 8, 9, and 11). The loss of these neurons is thought to cause the gradual deterioration of memory, language, and motor functions.

Of all individuals with Alzheimer's, 15 to 20% have inherited the disorder. Research indicates that the genetic form of this disease follows an autosomal dominant pattern of inheritance and that offspring have a 50% chance of inheriting the disease from a parent.

Environmental factors also may play a role in Alzheimer's disease. Studies have shown abnormally high concentrations of aluminum in degenerating neurons of Alzheimer's patients; however, the role of aluminum still needs to be clarified. In addition, infectious agents such as viruses also may be involved. Slow-acting viruses have been found to cause other forms of dementia in humans. At this time, there is no treatment that can prevent or slow the progression of Alzheimer's dis-

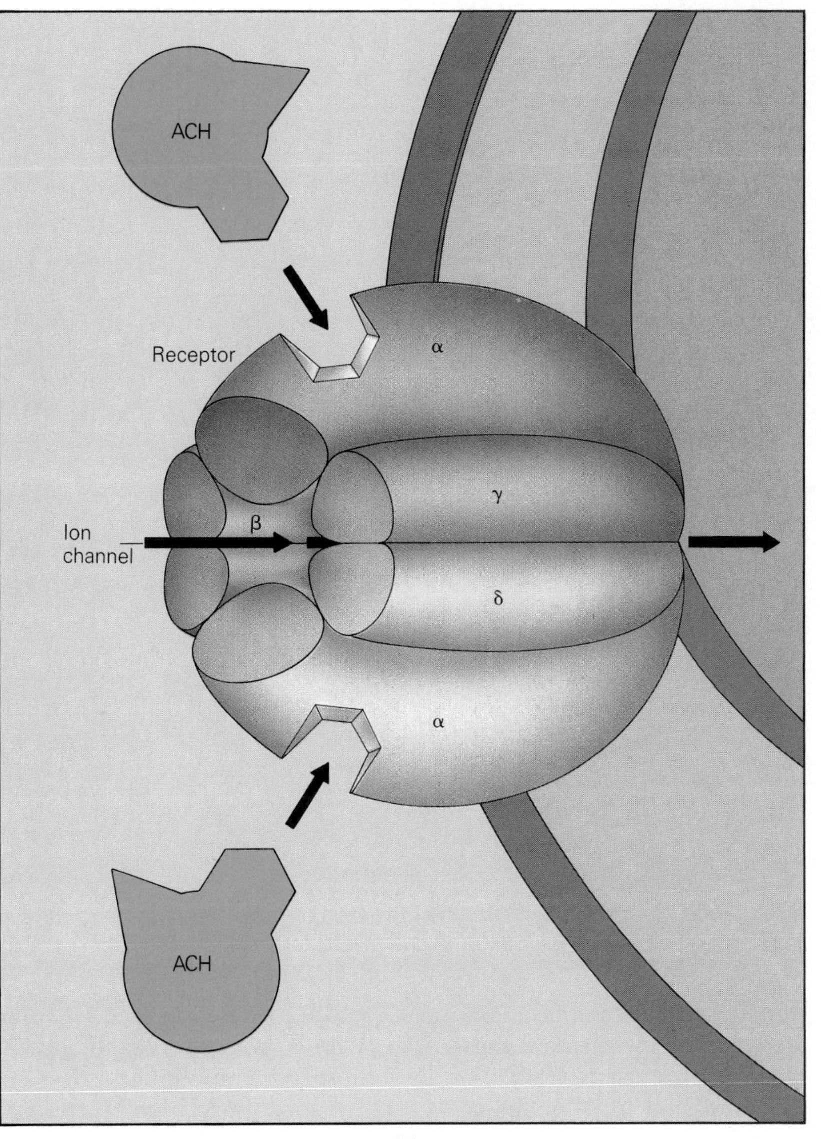

(b)

Figure 7–29 (continued) (*b*) The nicotinic acetylcholine receptor consists of five subunits. When acetylcholine binds to two of the subunits, the protein opens to form an ion channel into the postsynaptic cell.

ease. Drugs that increase the amount of acetylcholine have been found to temporarily improve memory in Alzheimer's patients. This approach, however, is useful only when there are still sufficient number of neurons that produce acetylcholine.

As the median age of our population increases, more people will succumb to this age-related disorder. Within the next 50 years, nearly 20% of our population will be older than 65 years. Up to 6% of this group will be afflicted with Alzheimer's disease, while approximately 15% will suffer from a milder form of this disorder. The cost of caring for these individuals currently exceeds $10 billion per year and is expected to exceed $40 billion within the next few decades.

hydroxylase is the controlling enzyme that regulates the overall synthesis of dopamine. DOPA, in turn, is converted to dopamine by the enzyme DOPA decarboxylase (Fig. 7–30).

The binding of dopamine to its receptor has been shown to hyperpolarize the postsynaptic potential by activating chemically sensitive K^+ channels. After uncoupling with the receptor, dopamine is transported back into the presynaptic terminal by a re-uptake pumping mechanism to be repackaged into synaptic vesicles. Eighty percent of the dopamine within the synaptic cleft is transported back to the presynaptic terminal. Thus, the re-uptake process is the main mechanism by which dopamine

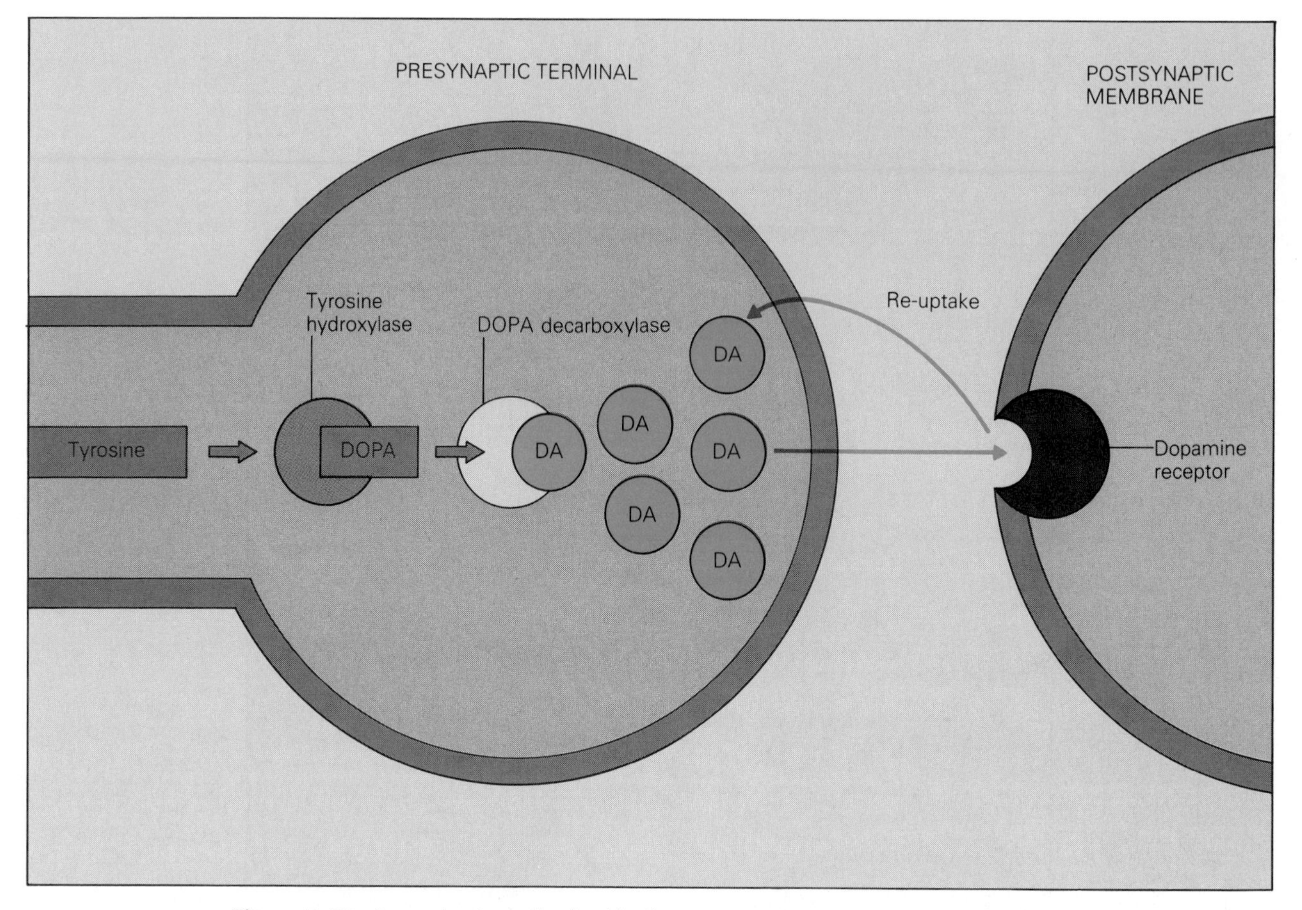

Figure 7–30 Dopamine is synthesized in the axon terminal and released to the synaptic cleft, to be pumped back into the presynaptic terminal.

transmission is terminated. The remaining 20% of dopamine is degraded within the cleft by the enzyme catechol-O-methyl transferase (COMT).

The synaptic transmission of dopamine is greatly affected by commonly used illegal drugs. Cocaine inhibits the re-uptake of dopamine into the presynaptic terminal, and amphetamine increases the release of dopamine into the synaptic cleft. Both drugs effectively increase the levels of dopamine within the cleft to activate dopamine receptors.

Norepinephrine. Norepinephrine (NE) is another member of the catecholamine family that is found throughout the central nervous system and also at the junction between nerves and smooth muscles in the autonomic nervous system. Norepinephrine is synthesized from dopamine by the enzyme dopamine-beta-hydroxylase (DBH) (Fig. 7–31*a*).

The formation of NE is quite labile and is regulated by mechanisms that can change the enzymatic activity associated with NE synthesis over either a short or long period of time. The short-term changes involve the modulation of tyrosine hydroxylase. Typically the activity of TH is regulated by both dopamine and norepinephrine by the process of end-product inhibition (see Chapter 5). When the enzyme is phosphorylated, however, the process of end-product inhibition is less effective. In addition, the phosphorylation of tyrosine hydroxylase increases the affinity of the enzyme to the tyrosine substrate. Thus the phosphorylation of the tyrosine hydroxylase can increase its enzymatic activity. The phosphorylation of tyrosine hydroxylase has been shown to occur by cAMP-dependent protein kinase and Ca^{2+}/calmodulin mechanisms (see Chapter 5). The short-term increases in the synthesis of NE occur within minutes and are easily reversible.

Long-term increases in NE can be caused by stress, since stress can lead to increases in tyrosine hydroxylase and dopamine beta-hydroxylase enzymes. Environmental factors can therefore play a significant role in modifying nerve cell function on a long-term basis.

The binding of NE to its receptor in the central nervous system closes K^+ channels and causes a depolarization of the membrane potential. NE-producing cells in the central nervous system act through cAMP to close K^+ channels, resulting in the depolarization of the postsynaptic cell.

In the peripheral nervous system, two types of

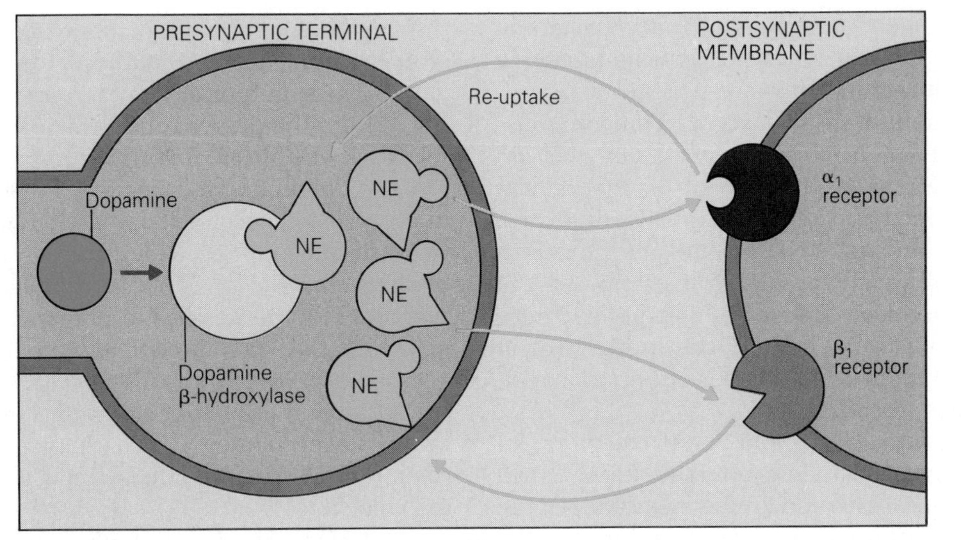

(a)

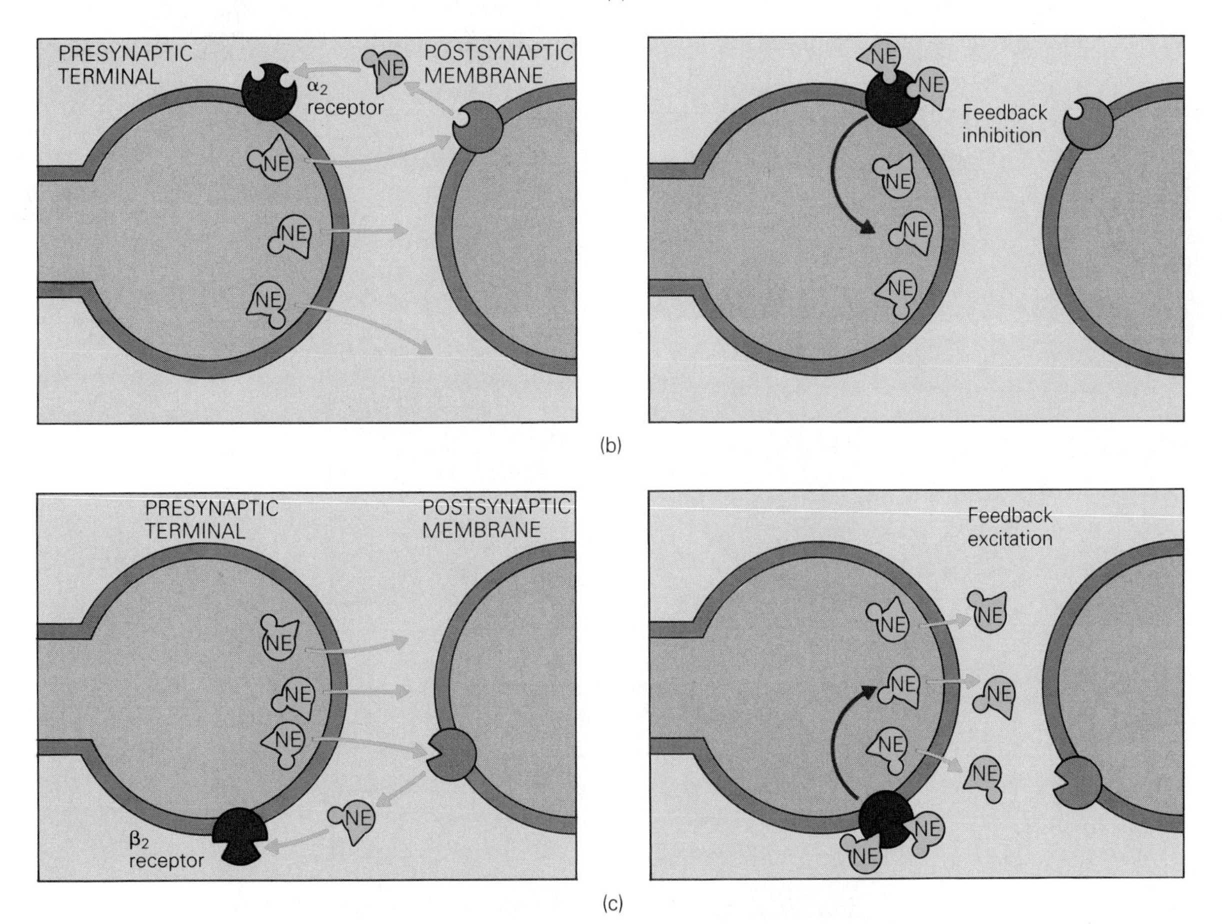

(c)

Figure 7–31 *(a)* Norepinephrine (NE) binds to two types of receptors on the postsynaptic membrane: α_1 and β_1. On the presynaptic membrane, binding of NE to α_2 receptors causes norepinephrine release to be inhibited *(b)*, while binding of NE to β_2 receptors causes more NE to be released *(c)*.

norepinephrine receptors have been identified: alpha receptors and beta receptors. Alpha-1 receptors are found on the smooth muscles of blood vessels. When they are activated, an increase in Ca^{2+} ion influx occurs, which in turn causes the contraction of smooth muscles. Alpha receptors also stimulate the hydrolysis of phosphatidylinositol, a lipid in the postsynaptic membrane that activates the intracellular messenger diacylglycerol. Diacylglycerol in turn activates protein kinase C, which initiates a variety of cellular functions (see Chapter 5).

Beta-1 receptors are also found on smooth muscles. Their activation, however, results in a relaxation of smooth muscles by mechanisms that remain to be determined. In addition to activating ionic channels, the binding of NE to beta-1 receptors activates metabolic pathways by converting ATP to cAMP. As previously discussed, this second messenger in turn activates a cAMP-dependent protein kinase, which regulates a number of important cellular functions (see Chapter 5).

In addition to alpha-1 and beta-1 receptors, located on the postsynaptic membrane, alpha-2 and beta-2 receptors are located on the membrane of the presynaptic terminal and appear to regulate the amount of NE that is released (Fig. 7–31*b*). These receptors are often referred to as **autoreceptors.** As the amount of NE within the synaptic cleft is increased, more alpha-2 autoreceptors become activated. These autoreceptors then inhibit the release of NE from the presynaptic terminal in a process of **feedback inhibition.** The activation of beta-2 receptors, on the other hand, increases the release of NE in a process called **feedback excitation.**

After uncoupling with its receptor, NE is taken back up into the presynaptic terminal, where it is re-packaged into vesicles in preparation for release again into the synaptic cleft. As with dopamine, this re-uptake process removes approximately 80% of the NE from the synaptic cleft and is the main mechanism the terminates NE synaptic transmission. The remaining NE is degraded within the cleft by the enzyme COMT.

Serotonin. Another common biogenic amine is **5-hydroxytryptamine (5-HT),** or **serotonin.** It is

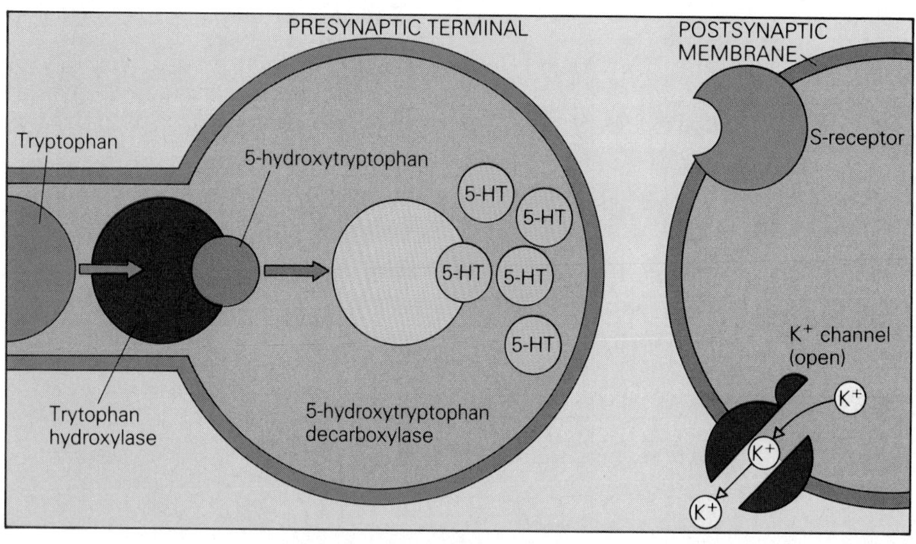

(a)

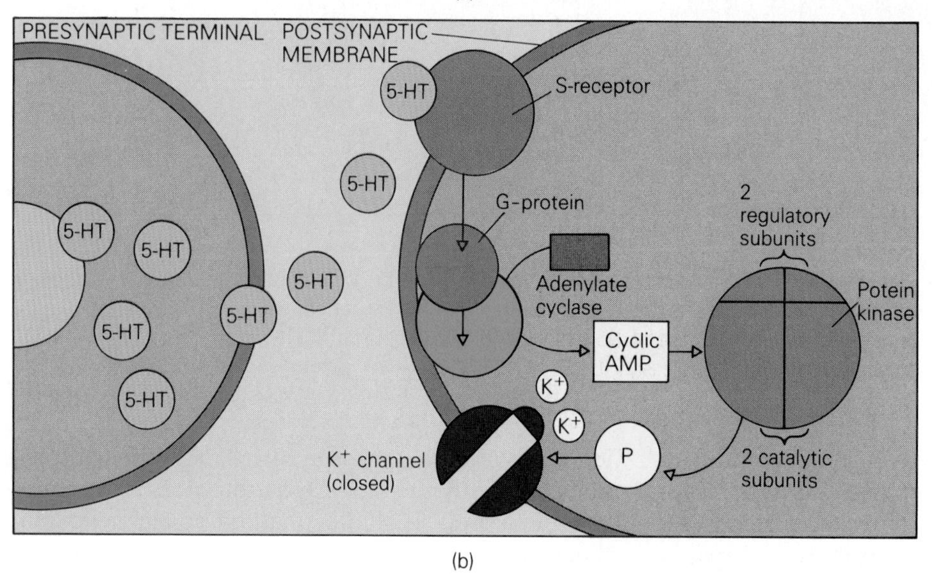

(b)

Figure 7–32 Through a sequence of enzymatic reactions culminating in a phosphorylation of a K^+ channel in the postsynaptic membrane, serotonin (5-HT) causes the K^+ channel to close, inhibiting K^+ diffusion out of the cell.

found throughout the brain, but is primarily synthesized in the region of the brain stem. The synthesis of 5-HT begins with the amino acid tryptophan, which is converted to 5-hydroxytryptophan (5-HTP) by the enzyme tryptophan hydroxylase (Fig. 7–32*a*). In turn, 5-HTP is converted to serotonin by 5-HTP decarboxylase.

After its release into the synaptic cleft, serotonin can interact with several types of serotonin receptors. Most is known about the so-called **S-receptor,** which closes gates to K^+ ions when 5-HT is bound to it. The S-receptor itself is not part of an ion channel, but part of a sequence of enzymatic reactions (Fig. 7–32*b*). Activation of the S-receptor causes a coupling protein (the G-protein) to activate the enzyme adenylate cyclase. This enzyme converts ATP to cAMP, which then activates cAMP-dependent protein kinase. This enzyme then phosphorylates a protein that acts on the K^+ channel to close it.

After uncoupling from its receptor, serotonin is transported back into the presynaptic terminal. As with NE and DA, approximately 80% of the serotonin within the synaptic cleft is removed by this re-uptake process. The remainder of the serotonin is degraded by an enzyme called monoamine oxidase (MAO).

Amino Acids

A variety of amino acids also satisfy the conditions necessary to be classified as neurotransmitters. We will consider only glutamate and gamma amino butyric acid, since they are found in great abundance throughout the nervous system.

Glutamate. **Glutamate** is synthesized from alpha-ketoglutarate by way of the citric acid cycle (Fig. 7–33). It is one of the most potent excitatory neurotransmitters in the nervous system. The binding of glutamate to its receptor opens ion channels that permit the passage of Na^+ into the nerve cell. As a result, the membrane potential depolarizes, and, if a sufficient number of glutamate receptors are activated, the membrane potential of the postsynaptic neuron is brought to threshold to generate an action potential.

Synaptic transmission by glutamate is terminated in part by its re-uptake into the presynaptic terminal. Much of the glutamate within the synaptic

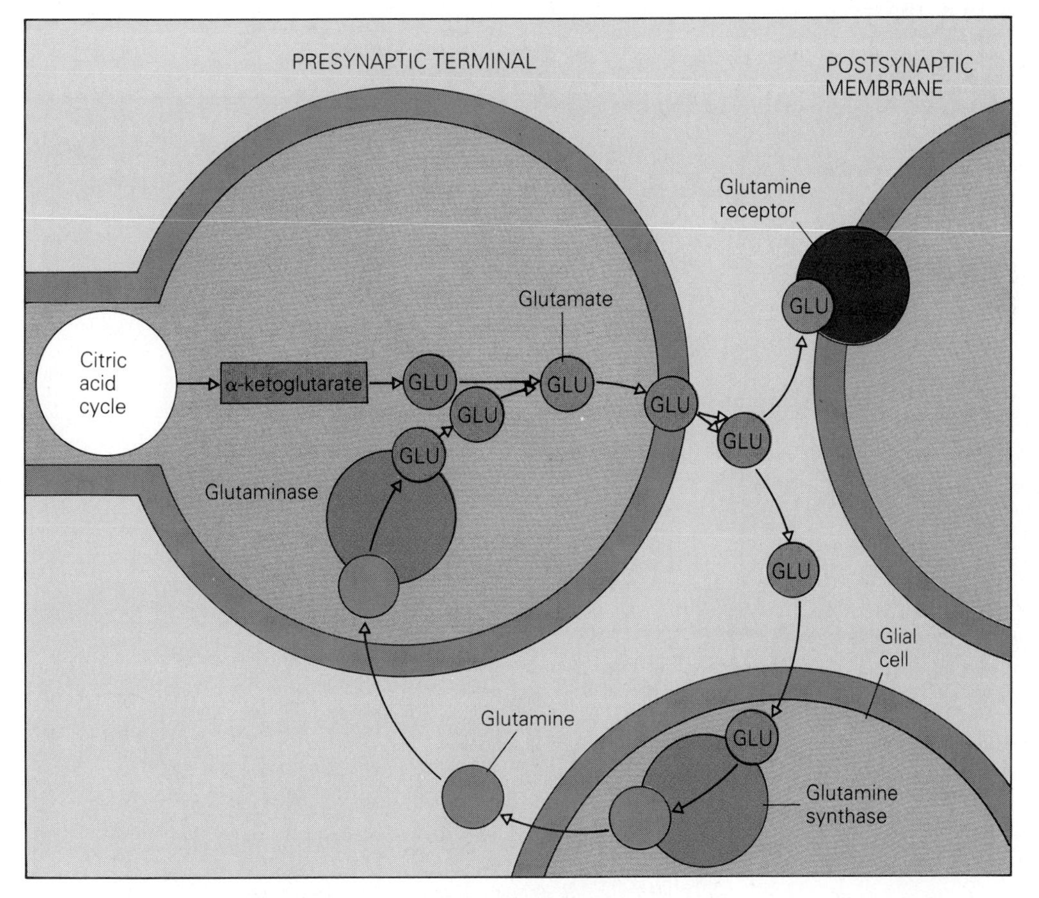

Figure 7–33 Glutamate is synthesized from a product of the citric acid cycle and is converted to glutamine in a glial cell before re-uptake.

cleft, however, is transported into glial cells. There, it is converted to glutamine by the enzyme glutamine synthase. The glutamine in turn is transported back into the presynaptic terminal, where it is then reconverted to free glutamate and repackaged into vesicles.

Gamma Amino Butyric Acid. Another amino acid neurotransmitter that is found throughout the central nervous system is **gamma amino butyric acid (GABA).** GABA is a potent inhibitory neurotransmitter that is synthesized from glutamate by the enzyme glutamic acid decarboxylase (GAD) (Fig. 7–34).

The binding of GABA to its receptor causes influx of Cl^- ions into the nerve cell, which hyperpolarizes the membrane potential and inhibits the postsynaptic neuron. The synaptic transmission of GABA is terminated by its re-uptake into the presynaptic terminal and by its transport into glial cells. The mitochondria within glial cells convert GABA into succinic semialdehyde by the enzyme GABA-T. At the same time, this enzyme is coupled to the conversion of alpha-ketoglutarate to glutamate. Glutamate in turn is converted to glutamine by glutamine synthase and is then transported to the presynaptic terminal. Within the presynaptic terminal, glutamine is converted into glutamate and subsequently to GABA to be packaged into synaptic vesicles.

Neuropeptides: Products of the Soma

Neuropeptides are chemical transmitters that consist of chains of amino acids. The processing of neuropeptides differs considerably from that of low-molecular-weight transmitters. Neuropeptides are synthesized in the soma of neurons rather than the synaptic terminal. In addition, neuropeptides are created when large proteins, or **polyproteins,** are broken down. An example of a large polyprotein is **beta-lipotropin,** which breaks down into a number of opiates that can inhibit the sensation of pain (Fig. 7–35). The various peptides are packaged within secretory vesicles and carried to the terminal area by mechanisms of fast axonal transport. Within the synaptic terminal, vesicles containing peptides are found to co-exist with vesicles containing low-molecular-weight transmitters. Table 7–3 lists the low-molecular-weight and neuropeptide transmitters that have been found within the same synaptic terminals.

At the terminal ending, the process of synaptic transmission of peptides is different from that of

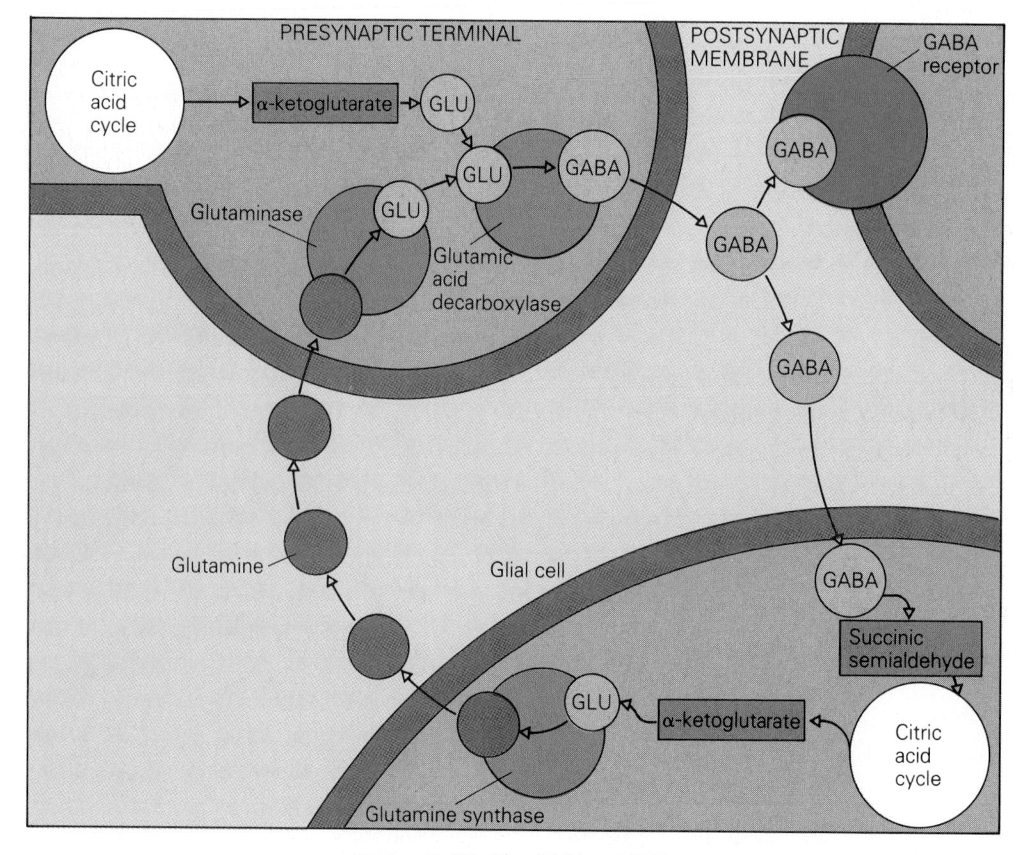

Figure 7–34 The GABA synapse.

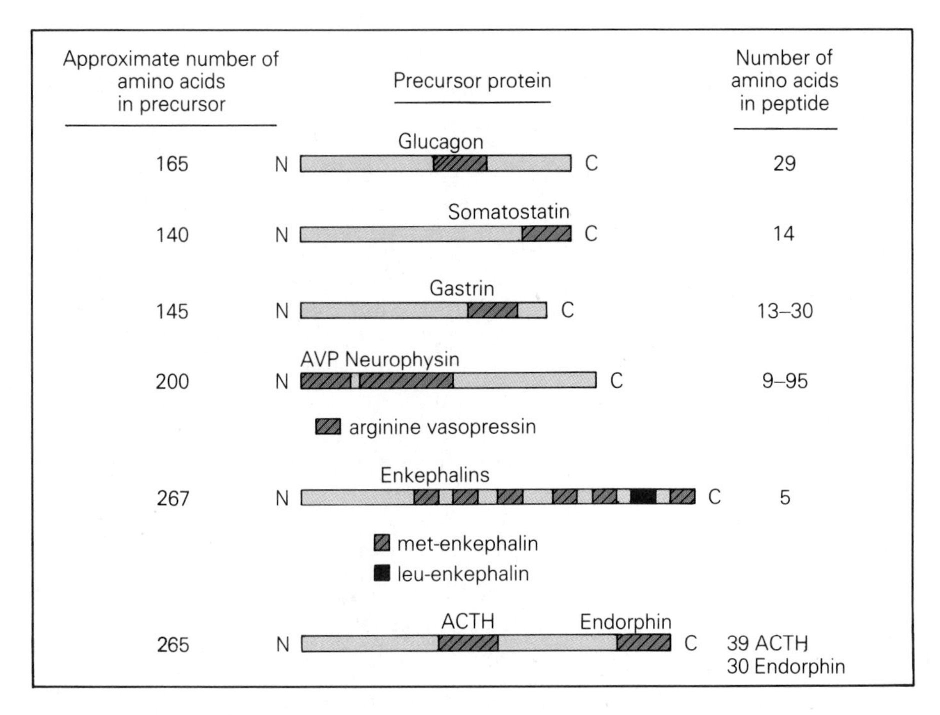

Figure 7–35 Precursors of neuropeptides, chemical transmitters synthesized in the soma.

low-molecular-weight transmitters. Once they are released into the synaptic cleft, there are no re-uptake mechanisms to recycle the neuropeptides. Therefore the process of peptide transmission cannot be sustained as it is for the low-molecular-weight transmitters.

Neuropeptides are classified in one of several families of peptides based on their amino acid sequence and function (Table 7–4). The family of **opioids** are structurally similar to opiates that have been shown to modulate the perception of pain. **Neurohypophyseal peptides** are structurally similar to those found in the posterior pituitary and function to regulate plasma osmolarity and lactation. **Secretins** and **gastrins** are peptides that are structurally similar to the peptides and hormones found in the gastrointestinal system (see Chapter 26). The family of **insulins** is similar in structure to the insuline hormone and are responsible for the growth and maintenance of nerve cells, and the **somatostatins** are structurally similar to growth hormone.

From our discussion of chemical neurotransmitters, it can be seen that nerve cells are capable of producing a variety of transmitters. Individual nerve cells, however, are able to synthesize only

Table 7–3
Colocalization of Low-Molecular-Weight Transmitters and Neuropeptides

Low-Molecular-Weight Transmitter	Neuropeptide
Acetylcholine	Vasoactive intestinal peptide
Norepinephrine	Somatostatin Enkephalin Neurotensin
Dopamine	Cholecystokinin Enkephalin
Adrenalin	Enkephalin
Serotonin	Substance P Thyrotropin-releasing hormone

Table 7–4
Families of Peptides

Opioid: opiocortins, enkephalins, dynorphin, FMRF amide

Neurohypophyseal: vasopressin, oxytocin, neurophysins

Tachykinins: substance P, physalaemin, kassinin, uperolein, eledoisin

Secretins: secretin, glucagon, vasoactive intestinal peptide, gastric inhibitory peptide, growth hormone–releasing factor, peptide histidine isoleucineamide

Insulins: insulin, somatomedins, relaxin, nerve growth factor

Somatostatins: somatostatins, pancreatic polypeptide

Gastrins: gastrin, cholecystokinins

specific combinations of low-molecular-weight and peptide transmitters based on the enzymes they have available. This defined set of chemical transmitters is used by a neuron at all of its synapses to transduce the action potentials it generates into chemical signals that are detected by target cells.

Neuronal Integration: Temporal and Spatial Summation

The dendrites and somata of nerve cells have many different types of receptors embedded within their membranes. The synaptic activation of these receptors can simultaneously produce excitatory and inhibitory responses. If a nerve cell receives both excitatory and inhibitory information from many synaptic inputs, how does it decide, so to speak, whether or not to initiate an action potential? The decision is made by adding all of the postsynaptic potentials by two different processes: temporal summation and spatial summation. If the addition of all of the synaptic potentials results in a depolarization of the membrane potential to approximately −45 mV near the initial segment or trigger zone of the axon, then an action potential will be generated.

Temporal summation involves the addition of postsynaptic potentials that arise from the activation at one synaptic input (Fig. 7–36). The first time that the synapse is activated, it may produce a depolarization that is not sufficient to cause the membrane potential to reach threshold. This depolarization by itself would decay back to the resting level. If, however, a second action potential invades the terminal, the release of neurotransmitters causes the postsynaptic membrane to again depolarize. Moreover, if the time between the first and second action potentials is short enough, the two postsynaptic membrane potentials together will produce a depolarization that may be sufficient to reach the threshold voltage and discharge an action potential.

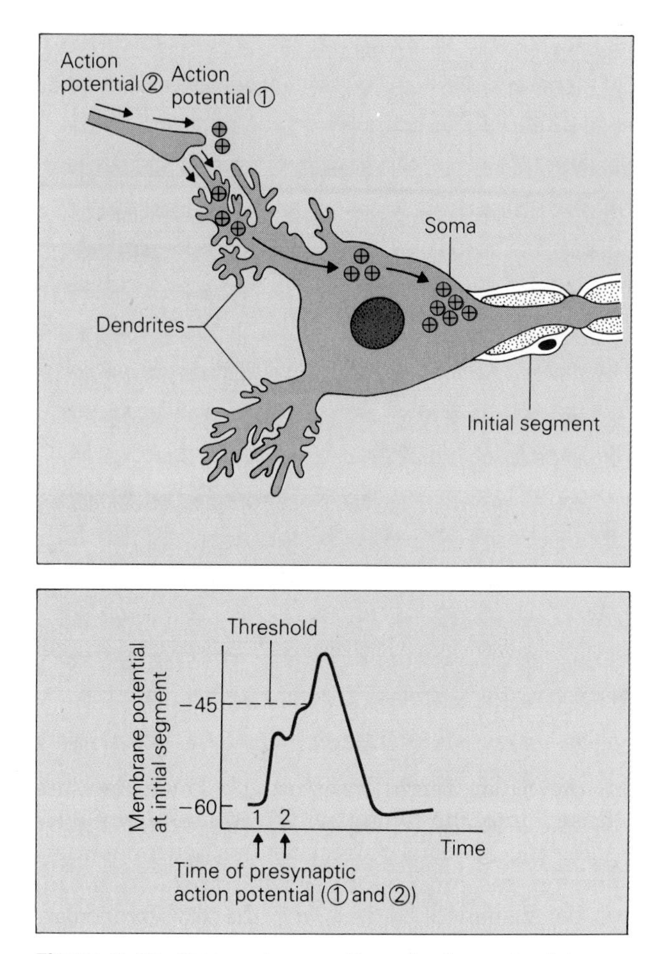

Figure 7–36 Temporal summation of action potentials.

Another summation process that involves the activation of more than one synaptic input is the process of **spatial summation.** In this case, two synaptic inputs, which alone are insufficient to cause the membrane potential to reach threshold (Fig. 7–37), are activated simultaneously, thus resulting in a depolarization sufficient to achieve a threshold value and discharge an action potential. The action potential discharge by a nerve cell, therefore, is based upon the summation of excitatory and inhibitory postsynaptic potentials that occur in time over the space of the neuron's receptive surface.

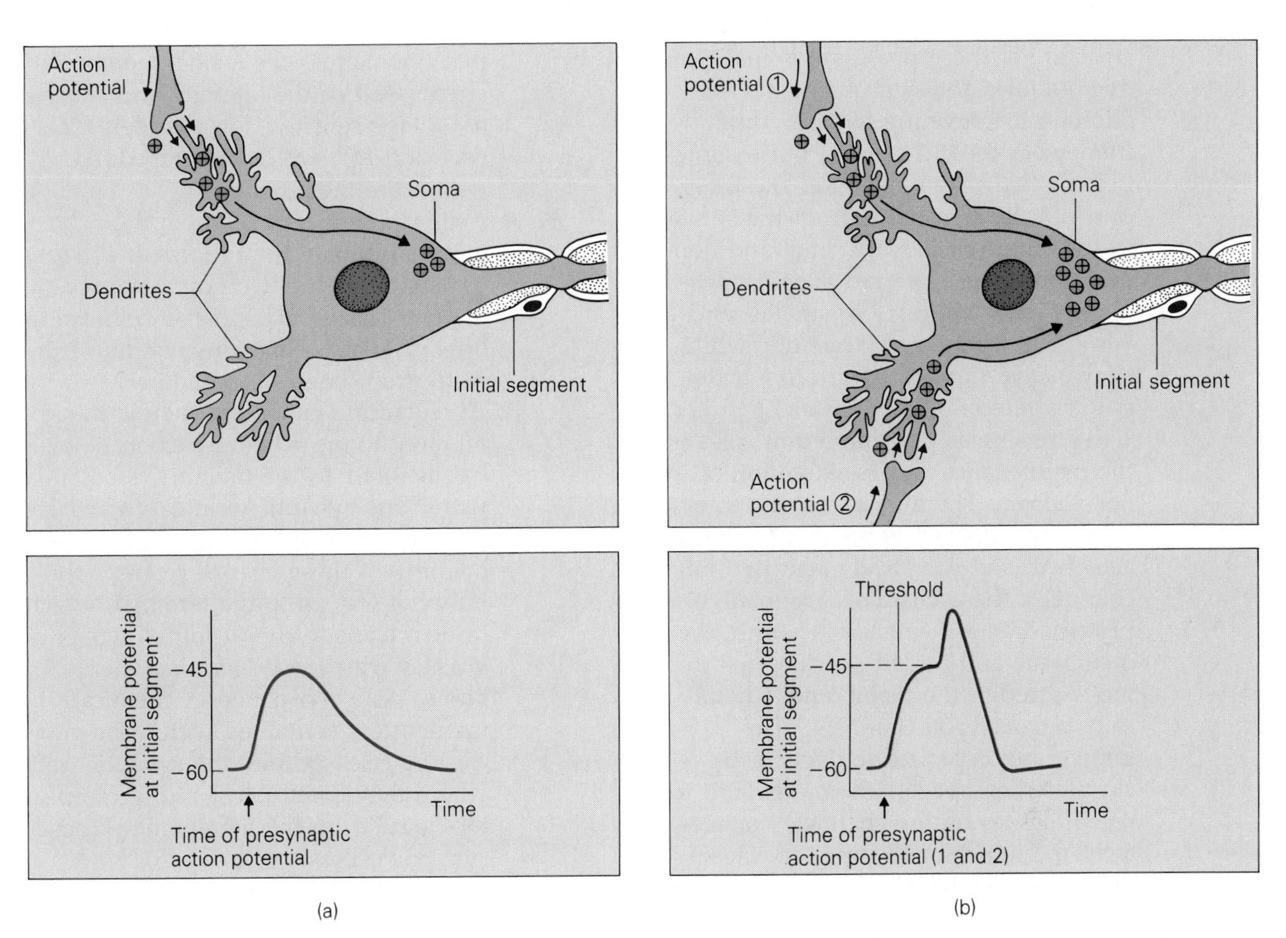

Figure 7–37 Spatial summation of action potentials.

Summary

I. The nervous system is composed of the brain, brain stem, spinal cord, and peripheral nerves. The first three make up the *central nervous system (CNS)* whereas the latter constitute the *peripheral nervous system (PNS)*. The brain is subdivided into functional regions that are involved in either sensory input from the skin, eye, ears, and other sensory organs. Other areas are involved with locomotor activity as well as other functions. This functional partitioning of brain areas is found throughout the nervous system. A general organizational theme of the nervous system is that the brain receives sensory input and provides the appropriate output for such functions as speech, movement, or the control of internal organs.

II. A. Glia, or glial cells, aid in maintaining the environment of the brain. With blood vessels, they form the blood-brain barrier, which prevents the passage of unwanted material from the circulatory system to the brain. They also act as scavengers in the brain to remove cellular debris. Another important function of glial cells is the formation of myelin around the axon of nerve cells to increase the conduction and transmission of electrical impulses. *Neurons* are the basic cellular elements in the nervous system involved in the transmission of information. The membranes of neurons have structural and functional properties that are specialized for the integration and rapid transmission of information from one nerve cell to another.

B. Neurons have extensive processes called dendrites and axons that are involved in the reception and transmission of information. The cytoskeletal structure of these processes is made of microtubules and neurofilaments. In addition to providing the structural framework for the neuron, these components are also involved in the transport of cellular material from the somal to the outer reaches of axonal and dendritic processes. The fast axonal transport process will carry organelles and vesicles at a rate of 400 mm/day while slow axonal transport will carry material at a rate of 3.0 mm or less per day. These processes play important roles in the maintenance and regeneration of nerve fibers. The membrane of nerve cells are composed of passive, voltage-activated, and chemically activated ion channels. These channels are made of proteins that are embedded within the membrane and permit specific ions to move through the membrane. Chemically activated ion channels, also known as receptors, are opened (or closed) when specific chemicals bind to receptor sites on the channel. Voltage-activated ion channels open (or close) when the membrane is depolarized. Passive ion channels remain in the open state and play a role in establishing the membrane potential.

III. A. A difference in the number of positive and negative ions between the inside and outside of a cell produces the membrane potential. In general, the inside of a cell is more negatively charged than the extracellular space, and thus establishes an electrical driving force for the movement of ions across the membrane. Differences in the concentration of ions between the inside and outside of cells establish concentration gradients, which act as chemical driving forces for the movement of ions across the membrane. The summated effects of electrical and chemical driving forces determine whether an ion will move through the membrane in an inward or outward direction.

B. The axon of nerve cells transmits information by generating action potentials. Action potentials are initiated and propagated along the axon as a result of the influx and efflux of Na^+ and K^+ ion, respectively. These ionic currents that flow across the axonal membrane are a result of the opening and closing of voltage-sensitive ion channels for Na^+ and K^+, which are embedded within the membrane.

IV. A. Action potentials that reach the axon terminal initiate the process of synaptic transmission. At electrical synapses, the cytoplasm between two cells are in direct contact. Thus ions will move directly from one cell to another.

B. At chemical synapses, synaptic transmission occurs when an action potential depolarizes the membrane potential within the synaptic terminal to activate voltage-sensitive Ca^{2+} channels. The opening of these channels causes the influx of Ca^{2+} into the terminal, which in turn leads to the fusion of synaptic vesicles with the preterminal membrane. As a consequence, chemical transmitters contained within the vesicles are released into the synaptic cleft where they eventually bind to receptors located on the membrane of postsynaptic cells.

V. A. The binding of chemicals called neurotransmitters with their receptors affects the ionic conductances of the target cells. These receptors are chemically activated ion channels that permit the passage of specific ions. Well-characterized neurotransmitters include acetylcholine, dopamine, norepinephrine, serotonin, glutamate, and gamma amino butyric acid.

B. Less well understood transmitters known as neuropeptides are synthesized in the soma.

VI. The influx and efflux of ions through the chemically activated channels produce excitatory or inhibitory postsynaptic potentials. These responses are integrated by the postsynaptic cell to produce a summated response. If the summation of these alterations in the postsynaptic membrane potential causes a depolarization that reaches threshold levels, an action potential will be generated in the postsynaptic nerve cell. The transmission of information between nerve cells, therefore, is an electrochemical process that features the conduction of an electrical impulse coupled to the secretion of chemical transmitters.

Review Questions

1. What are the differences between neurons and glia?
2. What are the different types of glial cells and their functions?
3. What are the specialized regions of the nerve cell and their general functions?
4. What are the differences between anterograde and retrograde axonal transport?
5. What are the two types of slow axonal transport process and what types of cellular material and organelles do they carry?
6. What are the functional differences between nerve and glial cells?
7. What are the functional characteristics of the three types of ion channels?
8. What are the differences and similarities between passive, chemically activated, and voltage-activated ion channels?
9. What are the functions of voltage-activated Na^+ and K^+ channels in the generation of an action potential?
10. What is the concept of the equilibrium potential?
11. What is the significance of Ohm's Law in relation to a single ion channel?
12. What are the factors that affect the conductance of an ion through its channel?
13. How are action potentials conducted along a myelinated axon?
14. What are the changes in membrane conductance in relation to various phases of the action potential?
15. What are the presynaptic mechanisms involved in the process of synaptic transmission?
16. What is the relationship between the synaptic vesicle and a unitary postsynaptic potential?
17. What are the mechanisms that are capable of producing inhibitory postsynaptic potentials?
18. What are the mechanisms that are capable of producing excitatory postsynaptic potentials?
19. Compare and contrast nicotinic versus muscarinic acetylcholine receptors.
20. Compare and contrast alpha and beta norepinephrine receptors.
21. Compare and contrast low-molecular-weight versus neuropeptide transmitters.
22. Discuss the synthesis, release, and degradation of acetylcholine as a neurotransmitter.
23. Discuss the synthesis, release, uptake, and recycling of dopamine as a neurotransmitter.
24. Discuss the synthesis, release, uptake, and recycling of norepinephrine as a neurotransmitter.
25. Discuss the synthesis, release, uptake, and recycling of serotonin as a neurotransmitter.
26. Discuss the synthesis, release, uptake, and recycling of glutamate as a neurotransmitter.
27. Discuss the synthesis, release, uptake, and recycling of GABA as a neurotransmitter.
28. Compare and contrast chemically activated receptors that open versus close when bound to a transmitter.
29. Compare and contrast temporal and spatial summation.

Suggested Readings

Siegel, G.J. and Albers, R.W. *Basic Neurochemistry.* Boston: Little Brown, 1981.

Jack, J.J.B. and Noble, D. *Electric Current Flow in Excitable Cells.* Oxford, England: Oxford University Press, 1983.

Kandel, E.R. and Schwartz, J. *Principles of Neural Science.* Amsterdam: Elsevier Press, 1985.

Kuffler, S.W. and Nicholls, J.G. *From Neuron to Brain.* Sunderland, Mass.: Sinauer Associates, 1984.

8

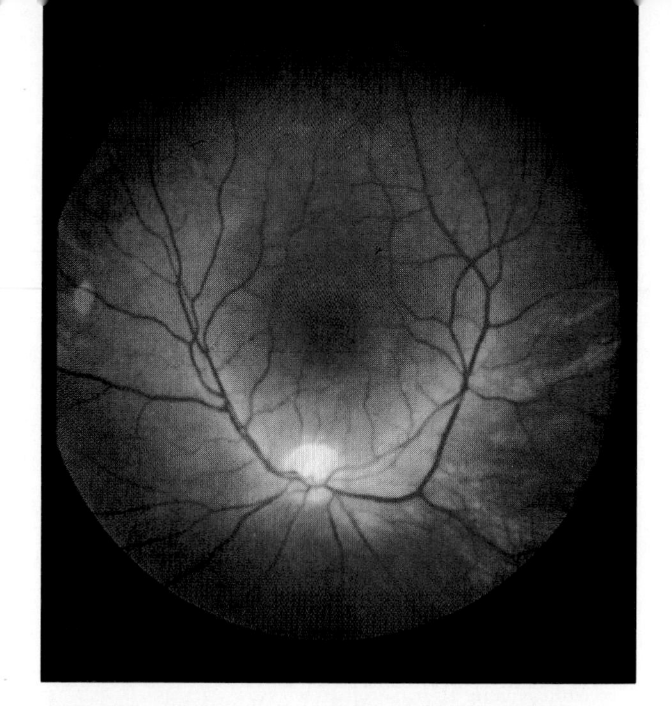

Sensory Systems

From the moment we awaken in the morning until we fall asleep at night, our bodies are bombarded with information from the outside world. Most of us awake to the sound of an alarm clock that is detected by our **auditory system.** Our eyes focus on the clock to send information to the **visual system** and confirm that it is time to get up. We then climb out of bed into the warmth of a hot shower that is sensed by nerve endings of the **somatic sensory system.** At the breakfast table the smell of toast and coffee comes wafting to our noses and **olfactory system,** and the flavor of a rich cup of coffee stimulates the taste buds of our **gustatory system.** Throughout the day we are continually exposed to information from the external environment that must reach the brain so that we can respond accordingly. The detection and transmission of this information is the function of our **sensory systems.**

Information about our physical environment is detected by **sensory receptor cells** located within various **sense organs** (see Table 8–1). Sensory information is transmitted to the neocortex by different pathways depending on the type of sensation (Fig. 8–1). Somatic sensation from the body and limbs, for example, is transmitted from receptor cells to the spinal cord, where it then ascends to the higher levels of the nervous system. Auditory information detected by the ears, on the other hand, is transmitted from sensory receptor cells to the brain stem, where it ascends to eventually reach the neocortex. In the visual system, sensory receptors are located in the back of the eye. These sensory cells detect

Table 8–1
Sensory Modalities and Receptor Cells

Sensory Mode	Receptor	Sense Organ
Vision	Rods and cones	Eye
Hearing	Hair cells	Ear (Organ of Corti)
Rotational acceleration	Hair cells	Ear (semicircular canals)
Linear acceleration	Hair cells	Ear (utricle and saccule)
Smell	Olfactory neurons	Olfactory mucous membrane
Taste	Taste receptor cells	Taste buds
Touch—pressure	Nerve endings	Skin
Warmth	Nerve endings	Skin
Cold	Nerve endings	Skin
Pain	Naked nerve endings	Skin
Joint movement and position	Nerve endings	Various
Muscle length	Nerve endings	Muscle spindle
Muscle tension	Nerve endings	Golgi tendon organ
Arterial blood pressure	Nerve endings	Stretch receptors in carotid sinus and aortic arch
Central venous pressure	Nerve endings	Stretch receptors in walls of great veins, atria
Inflation of lung	Nerve endings	Stretch receptors in lung parenchyma
Blood temperature in head	Neurons in hypothalamus	—
Arterial Po_2	Nerve endings?	Carotid and aortic bodies
pH of CSF	Receptors on ventral surface of medulla oblongata	—
Osmotic pressure of plasma	Cells in anterior hypothalamus	—
Arteriovenous blood glucose difference	Cells in hypothalamus	—

various wavelengths of light, corresponding to different colors, and **transduce** this information into the electrical impulses that nerve cells use to transmit information to the thalamus. Sensory receptors use several common mechanisms to transduce information from the external environment.

General Principles of Sensory Transduction

Sensory receptors are activated when they detect a specific stimulus. This specific stimulus is called an **adequate stimulus** (Fig. 8–2) and is unique to each sensory receptor. In the visual system, for example, the photoreceptors detect colors of light but are insensitive to frequencies of sound. Likewise, the auditory receptors in the ear respond only to sound and are insensitive to light.

An adequate stimulus will produce a change in the membrane potential of the sensory receptor cell. This change in the membrane potential is called a **generator potential.** In some sensory receptors, such as somatic sensory receptors, the generator potential is a **depolarization** of the membrane. In others, such as the photoreceptors of the eye, it is a **hyperpolarization** of the membrane. The generator potential, in turn, produces an action potential or a series of action potentials. The action potentials can be generated by the receptor cell itself or by a neuron connected to the receptor cell. These action potentials transmit information about the nature of the stimulus to the central nervous system.

The transduction mechanisms of all sensory systems, therefore, involve (1) an adequate stimu-

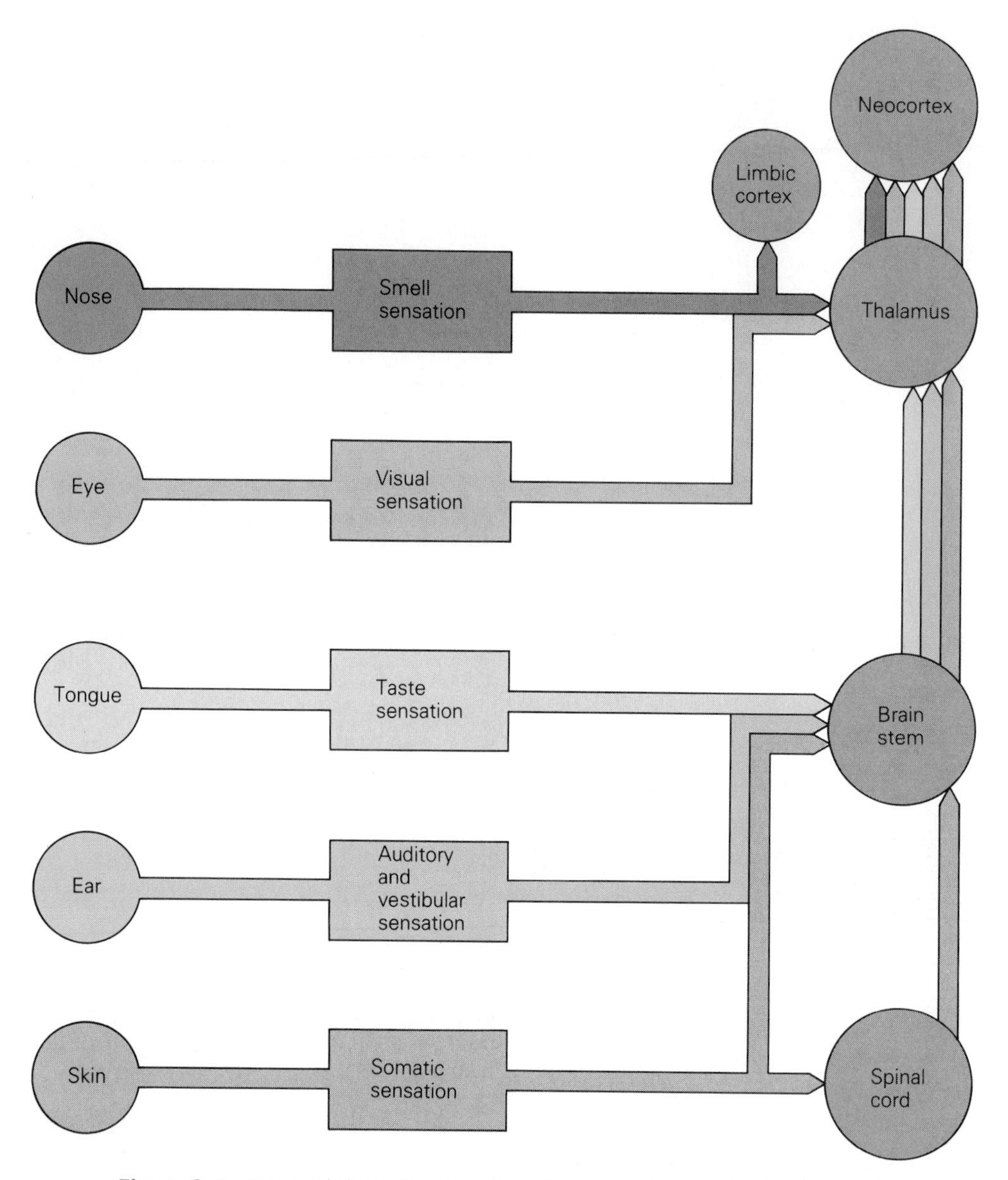

Figure 8–1 Sensory information travels to the brain via specialized pathways for each type of sensation.

lus, (2) a generator potential, and (3) the initiation of action potentials. In order to examine these mechanisms in greater detail, let us consider a somatic sensory receptor called a **Pacinian corpuscle** that responds to touch, or **tactile,** stimuli. This type of receptor is located beneath the skin and consists of a free nerve ending that is encapsulated by layers of connective tissue (Fig. 8–3). The nerve ending is wrapped with myelin along the length of the fiber. The fiber itself extends into the spinal cord, where it forms synapses with other nerve cells.

The adequate stimulus for the Pacinian corpus-

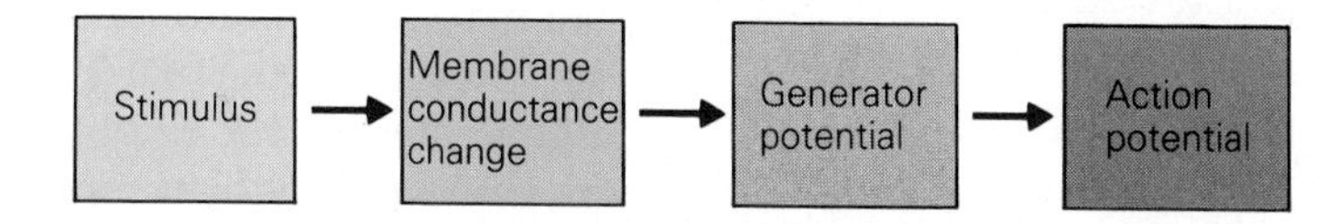

Figure 8–2 All sensory transductions involve an adequate stimulus, a generator potential resulting from a membrane conductance change, and one or more action potentials. The red "spikes" represent action potentials that have been compressed in time to reveal a pattern or frequency of occurrence.

cle is **pressure** applied to the skin. This pressure causes the layers of connective tissue and free nerve ending to compress. The compression of the free nerve ending causes an opening of Na^+ channels embedded within the nerve membrane (see Fig. 8–3*b*). The resulting influx of Na^+ ions depolarizes the membrane and produces a generator potential. If the generator potential is of sufficient amplitude, it will then depolarize the membrane in the region of the first node of Ranvier to threshold and initiate an action potential (Fig. 8–3*c*).

The Intensity of a Stimulus: How Nerves Inform the Brain

The number of action potentials that are generated per unit time is a function of the intensity of the pressure. Greater pressures produce a greater number of action potentials in a single fiber. This is known as the **frequency code of stimulus intensity** (Fig. 8–4*a*). The increase in the number of action potentials generated with greater pressure is due to the continued opening of the Na^+ channels and continued depolarization of the generator potential above the threshold value. As a consequence, after the repolarization of one action potential, another is generated.

A second method of informing the brain about the intensity of the stimulus is a **population code.** With this method of coding, as the pressure becomes greater more sensory receptors of the Pacinian corpuscle type are activated. This occurs because the increased pressure affects a greater area beneath the skin (Fig. 8–4*b*).

The Quality of a Stimulus: How Nerves Send Clear Messages

The type of information that is sent to the brain is also coded in the way that nerve fiber pathways that

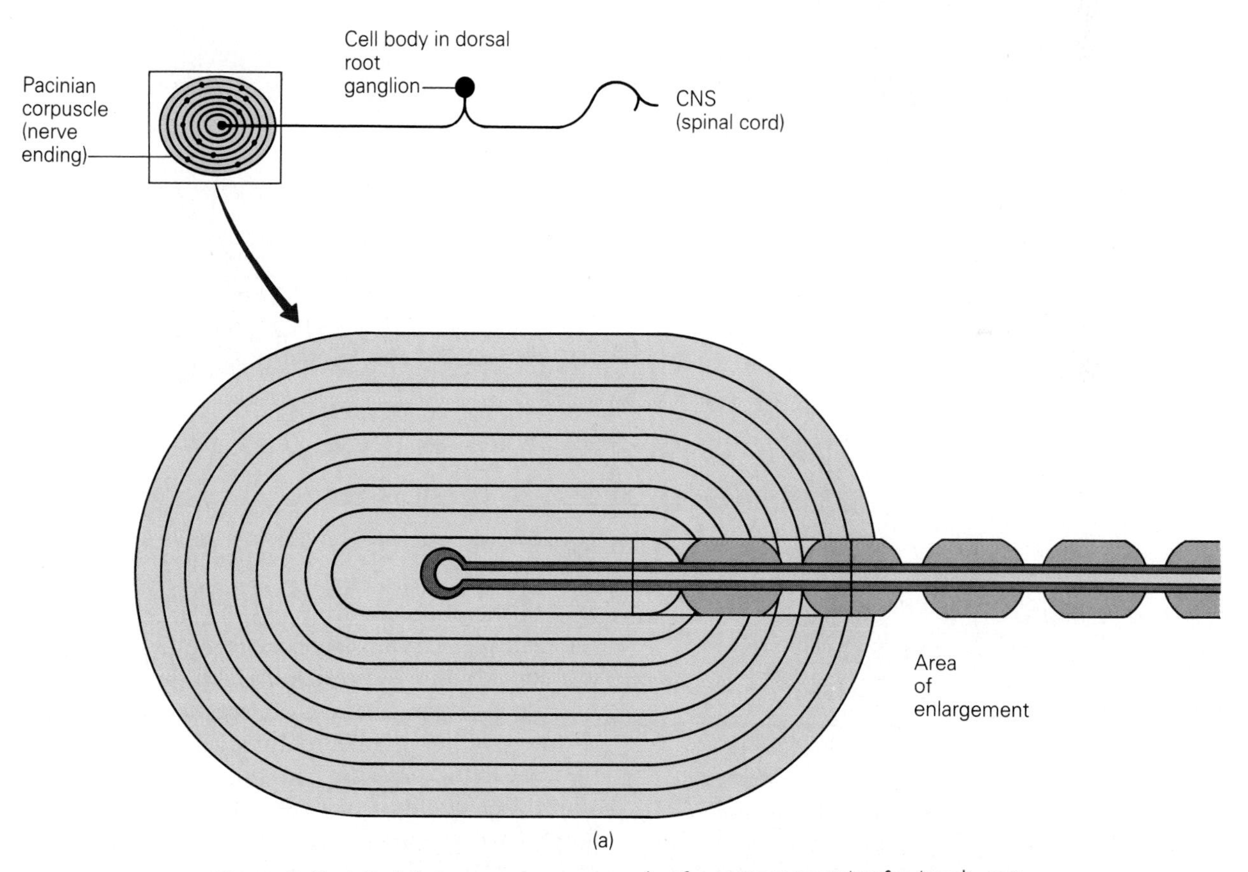

Figure 8–3 A Pacinian corpuscle, an example of a sensory receptor for touch, consists of a neuron wrapped in layers of connective tissue, with a myelinated nerve ending (axon). A tactile (pressure) stimulus causes the neuron within the connective tissue to change shape, allowing Na^+ to enter the cell. This produces a depolarization and hence an action potential that travels the length of the axon and represents outgoing sensory information.

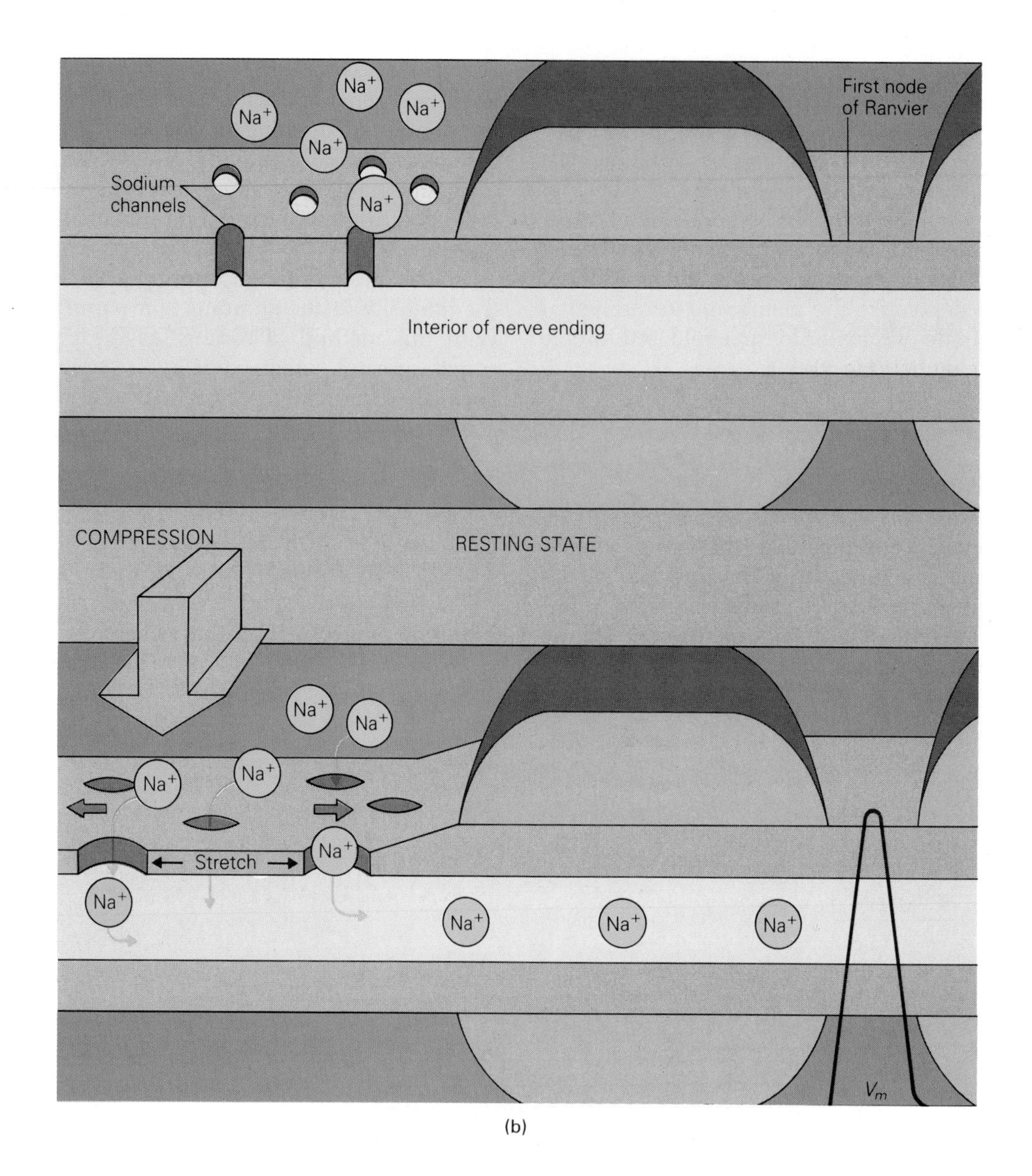

(b)

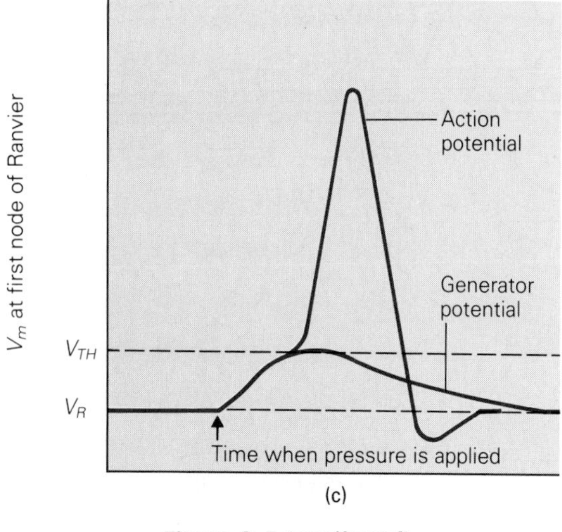

(c)

Figure 8–3 (continued)

lead to the brain are arranged physically. The skin, for example, contains sensory receptors for temperature in addition to pressure. The information about the temperature of the skin reaches the brain by a different nerve fiber pathway than does information about pressure. In this way, information about the quality, or type, of stimulus is maintained within each pathway without becoming mixed with other types of stimuli. This aspect of the nervous system, the mechanisms of coding for the **type** of stimulus detected by a sensory receptor, is called the **labeled-line code of stimulus quality** (Fig. 8–5).

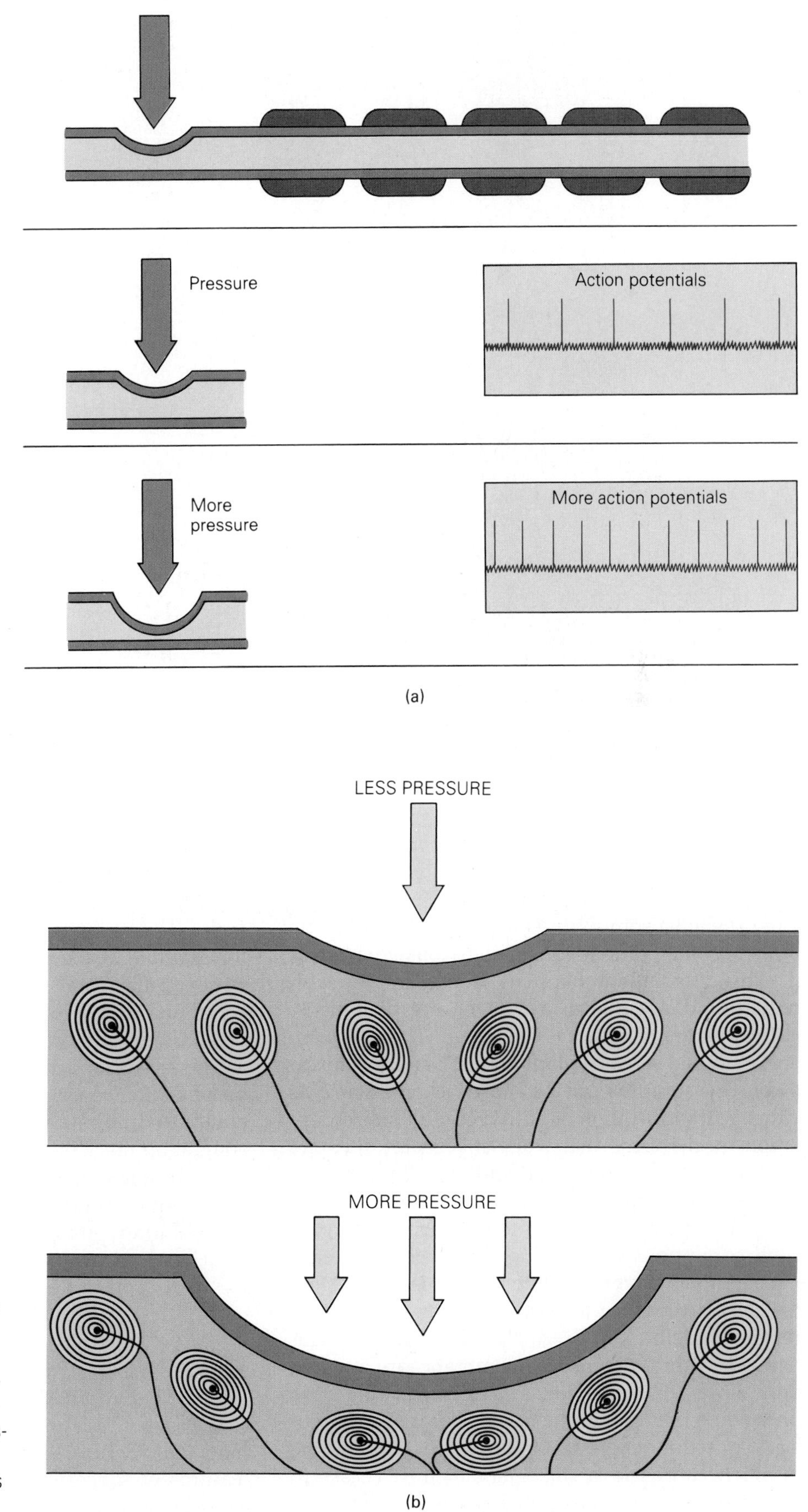

Figure 8–4 (*a*) According to the frequency code of stimulus intensity, more pressure produces more action potentials from the same receptor. (*b*) According to the population code of stimulus intensity, more pressure produces more action potentials because more receptors are affected.

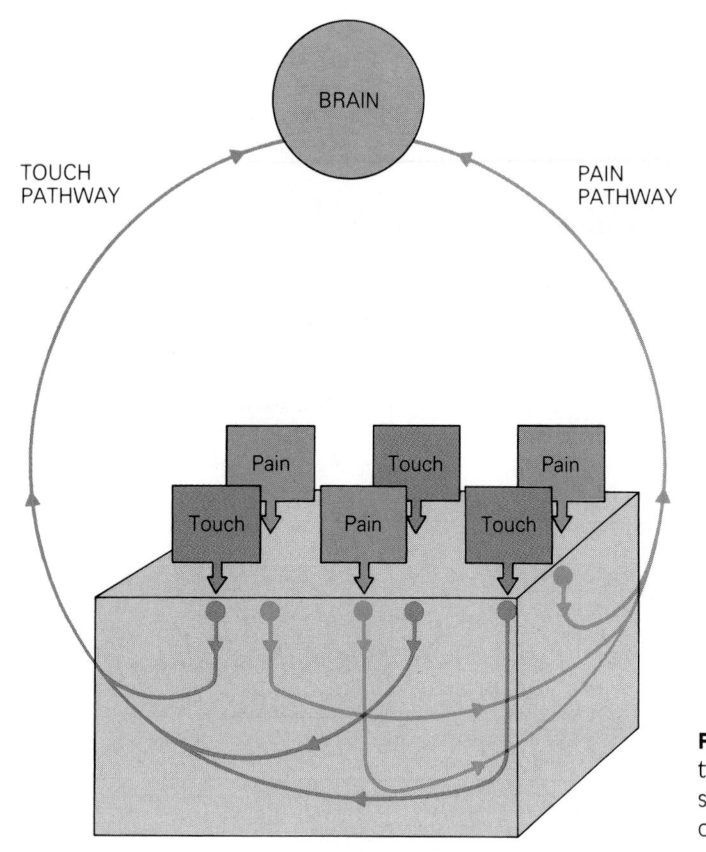

Figure 8–5 The labeled line code of stimulus quality. Although different stimuli affect the skin, each type of stimulus has its own pathway to the brain, ensuring clarity of sensory messages.

Sensory Adaptation: How Nerves Adjust to Repetitive Stimuli

Often, when a stimulus is continuously applied, the brain at some point no longer consciously perceives it. This adjustment happens, for example, with background noise such as the ticking of a clock. After a period of time it is unnoticed. This phenomenon is called **sensory adaptation.** The adaptation to a sensory stimulus can be caused by mechanisms either within the brain or at the receptor site. Adaptation mechanisms that work at receptor sites are the most clearly understood and best illustrated by the Pacinian corpuscle.

When pressure applied to the Pacinian corpuscle is continuously maintained, the free nerve ending eventually reverts back to its original shape (Fig. 8–6) even though the layers of connective tissue remained deformed. Since the Na^+ channels embedded within the nerve membrane are opened only when the nerve fiber is deformed, they close only when the form of the nerve fiber is again circular. Consequently, even though pressure on the Pacinian corpuscle is still maintained, it ceases to produce a generator potential when it resumes its normal shape and thus no longer transmits information about the pressure of the stimulus.

Mechanisms of adaptation are also found at the molecular level. Chemical receptors on membranes, for example, are internalized (see Chapter 12) and removed from the surface of the membrane after continued exposure to drugs. With the removal of these chemical receptors, higher levels of the drug are necessary to achieve the same effect. This may be why, after repeated use, the body develops a tolerance to drugs such as heroin. A similar molecular adaptation mechanism is found in receptors for vision. During sunny days, our eyes quickly adapt to the brightness of the outdoors by the "bleaching" of the photoreceptors. The underlying basis of this bleaching process is the removal of molecular receptors that capture light.

Sensory Systems and Their Processes

Now that we have an understanding of the general features of sensory systems, let us consider each sensory system in more detail.

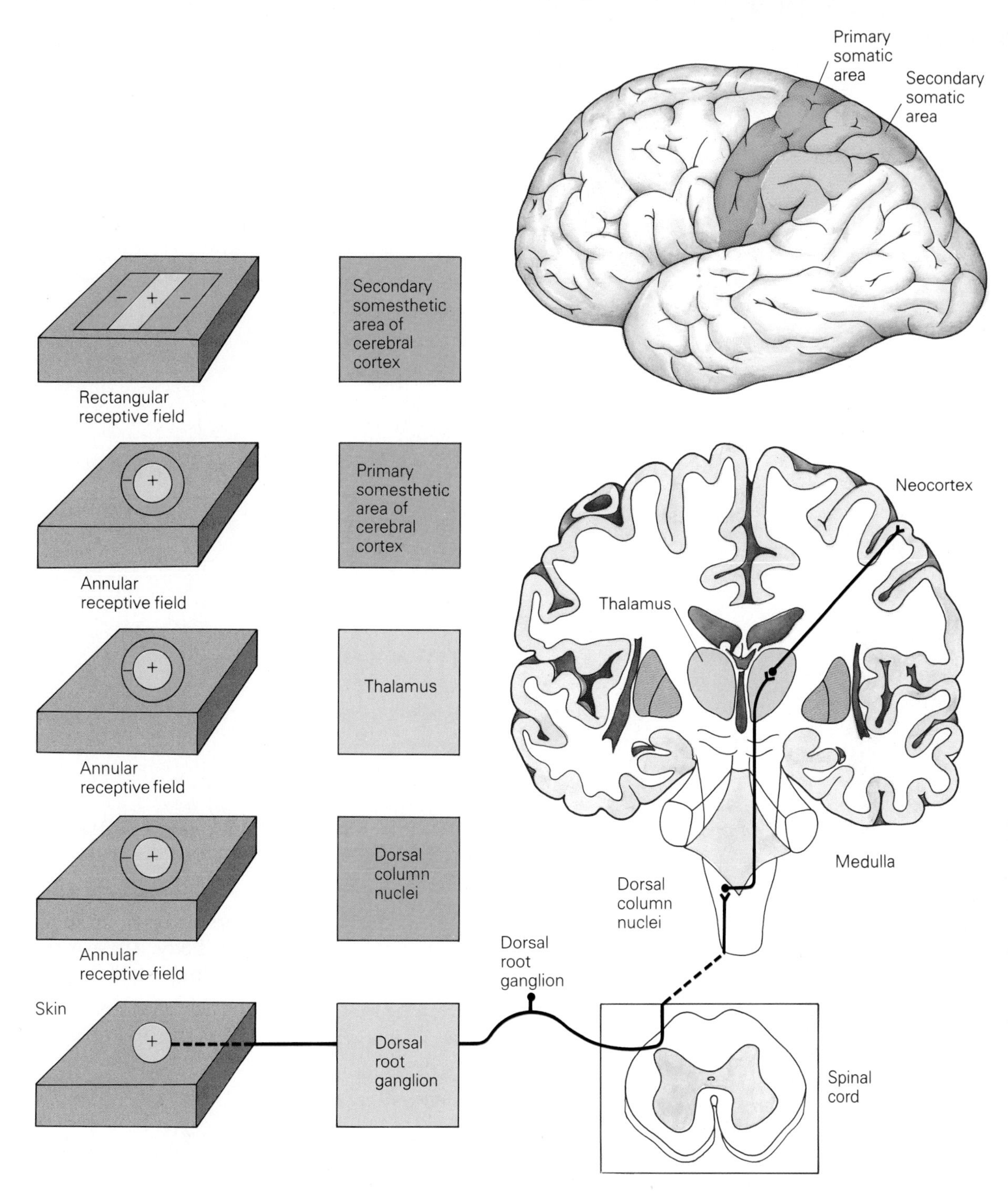

Figure 8–18 The different types of receptive fields for cells of different nuclei.

areas of the skin that, when pressed, will inhibit the response of dorsal column nerve cell #1 (see Fig. 8–17*b*). When the excitatory and inhibitory areas on the skin are mapped for nerve cell #1, a doughnut-shaped receptive field is revealed with an excitatory center and an inhibitory ring (Fig. 8–17*c*).

This doughnut-shaped annular receptive field is also found for nerve cells in the thalamus and somatosensory cortex. The cells in the somatosensory cortex that exhibit the annular receptive field are all located in the so-called **primary somatosensory area** (see Fig. 8–15). This is a narrow strip of cortical tissue that is also called **Brodmann area 3** (named for the neuroanatomical studies by K. Brodmann in 1909). In the **secondary** or **associational somatosensory areas** (Brodmann areas 1 and 2), the receptive fields are no longer annular. The stimulus that most effectively activates nerve cells in areas 1 and 2 is pressure applied to the skin in the form of a rectangle (Fig. 8–18) rather than a ring. The effectiveness of a rectangular stimulus is the result of nerve cells in area 3 that have overlapping receptive fields. The output fibers of these cells converge upon a nerve cell on area 2 (Fig. 8–19). Conse-

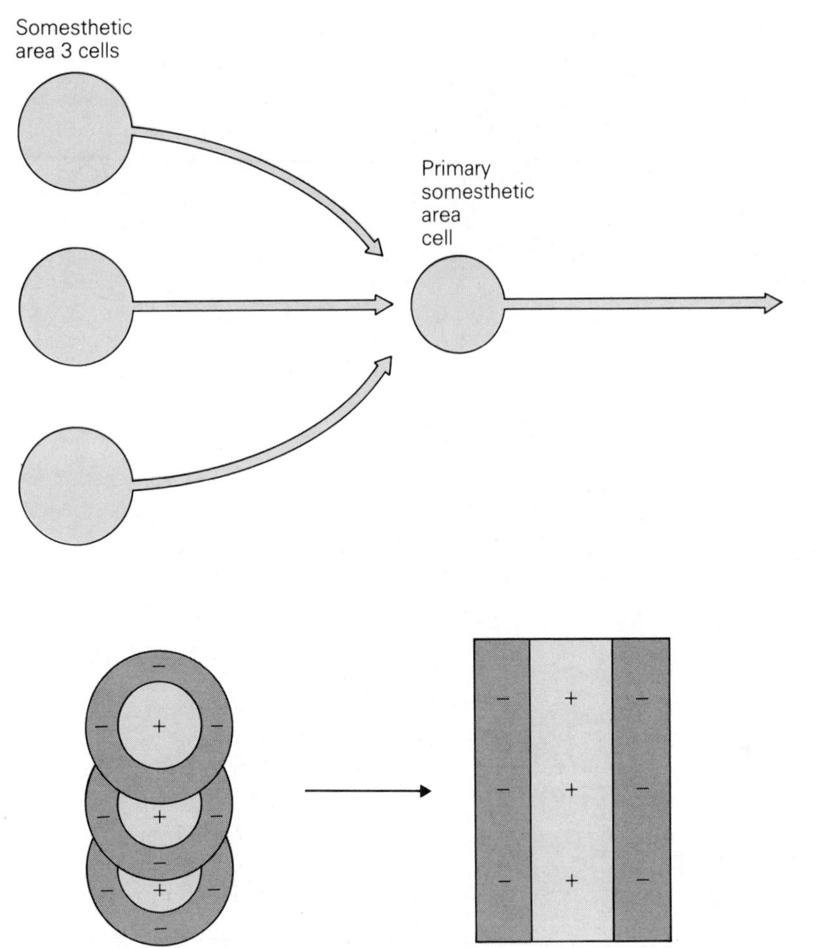

Figure 8–19 The receptive fields in the secondary somatosensory areas are rectangular.

quently, the simultaneous activation of all three input cells are needed in order to excite the one cell in area 1. This excitation occurs by the process of spatial summation as described in Chapter 7.

The most effective shapes of stimuli needed to activate nerve cells in area 2 can become quite complex. In essence, the physical features of an object in the environment are first broken down into its fundamental features by the peripheral sensory receptors. This information is then assembled to reconstruct the holistic features of the object that are capable of activating nerve cells in higher centers of the brain. In our discussion of the visual system, we will see that a similar synthesis and reconstruction of the external environment takes place for visual information.

The somatic sensory cortex is also organized so that all of the cells that respond to one type of sensation are located together within vertical columns. Each nerve cell within a vertical column of the somatic sensory cortex, for example, is activated by Pacinian corpuscles in one fingertip (Fig. 8–20). The location of the receptive fields for cells within a vertical column are also similar. Thus, the vertical columns represent basic units of sensation-specific and location-specific function.

The somatosensory cortex has four independent somatotopic maps of the body. These maps are found in regions called 3a, 3b, 1, and 2 (see Fig.

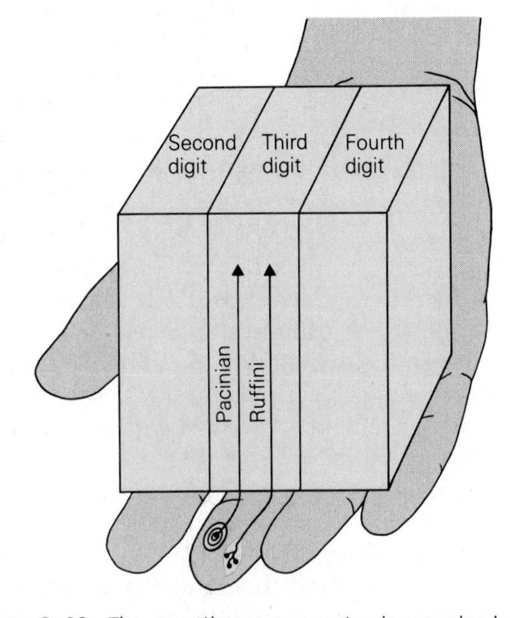

Figure 8–20 The somatic sensory cortex is organized so that all of the cells specialized for one sensation are grouped in a vertical column.

8–14). In any one Brodmann area of the somatosensory cortex, information from one type of somatic sensation tends to predominate. The nerve cells in Brodmann area 1, for example, respond to rapidly adapting cutaneous receptors; those in area 2 respond to deep pressure; those in area 3a respond to stretch receptors in skeletal muscle; and those in area 3b respond to rapidly and slowly adapting cutaneous receptors. Furthermore, the responses of nerve cells in the somatic sensory cortex are similar to those of the receptors in the periphery. This is also true for neurons in the thalamus, brain stem, and spinal cord. In this manner, the response characteristics of a sensory receptor are maintained while being transmitted to the neocortex.

The Anterolateral Pain and Temperature Pathway

As do the nerve fibers that transmit tactile sensation, the fibers that send pain and temperature information enter the dorsal portion of the spinal cord (Fig. 8-21). After entering the spinal cord, however, the majority of them form synaptic connections with nerve cells located in the dorsal horn. This group of nerve cells is collectively called the **substantia gelatinosa.** These cells, in turn, cross the midline and ascend toward the brain as a bundle of fibers along the anterior and lateral portion of the spinal cord and form the **anterolateral pathway.** These fibers have inputs to the reticular formation of the brain stem and the thalamus. The reticular formation serves as an alerting system and activates other areas of the brain and will be discussed in Chapter 11.

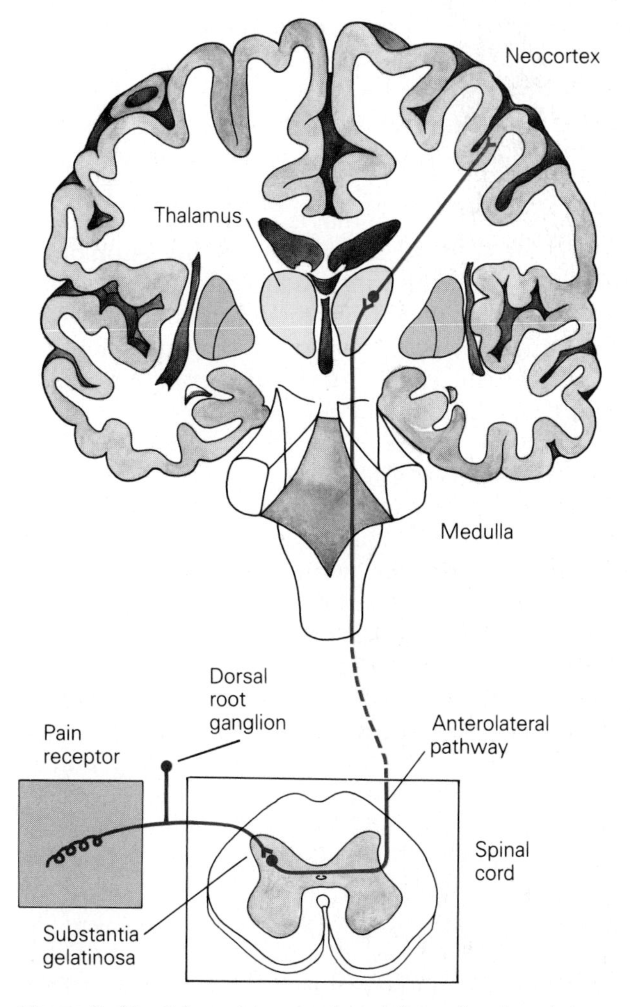

Figure 8–21 Pain and temperature information travels to the brain via the anterolateral pathway.

Inputs to the thalamus terminate in the **ventrobasal (VB)** nuclei. This information in turn is sent to the primary somatic sensory cortex in a somatotopic fashion and to the frontal cortex. The pain information that reaches the primary sensory cortex is associated with acute pain, whereas the information that reaches the frontal cortex is associated with chronic pain.

The perception of pain can be altered by other sensory inputs from the periphery or controlled by descending inputs from the brain stem and other higher brain centers. This gating of pain information is carried out by an inhibitory interneuron that is located in the dorsal horn of the spinal cord (Fig. 8–22). As previously discussed, the pain fibers from the periphery release the peptide called substance P as the neurotransmitter. In the absence of substance P release, pain cannot be detected by the brain. The interneuron acts to inhibit the release of substance P by a mechanism called **pre-synaptic inhibition.** The axon terminal of the interneuron forms a synapse upon the terminal of pain fibers. The terminals of the interneurons release the peptide called **enkephalin,** which in turn inhibits the release of substance P from the terminal of the pain fibers.

Thus, the interneuron is the key element in altering the sensation of pain. It can be activated by sensory fibers associated with touch, pressure, and vibration sense or by fibers from higher brain centers. The influence of other tactile inputs is seen in the reduction of pain by rubbing the skin surrounding a region of injury; this causes the pain to subside more quickly. The influence of higher brain centers on pain perception is seen in cases in which individuals who are involved in emergency situations do not realize that they have been injured until much later. The mechanisms that prevent the perception of pain under these highly stressful conditions are part of an autonomic "fight or flight" response that will be discussed in Chapter 10.

Another aspect of pain sensation is the perception of pain coming from the surface of the body that is actually caused by the distress of internal organs. This is often seen in cases of heart attacks where the area of the chest and left arm become

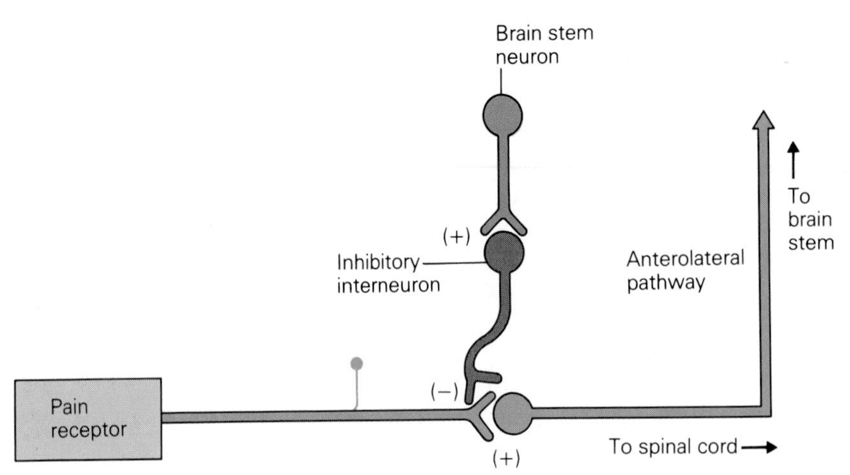

Figure 8–22 The perception of pain can be inhibited by the action of an interneuron.

painful. This type of **referred pain** will also be considered in Chapter 10.

The Visual System

The visual system allows us to determine the shape and color of objects such as a book, and the form of the letters printed within its pages. Parts of the visual system also allow us to recognize the grouping of letters in the form of words, and the grouping of words to represent ideas. These latter functions will be discussed in Chapter 11 when we consider the localization of language centers in the brain. In this section, we will consider how the visual system detects the shape and color of objects and the movement of objects in the external environment. The features of shape, color, and movement are processed by different cell groups within the visual system beginning with those found in the eye.

The eye is the sensory organ of the visual system that transduces light into electrical impulses, which the nervous system then uses to transmit information to the brain. The light perceived is part of the continuum of **electromagnetic radiation** that ranges from radio waves to X-rays (Fig. 8–23). Each type of electromagnetic radiation is characterized by a **wavelength.** Radio waves have long wavelengths whereas X-rays have very short wavelengths. The colors of light that we see have wavelengths that are in between radio waves and X-rays and range from 400 to 700 nm. These wavelengths correspond to the range of colors seen in a rainbow.

In addition to its wave-like nature, light can also be thought of as a composite of particles called **photons.** These photons have wavelengths and energies associated with different colors. All objects have the capability of selectively absorbing certain photons of light, and those photons that are not absorbed are reflected back into the environment. The reflected photons are those that give an object its color and that interact with the eye.

Anatomical Organization of the Eye

Figure 8–23 is an illustration of a cross-section of the eye. It shows the cornea, iris, lens, vitreous humor, and retina. The **retina** is the part of the eye that contains receptor cells called **photoreceptors** that absorb photons. As the illustration shows, photons must first pass through the cornea and lens before they reach the retina. The most sensitive part of the retina is a region called the **fovea;** consequently, the lens attempts to focus the image of an object onto the fovea. The image that is projected onto the retina, however, is upside down. The **inversion** of the image is similar to that produced by a camera lens on to a roll of film. Even though the retina detects the form of objects in an upside-down fashion, the brain reorganizes this information so that we interpret the outside world as right side up.

The Nerve Cells of the Retina. We can best understand how the eye operates to transduce the form of an object by first examining the functional organization of the nerve cells within the retina. In addition to photoreceptors, the retina contains other types of nerve cells called **ganglion cells, bipolar cells,** and **horizontal cells** (see Fig. 8–23). Photoreceptors, either **rods** or **cones,** are responsible for absorbing the photons and are situated at the back portion of the retina, behind these other types of cells. Photons must therefore pass through several layers of cells before they reach the layer of photoreceptors.

Figure 8–23 *(a)* The general structure of the eye and the portion of the electromagnetic spectrum visible to it. *(b)* The organization of rods and cones and of horizontal, bipolar, and ganglion cells within the retina.

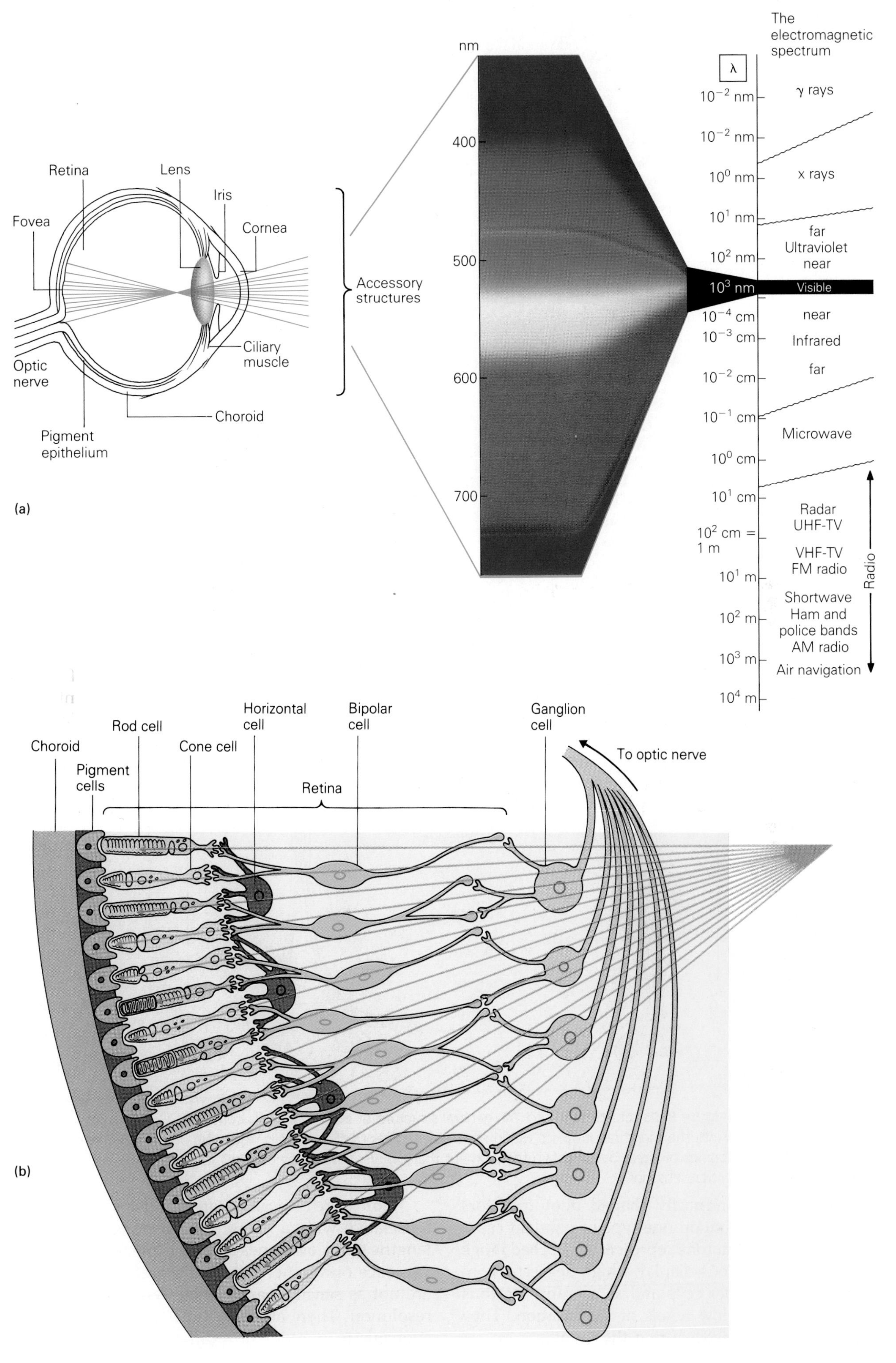

Retina
Lens
Iris
Fovea
Cornea
Optic nerve
Ciliary muscle
Choroid
Pigment epithelium
(a)
Accessory structures
nm
400
500
600
700
The electromagnetic spectrum
λ
10⁻² nm
γ rays
10⁻² nm
10⁰ nm
x rays
10¹ nm
far
Ultraviolet
near
10² nm
10³ nm
Visible
10⁻⁴ cm
near
10⁻³ cm
Infrared
far
10⁻² cm
10⁻¹ cm
Microwave
10⁰ cm
10¹ cm
Radar
UHF-TV
10² cm = 1 m
VHF-TV
FM radio
Radio
10¹ m
Shortwave
Ham and
police bands
AM radio
10² m
10³ m
Air navigation
10⁴ m
Horizontal cell
Bipolar cell
Ganglion cell
Rod cell
Choroid
Cone cell
To optic nerve
Pigment cells
Retina
(b)

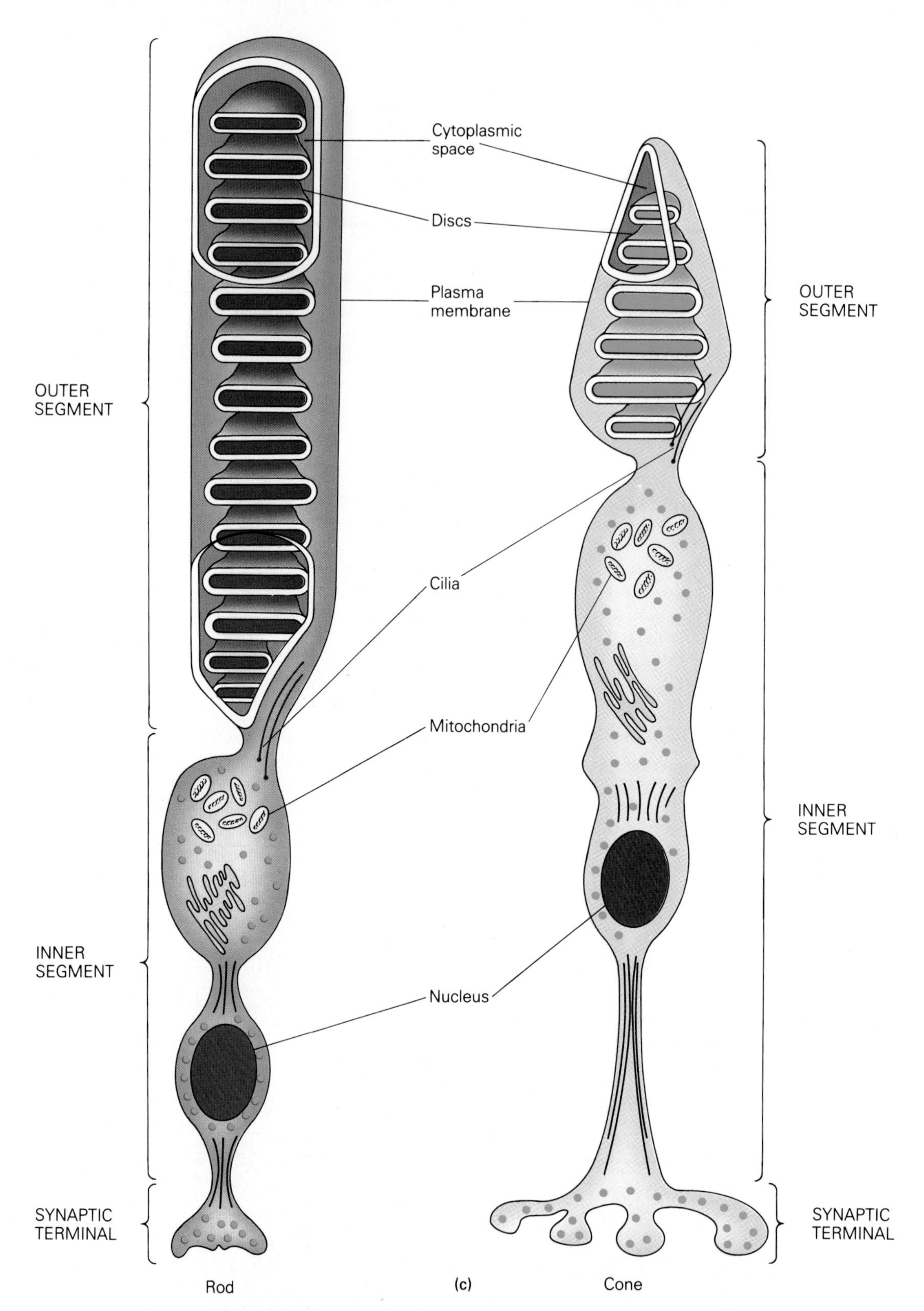

Figure 8–23 (continued) (*c*) The general structure of a rod cell and a cone cell. Note the membranous discs within the outer segments of both cells. The membranes of these discs contain the pigment that, when activated by photons, produce visual sensation.

Rods are cylindrically shaped photoreceptors (Fig. 8–23*c*) that contain one type of pigment capable of absorbing photons representing a broad range of wavelengths. Rods display a high degree of convergence onto other cells and as a group are thus very sensitive to low levels of illumination. They therefore operate well during the night.

Cones are conically shaped and absorb photons associated with a more narrow range of wavelengths for color. Cones display a low degree of convergence onto other cells and therefore as a group are not as sensitive as rods but have better spatial resolution. There are three types of cones, each responsive to photons of a different range of wavelengths. These cones form the basis of color perception and are used for day vision.

Photoreceptors and Phototransduction. The photoreceptor pigment that absorbs photons is called **visual pigment.** This pigment is embedded within the membranes of **discs,** lamellar structures in the **outer segment** of the photoreceptor. Also present in the disc membrane is a protein called **transducin** and an enzyme called **phosphodiesterase** (Fig. 8–24*a*). The pigment in a rod cell disc is **rhodopsin.** When the rhodopsin molecule absorbs a photon, its three-dimensional form changes. In its new form, rhodopsin is now capable of activating transducin. Transducin, in turn, is now able to activate phosphodiesterase, which hydrolyzes cyclic GMP (cGMP) to GDP. In this process, the net effect is the reduction in the amount of intracellular cGMP.

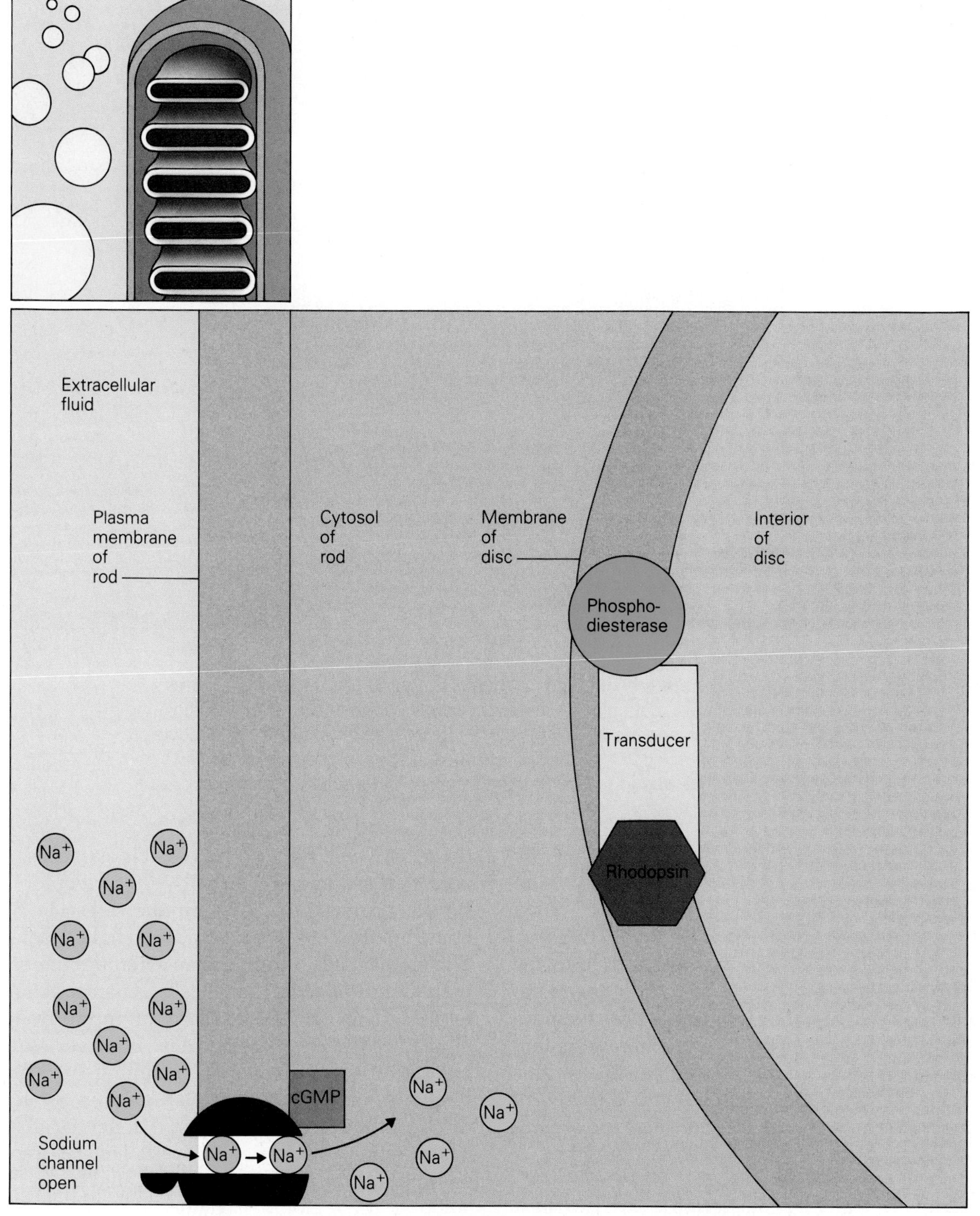

Figure 8–24 (*a*) In the absence of light, sodium channels in the plasma membrane of rod cells are kept open by the presence of cGMP, and sodium ions enter the cell.

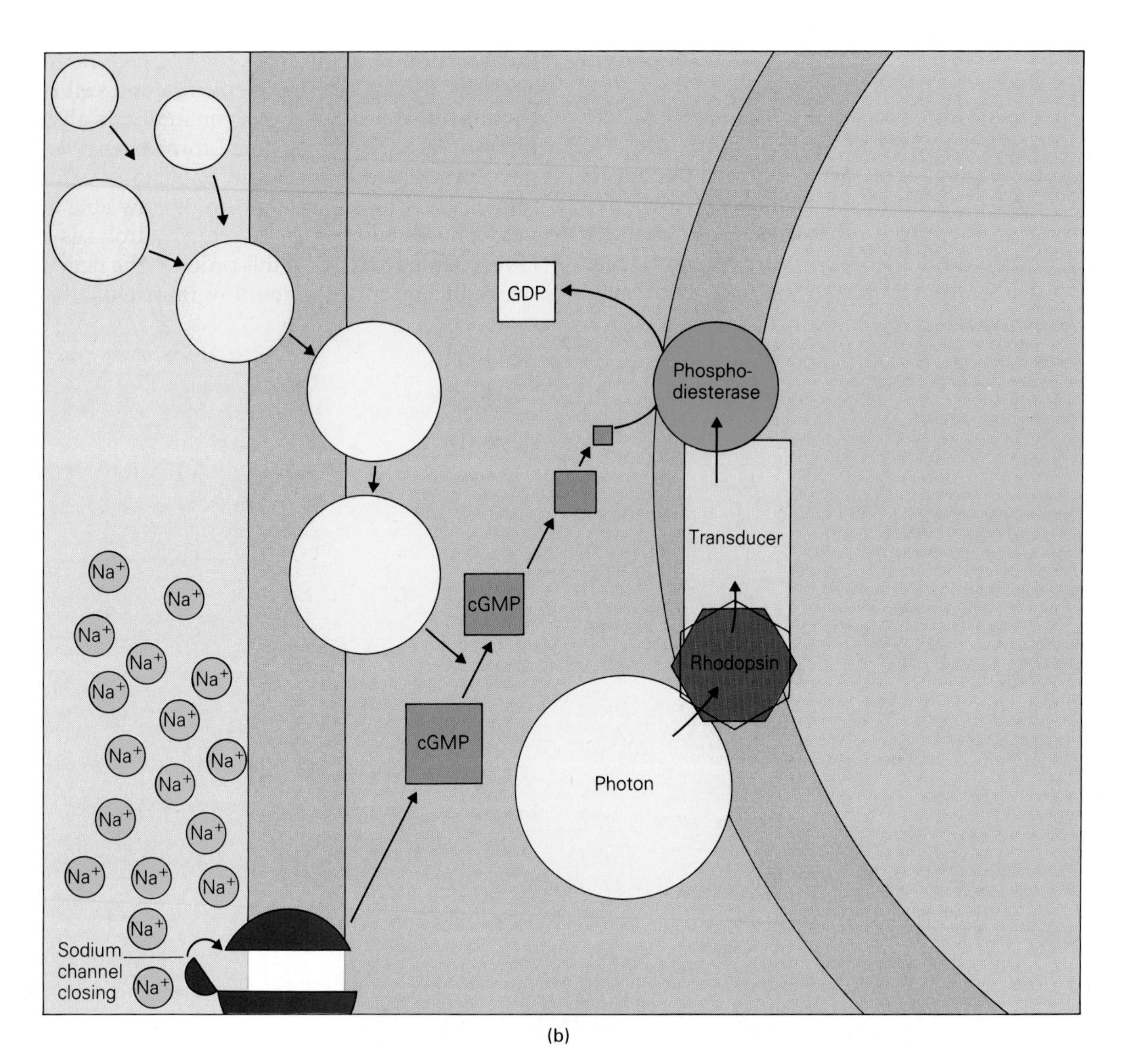

(b)

Figure 8–24 (continued) (*b*) When a photon of light enters a rod cell, it causes a rhodopsin (pigment) molecule to activate transducin, which in turn activates phosphodiesterase. This enzyme hydrolyzes cGMP cytosol of the rod to GDP. The reduction in the amount of cGMP causes Na^+ channels in the rod plasma membrane to close, reducing the amount of Na^+ entering the cell and producing a hyperpolarization (*b*).

What is the significance of decreasing the cGMP concentration? Normally, cGMP opens Na^+ channels in the photoreceptor plasma membrane. When cGMP is decreased, Na^+ channels begin to close, thus reducing the influx of positive ions into the cell. This results in a hyperpolarization of the membrane. Thus, the photoreceptor transduces the photon of light into a more negative membrane potential (Fig. 8–25). In order to understand how this hyperpolarization is transformed into a series of action potentials, we must first consider the organization and response of nerve cells in the retina.

Interaction of Cells in the Retina. We can best understand the way in which visual information is processed in the retina by considering the interactions between photoreceptors, bipolar cells, horizontal cells, and ganglion cells. Photoreceptors form synaptic connections with **bipolar cells** (Fig. 8–26). Some bipolar cells depolarize when light is detected. The bipolar cells, in turn, form synaptic connections with **ganglion cells.** These cells generate action potentials even in the absence of input from the bipolar cells. The bipolar cells, however, can depolarize the ganglion cells, resulting in a greater frequency of action potentials. Ganglion cell activity can be inhibited by a fourth type of nerve cell found in the retina, the **horizontal cell.** This cell receives its inputs from photoreceptors located in the retina lateral to the region of excitation (Fig. 8–26*b*). When these photoreceptors detect light, they inhibit the horizontal cell.

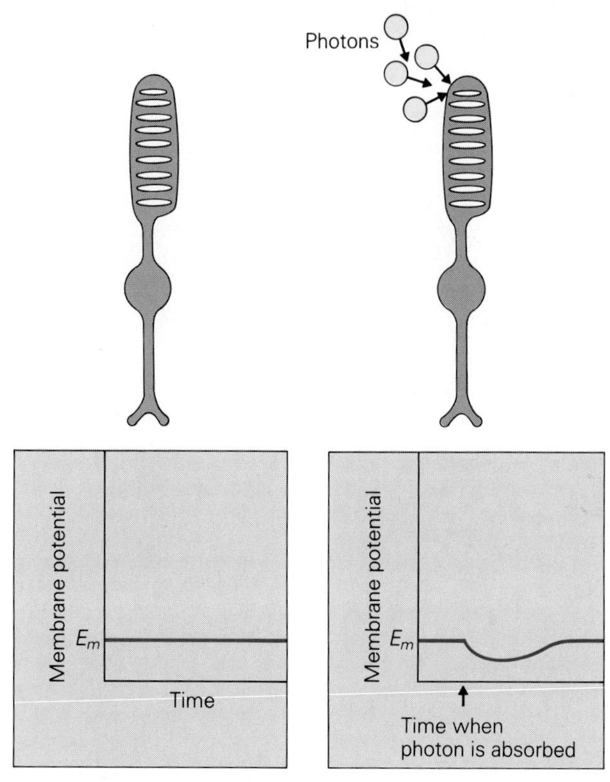

Figure 8–25 The hyperpolarization of a rod cell membrane caused by a photon is eventually transduced into action potentials that travel to the optic nerve.

The inhibition of the horizontal cell, in turn, causes a depolarization of adjacent photoreceptors (most likely by preventing the release of an inhibitory transmitter) and the subsequent inhibition of ganglion activity. Depending on where light impinges on the retina, ganglion cells can thus be activated or inhibited.

The areas of the retina that excite or inhibit a ganglion cell make up the receptive field for that ganglion cell. Therefore—as in the case of the dorsal column nuclei cells of the somatosensory pathway—the retinal ganglion cells have receptive fields that are shaped like an annulus. For some ganglion cells, the center of the annulus is excitatory, while the periphery is inhibitory (Fig. 8–27*d*). These neurons are called **"on" center and "off" surround cells.** Other ganglion cells have **"off" centers** and **"on" surrounds.**

The spatial pattern of activated ganglion cells in the two-dimensional plane of the retina represents the form of an object whose image is projected onto

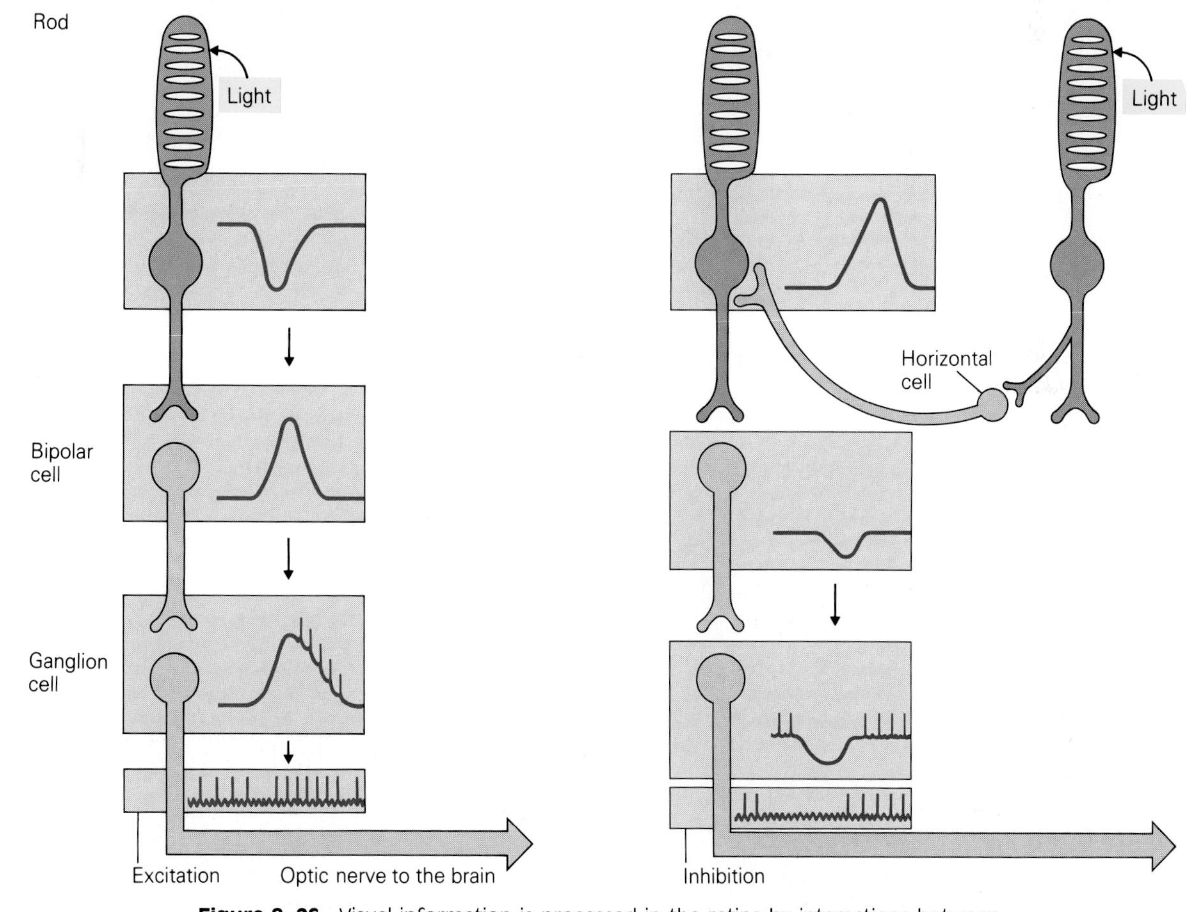

Figure 8–26 Visual information is processed in the retina by interactions between photoreceptors, bipolar cells, horizontal cells, and ganglion cells. Photoreceptors synapse with bipolar cells, causing bipolar cells to depolarize when the photoreceptors detect light. The bipolar cells can depolarize ganglion cells. Ganglion cell depolarization can be inhibited by a horizontal cell.

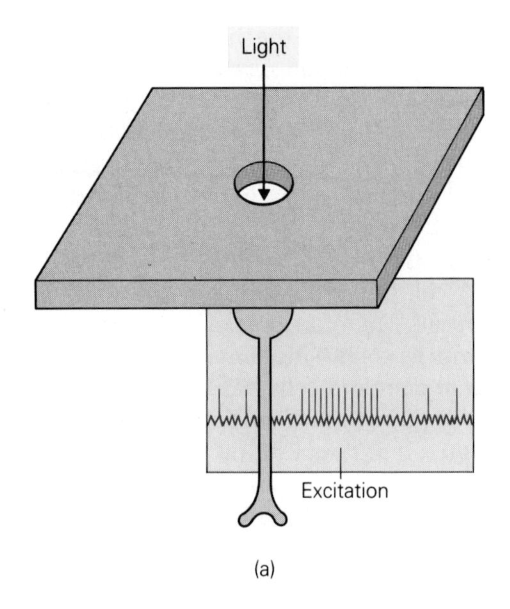
Light
Excitation

(a)

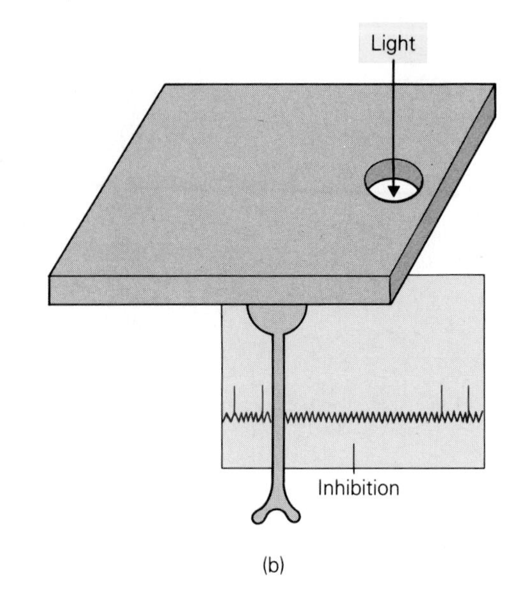
Light
Inhibition

(b)

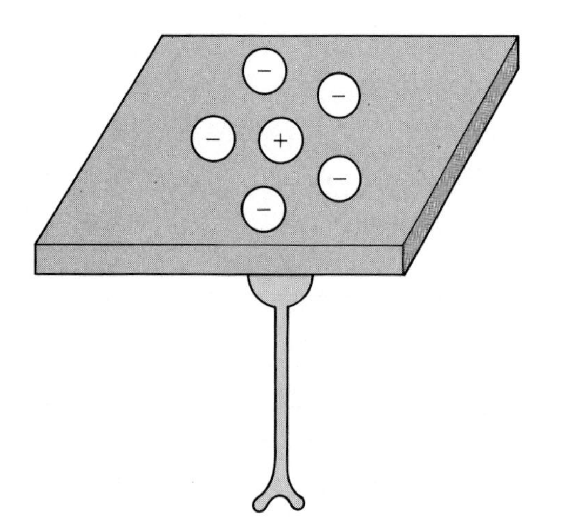
–
–
–
+
–
–
–

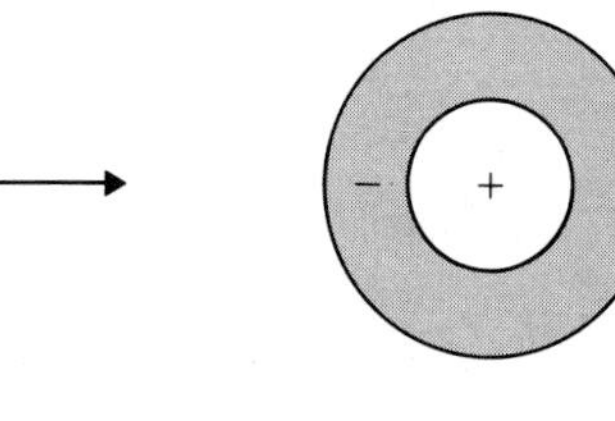
–
+

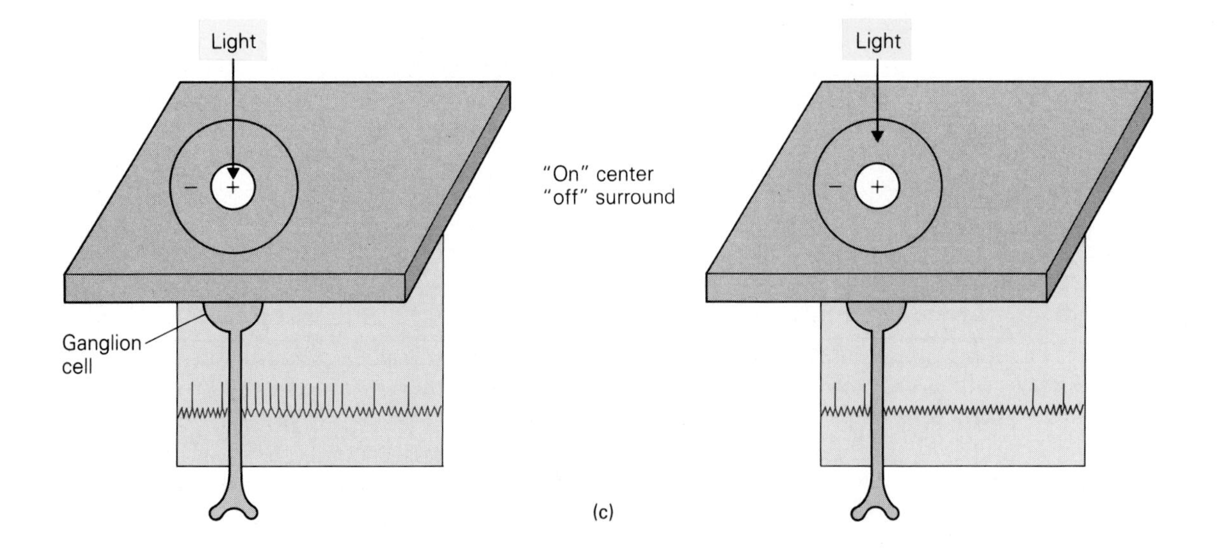
Light
–
+
Ganglion cell
"On" center
"off" surround
Light
–
+

(c)

20. What are ocular dominance and visual orientation columns?

21. What photoreceptors are responsible for color vision?

22. How are the broad-band, single-opponent, and double-opponent cells used in color vision alike and different?

23. What visual field defect would result from a lesion of the optic chiasm?

24. How are the middle and inner ear functionally organized? How do they transduce sound?

25. How do low and high frequencies of sound affect the basilar membrane?

26. What are the cellular mechanisms of sound transduction?

27. How are the functional properties of photoreceptors in the retina and the hair cells of the cochlea alike and different?

28. What is the tonotopic organization of the auditory nerves?

29. How are the taste receptors for sweet, salty, sour, and bitter substances organized in the tongue?

30. How do the functional properties of nerve fibers that innervate the taste buds differ from those that innervate the cochlear hair cells?

31. What are the types of primary odors that are sensed by olfactory receptors in the nose?

Suggested Readings

Beidler, L.M. *Handbook of Sensory Physiology, Vol. IV. Chemical Sense.* Heidelberg: Springer-Verlag, 1972.

Beidler, L.M. *Handbook of Sensory Physiology, Vol. V. Auditory System.* Heidelberg: Springer-Verlag, 1972.

Iggo, A. *Handbook of Sensory Physiology, Vol. II. Somatosensory System.* Heidelberg: Springer-Verlag, 1973.

Jung, R. *Handbook of Sensory Physiology, Vol. VII. Central Processing of Visual Information.* Heidelberg: Springer-Verlag, 1973.

Lowenstein, W.R. *Handbook of Sensory Physiology, Vol. I. Principles of Receptor Physiology.* Heidelberg: Springer-Verlag, 1971.

Somjen, G. *Sensory Coding in the Mammalian Nervous System.* New York: Plenum Press, 1975

9

Motor Systems

The central and peripheral nervous systems cannot act independently: their function is to collect information, process it, and give the signal for some appropriate action to be taken by the muscles of the body. Muscle translates the impulses of the nervous system into useful action. This conversion process, which takes place through the **motor system,** enables the performance of both voluntary (conscious) movements such as walking and talking and automatic (unconscious) ongoing internal movements such as the pumping of the heart and digestion of food.

I. Muscle as an Effector Organ

Recall from Chapter 1 that most of the control systems of the body employ mechanisms called **sensors** to detect changes in bodily conditions and effectors to adjust those conditions if necessary. While the sense organs are the sensors of the nervous system, the muscles are its effectors, because they *effect* changes (or cause changes to happen) in the body in response to sensory stimuli.

Muscles also help regulate the stimuli that enter the central nervous system (CNS). They help to adjust the sensitivity of the ears (particularly in very noisy environments), to point us in the proper direction to receive (or ignore) information, and to avoid stimuli that may be dangerous or unpleasant. Some muscles are very strictly controlled by the nervous system, whereas others have a large degree of independent function.

In the previous chapter, we examined the ways in which sensory systems detect stimuli from the external environment and transmit information about those stimuli to the brain. These **afferent** systems provide input to the brain, and we respond to this input—in part by making appropriate movements. The **efferent** flow of neural information that initiates and controls movement can be viewed as the output from the brain. The areas of the central and peripheral nervous systems that control movement through this output of information are collectively called the **motor system** and are the subject of this chapter. We will focus on the role of the motor system in the control of locomotion, posture, and fine movement.

The repertoire of motor responses ranges from **reflex** (automatic, involuntary) **actions** such as the knee jerk to complex, **voluntary** activities such as playing the piano. Reflex responses are controlled in the spinal cord, while more complex motor behaviors are organized in other areas of the brain. The range of complexity of motor behaviors reflects the **hierarchical** (or **serial**) **organization** of the motor system. The spinal cord is the lowest level of the hierarchy; it is followed in order by the **brain stem, motor cortex (Brodmann area 4),** and the **supplemental and premotor areas** of the cortex (Brodmann area 6). The premotor area is where the plans for movement originate. In this hierarchical organization, the highest motor levels influence the levels below them. For instance, the premotor area has inputs to the motor cortex; the motor cortex, in turn, provides inputs to the brain stem; and the brain stem provides inputs to the spinal cord (Fig. 9–1*a*). This hierarchical arrangement connects different areas of brain together serially in a descending fashion.

In addition to this serial organization, the motor system also has a **parallel organization.** In this arrangement, information from the motor cortex and brain stem, for example, can simultaneously and independently affect the spinal cord (see Fig. 9–1*b*). With both the parallel and serial organization of the motor system, the targets of the efferent (outgoing) information are the **motor neurons** of the spinal cord. These motor neurons are the nerve cells that directly activate or inhibit the skeletal muscles responsible for movement.

The body contains three major types of muscle: skeletal muscle, cardiac muscle, and smooth muscles. **Cardiac muscles** are the muscles of the heart; they pump blood through the circulatory system (see Chapter 18). **Smooth muscles** surround certain blood vessels to control blood pressure and are also associated with a variety of other organ systems. In general, they maintain the homeostatic state of the

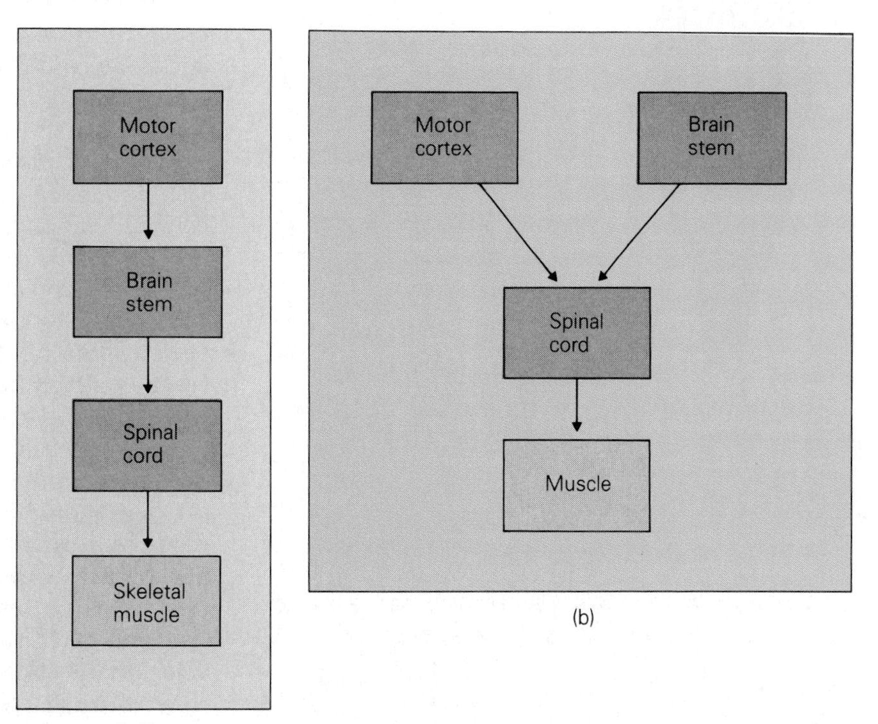

Figure 9–1 The motor system is organized in both hierarchical and parallel fashions.

body (see Chapter 19). In this chapter, we will focus on the areas of brain that control the **skeletal muscles,** also called **striate** or **striated muscles,** which produce voluntary movements of body parts.

Overview of the Skeletal Muscle System

Although the skeletal muscular system is discussed in detail in Chapter 16, we will review some general aspects of muscle structure and function here in order to better understand the interactions of muscles and neurons to be discussed shortly.

General Structure of Skeletal Muscle

A skeletal muscle is made up of **muscle cells,** which, because of their elongated shape, are called **muscle fibers.** Muscle cells are **multinucleate;** one muscle cell contains many nuclei. A **muscle** comprises many muscle fibers arranged in a parallel way. Two types of muscle fibers make up a skeletal muscle: **extrafusal fibers** and **intrafusal fibers** (Fig. 9–2*a*). Extrafusal fibers are the strongest and most common fibers and form the exterior of the muscle. Inside the muscle are the intrafusal fibers, arranged parallel to the extrafusal fibers. Both extra- and intrafusal fibers are attached to the tendon at either end of the muscle. By a process to be discussed in more detail in Chapter 16, muscle fibers shorten in response to certain stimuli, resulting in contraction of the entire muscle and producing movement of a bone.

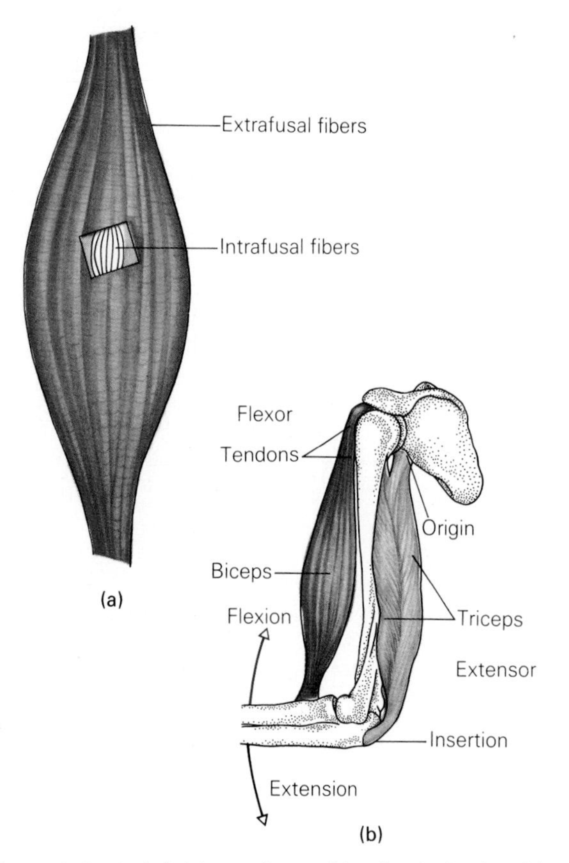

Figure 9–2 A skeletal muscle consists of extrafusal and intrafusal muscle fibers attached to tendons, which connect the muscle to a bone. Skeletal muscles work in antagonistic pairs of flexors and extensors.

Types and Actions of Skeletal Muscle

Most skeletal muscles are attached to bones at two points: an **origin** and an **insertion** (Fig. 9–2*b*), by strong connective tissue structures called **tendons.** As skeletal muscles **contract**—typically meaning that they shorten, although sometimes they remain the same length—they move bones, frequently causing joints to open or close. Skeletal muscles are usually arranged in **antagonistic** pairs or groups such that contraction of one muscle or muscle group will cause lengthening of another. In fact, all movements are controlled by antagonistic muscles working in opposite ways.

Antagonistic muscles are arranged in parallel fashion across a joint where two bones meet. The muscle that, by contracting, causes the joint to stretch out is called an **extensor,** while the muscle that, by contracting, causes the joint to close up is called the **flexor.** A flexor for a joint is said to be **synergistic** to another flexor for the same joint, while a flexor and an extensor for the same joint are said to be **antagonistic.**

Skeletal muscles produce movements of the bones of the **axial skeleton,** consisting of the skull, backbone, and ribs. These muscles are important in maintaining posture. Other skeletal muscles move the bones of the **distal skeleton,** primarily the limbs.

Skeletal muscles can be categorized as either fast muscles or slow muscles. **Fast skeletal muscles** contract rapidly. They are poorly vascularized (poorly supplied with blood vessels). Operating under anaerobic conditions, they fatigue rapidly. These muscles are best suited for physical activities such a sprinting. In contrast, **slow skeletal muscles** contract slowly and are more resistant to fatigue. They operate under aerobic metabolism. These muscles are best suited for activities such as long distance running and maintaining posture.

The motor neurons that innervate these two types of muscles have different functional properties. Motor neurons that innervate fast skeletal muscles discharge action potentials with high frequencies and have rapid action potential conduction velocities. In contrast, motor neurons that innervate slow skeletal muscles have low rates of action potential discharge and exhibit slow conduction velocities.

The properties of fast and slow skeletal muscles can be altered if the type of motor neuron that innervates the muscle is altered. For instance, if the slow motor neuron that innervates a slow skeletal muscle is replaced with a fast motor neuron, the muscle eventually develops into a fast skeletal muscle. Skeletal muscles therefore appear to be under **trophic** control by the type of nerve fibers that innervate them. Whether a muscle fiber can change from one type to another depending on the type of training (sprinting versus marathon running) remains to be determined. The mechanical properties of these skeletal muscles are discussed in Chapter 16.

Innervation of Skeletal Muscle

Skeletal muscle depends upon the somatic division of the CNS for control of its contraction. Both sensory neurons and motor neurons are involved in the control of skeletal muscle. Somatic motor neurons form **myoneural junctions** with both extrafusal and intrafusal muscle fibers, while somatic sensory neurons form **muscle spindles** by connecting with intrafusal muscle fibers and **Golgi tendon organs** by connecting with tendons. Figure 9–3 provides an overview of the nervous system components important in the innervation of skeletal muscle.

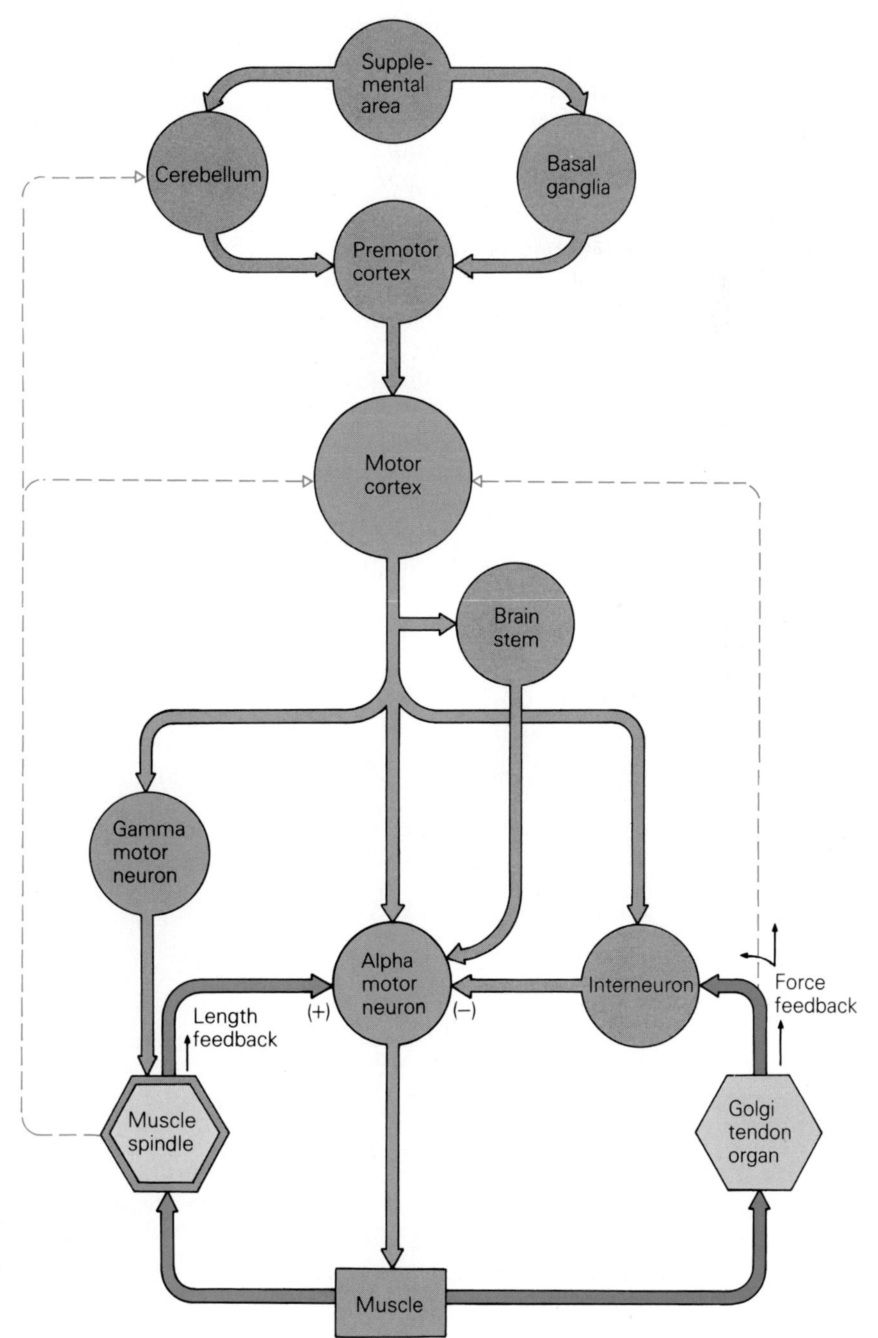

Figure 9–3 An overview of the motor system and the flow of information among the various components.

Motor Neurons in Skeletal Muscle

The Myoneural Junction: A Special Kind of Synapse. Under normal conditions, skeletal muscle will not contract unless stimulated by a **motor neuron,** an efferent neuron that originates in the CNS. The connection between a muscle fiber and the myelinated axon of a motor neuron is called a **myoneural junction,** or a **motor end-plate** (Fig. 9–4).

The myoneural junction can be thought of as a special type of synapse. As in most synapses in the CNS, an impulse is transmitted across a myoneural junction by a chemical transmitter released from the presynaptic cell (the motor neuron). Motor neurons that synapse with extrafusal muscle fibers are called **alpha motor neurons,** while those that synapse with intrafusal muscle fibers are called **gamma motor neurons.** Specific transmitters and processes involved in various myoneural junctions are discussed in Chapter 16.

The Motor Unit: The Myoneural Junctions of a Single Motor Neuron. The motor neurons that innervate skeletal muscles and cause them to contract have cell bodies that are located in the ventral horn of the spinal cord. Their axons leave the spinal cord through the ventral root and find their way to the appropriate skeletal muscle.

One motor neuron can form synaptic connections with many muscle fibers. An individual motor neuron and all the muscle fibers that it innervates forms a **motor unit** (see Fig. 9–4). The motor unit is the functional unit of the **neuromuscular system.**

Motor units come in various sizes. A small motor unit consists of a motor neuron that innervates few muscle fibers. This type of motor unit is responsible for the precise control of very fine movements such as the contraction of a finger or darting movements of the eyes. A large motor unit consists of a motor neuron that innervates many muscle fibers. Such a unit is responsible for gross movements such as the contraction of the legs or the maintenance of posture. The concept of motor unit size parallels that of the size of sensory receptive fields. Just as small receptive fields are associated with greater resolution of stimuli, small motor units are associated with greater precision of fine movement.

The Coding of Contractile Force. The amount of force that the muscle generates during contraction is controlled by the nervous system in two ways. First, a motor neuron can control the tension developed

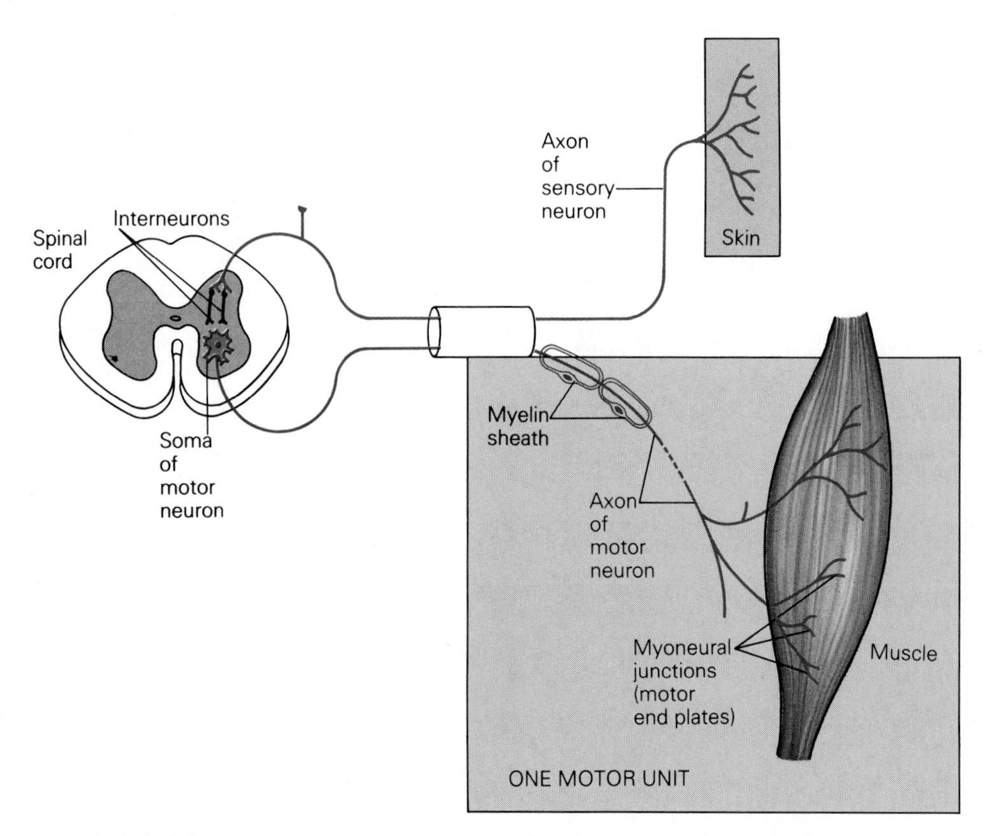

Figure 9–4 The motor unit consists of a motor neuron and all the muscle fibers it innervates through myoneural junctions.

by a single muscle fiber by the frequency of action potentials it generates (Fig. 9–5). This code is the nervous system's **frequency code** of contractile force. Secondly, more motor neurons and therefore more motor units can be activated to increase a muscle's force of contraction. This code is the nervous system's **population code** of contractile force.

Sensory Neurons in Skeletal Muscle

Sensory neurons also occur in the skeletal muscle system. They transmit information about the state of contraction and tension in muscle fibers to the CNS so that it can revise its instructions as necessary to the motor neurons producing the contraction or tension. Sensory neurons that innervate intrafusal muscle fibers form what is known as a muscle spindle, while sensory neurons that innervate tendons form Golgi tendon organs.

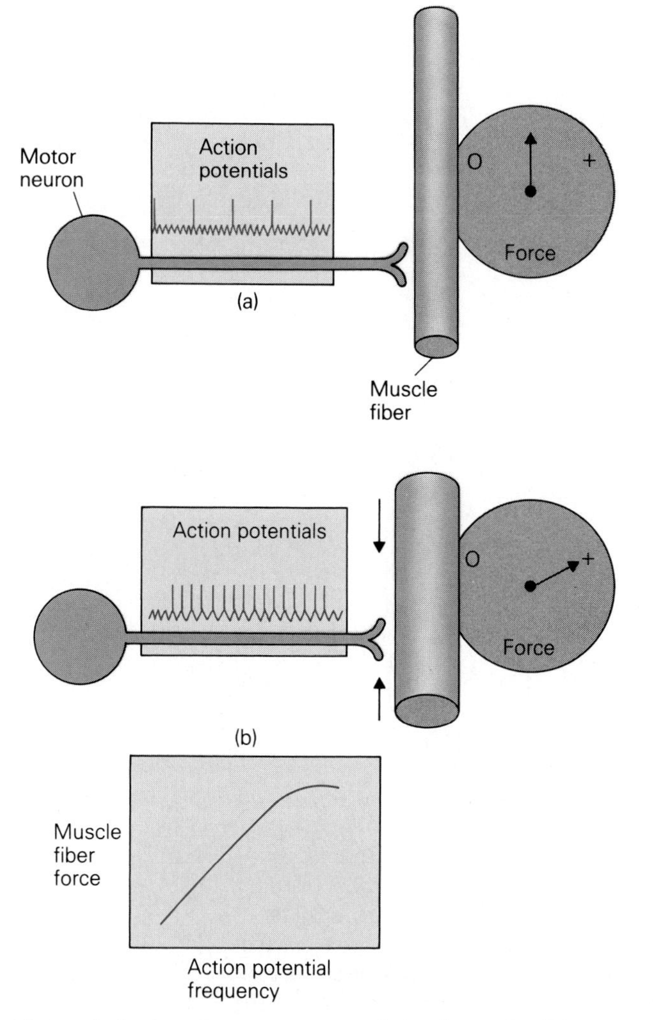

Figure 9–5 A motor neuron generates action potentials that cause a muscle fiber to contract and exert force.

The Muscle Spindle: A Length Detector. The sensory receptors that detect the length of the muscle and its velocity of contraction are called **muscle spindles** (Fig. 9–6). These structures are formed by sensory neurons that entwine intrafusal muscle fibers.

The muscle spindle is composed of two types of intrafusal fibers: **nuclear bag fibers** and **nuclear chain fibers** (see Fig. 9–6*a*). The nuclear bag fiber is innervated by **Type Ia nerve fibers,** which transmit information about muscle length and velocity of contraction to the CNS. Nuclear chain fibers are innervated by **Type II nerve fibers,** which transmit information about muscle length.

How are spindle receptors activated? The muscle spindle acts as a **stretch receptor**—that is, it increases its discharge of action potentials when the intrafusal fibers stretch. As a muscle contracts, both the extrafusal and intrafusal muscle fibers shorten. The shortening of the intrafusal muscle fiber causes a decrease in the action potential discharge of the Ia and II sensory nerve fibers (see Fig. 9–6*c*). As the extrafusal muscle fiber continues to shorten, the muscle spindle further reduces its discharge of action potentials. Hypothetically, at very short muscle lengths, the muscle spindle would no longer discharge action potentials. In this situation, the spindle would no longer be sensitive to further decreases in length. Muscle spindle sensitivity, however, is maintained even at short muscle lengths by gamma motor neurons.

Gamma motor neurons specifically innervate contractile elements at the poles of the muscle spindles (see Fig. 9–6*a*). The activation of these contractile elements causes the central region of the muscle spindle to stretch as a rubber band would. In this manner, the gamma motor neurons can regulate the sensitivity of the Type Ia sensory nerves of the muscle spindles. Figure 9–6*c* shows the action potential activity of Type Ia muscle spindle sensory nerves as a function of the length of the extrafusal muscle fiber. In the absence of gamma motor neuron activation, a change in the length of the extrafusal muscle fibers results in a slight change in the frequency of action potentials generated by the sensory neurons. With gamma motor neuron activation, however, a slight change in the length of the extrafusal muscle fibers causes a great change in the frequency of generated action potentials. In this way, the activation of gamma motor neurons maintains the sensitivity of the muscle spindles even as the extrafusal muscle fibers shorten during contraction.

Golgi Tendon Organs: Tension Receptors. A second type of muscle receptor is the **Golgi tendon**

organ. This type of receptor is located in series with the extrafusal muscle fiber (see Fig. 9–6*b*) and transmits information about the force or tension produced by the contraction of the muscle. As greater tension develops when the muscle contracts, the Golgi tendon organ generates more action poten-

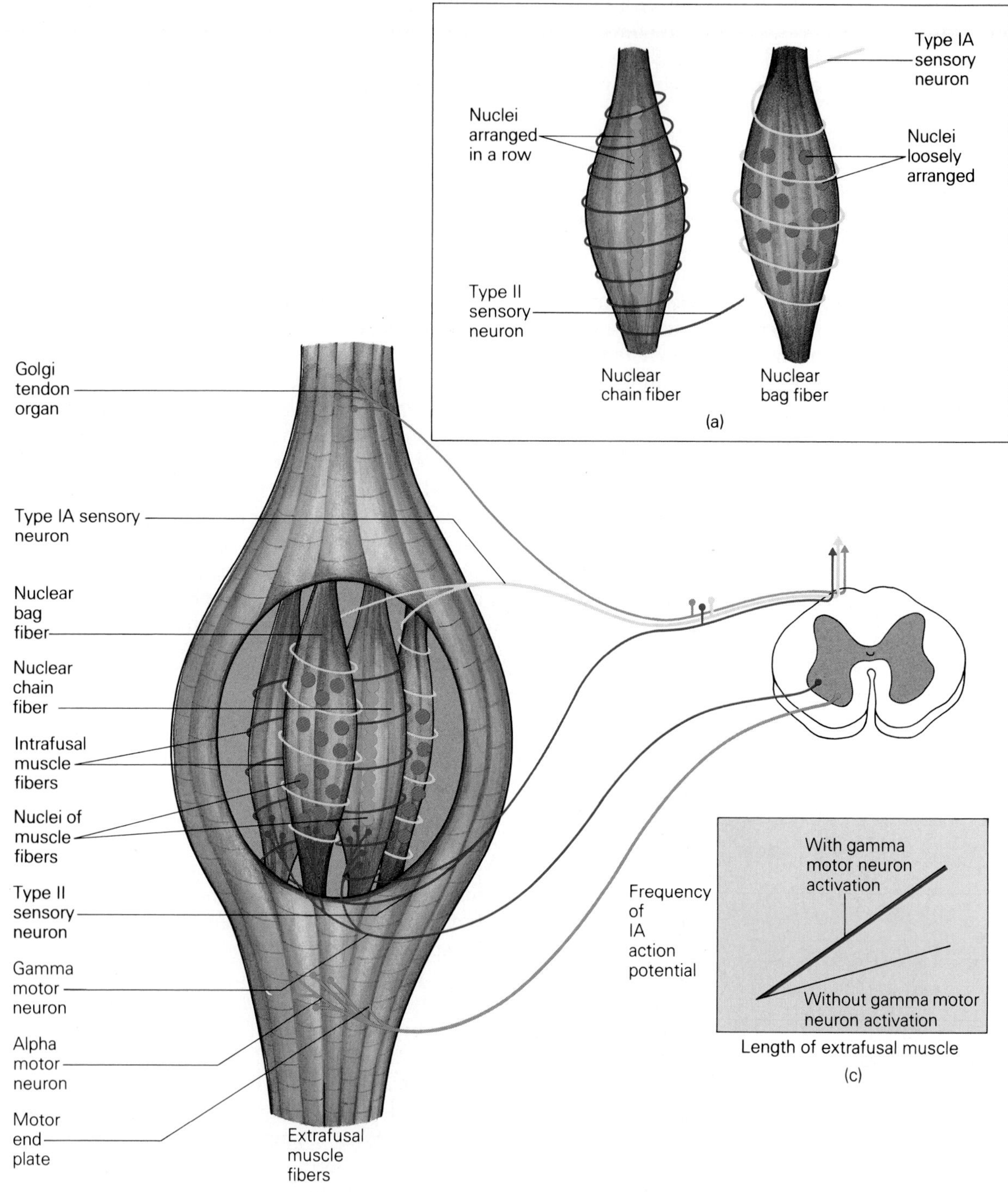

Figure 9–6 (*a*) Intrafusal muscle fibers consist of nuclear bag fibers and nuclear chain fibers. (*b*) These fibers and the sensory neurons that innervate them make up a muscle spindle, a receptor for muscle length and state of contraction. Golgi tendon receptors innervate tendons and transmit information about the force exerted by the muscle. Gamma motor neurons activate intrafusal fibers, while alpha motor neurons activate extrafusal muscle fibers. (*c*) When activated by gamma motor neurons, Type Ia sensory neurons generate action potentials in response to changes in extrafusal muscle length more frequently than in the absence of gamma motor neurons.

tials. Since the Golgi tendon organ and muscle spindle are somatic sensory receptors, they transmit their information about muscle force, velocity, and length to the spinal cord, where it is then forwarded to the somatic sensory cortex or used within the spinal cord for reflex action.

The Spinal Cord in the Motor Response System

The spinal cord plays a crucial role in the neuromuscular system. Motor neurons all originate in the gray matter of the organ. Some automatic body movements, called reflex movements, are controlled solely by the spinal cord and have no control from higher structures. Locomotion, or walking, for example, can be initiated in the spinal cord.

Organization of Motor Neurons in the Spinal Cord

Motor neurons are organized in the spinal cord in two ways: (1) the motor neurons in the dorsal area of the ventral horn are responsible for flexor movements while cells in the ventral region are responsible for extensor movements (Fig. 9–7); and (2) those located in the dorsolateral region of the ventral horn innervate muscles in the extremities while those located in the ventromedial region of the ventral horn innervate the axial muscles of the body to maintain posture. Motor neurons that innervate a single muscle are functionally grouped in the spinal cord in **motor neuron pools.** Motor neurons within a single pool are located within several adjacent segments of the spinal cord (Fig. 9–8). The activation of a motor neuron pool thus coordinates the contractile action of the muscle. Motor neurons within a pool are those that are recruited to increase contractile force by the population code.

Interneurons of the Spinal Cord

The interneurons of the spinal cord (see Fig. 9–4) also play an important role in movement. These cells are located in the intermediate zone of the spinal cord. Those located in the lateral part of this zone have axons that synapse **ipsilaterally** (on the same side of the body) with motor neurons that in-

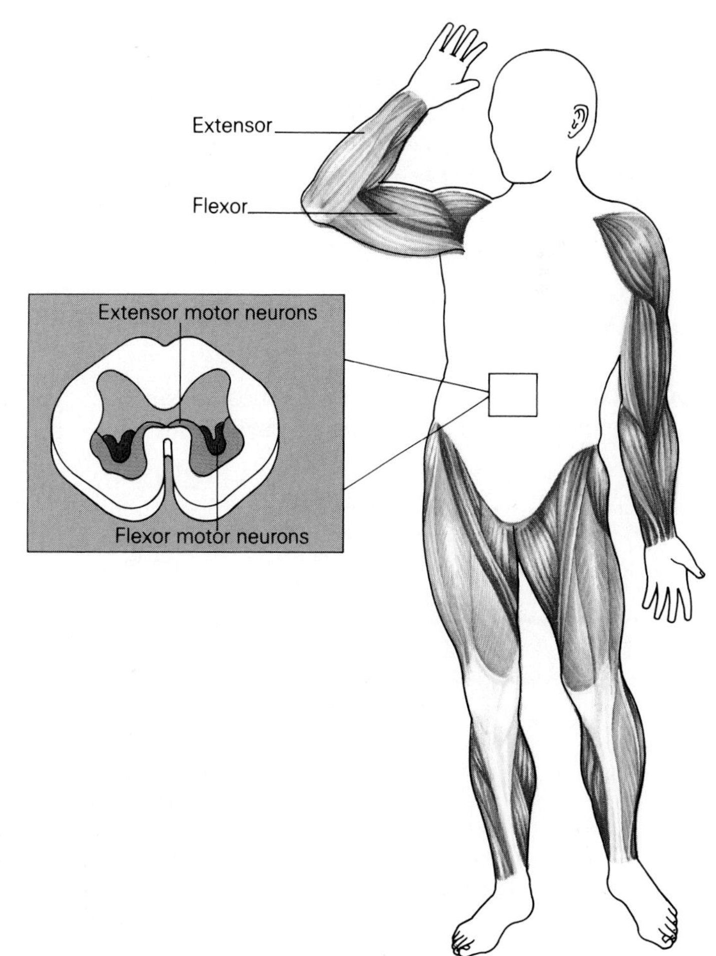

Figure 9–7 Motor neurons that activate flexors originate in the dorsal area of the ventral horn of the spinal cord, while those that activate extensors originate in the ventral region of the ventral horn.

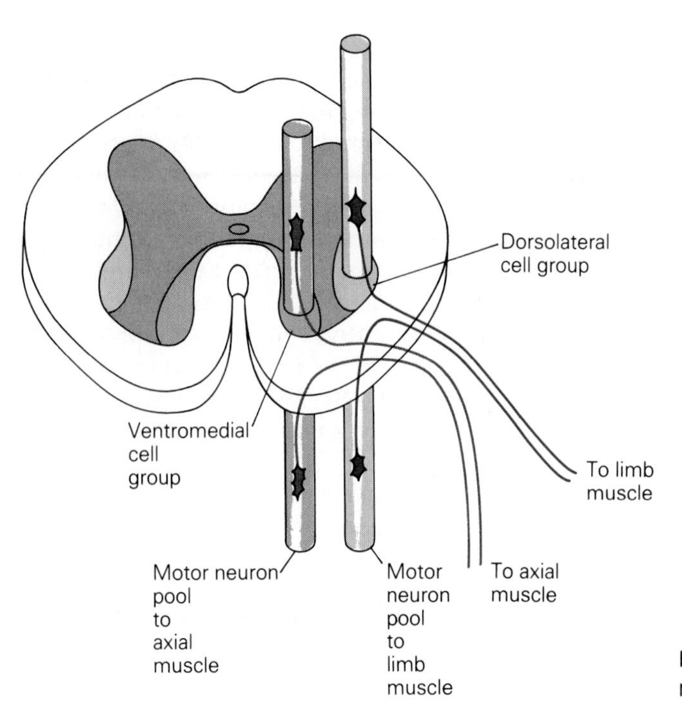

Figure 9–8 A motor neuron pool is a group of motor neurons that innervate a single muscle.

nervate distal limb muscles. Interneurons that lie closer to the midline have axons that synapse with motor neurons on both sides of the spinal cord that control muscles for posture.

Interneurons also send their axons up and down the spinal cord to form synaptic connections in the regions several cord segments away from where their cell bodies are located. They connect with motor neurons that control the contraction of axial muscles. In this way, the activity of several muscle groups are coordinated in maintaining posture.

Spinal Cord Reflexes

Spinal cord reflexes represent the most basic of motor responses. These reflexes are carried out entirely within the spinal cord and are modified by inputs from high centers to generate complex movements. They are also used to help diagnose disorders of the motor system.

The Stretch Reflex

The **stretch reflex** is commonly called the **knee jerk reflex.** A stretch reflex, for example, is activated by tapping the **patellar tendon** below the knee to stretch the muscle spindles. Action potentials conducted along the muscle spindle sensory neurons enter the spinal cord via the dorsal roots (Fig. 9–9). The fiber from the muscle spindle branches after entering the spinal cord, with the ascending branch joining the dorsal column pathway. The other branch forms synaptic connections with motor neurons in the ventral horn.

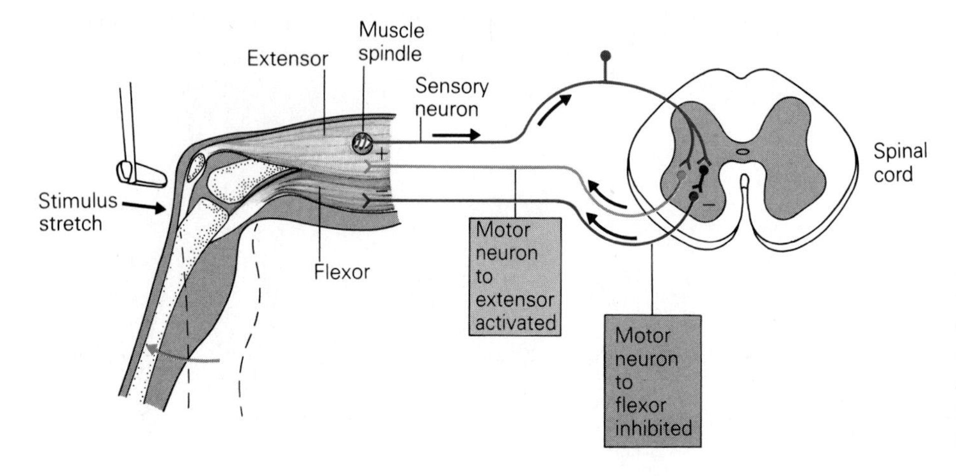

Figure 9–9 The stretch reflex occurs when a muscle spindle detects a tap on the knee and informs the spinal cord, which activates the leg extensor and inhibits the flexor, causing the knee joint to open up.

These motor neurons activate muscle fibers, causing the leg to extend quickly or kick out. In order for the leg to extend, however, the antagonistic muscles that would cause the leg to flex must be inhibited. This inhibition comes about by the activation of inhibitory interneurons within the ventral horn. These interneurons receive information from muscle spindle sensory neurons and in turn form inhibitory synaptic connections with motor neurons of flexor muscles. Consequently, the activation of these interneurons prevents the contraction of the flexors. By inhibiting flexors and activating extensors, the interconnections of nerve cells within the spinal cord produce well-coordinated motor responses.

The Inverse Myotatic Reflex

A reflex that involves the Golgi tendon organ is the **inverse myotatic reflex.** This reflex is seen when a person attempts to lift more weight than he or she can actually carry. Weightlifters, for example, often add increasing amounts of weight to bar bells when lifting the weights in a maneuver called the "curl." In this maneuver, a person lifts the bar bells by contracting the bicep muscles. If the weight is too great for the biceps to lift, however, they suddenly relax, and the bar bells are dropped.

The inhibition of the biceps occurs because the Golgi tendon organ detects excessive force, which might damage the muscle. With increasing force, the Golgi tendon organ begins to discharge action potentials (Fig. 9–10). As with the muscle spindles, these action potentials are transmitted to the dorsal column nuclei and subsequently to the somatic sensory cortex. In the spinal cord, however, information about excessive muscle tension is transmitted to inhibitory interneurons that turn off the motor neurons innervating the biceps. At the same time, interneurons turn off any muscles (flexors) synergistic to the biceps and turn on any muscles (extensors) that are antagonistic to the biceps, so that the arm will extend and drop the barbells. This reflex thus protects the muscle from damage that might occur when a person attempts to lift too heavy a weight.

In addition to regulating muscle tension, this reflex is also thought to involve the smooth onset and termination of muscular contraction in walking. In the motion of stepping, the flexors of one leg begin to contract while its extensors relax. The flexion of one leg needs to be coordinated with the extension of the opposite leg. To achieve this, sensory information from the Golgi tendon organ from the leg being flexed is transmitted to the other side of the spinal cord to activate the extensors and inhibit the flexors of the opposite leg. Although sensory inputs play an important role in the coordination of walking, the motor "program" for walking is contained within the spinal cord. As we shall see, this program can operate in the absence of sensory input.

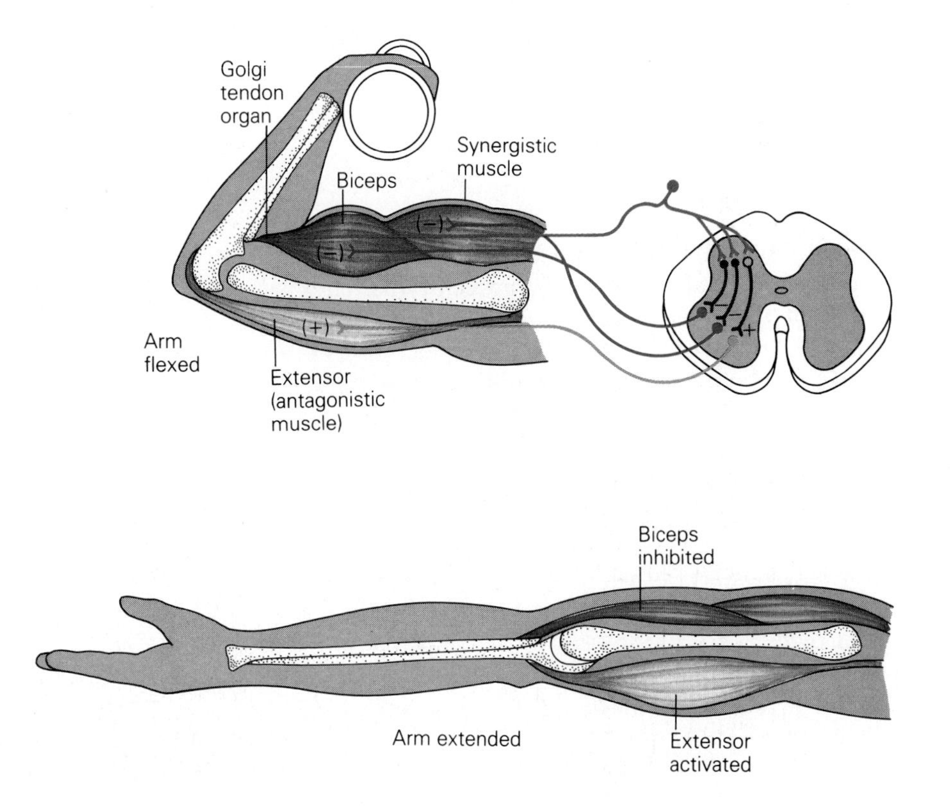

Figure 9–10 The inverse myotatic reflex causes a flexed bicep muscle to relax and an extensor in the arm to contract so that the arm extends and drops a weight that might be harmful.

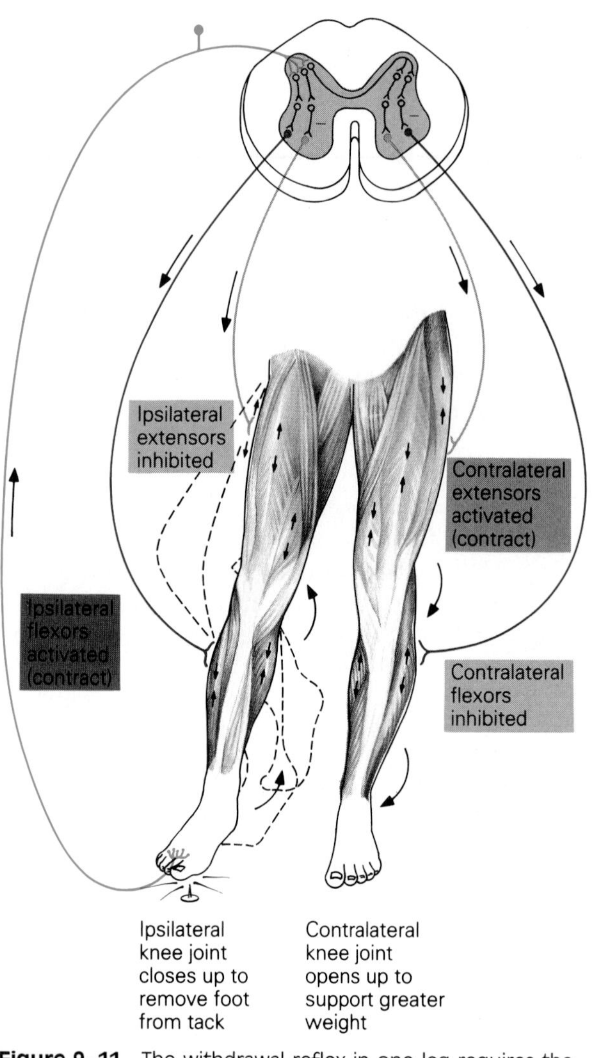

Figure 9–11 The withdrawal reflex in one leg requires the cooperation of the opposite leg, arranged by interneurons in the spinal cord.

The Flexor Withdrawal Reflex

A reflex that involves cutaneous receptors is the **flexor withdrawal reflex.** An example of this reflex is activated by pain receptors on the bottom of the foot (Fig. 9–11). When you step on a tack, for instance, pain information is transmitted to the spinal cord and causes a contraction of the flexors to remove the foot from the tack. At the same time, extensors of the leg that would normally keep the foot on the tack are inhibited. In order to support the rest of the body, however, the extensor muscles in the other leg are activated while its flexors are inhibited.

Control of Locomotion

The act of walking requires the coordinated contraction and relaxation of flexor and extensor muscles of the legs. This rhythmic alternation between flexion and extension is produced entirely by the interconnections of neurons in the spinal cord. These neurons and their interconnections comprise the **locomotor generators** for walking. Animals whose spinal cords have been severed can walk without inputs from higher centers. This phenomenon is seen in chickens, for example, which remain capable of running for a short period after decapitation. Experimental animals with spinal cord transections are also capable of walking on treadmills. The speeds at which they walk are dictated by the speed of the treadmill. In these experimental animals, if one limb is prevented from walking, the other limb will continue to walk. The locomotor generator for each limb, therefore, does not require activity from the other limb. However, when all limbs are active, the generators for limb movement are coupled to produce a coordinated response (Fig. 9–12).

The locomotor generators in the spinal cord are under the control of a "locomotor command center" in the brain stem. Electrical stimulation of this command center causes animals with spinal cord lesions to walk on a treadmill. Weak stimulation causes walking while stronger stimulation can cause the animal to run.

The locomotor generators coordinate the flexion and extension of leg muscles to carry out the **stance phase** and the **swing phase** of walking. In the

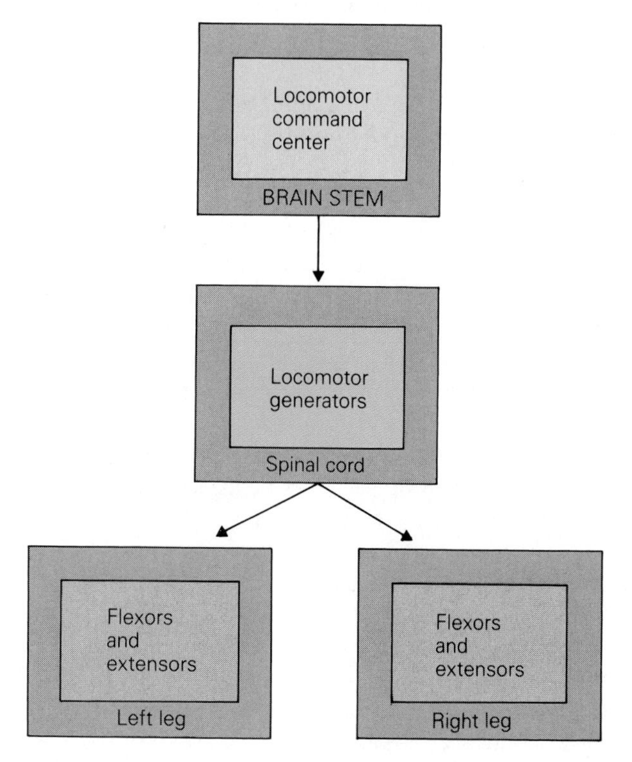

Figure 9–12 Locomotor generators and the locomotor command center.

stance phase, the foot is on the ground and supporting the weight of the body. In the swing phase, the foot is off the ground and swinging forward. If sensory feedback from muscle spindles and Golgi tendon organs is prevented from reaching the spinal cord during the stance phase, the step cycle can be stopped during the extension of the leg. In other phases of the walk cycle, however, the absence of sensory information does not inhibit the process of locomotion.

Sensory information can also change the activation of one motor response to another. The activation of tactile receptors on the top of the foot during the swing phase of walking, for example, would result in a flexion response that is appropriate for stepping over an object. Sensory information to the spinal cord is therefore capable of changing ongoing motor programs and reflexes.

The Brain in the Motor Response System

Several important structures in the brain represent the highest control centers for the motor system (Fig. 9–3). The brain stem controls the spinal cord and the posture and balance of the body. Various areas of the neocortex organize and interpret complex sensory input and program the responding movements. The cerebellum and the basal ganglia also plan and program specific types of motor responses.

These components of the motor system interact with each other to carry out specific functions. The desire to move activates the supplemental motor cortex, basal ganglia, and cerebellar components of the motor system. Information from these areas is transmitted to the premotor cortex, where movements are planned and programmed. From the premotor cortex, information about the planned movement is transmitted to the motor cortex. The motor program is then carried out when neurons in the motor cortex instruct those in the spinal cord to activate the skeletal muscles. The force and velocity of contraction, and the length of muscle shortening, is monitored by Golgi tendon organs and muscle spindles. This sensory information is fed back to the motor neurons at the spinal level and also transmitted back to the motor cortex and cerebellum.

Comparisons are made between the intended movement, as designed by the motor program, and the actual movement. Deviations from the intended movement are then corrected by transmitting another set of instructions to the motor neurons in the spinal cord.

The Brain Stem: Control of Posture and the Spinal Cord

One of the primary roles of the brain stem is to maintain the body's posture and balance. Nerve cells within different clusters in the brain stem send axons that terminate in the spinal cord. Three pathways from the brain stem provide input to the motor neurons of the spinal cord: (1) the ventromedial pathway, (2) the lateral reticulospinal pathway, and (3) the rubrospinal pathway.

The Ventromedial Pathway

The **ventromedial pathway** impinges on motor neurons that control axial muscles. This pathway has three major components, all of which descend along the ventral and medial region of the spinal cord (Fig. 9–13*a*). The first component is the **vestibulospinal tract,** which originates in the vestibular nucleus and carries information for the reflex control of equilibrium (discussed in a later section)]. The second component of the ventromedial pathway is the **tectospinal tract,** which originates in the **tectum,** a structure involved in the coordinated control of head and eye movements. The third component is the **medial reticulospinal tract,** which originates in the **reticular formation,** a structure involved in maintaining posture by the activation of extensor muscles.

The Lateral Reticulospinal Tract

The second major descending pathway from the brain stem to the spinal cord is the **lateral reticulospinal tract.** Nerve fibers in this pathway are derived from the lateral reticular nucleus and descend the spinal cord in the lateral region of the cord (see Fig. 9–13*b*). These fibers innervate flexors in the control of posture.

The Rubrospinal Tracts

The rubrospinal tracts have fibers that originate in the red nucleus of the brain stem. These fibers descend along the dorsal and lateral border of the cord to innervate motor neurons that control distal flexor muscles (see Fig. 9–13*c*). The effect of the major descending brain stem pathways on flexor and extensor muscles in the maintenance of posture and equilibrium are summarized in Figure 9–14.

The Motor Cortex and Fine Voluntary Movement

The fine control of voluntary movement is due to instructions transmitted via descending path-

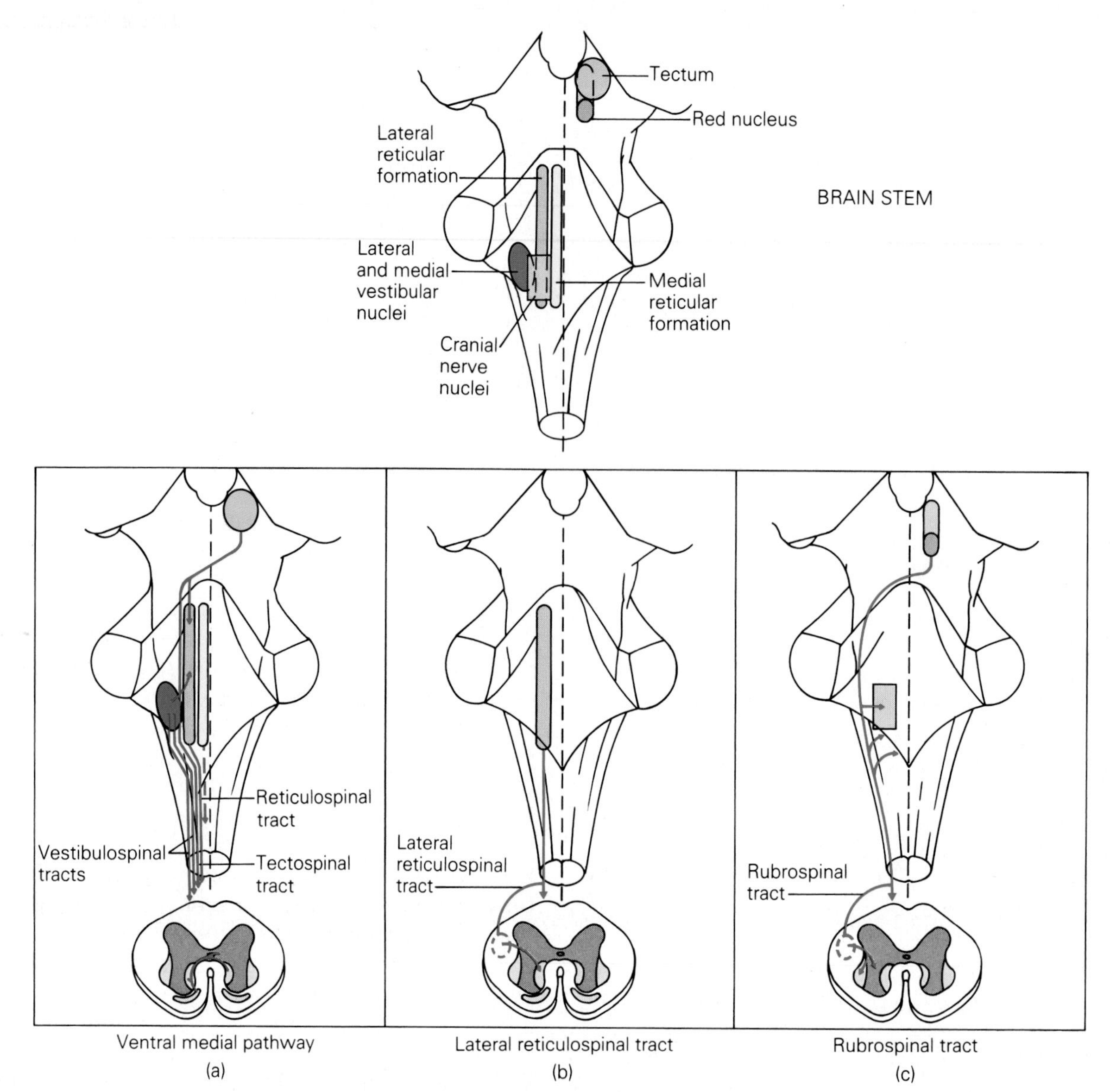

Figure 9–13 Pathways from the brain stem to the spinal cord: (*a*) the ventromedial pathway, (*b*) the lateral reticulospinal tract, and (*c*) the rubrospinal tracts.

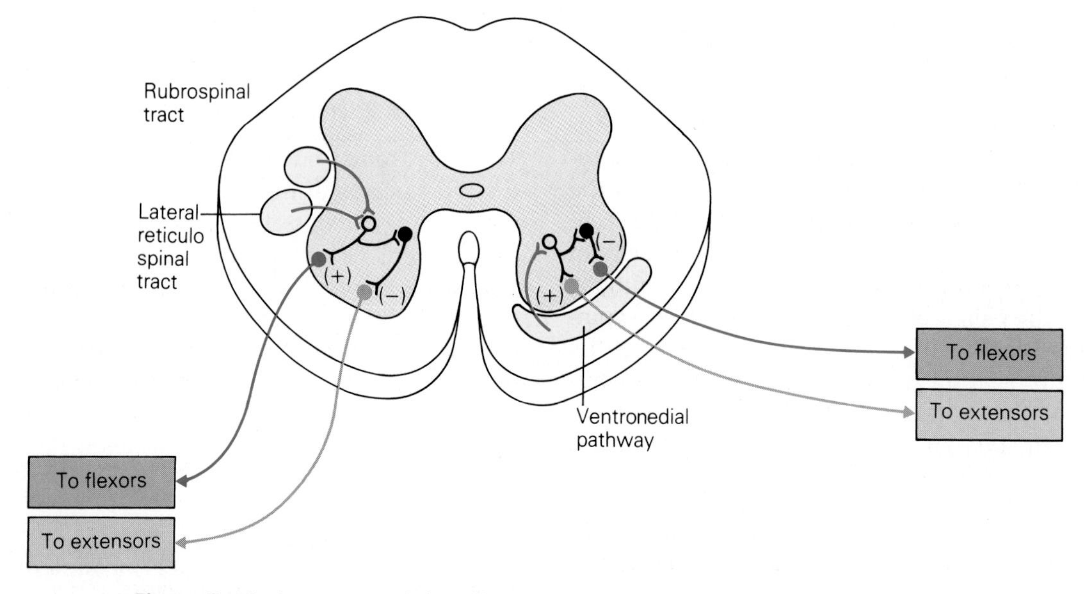

Figure 9–14 A summary of the effects of the brain stem on spinal motor neurons.

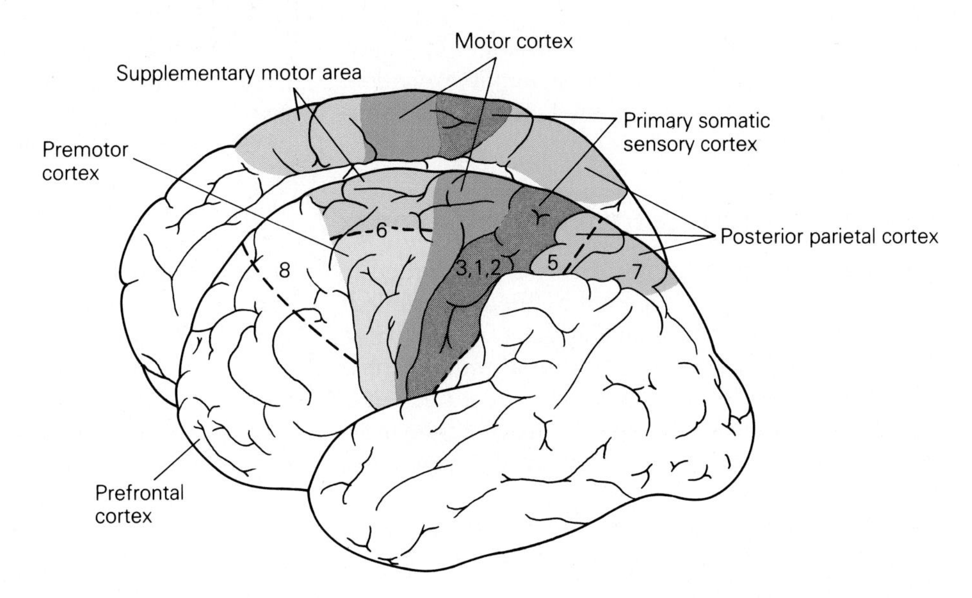

Figure 9–15 The functional areas of the cerebral cortex involved in motor control.

ways to the spinal cord from the **motor cortex.** The motor cortex occupies a cortical region rostral to the somatic sensory cortex (Fig. 9–15). The cells in the motor cortex are organized in a somatotopic manner similar to that of the somatic sensory cortex (Fig. 9–16). Cells in the medial region of the motor cortex cause the contraction of muscles in the leg. Nerve cells located in more lateral regions of the motor cortex activate the muscles of the torso, arm, hand, and face.

The parts of the body that are involved with fine movement, such as the fingers, occupy more space in the motor cortex than parts of the body involved in gross movement, such as the torso. This relationship is in keeping with the general principle that the amount of cortical space that is devoted to different parts of the body is in proportion to its sensory sensitivity or degree of fine motor control.

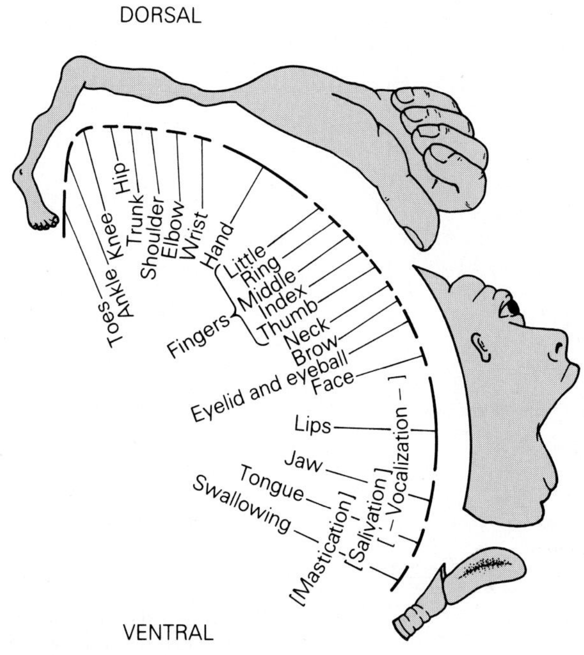

Figure 9–16 The somatotopic organization of the motor cortex.

Organization of the Motor Cortex

The cells of the motor cortex are organized in functional columns called **cortical efferent zones (CEZ).** All of the cells within one efferent zone are involved in the contraction of a given muscle. Neurons in the motor cortex are also organized horizontally into six different layers (Fig. 9–17). Those located in layer V provide the output from the motor cortex. Pyramid-shaped nerve cells in this layer send axons to the spinal cord to synapse with spinal motor neurons. These cortical neurons are involved primarily in controlling distal muscles. They excite both alpha and gamma motor neurons and thus allow the mus-

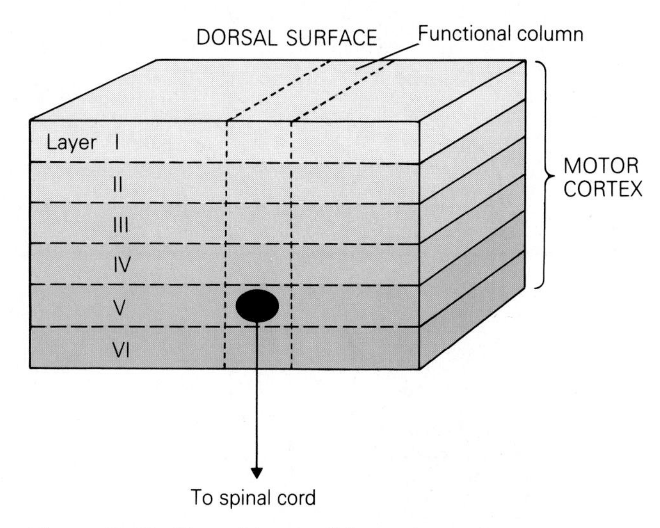

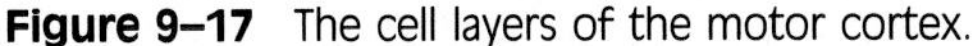

Figure 9–17 The cell layers of the motor cortex.

cle spindles to remain sensitive to changes in muscle length. In addition to the monosynaptic excitation, these cortical neurons also inhibit antagonistic muscles to coordinate movement.

Descending Pathways from the Motor Cortex

Information from the motor cortex is transmitted to the spinal cord and brain stem by the **corticospinal pathway** and **corticobulbar pathway,** respectively. Inputs to the brain stem control axial muscles near the midline for the maintenance of posture, while inputs to the spinal cord control the distal limb muscles. Approximately 30% of the corticospinal and corticobulbar fibers are from neurons in area 4; and another 30% originate from the premotor cortex (area 6). The remaining fibers are derived from the somatic sensory cortex.

As these axons descend from the motor cortex, they form the **pyramidal tract.** When the pyramidal tract reaches the level of the brain stem, the majority of the fibers that continue to the spinal cord cross the midline of the body and continue their descent along the **lateral corticospinal tract** (Fig. 9–18). Nerve fibers exit from the lateral corticospinal tract at various levels of the spinal cord. Fibers from neurons in the motor cortex innervate motor neuron pools that control distal limb muscles. Fibers from neurons in the somatic sensory cortex synapse with neurons in the dorsal horn of the spinal cord and most likely modulate sensory information. The direct connections from motor cortex to spinal cord permit the independent control of individual muscles. This ability is lost following lesions of the pyramidal tract.

In addition to the lateral corticospinal tract, a

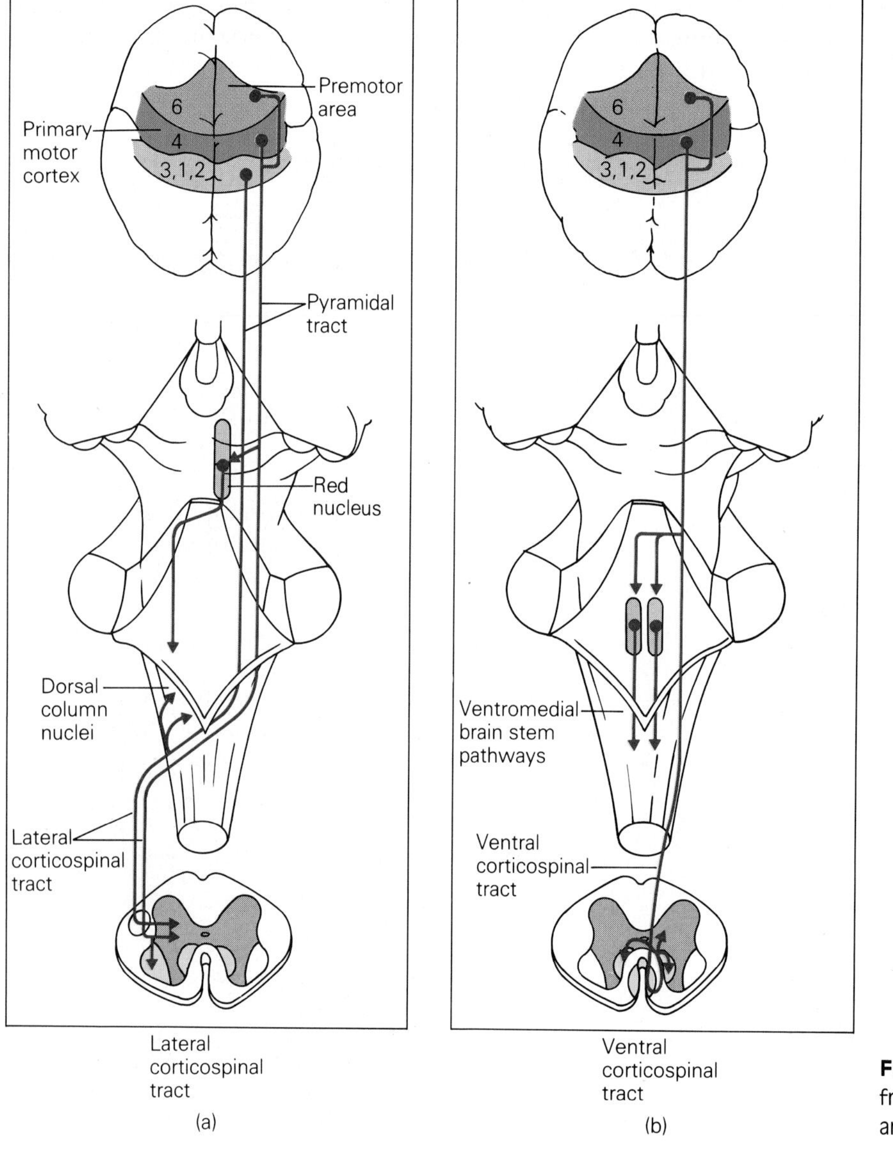

Figure 9–18 Descending pathways from the motor cortex: the lateral and ventral corticospinal tracts.

minority of the fibers from the motor cortex descend to the spinal cord without crossing the midline of the body. These fibers form the **ventral corticospinal tract** and primarily innervate motor neurons in the medial region of the ventral horn associated with axial muscles of the body (see Fig. 9–18).

Sensory Feedback to the Motor Cortex

Neurons in the motor cortex are informed of the consequences of movement through sensory feedback pathways. They receive input from either the muscles they project to or from areas of skin surrounding the muscle. This long loop of sensory feedback results in the alteration of information from the motor cortex to the spinal cord to correct any deviations from the intended movement. Sensory feedback to the motor cortex is by way of the somatic sensory cortex (Fig. 9–19).

The nerve cells in the sensory cortex are connected to those in the motor cortex in a topographic manner. Cells in the sensory cortex receiving proprioceptive input from muscles in the thumb, for example, are connected with cells in the cortical efferent zone responsible for the contraction of these same muscles. In this way, feedback is provided by the sensory system to inform the cells in the motor cortex whether the instructions they have transmitted for the muscular contraction have been faithfully executed. If not, then the sensory information that is sent back to the motor cortex causes the cells in the CEZ to modify their activity and thus modify the activity of the motor neurons.

Cortical Coding of Contractile Force

Since the activation of a muscle is associated with velocity, length, and force of contraction, the question arises as to what message a pyramidal cell in the motor cortex is trying to convey to the motor neuron in the spinal cord. Does the pyramidal cell tell the motor neuron to activate the muscle to contract with a certain force or velocity or to contract to a certain length?

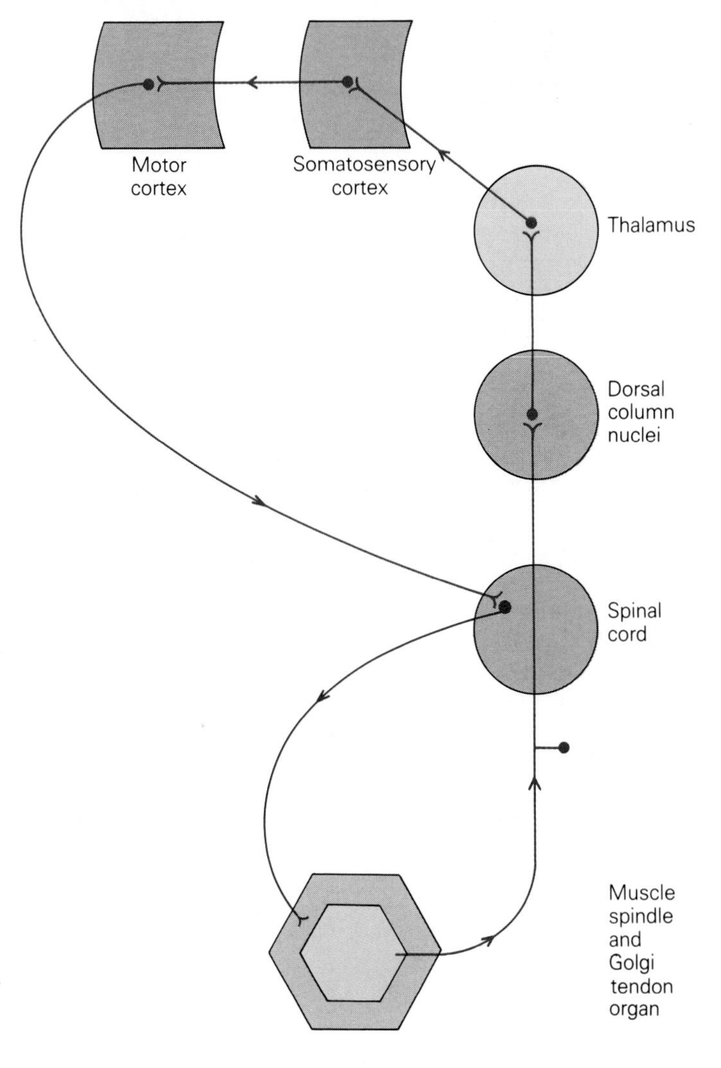

Figure 9–19 Sensory feedback pathways to the motor cortex.

Experiments with laboratory animals have shown that the message transmitted by the pyramidal cell in the motor cortex to the motor neuron in the spinal cord is to activate the muscle to contract with a certain force. Electrical recordings from pyramidal cells made while monkeys move their wrists have demonstrated that these cells generate action potentials at frequencies directly related to the amount of weight that was lifted in moving the wrist from an extensor to a flexor position (Fig. 9–20).

Coding of contractile force is carried out by two types of neurons. First, **dynamic neurons** code for the rate of force development. These neurons change their activity when there is an increase in the level of applied force. A second type of neuron codes for steady-state force. These neurons discharge action potentials throughout the entire period during which force is maintained.

The pyramidal cells in the motor cortex begin generating action potentials approximately 50 milliseconds before a movement. Nerve cells in other areas of the brain involved in motor control become active much earlier than the motor cortex. These areas of the brain, called the **supplemental and premotor cortex, posterior parietal cortex, cerebellum,** and **basal ganglia** are involved in the planning and programming of movement.

The Supplemental and Premotor Areas: Programming Movement

The motivational factors that cause specific movements most likely originate in subcortical areas of the brain such as the **hypothalamus.** This area of the brain is involved with processes such as hunger, thirst, and sex, and will be discussed in Chapter 11. The motivational areas of the brain that respond to the need to accomplish a particular task transmit this information to the **supplemental and premotor cortex** (Brodmann area 6), which designs the necessary movements to accomplish the desired task.

In the act of reaching for a glass of water to bring to the lips, the nervous system must determine which motor program to use to activate certain muscles, and when and how much they need to be contracted. The components of the motor program are thought to be developed by supplemental and premotor cortices (area 6).

Stimulation of the supplemental and premotor areas produces movements that are more complex than those produced by stimulation of the motor cortex. Activation of the premotor area causes movements of the torso or the opening and closing of the hand. In contrast, stimulating the motor cortex will produce small, twitching movements. The supplemental area has direct inputs to the motor

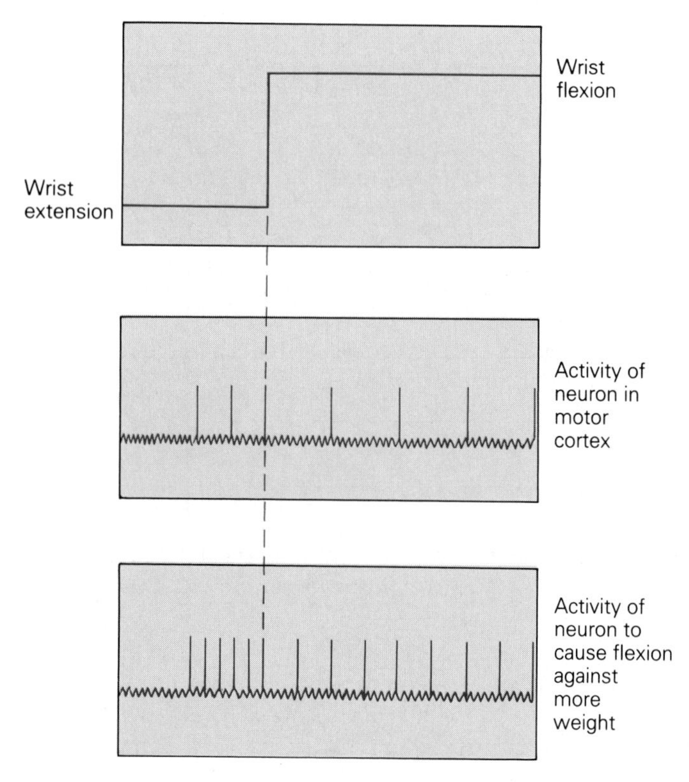

Figure 9–20 Cortical coding of contractile force by neurons in the motor cortex.

cortex for the control of distal limb muscles. However, it also has inputs to motor neurons in the spinal cord to control axial muscles involved with the maintenance of posture.

An example of the role of the supplemental area in the programming of a motor behavior is seen in studies of simple and complex movements of the hand. In these studies, blood flow to certain brain regions is monitored. The amount of blood flow is a measure of the activity of nerves cells in each area of the brain. With simple movements, blood flow increases in the motor and somatic sensory cortices, but no increase is found in the supplemental motor area (Fig. 9–21). With more complex movements, blood flow also increases in the supplemental motor area. Interestingly, when patients are told to rehearse a movement in their minds, blood flow increases only in the supplemental motor area and not in the motor and somatic sensory cortices. These studies suggest that the supplemental motor area is involved in the programming of complex movements.

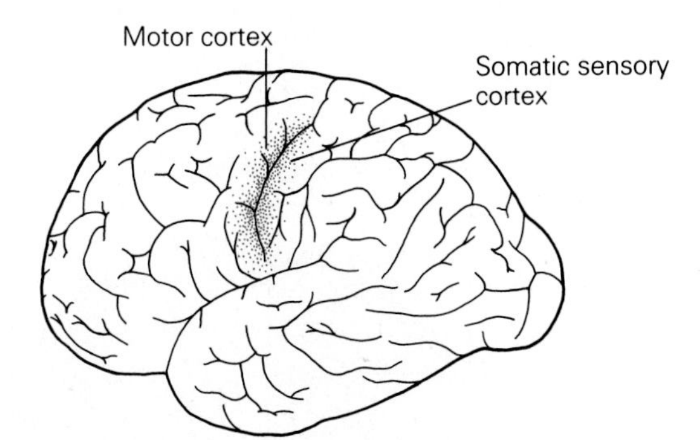

A Simple finger flexion (performance)

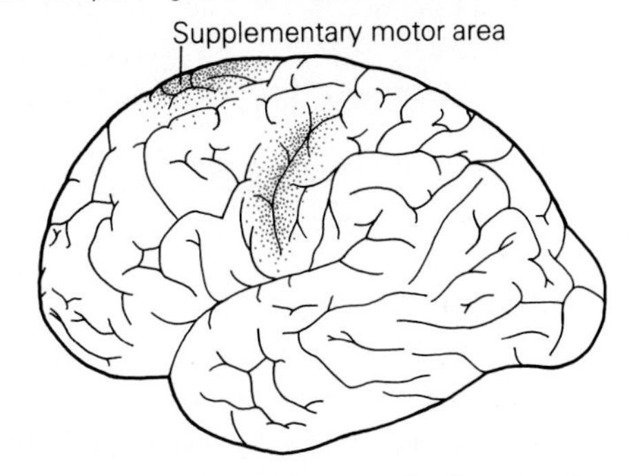

B Finger movement sequence (performance)

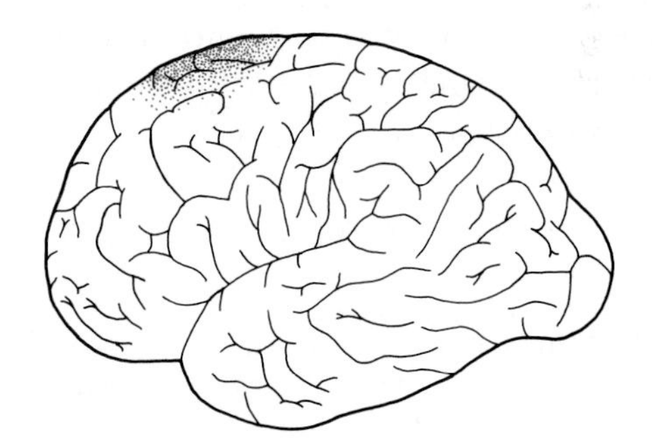

C Finger movement sequence (mental rehearsal)

Figure 9–21 The supplemental cortex is involved in the planning of complex movements. (*a*) During simple finger flexion, blood flow (indicated by red stippling) increases in the motor cortex, while (*b*) during a complex sequence of finger movements, blood flow is also increased in the supplemental cortex. (*c*) When the same sequence is mentally rehearsed, blood flow increases in the supplementary cortex but not in the primary motor cortex.

The Posterior Parietal Cortex: Sensory Stimuli and Purposeful Movement

The **posterior parietal cortex** is necessary for the processing of sensory stimuli leading to purposeful movement. It is located immediately behind the somatic sensory cortex (see Fig. 9–15). This cortex receives both somatic and visual sensory information and transmits information to the supplemental and premotor areas.

Three types of neurons have been identified in the posterior parietal cortex. **Arm projection neurons** generate action potentials when the arm reaches for an object. **Hand-eye coordination neurons** are most active when the eye fixates on an object that is being touched; **hand manipulation neurons** increase their firing rate when the hand explores an object. These types of neurons thus become active only with very specific behavioral motor responses.

Patients with lesions of this cortex caused by strokes or trauma are unable to perform previously learned movements in the appropriate sequence. They appear to synthesize the spatial coordinates of objects in abnormal ways and behave as if their movements are not in accord with the coordinates of the objects in space. For example, when drawing a clock, a patient with a posterior parietal lesion places all of the numbers on one half of its face.

The Cerebellum and the Basal Ganglia

The basal ganglia and cerebellum (Fig. 9–22) are the major subcortical components of the motor system. They both receive inputs from the neocortex and transmit information back to the cortex by way of the thalamus. The inputs to the basal ganglia, however, are from the entire cortex while those of the cerebellum are primarily from sensory and motor

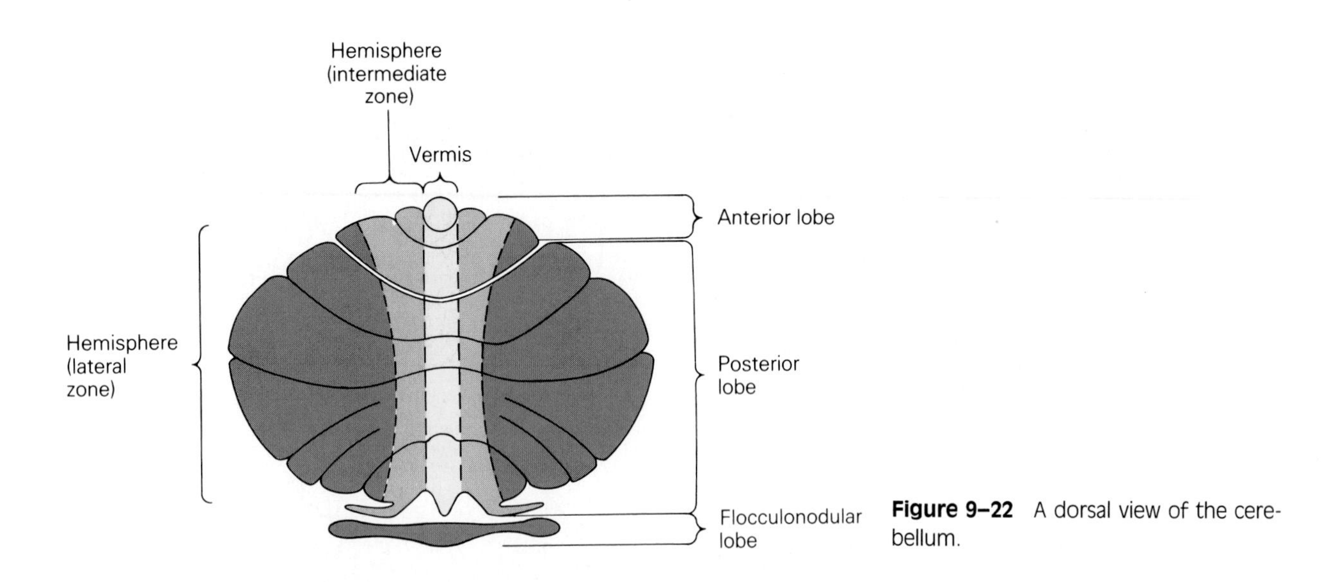

Figure 9–22 A dorsal view of the cerebellum.

areas. In addition, the basal ganglia do not receive direct sensory information from somatic receptors, nor do they transmit descending information directly back to the spinal cord as does the cerebellum. These differences suggest that the basal ganglia are involved in more complex motor functions, while the cerebellum is more involved with the control of movement that requires constant monitoring by sensory feedback.

The Cerebellum: Planning, Coordination, and Posture

The cerebellum ("little brain" in Latin) has three basic functions: (1) the planning of a movement, (2) the control of posture and equilibrium, and (3) the control of limb movement. The cerebellum accomplishes the latter two functions by comparing information concerning an intended movement with sensory feedback about the actual movement and adjusting its output to compensate for differences between the two. It participates in the planning of a movement by receiving information from motor (supplemental and premotor) and parietal cortices and then uses these inputs to initiate a planned movement.

These functions of the cerebellum are associated with specific anatomical locations. The cerebellum has three major lobes: the **anterior lobe,** the **posterior lobe,** and the **flocculonodular lobe** (see Fig. 9–22). The flocculonodular lobe is involved with the maintenance of equilibrium and posture, while the midline and intermediate regions of the anterior and posterior lobes are involved in limb movement. The planning and initiation of motor programs is the responsibility of the lateral region of the anterior and posterior lobes.

The most prominent nerve cell in the cerebellum is the **Purkinje cell.** These cells are found throughout the cerebellum and are packed tightly into a single cellular layer (Fig. 9–23). The axons of Purkinje cells form synaptic connections with neurons in one of three deep cerebellar nuclei, the **dentate nucleus,** the **interpositus nucleus,** or the **fastigial nucleus.** At the synapses between Purkinje cells and the neurons of the deep cerebellar nuclei, the transmitter GABA is released, causing a hyperpolarization of the postsynaptic neurons. The effect of Purkinje cells in the deep cerebellar nuclei is therefore inhibitory.

The deep nuclei receive information from Purkinje cells located in different regions of the cerebellum. The fastigial nucleus receives information from Purkinje cells in the flocculonodular lobe and is therefore involved in the maintenance of balance and posture. The fastigial nucleus transmits its information to the vestibular nucleus which in turn controls motor neurons that innervate axial muscles (see Fig. 9–23).

The interpositus nucleus receives information from Purkinje cells in the medial and intermediate regions of the anterior and posterior lobes of the cerebellum and is therefore involved in the control of limb muscles. The interpositus sends its information to the red nucleus of the brain stem, which in turn controls motor neurons innervating distal limb muscles.

The dentate receives information from Purkinje cells in the lateral regions of the anterior and posterior lobes of the cerebellum and is therefore involved in the planning and initiation of movement. The dentate transmits its information to the thalamus, which in turn forwards information to the motor and premotor cortex.

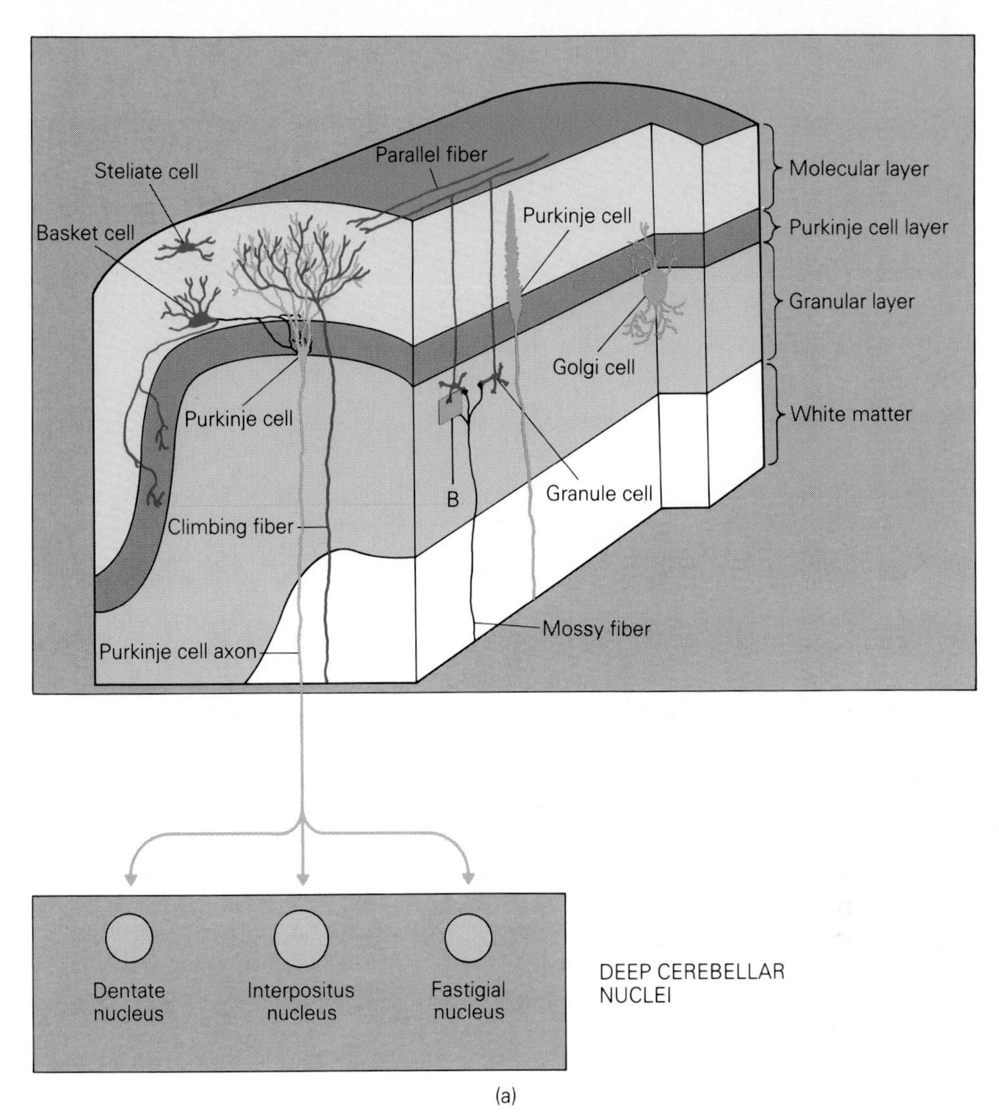

(a)

Figure 9–23 Cell types and circuits in the cerebellum.

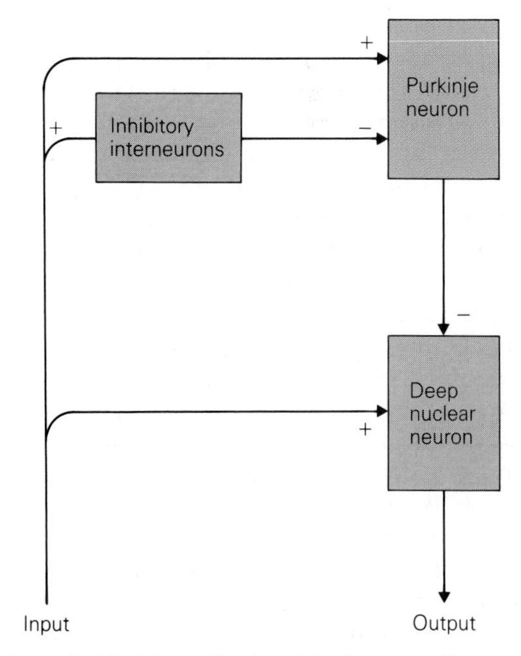

Figure 9–24 Information input to the cerebellum is simultaneously transmitted to the deep cerebellar nuclei.

Since the outflow of information from the cerebellum must eventually pass by way of the deep cerebellar nuclei, these nuclei constitute the output of the cerebellum. Research indicates that the input of information to the cerebellum is simultaneously transmitted to the deep cerebellar nuclei (Fig. 9–24). After the incoming information is processed by the cerebellum, it is transmitted by way of the Purkinje cells to the deep nuclei. The outputs from the deep nuclei are thus modulated by the inhibitory actions of the Purkinje cells in the comparison of intended and actual movements. Neurons of the deep nuclei then transmit information to correct for any deviation from the intended motion.

The inputs to the Purkinje cells of the cerebellum traverse two routes: the climbing fiber and parallel fiber pathways. Climbing fibers originate from the inferior olivary nucleus (Fig. 9–25*a*). Axons from

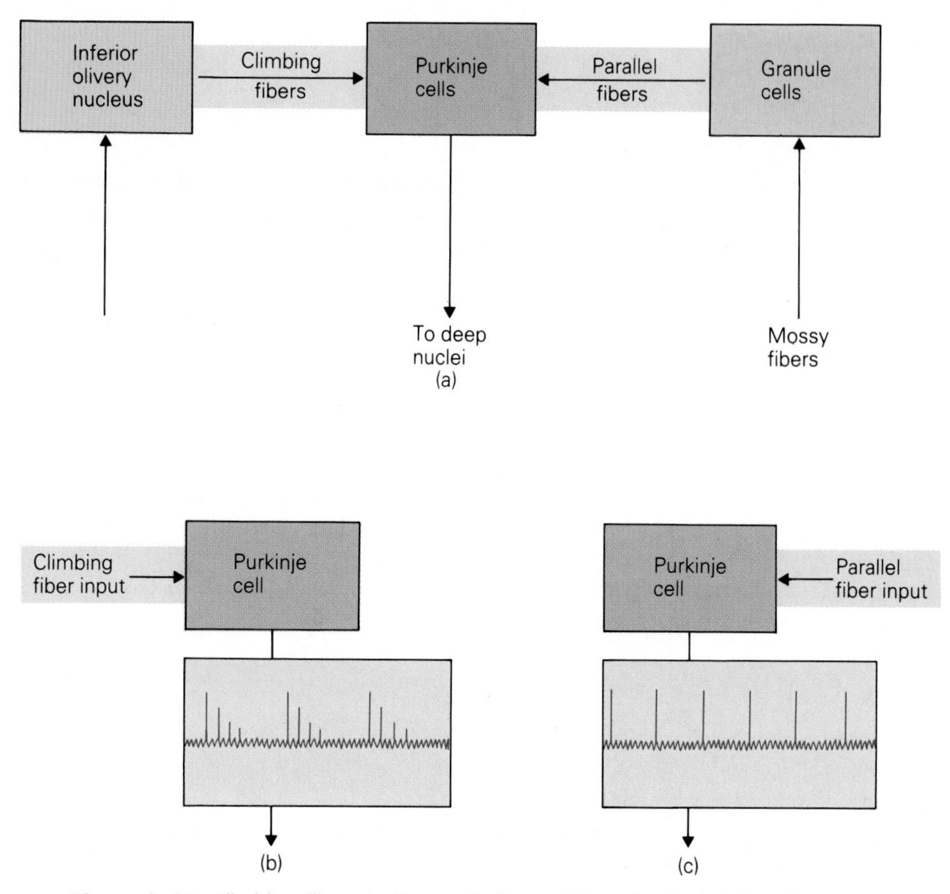

Figure 9–25 Climbing fibers to the cerebellum originate in the inferior olivary nucleus, while parallel fibers originate in the granule cells (*a*). Activation of climbing fibers generates complex pattern of action potentials (*b*), while activation of parallel fibers generates simple action potential patterns (*c*).

neurons in this nucleus form synaptic connections around the soma and dendrites of Purkinje cells. Each Purkinje cell receives synaptic connections from only one climbing fiber. The activation of this input causes the Purkinje cell to generate complex patterns of action potentials (Fig. 9–25*b*).

The parallel fiber input to the Purkinje cell originates from **granule cells.** These fibers synapse with the dendrites of Purkinje cells. Each Purkinje cell receives inputs from many granule cells, and the activation of the parallel fibers causes the Purkinje cells to generate a simple pattern of action potentials.

Inputs to the inferior olivary nucleus come from pathways originating in supplemental and premotor areas of the cortex as well as from the spinal cord, while inputs to granule cells are derived from so called **mossy fibers.** These mossy fibers also originate in the neocortex as well as in the spinal cord. The mossy fibers and olivary climbing fibers, however, respond quite differently during movement. Sensory stimuli and voluntary movements enhance the activity of mossy fibers but have little effect on climbing fiber activity. Climbing fibers are therefore thought to modulate the responsiveness of Purkinje cells to mossy fiber and hence to parallel fiber inputs. This is seen most clearly during the learning of motor behaviors.

When monkeys are trained to maintain a lever in one position with their hands, Purkinje cells generate a simple pattern of action potentials (Fig. 9–26). The pattern is characteristic of that produced by Purkinje cells activated by parallel fibers. When different weights are placed on the lever to change its position, the animal must compensate for this added force in order to maintain its original position. When more weight is added, Purkinje cells begin to produce complex (rather than simple) patterns of action potentials. These complex patterns of Purkinje cell action potentials are due to their activation by climbing fibers. As the animal learns to adapt to the new weight, the complex pattern of action potentials diminishes and the simple pattern re-emerges. Thus, the Purkinje cells are capable of changing their pattern of activity during the adaptation of new motor responses.

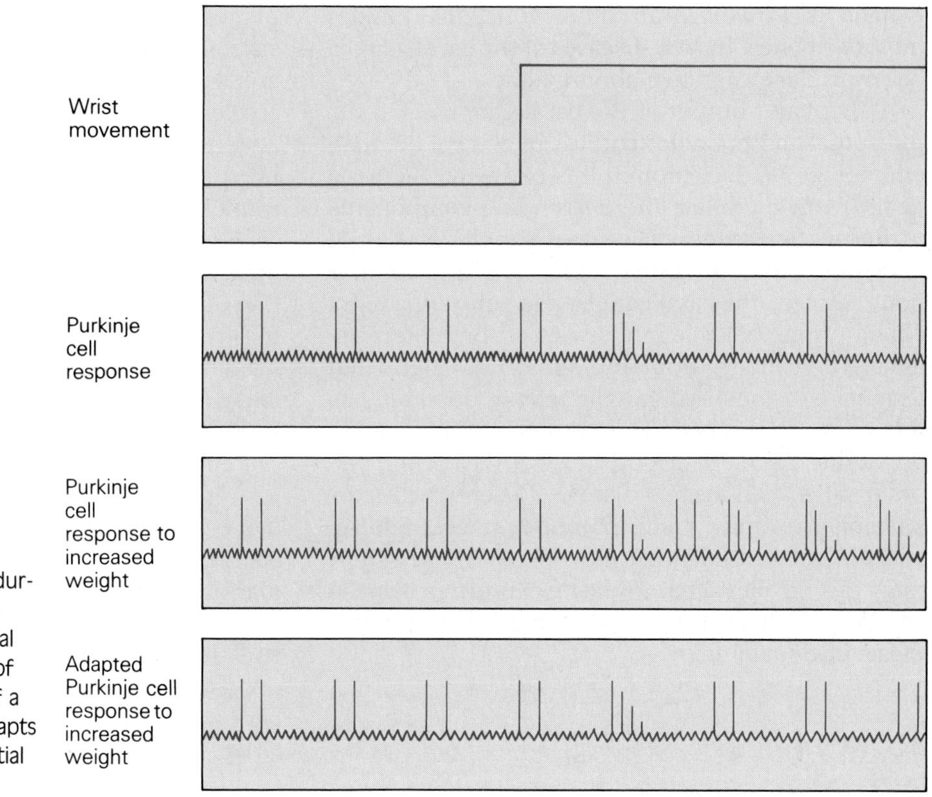

Figure 9–26 Purkinje cells can change their pattern of activity during adaptation to new motor responses. Complex action potential patterns show the involvement of climbing fibers in the learning of a new motor skill. As an animal adapts to the skill, simpler action potential patterns emerge.

The Basal Ganglia: Planning of Movements

Other areas of the brain involved in the programming of motor patterns are the basal ganglia. The basal ganglia include the **caudate, putamen,** and **globus pallidus** (Fig. 9–27). The caudate and putamen receive inputs to the basal ganglia, while the globus pallidus provides the output. Inputs to the basal ganglia are from the entire neocortex, thalamus, and substantia nigra of the brain stem. The primary input, however, is from the neocortex. The extensiveness of this input suggests that the basal

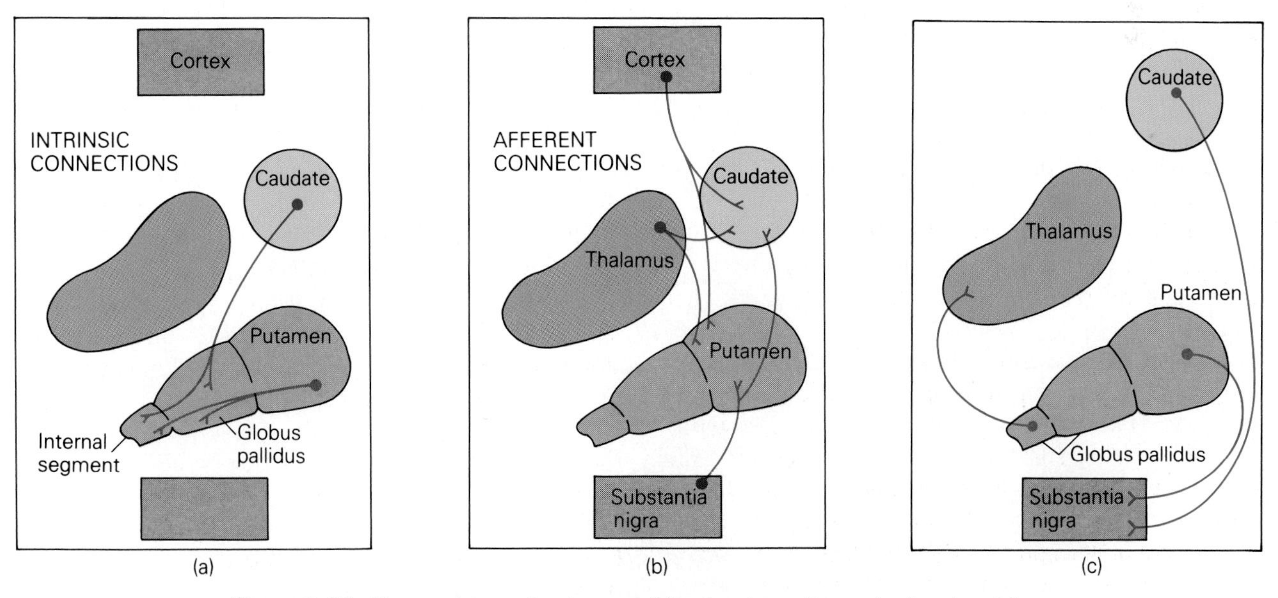

Figure 9–27 The caudate and putamen of the basal ganglia receive inputs, while the globus pallidus provides the output. The basal ganglia receive inputs from the neocortex, thalamus, and substantia nigra.

ganglia are involved in other functions besides motor activities. In fact, diseases of the basal ganglia often produce cognitive abnormalities.

The major output of the basal ganglia is to the prefrontal and premotor cortices by way of the thalamus (Fig. 9–27*c*). Through this pathway, the basal ganglia can modulate the descending components of the motor system. The basal ganglia also have outputs to the substantia nigra. The interconnections between the basal ganglia and substantia nigra play a prominent role in diseases of the motor system. Nerve fibers from the substantia nigra that terminate in the basal ganglia release dopamine as the neurotransmitter. The degeneration of these dopamine fibers is responsible for the motor disorder called Parkinson's disease. Patients with parkinsonism exhibit a rhythmic tremor at rest, rigidity, and difficulty in initiating movement. These symptoms can be alleviated, in part, with drugs such as L-dopa, which act as precursors to increase the synthesis of dopamine.

Vestibular Control of Equilibrium and Balance

The vestibular system aids in maintaining the body's balance by detecting the position and motion of the head in space. The sensory organs of the vestibular system are the **semicircular canals** (Fig. 9–28). Three semicircular canals make up the vestibular apparatus: the **superior, inferior,** and **horizontal semicircular canals.** Each semicircular canal is situated so that it is perpendicular to the other two. With this arrangement, each semicircular canal detects the angular acceleration in one of three planes in space. The vestibular apparatus also consists of the **utricle** and **saccule** (see Fig. 9–30). These are so-called **otolithic organs;** they detect linear acceleration.

The semicircular canals and otolithic organs are filled with the same endolymph fluid that surrounds the hair cells of the cochlea. As does the cochlea, the semicircular canals and otolithic organs have hair cells that are responsible for the transduction of sensory stimuli. The appropriate stimulus for these vestibular hair cells is the acceleration of the head. As the head begins to move, the inertia of the endolymph fluid causes the stereocilia to bend (Fig. 9–29). This bending of the stereocilia causes the hair cell either to depolarize or to hyperpolarize.

The type of membrane potential change in the hair cell depends on the direction in which the stereocilia bend. The stereocilia of hair cells in the vestibular apparatus are aligned according to size, with the smallest cilium placed at one end of the group and the largest cilium, called the *kinocilium,* at the opposite end (see Fig. 9–29). If the stereocilia are bent toward the kinocilium, the hair cell will depolarize. If the stereocilia are bent toward the smallest cilium, the hair cell will hyperpolarize.

These hair cells, as with those in the cochlear nucleus, are not capable of generating action potentials. They do, however, excite or inhibit the **vestibular nerve fibers** that innervate them. As the hair cell depolarizes, it releases a neurotransmitter that excites the innervating vestibular nerve (see Fig. 9–29). The hyperpolarization of a hair cell causes the vestibular nerve to produce fewer action potentials. This may be due to a decrease in the amount of an excitatory transmitter that is tonically being released from the hair cell.

How do these characteristics of vestibular hair

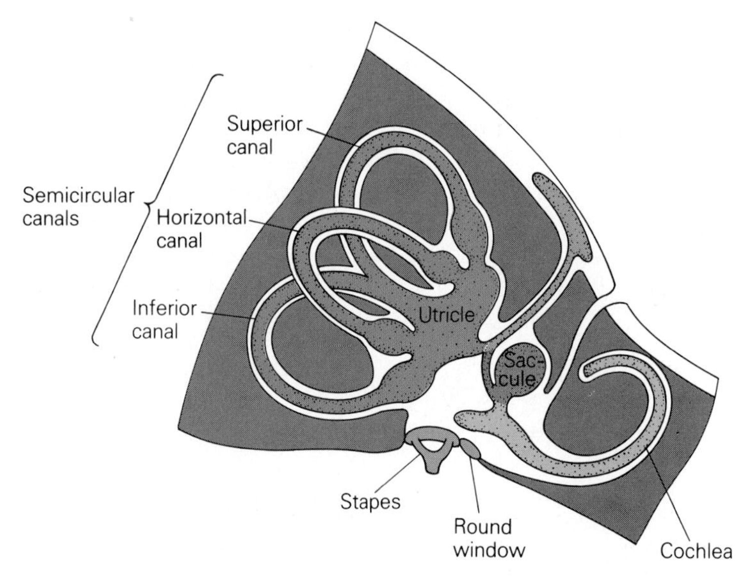

Figure 9–28 The vestibular apparatus consists of the semicircular canals, including the inferior, superior, and horizontal canals.

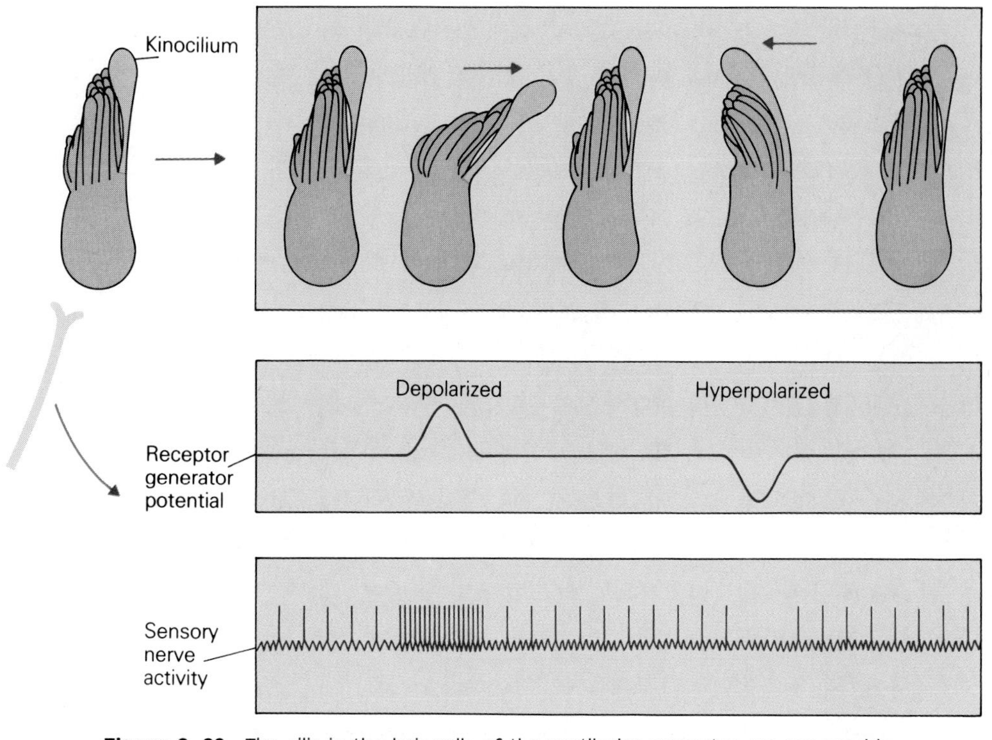

Figure 9–29 The cilia in the hair cells of the vestibular apparatus are arranged in order of increasing length, with the largest cilium, the kinocilium, at one end. When the cilia are bent in the direction of the kinocilium, the hair cell membrane becomes depolarized; when the kinocilium bends the other cilia toward the smallest cilium, the membrane becomes hyperpolarized.

cells allow them to detect the accelerated movement of the head? The hair cells are arranged in a polarized manner on both sides of the brain. In the horizontal semicircular canals, for example, the stereocilia on each hair cell are positioned so that the kinocilium is nearest to the nose or front of the head while the smallest cilium is nearest the back of the head (Fig. 9–30). As the head rotates from right to left (counterclockwise), the inertia of the endolymph makes it appear as though it is moving from left to right (clockwise) with respect to the stereocilia. The fluid's relative motion causes the stereocilia on the left side of the brain to depolarize the hair cells, while the stereocilia on the right side of the brain hyperpolarize the hair cells. Correspondingly, the vestibular nerves on the left side of the brain increase its rate of action potential generation while the nerves on the right side of the brain decrease their rates of action potential generation. This information is thus transmitted to inform the brain that the head is rotating clockwise.

The information from the vestibular nerves is transmitted to the vestibular nucleus located within the brain stem. From there, it is sent down to the spinal cord to act on nerve cells that regulate movement, and up to the cerebellum and somatosensory cortex. The cerebellum and somatosensory cortex use this information to coordinate muscular activity in order to maintain balance.

In addition to the transduction of information about the movement of the head, the vestibular apparatus also transduces information about the stationary position of the head in space. In addition to conveying information about linear acceleration, the hair cells of the utricle can transmit information about the position of the head with respect to gravity. The stereocilia of the hair cells in the utricle are embedded in an **otolithic membrane** that contains calcium carbonate crystals called **statoconia** (Fig. 9–31). The statoconia give the otolithic membrane mass, so that when the head is tilted to the left or the right, the gravitational force causes the weighted otolithic membrane to bend the stereocilia. The stereocilia are arranged on the hair cells so that when the head is tilted to the left, the hair cells depolarize; and when the head is tilted to the right, the hair cells hyperpolarize.

Disorders of the Motor System

Because the different parts of the motor system have such distinctive roles, lesions in these areas result in

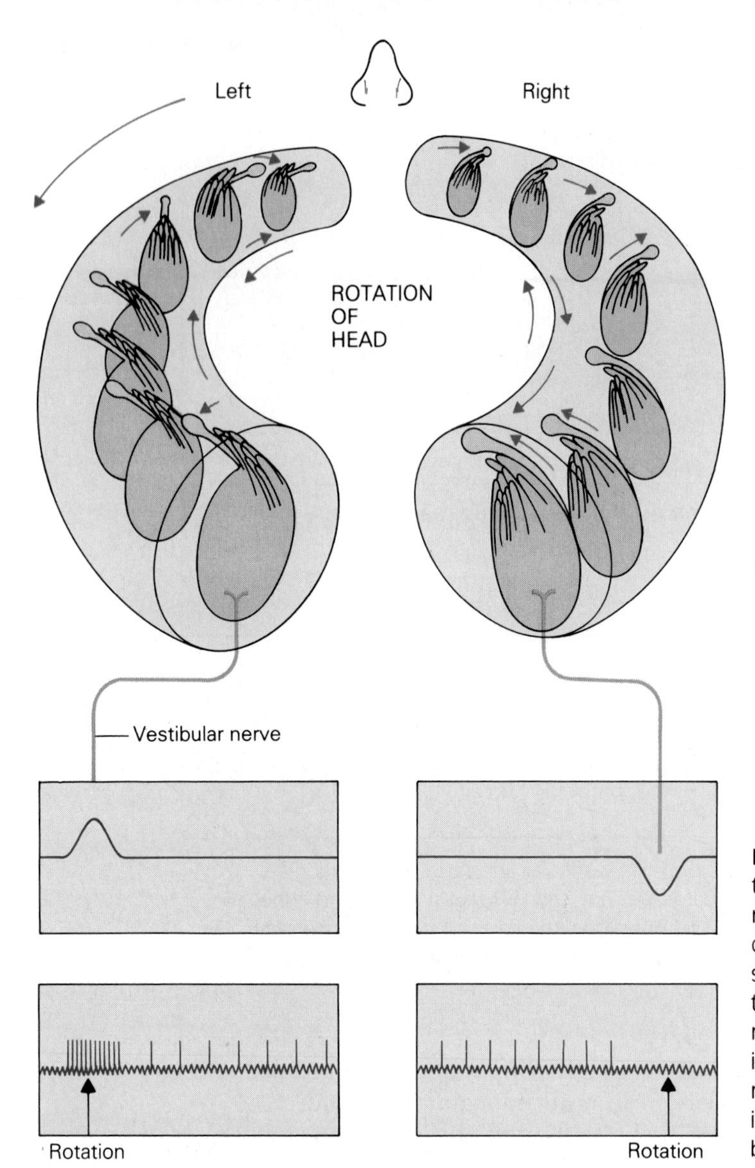

Figure 9–30 In the horizontal semicircular canal, the kinocilia of the hair cells are oriented toward the nose. When the head rotates in a counterclockwise direction, the endolymph in effect flows in the opposite direction relative to the cilia. On the left side of the brain, the hair cells depolarize because the fluid motion causes the cilia to bend toward the kinocilium, while the opposite effect takes place on the right side. The vestibular nerves detect the depolarization and hyperpolarization and thereby inform the brain about the direction of rotation.

Focus on Parkinson's Disease and Brain Cell Transplants

Parkinson's disease is a disorder of the motor system characterized by a slowness of movement, tremor of the fingers and hands (most prominent at rest), absence of facial expression, slowness of speech, stooped body posture, and a shuffling gait. In extreme cases, patients are bedridden and incapable of feeding and caring for themselves.

This disorder frequently appears in persons between the ages of 50 and 65 years. Currently in the United States, approximately 1 million patients are afflicted with parkinsonism, with 50,000 new cases reported each year.

Examination of brain tissue from Parkinson's patients has revealed a loss of dopamine-producing neurons that are located in the brain stem nucleus called the substantia nigra. These neurons send axons that form connections with other neurons in the neostriatum, an area of brain involved in locomotor activity. The resulting depletion of dopamine in the neostriatum appears to be the major cause of this disorder.

The reason for the degeneration of dopamine cells in the substantia nigra is unknown. A form of juvenile parkinsonism is thought to be genetically inherited. Evidence for a genetic component for adult parkinsonism is more controversial. Epidemiological studies reveal a strong correlation between parkinsonism and agricultural communities. It is thought that the use of pesticides may be a contributing factor in the development of Parkinson's disease.

One form of treatment for Parkinson's disease—the administration of drugs such as L-dopa that become precursors for the synthesis of dopamine—has been found to be an effective form of therapy for many patients. Some patients, however, do not respond to drug therapies or have debilitating side effects such as hallucinations.

10

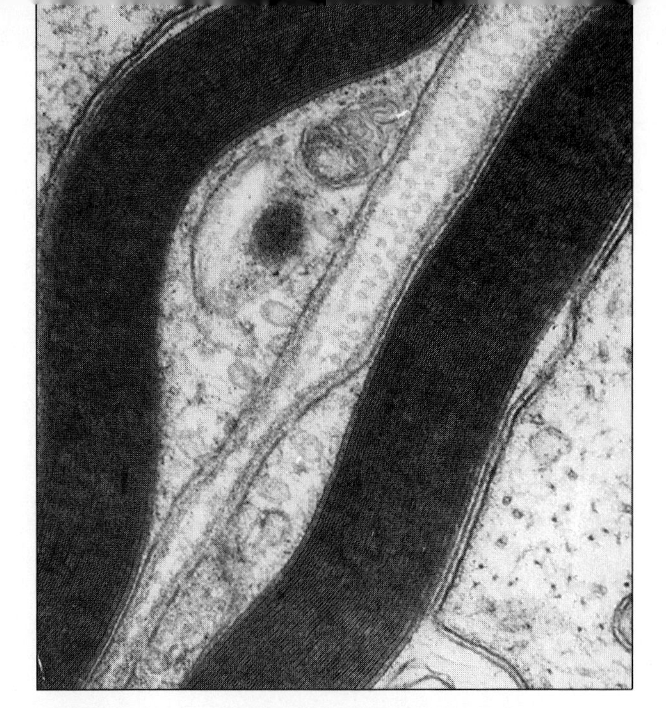

The Autonomic Nervous System

The **autonomic nervous system (ANS)** coordinates bodily functions that are needed for survival. It senses the need for food and activates the digestive processes; it senses the need for water and activates processes for the retention and consumption of fluids. The ANS aids in removing the waste products from the body, and it prepares the body for the stress of life-threatening conditions.

The ANS is also a regulatory system that maintains an environment necessary for the cells in the body to function properly. It maintains the body's temperature at 37°C, mean blood pressure at 90 mm Hg, pH at 7.4, and blood plasma osmolarity at 300 mOsm. These are the optimal conditions of the **internal environment** for the functioning of the organs and their cells.

The actions of the ANS are not under direct voluntary control. Unlike the **somatic nervous system,** by which we can consciously command skeletal muscles to move (e.g., the muscles in the fingers of the hand to pick up a pencil), the actions of the ANS are all subconscious. It is difficult to consciously will your heart to beat faster or command your blood pressure to increase. Instead, autonomic processes transmit sensory information from "visceral" organs to the central ANS, which in turn sends instructions to the smooth muscles and other cells of those organs, producing an appropriate automatic response.

This chapter examines the functional organization of the ANS, and the ways in which the ANS controls other organ systems to supply the necessary nutrients to the body, to remove waste products of metabolism, to coordinate the response to stress, and to regulate the internal environment.

Organization of the Autonomic Nervous System

The ANS has both central and peripheral components. The central components of the ANS are the **hypothalamus,** the **brain stem,** and **spinal cord** (Fig. 10–1). The peripheral components consist of the nerves that innervate the organs of the body,

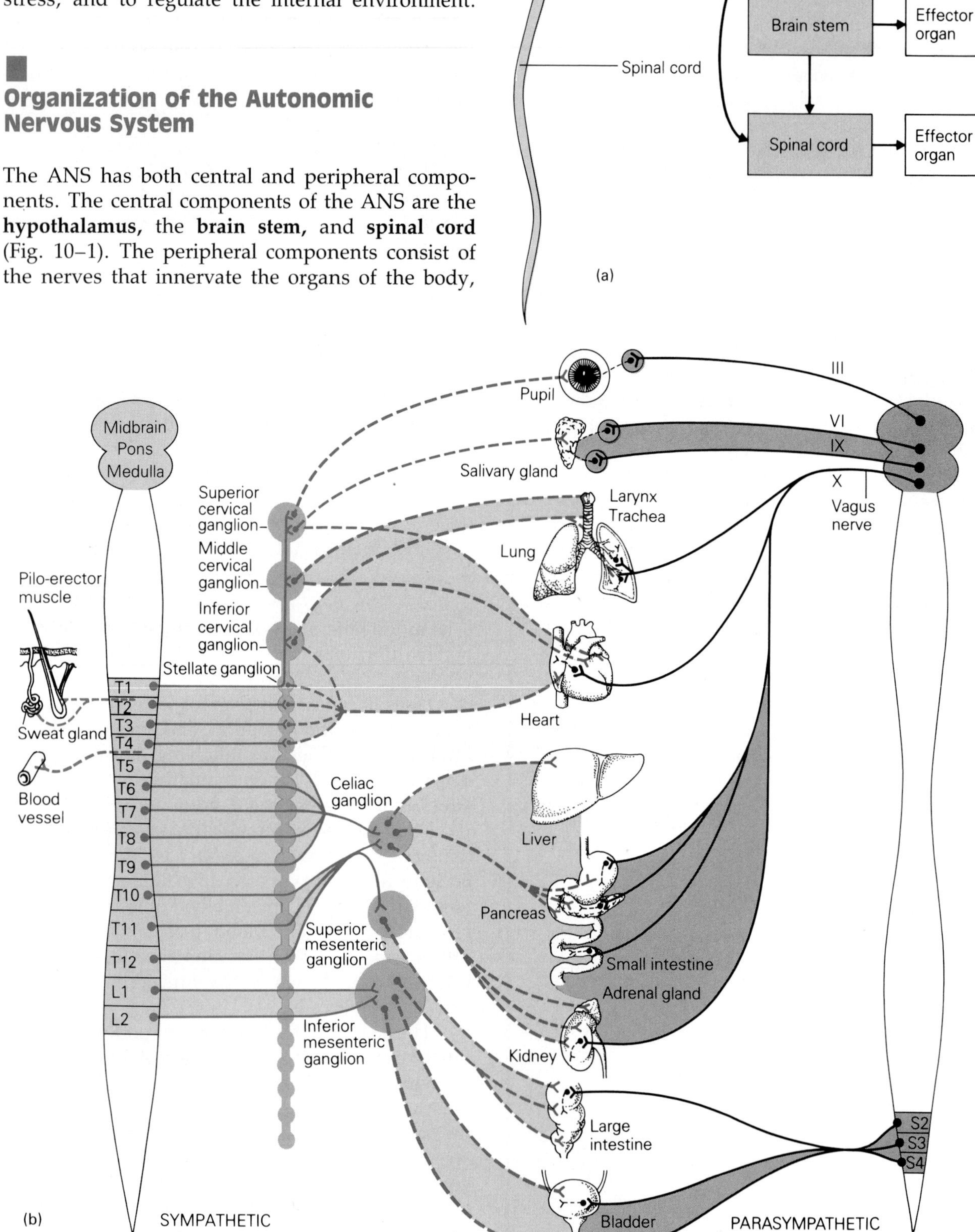

Figure 10–1 The principal components of the autonomic nervous system (ANS) are the hypothalamus, brain stem, and spinal cord (*a*). The ANS comprises the sympathetic and parasympathetic nervous systems (*b*). Sympathetic pathways are shown in orange, parasympathetic pathways in black. Dotted lines indicate postganglionic nerves.

classified as either parasympathetic or sympathetic nerves. Autonomic nerve fibers include both afferent (sensory) and efferent (motor) neurons.

The Peripheral Autonomic Nervous System

The peripheral ANS is functionally composed of the **sympathetic nervous system** and the **parasympathetic nervous system.** The sympathetic division of the ANS coordinates the body's response to stress, while the parasympathetic division coordinates the body's more "vegetative" activities such as digestion. The sympathetic and parasympathetic nervous systems not only have distinct anatomical differences, but also release different neurotransmitters at their target sites. The sympathetic nervous system simultaneously activates more organ systems than its parasympathetic counterpart. It plays an important role during stress responses where there is a need to coordinate increases in heart rate, blood pressure, and rate of respiration. In parasympathetic responses, in contrast, different organ systems are activated more independently.

Nerve fibers of the sympathetic nervous system emerge from the spinal cord at the thoracic and lumbar levels. The cell bodies of these fibers are located in the **interomedial lateral nuclei (IML)** of the spinal cord (Fig. 10–2). These cells are called **preganglionic nerve cells** and have short axons that innervate cells in **ganglia** located near the spinal cord. The cells within the ganglia are **postganglionic nerve cells.** The postganglionic cells have long axons and synapse with cells in other organ systems. They release **norepinephrine** as their neurotransmitter.

Fibers from the parasympathetic nervous system exit from the brain stem and also leave the spinal cord at the level of the sacrum. These preganglionic fibers have long myelinated and unmyelinated axons and innervate postganglionic nerve cells that are clustered near or in the target organ. The postganglionic parasympathetic nerve cells have very short axons that release **acetylcholine** at synapses.

As Figure 10–1*b* illustrates, most organs are innervated by both sympathetic and parasympathetic nerve fibers. The effects of each system on the target organ, however, are quite different. One system increases the activity of the target organ, while the opposing system decreases its activity. For example, the sympathetic innervation of the heart increases the heart rate, while parasympathetic innervation causes the heart rate to decrease. In contrast, sympathetic innervation of the small intestines decreases intestinal motility, while parasympathetic innervation causes it to increase. From this description, the actions of the sympathetic and parasympathetic nervous systems seem to oppose each other. However, the sympathetic and parasympathetic inputs to a target organ are not active at the same time; as one system is activated, the activity of the other is diminished. Later in this chapter, we will examine the mechanisms that coordinate this type of activity.

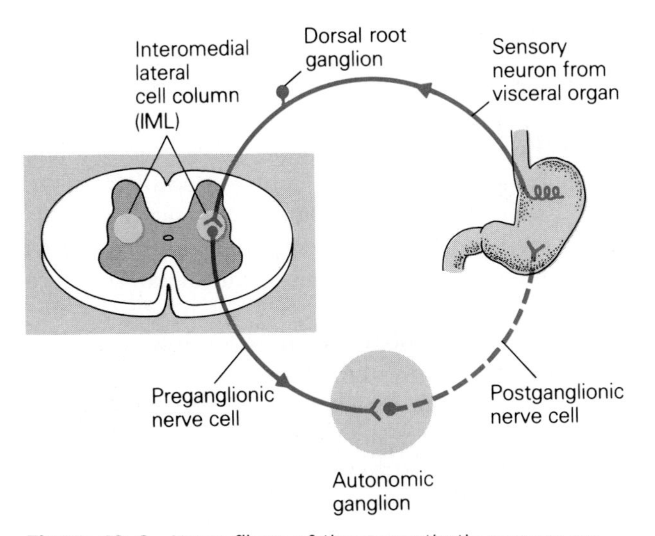

Figure 10–2 Nerve fibers of the sympathetic nervous system originate in the interomedial lateral nuclei (IML) of the spinal cord. These preganglionic nerve cells innervate postganglionic nerve cells located within ganglia near the spinal cord.

The Central Autonomic Nervous System

The organization of the ANS in the spinal cord is somewhat similar to that of the somatic nervous system. The preganglionic nerve cells are located in the interomedial lateral cell column and have fibers that leave the spinal cord from the ventral roots, while sensory fibers that carry information from the target organ enter the spinal cord through the dorsal roots. The cell bodies of these sensory fibers are located in the dorsal root ganglia. This spinal organization is identical for the sympathetic and the parasympathetic components of the autonomic nervous system. Sympathetic fibers, however, emerge from the thoracic and lumbar segments of the spinal cord, while parasympathetic fibers emerge from sacral segments of the cord.

The organization of the autonomic nervous system in the brain stem is different from that in the spinal cord. The autonomic sensory fibers coming into the brain stem are segregated from the somatic sensory fibers, which enter more laterally than autonomic sensory fibers (Fig. 10–3). On the other hand, autonomic efferent nerves leave the brain stem in more lateral regions than somatic efferents.

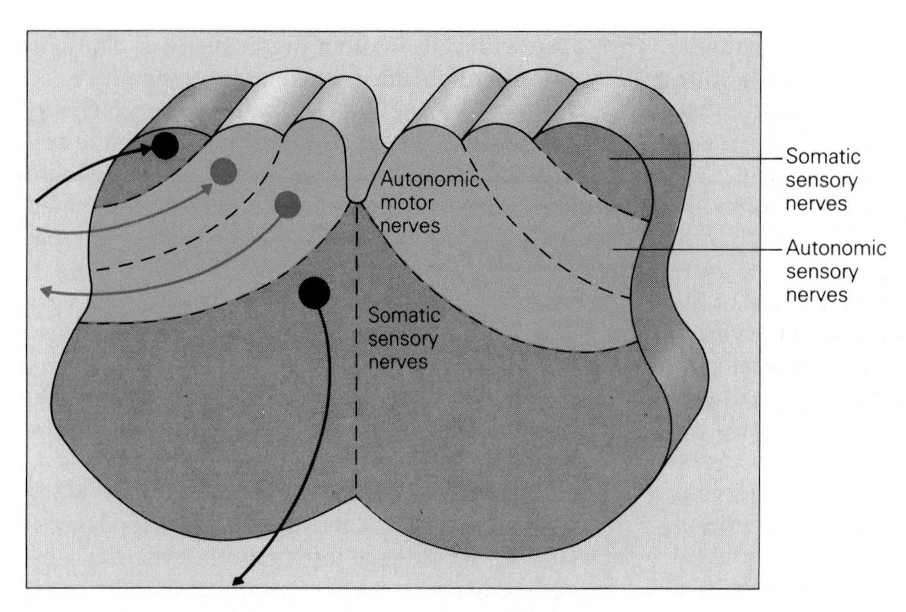

Figure 10–3 Autonomic sensory nerves enter and leave the brain stem from different regions than somatic sensory nerves do.

Neurotransmitters in the Autonomic Nervous System

The two major neurotransmitters in the autonomic nervous system, already mentioned, are **acetylcholine** and **norepinephrine.** Acetylcholine is released at the synaptic terminals of all preganglionic neurons (Fig. 10–4). In postganglionic nerve cells, norepinephrine is found in the terminals of sympathetic fibers, while acetylcholine is found in the terminals of parasympathetic fibers. Nerve fibers that transmit acetylcholine are sometimes called **cholinergic** fibers; those that transmit norepinephrine are sometimes referred to as **adrenergic** fibers.

The acetylcholine receptor on the target cell of the sympathetic nervous system, the postganglionic neuron, is different from the acetylcholine receptor on target cells of the parasympathetic nervous system. Cholinergic receptors are classified as either **nicotinic** or **muscarinic** acetylcholine receptors based on whether they are activated by either the drugs nicotine or muscarine. The acetylcholine receptor on the postganglionic neuron can be activated by nicotine and is thus a nicotinic receptor, while the acetylcholine receptor on target cells such as smooth muscle is muscarinic. Norepinephrine activates two types of receptors, **alpha receptors** and **beta receptors.** Beta receptors are distinguished from alpha receptors in that they are more sensitive to isoproterenol, a drug similar to norepinephrine. Table 10–1 lists different organs and the effects of acetylcholine and norepinephrine on them.

Control of Bodily Organs by the Sympathetic and Parasympathetic Systems

The actions of the sympathetic and parasympathetic inputs to an organ have opposite effects. These actions operate to control various parameters in the body. For instance, when the blood pressure becomes too high, the parasympathetic inputs are activated to a decrease the heart rate. They accomplish this by releasing acetylcholine. When acetylcholine binds to muscarinic receptors located on **pacemaker cells,** which are cardiac cells that regulate heart rate, these cells decrease their rhythmic rate of action potential discharge. This occurs because the muscarinic receptors, once acetylcholine binds to them, allow the K^+ ions to escape from the cells, causing them to become hyperpolarized. This slows their pacemaker activity, decreases the heart rate, and ultimately decreases the blood pressure (see Chapter 18).

When the blood pressure is too low, the sympathetic nervous system steps in. The sympathetic inputs to the heart release norepinephrine. The binding of norepinephrine to beta receptors on cardiac cells causes these cells to depolarize faster, resulting in an increase in heart rate. In addition, norepinephrine also causes the heart to contract with greater force. These mechanisms result in a compensatory increase in blood pressure (see Chapter 19).

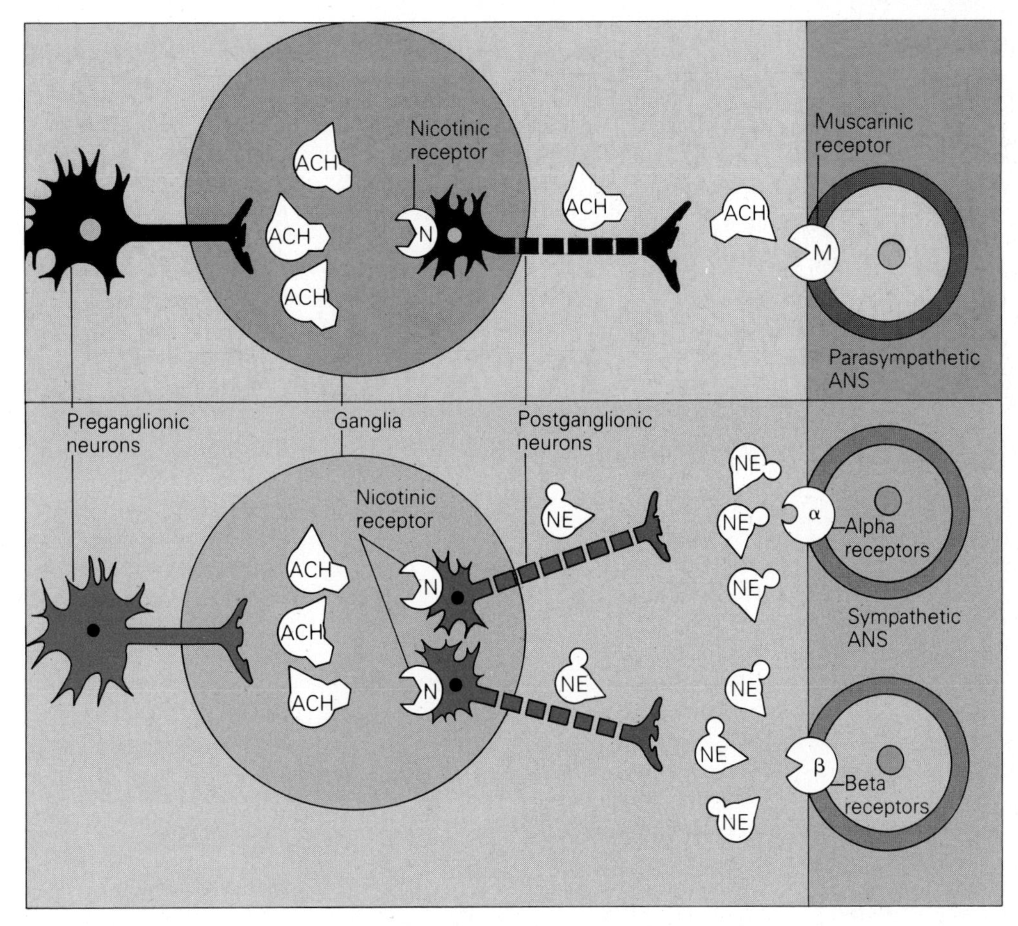

Figure 10–4 All preganglionic neurons release acetylcholine (ACH). Postganglionic sympathetic nerve cells release norepinephrine (NE), while postganglionic parasympathetic nerve cells release ACH. Receptors for ACH are either muscarinic (M) or nicotinic (N); receptors for NE are either alpha or beta receptors.

Table 10–1
Summary of Cholinergic and Adrenergic Influences on Effector Organs*

Effector Organs	Cholinergic Impulses	Adrenergic Impulses	
	Response	Receptor type†	Response
Eye			
Radial muscle of iris	—	α	Contraction (mydriasis)
Sphincter muscle of iris	Contraction (miosis)		—
Ciliary muscle	Contraction for near vision	β	Relaxation for far vision
Heart			
S–A node	Decrease in heart rate	β	Increase in heart rate
Atria	Decrease in contractility, and (usually) increase in conduction velocity	β	Increase in contractility and conduction velocity
A–V node and conduction system	Decrease in conduction velocity		Increase in conduction velocity
Ventricles	—	β	Increase in contractility, conduction velocity, automaticity, and rate of idiopathic pacemakers
Blood vessels			
Coronary	Dilation	α	Constriction
		β	Dilation

Table 10–1 (continued)

Effector Organs	Cholinergic Impulses	Adrenergic Impulses	
	Response	Receptor type†	Response
Skin and mucosa	—	α	Constriction
Skeletal muscle	Dilation	α	Constriction
		β	Dilation
Cerebral	—	α	Constriction (slight)
Pulmonary	—	α	Constriction
Abdominal organs	—	α	Constriction
	—	β	Dilation
Kidney	—	α	Constriction
Salivary glands	Dilation	α	Constriction
Lung			
Bronchial muscle	Contraction	β	Relaxation
Bronchial glands	Stimulation		Inhibition (?)
Stomach			
Motility and tone	Increase	β	Decrease (usually)
Sphincters	Relaxation (usually)	α	Contraction (usually)
Secretion	Stimulation		Inhibition (?)
Intestine			
Motility and tone	Increase	α, β	Decrease
Sphincters	Relaxation (usually)	α	Contraction (usually)
Secretion	Stimulation		Inhibition (?)
Gallbladder and ducts	Contraction		Relaxation
Urinary bladder			
Detrusor	Contraction	β	Relaxation (usually)
Trigone and sphincter	Relaxation	α	Contraction
Ureter			
Motility and tone	Increase (?)		Increase (usually)
Uterus	Variable‡	α, β	Variable‡
Male sex organs	Erection		Ejaculation
Skin			
Pilomotor muscles	—	α	Contraction
Sweat glands	Generalized secretion	α	Slight, localized secretion§
Spleen capsule	—	α	Contraction
Adrenal medulla	Secretion of epinephrine and norepinephrine		—
Liver	—	β	Glycogenolysis
Pancreas			
Acini	Secretion		—
Islets	Insulin secretion	α	Inhibition of insulin secretion
		β	Insulin secretion
Salivary glands	Profuse, watery secretion	α	Thick, viscous secretion
Lacrimal glands	Secretion		—
Nasopharyngeal glands	Secretion		—
Adipose tissue	—	β	Lipolysis
Juxtaglomerular cells	—	β	Renin secretion

*Modified from Goodman & Gilman, *The Pharmacological Basis of Therapeutics*. 4th ed. New York: Macmillan, 1970.
†Where adrenergic receptor type has been established.
‡Depends on stage of menstrual cycle, amount of circulating estrogen and progesterone, and other factors. Responses of pregnant uterus different from those of nonpregnant
§On palms of hands and in some other locations ("adrenergic sweating").

Blood pressure is also controlled by the contraction and relaxation of smooth muscles that surround blood vessels (Fig. 10–5). These smooth muscles respond to both norepinephrine and acetylcholine. Alpha-1 norepinephrine receptors cause contraction of smooth muscles, which increases blood pressure (see Chapter 19). Norepinephrine beta-1 receptors, on the other hand, cause relaxation of the smooth

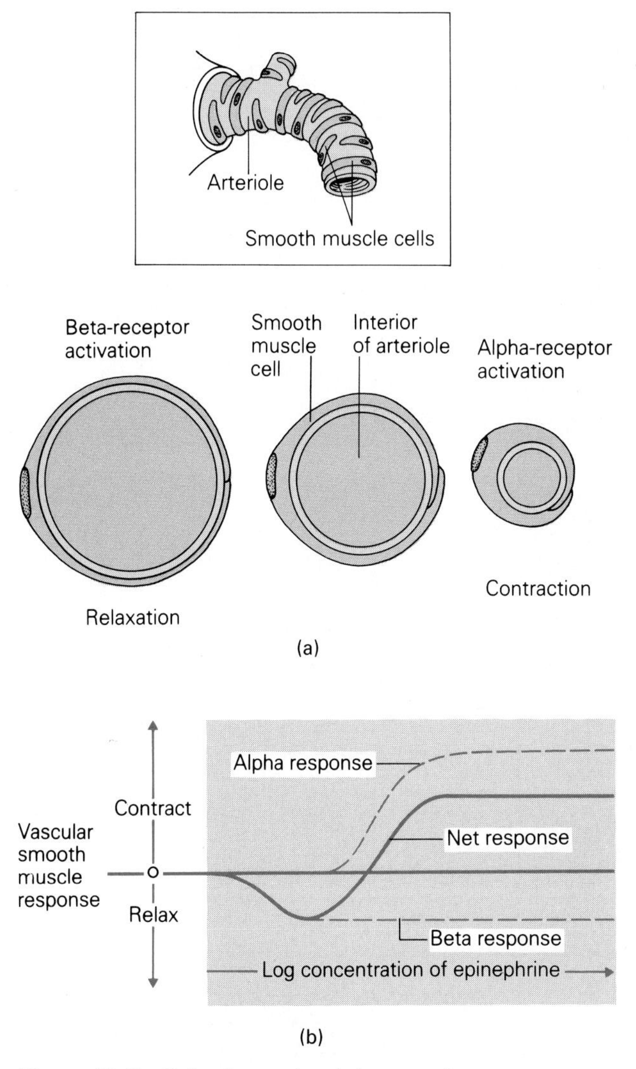

Figure 10–5 Alpha-1 norepinephrine receptors cause smooth muscle cells that line blood vessels to contract, reducing the inner space in the vessel and thereby increasing blood pressure. Beta-1 norepinephrine receptors cause smooth muscle cells to relax, thus widening vessels and decreasing blood pressure (*a*). The net response, however, is a contraction of smooth muscle cells (*b*).

muscle and decreased blood pressure. The relaxation of some blood vessels in smooth muscle can also occur by the activation of muscarinic acetylcholine receptors of the parasympathetic nervous system. Thus, smooth muscles are relaxed by the activation of either muscarinic or beta-1 receptors. Reduction of alpha-1 receptor activation can also produce relaxation. This is the mechanism typically found in the increase in blood flow caused by a decrease in the vasoconstriction of blood vessels.

Receptors for norepinephrine and acetylcholine also exist in the membranes of the presynaptic terminals. The alpha-2 receptors are found on synaptic terminals that release acetylcholine. These receptors inhibit the release of acetylcholine. Muscarinic receptors are found on synaptic terminals that release norepinephrine. Activation of these receptors inhibit the release of norepinephrine (Fig. 10–6). These receptors therefore allow the ANS to enhance the effect of either the sympathetic or parasympathetic actions while inhibiting the opposite action by local mechanisms at target organs. For example, when sympathetic inputs to the heart release norepinephrine, causing the heart to beat faster, the norepinephrine also binds to presynaptic alpha-2 receptors located on acetylcholine terminals. The activation of the alpha-2 receptors inhibits the release of acetylcholine, which would normally decrease the heart rate. Likewise, when parasympathetic inputs to the heart release acetylcholine and reduce the heart rate, acetylcholine also binds to muscarinic receptors located on norepinephrine terminals. The activation of these muscarinic receptors decreases the amount of norepinephrine released, which would typically increase the heart rate. Thus, the antagonistic actions of the sympathetic and parasympathetic nervous systems are controlled at the site of the target organ.

A more complex coordination of sympathetic and parasympathetic responses is found at the central levels of the ANS. At the level of the spinal cord and brain stem, these responses are viewed as **autonomic reflexes,** while at the hypothalamic level the responses are viewed as **autonomic control systems.**

Reflexes Governed by the Autonomic Nervous System

The Spinal Autonomic Reflex Arc

Earlier in this chapter, we discussed the sensory connections from the target organ to the spinal cord. This pathway transmits information about the status of the target organ to the spinal cord. Within the spinal cord, this information is then transmitted up to the brain and to neurons within the IML of the spinal cord. Since the cells in the IML provide the output from the ANS to the target organs, a loop of nerve cell activity is therefore established; this forms the **spinal autonomic reflex arc** (Fig. 10–7). The formation of this neuronal circuit starts with the detection of visceral information by autonomic sensory receptors in the organ. Sensory fibers in turn activate interneurons within the spinal cord. These neurons synapse with the cells of the IML. The IML cells integrate incoming information from many sensory neurons that enter the spinal cord within one spinal cord segment and discharge to activate

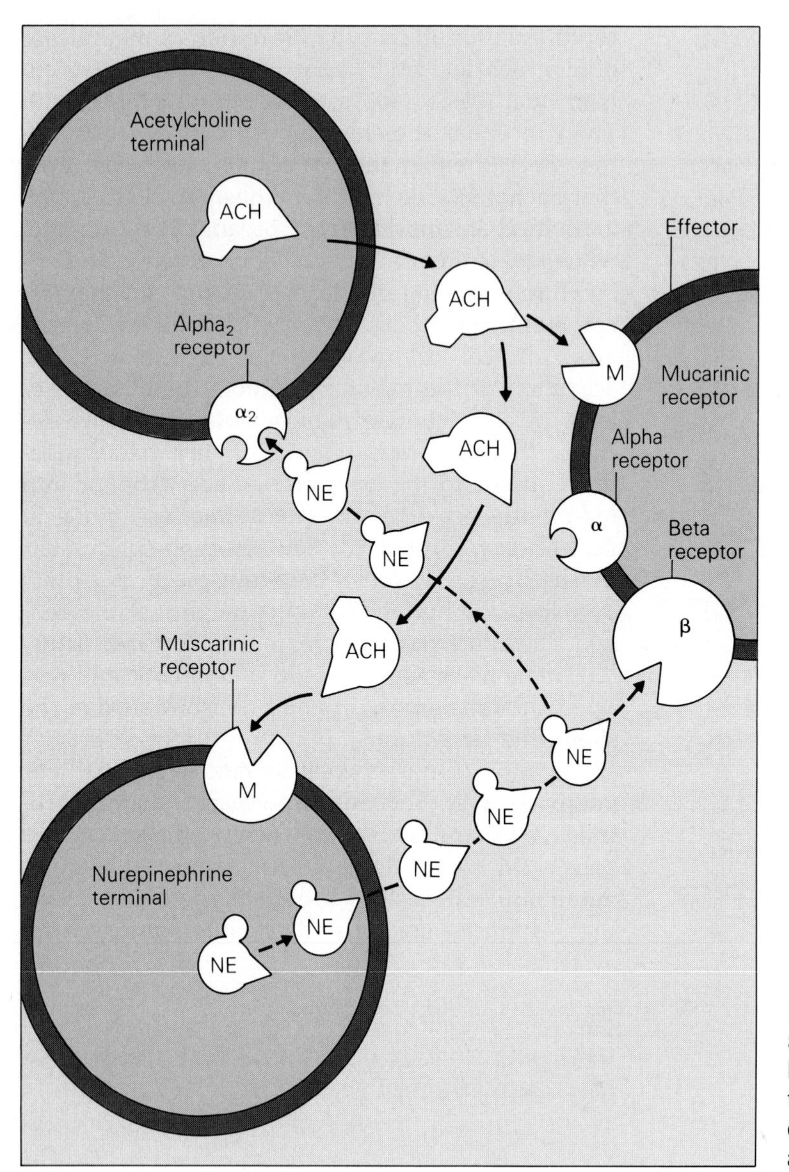

Figure 10–6 Receptors for norepinephrine and acetylcholine located in the presynaptic membranes can inhibit the release of these neurotransmitters, allowing the ANS to enhance the effect of a sympathetic or parasympathetic action.

postganglionic neurons. Finally, the postganglionic cells convey information to the target organ, causing it to make the appropriate responses. This reflex arc is quite similar to other reflex circuits such as the knee-jerk reflex of the somatic nervous system (see Chapter 9).

The particular response in the autonomic reflex system varies with the target organ. In the bladder, this reflex is responsible for urination; in the colon and rectum, this reflex is responsible for defecation.

The Urinary Bladder Reflex

The bladder stores urine that is produced by the kidneys. It fills at a rate of approximately 50 ml per hour and can store about 150 to 200 ml before the urge to urinate occurs. The ability to keep urine in

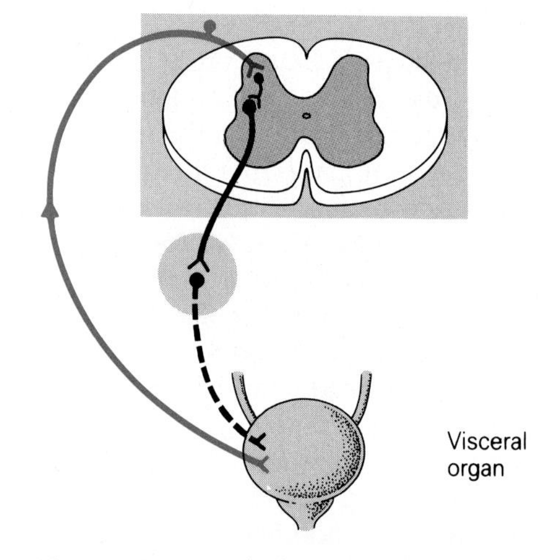

Figure 10–7 A spinal autonomic reflex arc.

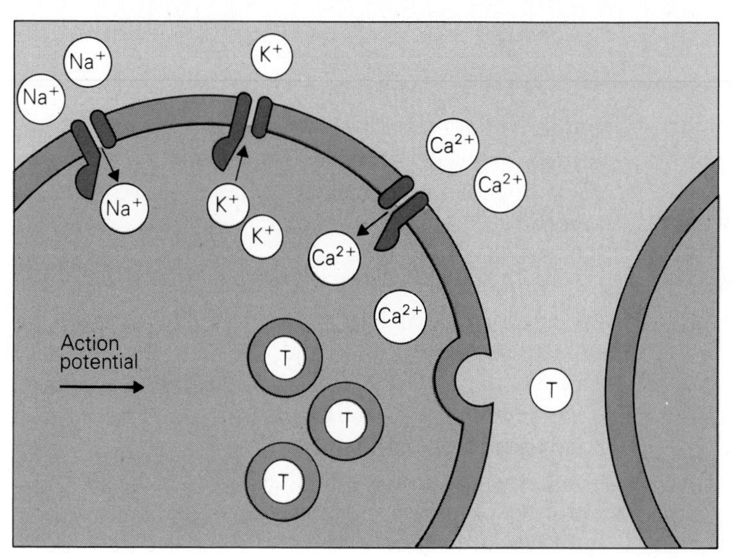

(a)

(b)

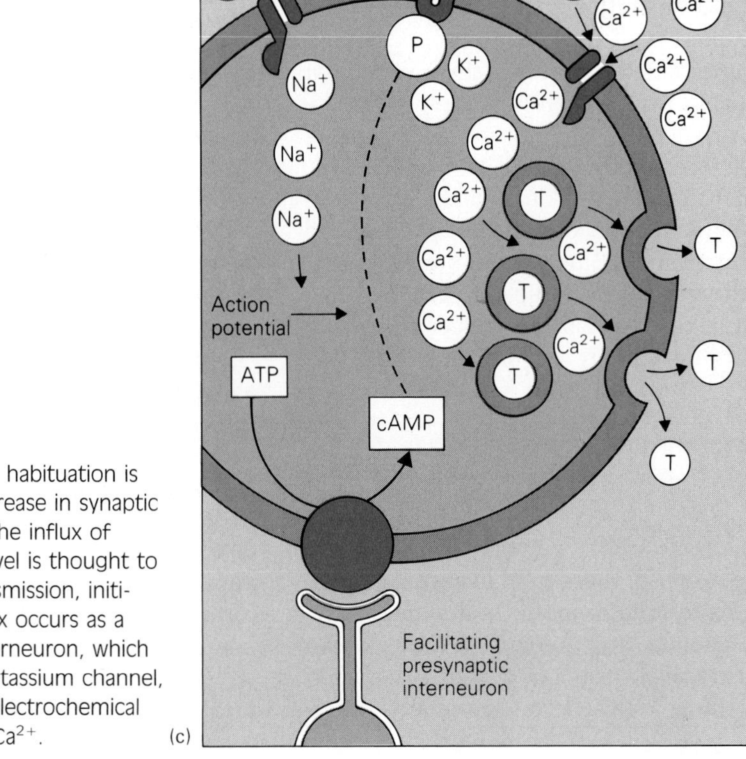

(c)

Figure 11–15 At the cellular level, habituation is thought to occur because of a decrease in synaptic transmission due to a decrease in the influx of Ca^{2+}. Sensitization at the cellular level is thought to involve an increase in synaptic transmission, initiated by an influx of Ca^{2+}. This influx occurs as a result of action by a facilitating interneuron, which causes cAMP to phosphorylate a potassium channel, thereby closing it and creating an electrochemical gradient favorable to the influx of Ca^{2+}.

flux effectively prolongs the action potential and as a consequence allows the Ca^{2+} channels to stay open longer. With an increase in Ca^{2+} influx, more vesicles release their transmitter contents into the synaptic cleft. The sensitization process is also associated with an increase in the number of synapses with active zones and a doubling in the area of the active zones. These changes, in conjunction with the enhancement in Ca^{2+} influx, suggest that an overall increase in transmitter release is the basis of the sensitization process.

Classical Conditioning: A Sensitization Process

Although there may be a variety of different cellular mechanisms of classical conditioning for different memory circuits, one mechanism has been found to be similar to that of the sensitization process. This mechanism leads to an enhanced response to a conditioned stimulus. The activation of a conditioned stimulus by this mechanism not only excites the postsynaptic neuron, but also temporarily raises the intracellular level of Ca^{2+} (Fig. 11–16). As we saw in Chapter 5, an increase in intracellular Ca^{2+} levels can activate a calmodulin-mediated increase in adenylate cyclase activity. As a consequence, cAMP levels are amplified, leading to the phosphorylation and inactivation of K^{+} channels and a prolongation of the action potential. As in the sensitization response, the prolongation of the action potential results in a greater Ca^{2+} influx at the terminals and an enhancement in transmitter release.

Thus, the pairing of the unconditioned stimulus and the conditioned stimulus activates cellular processes that alter the excitability of the postsynaptic cell. As a consequence, the CS input alone is ultimately capable of activating the cell.

The cellular mechanisms of the classically conditioned eye-blink response previously described remain to be determined. Since the response is associated with a decrease in Purkinje cell activity, however, the cellular mechanism of this conditioning process would most likely be similar to that of habituation.

The Effects of Use and Nonuse of Neural Pathways

During muscle exercise, there is an eventual change in muscle tone and muscle mass. In a similar way, the use and nonuse of neural pathways can strengthen or weaken the connections between nerve cells. Animals raised in "enriched" environments have nerve cells that form more synaptic connections with other nerve cells than would normally appear. In contrast, animals raised in "impoverished" environments develop fewer synapses than normal.

The use and nonuse of pathways can also alter topographical maps in the cortex. The map of the somatic sensory cortex and the motor cortex, for example, can be modified when an animal is trained to use one finger more than the others. The amount of cortex representing the function of the exercised finger expands beyond that which is normally found. In contrast, if the nerve fibers innervating the finger are severed, the cortical space devoted to that finger is gradually eliminated and eventually incorporated into the maps of other fingers.

The results of these studies indicate that the brain is a dynamic organ system, capable of changing the strength of existing synaptic connections and in some cases forming new synapses as part of the process of learning and memory. Many of these alterations in synaptic function can take place only during certain critical periods of development, while other modifications can occur at all ages. These observations argue for continued learning and exposure to new challenges, regardless of age.

Language Systems

Language is a tool that allows the communication of thoughts, knowledge, feelings, and emotions in both written and spoken form. The deciphering of words from complex sounds or from complex visual symbols and the extraction of concepts from these words represent activities at the highest level of sensory integration. Written language allows us to transfer knowledge from generation to generation so that each generation can build upon the achievements of the past.

The development of language has thus played a critical role in the progress of humankind. It is difficult to determine when language was first used; however, fossil evidence suggests that the areas of the brain that are responsible for language existed over 500,000 years ago. Spoken language obviously developed before written language. Consequently, another leap in human evolution may have been the formation of connections between the visual and language areas of the brain, allowing people to develop symbolic writing as a method of communication.

ripherally transformed hormones are **angiotensin II** and **1,25 dihydroxyvitamin D_3,** both of which are formed by sequences of conversion steps involving several different tissues.

The concentration of any one hormone circulating in the blood is determined both by its rate of secretion and also by the rate at which it is removed or metabolized to an inactive form. For most hormones, the major pathways for removal or degradation occur in the liver or kidneys.

Steroid hormones are primarily degraded by the liver and the metabolites secreted by the liver into the bile, eventually being eliminated in the feces. The liver first converts the steroid into a relatively less active form and then, to make the molecule more water soluble, couples the steroid to a polar sulfate or glucuronide group in a process known as **conjugation.** The conjugated steroids are then primarily secreted by the liver into the bile, with smaller amounts being excreted in the urine. Other hormones, such as epinephrine, are inactivated by specific degrading enzymes circulating in the blood and are then rapidly excreted in the urine. Determination of the rates of urinary excretion of a variety of steroid hormones or epinephrine metabolites has proven to be a useful indirect index of the rate of secretion of the active hormone and has provided physicians with an important diagnostic tool.

Many of the larger peptide hormones, such as insulin and prolactin, are taken into their target tissue cells by a process known as **receptor-mediated endocytosis.** After the hormone binds to its receptor on the surface of target cells, the hormone-receptor complex is internalized or taken into the cell. Once internalized, the hormone is separated from its receptor. In most cases, the hormone molecule is then proteolytically degraded, and the majority of the internalized receptors are recycled back to the cell surface. Many peptide hormones are taken up and degraded by the kidneys, although the exact contribution made by the kidneys can vary considerably depending on the hormone.

Measurement of Hormone Concentrations in the Blood

As indicated earlier, hormones are present in extremely low concentrations in the blood, usually in the range of 10^{-9} to 10^{-12} M. To put these low concentrations in perspective, consider that a hormone concentration of 10^{-9} M corresponds roughly to one hormone molecule in the blood per 50 billion water molecules. Thus, exceedingly sensitive techniques are needed to allow the measurement of circulating hormone concentrations. The measurement of circulating hormone concentrations, either in the basal (resting) state or after an appropriate stimulus has been given, is an important step in the diagnosis of many different endocrine disorders. Because of the similarity in structure and activity of many of the hormones, hormone assays (measurements) must be highly specific if meaningful results are to be obtained.

In the late 1950s, two American scientists, Solomon Berson and Rosalyn Yalow, developed an assay technique that is both extremely sensitive and highly specific. The technique can be performed quite rapidly and inexpensively. The assay that Berson and Yalow developed is known as the **radioimmunoassay (RIA).** The basic technique, described in Chapter 3, has been modified and adapted in a number of different ways, but these newer assays are still based on the original principal of **competition binding,** in which the amount of radioactive hormone that binds to an antibody reveals how much nonradioactive hormone is present. This assay technique can be used to measure the concentrations of hormones in the plasma, and also for the measurement of a variety of drugs and even vitamins. As a result, the radioimmunoassay has revolutionized the field of endocrinology both clinically as well as in the area of basic research. For their efforts in the development of the technique, Berson and Yalow were awarded the Nobel Prize in Physiology and Medicine in 1977.

Mechanisms of Hormone Action

For convenience, hormones can be considered as producing their effect by one of two different general mechanisms of action. One mechanism, used by amine and peptide hormones, involves membrane receptors and the generation of an intracellular signal or "second messenger" (see Chapter 5). These hormones take effect by altering (activating) proteins that already exist in the cell. The other mechanism, used by steroid hormones and thyroxine, uses intracellular receptors and involves alterations in the expression of particular genes within the nucleus of the cell. These hormones work by initiating production of new proteins in the cell. These two models of hormone action will be considered in more detail shortly. First, let us look at the nature of hormone receptors.

Properties of Hormone Receptors

As was described earlier, a receptor is a molecule inside the cell or in the plasma membrane that specifically recognizes and binds one particular hormone. Binding of the hormone to the receptor initiates a sequence of events that ultimately results in the biological effects typical of that hormone. Therefore, specificity within the endocrine system is determined at the level of the hormone receptor.

Receptors are themselves proteins, or in some cases glycoproteins, and thus have definite three-dimensional structures. As illustrated in Figure 12–3, only those hormones that have a structure complementary to the structure of a specific receptor can bind to and activate the receptor. A very simple key and lock analogy works here: the key functions like a hormone, and a lock is analogous to the receptor.

The ability of receptors to discriminate between hormones is not absolute, however. Similarities in structure between hormones can result in an overlap of hormonal activities. For example, cortisol primarily produces effects typical of a glucocorticoid, but it also interacts weakly with aldosterone receptors and therefore is considered a weak mineralocorticoid. This type of overlapping activity is fairly common among the steroids, owing to the numerous similarities in structure among these hormones. Similar, although fewer, examples of overlapping activities can be found among the peptide hormones. Growth hormone and prolactin have similar structures, and each displays a weak degree of the other's activity. Thus, while receptors are responsible for the specificity that exists within the endocrine system, the inability of the receptor to discriminate absolutely between hormones having similar structures leads to some overlap of activities. Normally, these weak secondary activities are of little or no physiological consequence. However, in an abnormal situation in which the circulating concentration of a particular hormone is markedly elevated, these secondary activities of the hormone may be strong enough to produce noticeable biological effects.

The binding of a hormone to its receptor involves weak chemical forces and is usually readily reversible. When a hormone binds to its receptor, the three-dimensional conformation of the receptor is altered. In this regard, a hormone functions in a manner similar to that of an allosteric modifier of an enzyme. The alteration in structure of the receptor is the first step in the sequence of events leading to the biological effects of the hormone. The magnitude of the biological response is therefore proportional to the number of receptors occupied by hormone molecules. However, receptors for any particular hormone are present only in a finite or limited number in their target tissue cells. It follows, therefore, that once a certain hormone concentration is reached at which all receptors are occupied, no greater response can be produced by further increases in the hormone concentration. Thus, for each biological effect of a hormone, there is a maximal response that can be produced by that hormone.

Current evidence indicates that all the biological effects of hormones are the result of interactions with their respective receptors. The hormones themselves do not directly affect other cellular constituents, such as enzymes, but do so only indirectly, using their receptors as intermediaries to produce the desired effects. The receptor therefore serves as a **signal transducer,** converting a specific extracellular hormonal signal into an intracellular signal that will redirect or alter cell function.

Given the role of the receptor in signal transduction, what is the nature of the intracellular signal generated by the receptor, and how does a single message or signal within the cell lead to a variety of different effects of the hormone?

The "Second Messenger" Model of Hormone Action

Peptide hormones are generally hydrophilic ("water-loving") and therefore are not soluble in the lipid bilayer of the plasma membrane. As a result, they are unable to readily penetrate the cell membrane and gain access to the interior of the cell. These hormones must therefore use mechanisms involving intracellular signals, or second messengers, to produce their effects. For this class of hormone, the receptor is an intrinsic membrane protein with its hormone-binding component exposed to the extracellular fluid. As illustrated in Figure 12–12, binding of the hormone to that portion of its receptor exposed to the outside of the cell results in generation of a second messenger. Many of the membrane receptors are glycoproteins oriented such that the carbohydrate portion of the molecule is exposed to the outside of the cell. As discussed in Chapter 5, several second messengers are known. Details of the mechanisms involved in the formation of each and the role they play in hormone action will be discussed shortly.

These surface receptors are not rigidly fixed in place in the membrane, as was once thought, but instead show considerable lateral mobility. In addition, as discussed in the earlier section on hormone degradation, a number of peptide hormones have been shown to undergo receptor-mediated endocytosis, in which both the receptor and the hormone are taken into the cell in an endocytotic vesicle. Binding of several peptide hormones to their receptors has also been shown to induce a clustering of

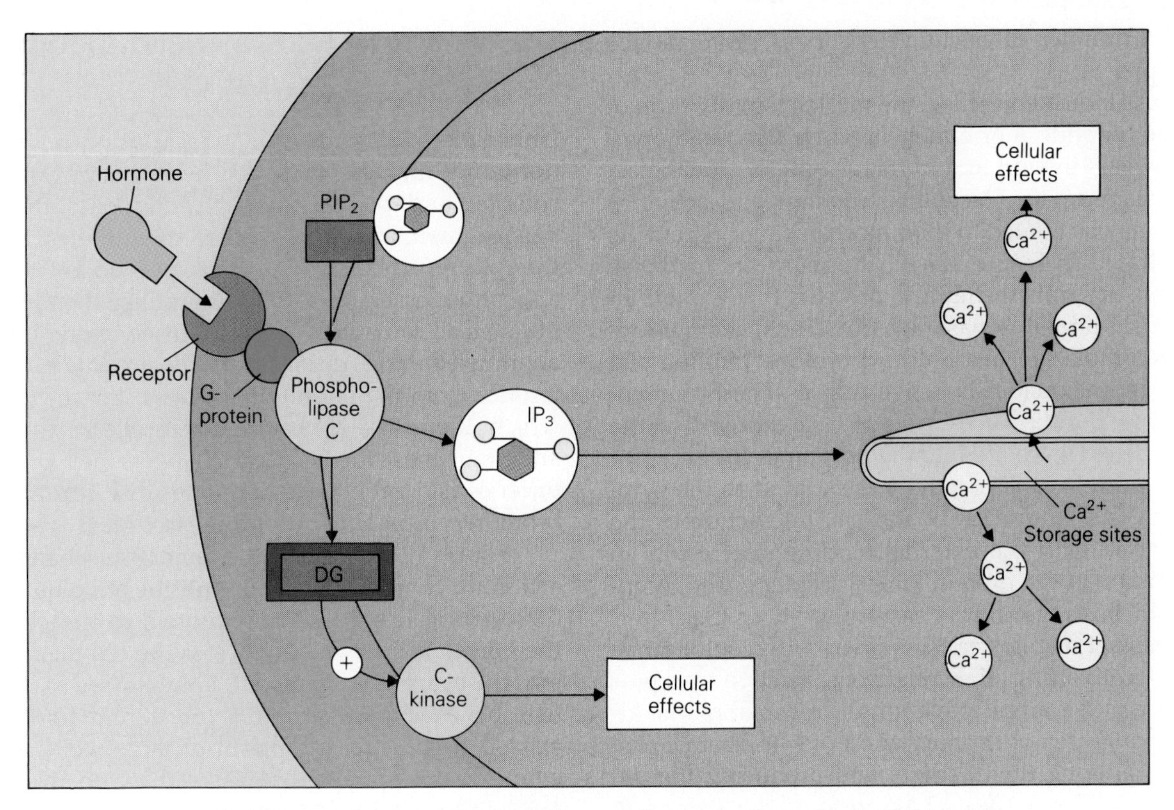

Figure 12–17 Diagram of the formation and the mechanisms of action of inositol trisphosphate (IP_3) and diacylglycerol (DG) as intracellular second messengers.

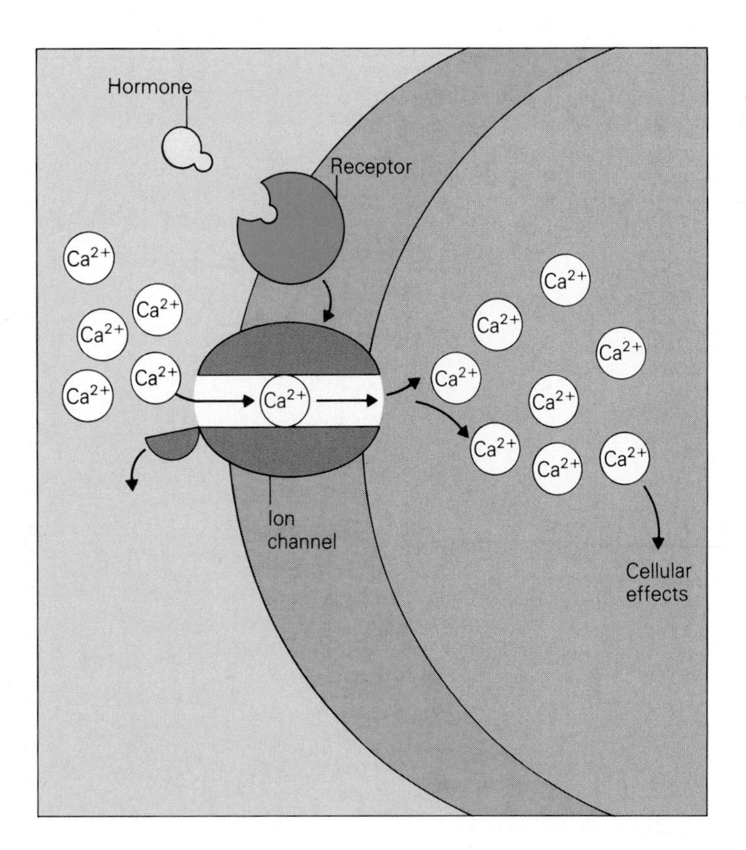

Figure 12–18 Coupling of membrane receptors to membrane calcium channels.

ing a number of calcium-dependent proteins (see Chapter 5).

As indicated earlier, the mechanism of action of some peptide hormones is currently unknown. However, insulin and certain peptide growth factors have unique receptor structures and activities that may be related to their mechanism of action. As illustrated in Figure 12–19, the receptors for these agents are transmembrane proteins that exhibit an intrinsic protein kinase activity. Thus, a single receptor molecule has both a hormone binding site and an enzymatically active region. The hormone-binding portion of the receptor is exposed to the outside of the cell, while the portion of the receptor with protein kinase activity is exposed to the cytoplasm. As Figure 12–19 depicts, interaction of the hormone with the external binding site results in activation of the protein kinase activity of the receptor. In this regard, it is easy to see how these hormones act as allosteric modifiers of receptor function, converting the relatively inactive receptor kinase into a more active form. Presumably, the kinase catalyzes phosphorylation of cytoplasmic proteins, altering their activity and producing the desired biological effects. At present, the degree to which these events are responsible for the biological effects of insulin and the growth factors is not completely understood.

The Gene Expression Model of Hormone Action

Unlike the hydrophilic peptide hormones, steroid hormones and thyroxine are hydrophobic, or lipid soluble, and therefore can readily pass through plasma membranes and other membranes within the cell. As a result, these hormones do not require a second messenger system to produce their effects, but instead can enter the cell themselves and bind to an intracellular receptor. In addition to this basic difference from the peptide hormones, the manner in which these hormones regulate cellular function and the time course of the changes they produce differ considerably from hormones that utilize membrane receptors and second messengers.

Figure 12–20 depicts the general mechanism of action for thyroid hormones and the steroids. After dissociating from its carrier protein in the plasma, the hormone freely moves across the cell membrane into the cytoplasm of the cell. For many years, scientists believed that receptors for steroid hormones existed free in the cytoplasm of target cells. However, recent evidence suggests that the receptors exist either on the nuclear envelope or within the nucleus of the cell. Binding of the hormone to the receptor causes a conformational change in the receptor such that the hormone/receptor complex

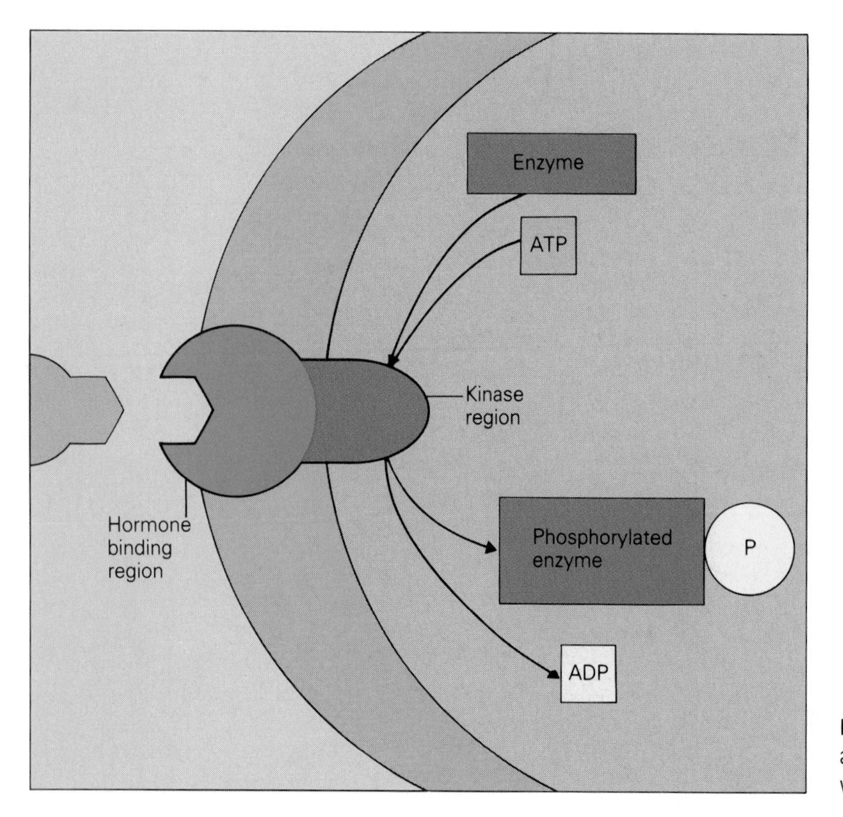

Figure 12–19 Possible mechanism of action of hormones that have receptors with intrinsic protein kinase activity.

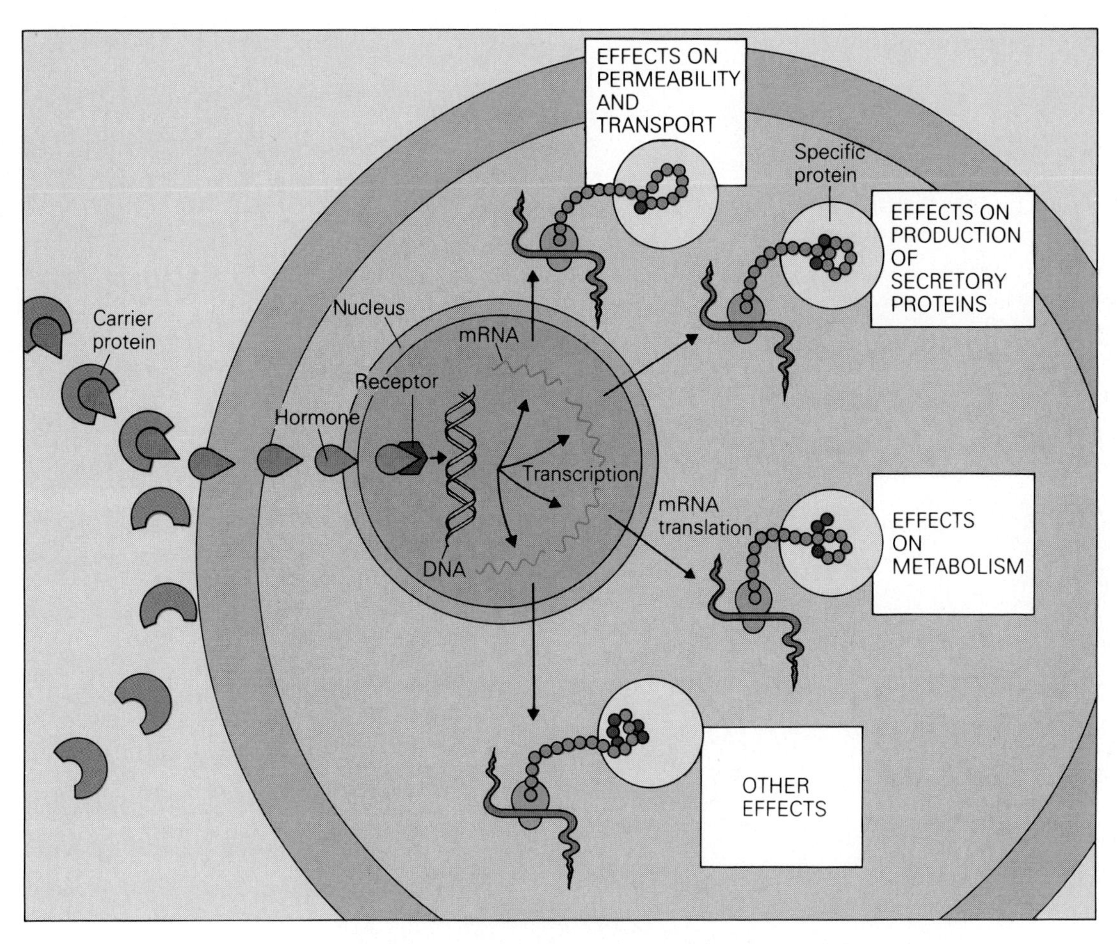

Figure 12–20 Mechanism of action of hormones that have intracellular receptors, indicating the effects on gene expression and the possible ways in which cell function might be altered.

then interacts with specific "acceptor sites" on the nuclear DNA. This interaction initiates transcription of mRNA molecules that code for specific proteins. After transcription, the mRNA precursors are processed within the nucleus before moving into the cytoplasm of the cell (see Chapter 5). Because of the increased abundance of these particular mRNA molecules, the synthesis of a few specific proteins is increased. Increased synthesis of these particular proteins leads to the altered cellular function characteristically produced by the particular hormone. As indicated in Figure 12–20, these newly synthesized proteins may be involved in a variety of different cellular functions.

The cellular response to a particular hormone acting via these mechanisms usually results from increased synthesis of several proteins. Although each of these proteins may serve a different role in the cell, each contributes in some way to producing the final coordinated biological effect. For example, aldosterone, a mineralocorticoid produced by the adrenal cortex, has the effect of increasing sodium transport across tubule cells of the kidney. As shown in Figure 12–21, sodium first enters the cell on the tubular side by diffusion through specific sodium channels and is then pumped out on the opposite side of the cell by an energy-requiring sodium pump. Aldosterone both increases the synthesis of several mitochondrial citric-acid-cycle enzymes involved in ATP production, making more energy available to power the sodium pump, and it also stimulates synthesis of sodium channel proteins, making more sodium available to the sodium pump. There is some evidence that aldosterone may also increase the synthesis of sodium pump proteins themselves. The overall effect is to increase the transfer of sodium across kidney tubule cells. Thus, as is typical of other steroid or thyroid hormones that take effect by regulating events within the nucleus, aldosterone alters the function of its target cells by changing the rate of synthesis and therefore the amount of several key proteins in the cell. Recall that, by contrast, hormones that act through membrane receptors using second messengers generally

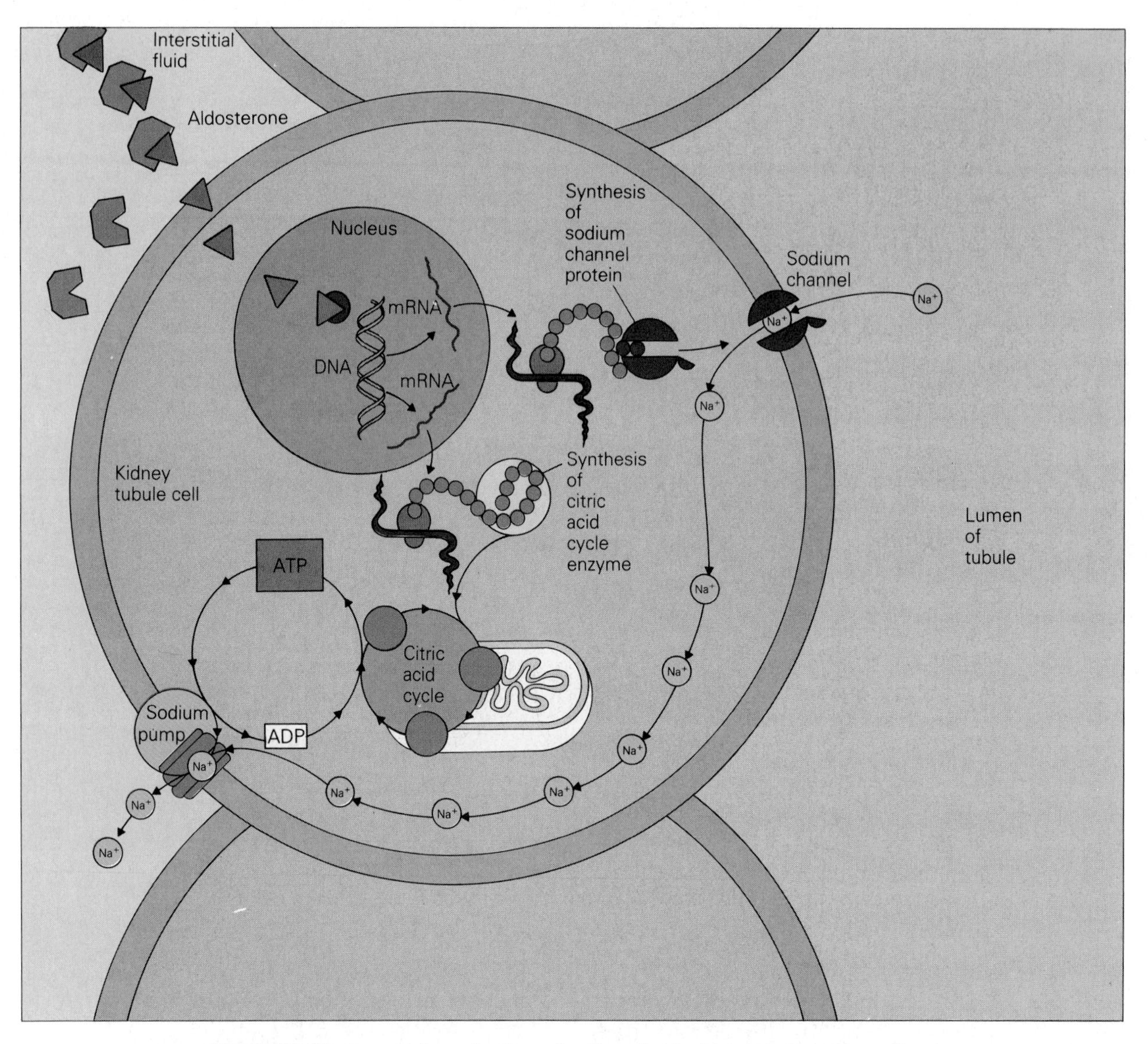

Figure 12–21 Apparent mechanism of action of aldosterone to increase sodium transport across kidney tubule cells.

do so by modifying the activity of pre-existing proteins, using such mechanisms as phosphorylation or dephosphorylation.

Another basic difference between the cell-surface and intracellular receptor systems involves the time course over which responses occur. In general, responses to hormones that use the second-messenger system can be observed in a matter of seconds or minutes. In addition, once the hormonal stimulus is withdrawn, cell function rapidly returns to its basal (resting) state. In contrast, responses to steroid or thyroid hormones are usually not evident for many minutes or several hours. In some cases, it may even take days for the response to develop. Likewise, after removal of the hormonal signal, it may take hours or days for the effect to diminish and the cells to return to their original state. With these two mechanisms, the endocrine system is able to closely regulate a number of cellular processes on both a short-term (acute) and a long-term (chronic) basis.

Amplification of Hormone Signals

One characteristic feature within the endocrine system is the amplification or increase in signal strength. As a result of this amplification, binding of just a few hormone molecules to their receptors can result in a dramatic change in cell function. Signal amplification occurs at nearly every step in hormone action, from binding of the hormone to its receptor to the final biological effect produced by the hormone. An example of signal amplification is illustrated in Figure 12–22. It is easy to see that after just a few steps, the original signal consisting of one hor-

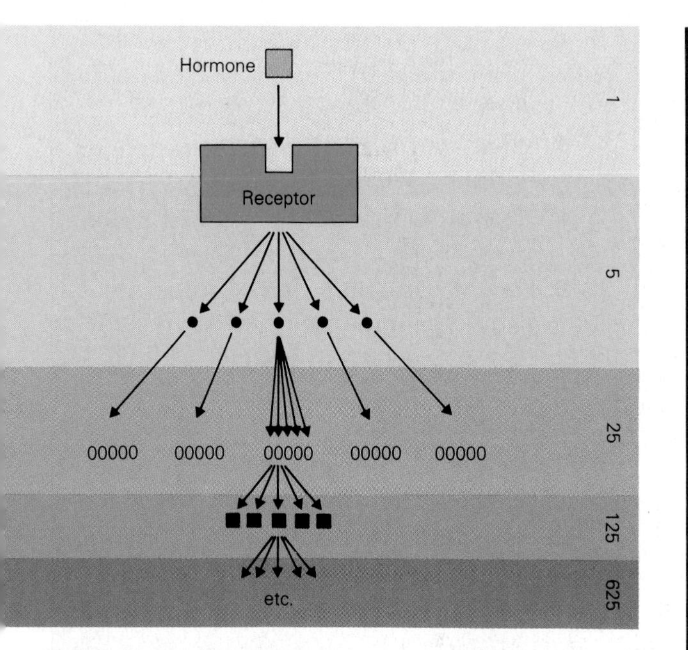

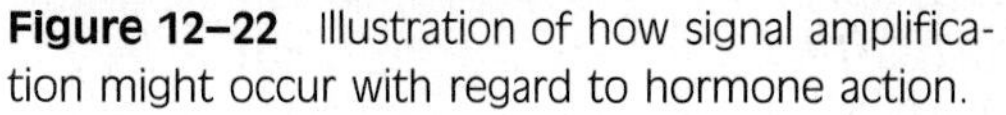

Figure 12–22 Illustration of how signal amplification might occur with regard to hormone action.

mone molecule has now been multiplied many times over. The amplification occurs because each step in the sequence involves enzymes that, when active, are responsible for forming not just one but many single "signal" molecules. For example, when glucagon binds to its receptor, it activates adenylate cyclase. Adenylate cyclase catalyzes the formation of many cAMP molecules, each of which in turn activates a particular protein kinase. Each cAMP-activated protein kinase will now catalyze the phosphorylation of many copies of a particular enzyme, leading to an alteration in the activity of many enzyme molecules. Thus, just one glucagon molecule can result in the activation of many individual cell proteins. This amplification allows hormones to be effective regulators of cell function even though they are present in the blood in extremely small amounts.

Summary

I. A. *Endocrinology* is the study of hormones and the glands that produce them, including the synthesis, secretion, and action of hormones and the clinical abnormalities produced by hormone dysfunction. Most substances considered to be hormones are secreted in small amounts into the blood and are then carried to other parts of the body, where they interact with receptors and produce a distinct biological response in the cells of their target tissues.

B. Neuroendocrine cells link between the *nervous* and *endocrine* systems by converting electrochemical signals within the nervous system into hormonal signals in the blood.

C. Feedback regulation in the endocrine system usually involves *negative feedback,* although positive feedback can also occur.

II. A. In general, hormones regulate processes related to either energy metabolism, salt and water metabolism, growth and development, or reproduction. Most hormones act on several tissues and produce several different specific effects.

III. A. Chemically, hormones generally fit into one of three categories: amino acid derivatives, peptide or protein hormones, and steroid hormones, which are derived from cholesterol. Steroids can be further subdivided into glucocorticoids, mineralocorticoids, androgens, estrogens, and progestins.

B. Steroid hormones and thyroid hormones travel through the blood by binding to specific carrier proteins and to albumin and other plasma proteins. Peptide and protein hormones do not require carrier proteins for transport.

C. Most hormones are secreted in a biologically active form. Steroid hormones are primarily degraded by the liver. Many of the peptide hormones are proteolytically degraded by their target tissues. The kidneys also contribute to the proteolytic degradation of several peptide hormones.

D. The radioimmunoassay allows for the measurement of hormone concentrations in the blood. Development of the assay was an extremely important advancement in the field of endocrinology.

IV. A. The receptor is the molecule in the cell that recognizes the hormone and initiates its biological effects. Receptors are responsible for the *specificity* that exists within the endocrine system, although the specificity is not absolute in every case.

B. Peptide and protein hormones generally act via plasma membrane receptors present on the cell surface. Binding of a hormone to its receptor on the cell surface results in the generation of an intracellular second messenger, which then alters, or regulates, cell function. Second messengers include cyclic AMP, inositol trisphosphate, diacylglycerol (DG), and ions, notably calcium.

C. *Steroid hormones* generally exert their effects via intracellular receptors located on or in the nucleus. The hormone/receptor complex interacts with the DNA to alter transcription of messenger RNA molecules coding for specific proteins. Alterations in the synthesis of these specific proteins leads to the particular cellular response characteristically produced by the hormone.

D. Signal amplification occurs at nearly every step in hormone action, allowing hormones to be effective regulators of cell function even though they are present in the blood in extremely small amounts.

Review Questions

1. What are the key features of a hormone? How are the endocrine, autocrine, and paracrine systems different?
2. How do neural, endocrine, and neuroendocrine cells receive and transfer information?
3. What is the role of the receptor in hormone action?
4. How do simple and multilevel negative feedback endocrine systems work?
5. What are three different chemical classifications of hormones? Provide examples of each.
6. What are the five different types of steroid hormones? Give at least one example of each.
7. What are the primary steps in steroid hormone synthesis?
8. How are hormones transported in the bloodstream?
9. How do receptors convey specificity within the endocrine system? What is the role of receptors in hormone action?
10. What are three second messenger models of hormone action? How do they work?
11. What is the general mechanism of action of steroid and thyroid hormones?
12. How does signal amplification occur as part of hormone action?

Suggested Readings

Berridge, Michael J. The molecular basis of communication within the cell. *Scientific American* (1986, 253:142–152).

Carafoli, Ernesto, and Pennington, John T. The calcium signal. *Scientific American* (1986, 253:70–78).

Fregley, Melvin J., and Luttge, William G. *Human Endocrinology: An Interactive Text.* New York: Elsevier Biomedical, 1982.

Hedge, George A., Colby, Howard C., and Goodman, Robert L. *Clinical Endocrine Physiology.* Philadelphia: W. B. Saunders Company, 1987.

Smith, Emil L., Hill, Robert L., Lehman, I. Robert, Lefkowitz, Robert J., Handler, Philip, and White, Abraham. *Principles of Biochemistry: Mammalian Biochemistry.* 7th ed. New York: McGraw-Hill Book Company, 1983.

Snyder, Solomon H. The molecular basis of communication between cells. *Scientific American* (1986, 253:132–141).

Tepperman, Jay, and Tepperman, Helen M. *Metabolic and Endocrine Physiology.* 5th ed. Chicago: Year Book Medical Publishers, Inc., 1987.

Walters, Marian R. Steroid hormone receptors and the nucleus. *Endocrine Reviews* (1985, 6:512–543).

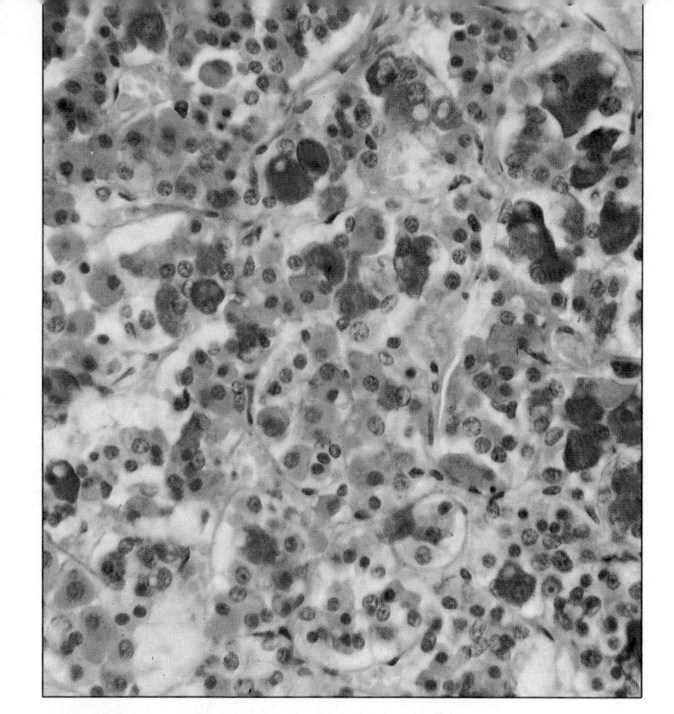

13 The Pituitary Gland

The pituitary gland plays a central role in the hormonal regulation of a wide variety of processes, and thus it has been one of the most extensively studied of all endocrine glands. As we will show, pituitary function is under the control of the hypothalamus. The interaction of the pituitary and hypothalamus is therefore an excellent example of the interaction between the nervous and endocrine systems.

We begin with a discussion of pituitary hormones and the regulation of pituitary function by the hypothalamus. Later sections of this chapter describe the actions of two pituitary hormones, growth hormone and thyroid-stimulating hormone.

Anatomy of the Pituitary and Its Relation to the Hypothalamus

Gross Anatomy of the Pituitary Gland

In humans, the pituitary measures about 1 cm (1/2″) in diameter and is located just below the hypothalamus at the base of the brain (Fig. 13–1). (As described in Chapter 10, the hypothalamus serves as an integrative center for a variety of different bodily processes. The hypothalamus also controls pituitary function and therefore serves as an important regu-

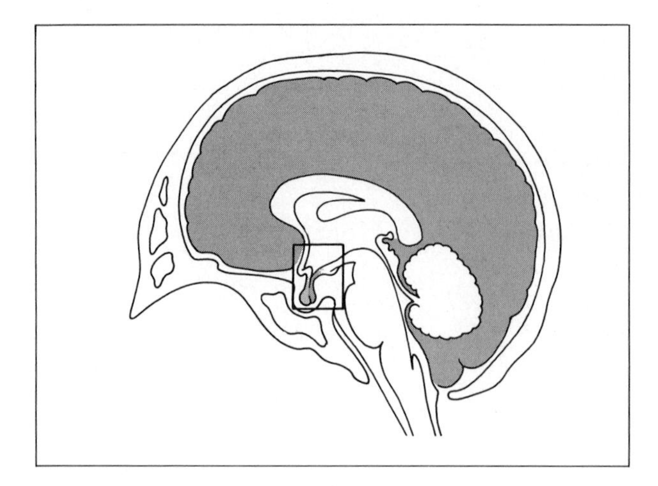

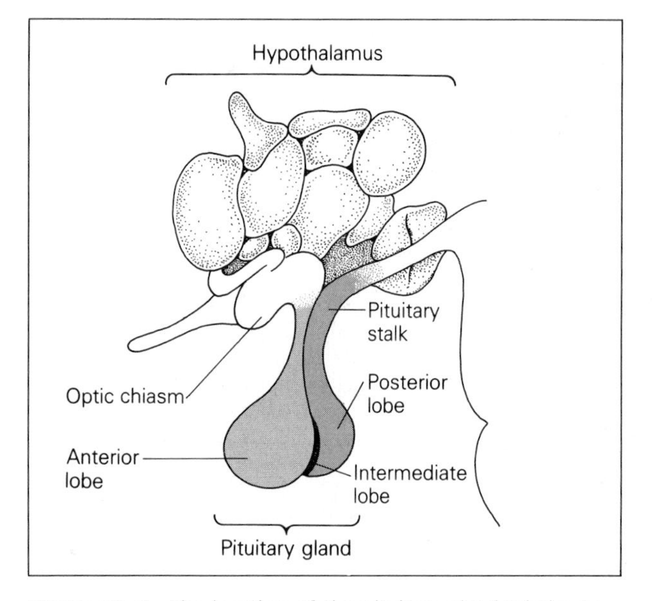

Figure 13–1 The location of the pituitary gland relative to the brain and hypothalamus.

lator for the endocrine system.) The pituitary is not a single gland, but consists of three separate glands, or lobes: the **anterior lobe,** the **intermediate lobe,** and the **posterior lobe.** The pituitary is connected to the hypothalamus by a thin segment of tissue known as the **pituitary stalk.** The hormones secreted by each of the three different lobes of the pituitary are listed in Table 13–1. As this table indicates, the pituitary gland secretes at least nine different hormones. However, as will be described later, the pituitary does not itself synthesize all nine hormones.

Regulation of hormone secretion by each of the lobes of the pituitary involves different mechanisms. Cells of the anterior pituitary function as true endocrine cells (see Fig. 12–1) in that they receive signals or information via the blood and in turn release their own hormones directly into the blood.

Table 13–1
Hormones Secreted by the Pituitary Gland

Hormones secreted from the anterior lobe
- Luteinizing hormone (LH)
- Follicle-stimulating hormone (FSH)
- Prolactin (PRL)
- Adrenocorticotropic hormone (ACTH)
- Growth hormone (GH)
- Thyroid-stimulating hormone (TSH)

Hormones secreted from the intermediate lobe
- Melanocyte-stimulating hormone (MSH)

Hormones secreted from the posterior lobe
- Oxytocin
- Antidiuretic hormone (ADH, also known as vasopressin)

The posterior pituitary, in contrast, is really an extension of the nervous system and functions like typical neuroendocrine cells. In humans, the intermediate lobe is rudimentary and does not exist as a distinct lobe per se, but consists of only a few cells interspersed along the borders of the anterior and posterior lobes. Although the intermediate lobe serves an important role in a number of lower vertebrates, it has little or no endocrine function in humans.

Functional Anatomy of the Pituitary Gland

An examination of the embryologic development of the pituitary illustrates how both the endocrine and neuroendocrine portions of the gland are formed. As Figure 13–2 indicates, during early fetal development a pocket of cells known as **Rathke's pouch** pinches off and moves upward from the roof of the primitive oral cavity. At about the same time, a finger-like projection of neural tissue extends downward from the base of the hypothalamus. Rathke's pouch eventually meets this neural protrusion of the hypothalamus, and the two structures fuse to form the anterior and posterior lobes of the pituitary, respectively.

Neural connections between the hypothalamus and the posterior lobe of the pituitary are maintained during development. As Figure 13–3 shows, the posterior pituitary contains the axons and terminals of neurons that have their cell bodies within the hypothalamus. These neurons terminate close to numerous capillaries located throughout the posterior pituitary. As indicated in Figure 13–3, the cell bodies of these neurons are found within two dis-

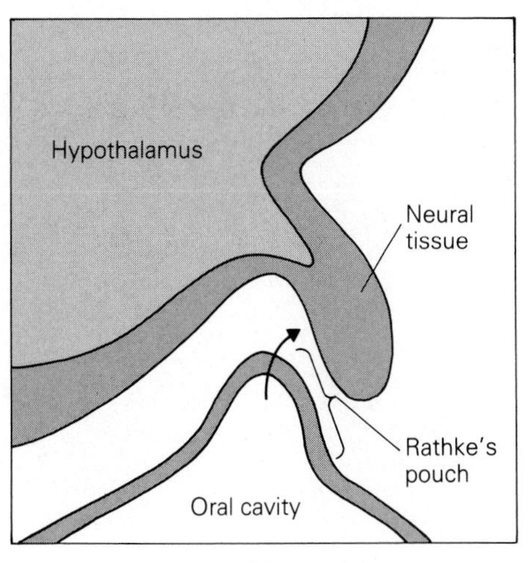

(a)

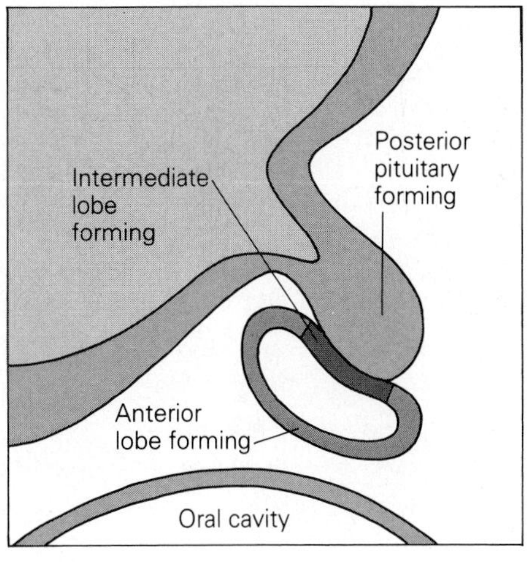

(b)

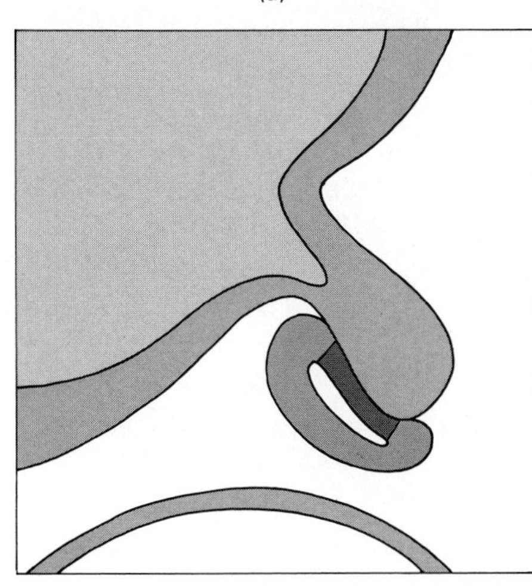

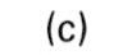
(c)

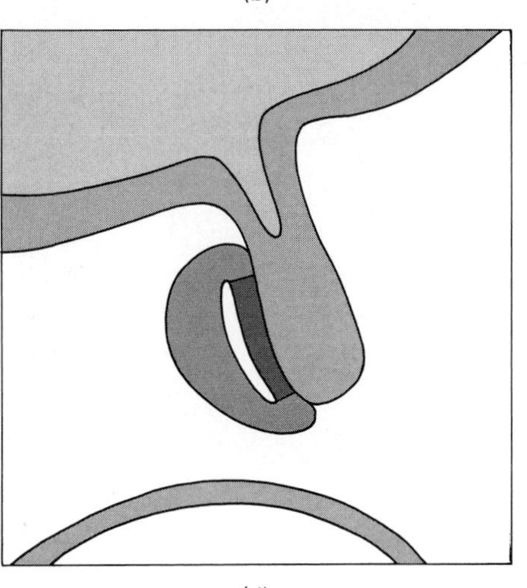

(d)

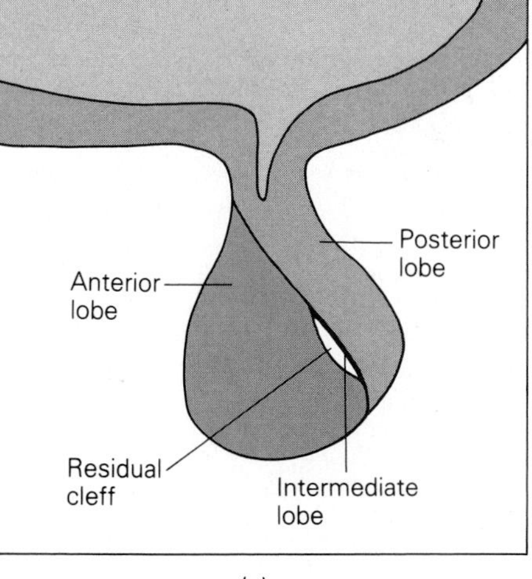

(e)

Figure 13–2 The formation of the anterior, posterior, and intermediate lobes of the pituitary during embryological development. The normal sequence of development is from (*a*) through (*e*). (*a*) and (*b*) Note that the posterior lobe arises from neural tissue, while the anterior lobe is formed from Rathke's pouch. The intermediate lobe forms from that portion of Rathke's pouch that makes contact with the neural tissue. The cleft (*e*) is the remnant of what was once the hollow interior of Rathke's pouch.

tinct areas of the hypothalamus, the **supraoptic nucleus** and the **paraventricular nucleus.** The significance of these anatomical features will become evident in the discussion of the regulation of posterior pituitary hormone secretion that follows.

In contrast to the posterior pituitary, the function of the anterior pituitary is not regulated by direct innervation but instead by vascular (blood vessel) connections with the hypothalamus. As shown in Figure 13–3, arterial blood reaching the hypothalamus enters a specialized region known as the **median eminence,** where vessels branch into a network of primary capillaries. From these primary capillaries, the blood enters the long portal veins

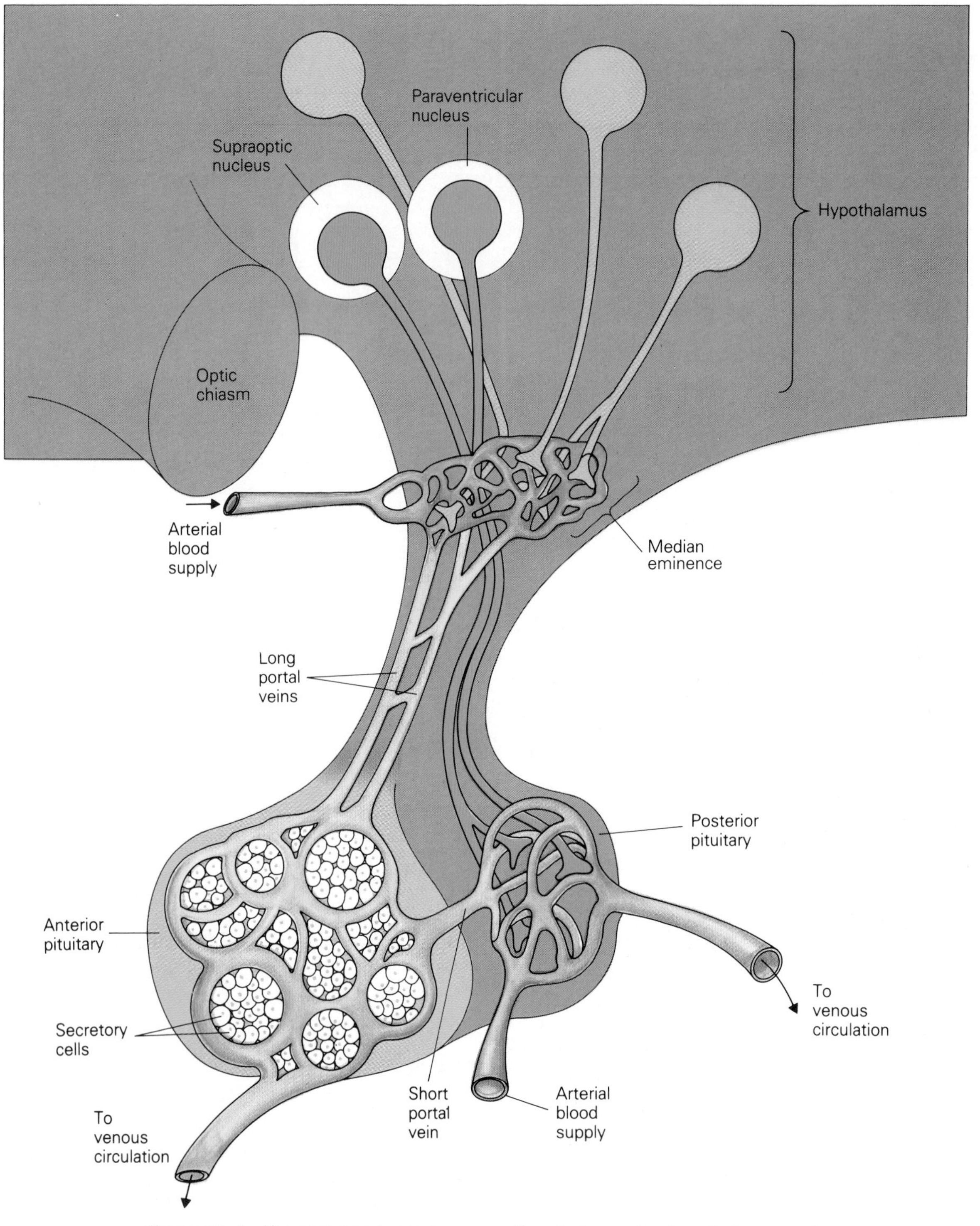

Figure 13–3 The neural and vascular connections between the hypothalamus and the anterior and posterior lobes of the pituitary.

and is carried down through the pituitary stalk to the anterior lobe of the pituitary, where a second capillary network is located. From this secondary capillary network in the anterior pituitary, blood leaves the gland through veins and then mixes with the systemic venous blood.

This complex of blood vessels, which consists of the primary capillary network, the secondary capillary network, and the vessels that connect them, is known as the **hypothalamic–pituitary portal system.** As will be described later, the hypothalamic–pituitary portal system carries factors that regulate the secretion of anterior pituitary hormones from the hypothalamus to the anterior pituitary.

In addition to the long portal veins that connect the hypothalamus and anterior pituitary, a less prominent portal system connects the posterior and anterior lobes of the pituitary. This vascular system is referred to as the **short portal system.** Some evidence suggests that the short portal veins carry substances secreted by the posterior lobe to the anterior lobe, where they regulate hormone secretion from the anterior pituitary. The exact degree to which this system is involved in regulating anterior pituitary function is not well understood at this time, however.

Regulation of Pituitary Hormone Secretion by the Hypothalamus

The mechanisms by which the hypothalamus regulates the secretion of hormones from both the posterior and anterior lobes of the pituitary are shown in diagrammatic form in Figure 13–4. This figure illustrates the functional significance of the neural and vascular connections between the hypothalamus and the pituitary shown in Figure 13–3.

The Posterior Pituitary: A Neuroendocrine System

Neural connections between the hypothalamus and posterior pituitary are shown on the right side of Figure 13–4. As indicated previously, the cell bodies of the neurons that send projections to the posterior pituitary are located primarily in the supraoptic and paraventricular nuclei of the hypothalamus. As indicated in Table 13–1, the posterior pituitary secretes two hormones: **oxytocin** and **antidiuretic hormone** (**ADH,** or vasopressin). These two hormones are actually synthesized within the cell bodies of the neurons in the supraoptic and paraventricular nuclei of the hypothalamus. Oxytocin and ADH are then transported to the posterior pituitary through the axons of the same neurons that produce them. Generation of action potentials in these nerves results in the release of oxytocin or vasopressin from the nerve endings that are adjacent to capillary beds in the posterior pituitary. The hormones enter the blood vessels and are then carried throughout the body. The cells that produce oxytocin and ADH are therefore good examples of neuroendocrine cells.

Neural inputs to the supraoptic and paraventricular nuclei from other areas of the hypothalamus or from other areas of the brain modulate the electrical activity of the neuroendocrine cells in these two nuclei. In this manner, a variety of stimuli are capable of regulating the secretion of oxytocin and ADH.

The Anterior Pituitary: An Endocrine System

As Table 13–1 indicates, cells of the anterior pituitary synthesize six different hormones. Secretion of these hormones is under the control of several different **releasing/inhibiting hormones** produced in the hypothalamus. All of the releasing/inhibiting hormones that have been chemically characterized thus far are either small peptides or amino acid derivatives. As illustrated on the left side of Figure 13–4, these hormones are synthesized within the cell bodies of neuroendocrine cells located in a number of different sites in the hypothalamus. The axons of these neuroendocrine cells converge in the median eminence of the hypothalamus, where they terminate on or near the primary capillary network of the hypothalamic–pituitary portal system. Generation of action potentials in these nerves causes secretion of particular releasing/inhibiting hormones from the nerve endings. After release from the nerve endings, the releasing/inhibiting hormones then enter the primary capillaries and travel via the long portal vessels to the anterior pituitary. The releasing/inhibiting hormones then exit from the secondary capillaries and act on the endocrine cells of the anterior pituitary, where they regulate (stimulate or inhibit) secretion of hormones.

Table 13–2 lists the hypothalamic releasing/inhibiting hormones involved in regulating anterior pituitary secretion. Note that some stimulate secretion of a particular hormone, while others inhibit hormone secretion. As a group, they are therefore referred to as "releasing/inhibiting hormones." The system would be relatively easy to understand if there were a simple one-to-one relationship between each anterior pituitary hormone and a single releasing hormone or a pair of releasing and inhibiting hormones. However, a single hypothalamic hormone may influence the secretion of several anterior pituitary hormones. For example, **somatostatin** regulates the secretion not only of growth hormone but also of **prolactin** and **TSH.** Similarly, GnRH regulates the secretion of both FSH and LH. The degree to which multiple activities of these hypothalamic releasing/inhibiting hormones regulates hormone secretion from the anterior pituitary has not yet been fully established. The general belief, however, is that secretion of any particular hormone from the

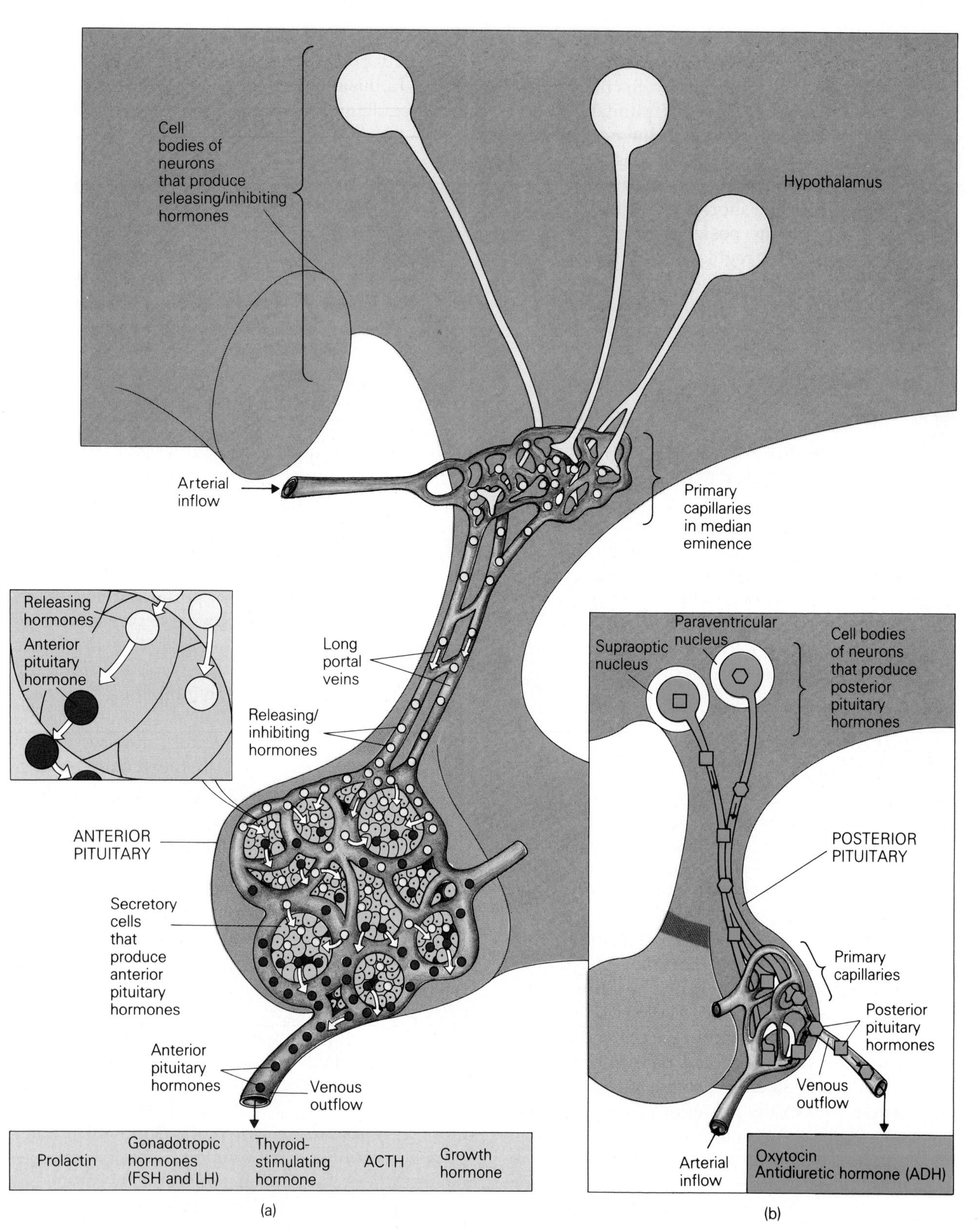

Figure 13–4 An illustration of the mechanisms involved in the regulation of anterior and posterior pituitary function. (*a*) Neuroendocrine cells of the hypothalamus transmit releasing/inhibiting hormones to the capillaries in the median eminence. These hormones travel via the long portal veins to the secretory cells of the anterior lobe, where they stimulate or inhibit the release of anterior pituitary hormones into the blood. (*b*) Neurons that originate in the supraoptic and paraventricular nuclei of the hypothalamus synthesize oxytocin and ADH and transmit these through axons that release the hormones into capillaries in the posterior pituitary. The hormones are then funneled into the circulation.

Table 13–2
Releasing Hormones Involved in Regulating Anterior Pituitary Secretion

Hypothalamic-releasing Hormone	Primary Effect on Anterior Pituitary*
Corticotropin-releasing hormone (CRH)	Stimulates ACTH secretion
Thyrotropin-releasing hormone (TRH)	Stimulates TSH secretion
Gonadotropin-releasing hormone (GnRH)	Stimulates LH and FSH secretion
Somatostatin	Inhibits GH secretion
Growth-hormone-releasing hormone (GHRH)	Stimulates GH secretion
Prolactin-releasing hormone (PRH)	Stimulates prolactin secretion
Prolactin-inhibitory hormone (PIH)	Inhibits prolactin secretion

*As described in the text, several of the releasing hormones influence the secretion of more than one anterior pituitary hormone. For simplicity, only the primary effects of the releasing hormones are presented here.

anterior pituitary depends upon the relative amounts of the various hypothalamic releasing/inhibiting hormones that reach the anterior pituitary.

Cell bodies of the neurons that synthesize the releasing/inhibiting hormones are not randomly distributed throughout the hypothalamus, but instead each releasing factor is synthesized by neurons located in different, discrete nuclei of the hypothalamus. For example, GnRH is produced primarily in one nucleus of the hypothalamus, while CRH is produced in another. Although they are synthesized in different regions of the hypothalamus, both hormones are ultimately released into the capillary network in the median eminence. Secretion of releasing/inhibiting hormones from these hypothalamic neurons depends upon the input they receive from neurons located in other areas of the hypothalamus or in higher brain centers.

Actions and Effects of the Pituitary Hormones

Now that we have examined the anatomical and physiological bases for regulating hormone secretion from the anterior and posterior lobes of the pituitary, a logical question might be, "What are the actions of the pituitary hormones?" The following sections describe the specific biological effects of the pituitary hormones. Note the wide variety of physiological processes that the pituitary hormones regulate.

Hormones of the Posterior Pituitary

Antidiuretic Hormone (ADH): An Increase in Water Retention

Antidiuretic hormone, or **ADH,** is one of two small peptide hormones secreted by the posterior pituitary. It is eight amino acids in length and is synthesized primarily in the supraoptic nucleus of the hypothalamus, although smaller amounts are produced in the paraventricular nucleus as well.

The peripheral mechanisms of action of ADH will be discussed in detail in Chapters 24 and 25 when the regulation of fluid balance by the kidney is described. Briefly, however, the primary effect of ADH is to increase water retention by the kidney. The net result of this effect is a decrease in urine volume and an increase in extracellular fluid volume. ADH can also act to constrict blood vessels. The other name for ADH, **"vasopressin,"** stems from this vasoconstrictor effect, but this response occurs only when the hormone is secreted in relatively large amounts.

The two primary physiological stimulators of ADH release are (1) an increase in osmolality of the blood or extracellular fluid and (2) a large decrease in blood volume, as in a hemorrhage, or heavy bleeding. How do changes in osmolality or blood volume trigger ADH release? Specialized cells within the hypothalamus known as **osmoreceptors** are very sensitive to increases in osmolality of the extracellular fluid. Even a slight increase in osmolality will stimulate the osmoreceptor cells. The osmoreceptors in turn trigger ADH release from the posterior pituitary. As a result of its effect to promote water retention by the kidney, ADH causes a dilution of the extracellular fluid, which returns plasma osmolality to normal.

A decrease in blood volume also stimulates ADH release. However, stimulation of ADH release requires a 10% or greater blood volume decrease, which is a severe loss of blood. As a result of its promotion of fluid retention by the kidneys, ADH acts to restore blood volume to normal. In addition, a large loss of blood can result in sufficient ADH secretion to cause constriction of blood vessels. By constricting blood vessels, this effect serves to lessen the decrease in blood pressure that would occur with severe hemorrhage.

In addition to changes in osmolality and blood volume, several other factors are known to influence ADH release. For example, alcohol is a strong inhibitor of ADH release. When ADH secretion is inhibited, urine volume increases, and excess fluid is lost from the body. This accounts, in part, for the increase in urine volume and the thirst experienced following the consumption of alcoholic beverages. Nicotine, barbiturates, and certain anesthetics are known to stimulate ADH release. The overproduction of ADH caused by these agents results in excess fluid retention.

In the absence of ADH secretion, which can occur as a result of either a genetic abnormality or damage to the posterior pituitary, the kidneys are unable to conserve water, and large quantities of fluid are lost in the urine (a process called *diuresis*). This condition is known as **diabetes insipidus.** Diabetes insipidus is distinctly different from what is commonly termed "diabetes," which is actually diabetes mellitus and will be discussed in detail in Chapter 15.

Oxytocin: Regulation of Breast Milk Release

Oxytocin, the other hormone secreted by the posterior pituitary, is also a small peptide hormone (eight amino acids in length) and is very similar in structure to ADH. Like ADH, oxytocin is synthesized in the cell bodies of neurons in both the supraoptic and paraventricular nuclei of the hypothalamus, but its primary site of synthesis is in the paraventricular nucleus.

The principal site of action of oxytocin is the female breast. In the breast, oxytocin stimulates contraction of specialized smooth muscle cells, which results in the transfer of milk from its site of synthesis in structures known as **alveoli** into the larger ducts of the breast. The net result of this effect is that the milk is made available for a nursing infant. This action of oxytocin will be described in Chapter 31, when the hormonal control of lactation is discussed. In addition to its effects on the breast, oxytocin can also stimulate contraction of smooth muscle in the uterus.

The primary stimulus for oxytocin secretion in a female is suckling by a nursing infant. When an infant nurses, touch sensors in the mother's nipple are activated and neural signals are transmitted to her brain. This information is processed in the brain, and the oxytocin-producing cells in the paraventricular nucleus are stimulated, resulting in oxytocin release from the posterior pituitary.

A variety of psychological stimuli can also influence oxytocin release. The mere sound of a baby crying can trigger oxytocin release in nursing mothers, while fear or apprehension can strongly inhibit secretion of this hormone. Thus, in addition to the neural mechanism that is triggered by suckling, input from other areas of the brain are also important in regulating oxytocin secretion.

Hormones of the Anterior Pituitary

Five different hormone-secreting cell types are found in the anterior pituitary. These are listed in Table 13–3 along with the hormone(s) produced by each cell type. The **somatotrophs,** which produce growth hormone, are the most numerous cells of the anterior pituitary. Normally, about 50% of the cells present in the anterior pituitary are somatotrophs. **Corticotrophs,** the ACTH-secreting cells, comprise about 20% of the cells, and the TSH-producing cells, the **thyrotrophs,** about 5%. The **lactotrophs** and **gonadotrophs** are the other two cell types in the anterior pituitary.

The size and activity of these cells varies under different physiological conditions. In lactating (nursing) females, the lactotrophs, which secrete prolactin, are large and numerous, reflecting the high rate of prolactin secretion in these individuals. In males or nonlactating females, fewer lactotrophs are present, and they are smaller as well. The gonadotrophs, which secrete both LH and FSH, account for only about 5% of the cell population of the anterior pituitary of males. In females, their size varies in accordance with the cyclical changes in LH and FSH secretion that occur during the monthly menstrual cycle.

As Table 13–3 indicates, these five cell types are responsible for the production of six hormones in the anterior pituitary. Gonadotrophs are responsible for the synthesis and secretion of *both* LH and FSH. As Chapter 30 will describe, the concentrations of LH and FSH in the blood do not always change in parallel during the monthly ovarian cycle. Thus, although they are produced within the same cell, their secretion is regulated independently. At present, the mechanisms by which a single releasing hormone (GnRH) is able to differentially regulate the secretion of two hormones (LH and FSH) are not well understood.

The following sections describe the secretion and actions of each of the anterior hormones. The physiology of LH, FSH, and prolactin will be presented in detail in Chapters 30 and 31. Similarly, the physiology of ACTH will be presented primarily in Chapter 14. Therefore, these hormones will only be discussed briefly here. Instead, we will emphasize the remaining two hormones, growth hormone and TSH.

Table 13–3
Hormone-Producing Cells of the Anterior Pituitary

Name of Cell Type	Principal Hormone Produced
Somatotrophs	Growth hormone (GH)
Corticotrophs	ACTH
Thyrotrophs	TSH
Lactotrophs	Prolactin
Gonadotrophs	LH and FSH

Luteinizing Hormone (LH) and Follicle-Stimulating Hormone (FSH): Regulation of Reproductive Function

LH and FSH are collectively referred to as the **gonadotropins.** This name reflects the fact that the target tissue of these hormones is the **gonads,** or the ovaries and testes in females and males, respectively. In general, the gonadotropins have two primary effects: (1) to promote the development and maturation of sperm and egg, and (2) to stimulate production of sex steroid hormones by the gonads. The principal sex steroids in males and females are **testosterone** and **estradiol,** respectively. As Table 13–2 indicates, secretion of gonadotropins is under the control of GnRH from the hypothalamus.

Prolactin: Regulation of Milk Synthesis

Prolactin has no known function in males. In females, the primary effect of prolactin is to stimulate milk production by the breast. The target cells on which prolactin has its effects are the alveolar cells, which are the milk-producing cells of the breast. (Specific mechanisms by which prolactin stimulates milk formation will be presented in Chapter 31.) Prolactin secretion is under the dual control of a **prolactin-releasing hormone (PRH)** and a **prolactin release-inhibiting hormone (PIH).** As in the situation with oxytocin described previously, prolactin release increases in response to stimulation of the mother's nipple by a suckling infant.

Adrenocorticotropic Hormone (ACTH): Regulation of Cortisol

The primary effect of **adrenocorticotropic hormone (ACTH)** is to regulate the synthesis and secretion of the steroid hormone **cortisol** from the adrenal cortex (a portion of the adrenal gland, an endocrine gland located near the kidney). The cortisol produced by the adrenal gland in turn has numerous effects upon intermediary metabolism in a variety of different tissues. As the name indicates, ACTH also has **trophic** or "growth-promoting" effects on cells of the adrenal cortex. As Table 13–2 indicates, the secretion of ACTH is regulated by **corticotropin-releasing hormone (CRH),** which is produced in the hypothalamus. The specific effects of ACTH on the adrenal cortex and the effects of adrenal steroids on metabolism will be covered in detail in the following chapter.

ACTH is actually synthesized as part of a much larger precursor protein that contains not only ACTH but also several other biologically active peptides. This precursor has been named **pro-opiomelanocortin peptide,** or **POMC peptide,** to reflect the fact that it contains the sequences for endogenous (internally synthesized) opioid compounds (see Chapter 7), the sequence for **melanocyte-stimulating hormone (MSH),** and the sequence for ACTH.

Growth Hormone (GH): Body Growth and Metabolism

As was indicated in Figure 12–9 in the previous chapter, **human growth hormone (hGH** or **GH)** is a peptide hormone consisting of 191 amino acids. The structure of GH is remarkably similar to that of a peptide hormone secreted by the placenta during pregnancy, **human chorionic somatomammotropin (HCS).** Growth hormone is also somewhat similar to prolactin, although to a much lesser extent. GH, HCS, and prolactin comprise a family of related peptide hormones collectively referred to as the **somatomammotropic hormones,** named to indicate their effects on body growth and breast development. The sections that follow focus on the effects of GH on body growth and metabolism. The physiology of HCS and prolactin is presented in later chapters.

Regulation of Growth Hormone Secretion. As Tables 13–2 and 13–3 indicate, GH secretion from the somatotrophs of the anterior pituitary is under the dual control of growth-hormone-releasing hormone (GHRH) and somatostatin, both of which are produced by the hypothalamus. As indicated previously in Table 12–1, somatostatin is also produced in the pancreas, where it is thought to play a local role in regulating pancreatic hormone secretion. This particular action of somatostatin will be described further in Chapter 15. At this point we should emphasize, however, that the somatostatin produced by the hypothalamus and that produced by the pancreas serve different physiological functions and that the secretion of somatostatin from each site is also independently regulated.

As Figure 13–5 shows, growth hormone is not secreted in a steady continuous fashion throughout

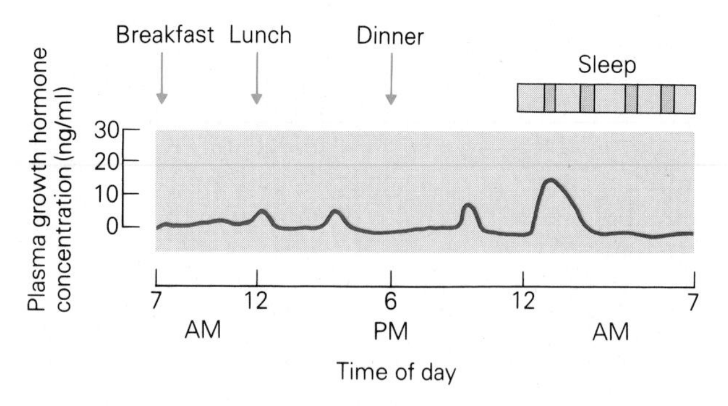

Figure 13–5 Typical 24-hr pattern of growth hormone secretion in a normal individual. As indicated, the person slept from 11:30 PM until 7:00 AM. The hatched bars indicate periods of REM sleep. Note the sleep-induced increase in the plasma growth hormone concentration that occurs between midnight and 1:00 AM. Note also that with the onset of REM sleep, growth hormone secretion returns to the basal level.

the day, but instead in a somewhat pulsatile manner. The most consistent period of GH secretion occurs about one hour after the onset of deep sleep (Fig. 13–5). This can occur during the normal sleep period at night or during a daytime nap. The phase of sleep associated with dreams, the phase known as REM sleep (see Chapter 11), initiates the return of GH secretion to the basal pre-sleep levels. The exact significance of this sleep-related surge in GH secretion is not known, but it has been suggested that it may be important in stimulating processes of tissue growth and repair during sleep.

In addition to the sleep-induced increase in GH secretion, a series of pulses often occurs anywhere from two to four hours after a meal (see Fig. 13–5). The relative frequency and the size of these pulses tends to increase about the time of puberty. Some of the other agents or conditions known to influence GH secretion are listed in Table 13–4. Hypoglycemia (low blood glucose) stimulates GH secretion, while high blood glucose inhibits GH secretion. A useful test that physicians can perform to evaluate whether a patient's pituitary is capable of secreting GH involves the administration of a dose of insulin, followed by the monitoring of GH concentrations in the blood. As Chapter 15 will describe, insulin decreases the blood glucose concentration. In a normal individual in this test situation, the hypoglycemia caused by insulin will trigger a pronounced increase in GH secretion. In contrast, the absence of increased GH secretion in response to insulin indicates to the physician that pituitary GH secretion is impaired. As Table 13–4 also shows, a variety of types of stress, both physical and emotional, stimulate GH secretion. Several amino acids, especially arginine, also stimulate GH secretion.

How are such a wide variety of stimuli such as those listed in Table 13–4 able to regulate GH secretion? As Figure 13–6 illustrates, GHRH and somatostatin serve as the final common pathway for regulating GH secretion. Inputs from various areas of the brain and hypothalamus regulate GHRH and

Table 13–4
Partial Listing of Factors or Conditions Known to Influence Growth Hormone (GH) Secretion

Stimulators
- Deep sleep*
- Low blood glucose concentration
- Stress
  - Physical trauma
  - Infection
  - Psychological stress
- Amino acids, especially arginine

Inhibitors
- REM sleep*
- High blood glucose concentration

*See Chapter 11 of this text.

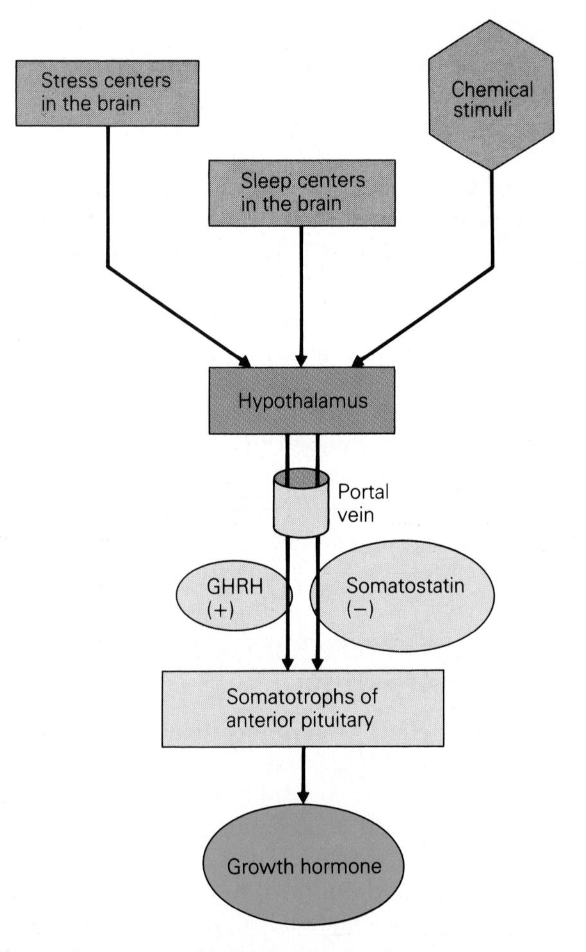

Figure 13–6 A variety of factors can influence growth hormone secretion by regulating the production of GHRH and somatostatin by the hypothalamus.

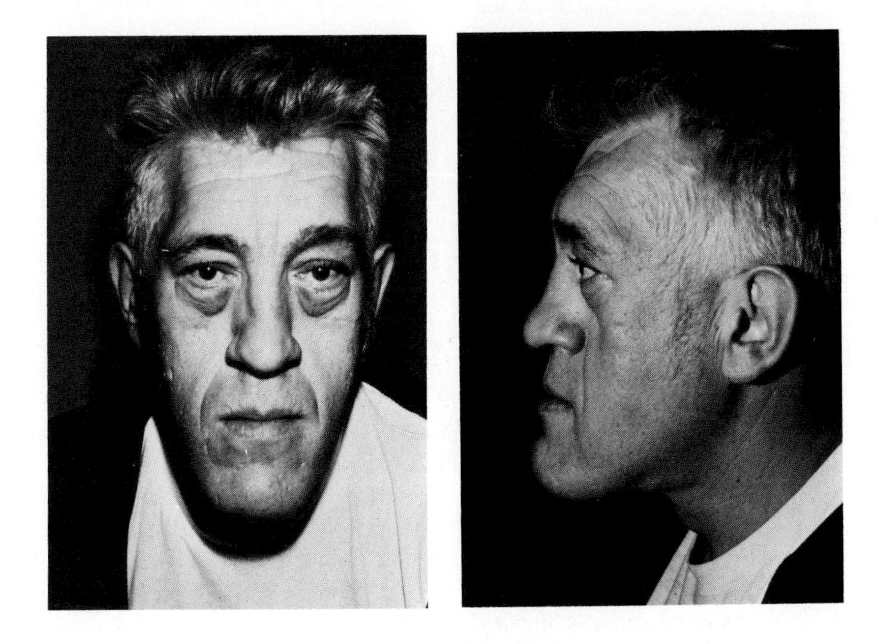

Figure 13–9 Typical characteristics of acromegaly. Note in particular the enlarged facial features and lower jawbone. (Courtesy of Dr. Gordon Williams.)

dent in the figure, the long bones of the body are stimulated to grow disproportionately, and as a result these persons have relatively long arms and long legs.

If, on the other hand, overproduction of GH begins after the time of puberty, acromegaly results. Around the time of puberty, the growth centers, or epiphyseal plates, of the long bones of the body "close," meaning that they become unresponsive to hormonal stimulation. Thus, bones of the arms and legs cease to grow at this time. However, the bones of the hands, feet, skull, and lower jaw do not "close" and can still be stimulated to grow. As a result of the oversecretion of GH, these bones continue to grow and enlarge, leading to the distinctive physical appearance of acromegaly shown in Figure 13–9. Typically, these individuals have an enlarged lower jaw, large hands and feet, and also have generally large facial features due to the stimulation of soft tissue growth that occurs as a result of increased somatomedin production. In extreme situations, the diabetogenic effects of GH can also lead to overt cases of diabetes mellitus in individuals suffering from acromegaly.

The Thyroid Gland and Thyroid-Stimulating Hormone (TSH)

As its name implies, the primary target tissue of thyroid stimulating hormone (TSH) is the thyroid gland. In the thyroid, TSH stimulates cell growth and the secretion of thyroid hormones. Thyroid hormones in turn affect a variety of metabolic processes in the body. Therefore, this section provides a description not only of the physiology of TSH, but also of the thyroid gland and the actions of thyroid hormones.

Anatomy of the Thyroid Gland. The gross anatomy of the thyroid gland is shown in Figure 13–10. The thyroid is one of the largest endocrine glands in the body, weighing approximately 20 grams in a normal adult. As Figure 13–10 indicates, the thyroid consists of two lobes that lie on either side of the trachea, just below the larynx. A thin band of tissue known as the **isthmus** connects the two lobes. The gland has an abundant blood supply and actually exhibits one of the highest blood flow rates of any tissue or organ in the body. The thyroid gland also has a tremendous capacity for growth. With the appropriate stimulus, the gland can become greatly enlarged, a condition known as **goiter.** An example of an extreme case of goiter is shown in Figure 13–11. In this particular case, the goiter was caused by a dietary deficiency in iodide. A later section of this chapter describes the mechanism by which iodide deficiency causes goiter.

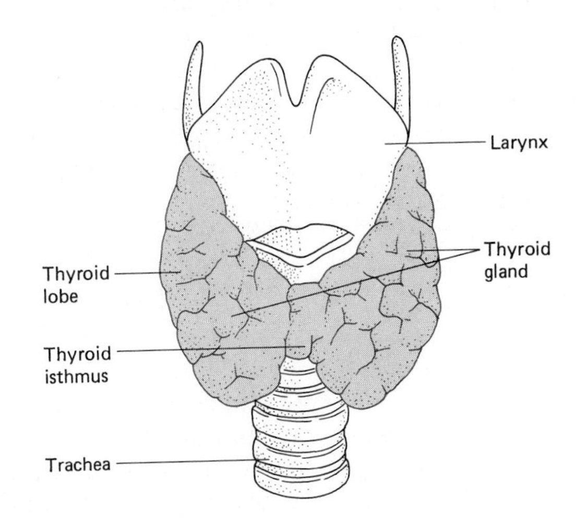

Figure 13–10 Gross anatomy of the thyroid gland showing its location relative to the trachea and the larynx.

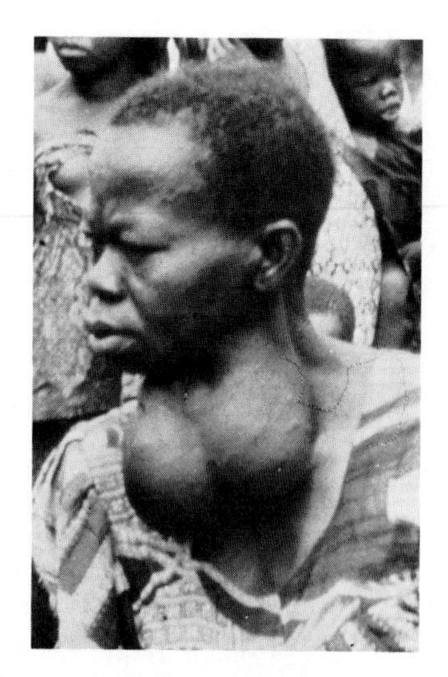

Figure 13–11 An extreme case of thyroid goiter.

The thyroid gland consists of cells that are arranged in closely packed follicles (Fig. 13–12). The thyroid of a normal healthy adult contains about 3 million of these follicles. The cells that form a follicle are usually cuboidal in shape, and the interior of the follicle is filled with a protein-containing material termed **colloid.** The major component of colloid is a protein known as **thyroglobulin,** which serves as the precursor of thyroid hormones. A considerable amount of colloid is usually present within each follicle, such that colloid is normally the major constituent of the total mass of the thyroid gland. An extensive capillary network surrounds each follicle.

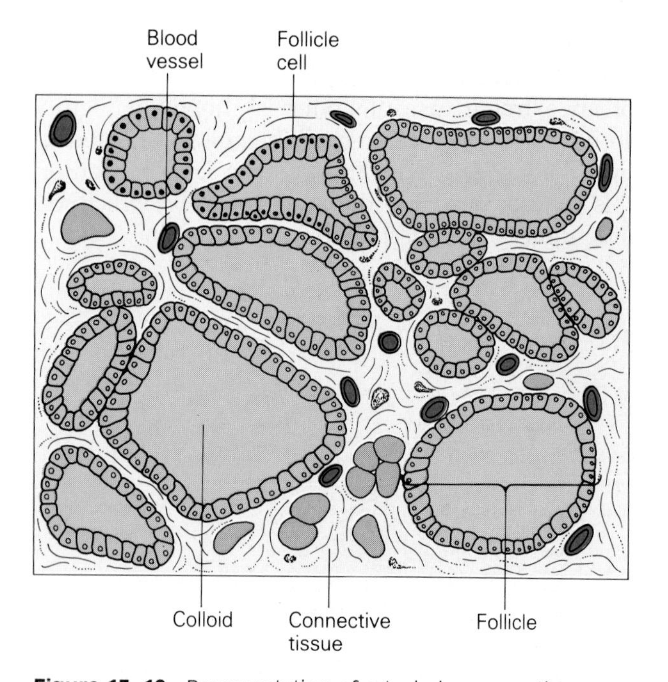

Figure 13–12 Representation of a typical cross-section through a portion of the thyroid gland. Note the numerous blood vessels and the relative abundance of colloid in the gland.

Synthesis and Secretion of Thyroid Hormones. The major compounds of interest with regards to thyroid hormone production and metabolism are shown in Figure 13–13. **Thyroxine (tetraiodothyronine, or T_4)** is the primary hormone product of the thyroid gland. **Triiodothyronine (T_3)** is also produced by the thyroid gland but in lesser amounts. (In target tissues, T_4 is converted into T_3, as will be discussed later.) The other two iodinated compounds shown in Figure 13–13, **monoiodotyrosine (MIT)** and **diiodotyrosine (DIT),** are found primarily within the thyroid follicle cell, although small amounts are also secreted into the blood.

Each of the compounds shown in Figure 13–13 contains iodine as an integral part of the molecule. In comparison to other hormones, thyroid hormones are unique in that their synthesis requires a constituent (iodine) that is not always present in the diet. As a result, mechanisms have evolved that permit the concentration of iodide in the gland and the storage of large amounts of thyroglobulin, the thyroid hormone precursor. Normally, enough thyroglobulin may be stored in the gland to sustain thyroid hormone secretion for 2 months in the absence of any further thyroglobulin synthesis.

Thyroglobulin itself is a large glycoprotein that is synthesized in the thyroid follicle cell (Fig. 13–14*a*). As with other glycoproteins that are secreted, the protein component of thyroglobulin is synthesized in the rough endoplasmic reticulum of the cell. The carbohydrate portion of the molecule is added in the Golgi complex. From the Golgi complex, thyroglobulin is then secreted on the apical side of the cell into the lumen of the follicle, where it is iodinated and stored as part of the colloid.

The first step in thyroid hormone formation involves the uptake of iodide (I^-, the ionic form of the element iodine) into the thyroid follicle cell across its basal membrane from the plasma (Fig. 13–14*b*). The uptake involves an active transport process that pumps iodide into the cells and can produce an intracellular iodide concentration some 30 times higher than that in the extracellular fluid. This transport system is so effective at pumping iodide into the follicle cell that the process is often referred to as the **"iodide trap."**

Once inside the cell, iodide diffuses down a concentration gradient toward the apical surface of the cell, where it then leaves the cell and enters the interior of the follicle. Soon after leaving the follicle cell, the iodide undergoes **oxidation** and **organification** reactions. Enzymes that rapidly oxidize the iodide and convert it to iodine (I) are present on the surface of the follicle cell membrane that faces the colloid. The iodine formed by these enzymes then serves as a substrate for other enzymes that

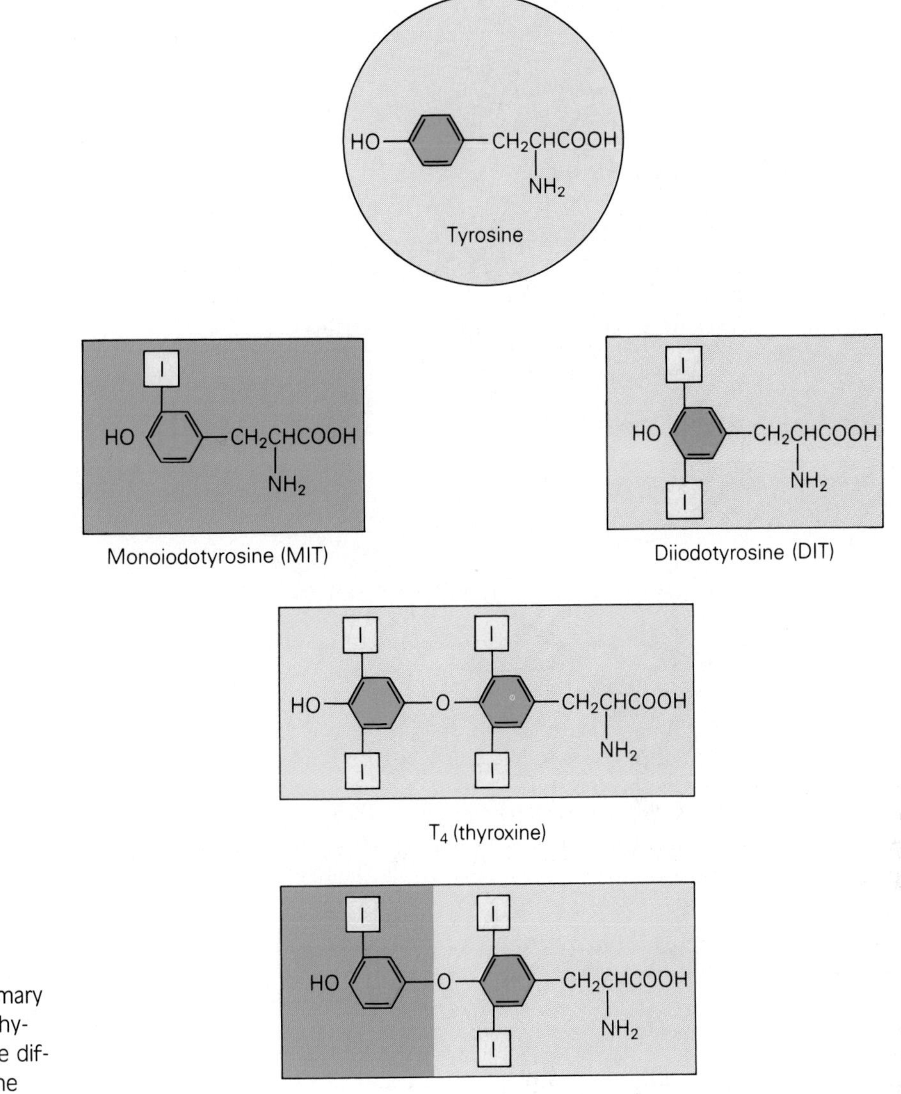

Figure 13–13 Structures of the primary iodinated compounds found in the thyroid gland and in the blood. Note the difference in number and location of the iodides between T_4 and T_3.

catalyze the addition of iodine to tyrosine residues within the thyroglobulin protein. Tyrosine residues containing either one or two iodine molecules can be formed, and thus at this point thyroglobulin contains both MIT and DIT as part of its protein structure.

The next major step in thyroid hormone formation involves the coupling of two of these iodinated tyrosines within the thyroglobulin molecule to form either T_3 or T_4. If two DIT residues are coupled, then T_4 will be formed (refer to Fig. 13–14*c*). If, however, DIT and MIT residues of thyroglobulin are coupled, then T_3 is produced. It should be emphasized that at this point the T_3 and T_4 are *not* free, but are still present as part of the thyroglobulin peptide. In later steps to be described, the thyroglobulin is broken down to release the free hormones.

When follicle cells are stimulated to produce thyroid hormones, numerous microvilli are formed on the apical surface of the cells. These microvilli extend out from the cell into the colloid and engulf a portion of colloid by pinocytosis (see Fig. 13–14*b*). The colloid droplet is then taken into the follicle cell and is moved toward the basal surface of the cell. During this movement, the colloid droplet meets with lysosomes, and the structures fuse to form a **phagolysosome** (see Fig. 13–14*b*). The phagolysosome then continues to move toward the basal end of the cell, and as it does, the thyroglobulin molecule is broken down by proteolytic enzymes to produce free T_4 and T_3. The T_3 and T_4 are secreted into the interstitial fluid, where they are picked up by nearby capillaries and carried by the circulation throughout the body.

The proteolytic breakdown of thyroglobulin also results in the release of small amounts of free MIT and DIT that are present due to incomplete coupling reactions. The MIT and DIT that are freed

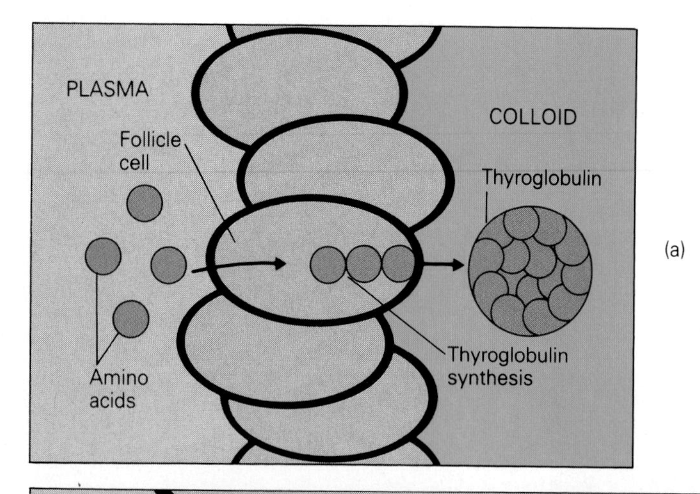

(a)

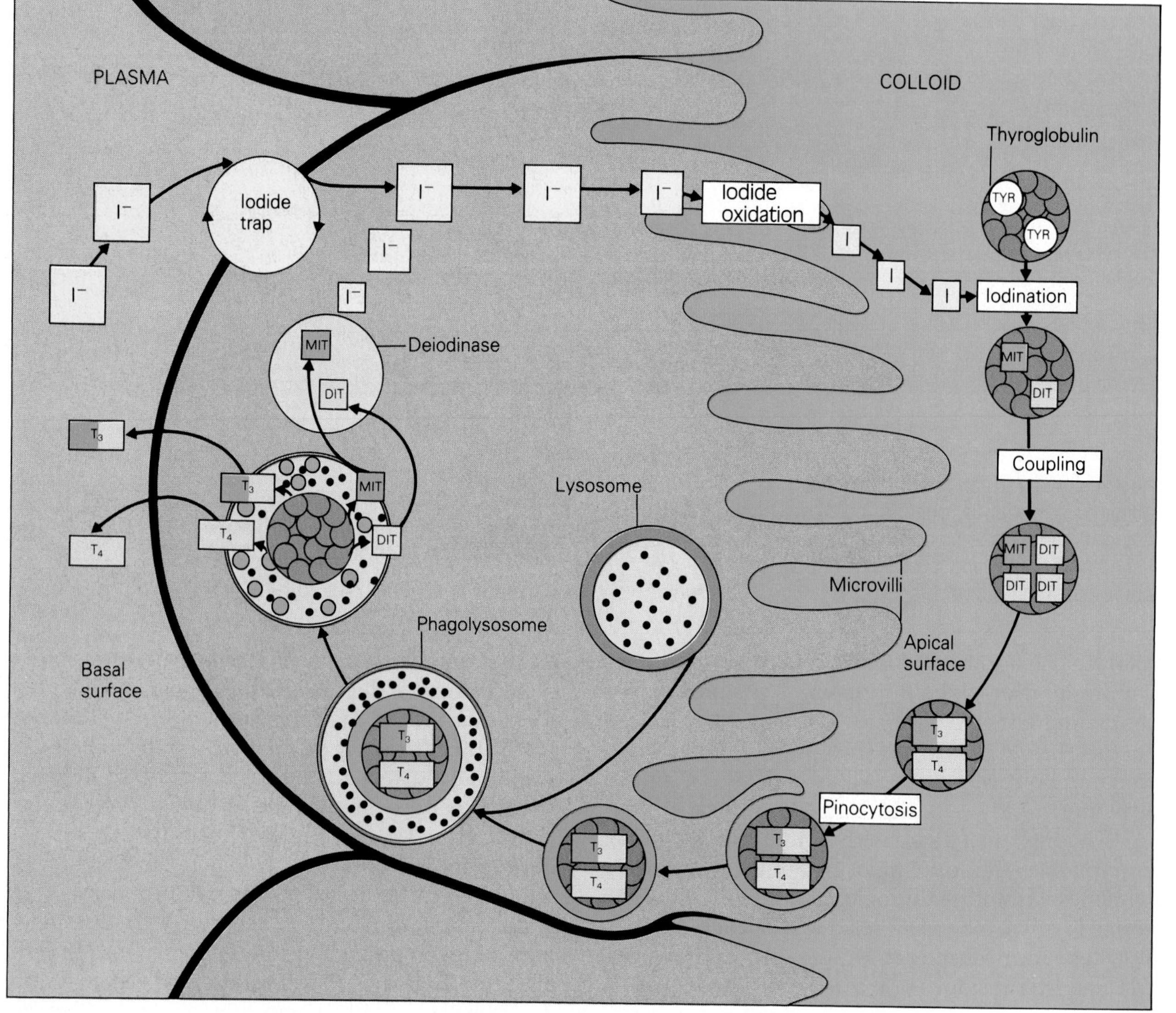

(b)

Figure 13–14 *(a)* The synthesis of thyroglobulin occurs in the follicle cell. *(b)* The major steps involved in thyroid hormone formation and release.

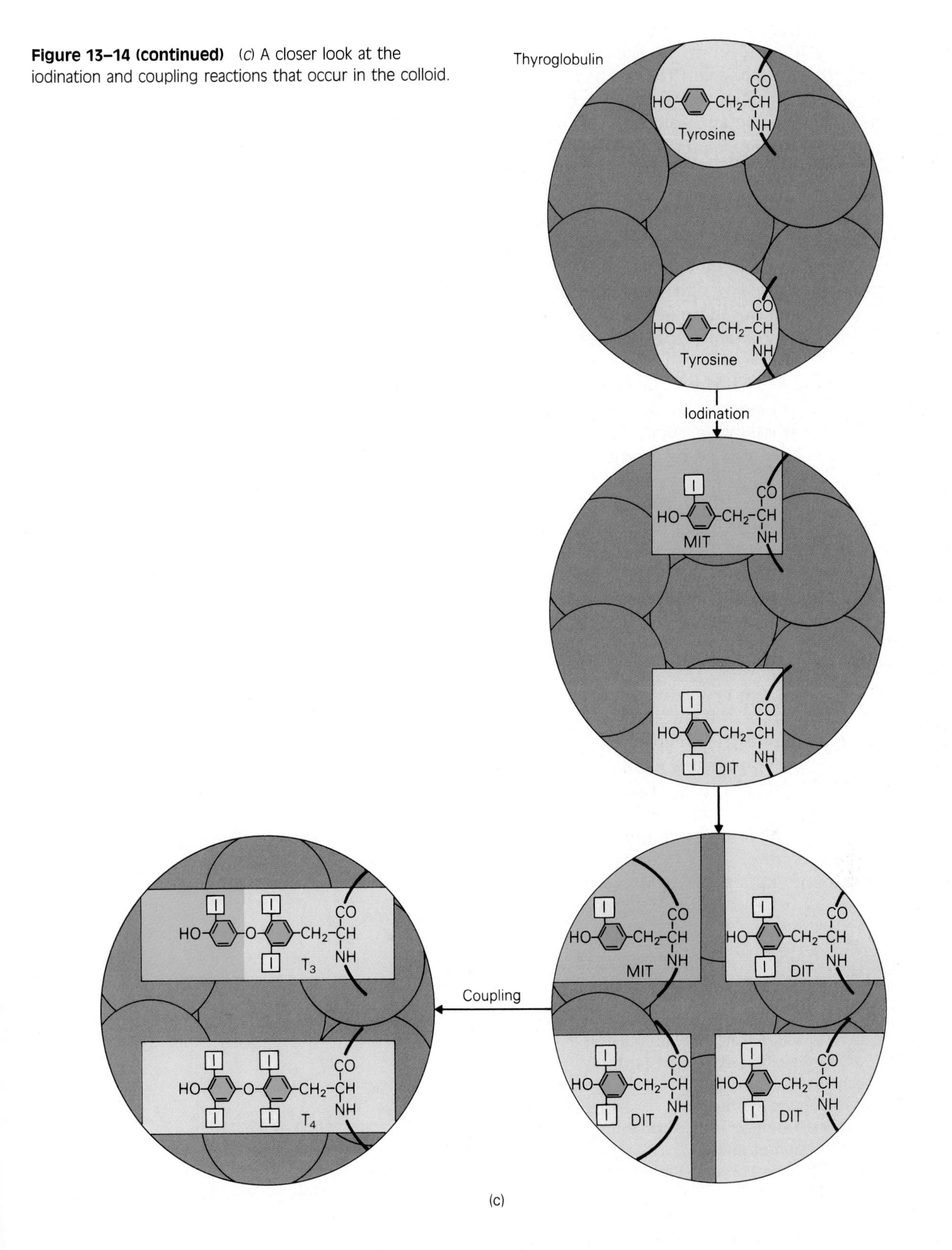

Figure 13–14 (continued) (c) A closer look at the iodination and coupling reactions that occur in the colloid.

are subject to metabolism within the follicle cell by very active **deiodinases** present in the cytoplasm. The deiodinases remove iodide from MIT and DIT to produce iodide and tyrosine. The iodide can be recycled for further hormone synthesis, thereby conserving the iodide. Very little free MIT or DIT is actually released from the thyroid gland.

Regulation of Thyroid Hormone Synthesis and Secretion. The primary regulator of thyroid follicle cell activity and thyroid hormone secretion is **thyroid-stimulating hormone,** or **TSH,** which is produced by the thyrotrophs of the anterior pituitary (see Table 13–3). Specific receptors for TSH are present on the basal surface of the follicle cell. These receptors are coupled to adenylate cyclase, and hence TSH binding to its receptor leads to an increase in the intracellular concentration of cAMP. The specific effects of TSH on thyroid hormone production are listed in Table 13–5. Note that TSH stimulates virtually every aspect of thyroid hormone formation. In addition to these specific effects on thyroid hormone formation, TSH also stimulates growth of the thyroid follicle cell. Thus, when TSH secretion is increased, the follicle cells are stimulated to grow and become elongated and more columnar in shape. Likewise, when TSH secretion is low, the cells regress and become somewhat flattened.

A diagram outlining the regulation of thyroid hormone secretion is presented in Figure 13–15. The relationship between the hypothalamus, anterior pituitary, and thyroid gland indicated in this figure is known as the **hypothalamic-pituitary-thyroid axis.** As described previously, thyroid hormone secretion is under the control of TSH from the pituitary. The secretion of TSH in turn is regulated primarily by TRH produced by the hypothalamus. The secretion of TRH can be influenced by a variety of factors, including inputs from higher centers within the CNS and from temperature regulatory centers in the hypothalamus. The temperature regulatory centers in turn receive information concerning changes in body temperature and changes in environmental temperature. Also indicated in Figure 13–15 is the important negative feedback effect that

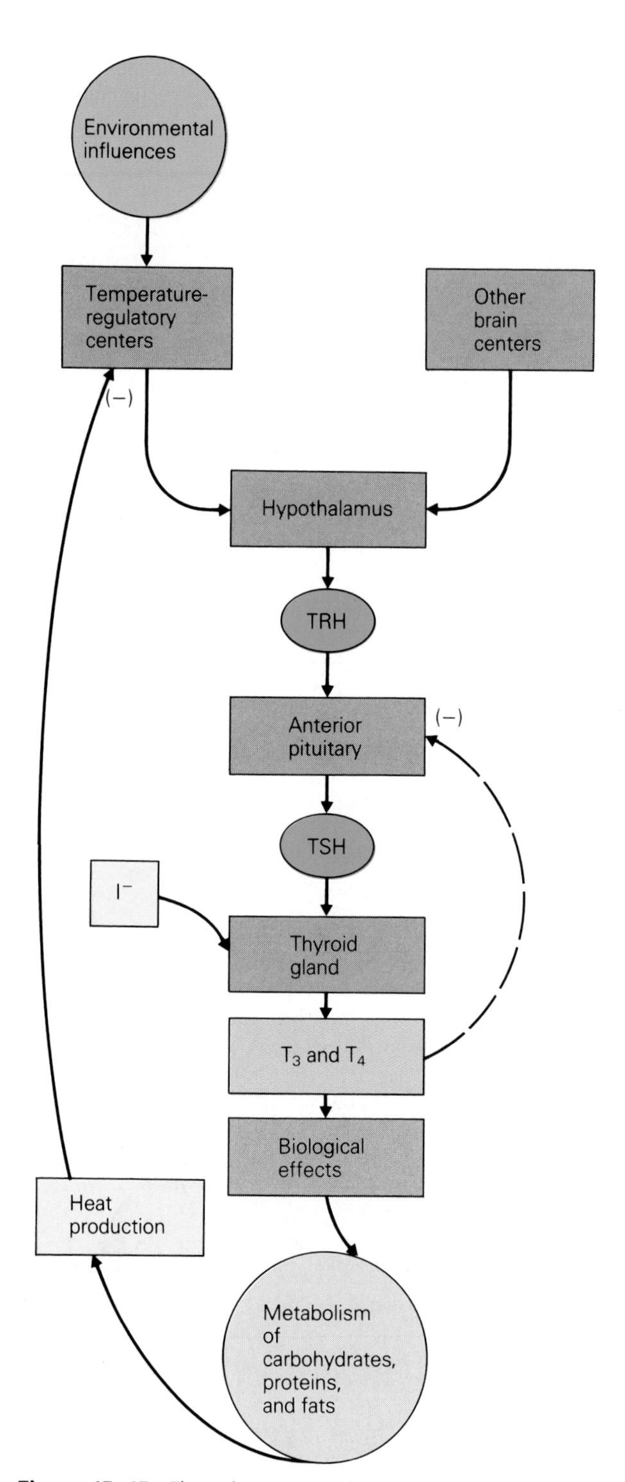

Figure 13–15 The primary steps involved in the regulation of thyroid hormone production.

thyroid hormones exert upon the anterior pituitary to limit TSH secretion. As a result, thyroid hormones exert negative feedback effects that limit their own production. In addition, some of the biological effects of thyroid hormones result in increased heat production in the body (see the following), which also has a negative feedback effect to limit further thyroid hormone formation.

Table 13–5
Effects of TSH on Thyroid Hormone Production

In the thyroid follicle cell, TSH stimulates:

1. Iodide uptake by active transport mechanisms
2. Thyroglobulin synthesis
3. Reactions resulting in the oxidation and organification of iodide
4. Microvilli formation and the engulfment of colloid at the apical cell surface
5. The rate of movement of lysosomes from the basal toward the apical surface of the follicle cell
6. The rate of movement of phagolysosomes from the apical toward the basal surface of the cell
7. The activity of the deiodinase enzymes

Prior to the introduction of iodized salt into the diet, the occurrence of goiter, or enlargement of the thyroid gland, was fairly common. Figure 13–15 illustrates how a deficiency of dietary iodide would lead to an enlargement of the thyroid gland. As described previously and indicated in Figure 13–15, iodide is a necessary component required for thyroid hormone formation. In the absence of iodide, thyroid hormones cannot be produced. In the absence of thyroid hormone secretion, the negative feedback effects of T_4 and T_3 on TSH release will also be absent (see Fig. 13–15). TRH and TSH secretion would therefore increase, and the growth-promoting effects of TSH on thyroid follicle cells would cause the gland to increase in size, despite the inability to actually synthesize active hormone. As a result, goiter would be produced with a prolonged absence of dietary iodine. A number of agents known as **goitrogens** inhibit iodide uptake or organification reactions, resulting in a blockage of thyroid hormone formation and the formation of a thyroid goiter. Vegetables of the turnip family are known to contain goitrogens. These vegetables can produce goiter if eaten raw in excessively large quantities; however, cooking destroys the activity of the goitrogens.

Thyroid Hormone Transport and Metabolism. After secretion from the thyroid gland, T_3 and T_4 are carried in the blood reversibly bound to a number of plasma proteins. The carrier proteins serve to increase solubility of the thyroid hormones and to buffer acute changes in secretion, as was discussed earlier in Chapter 12. Both T_4 and T_3 are associated primarily with **thyroxine-binding globulin (TBG),** but are also bound to a lesser extent to **albumin** (see Table 12–2). The relative affinity or strength with which the two hormones bind to these proteins differs greatly, however. T_4 has a much higher affinity and therefore binds more tightly to the plasma proteins than does T_3. As a result, T_3 is more rapidly degraded and removed from the plasma. Thus, not only is T_4 secreted from the thyroid gland in much greater amounts than is T_3, but it also is removed from the blood much more slowly than is T_3.

Target tissues for thyroid hormones have the capacity to convert T_4 into T_3. Although T_3 and T_4 are capable of producing exactly the same biological effects, their relative strengths or potencies differ considerably. Within a target cell, T_3 is much more active than is T_4. It is currently thought that most, if not all, of the biological effects of T_4 are due to its conversion to T_3 within the target tissue cell.

Table 13–6
Summary of the Effects of Thyroid Hormones

- Stimulate calorigenesis in most cells
- Increase cardiac output
  - Increase rate of cardiac contractions
  - Increase strength of cardiac contractions
- Increase oxygenation of blood
  - Increase rate of breathing
  - Increase number of red blood cells in the circulation
- Effects on carbohydrate metabolism
  - Promote glycogen formation in liver
  - Increase glucose uptake into adipose and muscle
- Effects on lipid turnover
  - Increased lipid synthesis
  - Increased lipid mobilization
  - Increased lipid oxidation
- Effects on protein metabolism
  - Stimulate protein synthesis
- Promote normal growth
  - Stimulate growth hormone (GH) secretion
  - Promote bone growth
  - Promote IGF-I production by liver
- Promote development and maturation of nervous system
  - Promote neural branching
  - Promote myelinization of nerves

Effects of Thyroid Hormones on Metabolic Processes. Thyroid hormones exert numerous effects on metabolic processes in a wide variety of different tissues and cells: these are summarized in Table 13–6. The effects of thyroid hormones are so widespread that virtually no tissue or organ system escapes the adverse effects that result from thyroid hormone deficiency or excess. In general, the effects of thyroid hormones are slow in onset and long-lasting, as compared to the effects of other hormones on metabolic processes. Thus, thyroid hormones exert relatively long-term regulatory influences over metabolism.

One of the primary effects of thyroid hormones is to stimulate **calorigenesis,** or heat production, in the body (see Table 13–6). This response occurs after a delay of several hours or days and is reflected by an increase in oxygen consumption. This particular effect is evident in most tissues, with brain, spleen, and testes being the most notable exceptions. To provide for an increase in oxygen consumption by the tissues, thyroid hormones also have effects on the cardiopulmonary system. In the heart, thyroid hormones increase cardiac output (the amount of blood being pumped) by increasing both the rate and the strength of cardiac contractions. The amount of oxygen carried by the blood is also enhanced by thyroid hormones as a result of increases in both the resting rate of breathing and the number of red blood cells present in the blood.

Thyroid hormones also influence several aspects of carbohydrate metabolism, although many of the effects that they produce depend on or are modified by other hormones. For example, normal concentrations of thyroid hormones increase glycogen formation in the liver (see Table 13–6), but this response occurs only when insulin is also present. Thyroid hormones also increase the uptake of glucose from the blood into adipose tissue and muscle and thereby act to potentiate the effect of insulin in this regard. In addition, many of the effects of epinephrine are enhanced by thyroid hormones as a result of an effect of the thyroid hormones that enhance the responsiveness of the adenylate cyclase/cAMP system.

Thyroid hormones also have a number of effects on lipid metabolism in both liver and adipose tissue (see Table 13–6). Thyroid hormones stimulate virtually all aspects of lipid metabolism, including the synthesis, mobilization, and oxidation of lipids. In general, however, the oxidation of lipids is affected more than is their synthesis, so that in conditions of hormone excess the net effect is a decrease in the size of most fat stores in the body.

Thyroid hormones also exert important regulatory effects on protein metabolism in the body. When present in normal physiological concentrations, thyroid hormones increase protein synthesis and promote an overall accumulation of protein in the body (see Table 13–6). However, when thyroid hormones are present in excess, they tend to cause a decrease in protein synthesis and an increase in protein breakdown. The result of these two effects is an overall loss of protein from the body. The effects of thyroid hormones on protein metabolism are therefore described as **biphasic.** Thyroid hormones have an additional effect on the pituitary that influences protein metabolism indirectly. In the anterior pituitary, thyroid hormones are known to stimulate the somatotrophs and increase GH secretion. As a result, thyroid hormones also affect protein metabolism owing to their influence on pituitary GH secretion.

In addition to specific effects on protein metabolism, thyroid hormones also play a role in regulating the overall growth and development of several tissues. In humans, thyroid hormone is required for normal skeletal growth. This is due in part to the stimulatory effects of thyroid hormones on GH production and in part to thyroid hormone stimulation of maturation of the epiphyseal growth centers in bone. In addition, as was noted in the earlier section describing somatomedins, thyroid hormones also influence the production of IGF-I by the liver, which further contributes to the promotion of normal growth. Thyroid hormones are also required for the normal development and maturation of the teeth, hair follicles, and skin.

Development and maturation of the CNS is markedly affected by thyroid hormones. This effect of thyroid hormones is particularly evident in their absence. The branching of axons and dendrites that normally occurs during fetal and early neonatal development occurs to only a limited degree in the absence of thyroid hormones. Normal myelinization of nerves is also impaired in the absence of thyroid hormones. As a result, severe mental retardation can occur if thyroid hormone is deficient during the period of fetal and early neonatal development. In adults, thyroid hormones also influence mental alertness and responsiveness to external stimuli. The velocity of conduction of action potentials in peripheral nerves can also be shown to vary in response to an excess or deficiency of thyroid hormone secretion.

Conditions of Abnormal Thyroid Hormone Secretion. Many of the specific effects of thyroid hormones that were discussed previously are readily evident in conditions of either thyroid hormone deficiency or thyroid hormone excess.

Hypothyroidism is the condition that exists when there is a deficiency in thyroid hormone production. In most instances, this occurs as a result of a defect in the thyroid gland itself. However, in some cases the defect may occur in either the hypothalamus or pituitary, resulting in a deficiency in TSH production. Because of a slower metabolic rate and a reduced rate of heat production, hypothyroid individuals cannot tolerate cold temperatures. Water tends to accumulate in the skin of these individuals, leading to a condition termed **myxedema,** in which the skin has a thickened puffy appearance. This change is most evident in the appearance of the facial features. The heart rate and the strength of cardiac contractions are reduced, both effects contributing to an overall reduction in the cardiac output. There is also a general slowing of all intellectual functions, leading to a feeling of lethargy and also possibly to some degree of speech impairment.

Excess secretion of thyroid hormones results in **hyperthyroidism.** When tissues are presented with excessive quantities of thyroid hormones, a complex of biochemical and physiological events occur. Hyperthyroid individuals exhibit such symptoms as a markedly increased heart rate, an insensitivity to heat, and considerable weight loss owing to the lipid-mobilizing and protein-catabolic effects of excess thyroid hormones. In contrast to the lethargy that occurs in hypothyroidism, hyperthyroid indi-

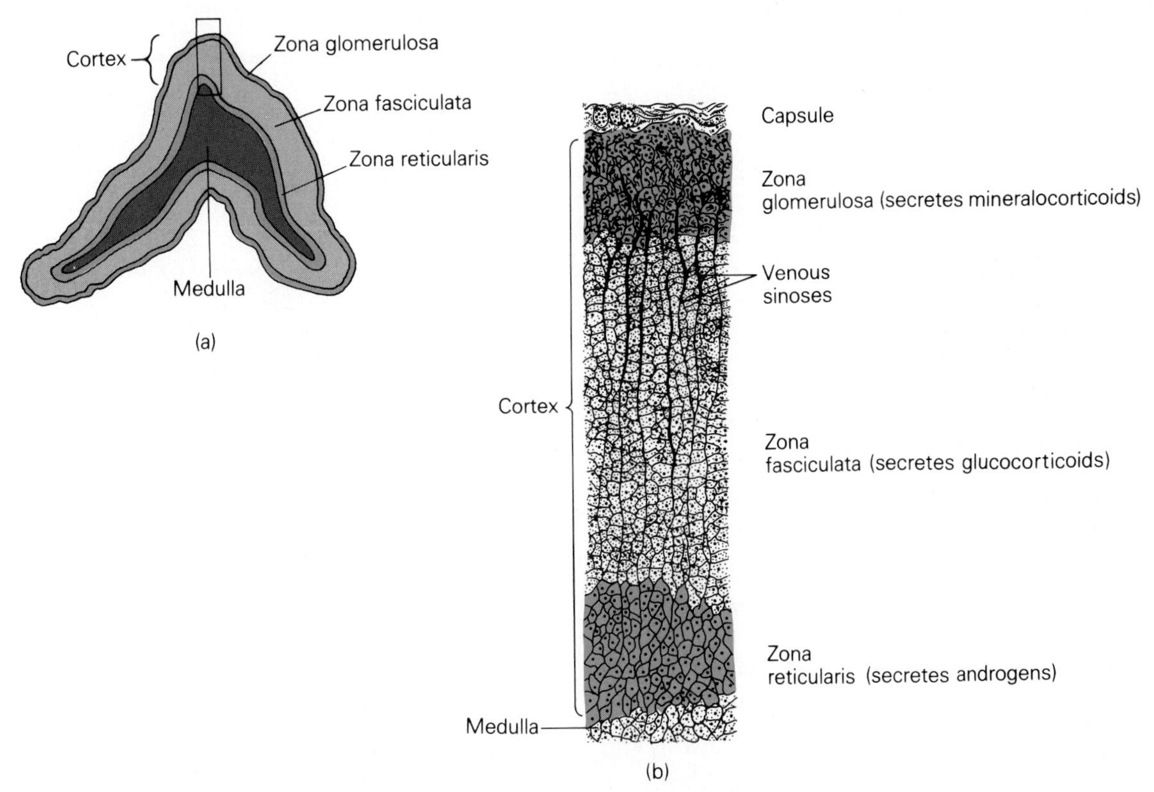

Figure 14–2 The three zones of the adrenal cortex.

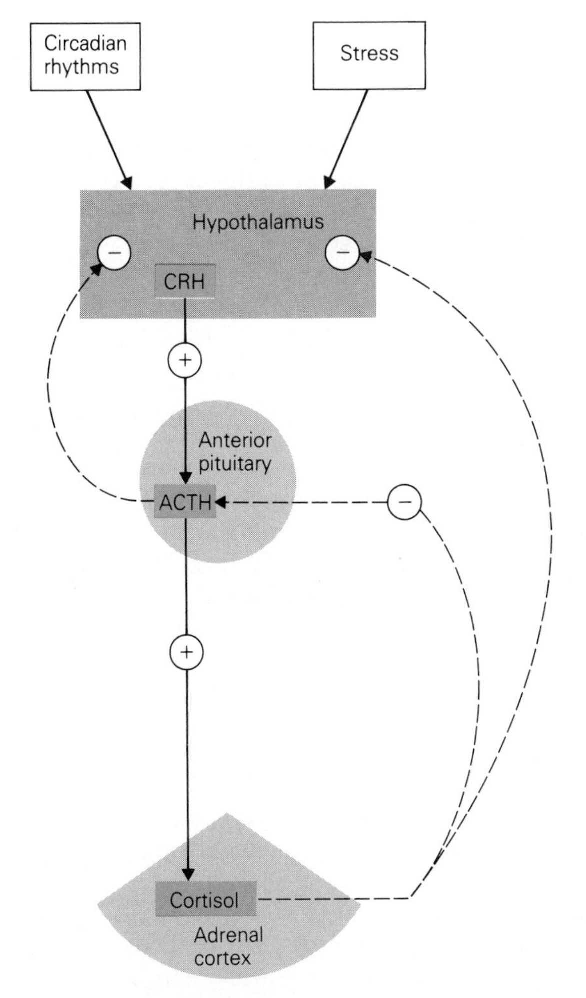

Figure 14–3 The hypothalamo-pituitary adrenal axis. Stimulatory effects are shown by the solid arrow; inhibitory effects by the dashed arrows.

the least amount. This chapter focuses on the physiology of these two steroids. The third, dehydroepiandrosterone (DHEA), is itself a rather weak androgen, but it can be converted in peripheral tissues to testosterone, a much more powerful androgen (see Chapter 30). In males, androgen production by the adrenal is of little or no physiological significance, since far greater amounts are produced by the testes. In females, however, androgens produced by the adrenal gland account for most of the androgens circulating in the blood.

The general pathways involved in steroid hormone biosynthesis were previously described in Chapter 12 (see Fig. 12–11). For reference, a simplified diagram of the pathways is also presented here (Fig. 14–5). Recall that steroid hormones are synthesized and secreted on demand and are not stored within the cells in which they are produced. Therefore, the regulation of steroid hormone secretion occurs at the level of steroid hormone synthesis. As was described in Chapter 12, the first step in the synthesis of all steroid hormones begins with the enzymatic conversion of cholesterol to pregnenolone. In most instances, this step in steroid synthesis is the slowest, or rate-limiting, step. Once pregnenolone is formed, the other steps proceed fairly rapidly. Therefore, the enzyme that catalyzes the metabolism of cholesterol to pregnenolone represents the primary site of regulation of steroid synthesis (see Step A, Fig. 14–6).

Increasing the supply of cholesterol inside the cells also enhances steroid synthesis. Most of the cholesterol that the cells of the adrenal cortex use to

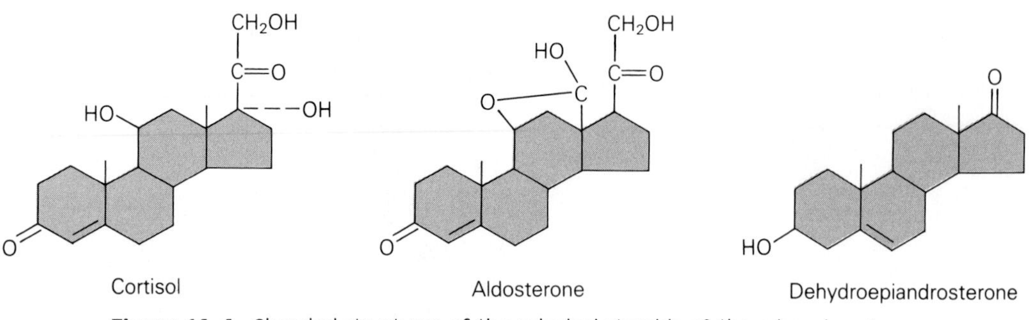

Figure 14–4 Chemical structures of the principal steroids of the adrenal cortex.

make steroids is taken up from the blood, where it is carried in the form of lipoproteins. Cells of the adrenal cortex can remove these lipoproteins from the blood and bring them inside the cell, making the cholesterol available for steroid synthesis (see Step B, Fig. 14–6). The cells store any cholesterol not immediately used for steroid synthesis in the form of lipid droplets. This stored cholesterol can later be rapidly mobilized when needed and used for steroid synthesis (see Step C, Fig. 14–6).

Cells within each of the three different zones of the adrenal cortex contain the appropriate enzymes specifically suited to the production of the steroid hormone that is characteristic of that zone. For example, only cells of the zona glomerulosa contain the enzymes required for the conversion of corticosterone to aldosterone, and thus only these cells can produce aldosterone (see Fig. 14–5). These cells lack the enzyme required for the conversion of pregnenolone to 17-OH-pregnenolone and so are unable to synthesize either cortisol or DHEA. Similarly, cells of the zona fasciculata contain the enzymes that favor the production of cortisol, and cells of the zona reticularis exhibit high levels of the enzymes favoring the synthesis of DHEA.

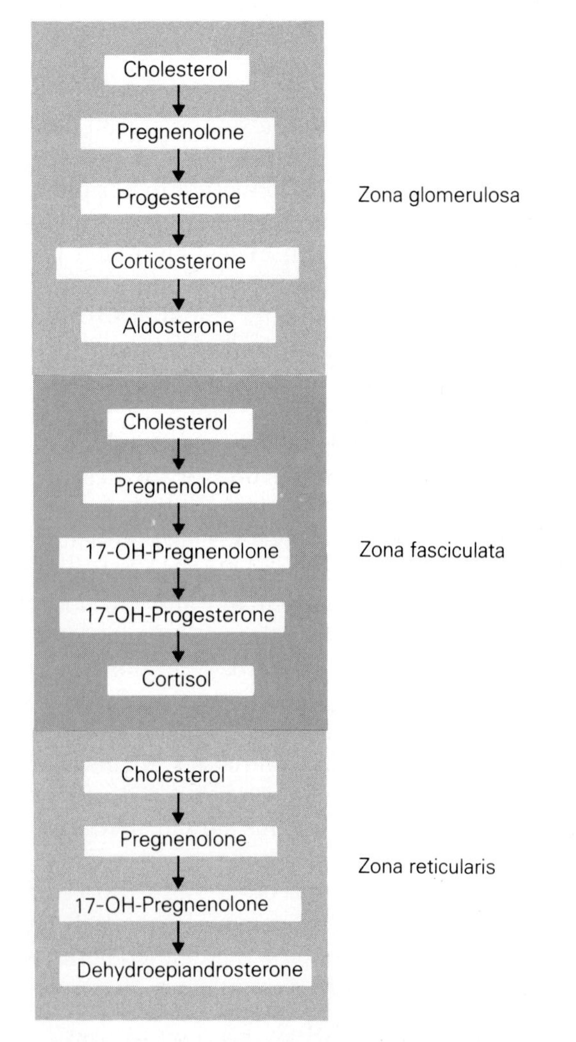

Figure 14–5 Pathways of steroid synthesis in the adrenal medulla.

Secretion and Actions of Glucocorticoids

Regulation of Glucocorticoid Secretion

As illustrated in Figure 14–3, secretion of glucocorticoids from cells of the adrenal cortex is regulated by a sequence of steps beginning with CRH secretion by the hypothalamus. CRH release by the hypothalamus is ultimately determined by inputs from a variety of higher brain centers to the hypothalamus. Therefore, much of the regulation of glucocorticoid secretion is due to these inputs to the hypothalamus. In general, glucocorticoid secretion from the adrenal glands is responsive to two types of stimuli. The first type involves a pattern of secretion that varies over the 24-hour period of a day, and the second category involves increased secretion in response to a variety of specific stimuli.

The Circadian Rhythm of Glucocorticoid Secretion. Figure 14–7 shows the typical changes in plasma cortisol concentrations that might occur over a 24-hour period in a normal individual. As the figure

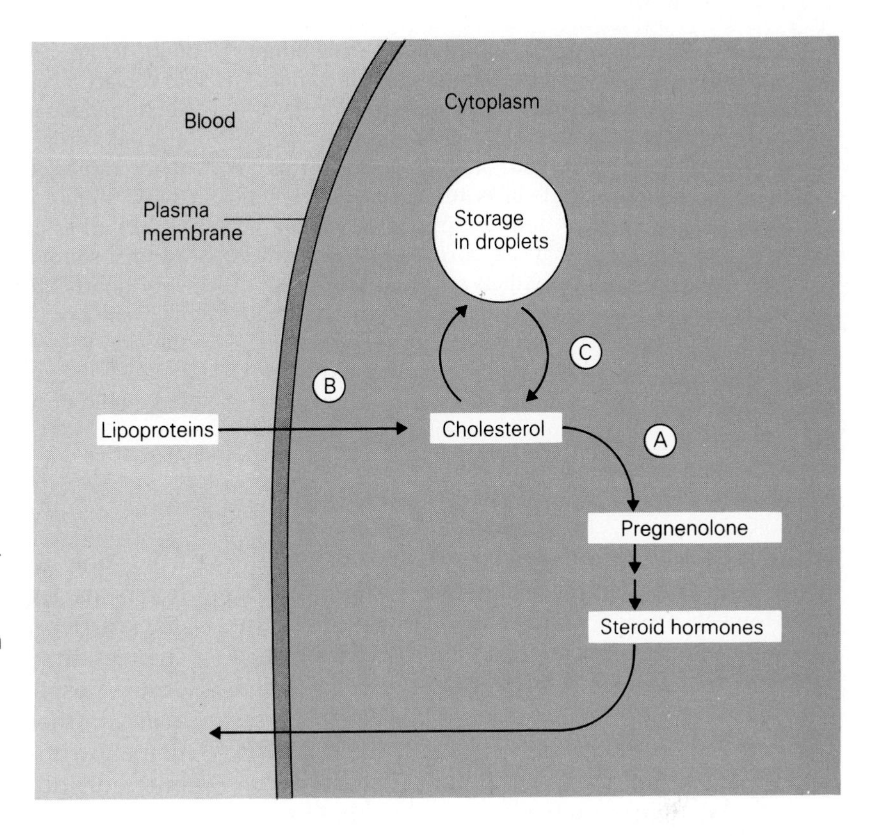

Figure 14–6 The regulatory steps in biosynthesis of adrenal cortex steroids. Step A, the conversion of cholesterol to pregnenolone, is the primary site of regulation of steroid synthesis. Steps B and C, the uptake of cholesterol from the bloodstream and its removal from intracellular storage sites, respectively, are secondary sites of regulation.

Focus on Clinical Use of Steroids

One of the first adrenal steroids to be isolated and synthesized was cortisone, which is structurally related to cortisol. When cortisone was discovered, it was hailed as being a potential wonder drug and was at first believed to have potential for widespread use in medicine. Although these initial expectations of cortisone did not materialize to the extent hoped, a variety of steroids have in fact been found to be extremely useful in medical practice.

Most commonly, corticosteroids are prescribed (1) to replace hormones in patients unable to secrete them in adequate quantities; (2) to suppress inflammatory reactions in situations in which inflammation would be undesirable; and (3) to minimize the immune response in cases of allergy or following organ transplantation. The requirements for steroid hormone action in each of these three cases differ, however, and dictate which type of corticosteroid should be used. Depending on the situation, a physician may want to treat a patient with a corticosteroid that has primarily glucocorticoid activity, while in another situation a corticosteroid that has primarily mineralocorticoid activity would be more beneficial.

The pharmaceutical industry has produced numerous steroid derivatives, each of which has differing degrees of glucocorticoid or mineralocorticoid activity. Organic chemists have synthesized many analogs of adrenal steroids, hoping to produce compounds that have certain desirable qualities exaggerated and other undesirable qualities reduced or eliminated.

The table shows the relative anti-inflammatory and mineralocorticoid potencies of four commonly prescribed steroid preparations. Dexamethasone and prednisone are synthetic steroids; cortisol and aldosterone are naturally occurring steroids that have been described in this chapter.

Steroid	Relative Anti-inflammatory Activity*	Relative Mineralocorticoid Activity*	Ratio of Anti-inflammatory Mineralocorticoid Activity
Dexamethasone	30	0	infinite
Prednisone	4	0.7	6
Cortisol	1	1	1
Aldosterone	0.1	400	0.00025

*Activities are expressed relative to that of cortisol, which is arbitrarily set equal to 1.0.

Note that the ratio of anti-inflammatory/mineralocorticoid potency of these four steroids varies considerably, from a nearly infinite ratio for dexamethasone to 0.00025 for aldosterone.

In addition to the four steroids listed in this table, literally hundreds of synthetic steroids have been produced. How does the physician choose which steroid to use? If the purpose of treatment is to reduce inflammation, then a steroid that has a high ratio of anti-inflammatory activity to mineralocorticoid activity, such as dexamethasone, should be chosen. Large doses of the steroid could then be given to suppress the inflammation while having little or no effect on stimulating sodium retention. If, however, the purpose of steroid therapy is to replace normal hormone levels in an individual with inadequate adrenal function, then a hormone such as cortisol, which has both glucocorticoid activity and a weak amount of mineralocorticoid activity, should be used.

In other cases, pharmaceutical derivatives of adrenal steroids have been formulated that are very poorly absorbed by the skin. These are useful in creams or ointments, which can be applied to the skin to produce a maximal localized effect, but because they are poorly absorbed, these steroids have little or no generalized effect on the body.

indicates, the plasma cortisol concentration is highest around the time of awakening in the morning and lowest around midnight. Because, like the sleep–wake cycle, this general pattern of secretion is repeated approximately every 24 hours, it is termed a circadian rhythm. These changes in cortisol secretion are due to similar circadian variations in CRH and ACTH secretion. This rhythm in cortisol secretion has its origin in higher brain centers that in turn influence the hypothalamo–pituitary system. These circadian rhythms are related primarily to the sleep–wake patterns of the individual rather than by environmental light–dark cycles. Changes in a person's sleep–wake cycle, such as working nights and sleeping days, will result in a temporal shift in their daily rhythm of cortisol secretion.

Cortisol secretion from the adrenal does not occur continuously but in irregular bursts throughout the day, as reflected in the somewhat jagged nature of the curve shown in Figure 14–7. The size and timing of these pulses can vary considerably from one individual to another. However, the bursts are largely buffered by the presence of specific carrier proteins in plasma, such that rapid changes in the free cortisol concentration do not occur. Despite the bursts of secretion and the variability in their timing, the general pattern of secretion shown in Figure 14–7 is fairly consistent from one individual to the next.

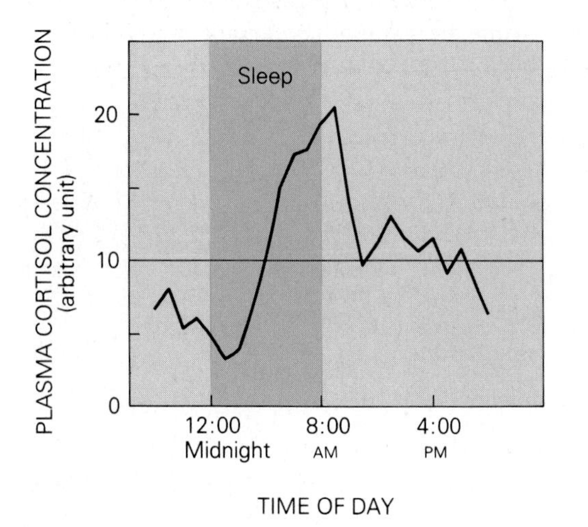

Figure 14–7 Graph of the daily rhythm of the plasma cortisol concentration.

Stress-Induced Glucocorticoid Secretion. A variety of different stress stimuli can also cause significant increases in cortisol synthesis and secretion (Fig. 14–3). Increased secretion in response to stress occurs in addition to the normal circadian rhythm of secretion described previously. Table 14–1 lists several of the different types of stress known to cause increased glucocorticoid secretion. Generally these stresses can be either physical or psychological in nature. Hypoglycemia (low blood glucose) is an important stimulator of cortisol secretion and provides for important adaptive changes that occur with fasting. Practically any type of physical trauma or injury, such as a broken bone, severe burn, infection, or surgery, will result in increased cortisol secretion. Strenuous exercise, such as in competitive athletics, also results in increased cortisol secretion.

Several types of psychological or emotional stress can also stimulate cortisol secretion. Acute anxiety is probably the strongest psychological stimulus for cortisol secretion. Examples of such situations might be the time just prior to a surgical operation or anticipation of final course exams in school. Encountering a new social situation often triggers increased cortisol secretion. In contrast, chronic anxiety is considerably less important as a stimulus for cortisol secretion.

Recent experiments have suggested that ACTH may play a role in promoting learning. A common denominator in many of the psychological stresses that trigger increased ACTH and cortisol secretion is

Complications of Diabetes

Several immediate complications can arise if the glucose concentration in the blood is not checked by insulin; these problems include hyperglycemia, ketoacidosis, and electrolyte imbalance. In addition, long-term complications of the condition include vascular problems, vision problems, and impairment of nerve function.

Acute Complications of Diabetes

As indicated previously, in the absence of either insulin secretion or insulin action, the blood glucose concentration increases markedly, a condition termed **hyperglycemia.** If the hyperglycemia becomes severe enough and the concentration of glucose in the blood exceeds the capacity of the kidneys to recapture the glucose by active transport, glucose will be excreted in the urine (a condition called **glucosuria**). Because of osmotic effects, the glucose that is present in the urine will also attract considerable amounts of water, which will be excreted along with the glucose. As a result, the urine volume and the frequency of urination will increase (a condition known as **polyuria**). As increased amounts of water are lost in the urine, **dehydration** of the individual can also occur.

In Type I (insulin-deficient) diabetics, the unopposed actions of glucagon result in increased ketone formation by the liver. The ketones are acids, and when they are produced in sufficiently large amounts, the normal acid/base balance in the body is considerably disturbed, resulting in the condition termed **ketoacidosis.** As with glucose, if ketones reach a high enough concentration in the blood, they will begin to "spill over" into the urine. As this spillover occurs, ketones also carry cations, such as sodium and potassium, with them into the urine. Thus, accompanying severe ketoacidosis is a loss of these ions, resulting in an **electrolyte imbalance** in the body. If left untreated, severe ketoacidosis accompanied by dehydration can rapidly result in coma and death.

Chronic Secondary Complications of Diabetes

In addition to the possible acute complications that may occur, several secondary complications usually accompany longstanding diabetes. These secondary complications usually involve gradual changes that develop over a period of years and are often of such a nature that they considerably shorten the life expectancy of these individuals. The most common secondary complications of diabetes are seen within the vascular system. Changes much

18 to 24 hours later. By tailoring the type and dose of different insulin preparations to an individual patient's needs, it is possible that only a few injections per day are required.

In the past few years, advancements in electronics and biotechnology have been applied to the field of insulin delivery as well. One example from this area is the development of an artificial pancreas. The essential elements to such a device include a glucose sensor to measure the blood glucose concentration, a pump and reservoir to deliver insulin to the bloodstream, and a computer to monitor information from the glucose sensor and make appropriate adjustments in the rate of insulin delivery by the pump. Initially, the equipment needed to perform these functions nearly filled a room. However, considerable progress has been made towards the miniaturization of this equipment. The hope is that someday an artificial pancreas can be produced that is small enough to be implanted and left in place for prolonged periods of time.

Another approach to the problem of insulin delivery that has shown considerable promise is that of islet transplantation. Although the method has been used only in experimental animals, there is considerable hope that the technique will one day be applied in humans, since in theory such a procedure could "cure" a person's diabetes. The basic technique of islet transplantation involves separating the islets of Langerhans from the exocrine pancreas and then giving the recipient animal an injection containing a suspension of the isolated and purified islets. The primary obstacle to the adoption of this technique, as is the case with many transplant procedures, is rejection of the transplanted tissue by the immune system. As techniques for immunosuppression improve, the likelihood that islet transplantation will be used in the treatment of human diabetics also increases.

Finally, yet another major advance has been the production of insulin by the application of recombinant DNA technology. To accomplish this, researchers insert the human proinsulin gene into the DNA of bacteria. These altered bacteria produce human proinsulin and can be grown in large quantities in fermentation vats. The proinsulin is harvested, purified from the bacteria, and then converted into human insulin by the proteolytic removal of the C-peptide. Using this approach, virtually limitless amounts of authentic human insulin can be produced, ensuring the widespread availability of insulin despite the growing numbers of diabetics in our population.

like those seen with atherosclerosis lead to the narrowing of larger blood vessels in the brain, heart, and lower extremities. The resulting reduction in circulation to these areas may result in stroke, heart attack, or loss of limb, respectively. Lesions in the microvasculature (small blood vessels and capillaries) are also common in diabetics and are most detrimental in the kidney and in the retina of the eye. These changes can result in kidney disease and in the condition termed **diabetic retinopathy.** Diabetic retinopathy is due to the deterioration of blood vessels that nourish the retina. As the blood vessels in the retina break down, scar tissue is formed that eventually interferes with the light-sensing function of the retina. Each year, several thousand diabetics become legally blind as a result of diabetic retinopathy.

Another common secondary complication of diabetes involves impairment in nerve function (**diabetic neuropathy**). Fibers of the ANS are often involved, frequently resulting in abnormalities in bladder or gastrointestinal tract function, or in impotence. Diabetic neuropathy also frequently involves peripheral sensory nerves, resulting in loss of feeling in the lower limbs in particular.

Summary

I. The functional units of the endocrine pancreas are groups of cells known as the *islets of Langerhans.* The islets comprise four major cell types: the alpha cells, which secrete glucagon; beta cells, which secrete insulin; delta cells, which secrete somatostatin; and F cells, which secrete pancreatic polypeptide.

II. A. Insulin consists of an A-chain and a B-chain held together by disulfide bonds. Insulin is first synthesized as proinsulin. The most important physiological stimulus for insulin secretion is an increased blood glucose concentration. Other factors that stimulate insulin secretion from the pancreas include amino acids, several of the GI hormones, acetylcholine, sulfonylureas, and glucagon. Inhibitors of insulin secretion include somatostatin, epinephrine, and norepinephrine.

B. Glucagon is a simple polypeptide first synthesized in the alpha cells as proglucagon. Glucagon secretion is stimulated by a decreased blood glucose concentration. Conversely, increased blood glucose inhibits glucagon secretion. Amino acids, fatty acids, and both branches of the ANS also stimulate glucagon secretion. Insulin and somatostatin both inhibit glucagon secretion.

C. Somatostatin is first synthesized in the delta cells as prosomatostatin. Somatostatin secretion is stimulated by increased blood glucose, glucagon, and amino acids.

III. A. Insulin promotes fuel storage and protein synthesis and therefore has anabolic effects. Insulin rapidly lowers the blood glucose concentration by increasing glucose uptake into muscle and adipose tissue by facilitated diffusion. Insulin also stimulates glycogen formation and glycolysis in its target cells and decreases glucose output by the liver. In liver and adipose tissue, insulin *stimulates fatty acid synthesis* and promotes the *storage* of fatty acids as *triglycerides*. In adipose tissue, insulin promotes the uptake of lipoproteins into fat cells from the blood and inhibits lipolysis. Insulin stimulates the accumulation of protein in most tissues of the body, promoting a positive nitrogen balance.

B. Glucagon, which raises the intracellular concentration of cAMP, has catabolic effects on metabolism. The primary target tissue of glucagon is the liver. Glucagon increases the blood glucose concentration by stimulating glycogenolysis and gluconeogenesis in the liver. Glucagon also stimulates lipolysis and stimulates ketogenesis in liver. Fatty acids and ketones are alternative energy sources for many tissues and thus their increased availability has a glucose-sparing effect. Glucagon provides increased substrate for gluconeogenesis in liver by stimulating protein degradation and amino acid transport.

C. Because of the opposing actions of insulin and glucagon, the status of nutrient flow and metabolism in the body is determined by the relative amounts of

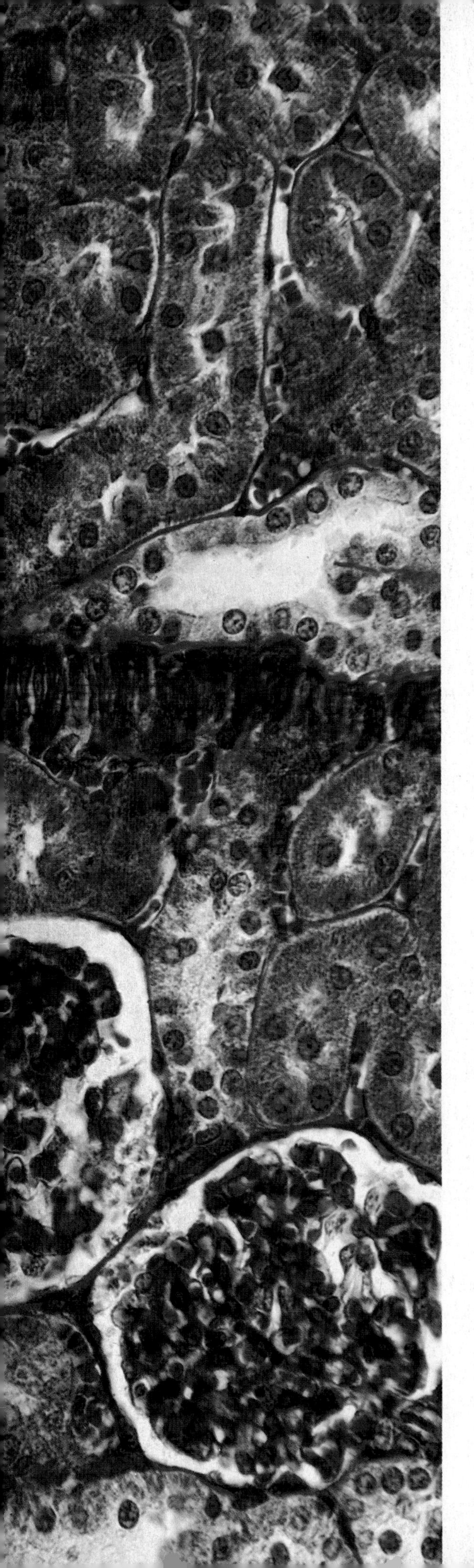

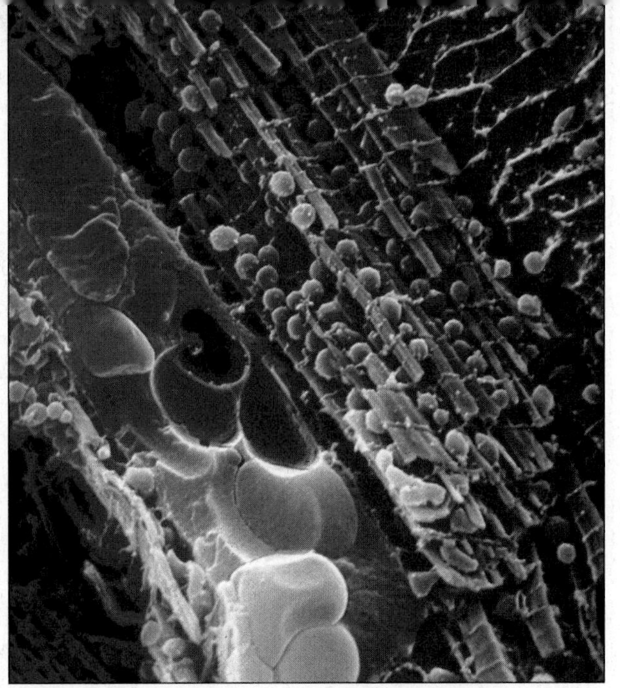

Integrative Organ Functions

III

16

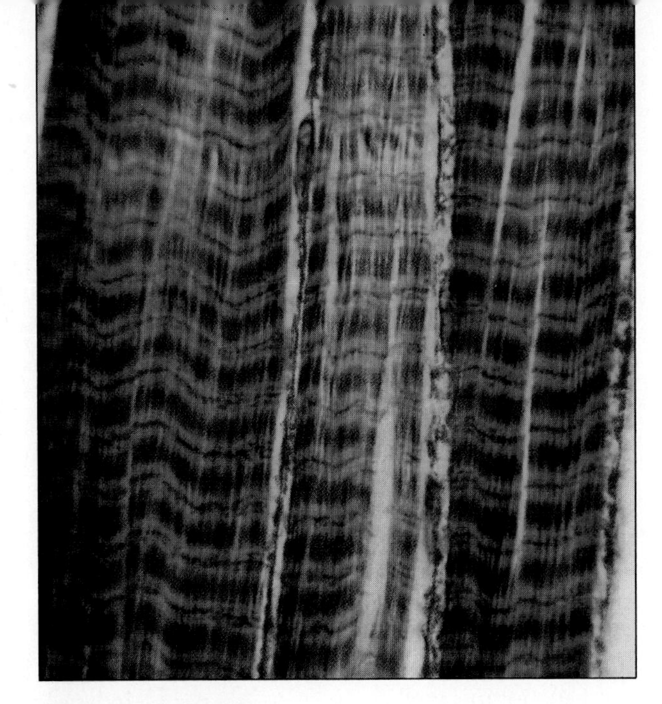

Muscle

There are many types of muscle in the body. While we are most familiar with those large muscles that are attached to bones and are responsible for our ability to move, breathe, and manipulate our environment, many other muscles are used for bodily activities of which we are usually unaware. All of our muscles have important features in common. They may be similar in structure, functions, and other aspects. This chapter covers all of these aspects of muscle, and the critical roles of muscle in the body economy will become apparent. Because skeletal muscle is the type of muscle best understood, most of the chapter will be devoted to it.

Overview: The Roles and Types of Muscle

The range of activities that muscles carry out in the body is extremely broad, so it is not surprising that muscles show a wide range of functional adaptations that specifically suit their many tasks in the overall function of the body.

Muscle as an Effector

As described in Chapter 10, our central and peripheral nervous systems function by gathering information and instructing muscles to take useful action in response. This coordination between nerve and muscle enables us to walk, talk, digest our food, defend our bodies, propagate our species, and do almost everything else that we do. In addition to effecting changes prescribed by the nervous system, muscles also help to regulate what goes into that system by adjusting the sensitivity of our sense organs. Some muscles are under neural control, while others are more independent. Very few muscles, however, are totally independent. The degree of neural control is a feature of muscle that serves to adapt various muscles to their special roles. Some of the many functions of muscle in relation to the nervous system are shown in Figure 16–1.

Our other important control system, the endocrine system, also has muscle as one of its important effectors, or means of expression, although this function may not be very obvious when our bodies are functioning well.

Muscle as a Motor

Muscles can do what they do because they are a sort of **biological motor.** Like mechanical or electrical motors, they consume fuel, do useful work, and waste some of the energy that they consume as heat. Muscles produce extra heat because they, like other motors, are not completely efficient. That is to say, they cannot turn all of the energy that they consume into mechanical work. The heat that they produce as a result of this inefficiency may at times be very important in overall bodily function. Under most conditions of activity, most of our body heat is produced as a result of the contraction of muscles. When the body temperature falls, brisk physical exercise (which may be relatively efficient) or shivering (which is very inefficient from a mechanical standpoint) provides the necessary warmth. During heavy exercise or under the stress of high environmental temperatures, our problem may involve getting rid of the excess heat produced by muscular contraction. Many of our conscious "cooling" mechanisms themselves also involve the use of muscle, as may become apparent when the work of fanning to cool off produces more heating than cooling.

Muscle as a Regulator

Muscles also help to regulate many important body functions. Muscles control the movement of substances through the tubular structures (such as blood vessels and intestines) of the body and expel those substances from the body at the proper time. For example, the regulation of blood pressure involves a complex interaction between the heart muscle, which pumps the blood, and other muscles that control the diameter of the blood vessels. Other

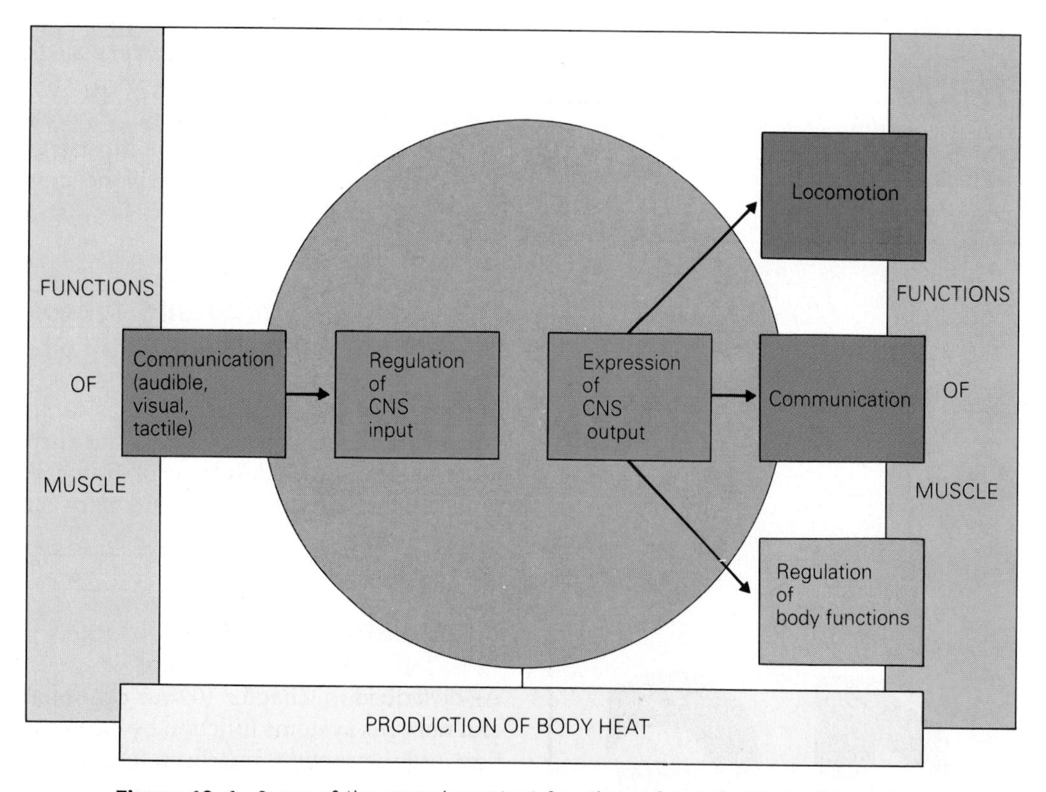

Figure 16–1 Some of the more important functions of muscle. Most of the categories shown could be divided into even more functions.

regulatory mechanisms in which muscles play a central role include the maintenance of an upright posture and the control of the body temperature.

Classification of Muscle Types

Several different sets of criteria are used to place muscles in separate categories relating to location and function, microscopic structure, or mode of control and action. While these topics serve as convenient ways to describe and discuss particular aspects of muscle, as with any attempt at classification, some exceptions will have to be noted. The most usual sort of classification is shown in Figure 16–2.

Classification of Muscles by Location and Function

The most obvious muscles in the human body are called **skeletal muscles.** These are the large muscles that are used, among other things, for locomotion and for maintaining body posture. They usually attach to the skeleton and move the jointed bones with respect to one another. In some instances, the skeletal attachments may be indirect or may exist at only one end of the muscle (in the case of the tongue, which is a skeletal muscle), or they may not exist at all (as is the case for the upper end of the esophagus). The bulk of skeletal muscles, however, are attached to bones at both ends.

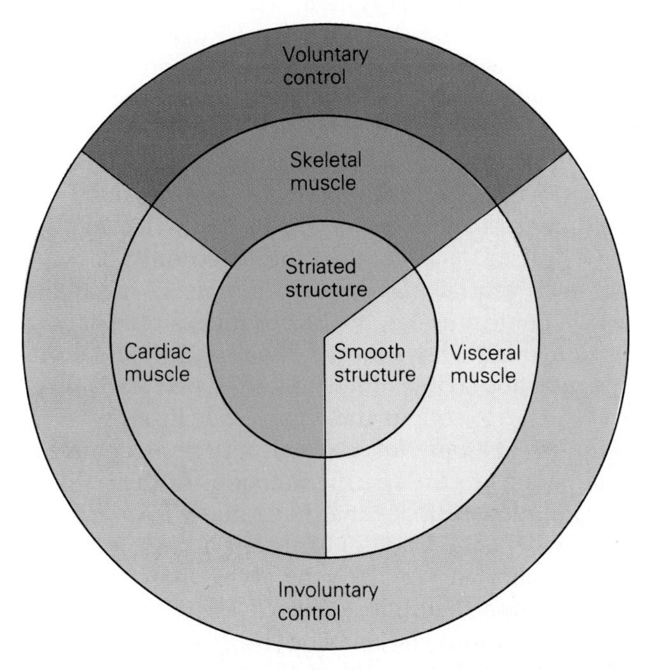

Figure 16–2 A simplified view of the classification of muscle. Although there may be some difficulties with each way of classification, muscle may be classified by three different means: by location (skeletal, cardiac, or visceral); by structure (smooth or striated); and by mode of control (voluntary or involuntary). In reality, some categories and functions overlap.

Muscles that line the walls of the internal organs (the viscera) are called **visceral muscles.** Muscles of this type are also located in nonvisceral organs such as blood vessels and in sensory organs such as the eye, where they aid in focusing and adapting to different levels of lighting.

The location of **cardiac muscle** is obvious from its name; it makes up the pumping muscle mass of the heart. Cardiac muscle is found only in the heart (although it may extend a little way into the large vessels that enter and leave the organ). While it is confined to a relatively small anatomical region, its critical role in body function in health and disease gives great importance to its understanding and study.

Classification of Muscles by Structure

When viewed at the level of microscopic structure, muscle falls into two general categories: striated muscle and smooth muscle. The term *striated* means "striped" and refers to the regular cross-striped appearance of the muscle cells under a microscope. This appearance is an important clue to the way in which these muscles function on a molecular level, a topic that will be treated in some detail later in this chapter.

Skeletal muscle is **striated muscle.** Cardiac muscle is also a striated muscle. While its cells are very much smaller than those of skeletal muscle, its molecular function is similar in many respects. Many of the details of the function of skeletal muscle have been found to apply to cardiac muscle also. In addition, cardiac muscle also shares some functional similarities with visceral muscle.

Visceral muscle cells lack the striations found in the other muscle types and are therefore referred to as **smooth muscle.** Their structure provides fewer clues to their function than in the case of skeletal muscle, although many aspects of their internal processes are becoming better understood as research in this field progresses.

Classification of Muscles by Mode of Action and Control

Because most movements of skeletal (striated) muscle occur in response to a conscious (willed) effort, these muscles may be called **voluntary muscles.** However, many actions of skeletal muscle, such as the acts of breathing or walking, are almost automatic and thus are in a sense involuntary. The strictest basis for this type of classification is that skeletal muscles, in order to function, must receive a specific signal or signals from the central nervous system (CNS). In the event of an accident or illness that disconnects a skeletal muscle from its nerve supply, paralysis results.

On the other hand, the actions of most smooth (visceral) muscles are **involuntary.** In many cases these involuntary actions are initiated by the **autonomic nervous system (ANS)** in response to some internal reflex adjustment that may take place automatically. Smooth muscle is also capable of acting on its own, sometimes in response to hormonal or environmental factors. Voluntary control of some smooth muscle functions is also possible; for instance, specially trained persons using the techniques of "biofeedback" can cause blood vessels to dilate or constrict. In the majority of situations, however, the function of smooth muscle is completely involuntary.

Cardiac muscle is also an involuntary muscle, and no signals from the nervous system are required to start or maintain its contraction. The heart will continue to beat indefinitely after all nerve connections to it are removed, although the rate and strength of the heartbeat are no longer under external control. Because of the special nature of the cardiac muscle, it is not usually described as being either voluntary or involuntary.

Smooth muscle is itself subject to some further classification in the area of its mode of control. Some smooth muscle is called **multiunit smooth muscle;** this muscle is closely controlled by the nervous system. Since the nervous elements involved in its control are part of the ANS, it is not subject to voluntary control. Other smooth muscle, that which makes up the bulk of our visceral organs, is called **unitary smooth muscle.** This sort of muscle is less strictly controlled by the nervous system, and large numbers of its cells function together as a single unit. The distinctions mentioned here should become clearer later in this chapter.

A summary of the three classifications is provided in Figure 16–2. When allowance is made for the exceptions noted previously, the classifications are useful for discussing and comparing the various muscle types. Within each type are important variations, and these will be taken up as the specific details of each muscle type are discussed.

The Structure of Muscle

The study of muscle structure can be done at two different levels. At one level, large-scale structural aspects of muscle can be observed and described without microscopic aid. The position, size, shape, and attachment points of a skeletal muscle give important information about its function. At another level, light microscopes and electron microscopes reveal details of the fine structure responsible for the actual mechanism of contraction. Muscle is an excellent example of the way in which the structure and function of a biological tissue are intimately related. For this reason we will discuss in detail the structure of striated muscle before going on to other muscle types. Striated muscle can serve as a basis of comparison for the description of other muscle types.

The Structure of Skeletal Muscle

Contraction, the major function of a skeletal muscle, involves its shortening and the resulting movement of parts of the skeleton. The large-scale structure of skeletal muscle reflects this function, and the organization of the cells and tissues reflects the structural and metabolic adaptations necessary for functions in which large forces are generated and large amounts of energy are expended.

Gross Structure of Skeletal Muscle

In mechanical terms, the skeleton works with the muscles as a **lever system.** The term *lever system* refers to the fact that groups of muscles working together with bones permit a larger range of movement than could muscles acting on their own. Muscles attach to bones by means of very strong connective tissue structures called **tendons** (Fig. 16–3). The more stationary (and **proximal**) attachment is called the **origin** of a muscle, and the other **(distal)** end is called the **insertion.** The wide or thick central portion of a muscle is called the **belly.** Some muscles, the **biceps** for example, have a double origin (*biceps* means "two heads") and a single insertion, while others (e.g., the **pectoralis**) may have a narrow origin and a very wide insertion. The variable anatomy of skeletal muscles represents adaptations to the specific functions of the individual muscles. Muscles that have many fibers placed in parallel (side-by-side) can exert a large force but cannot shorten rapidly or by a very great amount. On the other hand, long muscles with many fibers in series (end-to-end) may be able to shorten rapidly but may develop less force.

Because most skeletal muscles are attached to bones at either end, the range of motion of a muscle itself may be limited by the skeletal system. The needed range of motion must then be multiplied by the lever system that the skeleton provides. This

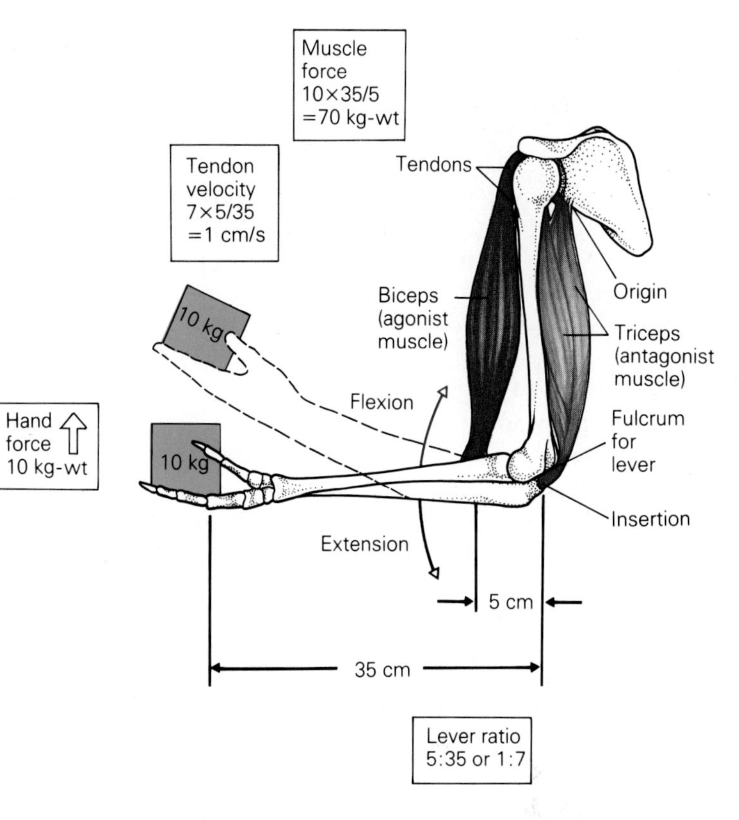

Figure 16–3 Antagonistic muscle arrangements and the skeletal lever system. The major muscles that move the upper arm work in opposition to each other. The biceps flexes the arm, while the triceps extends it. The tendon of the biceps attaches at a point approximately 5 cm away from the fulcrum of the lever system (the elbow joint), and the hand is 35 cm away. Therefore, the upward force exerted by the muscle must be seven times the downward force supplied by a weight in the hand. However, any shortening of the muscle is multiplied by seven, meaning that the hand will move seven times as fast and as far as does the lower end of the biceps muscle.

arrangement has some important consequences. Large amounts of muscle shortening, as we shall see, can significantly affect the amount of force a muscle can exert, and limiting the range of muscle shortening helps to compensate for these effects. However, this arrangement also means that the muscle must exert a much greater force at its insertion than is actually produced at the end of the limb. For instance, in the human forearm, the lever is hinged at the elbow. The hand is approximately seven times as far from the joint as is the insertion of the biceps. This means that if the hand is holding an object with a mass of 10 kilograms, the tendons of the biceps must exert seven times as much force (see Fig. 16–3). This situation can lead to serious muscle and tendon injury in athletes and in people who do heavy manual labor. On the other hand, it means that the hand can move seven times as fast as the biceps can shorten. Such an increase in speed is often valuable. For instance, it allows the leg muscles to propel the body at a high rate of speed, or to cause a person to jump up in the air for a distance far larger than the distance over which the muscles themselves can shorten.

As mentioned in Chapter 10, in most situations, especially in the limbs, skeletal muscles are arranged in so-called **antagonistic pairs.** When a muscle that moves a joint in one direction (e.g., **flexion,** such that a joint closes and a limb assumes a more withdrawn position), another muscle or muscles move the joint in the opposite direction, toward a more extended position (i.e., **extension**). This arrangement is necessary because muscles can only *pull;* they cannot *push.*

Skeletal **muscle cells** (also called **muscle fibers**) are quite large compared to other cells. Typical diameters are on the order of 100 micrometers, while cells from some large muscles may be many centimeters in length. The way in which muscle cells are assembled into the complete muscle is diagrammed in Figure 16–4. In an intact muscle, shown in cross section in Figure 16–4*a*, the individual muscle fibers are surrounded by a delicate connective tissue sheet called the **endomysium.** Bundles of muscle fibers are grouped into **fasciculi** (shown in longitudinal section in Fig. 16–4*b*) by a connective tissue called the **perimysium,** and the entire muscle is covered by a connective tissue sheet called the **epimysium.** During physical exercise, muscle requires a plentiful supply of oxygen and nutrients, and it must rapidly get rid of waste products. Accordingly, muscle is supplied with a plentiful network of blood vessels arranged among the fibers so that no muscle cell is very far away from a blood vessel.

Both the connective tissue sheets and the muscle fibers themselves are connected at either end into a very tough connective tissue extension of the muscle that makes up the tendon. Some tendons are thick and short, like the one that provides the inser-

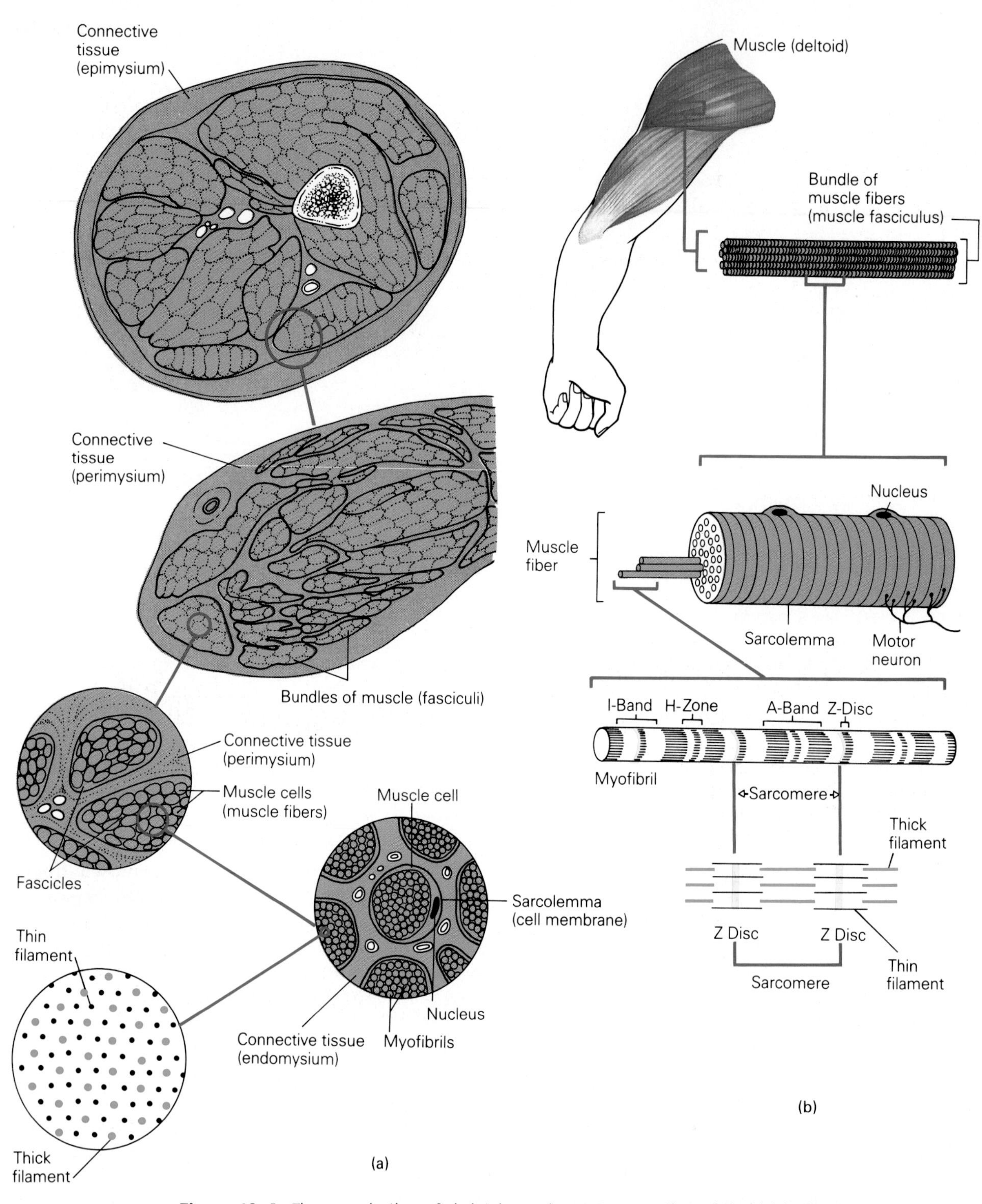

Figure 16–4 The organization of skeletal muscle. (*a*) Cross section of the thigh dissected into smaller and smaller parts. (*b*) Longitudinal section of the deltoid muscle, showing the bundles of fibers (fasciculi) that make up a whole muscle. The individual muscle fibers, or cells, contain long slender structures called myofibrils. These are made up of myofilaments, which are organized into sarcomeres.

tion of the biceps. Some tendons are very long; the muscles that move the fingers are actually located in the forearm, and their tendons run across the wrist joint and the back and palm of the hand to connect the muscles to the bones of the fingers.

Cellular and Molecular Structure of Skeletal Muscle

At the cellular level of organization, there is a remarkably clear relationship between the structure and function of muscle. The molecules that make up the contractile substance of muscle have a dual role. Not only are they the major structural elements of the muscle cell; they are also the enzymes that convert the biochemical energy (derived ultimately from the food we eat) into mechanical energy.

When viewed through a light microscope, skeletal muscle cells show a repeating pattern of dark and light crossbanding. This striped, or striated, pattern gives skeletal muscle one of its names; more importantly, it provides a clue as to how striated muscle actually works. Analysis using electron microscopy and other methods has revealed that the dark and light crossbands, or striations, are composed of two types of protein filaments, with each type being arranged in sets side by side across the cell. These protein filaments are called **myofilaments,** and the two types differ from each other in their size and internal structure. The two sets overlap each other in a characteristic way that changes as the muscle is lengthened or shortened.

The Structure of a Sarcomere. The two sets of myofilaments are organized into a structure called a **sarcomere;** this is the basic unit of the contractile machinery of the skeletal muscle. Figure 16–5 shows both a cross-section and longitudinal section view of the microscopic structure of skeletal muscle. The center portion of the sarcomere (the **A-band**) is made up of **thick filaments,** and the lateral regions are made up of **thin filaments.** A structure called the **M-line** runs across the middle of the A-band

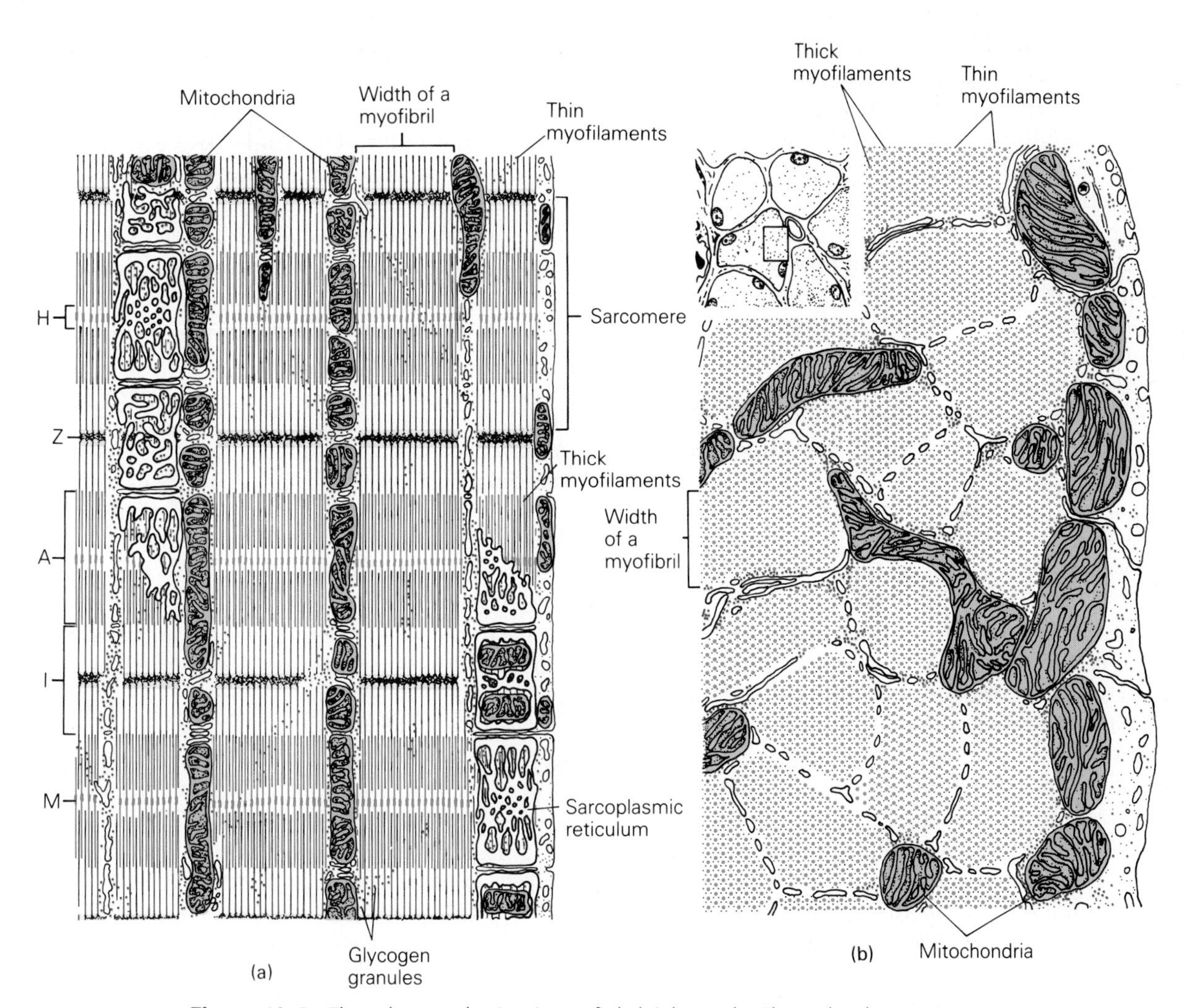

Figure 16–5 The microscopic structure of skeletal muscle. These drawings were made from electron micrographs of skeletal muscle. (*a*) A longitudinal section showing the arrangement of the sarcomeres. The letters H, Z, A, I, and M refer to the parts of the sarcomere diagrammed in Figure 16–6. (*b*) A cross section of the same tissue. Several cells are shown in the inset, at the upper left, and the area in the rectangle is magnified in the rest of the figure. (Modified from Krstic, R.V., *Ultrastruktur der Saugetierzelle.* Springer-Verlag, Berlin, 1976.)

and holds the thick filaments in side-by-side alignment. The thin filament array interdigitates (overlaps) with the thick filament array, and the depth of its penetration depends on the overall length of the muscle. The region where only thin filaments are present is called the **I-band.**

Both sets of filaments are packed in such a way that a cross-sectional cut through a muscle fiber shows a hexagonal pattern; this is the most compact way in which the filaments could be packed (Fig. 16–6*a*). If a cut is made in a region in which sets of thick and thin myofilaments overlap, then the spaces provided by the hexagonal array of the thick filaments will be occupied by an interlocking hexagonal array of the thin filaments. "Crossbridges" from the thick filaments will be seen to reach out directly toward the thin filaments.

The usual names given to the parts of the sarcomere are shown in Figure 16–6*b*. The center of each I-band is marked by a structure called the **Z-line,** which serves to mark the boundary of the sarcomere. (Sometimes this structure is called the **Z-disc** in order to emphasize its two-dimensional nature.) The function of the Z-line is to connect the ends of the thin filaments from one sarcomere to those of the next one in line. The Z-lines also have cross-connections that keep the thin filaments in register with each other in a lateral direction. In the center region of the A-band where the thin filaments of the I-band have not penetrated is a somewhat lighter region called the **H-zone** (the M-line runs through the center of the H-zone, since the filament overlap is symmetrical).

During the process of contraction, in which the muscle may shorten considerably, *neither the thick nor the thin filaments change their length.* All of the change in the length of the muscle is accounted for by a greater or lesser degree of overlap between the thick and thin filaments. This means that while the muscle shortens, the length of the A-band will stay the same while that of the I-band will decrease (see Fig. 16–6*c*). The thin filaments themselves will not change in length. Shortening of the muscle is brought about by a cyclic "rowing" motion of the crossbridges, in which they successively attach to the closest available thin filament site, change their shape, then let go of the actin and re-attach further along. This process will be discussed in more detail.

Hundreds of sarcomeres connected in series (end-to-end) form a functional unit called a **myofibril,** which runs for the length of the cell. Most muscle cells contain a large number of parallel myofibrils, and the intracellular spaces between them are occupied by **mitochondria,** and by molecules of **glycogen,** a source of chemical energy. Space between the myofibrils is also occupied by the **internal membrane system.**

Structure of the Myofilaments. **Thick filaments** are made up of a very large protein called **myosin** (Fig. 16–7). A thick filament is composed of about 300 to 400 myosin molecules. Each myosin molecule has a long rod-shaped section (a structural portion sometimes called the **tail region**) and a double globular head at one end. The head portion of the molecule has the biochemical and enzymatic properties that participate in the actual contractile function. On the head portion are protein chains called "light chains" because of their molecular weights. These will be discussed in detail later.

When myosin molecules are packed together to form a thick filament, the tail regions of the molecules make up the structure of the filament itself, while the head portions project from the sides of the filament. The projections of the myosin heads are what form the structures called **crossbridges.** Because of the way the molecules are packed, the heads project from the filament in a radial fashion, and the projection makes a full turn for every six myosin molecules. The angle between the successive myosin heads is thus 60 degrees, and they project out of the filament every 14.3 nanometers. In this way, they form a helical pattern (like a spiral staircase) that winds down the length of the thick filament.

The tail regions of molecules at either end of the filament point toward its center, making the thick filament **symmetrical** about its center and **bi-directional.** It also means that there is a **bare zone** in the middle of each thick filament; this is an area containing only the tail portions of myosin molecules. These arrangements are critical to the way in which skeletal muscle functions.

A globular protein called **actin** makes up the **thin filaments.** The slightly oblong actin molecules are each about 5.5 nanometers in diameter and are joined end-to-end to form a long structure that re-

Figure 16–6 The structure of a sarcomere. (*a*) Diagram → showing the relationship of the longitudinal and cross-sectional arrangement of the myofilaments. (*b*) The A-band is made of thick filaments arranged side by side, and the thin filaments extend from the A-line into the A-band. The I-bands are made up of the portions of thin filaments where there is no overlap between thick and thin filaments, and the H-zone is the region of the A-b and into which the thin filaments do not extend. The M-line runs down the center of the A-band and appears to help in maintaining the uniform positions of the thick filament array. A sarcomere may also be defined as the distance between M-lines, with the Z-line at its center, since the sarcomeres repeat hundreds of times down the length of a myofibril. (*c*) During contraction, the width of the A-band stays constant, while the H-zones and the I-bands become narrower, and the Z-lines move closer together.

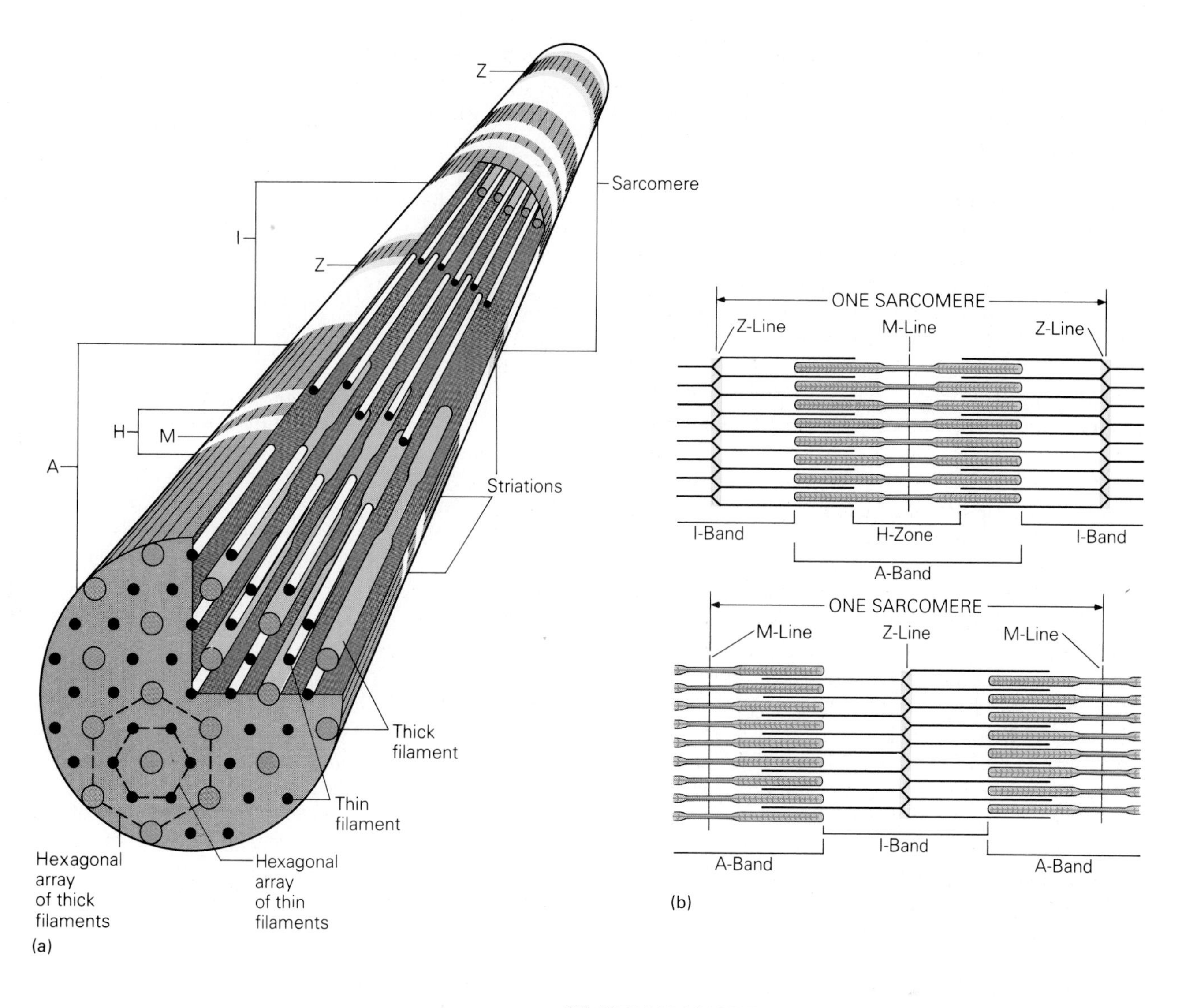
Z
Sarcomere
I
Z
H
M
A
Striations
Thick filament
Thin filament
Hexagonal array of thick filaments
Hexagonal array of thin filaments
(a)
ONE SARCOMERE
Z-Line
M-Line
Z-Line
I-Band
H-Zone
I-Band
A-Band
ONE SARCOMERE
M-Line
Z-Line
M-Line
I-Band
A-Band
A-Band
(b)

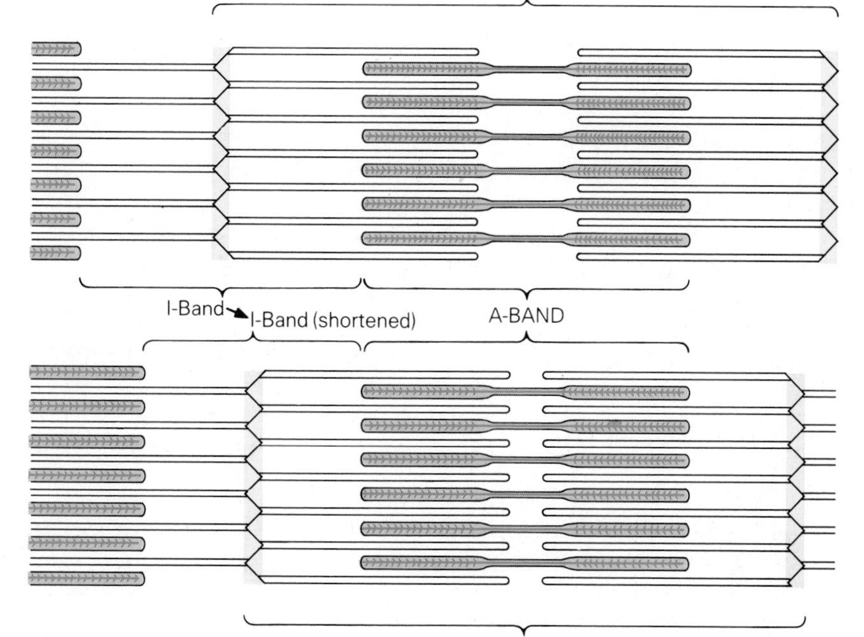
RELAXED SARCOMERE
I-Band
I-Band (shortened)
A-BAND
CONTRACTED SARCOMERE
(c)

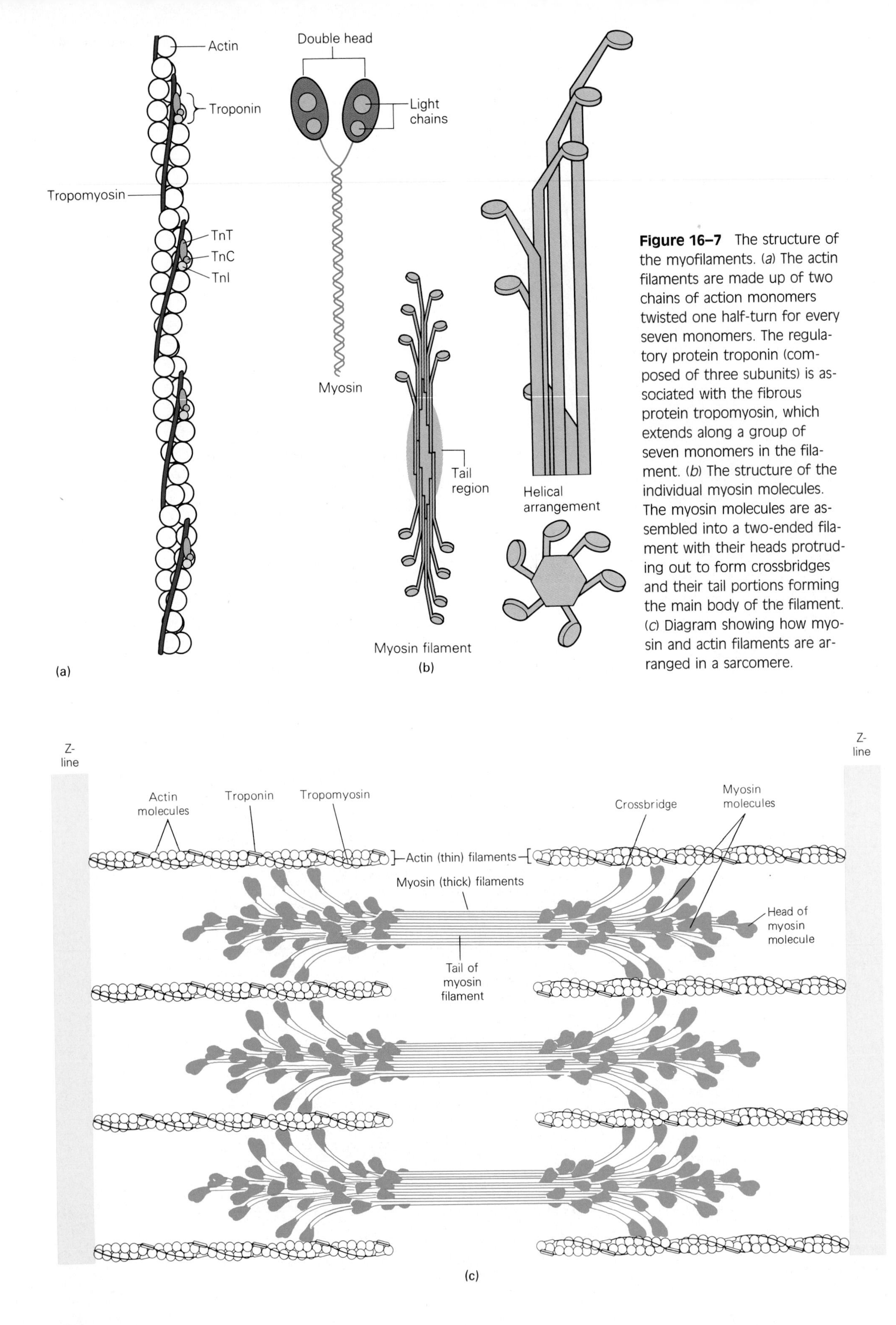

Figure 16–7 The structure of the myofilaments. (*a*) The actin filaments are made up of two chains of action monomers twisted one half-turn for every seven monomers. The regulatory protein troponin (composed of three subunits) is associated with the fibrous protein tropomyosin, which extends along a group of seven monomers in the filament. (*b*) The structure of the individual myosin molecules. The myosin molecules are assembled into a two-ended filament with their heads protruding out to form crossbridges and their tail portions forming the main body of the filament. (*c*) Diagram showing how myosin and actin filaments are arranged in a sarcomere.

sembles a bead-chain. Two such chains are wound about each other so that the pair makes a half-turn every seven actin monomers. This means that the thin filaments are also helical structures, with a half-pitch of approximately 35 to 37 nanometers (compared with the 42.9 nm pitch of the thick filaments). Each half-turn of seven actin molecules is also associated with a long fibrous protein molecule called **tropomyosin.** Each of these tropomyosin molecules bears a protein complex called **troponin.** The troponin is in turn composed of three subunits whose names are **troponin-I, troponin-T,** and **troponin-C.** Although it is the actin protein of the thin filaments that interacts with the myosin crossbridges of the thick filaments, the troponin-tropomyosin system has an important function in regulating the contraction of skeletal muscle.

The Internal Membranes of a Muscle Fiber. These membrane structures participate in the control of contraction and relaxation. See Figure 16–8 for their anatomical arrangements. The **plasma membrane** and a thin layer of connective tissue fibers form the outer covering of the cell, called the **sarcolemma.** Extending inward from the sarcolemma into the depths of the cell is a set of membrane-lined passageways, the **transverse tubules** (often called **t-tubules** for short). Depending on the type of mus-

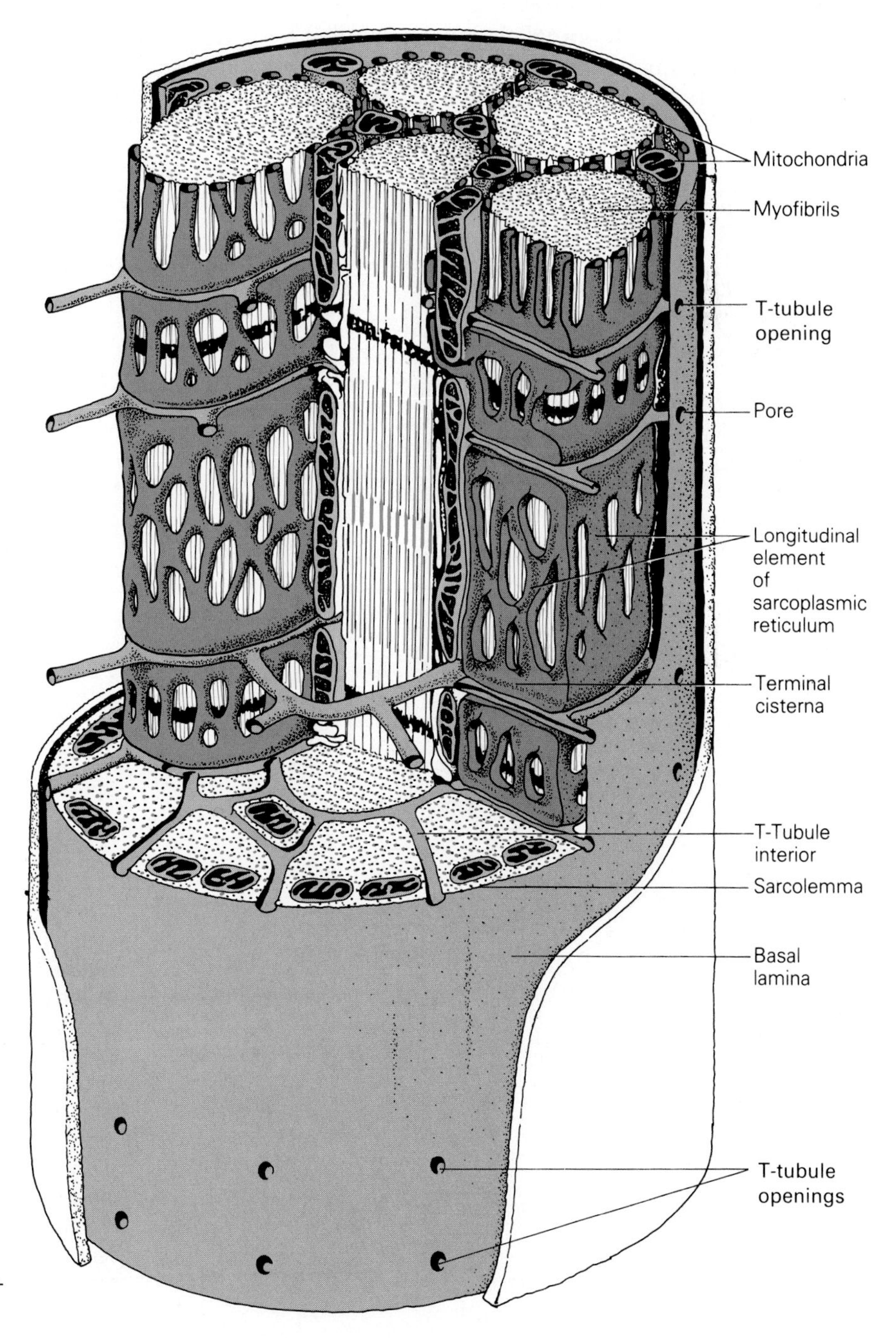

Figure 16–8 The internal membranes of skeletal muscle showing the structure of the t-tubule system and the sarcoplasmic reticulum. (Modified from Krstic, R.V. *Ultrastruktur der Saugetierzelle.* Berlin: Springer-Verlag, 1976.)

cle being considered, the t-tubules may enter the fiber either at the region where the A- and I- bands overlap, or at the level of the Z-discs. The t-tubules are an inward extension of the plasma membrane, and the interior of the t-tubule system is continuous with the extracellular space. Closely associated with the t-tubular system is another set of membranes that run at right angles to the t-tubules and down the length of the sarcomere. This is the **sarcoplasmic reticulum.** Unlike the t-tubules, the sarcoplasmic reticulum is not connected to the extracellular space; however, it does come into close contact with the t-tubules as they course across the fiber. This region of contact, which consists of the t-tubules and the enlarged end portions of the sarcoplasmic reticulum (called **lateral sacs** or **terminal cisternae**), is called a **triad** in skeletal muscle because it is made up of one t-tubule and the terminal cisternae from two adjacent sarcomeres. Periodic structures (sometimes called **"feet"**) appear to join the two membrane systems at the triad. These internal membrane systems function in the control of the contraction and relaxation of the muscle. (Details of this process and further aspects of the anatomy will be discussed later.) The function of the "feet" structures is not fully understood, but they are thought to play a role in communicating the signal for contraction from the t-system to the interior of the fiber.

The Structure of Cardiac Muscle

The cellular structure of heart muscle is similar in many ways to that of skeletal muscle, but there are also some very important differences (Fig. 16–9). The cells of heart muscle are small compared with

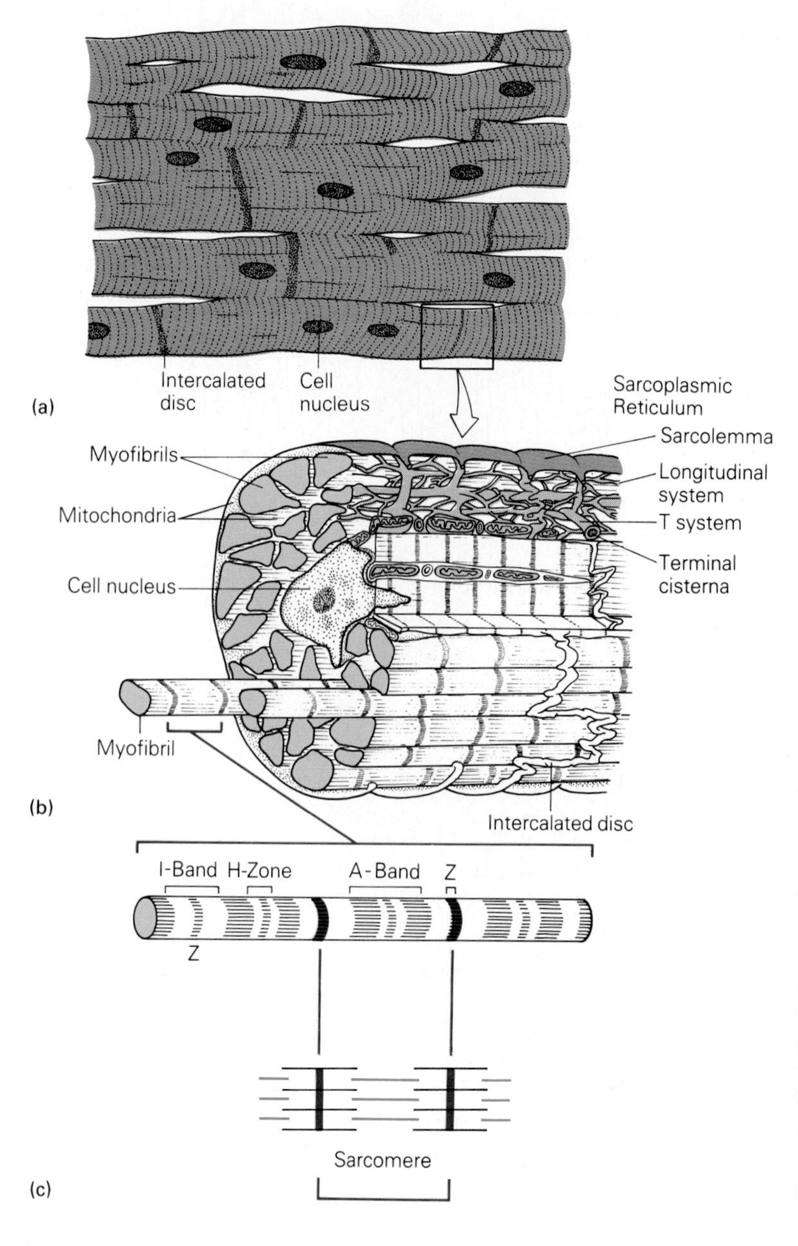

Figure 16–9 The structure of cardiac muscle. *(a)* The cells of the muscle are small, sometimes branched, and they attach end to end. *(b)* Seen at higher magnification, bundles of myofibrils run the length of the muscle, and a sarcoplasmic reticulum and t-system, which function much as they do in skeletal muscle, are present. *(c)* The sarcomere is organized in the same way that it is in skeletal muscle. (Modified from Braunwald, E., Ross, J. and Sonnenblick, E. H. *Mechanism of Contraction of the Normal and Failing Heart.* Boston: Little, Brown, 1968.)

those of skeletal muscle. They are from 5 to 15 micrometers in diameter and may be from 20 to 30 micrometers long. The cells are connected end-to-end and, to a lesser extent, side-by-side. Some of the cells may branch so that one end connects to two other cells. At the regions of connection between the cells is a specialized structure called the **intercalated disc.** This is an area of extremely close contact in which intercellular space is essentially absent; many projections from one cell will extend into another, and the other cell will have the mirror image of this arrangement. There is a strong mechanical connection between cells in this region as well as a close electrical connection. The mechanical connection is provided by a cellular structure called a **desmosome,** and the electrical connection is made through a **gap junction** (a structure found also between smooth muscle cells). The specialized contact structure of the intercalated disc allows a tissue made up of small cells to function in many ways as though it were a single large cell, and this forms the cellular basis of the coordinated action of the muscle in the beating heart.

The intracellular organization of cardiac muscle is the aspect in which cardiac muscle is most like skeletal muscle. The structure and arrangement of skeletal and cardiac myofilaments is quite similar, as is their protein composition. The similar contractile structure gives rise to the same sort of striated appearance that skeletal muscle has, and many of the mechanical responses are similar. Because the muscle of the heart has a largely **aerobic** metabolism, it contains numerous mitochondria. A transverse tubular system is present in muscle from many areas of the heart, although the tubules are larger in diameter and usually make close contact with only one portion of longitudinal sarcoplasmic reticulum. This results in a structure called a **dyad,** rather than a **triad,** as in skeletal muscle.

The Structure of Smooth Muscle

The Ultrastructure of Smooth Muscle

Although it is very much like skeletal and cardiac muscle in the molecular structure of its contractile proteins, smooth muscle has a very different cellular and tissue structure. Drawings made from electron micrographs are shown in Figure 16–10. Its cells are quite small, 5 to 15 micrometers in diameter and perhaps 200 to 500 micrometers long, depending upon which type of muscle is examined. The cells have a single nucleus, centrally located, and the ends taper off to a small diameter. In many smooth muscle tissues, numerous portions of cells are connected very closely together by a structure called a **nexus,** or **gap junction.** This structure involves a fusion of the outer layer of the membrane of each cell so that there is no extracellular space at this region between the cells. The interiors of the cells are still separated by the inner layers of the cell membranes, but the gap junction area has been shown to allow electrical current to pass more easily between adjacent cells. In many tissues, the gap junctions are not very stable structures, and they may form or disappear under the influence of hormonal action. These close cell-to-cell contacts in smooth muscle allow for coordination of smooth muscle activity (as in the heart, which also has gap junctions between cells).

A network of connective tissue fibers (collagen and elastin) surrounds the cells and helps to link them together. It also gives strength to the whole tissue, particularly in those organs that may be subject to a high degree of stretch. The force of contraction of the individual cells is transmitted to the whole tissue by strong cell-to-cell connections and by the connective tissue network.

Mechanical Arrangements of Smooth Muscle

Smooth muscle makes up much of the walls of organs such as the stomach, intestines, and uterus. It is generally arranged in layers. In the small intestine, for example, one layer (the inner one) is arranged in a circular fashion around the lumen, and the outer layer runs lengthwise (longitudinally). When the circular layer contracts, the intestine is reduced in diameter, and when the longitudinal layer contracts, the intestine becomes shorter. Unlike skeletal muscles, which usually have antagonistic muscles to stretch them out after contraction, smooth muscles must relax in a different way. An increase in the volume of the contents of the organ re-stretches the muscles in the walls. This usually comes about by the entry of new fluid (e.g., by newly formed urine in the bladder) or by movement of contents from elsewhere along a tubular structure (as in the intestine). Coordinated movements of smooth muscle organs such as the intestine often involve the alternate contraction and relaxation of the two muscle layers. In some saclike organs such as the uterus, muscle fibers are oriented in several directions, reducing the overall volume of the organ when the muscle contracts. The musculature of most blood vessels is simpler, usually consisting only of a circular layer of muscle. In very tiny blood vessels (e.g., an arteriole, which is pictured in Fig. 16–11), a single cell may wrap completely around the circumference of the vessel. Contraction of this muscle cell would lead to a narrowing of the blood vessel and reduce the amount of blood flow. Such

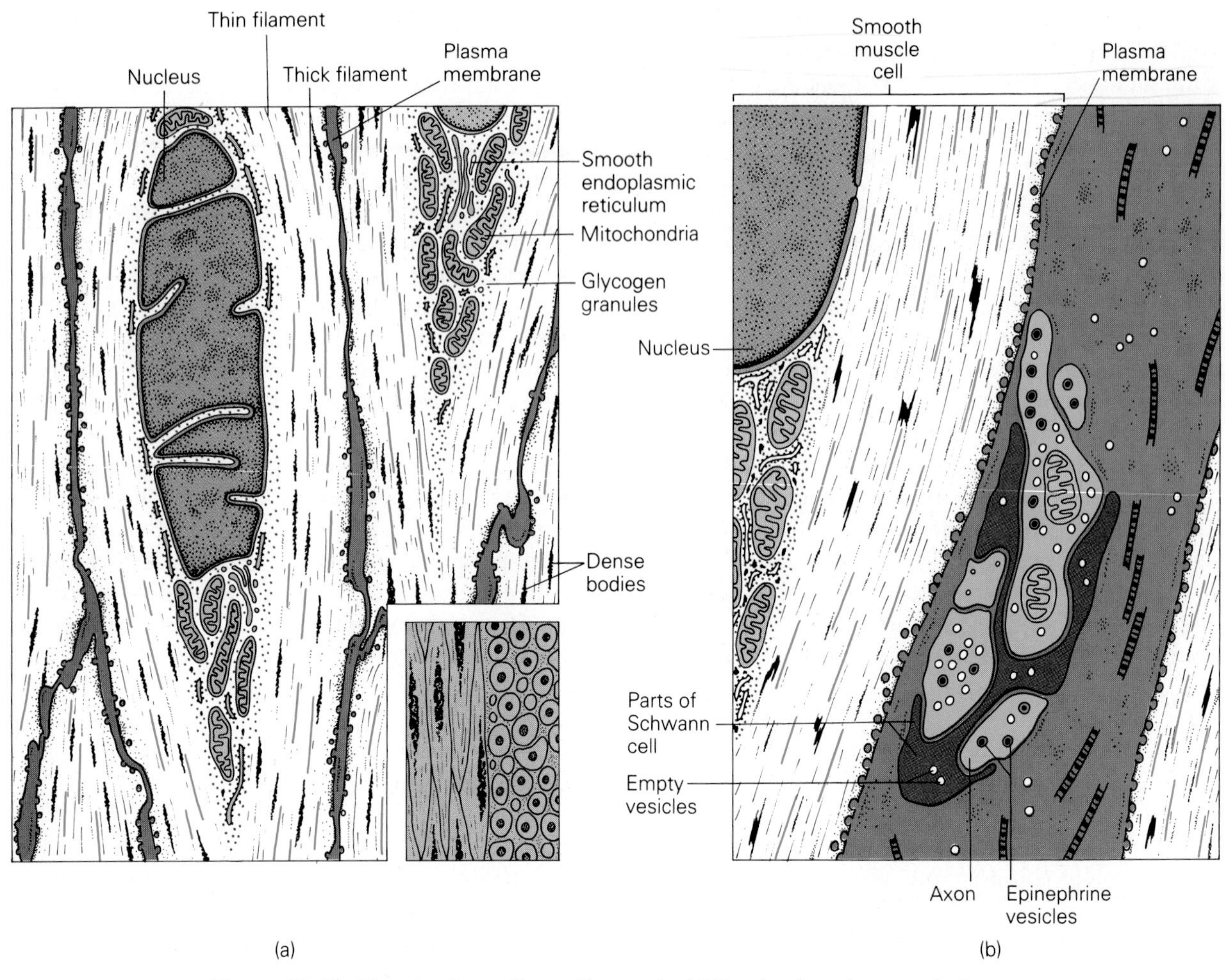

Figure 16–10 The structure of smooth muscle. (*a*) Drawing based on an electron micrograph of a longitudinal section of parts of several smooth muscle cells. It is the area enclosed in the rectangle in the inset portion at the lower right. Most of the cell organelles tend to cluster at the ends of the nucleus. (*b*) A nerve-muscle contact between a smooth muscle cell and a small bundle of autonomic nerve fibers. There is no specialized region of contact. (Modified from Krstic, R.V. *Ultrastruktur der Saugetierzelle.* Berlin: Springer-Verlag, 1976.)

smooth muscle plays a vital role in the regulation of blood flow and pressure.

The Function of Muscle

Most of what muscle does is obvious. We are aware that it shortens and develops force to lift an object, and there is some apparent relationship between how heavy a load is and how fast it can be lifted. These everyday impressions have been refined into a more precise set of concepts that are used to describe, quantify, and compare the way in which muscle functions. Before we approach this more exact analysis of muscle function, some basic definitions are necessary.

The Nature of Skeletal Muscle Contraction

First of all, the term **contraction** has a special meaning. In everyday usage, it means "to shorten"; in the terminology of muscle physiology, it means any form of muscle activity in response to stimulation, whether it involves shortening or not. For instance, in an **isometric contraction** (Fig. 16–12*a*) (*iso* = same; *metric* = measure), the muscle does not do any external shortening; instead, it develops **force,** or **tension,** while pulling against immovable attachments. In an **isotonic contraction** (*tonic* = tension), the force in the muscle remains constant while the muscle shortens. In an **auxotonic contraction,** the force continually increases as the muscle shortens (e.g., pulling back a bowstring or stretching a rub-

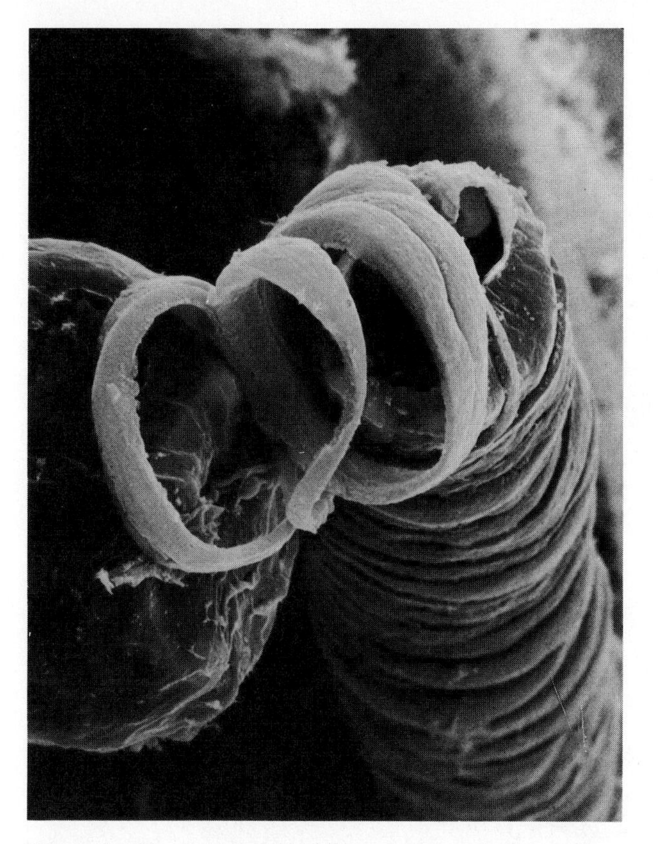

Figure 16–11 An example of the arrangement of smooth muscle cells in the wall of a blood vessel. This is a scanning electron micrograph of a small blood vessel called an arteriole. At the cut end of this vessel, some of the smooth muscle cells have been spread apart so that they may be seen better, while further down the vessel the cells are in their natural position. Contraction and shortening of the smooth muscle cells, which run in a circular direction, would cause the arteriole to become smaller in diameter. It is by this means that smooth muscle can regulate the flow of blood to the tissues. (Scanning electron micrograph courtesy of Dr. A. Evan.)

ber band involves this sort of contraction). A **meiotonic** contraction, however, is one in which the force *lessens* as the muscle shortens. Many useful mechanical devices, such as gearshift levers and typewriter keyboards, allow this type of contraction in order to provide a clear indication that an action has been performed. These defined conditions represent special circumstances, and realistic contractions usually involve some combination of them. For purposes of analysis, however, contraction conditions can be set up to emphasize the specific properties desired.

Isometric Contractions in the Laboratory: Clues to the Sliding-Filament Hypothesis of Contraction

Under laboratory conditions, an isolated skeletal muscle (and other muscle types as well) can be made to produce contractions conforming to any of these definitions. The relationships between muscle and its external mechanical environment are usually studied by using some "standard" experimental conditions. The simplest type of setup (as diagrammed in Fig. 16–12) can record **isometric contractions.** Here the muscle is securely attached to a device called a **force transducer,** which produces an electrical signal that varies in direct proportion to the force the muscle exerts. The other end of the muscle is attached rigidly to a fixed support (whose position the experimenter may adjust). All connections, and the force transducer itself, are made very stiff so as to prevent any shortening of the muscle. The force signal from the transducer is fed to a device such as a **chart recorder,** which produces an ink-on-paper graph of the force as it changes with time. (An **oscilloscope,** which produces a similar, although temporary, record on a **cathode ray tube,** may also be used.) The muscle is kept alive, by being immersed in a physiological saline solution containing all of the necessary ionic and organic substances (along with oxygen and an energy source) that it requires.

The Muscle Twitch. A single electrical stimulus to a skeletal muscle sets off a series of electrical and chemical events that result in a single, brief contraction called a **twitch.** On the graph drawn by the chart recorder, the force rises rapidly to a peak and then declines a bit more slowly back to the resting value. For a given stimulus sufficient to activate all of the muscle fibers, and under a constant set of temperature and metabolic conditions, all twitches made at the same muscle length will produce similar contraction records.

The Isometric Length–Tension Curve. The mechanical factor that has the most profound effect on an isometric contraction is the *length* at which the muscle is held. Figures 16–13 and 16–14 illustrate the relationships that will be described shortly. Making a series of isometric twitches, and changing the muscle length *between* (not during) the contractions, allows an important relationship to be observed. First the muscle is held at a very short length (shorter than it would ever be in the body) and then stimulated. If it is then progressively stretched out between contractions, the peak amount of twitch force will be seen to increase each time until some length is reached at which the twitch force is greatest. From this point on, further lengthening will result in two effects. First, force will begin to be measured in the muscle even before the stimulus is applied. This is called the **resting** (or **passive**) **force** and is due mostly to the connective tissue that holds the muscle together, acting much like a rubber band as it is stretched. Secondly, the

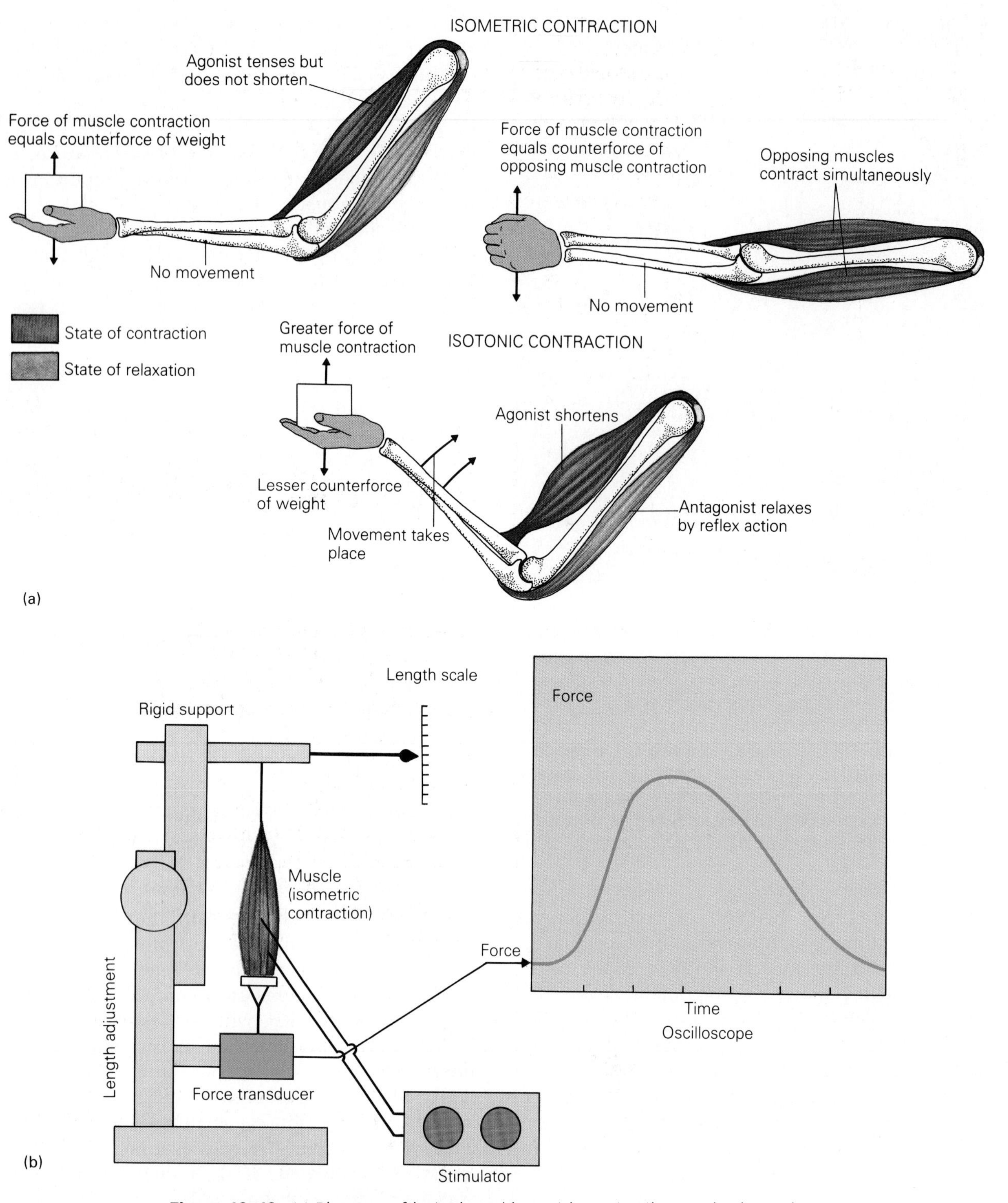

Figure 16–12 *(a)* Diagrams of isotonic and isometric contraction mechanisms. *(b)* Typical apparatus for measuring isometric contraction. The length of the muscle may be varied between contractions, and the record of the contraction is viewed on the face of an oscilloscope. The muscle is stimulated with an electric shock.

amount of force (the so-called active force) that the muscle can develop over and above the resting force now diminishes as the muscle is lengthened further. At an extreme degree of stretch (provided that the muscle has not torn), there will be a very high resting force and very little active force in response to stimulation.

This relationship, in which the force capability of the muscle is greatest at some intermediate length, is called the **isometric length–tension curve.**

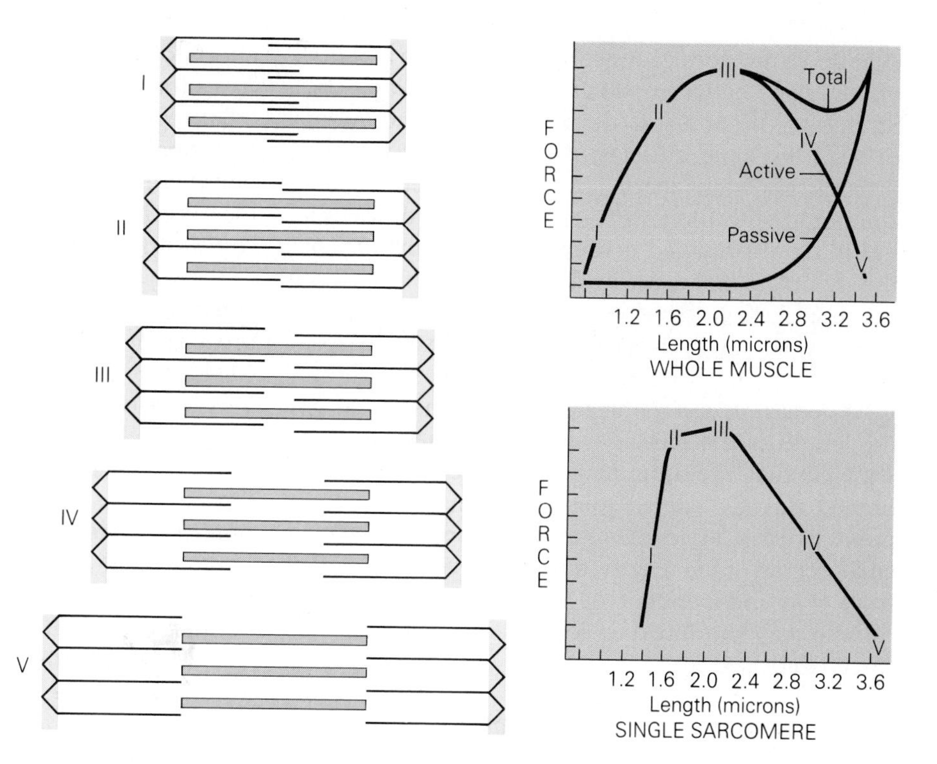

Figure 16–13 The length-tension curve. The upper portion shows how the resting length affects the tension (or force) produced by a muscle during an isometric contraction. The curve labeled "Total" represents the "Active" force plus the "Passive" force. The key to understanding the basis for this curve lies in the sarcomere diagrams to the left. They show the relative amounts of overlap in most of the sarcomeres of the muscle at the various resting lengths. This may be better seen in the lower graph, the length-tension curve of a single sarcomere. This curve is made up of straight lines that correlate very well with the overlap of myofilaments; this correlation is one of the main bits of evidence for the sliding-filament hypothesis. In the upper curve, the collective effect of millions of sarcomeres obscures the simple relationship.

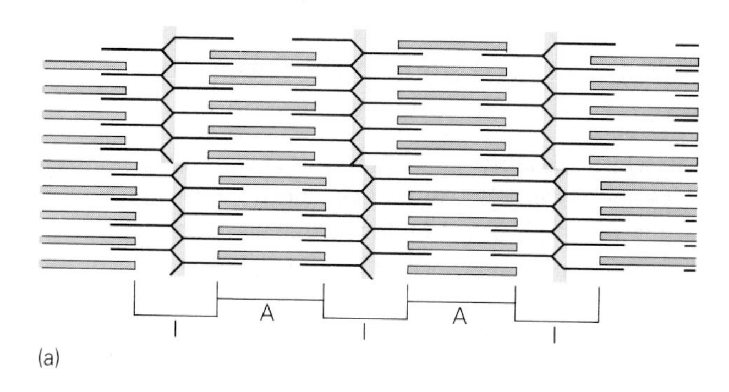

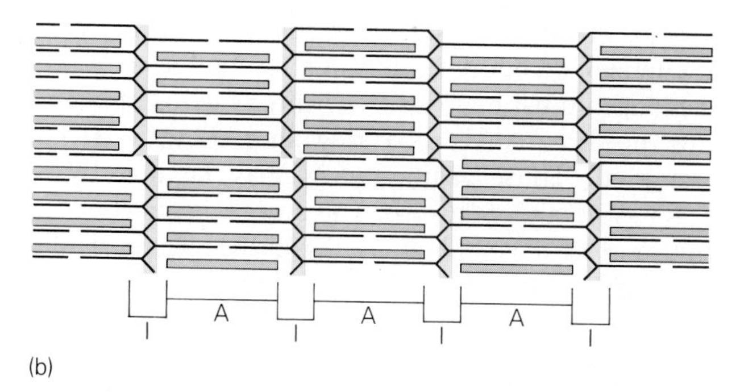

Figure 16–14 The shortening of sarcomeres in series. The upper portion represents a "snapshot" of two sarcomeres somewhere along a myofibril. There is about 50% overlap of thick and thin filaments at this length. The lower portion shows sarcomeres from a shortened muscle. Here the overlap is complete, and three sarcomeres now fit roughly in the space of two. The shortening of all of the sarcomeres in a muscle is added together, and the tiny amount of shortening of each sarcomere results in a very large shortening of the whole muscle.

Usually the maximum force capability occurs at a length that is close to the natural length of the muscle in the body. This length is often termed the **optimal length** or $\mathbf{L_o}$, while the peak force at this length is designated $\mathbf{P_o}$ or $\mathbf{F_o}$.

While this length–tension behavior is very important to an understanding of how muscle works, most skeletal muscles are limited in how far they can be stretched out by their attachments to the skeleton, and the length–tension curve is of relatively little importance in normal function. Heart muscle, however, does not have its length changes limited by skeletal attachments, and its resting length is set by the amount of blood that returns to it between beats. The length–tension curve thus allows the heart to make a more forceful contraction if it receives a larger filling of blood between contractions.

Optimal Length and the Overlap of Myofilaments. Why does the resting length have so important an effect on the contraction of the muscle? The answer to this question has provided some important clues to the basic physiology of skeletal muscle, clues that relate the structure of the muscle to its function. Careful measurements with the electron microscope have shown that the A-bands do not change in width as the muscle is lengthened or shortened, but the width of the I-bands increases as the muscle is stretched (and vice-versa). The spacing between the Z-lines also changes in proportion to the muscle length. Other measurements have shown that the length of the myofilaments themselves also does not change. What does change, however, is the amount of overlap between thick and thin filaments. Stretching the muscle decreases this amount of overlap; in some muscles, the sarcomeres may be pulled completely apart. These observations (when compared with the length–tension curve of the living muscle) provided evidence that the overlap of the myofilaments controls the amount of force that the muscle can develop. During delicate experiments with single fibers of skeletal muscle in specially controlled isometric contractions, the sarcomere length of the living muscle was measured as the length of the muscle was varied. These studies showed that the amount of force developed was indeed directly related to the amount of myofilament overlap at lengths greater than L_o. Over a small region of the length–tension curve near L_o, the force was independent of the length, but as the muscle was shortened to lengths less than L_o, the force decreased.

Myofilament Overlap and the Amount of Force. These findings are interpreted in the light of the structural evidence for the existence of crossbridges and the biochemical evidence that it is the head (crossbridge) portion of the myosin molecule that interacts with actin and catalyzes the release of energy from ATP. When the muscle is stretched out beyond L_o, there is incomplete overlap between the thick and thin filaments, and not all of the crossbridges on the thick filaments have the chance to attach to an actin filament. Therefore, the muscle produces force, and the amount of force is proportional to the myofilament overlap. In the region of the curve near L_o, overlap is complete, and force is at its greatest. Because of the **bare zone** at the center of the thick filaments (where there are only myosin "tails" and no crossbridges), shortening the sarcomere does not allow any more crossbridges to attach to the thin filament as it slides further in. This is why the force does not change with length in this region. As the sarcomere shortens further, myofilaments from opposite sides of the sarcomere begin to interfere with each other, and less force can be produced. At the extreme shortening, where force has virtually disappeared, the thick filaments have been pushed against the Z-lines, and there is no further opportunity for shortening or developing force.

It has also been shown that the processes that activate the myofilaments do not function as well at short lengths, and this factor also reduces the force. In experiments with whole muscles, the regions of the length–tension curve are not as clear-cut, because the millions of sarcomeres that make up the whole muscle are all at somewhat different places on this curve. For this reason, and because of the effects of the connective tissue, the length–tension curve for a whole muscle is a smooth curve without a flat region at its peak. Nevertheless, it is well established that the effects of length on muscle function do reflect a fundamental property of the molecular mechanisms that are responsible for muscle contraction.

The Sliding-Filament Model of Contraction. The account of muscle function just given is usually called the **sliding-filament hypothesis.** This is a very well-established body of knowledge that integrates much of what is known about skeletal muscle into a consistent framework. Because cardiac muscle is similar to skeletal muscle in structure and function, the same basic hypothesis appears to be valid here as well. Less is known about the relationship between structure and function in smooth muscle because of the different cell arrangement and lack of obvious structural regularity. However, what is known about the details of smooth muscle function and structure is at least consistent with a sliding-filament hypothesis, and as more information becomes available such a mechanism will probably be experimentally supported.

Isotonic Contractions in the Laboratory: Clues to the Work of Muscle as a Motor

Analysis of isotonic contractions can reveal other aspects of muscle contraction. Because muscles that are shortening at a constant force are doing work, it is during isotonic contraction that the functions of muscle as a **motor** are best emphasized.

The equipment necessary to produce isotonic contraction from an isolated muscle is somewhat more complex than that used in isometric experiments. A force transducer is still used to measure the tension produced, but now the changes in muscle length must be measured, and a means of keeping the force constant during shortening must be provided. These requirements may be met by a **lever system** (Fig. 16–15). The muscle is attached to one end of a very light (but rigid) lever, and the desired load (such as a brass weight) is hung from the other end. The muscle is anchored at its other end to the force transducer. A support is placed under the weight so that when the muscle is at rest it is not stretched out. The position of this support sets the resting length of the muscle. If the muscle length is appropriately set (see Fig. 16–15), there will be no passive tension (recall this definition from the previous section). The force that the weight will provide during contraction is called the **afterload,** because the muscle "feels" it only after beginning to contract. (If the muscle is stretched out by the at-brane system. This means that it is a structure that maintains a separate region of the cell interior. This enclosed region is used to store and release the actual chemical substance that controls the contractile activity of the muscle proteins. This substance is **ionized calcium** (calcium ions), and in the resting muscle it is highly concentrated within the SR. It is contained within the saclike areas called **terminal** (or **lateral**) **cisternae.** The longitudinal elements of the SR communicate with the terminal cisternae at each end of the sarcomere and form a highly branched network surrounding the myofilament bundles. The interior of the longitudinal elements is continuous with the terminal cisternae.

As this description suggests, each sarcomere has its own separate portion of sarcoplasmic reticulum. The junction between a t-tubule and the terminal cisternae of two adjacent sarcomeres (the **triad**) is the place where the electrical signal from the outside of the fiber actually is communicated with the interior of the fiber. Just how the signal passes from the t-tubule to the SR at this point is not well understood, although it is assumed that the local flow of electrical currents plays a role. It is well known, however, that when this signal has been passed, the SR is caused to release a large portion of its stored calcium ions into the cytoplasm very near the myofilaments. These calcium ions free the myofilaments from their resting state (by a process that will be discussed shortly), and contraction takes place. Relaxation is associated with the movement of calcium ions back into the longitudinal elements of the SR. This ionic movement takes place up a steep concentration gradient via an ATP-dependent calcium ion pump present in this region of the SR membrane. While this mechanism (diagrammed in Fig. 16–18) may seem complicated, it quite effectively overcomes the diffusion and conduction obstacles posed by the large size of skeletal muscle fibers. It is capable, in some muscles, of allowing activation to be repeated as many as 60 to 100 times per second. Muscles without such a mechanism would be limited to very slow rates of contraction.

Control of Muscle by Calcium Ions

The regulation of muscle contraction involves controlling the interaction of the actin and myosin myofilaments. In all types of muscle, this control is carried out through the action of calcium ions, though the actual process differs greatly between smooth muscle on one hand and cardiac and skeletal muscle on the other.

The Molecular Mechanisms of Muscle Contraction

When an action potential, through the sequence of events described previously, causes the SR of skeletal muscle to release a quantity of stored calcium ions, they diffuse to the region of the myofilaments. Here they bind to the **troponin-C (TnC)** subunit of the regulatory protein complex, troponin. The troponin complex is attached to the tropomyosin molecule on the thin (actin) myofilament. When calcium ions bind to troponin-C, the actin filament becomes able to interact with the myosin crossbridge. The contraction that results involves the repeated interaction of actin and myosin in a series of steps that are diagrammed in Figure 16–19. In resting skeletal muscle (with calcium ions not yet released), an ATP molecule is bound to each myosin molecule projection that forms a crossbridge. The ATP molecule has been "split," but its energy cannot be released until the actin is allowed to interact with the myosin. When the presence of calcium ions permits this interaction, the crossbridge attaches to the actin filament at an angle of approximately 90°. As the energy of the ATP molecule is released, the angle of attachment becomes closer to 45°. This change in attachment angle (the so-called **power stroke**) causes the actin filaments to be pulled in toward the center of the sarcomere, and a phosphate ion and ADP are given off. Only when a new ATP molecule binds to the myosin head can the crossbridge detach itself from the actin and become ready for a new cycle.

As long as enough ATP and calcium ions are present, the cyclic process of attachment, angle change, ATP binding and hydrolysis, and detachment can take place. Repeated crossbridge cycles result in the actin filaments being pulled further into the myosin filament array, and the whole muscle becomes shorter. As was discussed previously, the force that the muscle exerts is important in determining how fast the crossbridge cycle will run. If the ends of the muscle are held stationary (as in isometric conditions), the crossbridge cycles will still continue to some extent, since some internal movement is allowed by the elastic nature of the muscle. When stimulation of the muscle is stopped, no additional calcium ions are released from the SR, and calcium ions that diffuse free from the troponin or are free in the cytoplasm are taken back up by the SR. Crossbridges will not reattach, and relaxation takes place. If the supply of ATP becomes depleted, attached crossbridges will not be able to detach, and the muscle will become stiff and unable to relax. An extreme example of this situation is the condition known as

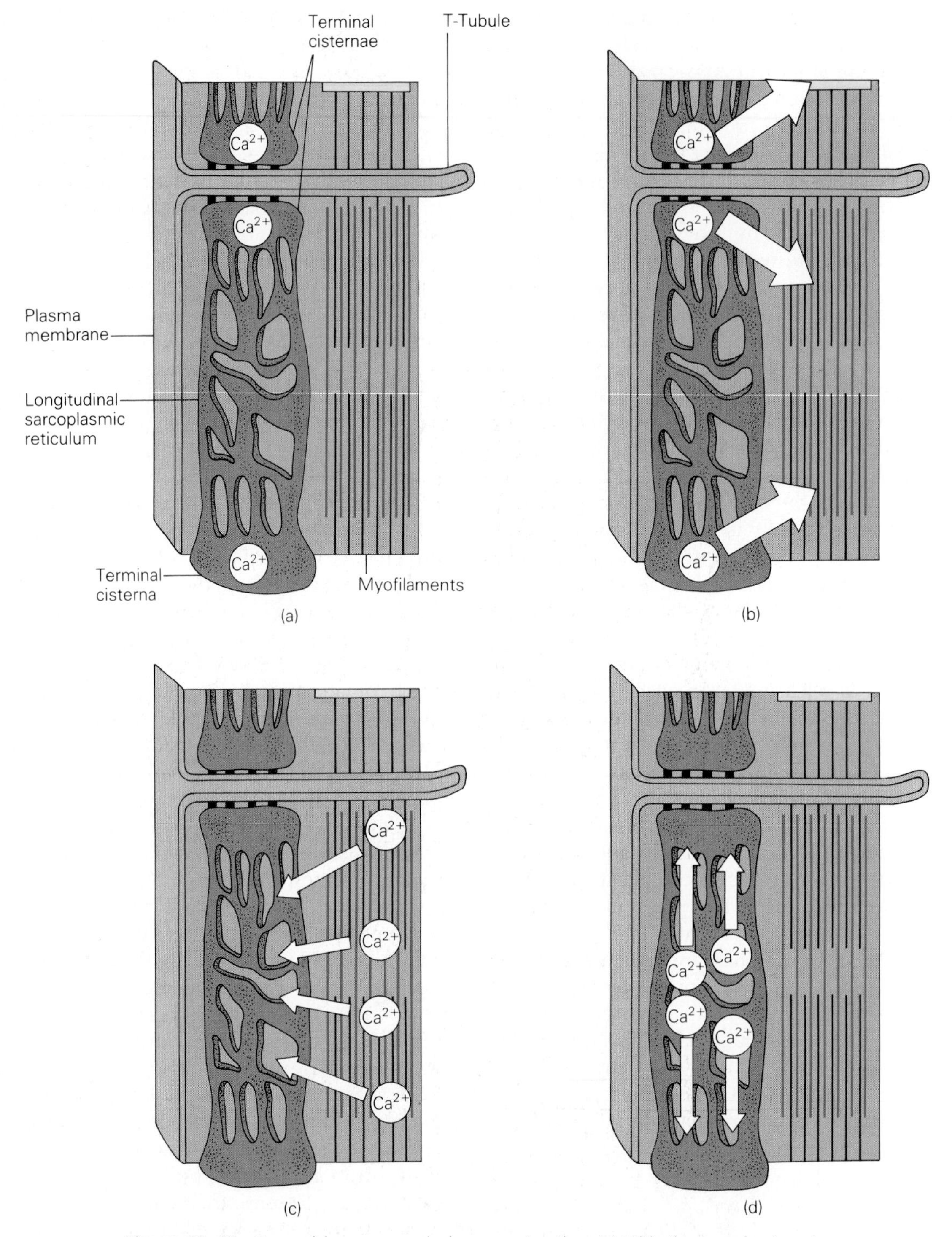

Figure 16–18 How calcium moves during a contraction. (*a*) With the muscle at rest, the calcium ions are stored in the terminal cisternae. (*b*) As activation begins, calcium is released from the sarcoplasmic reticulum into the region of the myofilaments. (*c*) Activation ends as calcium is taken back up into the longitudinal elements of the sarcoplasmic reticulum. (*d*) Immediately after the contraction, the calcium moves back into the terminal cisternae.

Phasic and Tonic Contractions: A Variety of Responses in Smooth Muscle

Smooth muscle can contract in a variety of different ways. Some smooth muscles can give twitchlike contractions in response to single stimuli or brief bursts of stimuli. These contractions are called **phasic** and are often found in smooth muscles that propel or move materials. Other smooth muscles give long and sustained contractions in response to single or continued stimulation by nerves, drugs, or hormones. These contractions are called **tonic** and are likely to be found in smooth muscle organs, such as sphincters or bladders, that must remain contracted for long periods of time. Many smooth muscles are almost never in a state of complete relaxation, and the steady force that they generate is called **tonus.** The level of tonus is often set by the concentration of a specific hormone.

In many cases, the contraction of smooth muscle is associated with electrical activity of the cell membranes. Most phasic contractions are associated with **action potentials.** In some tissues, such as intestinal muscle, that have a rhythmic contraction pattern, there is a kind of electrical activity called the **slow wave.** The slow waves, which are many seconds in duration, periodically depolarize the membrane to the threshold for action potential generation; this mechanism acts to produce periodic contractions. Other smooth muscles show action potential activity that may be caused by a pacemaker cell and conducted to adjacent cells by means of the gap junctions previously mentioned.

Control of Muscle by the Central Nervous System

Almost all of the muscle in the body is under the direct or indirect control of the CNS, although smooth muscle and cardiac muscle do have a degree of autonomous function that is suited to their specialized roles. Because neural control is so highly developed in the case of skeletal muscle, most of the general process can be understood by a study of how the action of a motor nerve regulates the function of a skeletal muscle.

Nervous Control of Skeletal Muscle

Skeletal muscle depends upon the somatic division of the CNS for the control of its contraction. As described in Chapter 10, under normal conditions, skeletal muscle will not contract unless it is stimulated by a **motor neuron,** which has its origin in the CNS. Connection between the motor nerve fibers, which are large myelinated axons, and the muscle fibers, takes place at a structure called the **myoneural junction.** Because in some cases this structure spreads out somewhat on the muscle fiber, it is also called a **motor end-plate.**

The Myoneural Junction and the Motor Unit. The myoneural junction is a special case of a **synapse.** As in the vast majority of synapses in the CNS, the transmission of the impulse from the nerve to the muscle is by chemical means. The particular transmitter substance at the myoneural junction is **acetylcholine (ACh).** Unlike some CNS synapses, only **excitation** of the postsynaptic membrane is possible at the myoneural junction. Because many of the myoneural junctions in a muscle lie at its surface, they have been extensively studied, and the process of transmission is well understood. In fact, much of what we know about synapses in general has been learned by the study of this highly specialized synapse.

The structure of the myoneural junction provides keys to an understanding of its function. As a motor axon from the motor nerve bundle nears the muscle, it branches to innervate a number of individual muscle fibers. The assemblage of a motor axon, together with all of the muscle fibers that it innervates, is called a **motor unit.** Since all of the muscle fibers served by a single branched axon are activated together when an impulse passes down the axon, the fineness of control of a muscle depends upon just how many muscle fibers a single axon controls. Muscles that are capable of fine and delicate movements will have only a few muscle fibers served by a single axon, while muscles used for rapid and relatively coarse movements will have many muscle fibers controlled by a single axon. Each muscle fiber normally receives innervation from only a single axon terminal, although some multiply-innervated muscle fibers do exist.

Synaptic Transmission at the Myonerual Junction. As the motor axon branches, the myelin sheath becomes reduced to the cytoplasm and cell membranes of a single **Schwann cell** (the type of cell that forms the myelin sheaths around motor axons), and this covers the axon as it makes its final approach to the muscle (Fig. 16–22). The terminal region of the axon lies in a "groove" on the surface of the muscle. This arrangement increases the amount of nerve and muscle cell portions that make up the junctional region. The motor axon terminal contains small vesicles that carry the transmitter substance. Also present in this presynaptic region is a large number of mitochondria. These provide the metabolic energy for the synthesis of the transmitter substance and

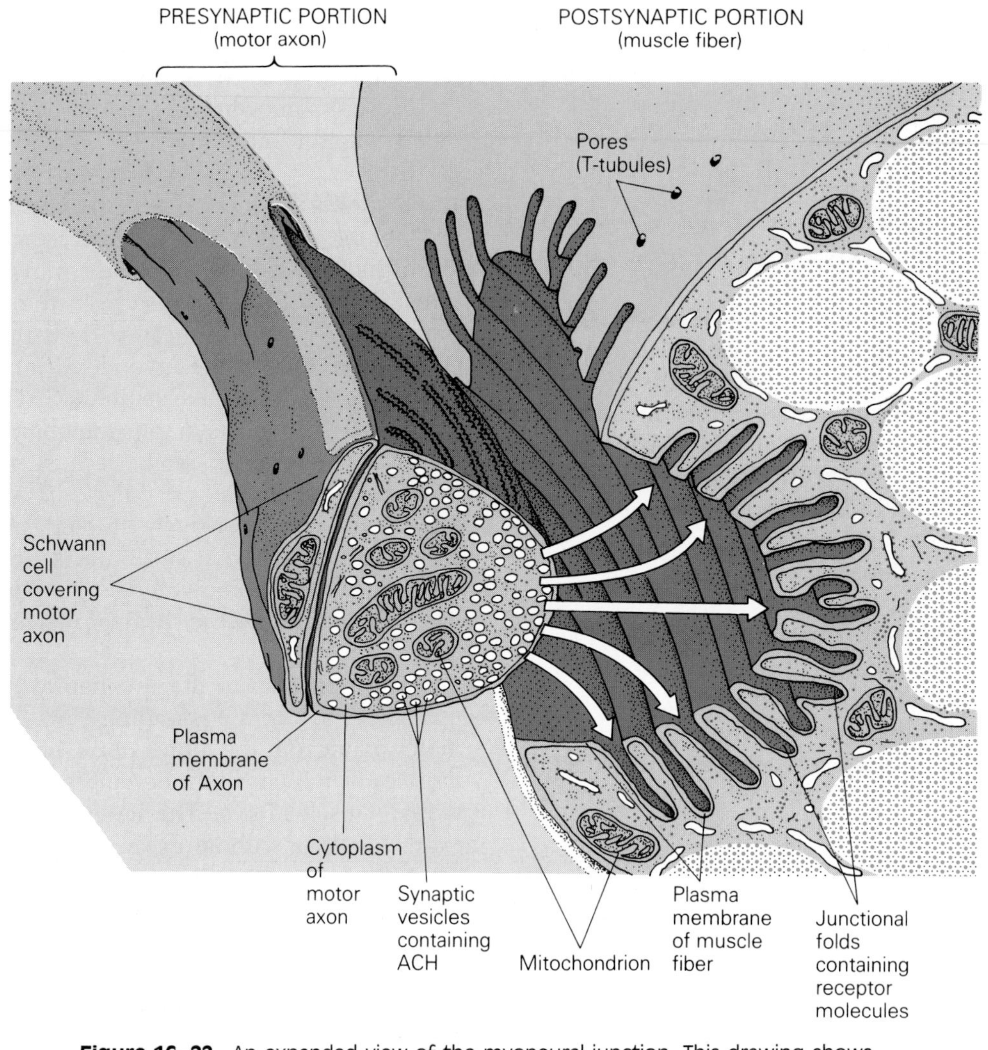

Figure 16–22 An expanded view of the myoneural junction. This drawing shows the essential features of the myoneural junction of skeletal muscle. (Modified from Krstic, R.V. *Ultrastruktur der Saugetierzelle.* Berlin: Springer-Verlag, 1976.)

the ionic pumping mechanisms that are necessary for the recovery processes that follow the passage of an impulse. The muscle plasma membrane that makes up the postsynaptic side of the junction is folded into a series of deep grooves called **junctional folds,** which are lined with the receptors for the ACh transmitter substance. The receptors are complex protein molecules that have a dual function. Each receptor has two binding sites for ACh and also functions as an ion channel. Normally, the receptor is not permeable to ions, but when the transmitter substance has attached to the binding sites, sodium and potassium ions can pass through the channel down their electrochemical gradients. When the ACh is no longer bound to the receptor, the channel again becomes impermeable to ions. The postsynaptic membrane also contains an enzyme called **acetylcholinesterase.** This enzyme is essential for breaking down the transmitter to an inactive form once it has done its job of causing the channel to open.

Between the time when a motor nerve is stimulated and when the muscle contracts, a series of many events takes place (Fig. 16–23). The action potential travels down the motor axon by the mechanisms previously described. When the action potential spreads into the terminal regions of the axon branches, membrane calcium channels open in response to the depolarization, allowing the entry of calcium ions into the neuron. This increase in the intracellular calcium ions causes the synaptic vesicles that contain the acetylcholine to move to the region of the presynaptic membrane. Here, the vesicle membranes fuse with the membrane. In the region of the fusion, the plasma membrane breaks down, and the contents of the vesicle are released into the synaptic cleft. They diffuse across the cleft to the postsynaptic region, where they attach to the

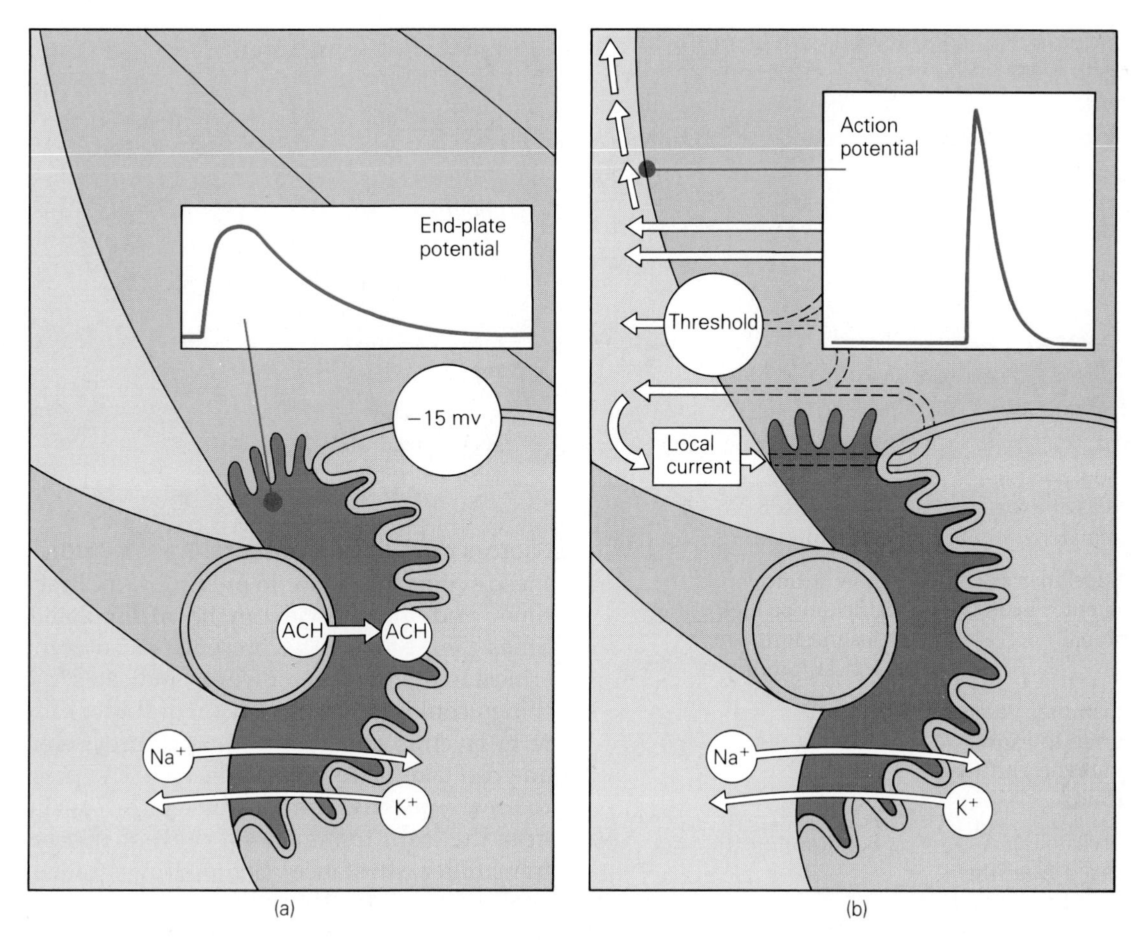

Figure 16–23 How the end-plate potential produces a muscle action potential. The region of membrane (the postsynaptic membrane) immediately opposite the axon terminal does not produce an action potential. Instead, its depolarization causes local currents to flow through the interior of the muscle cell and across the adjacent muscle cell membrane. This flow of current depolarizes the membrane to a threshold value, and it produces an action potential. This action potential then travels the length of the muscle cell membrane and enters via the t-tubule system to activate the contraction.

specific binding sites on the receptor molecules. When this binding takes place, the membrane channels associated with the receptors become permeable simultaneously to both sodium and potassium ions. This causes the membrane potential to move away from the resting potential (about −80 mV) and toward a new potential of about −15 mV, which is called the **end-plate potential.** (This is the potential expected from a membrane permeable to both sodium and potassium at the same time.) The postsynaptic membrane is unlike the rest of the plasma membrane in one important aspect. Since its channels are chemically activated and both permeability changes occur together, it cannot produce an action potential. What the end-plate potential does, however, is to cause a small local current (the **end-plate current**) to flow across the membrane at this region. This inward current flows into ıe fiber and begins to spread down its length. The corresponding outward current is forced to flow across adjacent resting areas of the general plasma membrane, and it depolarizes these regions so that they reach their threshold for action potential generation. The action potentials that the end-plate potential has produced then propagate down the length of the fiber and activate the contraction in ways that have already been described. After the passage of the action potential, the muscle membrane returns to its resting condition.

If the action of the transmitter substance were not terminated, the end-plate depolarization would continue. This would prevent the muscle plasma membrane from being stimulated again, and subsequent synaptic transmission would fail. Hydrolysis of the ACh by the postsynaptic acetylcholinesterase terminates ACh action and produces its components, acetate and choline. The choline is actively taken back up into the presynaptic terminal, where it is made back into ACh. The acetate fragment, which is a chemical common throughout the body, is not saved by any special mechanism. These events are summarized in Table 16–1.

Table 16–1
Sequence of Events During Neuromuscular Transmission

Event	Possible Blocking Agent
Action potential arrives at axon terminal.	Nerve-blocking agents such as procaine or tetrodotoxin
Extracellular calcium ions enter axon terminal.	Low extracellular calcium or high magnesium concentration
Transmitter vesicles migrate to axon membrane and fuse with it.	
Acetylcholine is released from the transmitter vesicles into synaptic cleft.	Botulinum toxin
Acetylcholine molecules diffuse across synaptic cleft.	
Acetylcholine molecules bind to receptor proteins in postsynaptic membrane on muscle.	Curare, succinylcholine
Ion channels in postsynaptic membrane open and allow the flow of sodium and potassium ions.	
End-plate region (postsynaptic membrane) becomes depolarized; end-plate potential is set up.	
Local end-plate currents cause adjacent muscle membrane to depolarize.	
Muscle action potential is triggered.	
Acetylcholine diffuses away from receptors and is hydrolyzed by cholinesterase enzyme.	Eserine or other cholinesterase inhibitors
Choline is taken up by presynaptic terminal (now repolarized).	Hemicholinium drugs
Acetylcholine re-synthesized and vesicles re-filled.	

Factors that Block Myoneural Transmission. All of these events take place in only a few milliseconds of time, and the muscle can be restimulated many times per second. The presence of so many steps critical to the process, however, indicates that many things could go wrong. **Myoneural blockade** is the term given to this transmission failure. Some blockade can take place presynaptically. The presence of too many magnesium ions or too few calcium ions near the axon terminal will prevent release of the transmitter substance. Some drugs, called **hemicholiniums,** interfere with uptake of choline, and the transmitter substance becomes depleted. **Botulinum toxin,** an extremely deadly poison produced by a bacterium *(Clostridium botulinum)* present in spoiled food, also causes a failure of transmitter release, and poisoning with this substance results in paralysis and death.

Postsynaptic blockade is also possible. The substance **curare** (used on blowgun darts by aboriginal hunters in South America) can bind to the acetylcholine receptors, in preference to acetylcholine. However, its binding does not produce a permeability change or depolarization. It is also resistant to breakdown by the acetylcholinesterase. When it is present in sufficient quantity, therefore, myoneural transmission cannot take place.

A close chemical relative of acetylcholine is the drug **succinylcholine,** which is used during surgical procedures to produce muscle relaxation. This drug is broken down only very slowly by acetylcholinesterase. Another possible site for blockade is the acetylcholinesterase enzyme. This enzyme is inhibited by drugs such as **physostigmine** and **eserine,** which are similar to the active (organophosphate) compounds in chemical warfare agents, the so-called nerve gases. When the action of cholinesterase is inhibited, the end-plate membranes remain depolarized, and the muscle action potential mechanism cannot be reset.

Focus on Myasthenia Gravis

Myasthenia gravis (meaning "serious muscle weakness") is an illness associated with muscular weakness and extreme fatigue following moderate exercise. It is related to a failure of neuromuscular transmission rather than with any defects in the muscle contraction mechanism itself. While no cure has been found for the disease, several kinds of therapy can give patients some symptomatic relief.

Scientists now know that myasthenia gravis is one of the so-called **autoimmune diseases.** For unknown reasons, some people produce antibodies that react with the **acetylcholinesterase receptors** in the postsynaptic (motor end-plate) plasma membrane. This immune reaction destroys many of the receptors, and adequate **end-plate potentials** can no longer be produced. This means that a muscle action potential will not result from a nerve impulse, and the muscle fails to contract. For a person afflicted with the condition, the result is that extreme effort may be required for even ordinary movements. In extreme cases, even breathing becomes very difficult.

Specific diagnosis of the condition involves study of the neuromuscular junction and immune system. Careful observation of the relationships between the nerve stimulation and muscle response can reveal delays in the transmission process even before it fails completely, and the presence of antibodies specific to the acetylcholine receptors may be detected in the blood of affected persons.

Several forms of treatment are possible. Drugs that inhibit the action of acetylcholinesterase allow the remaining postsynaptic receptors to function for a longer period of time, and adequate end-plate potentials can be produced. Unfortunately, the effect is only temporary, and the drug must be readministered repeatedly. This approach can lead to harmful side effects and, in any case, treats only a symptom of the underlying cause, with no effect on the general course of the disease.

Like other autoimmune diseases (such as rheumatoid arthritis), myasthenia gravis can undergo spontaneous remissions in its early stages, although the symptoms usually return. Removal of the thymus gland (which is known to play an important role in the immune system) results in substantial improvement of the condition, especially if it is done early in the course of the disease.

Other therapies, also directed at the immune system, may be applied. Administration of gamma globulin may produce temporary improvement. Steroid drugs can be used to reduce the severity of the problem for some periods of time, but there are harmful side effects in many cases. Immunosuppressant drugs reduce the function of the entire immune system and can produce an improvement in the symptoms, although the patient now is at risk from other infections and diseases. The technique of **plasmapheresis,** in which the proteins (including antibodies) are removed from the blood by a special transfusion method, can produce temporary improvement. But the procedure, though relatively safe, is time-consuming and expensive.

While an understanding of the effects of the disease can be obtained through study of the neuromuscular junction, its actual cause lies elsewhere. With immunological methods, myasthenia gravis can be produced in experimental animals, especially in rodents and monkeys. Since the animal condition very much resembles the human disease, studies involving animal immune responses offer hope for development of a more satisfactory treatment.

In all of the situations mentioned, the myoneural blockade would result in death from respiratory failure. The diaphragm, a skeletal muscle, would fail to respond to nerve impulses, and along with other skeletal muscles, it would be paralyzed. Carefully controlled doses of blockers such as succinylcholine can be used as surgical muscle relaxants as long as respiration is maintained artificially.

Other factors also affect the myoneural junction. As are other synapses, it is subject to **fatigue** due to depletion of the transmitter substance. However, the transmission normally has a large **safety factor.** That is, more transmitter is released than is necessary for transmission, and the end-plate potential is normally larger than necessary to stimulate the muscle membrane. In addition, the process of contraction in the muscle is subject to fatigue far sooner than is the process of myoneural transmission. An abnormal situation, though, is present in the disease known as **myasthenia gravis,** which means "severe weakness of the muscles." In this condition, it appears that an immune reaction against the acetylcholine receptors reduces their numbers, and the end-plate potentials may not be large enough to stimulate the muscle fibers. Weakness or paralysis results from this condition, but effective therapy, at least in the short term, is possible. Careful administration of cholinesterase inhibitors allows the acetylcholine to remain active in the synaptic cleft for a longer time, and what receptors there are may be bound to and activated several times in succession. This allows sufficient end-plate current to flow to permit near-normal muscle function.

Nervous Control of Smooth and Cardiac Muscle

The two general types of smooth muscle, multiunit and unitary, differ considerably in the way they are controlled by the nervous system. Multiunit smooth muscles are generally very densely innervated by autonomic nerve fibers and have rudimentary motor end-plates on each cell in the tissue. Myoneural transmission in these muscles is much like that found in skeletal muscles. The structures involving this muscle are quite small. They include some very small blood vessels, the **arterioles,** and the muscles that make our hair stand on end, the **piloerector muscles.** Because of the small size and the plentiful innervation of the tissues, the fact that the cells do not communicate with one another is of little practical consequence.

Most tissues composed of unitary smooth muscle are also innervated by postganglionic fibers of the autonomic nervous system. There is nothing approaching a one-to-one connection between the nerve supply and the muscle cells. Instead of a nerve-muscle connection such as a motor end-plate, the innervating fibers course among the smooth muscle cells. Periodically the nerves have swellings (called **varicosities**) that hold vesicles containing various transmitter substances (e.g., acetylcholine or norepinephrine). Action potentials traveling down the nerves cause the release of the transmitter substance, which can then diffuse across the intercellular spaces to the vicinity of the muscle cells. Because there are so many cells in the tissue, only a few of them are directly affected by the transmitter substance. The cells that depolarize and produce action potentials, however, can affect neighboring cells through gap junctions. The gap junction forms a low-resistance connection between adjacent cells; and depolarizing one cell can cause an adjacent cell to depolarize by means of **local current flow.** This has the effect of spreading the activity throughout the tissue, and even though most individual cells are not directly innervated, the tissue tends to function as though they were. This is the source of the term *unitary,* and a tissue that behaves in this way is called a **functional syncytium** (*syn* = same; *cytium* = cells). Many unitary smooth muscle tissues are spontaneously active and can contract rhythmically on their own. In these cases, the effect of the autonomic innervation may be to regulate (increase or decrease) the ongoing function.

Heart muscle also receives an extensive autonomic innervation, but the function of the nerves in heart muscle is not to initiate a contraction. The muscle of the heart does not contain motor end-plates, and its contractions begin directly in the muscle tissue, not in response to an external stimulus. As in some smooth muscles, the function of the nerves in the heart is to regulate the ongoing timing and strength of contraction. Cell-to-cell communication is also of very great importance in the function of heart muscle, because, as is smooth muscle, it is composed of cells that are very small in relation to the size of the tissue. The process of the control of heart muscle contraction is sufficiently specialized to be treated separately in Chapter 18.

The Metabolism of Muscle Contraction

Because muscle does work, it must consume fuel in the form of energy-yielding biochemical compounds. The fuel most directly consumed by the actin-myosin contractile system is the universal high-energy compound, ATP. While muscle cells, as do other cells, require energy for ion pumping, growth, and cell maintenance, their major function is that of contraction, and this function consumes more energy than does any other type of cell. Not surprisingly, the metabolic adaptations of muscle are specialized to provide an adequate and constant supply of ATP for the contractile process.

Energy Pathways in Muscle Contraction

As we have seen, ATP is broken down during the crossbridge cycle and provides the energy that is turned into mechanical work. If chemical inhibitors of the general cell energy metabolism are used to prevent replenishment of the ATP supply, a contracting muscle will shortly become exhausted and enter a state similar to **rigor mortis.** However, a muscle maintains activity for a longer period of time than simple chemical measurements of the ATP content would indicate. This is because of a reserve "energy pool" in the cell in the form of a compound called **creatine phosphate.** This high-energy compound is capable of transferring its energy to ATP very readily, so that as soon as the supply of ATP to the myofilaments begins to fall, it is rapidly replenished. The ADP from the contraction process is rephosphorylated to ATP, and the creatine phosphate becomes creatine. This creatine is itself rapidly rephosphorylated by ATP derived from the general metabolism of the cell. The creatine phosphate pool provides a "buffer" for the rapid supply of ATP for the work of contraction as well as a link to the cellular sources of ATP. Depending upon the type of muscle fiber, this cellular ATP is produced by one or both of two common biochemical pathways, as diagrammed in Figure 16–24.

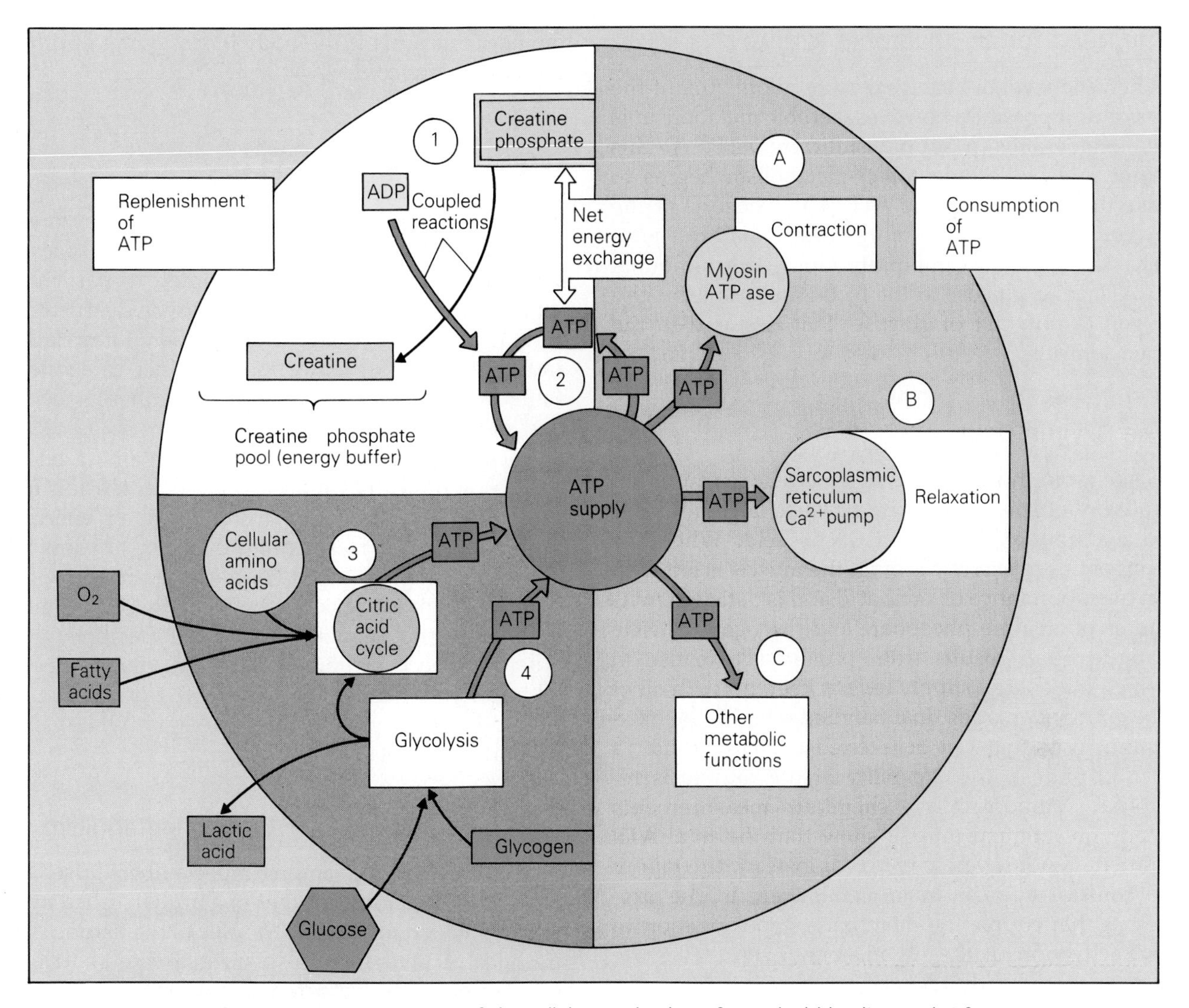

Figure 16–24 A summary of the cellular mechanisms for replenishing its supply of ATP, with the numbers indicating the order of replenishment. The cell used ATP for a variety of functions. The letters A, B, and C indicate the order in which cellular functions use up ATP.

Glycolysis: An Anaerobic Pathway

Glycolysis, the first of these pathways, yields a relatively small amount of ATP from the metabolic fuel (glucose) in a series of reactions that take place in the cytoplasm of the cell. Recall from Chapter 6 that since this reaction pathway takes place in the absence of oxygen, it is called **anaerobic.** For each molecule of glucose metabolized by this pathway, only two molecules of ADP are "recharged" to ATP.

The Citric Acid Cycle: Aerobic Metabolism

When sufficient oxygen is present, however, the end products of glycolysis (and those of fatty acid metabolism) can enter an **aerobic (oxidative)** reaction pathway, the citric acid cycle. These reactions, which take place in the mitochondria of the cell, yield an additional 36 molecules of ATP from a metabolized glucose molecule.

Give the much higher ATP production of aerobic metabolism, it is reasonable to ask why muscles would also have the less efficient anaerobic pathways. The answer lies in the rates at which energy is consumed and at which metabolic fuels can be supplied. Under conditions of normal activity, the blood supply to a muscle can bring the energy-supplying molecules (fatty acids and glucose) to a muscle at a rate that allows aerobic metabolism to supply almost all of the ATP the contractile system needs. During moderate exercise as well, the blood supply and the oxygen it delivers can keep up with the needs of the muscle.

The Need for an Anaerobic Pathway

When the level of activity reaches about 70% of the maximum possible, however, aerobic metabolism is no longer able to supply sufficient ATP. At this point, glycolytic (anaerobic) metabolism begins to take over to provide the ATP. The reactions of the glycolytic pathway run more quickly than the aerobic ones, and they can rapidly supply large amounts of ATP from glucose in the blood and from the glycogen (a polymer of glucose) that is stored in the muscle itself.

The Oxygen Debt

While rapid, however, these anaerobic reactions are not very efficient, and they produce large amounts of a temporary byproduct, **lactic acid.** When the bout of heavy exercise is over, the muscle is left with an overabundance of lactic acid and is deficient in its stores of creatine phosphate and glycogen. **Muscle fatigue** often results from prolonged exercise in which the energy supply fails to keep up with all of the energy demands. In a fatigued muscle, the maximum force that can be exerted decreases; after a period of rest, the capability of the muscle is restored. Paradoxically, chemical measurements made on fatigued muscle show that the total ATP content is not severely reduced; instead, the failure to contract appears to lie somewhere in the processes that couple the membrane depolarization to the activation of the myofilaments. Whatever the defect, it is quickly restored.

The metabolic imbalances that follow contraction are corrected by the aerobic metabolism that occurs in the resting muscle after the activity. The lactic acid is oxidized to compounds that can provide ATP, and the glycogen stores are resynthesized from glucose. These processes are responsible for the continued high oxygen consumption of a resting muscle after exercise. This oxygen is required to run these aerobic reactions until the original state of the muscle is restored. The burden of lactic acid and the glycogen deficit that occur during exercise make up the so-called **oxygen debt,** which must be "paid back" quantitatively by postcontraction oxygen consumption.

Muscle Efficiency and Heat Production

Another important byproduct of muscle metabolism is **heat.** The biochemical reactions of contraction are only about 20% efficient; the other 80% of the energy is degraded into heat. This is the heat that the body must get rid of during heavy exercise; it is also the heat that warms the body by shivering or other exercise in the cold.

Types of Muscle Fiber

In light of the many functions that muscles must perform, it is not surprising that there are specialized types of fibers that are adapted to particular sorts of tasks. These adaptations involve structural and biochemical specializations that are interrelated and that fall into some fairly distinct categories. Table 16–2 gives the characteristics of these specialized muscle types.

Some muscle tasks, such as maintaining body posture, require that muscles maintain tension for long periods with a low expenditure of energy. Other tasks, such as sprinting in a race or running after a bus, require rapid contractions that may be made at a high energy cost. Between these extremes are muscles that require a mixture of metabolic and structural specializations. In general, muscle fibers fall into two broad categories, **white fibers** and **red fibers.** (For example, consider the light (white) and the dark (red) meat from poultry.)

Red Muscle Fibers and Aerobic Metabolism

The color differences among muscle fibers arise because of the differing content of a protein called **myoglobin,** which imparts a red color to the tissue. The presence of myoglobin also gives a clue as to the most important distinction between the muscle classes. Myoglobin is a protein (similar to the protein **hemoglobin** found in red blood cells) that is capable of binding, storing, and releasing oxygen. It is found most abundantly in those muscle fibers that depend on oxidative (aerobic) metabolism, where it serves as an oxygen source in times of heavy demand.

Red fibers can be further classified as **fast twitch** and **slow twitch** types. They both have plentiful mitochondria and a rich blood supply. Their speeds of contraction are correlated with their rates of ATPase activity; that is, with the rate at which the actin-myosin contractile proteins can break down ATP and release its energy. The slow fibers are very resistant to fatigue when fulfilling their usual function of maintaining posture or working at lower rates. Attempts to use such muscles for rapid, sustained, tasks will cause them to fatigue rather quickly as their ATP consumption outstrips the ability of the citric acid cycle to replenish the ATP supply. The fast fibers of the red group have a higher ATPase activity; they contract

Table 16–2
Classification of Skeletal Muscle

Feature	Fast		Slow
	White	Red	Red
Mechanical			
Contraction speed	Fast	Fast	Slow
Force capability	High	Medium	Low
S.R. Ca^{2+} pumping	High	High	Moderate
Motor axon velocity	100 M/sec	100 M/sec	85 M/sec
Biochemical			
ATPase activity	High	High	Low
Source of ATP	Anaerobic glycolysis	Oxidative phosphorylation	Oxidative phosphorylation
Glycolytic enzymes	High	Moderate	Low
Number of mitochondria	Low	High	High
Myoglobin content	Low	High	High
Glycogen content	High	Moderate	Low
Rate of fatigue	Fast	Moderate	Slow
Structural			
Diffusion distance	Large	Small	Moderate
Fiber diameter	Large	Moderate	Small
Number of capillaries	Few	Many	Many
Sarcomere structure	Very regular	Less regular	Irregular Z-lines
Functional			
Role in body	Rapid, powerful movements	Medium endurance	Postural endurance
Example	*Latissimus dorsi*	Found in mixed-fiber muscles such as *Vastus lateralis*	*Soleus*

and relax more quickly and are suited to moderate endurance-type activities.

White Muscle Fibers and Anaerobic Metabolism

White muscle fibers, on the other hand, are adapted for fast and powerful contractions, but they will fatigue quickly. They contain the enzymes necessary for glycolytic (anaerobic) metabolism. Glycolysis, a process that operates faster than oxidative metabolism, can use the stored glycogen at a high rate to produce ATP. Since their immediate contractile function does not depend heavily on their oxygen supply, these muscles are not as well supplied with a dense capillary network as are the red types. The white muscles show a high ATPase activity and very rapid contraction and relaxation. The rapid contraction appears to be due to specializations in the myosin molecules that permit them to convert the energy in ATP more rapidly. The rapid relaxation is due in part to a more aggressive calcium-pumping mechanism present in the sarcoplasmic reticulum. As might be inferred from the anaerobic character of their metabolism, the cells contain fewer mitochondria than do the red types.

Patterns of Innervation and Muscle Growth

These various characteristics are summarized in Table 16–2. A close look at the features compared there will reveal many correlations among function and biochemical makeup. Note that the characteristics listed are for fibers, not whole muscles. Most whole muscles contain a mixture of these fiber types, so that they are not strictly confined to a single type of activity. Studies on athletes have shown that, in spite of particular types of training programs, the fiber composition of a given muscle remains relatively constant. For an individual, the specific mixture of fiber types begins to be established before birth and becomes set in childhood. In some newborn mammals, the fibers are predominantly the slow type. Patterns of innervation and usage during early development play an important role in deciding the ultimate distribution of fiber types. Experiments have been done with animals in which the nerve supply to fast and slow muscles was surgically reversed. These studies have shown that the re-innervated muscles take on characteristics related to the type of nerve supplying them. That is, formerly slow muscles begin to take on the contraction characteristics of fast muscles.

The Development of Muscle in Response to Use

It is well known that the pattern of use is important in determining the size and strength of a muscle. An increase in the mass of a muscle (without an increase in the actual number of cells) is called **hypertrophy.** Such hypertrophy is most readily brought about by exercises that emphasize isometric or slow isotonic contractions that are repeated many times. Muscles subjected to such routines will increase in their cross-sectional area and in the amount of contractile protein they contain. The relative distribution of fiber types will remain relatively constant. A negative aspect of this type of muscle adaptation (to a specific set of exercise conditions) is that the vascular supply to the muscle does not increase in proportion to the increase in muscle mass, nor is there a large improvement in the function of the heart. This means that the endurance of the hypertrophied muscle is not greatly increased, and its improved function is largely limited to the sort of exercise that produced the hypertrophy.

Heart muscle is also subject to hypertrophy in response to increased demands. This occurs normally during athletic conditioning but may also be a result of some disease process. For example, if the improper functioning of a heart valve makes the heart do extra work to maintain the blood pressure, (a so-called enlarged heart) will result. Such enlargement can at least partially compensate for the condition that caused it, but it also increases the metabolic demands of the heart and may make it more prone to disturbances in the rhythm of the heartbeat (see Chapter 18).

Exercise that emphasizes isotonic contractions at moderate workloads produces a different type of modification in the muscle. The popular "aerobic" exercise regimens produce less marked increases in strength than the isometric exercises mentioned previously, but do result in an increased vascular supply to the muscle. This increases the ability to produce increased performance for extended periods of time, such as for the duration of a cross-country race. Such exercises also provide conditioning for the heart muscle and can increase the efficiency of its pumping.

Adaptation in the opposite direction is also possible. Long-term immobilization of a muscle (such as might be produced by wearing a cast) causes **disuse atrophy,** in which the mass and strength of the muscle decrease. When a muscle is paralyzed by injury to its motor nerve and cannot contract on its own, it is also subject to atrophy. If there is a probability that nerve function may eventually be restored, periodic electrical stimulation of the muscle may prevent some of the atrophy.

Less severe atrophy occurs when a regular form of exercise is discontinued, and the muscle adapts to the lower level of use. This may occur during the "off-season" periods in the life of an athlete and is a factor in the "deconditioning" that is observed after long periods of weightlessness during space travel. Both of these situations can be at least partially reversed by a proper exercise program.

Summary

I. A. The many functions of muscle require that different types of muscle be specialized for different tasks. Muscle functions as an effector, a motor, and a regulator. It also regulates much of the input to the nervous system. The degree of neural control varies with the type of muscle.

B. Muscle may be classified in three ways: on the basis of its anatomical location (as skeletal, visceral, or cardiac); on the basis of its structure (as striated or smooth); and on the basis of its mode of control (as voluntary or involuntary). There is some overlap in these categories.

II. A. Most skeletal muscles are attached to bones at an origin and an insertion by tendons. As skeletal muscles shorten, they power the lever systems made up of the bones of the skeleton. Usually, skeletal muscles are arranged in antagonistic pairs or groups, and contraction of one muscle group will cause the lengthening of another. A whole skeletal muscle is made up of long and slender cells called fibers, which are bound up into the muscle by connective tissue. A skeletal muscle receives a plentiful blood supply.

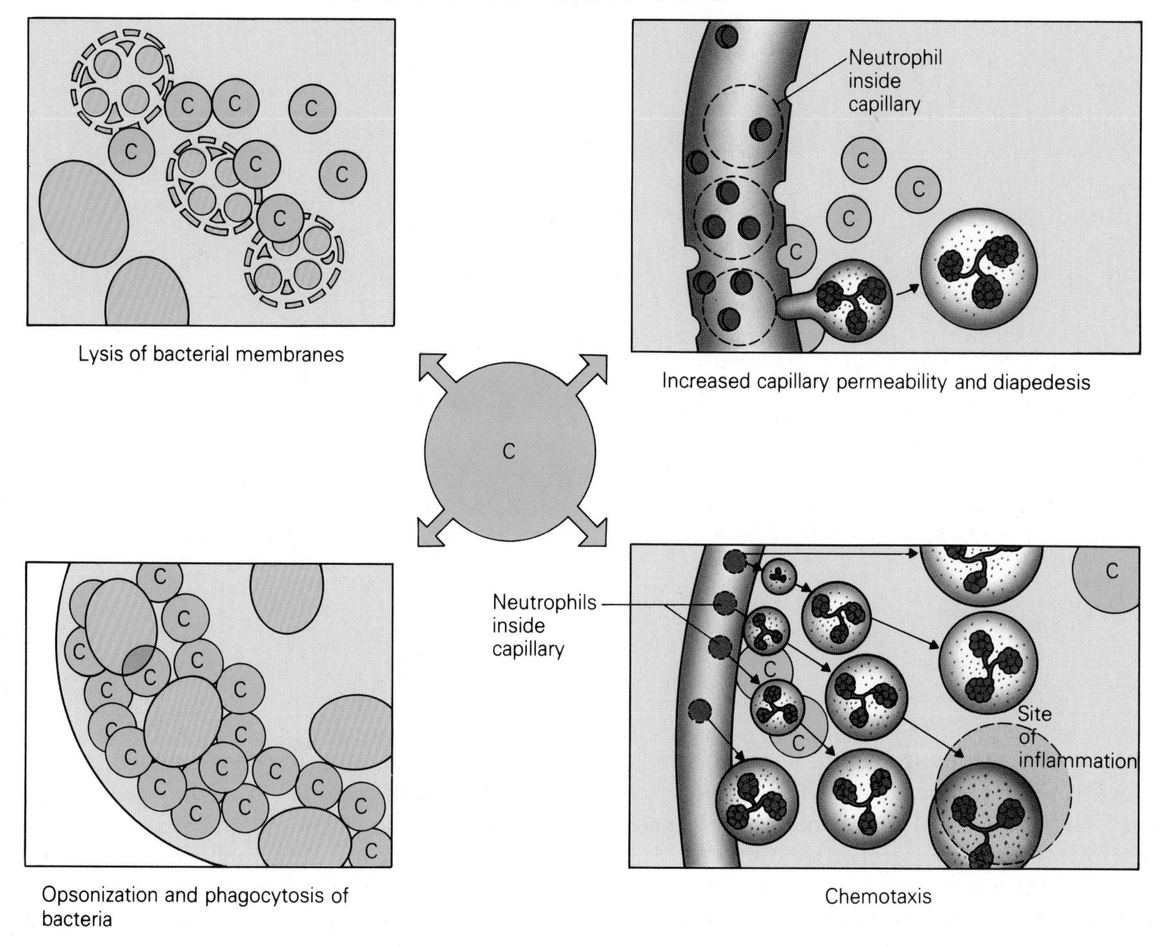

Figure 17–14 Four different ways in which complement proteins function.

trophils and macrophages, which in turn ingest and destroy the microbe. The latter process, similar to coating distasteful medicine with candy, is called **opsonization,** and the complement proteins that function in this matter are called **opsonins.**

Inflammation

Inflammation is usually a localized body response to tissue damage or injury such as occurs when pathogens invade tissues. It signifies the body's attempt to defend itself against the pathogen and prevent the pathogen from spreading and damaging other tissues (Fig. 17–15). The clinical signs of inflammation are **redness, heat, edema,** and **pain.** They are caused by the following inflammatory responses:

1. Injured cells, basophils, mast cells, and others release **histamine** and other chemicals that dilate blood vessels and increase the blood supply to the injured area. This process makes the skin look red and feel warm.
2. Histamine and other chemicals increase the permeability of local capillaries, which allows antibody proteins and other chemicals that mediate an immune response to pass out of the capillary and enter the tissues. The increased flow of materials out of the blood and into the tissues causes local swelling or edema (fluid accumulation), which in turn with chemicals released from damaged cells, causes pain.
3. Neutrophils, and to a lesser extent other leukocytes, migrate out of the capillaries (diapedesis) and into the surrounding tissues (Fig. 17–16). Neutrophils and tissue macrophages are attracted to the area of tissue damage by chemicals (chemotaxis) released from damaged cells, activated complement proteins, lymphokines (from lymphocytes), and other substances.

Phagocytosis

After a pathogen penetrates the body's epithelium, it encounters **phagocytes,** cells that ingest and destroy microorganisms by the processes of **phagocytosis.** Although there are several types of phago-

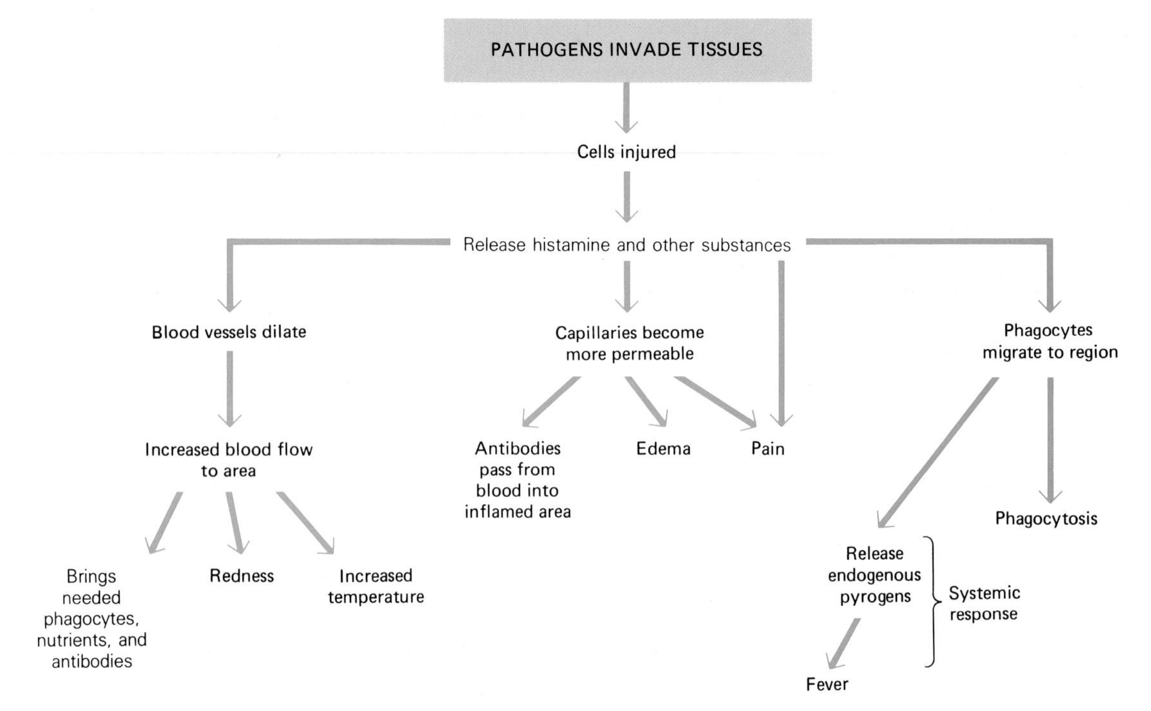

Figure 17–15 The inflammatory response.

cytes, they are all derived from stem cells in the bone marrow. Phagocytes are divided into two groups: The **neutrophils,** or **polymorphonuclear leukocytes,** and the **monocyte/macrophages.**

Neutrophils, as previously discussed, are the most numerous of the leukocytes, making up about 70% of the total. They are highly phagocytic, engulfing all kinds of particles (including pathogens) and destroying them. They are short-lived, existing for 1 to 5 days. They are continually being replaced and are the first phagocytes to arrive at an invasion site.

Monocytes migrate out of the blood and into

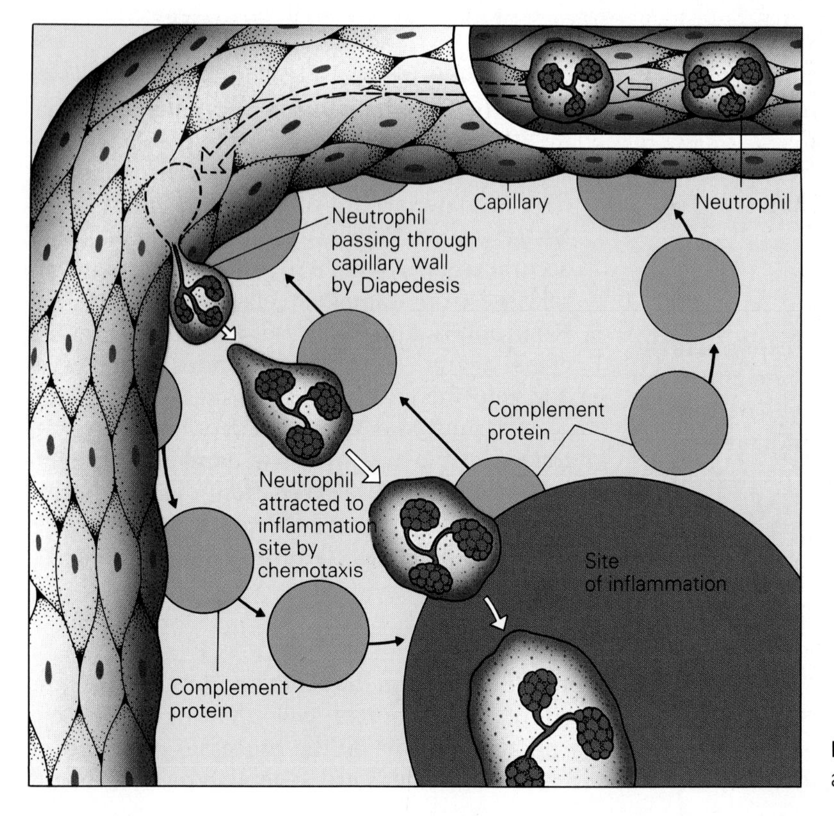

Figure 17–16 The processes of diapedesis and chemotaxis.

and skeletal muscles, action potentials in the heart involve additional membrane processes. Cells that are found in the ventricular region have most of the special features that we need to discuss, so a cell of this sort can serve as our typical example.

Figure 18–11 shows such a complex action potential. This is the type of electrical activity that is found in the **Purkinje fibers** of the ventricles (we will discuss the function of these fibers later). Notice first of all that this action potential lasts hundreds of times longer than one from a nerve. In order to identify the extra voltage changes present, we use numbers to designate the various parts of this action potential. **Phase 0** is the rapid upstroke. This is a very fast depolarization that **overshoots** beyond the zero millivolt level. As in nerve tissue, this rapid depolarization is due to a rapid increase in the membrane permeability to **sodium ions.** Associated with this is a reduction of the potassium permeability to a value lower than that at rest. As the high sodium permeability begins to decline, a small amount of repolarization takes place. This brief event is **Phase 1.** Shortly thereafter, the membrane potential becomes relatively steady at a value near zero millivolts. This is **Phase 2,** also called the **plateau** of the action potential. Here a slow inward membrane current due to the flow of **calcium ions** is balanced against the efflux (outward flow) of **potassium ions,** even though the potassium permeability is reduced at this time. The membrane channels through which calcium passes will also allow a small amount of sodium to pass inward. Late in the plateau, the calcium permeability declines, and with the falling membrane potential, the potassium permeability begins to increase. These factors produce a **repolarization,** which is called **Phase 3.** During this time, the membrane potential falls to its original value and remains steady until the next beat. This steady portion is **Phase 4,** which is equivalent to the resting potential of other muscle cells. This sequence of membrane potential changes lasts for about 300 milliseconds, depending on the heart rate.

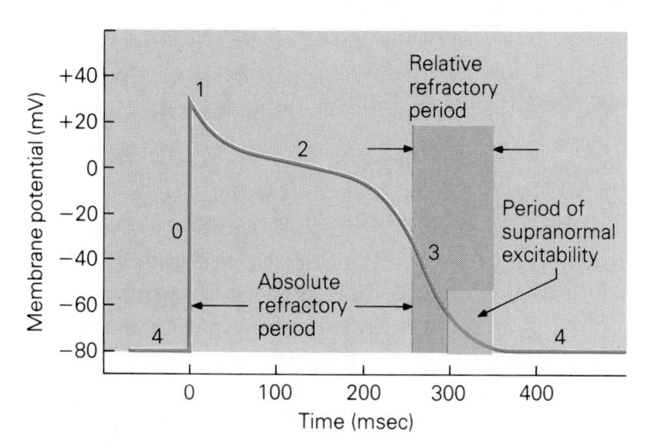

Figure 18–11 A sample action potential from heart muscle. This action potential is typical of those found in the fast conduction fibers of the ventricle, such as the Purkinje fibers. The numbers above the curve designate the phases that are described in the text, and the approximate durations of the refractory periods are shown.

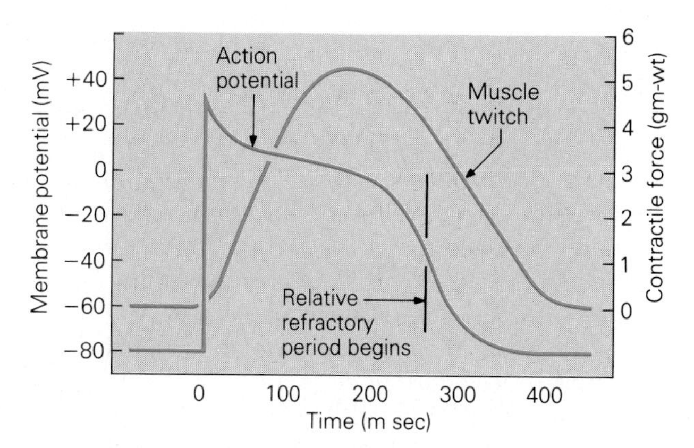

Figure 18–12 The relationship between the action potential and the twitch. The isometric twitch of a sample of heart muscle is shown in relation to the action potential that caused it. To some extent the duration of the twitch is determined by the duration of the action potential. Agents that prolong the plateau of the action potential make the twitch last longer, and the force of contraction may be increased because of the greater opportunity for calcium to enter the cell during the action potential.

As in nerve and other types of muscles, these phases of the action potential are associated with refractory periods. These are diagrammed in Figure 18–12. During the **absolute refractory period,** the muscle cannot be stimulated again. The **relative refractory period** then follows, and latest of all is the period of **supernormal excitability.** Here the muscle is actually easier to stimulate, because the potassium permeability is still a bit lower than normal. The action potential produced will be smaller than normal, because not all of the sodium channels from Phase 0 have had time to "re-set." Normally, the muscle is never stimulated during this time period. If such stimulation should occur, the result could be very dangerous, as will be discussed later in this chapter.

The Cardiac Action Potential and Contraction of Cardiac Muscle

Notice in Figure 18–12 that the absolute refractory period lasts almost as long as the isometric twitch. This relationship prevents the muscle from contracting again before it has relaxed, since the muscle will not respond to a second stimulus so soon after the first. This means that a tetanus is not possible in heart muscle; such a long-lasting contraction would prevent the heart from filling with blood for the next beat, and it would be a poor pump indeed.

Action potentials are necessary to make heart muscle contract. In skeletal muscle, action potentials are produced as a result of stimulation by a motor nerve. Each cell in a skeletal muscle is individually supplied by its own motor nerve terminal, and coordinated activity is due to the timing of impulses from the central nervous system (CNS).

The heart, in contrast, has no motor nerve supply, and its cells are very small. A separate nerve supply to each of the millions of muscle cells in the heart would make it a huge and complex organ. Instead, the cells of heart muscle communicate with one another directly, and muscle takes over the function of nerve. This means that the heart can be stimulated to contract at a single location, and the message will be passed along to the rest of the heart muscle. Within the individual cells of heart muscle, the process of activation is similar to that in skeletal muscle. The t-tubule system of the ventricular muscle cells aids in conducting the surface action potential inward, where it causes the release of stored calcium for the activation of the troponin-tropomyosin regulation system. Significant amounts of calcium enter the cell during the plateau of the action potential, and this calcium is added to the internal stores. The entering calcium is also directly responsible for causing some release of internal stored calcium from the sarcoplasmic reticulum. Because of this continual entry of calcium, cardiac muscle cells must (and do) have a way of getting rid of excess internal calcium by exchanging it for incoming sodium.

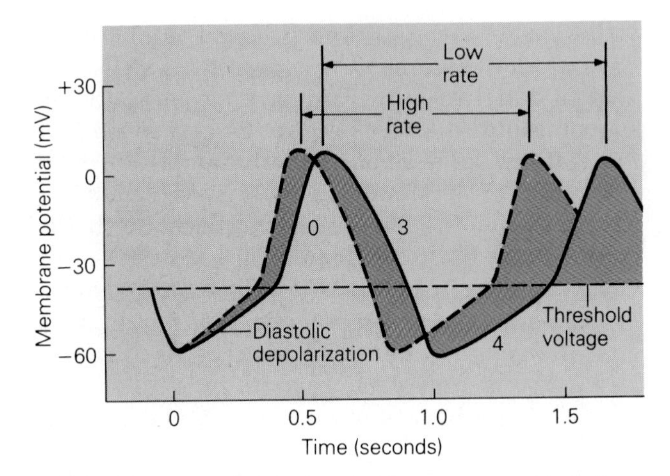

Figure 18–13 An action potential from a pacemaker cell. The slow diastolic depolarization (Phase 4) lets the membrane potential reach the threshold voltage shortly after the cell has repolarized, and another action potential results. The solid line shows a pacemaker potential at a low heart rate, which is determined by the spacing between successive action potentials. In this case the low rate corresponds to a heart rate of 65 beats per minute. The dotted line shows the rate that would result from a more rapid diastolic depolarization. The action potentials occur closer together, and the new spacing corresponds to a rate of 72 beats per minute.

Initiation of the Action Potential by Pacemaker Cells

The single location from which the stimulus for the heart action arises is called the **pacemaker.** Normally, the pacemaker region is the **sinoatrial node** (also called the **SA node**), a specialized region of muscle cells in the right atrium. A pacemaker action potential is shown in Figure 18–13. These cells do not have a steady resting potential. After each action potential, the cells repolarize to a value called the **diastolic potential.** Very soon after repolarization, the membrane begins a slow depolarization. This depolarization quickly brings the membrane potential to its **threshold,** and another action potential takes place. This action is repeated again and again, as many as 3 billion times in a 70-year lifetime. At resting heart rates, it is repeated approximately 70 times per minute, and it can reach as high as 160 times per minute during heavy exercise. The rate of the heart is regulated by the rate at which the diastolic potential reaches the threshold.

Normally, only one pacemaker is active at one time in the heart. This is because whichever pacemaker cell "fires" first will stimulate other cells that were about to produce an action potential. Therefore, the fastest pacemaker cell will predominate. The heart actually contains many cells that are capable of becoming pacemakers, but their activity is not usually expressed because they are inherently slower. Under some conditions, if these "hidden" pacemakers do not receive a stimulus from other cells, their activity will become apparent. As we will see later, this can be an important safety device.

The Sequence of Cardiac Activation

Activation of the Atria. The rhythmic activity of the SA node pacemaker is conducted first into the musculature of the right atrium and then to the left atrium (Fig. 18–14). This impulse conduction occurs when the action potential in one cell causes current to flow in an adjacent resting cell. This local current, which flows between the cells in the region of their intercalated disks (see Chapter 16), brings the membrane potential of the resting cell to its threshold. It then produces its own action potential, which in turn stimulates the next adjacent cell. This process occurs (with varying degrees of efficiency) in all regions of the heart. Conduction of action potentials between separate cells in heart muscle is similar

Focus on Cardiac Pacemakers

The **artificial cardiac pacemaker** is an electronic device that delivers an electrical stimulus to the heart to initiate contraction. Pacemakers are frequently used to treat patients who suffer from permanent or recurrent bradycardias (slow heart rates) that have undesirable symptoms (e.g., fainting). The first totally implantable pacemakers were reported in the 1950s, and today an estimated 1 million patients worldwide have implanted pacemakers. Implantable pacemakers are powered by lithium batteries, which may last for 6 to 10 years. Replacement of the battery requires the surgical implantation of a new pacemaker.

Early pacemakers were unable to sense the electrical activity of the heart and stimulated the heart (either the atrium or ventricle) at a fixed rate. Modern pacemakers can sense the heart's normal electrical activity and are activated only when they fail to detect appropriate spontaneous activity.

Such **demand pacemakers** are frequently programmable to respond in a specified manner when a particular arrhythmia is detected. For example, to treat **complete heart block,** a pacemaker is used that senses the depolarization of the atria and then delivers a stimulus to the ventricles after a delay equal to the normal atrioventricular conduction time. This method of pacing retains the normal temporal relationship between atrial and ventricular contraction, and therefore preserves the normal contribution of atrial contraction to ventricular filling. In addition, by sensing the normal rate of atrial depolarization, the pacemaker can respond appropriately to increases in heart rate associated with exercise and stress. Programmable pacemakers can be reprogrammed, without removing the pacemaker, so that the pacemaker operating parameters can be revised to fit the patient's changing needs.

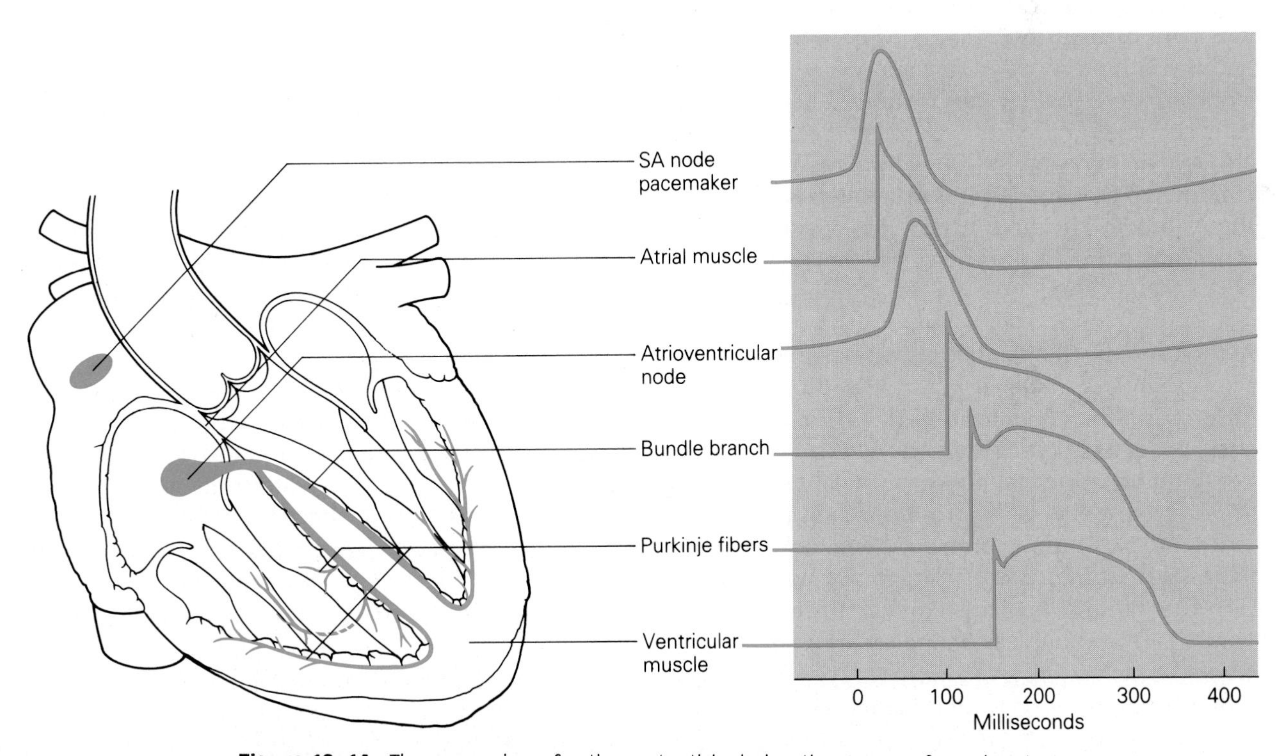

Figure 18–14 The succession of action potentials during the course of one heartbeat. The sequence starts with the *pacemaker potential* at the sinoatrial node. The activity spreads rapidly to the *atrial muscle.* The *atrioventricular* node action potential, similar in form to the SA node potential, is conducted slowly, and there is a longer space between this and the first ventricular potential shown, that of the *left bundle branch.* (This is similar to the slightly earlier action potential of the *Bundle of His,* which is not shown.) Action potentials of the *Purkinje fibers* and the *ventricular muscle* follow quickly in succession. The entire activation process takes approximately two-tenths of a second from the time the action potential begins in the SA node until the last ventricular muscle is activated.

in many ways to conduction along a nerve axon (a single cell), in which active regions stimulate the resting regions next to them. Cell-to-cell conduction in heart muscle is made possible by the low intercellular resistance associated with the intercalated disks.

Conduction through the atria is quite rapid. Atrial action potentials are conducted quickly because the muscle cells are relatively large and have a very negative resting potential. In addition, some specialized atrial conducting pathways conduct very rapidly. The rapid conduction and fast action potentials result in a very rapid Phase 0 depolarization with a large overshoot in the cells, which produces a large local current flow. These conditions favor the rapid conduction of the impulse throughout the tissue.

Conduction of the Action Potential from the Atria to the Ventricles. At the lower portion of the right atrium is another region of specialized cells called the **atrioventricular (AV node).** These cells are quite small and have resting potentials that are not as negative as those of the atrial cells. As action potentials pass through this region, they become weaker and travel more slowly. This slowed impulse travel is called **decremental conduction.** If the distance is great enough, or the impulse becomes weak enough, it may die out altogether. Under normal conditions, the impulse is slowed but not stopped at the AV node. The delay this causes ensures that the atria finish their contraction before the ventricles begin theirs.

Activation of the Ventricles. After the impulse crosses the AV node, it enters the **Bundle of His** (also called the **common bundle**). This conducting structure is composed of muscle cells specialized for very rapid conduction, and here the impulse travels at its greatest speed. The common bundle soon divides into two **bundle branches,** one for each of the ventricles. Conduction is very rapid here as well.

The bundle branches divide further in each ventricle, eventually becoming an array of **Purkinje fibers,** which form a distribution system over the endocardial surface and even penetrate slightly into the main muscle mass of the ventricles. The Purkinje fibers are also specialized muscle fibers with a high speed of conduction.

From the Purkinje fibers, the impulse passes into the muscle of the ventricle wall, stimulating contraction. The ordinary ventricular muscle fibers are somewhat slower conductors than the specialized conducting tissues, but by this time the activity is well distributed throughout the heart, and the distance remaining is relatively short. The arrangement of the specialized conducting system ensures a nearly simultaneous activation of all of the ventricular muscle.

The overall effect of the distribution of the stimulating impulse by the rapid conduction system is a wave of depolarization that sweeps over the heart from its base to its apex and from the endocardial to the epicardial surface. Because of the rapid conduction of the impulse and the fast rise of the action potential, the boundary between active and resting tissue is very sharp and moves very quickly. During repolarization, there is also an apparent movement of the active areas.

As active cells return to rest, a less-pronounced wave of repolarization sweeps over the heart in the opposite direction. However, the directionality of the repolarization is not determined by the conduction system, but by the fact that the cells that are activated last have the shortest action potentials and begin to repolarize first. This produces an apparent movement in the opposite direction from the wave of depolarization. Because this is not a conducted phenomenon, and because the repolarization is not nearly as rapid as the depolarization is, the speed of the repolarization wave is lower and the boundary between recovering and still-active cells is not very sharp.

Abnormal Activation of the Heart. The proper function of the heart as a pump depends on the correct sequence of activation, and this in turn depends on the normal function of the conducting system. For example, failure of the AV node to conduct the impulse will cause **heart block,** in which the ventricles are not stimulated by the atrial impulse. At such times, cells other than those of the SA node can adopt a pacemaker function and provide a source of stimulation, although the rate from such auxiliary pacemakers is usually lower than normal. These "emergency" pacemakers usually lie somewhere in the conduction system, and their activity is normally suppressed by the faster pacemaker rhythm of the SA node.

Regular muscle tissue rarely shows any spontaneous activity. If there is such a pacemaker in the ventricular region—and if for some reason it becomes active from time to time—its impulse will be conducted both in the normal direction and "backwards" toward the atria. When it stimulates the AV node, it will produce refractoriness in the nodal tissue. The next atrial impulse arriving may not be able to be conducted through the AV node, and the ventricles will "drop" a beat. This can be very apparent to the person in whom it happens, because the delay gives the ventricles a chance to become filled with extra blood, and the next heartbeat is larger than usual. Such "skipping a beat" is not rare and usually is of no concern.

However, a dangerous condition may arise if the conducting system of the ventricle is damaged so that conduction along parts of it is slowed. Such damage is not uncommon following a heart attack (a **myocardial infarction**), because areas of the heart muscle may have been deprived of oxygen for some time. This slowed conduction can cause cells that are in the process of normal repolarization to be stimulated by a late-arriving action potential, which may arrive during the period of supranormal excitability. The resulting stimulation can give rise to even more action potentials out of their proper time sequence, which are then conducted to normally resting cells. The final result may be an uncoordinated stimulation of the ventricular muscle called **ventricular fibrillation.** Such uncoordinated contractile activity is useless for pumping blood, and death will follow shortly. These conduction and rhythm disturbances, and many others related to them, are called **arrhythmias** and are often fatal if not treated immediately.

The Electrocardiogram. The action potentials and their conduction sequence as just described can be observed only by painstaking study of exposed or isolated hearts. While the knowledge gained is important in understanding the operation of the heart, it is obviously not a realistic approach to the study or treatment of human (or veterinary) patients. Fortunately, the electrical activity of the heart may be harmlessly detected at the body surface by a technique called **electrocardiography.**

The heart may be thought of as an electrical generator lying in a conducting medium made up of the body tissue and fluids. During the phases of the cardiac cycle in which depolarization or repolarization is sweeping over the heart muscle mass, some portions are positively charged and others negatively charged. This potential difference, or **voltage gradient** (as much as 100 millivolts), causes current to flow in the external medium between these regions of the heart. These currents are strongest close to the heart, and they become weaker at greater distances from the heart. The largest voltage measured at the body surface is only around 1 millivolt; however, this is sufficient to be detected by a sensitive instrument called an **electrocardiograph.** This device converts the minute surface currents to movements of a pen on paper or a spot of light on a cathode ray tube (CRT), producing the familiar **electrocardiogram** (**ECG** or **EKG**).

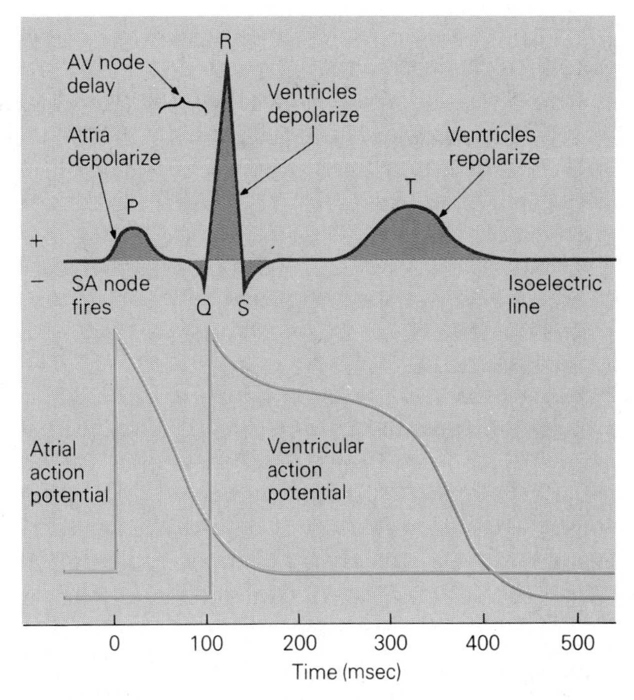

Figure 18–15 The time relationship between the action potentials and the electrocardiogram. All three traces share the same time axis, and the ECG recording may be used to infer when the actual electrical changes took place in the heart muscle.

Figure 18–15 is a diagram of a typical electrocardiographic recording. For convenience, the principal features of the recording have been designated by the letters P, Q, R, S, and T. The so-called **P-wave** is caused by the depolarization of the atrial muscle, and the **QRS complex** is caused by the depolarization of the ventricles. The **T-wave** is the result of the repolarization of the ventricles. Repolarization of the atria takes place during the ventricular depolarization, and its weak signal is lost. When the entire heart is either depolarized (active) or is resting, there is no potential difference to set up a current flow, and there is no deflection seen in the ECG. The line of no deflection is called the **isoelectric line.** Figure 18–15 also shows the relationship between cellular action potentials and the surface ECG.

It is important to realize what the ECG can and cannot show. First, only electrical events are shown, and one can only infer that a contraction of the muscle has taken place (sometimes it has not). The ECG is only a record of **voltage** and **time.** Secondly, only the electrical events that involve large amounts of muscle show up in the ECG record. Because of the small tissue mass involved, activity of the conducting system and nodal tissues can only be inferred, but these inferences can be very revealing. For example, if a P-wave is not followed by a QRS-

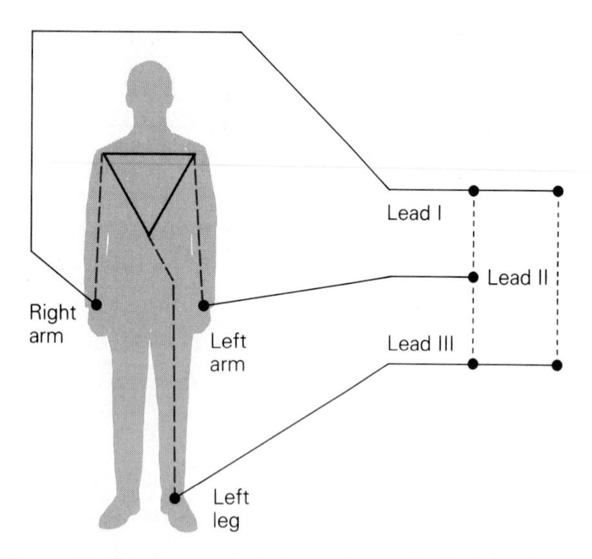

Figure 18–16 Standard electrocardiographic limb leads.

complex, then one can assume that conduction was blocked somewhere between the atria and the ventricular muscle. If the QRS complex is very broad or abnormally shaped, one can assume that some areas of the muscle of the ventricle were activated later than they should have been. This implies a defect in the ventricular conducting system. A very long delay between the P-wave and the QRS complex most probably means that the conduction in the AV node was more decremental than usual.

Further information can be gained about the direction of the spread of electrical activity by recording from electrodes located on different areas of the body. The portion of the electrical activity that the recording electrodes will intercept depends on their location and orientation. Clinical recording of the ECG assumes that the heart lies at the center of an equilateral triangle and that the recording connections are made at the vertices of this triangle (Fig. 18–16). In practice, the electrical connections to the right and left arms and to the left leg are considered to connect to these vertices, and the standard **limb leads** are recordings between any two of the connections. A recording made between the right and left arms is designated as **Lead I** and is most sensitive to activity spreading laterally across the heart. **Leads II** and **III** are recordings made between the left leg and the right and left arms, respectively. Records from these leads are most sensitive to activity proceeding from the top to the bottom (base to apex) of the heart. By combining the timing and voltage information recorded from a number of leads, someone skilled in the interpretation of the ECG can build up a very complete picture of the functioning of the heart.

The Ventricular Pump and the Cardiac Cycle

As discussed in the previous section, the electrical activation of the left and right hearts occurs almost simultaneously. As a result, the sequence of mechanical events that occur during a single beat or cycle of the heart are very similar for both hearts. Because of this similarity, portions of the following discussion of the cardiac cycle will focus on the left heart. Keep in mind that similar events are occurring at the same time in the right heart.

The Ventricular Pump

Let us first focus on the ventricles of the heart and examine how the ventricles function as a pump and how the heart valves ensure that blood flows in one direction. Figure 18–17 shows schematically the four basic steps in the pumping process. Beginning with Panel A **(ventricular filling),** we see that the ventricle is being filled with blood from the atrium. Contraction of the atrial muscle facilitates the movement of blood into the ventricle through the one-way AV valve. Notice that the semilunar valve, between the ventricle and the outflow tract, is closed and prevents the backflow of blood from the outflow tract into the ventricle.

Panel B **(ventricular contraction)** shows the beginning of the ventricular contraction phase of the cycle. Notice that as the ventricle contracts, pressure increases inside the ventricle. This initially causes a small backflow of blood into the atrium, which causes the AV valve to close.

As soon as ventricular pressure exceeds the pressure in the outflow tract (Panel C), the semilunar valve is forced open, and blood flows out of the ventricle **(ventricular ejection).** In order for the cycle to repeat, the ventricle now must relax and refill with blood. As the ventricle relaxes **(ventricular relaxation),** pressure inside the ventricle falls, and the semilunar valve is forced closed as blood attempts to flow from the outflow track back into the ventricle (Panel D). When ventricular pressure falls below that of the atrium, the AV valve opens and blood flows into the ventricle (Panel A). This completes the cycle.

The Cardiac Cycle

You will notice that the ventricle is the main pumping chamber of the heart. The atrium serves as a reservoir to receive the continuous flow of blood back to the heart. Although the atrium does contract, its contraction adds very little to the filling of

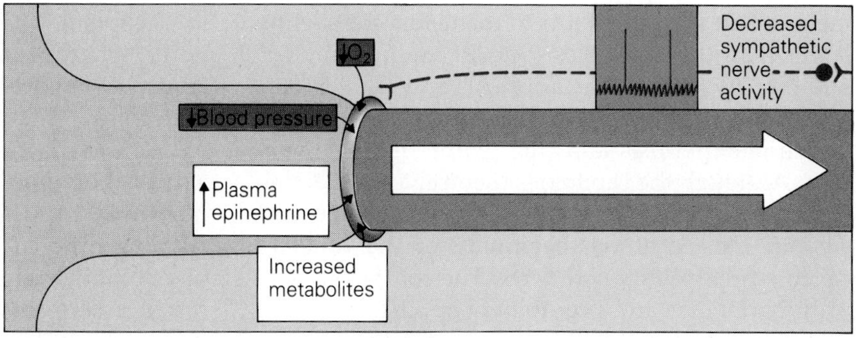

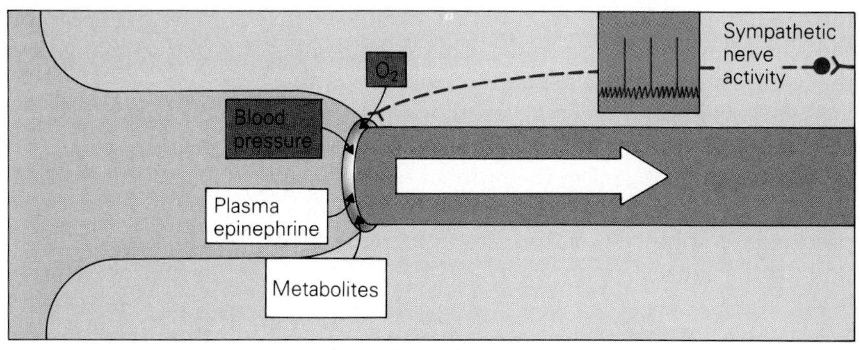

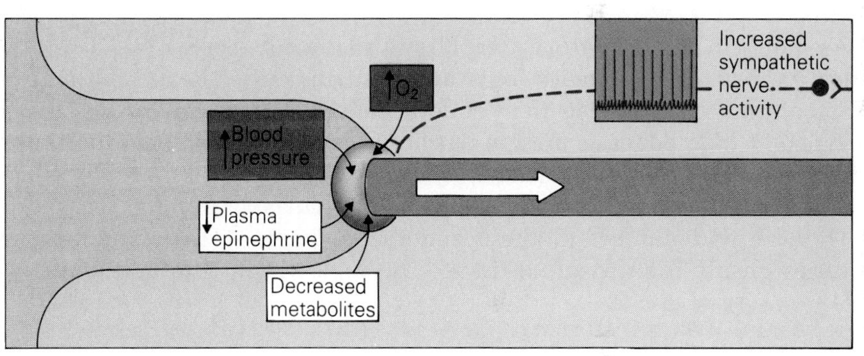

Figure 19–14 Summary of local tissue factors that produce either vasodilation or vasoconstriction of vascular smooth muscle.

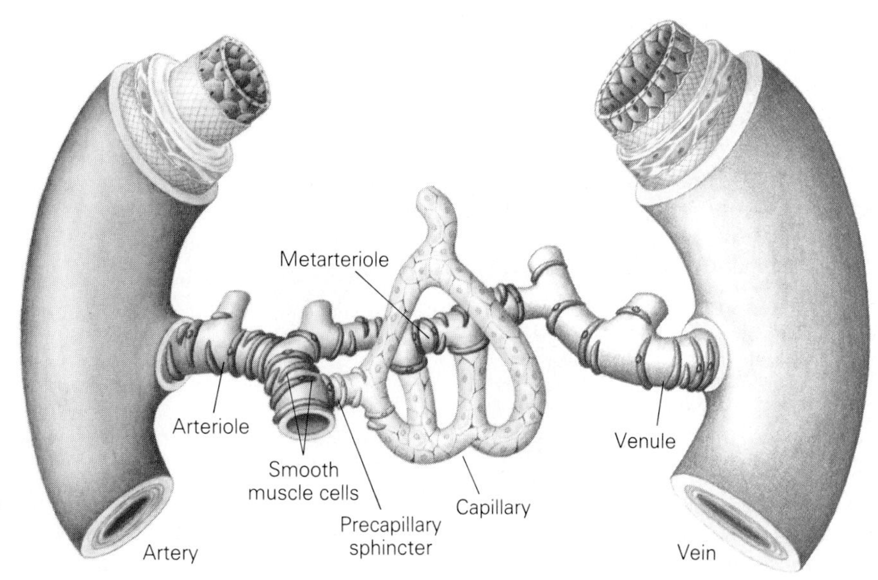

Figure 19–15 Diagrammatic representation of the microcirculation. The diameter of the capillaries is approximately 5 μm.

metarterioles (10 to 20 μm in diameter) are slightly larger than the capillaries (5 to 10 μm in diameter) and directly connect the arterioles and the venules. The capillaries branch out from either the arterioles or the metarterioles and then empty into the venules. Although the capillaries themselves contain no smooth muscle, in some cases there is a small cuff of smooth muscle at the beginning of the capillary called a **precapillary sphincter.** The contractile state of the arteriolar and precapillary sphincter smooth muscle determines blood flow through the capillaries. If we were to look at the flow of blood through the capillaries with a microscope, we would see that the blood flow is often sporadic. This is the result of the contraction and relaxation of the precapillary vessels **(vasomotion).** This vasomotion is heavily influenced by the metabolic state of the tissue. For example, during exercise, the number of open capillaries in skeletal muscle will increase several-fold as precapillary sphincters vasodilate to increase blood flow to the muscle.

The Capillary Wall and Endothelial Pores

As shown in Figure 19–16, the capillary wall is made up entirely of endothelial cells and contains no smooth muscle. The flat, thin, endothelial cells are joined at their edges as are the patches in a patchwork quilt. Small pores (clefts) between the endothelial cells allow for the passage of molecules. The size and number of these **endothelial pores** varies greatly from tissue to tissue. The capillaries in the brain, for example, may not contain any pores, whereas the capillaries in the kidney contain relatively large pores to allow the passage of large molecules.

A thin basement membrane surrounds the outside of the capillaries and may serve as a molecular filter to restrict the passage of very large molecules. Although the endothelial pores represent an obvious point of exchange, in most capillaries they represent a very small percentage of the capillary surface area. As we will see in a later section, direct diffusion of molecules across the endothelial cell membranes accounts for the major portion of molecular exchange.

Movement of Substances Across the Capillary Wall

There are three fundamentally different mechanisms by which substances can cross the capillary wall. These are **pinocytosis, diffusion,** and **bulk flow** (Fig. 19–17).

Pinocytosis

Relatively large molecules cross the capillary wall via vesicles that pinch off from the plasma membrane. Pinocytosis accounts for a very small portion of the total exchange and is probably important only for the transport of large, lipid-insoluble proteins that cannot cross the capillary wall by movement through pores.

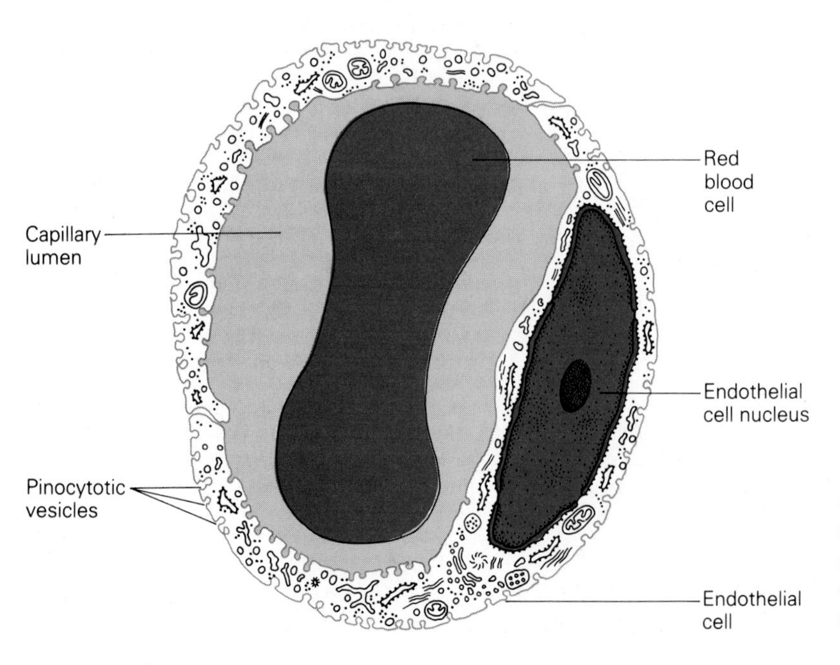

Figure 19–16 Cross-sectional view of a single capillary. The capillary wall is made up of a single layer of endothelial cells. (Adapted from *Human Physiology,* Vander, Sherman, and Luciano, 1985, page 342)

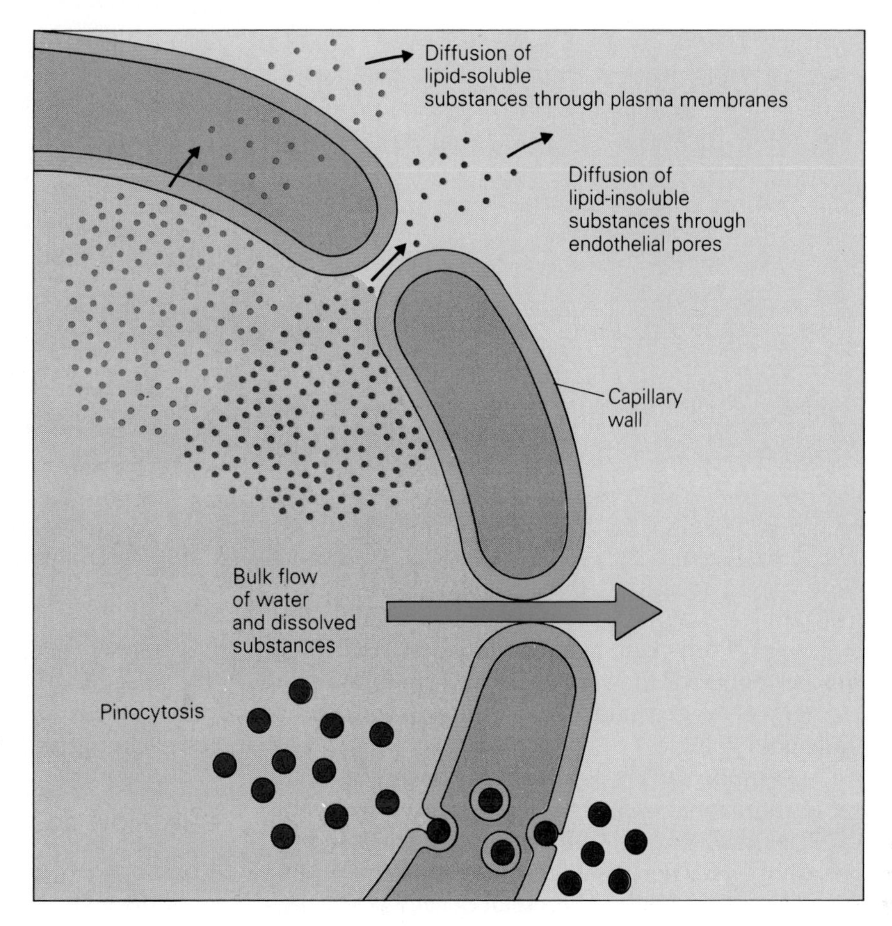

Figure 19–17 Diagrammatic summary of transport pathways across the capillary wall. Movement of substances is in response to either a hydrostatic pressure gradient (bulk flow) or a diffusional gradient.

Diffusion

By far the most important mechanism for exchange of water and dissolved substances is diffusion. We can get some idea of this exchange by comparing the rate of blood flow through a capillary with the rate of water exchange across the capillary wall by diffusion. The rate of exchange is as much as 100 times greater than the rate of blood flow down the length of the capillary. Since water is moving across the capillary wall in both directions, there is little net gain or loss of water from the circulation by diffusion; however, the exchange rate is very high.

Of course, many solute molecules that are dissolved in blood and interstitial fluid are also exchanged by diffusion. Blood that is entering the capillary has a relatively high concentration of oxygen and glucose and a relatively low concentration of carbon dioxide. The concentrations of oxygen and glucose in the interstitial fluid are lower because these substances are being continuously taken up by the cells. Metabolism of these substances results in the production of carbon dioxide, which diffuses out of the cells into the interstitial space. As a result of these concentration gradients, there is net movement of oxygen and glucose out of the capillary and net movement of carbon dioxide into the capillary (Fig. 19–18). Of course, similar concentration gradients for other nutrients, metabolites, and hormones result in net transfer of these substances between the interstitial fluid and the blood. In all cases, the direction of the concentration gradient determines the direction of exchange. Analogously, heat generated by metabolism is transferred by convection down its thermal gradient from the cell to the blood.

Molecules can diffuse across the capillary walls either through the water-filled pores or directly through the endothelial cells. The path a molecule takes is determined by its relative solubility in water and lipids. Molecules that are only slightly soluble in lipid (e.g., Na^+, K^+, Cl^-, and glucose) will more likely diffuse through pores. Molecules such as oxygen, carbon dioxide, and urea, which are more lipid soluble, can diffuse directly through the plasma membranes of the endothelial cells. Water molecules are able to cross either through pores or by diffusion through the endothelial cells.

These pathways for exchange are summarized in Figure 19–17. Because pores occupy less than 1% of the capillary wall area, the surface area available for diffusion of lipid-soluble substances is more than 100 times greater than that for lipid-insoluble sub-

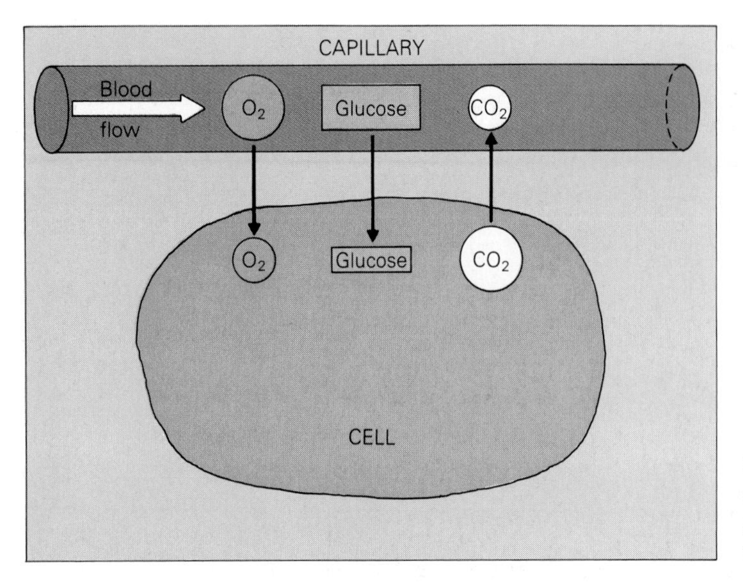

Figure 19–18 Transfer of substances between the cells and the capillary blood in response to diffusional gradients. The size of the label indicates the relative concentration of a substance.

stances. We can begin to appreciate how well suited this system is for the very rapid and continuous exchange of oxygen and carbon dioxide, which are both lipid soluble.

A number of physical factors contribute to the rapid diffusional exchange of substances across the capillary wall through either pores or the endothelial cell. First, the distances that molecules must move are very small. Most metabolically active cells are within 20 μm of a capillary. Secondly, the surface area of the capillary walls is extensive. Both diffusion distance and surface area influence the rate of net diffusion. The total surface area of all the capillary walls may be as high as 7,000 sq. ft. Imagine spreading the entire blood volume of the capillaries (approximately 250 ml or 1 cup) over an area the size of a football field: this gives you some idea of the ideal conditions for diffusion that exist in the capillaries.

A third factor is the relatively slow rate of blood flow that exists within the capillaries. The velocity of blood flow through the capillaries is greatly reduced because of the large total cross-sectional area of the capillary bed. The velocity of blood flow is inversely proportional to the cross-sectional area of the vessel(s) through which it is flowing. We see this principal in action whenever we observe the flow of water down a river. The velocity is greatest where the stream narrows (as in rapids) and slowest where the stream widens; total flow, however, is the same at both points. When a stream empties into a lake, the velocity of water movement in the lake is imperceptible because of the dramatic increase in cross-sectional area. For precisely the same reason, the velocity of blood flow in the aorta may be as much as 600 to 700 times greater than that in the capillaries. The average velocity of blood flow in a capillary is about 1 cm/sec (about 2 ft/min) compared to 40 cm/sec in the aorta. The dramatic reduction in flow velocity in the capillaries means that the time for exchange is substantially increased.

Bulk Flow and Oncotic Pressure

Exchange of water (and dissolved solutes) also occurs through endothelial pores by **bulk flow** in response to the pressure gradient between the inside and outside of the capillary (Fig. 19–19). As we will see, the pressure gradient is always from inside the capillary to outside, which causes water to flow out of the capillary. Although the total exchange of fluid by this means is relatively small (except in the kidney), it represents an extremely important mechanism for the maintenance of the circulating blood volume. This exchange is important because it occurs in response to a pressure gradient rather than a diffusional gradient.

The hydrostatic pressure gradient that drives water through the endothelial pores is simply the difference between the hydrostatic pressure inside the capillary (capillary blood pressure) and the hydrostatic pressure outside the capillary (interstitial hydrostatic pressure). The hydrostatic pressure outside the capillary is extremely difficult to measure, and there is some uncertainty about the actual value. Although it may in fact be slightly subatmospheric (negative), for purposes of our discussion we will assume that it is zero. Given this assumption, the hydrostatic pressure gradient (capillary hydrostatic pressure − interstitial hydrostatic pressure) is equal to the capillary hydrostatic pressure. When a blood vessel is cut (usually a superficial vein), blood flows out of the vessel in response to the pressure difference between the inside and the outside of

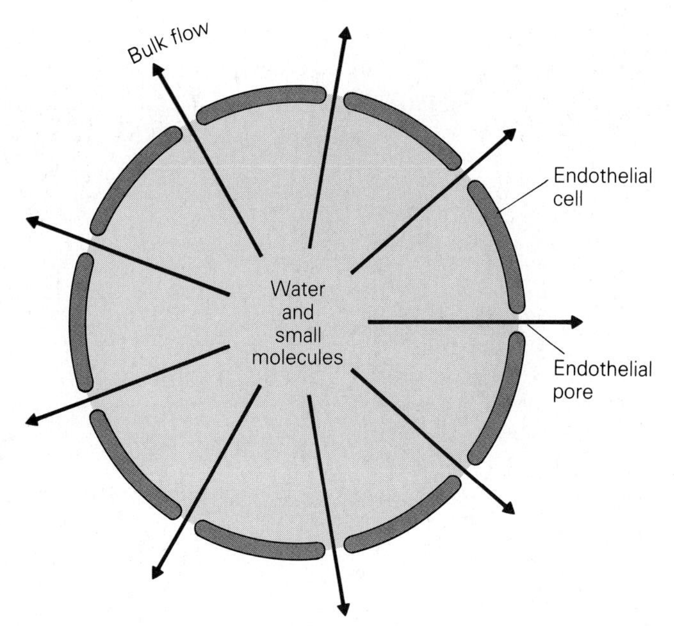

Figure 19–19 Bulk flow of water and dissolved substances out of the capillary through endothelial pores in response to a hydrostatic pressure gradient. Hydrostatic pressure inside the capillary is always greater than the pressure of the interstitial fluid surrounding the capillary.

the vessel. Since water can move through the capillary pores by bulk flow, we can reason that water and dissolved solutes should be forced out of the vessels. Clearly, if this were the whole story we would have a great deal of trouble maintaining a normal circulating blood volume. There is in fact an opposing force that minimizes the loss of fluid by bulk flow.

The opposing force that prevents this loss of vascular volume is the **osmotic pressure** due to the plasma proteins (proteins in the blood plasma). Although all dissolved solutes contribute to the **total osmotic pressure of blood,** we can ignore the contribution of all solutes that are freely exchanged across the capillary wall because these solutes are at diffusional equilibrium. Only the plasma proteins that do not freely cross the capillary wall contribute to the oncotic pressure gradient between the inside and outside of the capillary. The term **oncotic pressure** refers to that portion of the total osmotic pressure of blood that is due to the presence of plasma proteins. Recall from Chapter 4 that net diffusion of water across a selectively permeable membrane will occur in response to a concentration gradient for water. The capillary wall functions as a selectively permeable membrane because of its limited permeability to plasma proteins. Recall that the major plasma protein is albumin. Most plasma proteins cannot leave the capillary and therefore establish an oncotic pressure gradient that favors the net diffusion of water from the interstitial space to the intravascular space. You have probably guessed by now that the oncotic pressure exactly balances the hydrostatic pressure so that there is no net loss of water from the circulating blood.

Filtration and Absorption: Starling's Hypothesis

This balance of hydrostatic and oncotic pressures across the capillary endothelium was originally described by Starling in what is now called **Starling's hypothesis** (Fig. 19–20). Let's begin by looking at an idealized average capillary. (The situation in any particular capillary might differ significantly from this model, but the principles remain the same.) The oncotic pressure gradient is determined by the difference between the oncotic pressure of the capillary blood and that of the interstitial fluid. The oncotic pressure of blood is normally about 28 mm Hg. Although the capillaries are relatively impermeable to plasma proteins, some proteins do leak out of the capillaries. This small amount of protein in the interstitial fluid results in a tissue oncotic pressure of about 3 mm Hg. The oncotic pressure gradient ($28 - 3 = 25$ mm Hg) causes a net flow of water into the capillary. Remember that the osmotic flow of water is in the direction of increasing osmotic pressure. This is easy to remember if you recall that osmosis is the diffusion of water down its concentration gradient. Water is less concentrated in blood because of the presence of plasma proteins.

If we examine a typical capillary, we would measure a hydrostatic pressure of about 35 mm Hg at the arterial end of the capillary and about 15 mm Hg at the venous end. The pressure drop along the length of the capillary reflects the resistance to flow of blood through the capillary. The hydrostatic pressure causes fluid to move out of the capillary; this is opposed by the oncotic pressure that moves water back into the capillary (Fig. 19–20). At the arterial

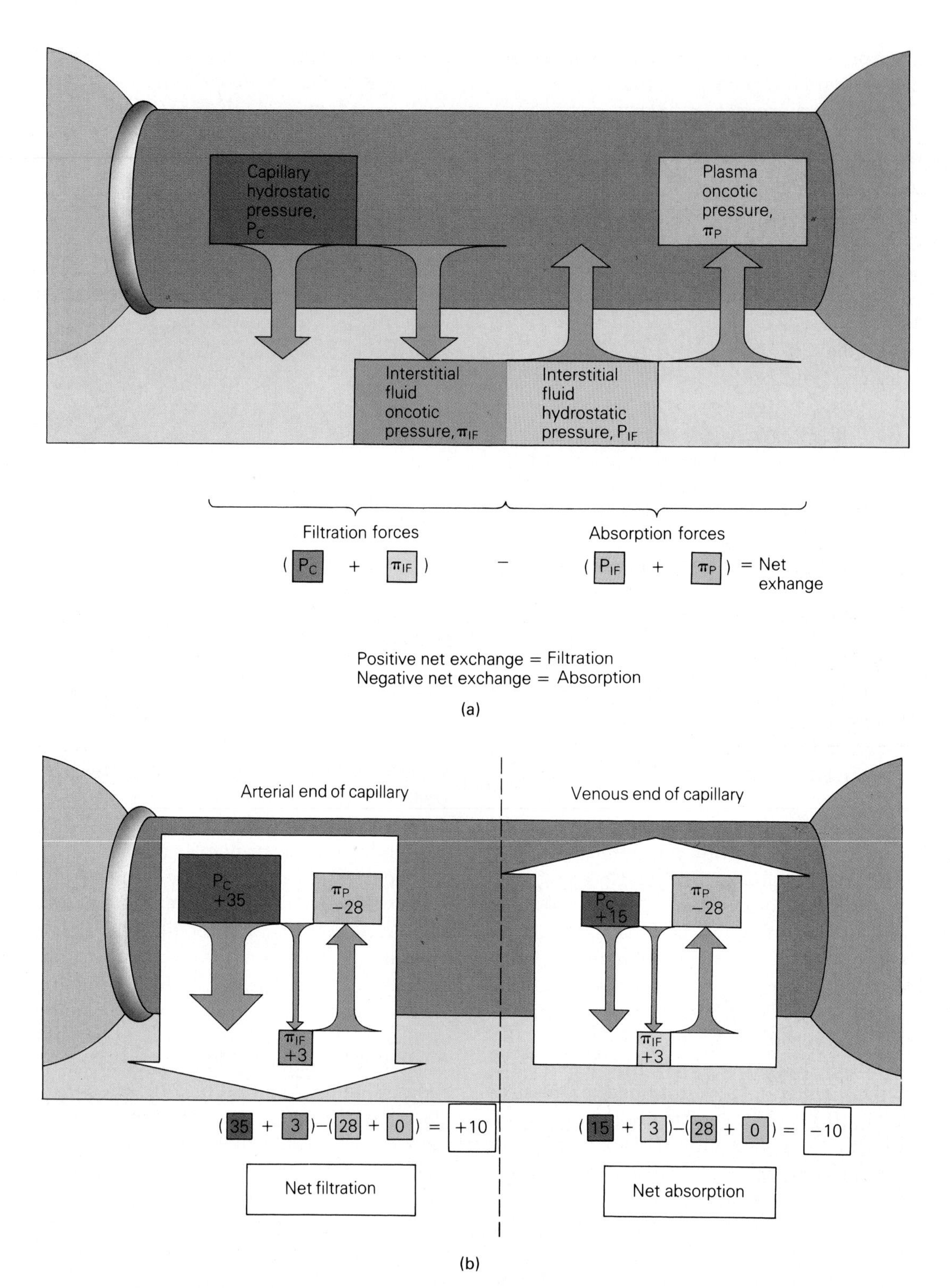

Figure 19–20 Summary of hydrostatic and osmotic forces causing movement of water across the capillary wall. The ideal capillary shown here represents the average of all capillaries in the body. Pressures within individual capillaries will vary considerably. (Modified from *Human Physiology*, 4th Ed, 1985, ed. Vander, Sherman, Luciano. Figure 11–53, p. 346).

end of the capillary, hydrostatic pressure (35 mm Hg) exceeds oncotic pressure (−25 mm Hg), leading to a net **filtration pressure** of 10 mm Hg. On the other hand, at the venous end of the capillary, hydrostatic pressure (15 mm Hg) is less than oncotic pressure (−25 mm Hg), leading to a net **absorption pressure** of −10 mm Hg. At the middle of the capillary, a hydrostatic pressure of 25 mm Hg is exactly balanced by an opposing oncotic pressure of −25 mm Hg. In this example, filtration exactly equals absorption, which means there will be no change in the overall capillary volume.

We stated earlier that capillary flow often changes dramatically owing to the contraction of smooth muscle in the terminal arterioles and precapillary sphincters. These changes in flow are associated with similarly large changes in intracapillary pressure. As shown in Figure 19–21, when precapillary smooth muscles are fully dilated, the associated capillaries would show filtration along their entire length. Fully contracted precapillary muscles would lower the capillary pressure enough to cause absorption of fluid along the entire length of the capillary. Changes in arteriolar pressure could also lead to changes in capillary pressure to favor either filtration or absorption. We can get some idea of what the "average" state of the capillaries is by looking at the net loss or gain of fluid from the cardiovascular system. Under normal conditions, approximately 2 L of fluid is returned from the interstitial space to the circulation by the lymphatic system during a single 24-hour period. Thus, the average situation slightly favors filtration over absorption.

Filtration and absorption provides an important mechanism for the dynamic maintenance of the circulating blood volume. Although the cardiovascular system is a closed system at the "macro" level, at the level of the microcirculation, it is an open system. The constancy of the circulating blood volume in fact represents a dynamic balance between forces

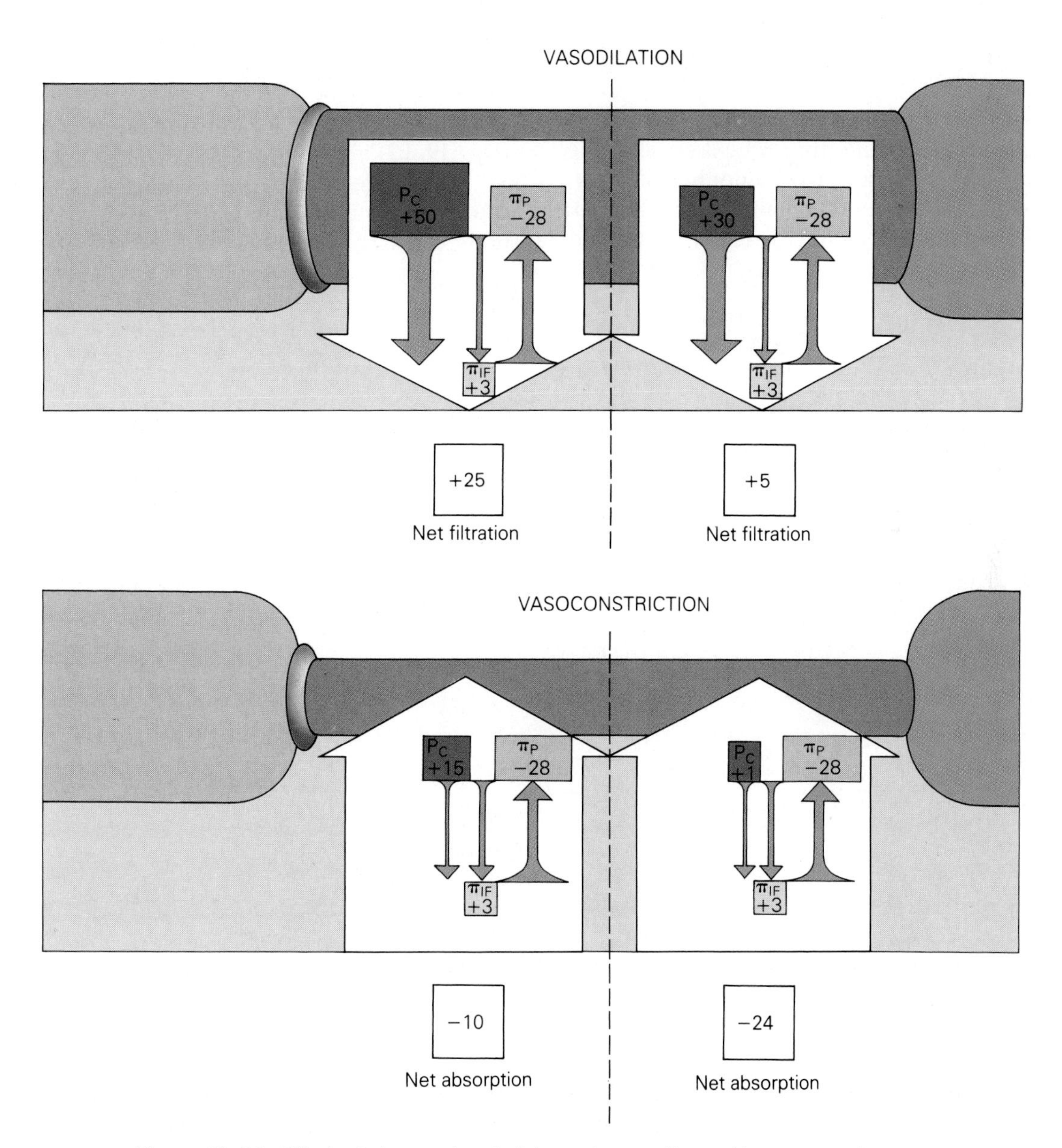

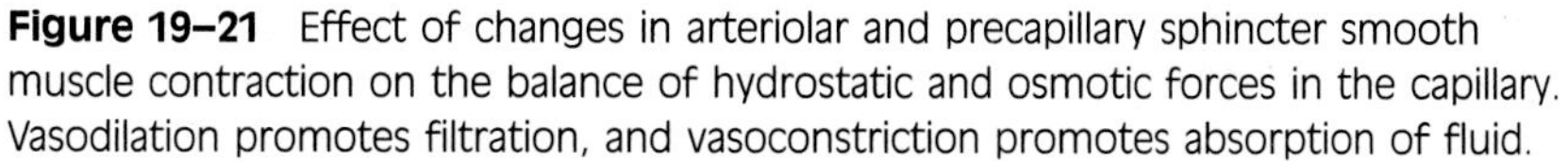
Figure 19–21 Effect of changes in arteriolar and precapillary sphincter smooth muscle contraction on the balance of hydrostatic and osmotic forces in the capillary. Vasodilation promotes filtration, and vasoconstriction promotes absorption of fluid.

tending to move water into and out of the capillary. Recall that cardiac output is determined in part by filling of the heart (end-diastolic volume). Cardiac filling, in turn, is greatly influenced by the total volume of circulating blood. Any significant reduction in blood volume would lead to a reduction in cardiac output. As we will see later, transfer of fluid from the extravascular space into the capillary represents an important compensatory response to blood loss (hemorrhage). The distribution of fluid can also be affected by disease. Heart failure will ultimately lead to an increase in capillary pressure, causing the accumulation of fluid in the interstitial space (edema). Other diseases may also lead to edema, because of a decrease in plasma protein concentration.

The Venous System: Returning Blood to the Heart

The capillaries empty into thin-walled vessels called **venules,** which represent the beginning of the venous system. The venules contain endothelial clefts as do the capillaries. Although fewer in number, these allow for additional exchange of substances. The capillaries and venules together are often referred to as the **exchange vessels.** The venules then converge to form progressively larger veins.

Venous Capacitance

The physical nature of the veins is considerably different from that of the arteries and arterioles. The veins are to the arteries as a paper bag is to a balloon. The veins, like a paper bag, can accommodate large changes in volume with little change in internal pressure. In addition, the total volume of the venous system is several times greater than that of the arteriolar system. These characteristics allow the veins to function as a low-pressure reservoir for blood; the veins are often referred to as the **capacitance vessels.** At any one time, approximately 75% of the total blood volume is contained within the veins as compared to 20% in the arterial system (Fig. 19–22).

This large capacity in and of itself is not significant. What is important is that the veins can reduce their capacity through contraction of the smooth

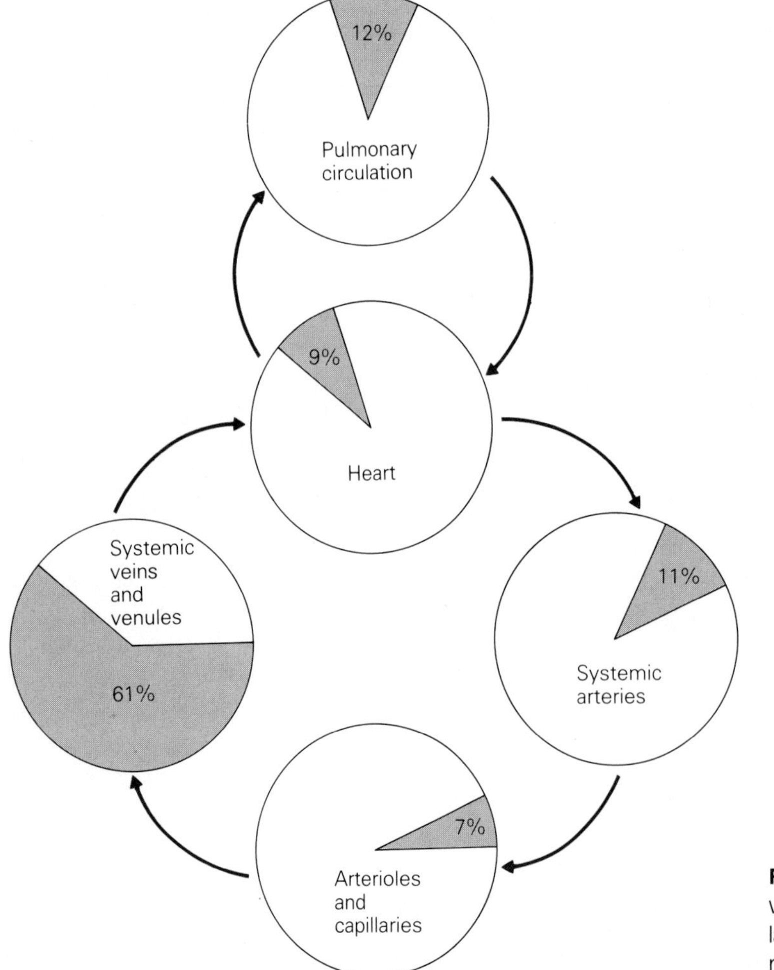

Figure 19–22 Distribution of the total blood volume among the various regions of the circulatory system. The total blood flow through each region is the same (see Figure 18–3).

muscle in their walls. This occurs in response to increased sympathetic nerve activity. As the venous smooth muscle contracts, blood is transferred from the veins to the heart and the arterial system. As we will see in a later section, this ability of the veins to dynamically alter their capacity provides a mechanism by which the body can maintain end-diastolic volume of the heart. Recall that the distribution of blood volume can change dynamically as a result of changes in arteriolar resistance. For example, during exercise, the volume of blood in muscle increases as capillary beds open up in response to dilation of precapillary resistance vessels. Other factors such as postural changes (e.g., standing up from a recumbent position) can result in passive redistributions of blood volume in response to gravity. Contraction of the veins can compensate for these changes. We will say more about this in a later section when we discuss the overall regulation of cardiovascular function.

Venous Return

The pressure in the venules and small veins is only about 10 mm Hg. Since the pressure at the right atrium is essentially zero, there is a pressure drop of only 10 mm Hg between the venules and the heart, indicating that the resistance to venous blood flow is low. This low resistance is due to the large cross-sectional area of the veins.

Several other factors also contribute to the flow of blood back to the heart. Some veins contain one-way valves, which serve a function similar to that of valves in the heart (Figure 19–23). These valves serve two major functions. First, by preventing backflow of blood away from the heart, they tend to minimize the pooling of blood in the feet and legs upon standing. Secondly, these valves allow the contractile activity of the muscles that surround the veins to contribute to the flow of blood back to the heart. Contraction of the surrounding muscle tends to squeeze the veins. Blood is forced out of the locally constricted vein and, because of the valves, is constrained to flow toward the heart. So you can see that the combination of skeletal muscle contraction and the venous valves constitutes a pump that assists the heart to maintain the unidirectional flow of blood through the circulatory system.

An additional pumping activity further facilitates the flow of blood back to the heart. The muscular activity associated with normal respiration acts through the venous system to pump blood back to the heart. When the diaphragm descends during inspiration, the contents of the abdomen are compressed along with the abdominal veins. As before, this compression of the veins forces blood back toward the heart. During inspiration, the intrathoracic pressure decreases; this provides a negative pressure that draws air into the lungs. Since the right atrium is located within the thoracic cavity, this negative pressure also increases the flow of blood from the veins into the right heart.

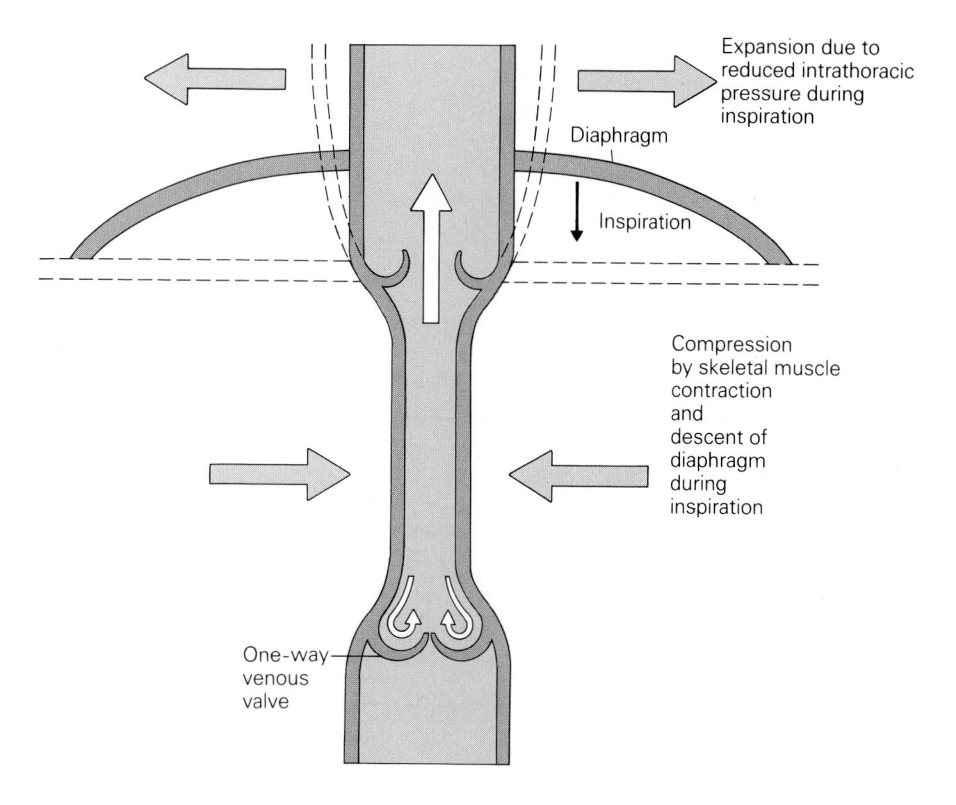

Figure 19–23 Factors promoting venous return to the heart. The combination of one-way venous valves and the compression of veins by skeletal muscle contraction and respiration enhance the flow of blood from the veins back to the heart.

The Lymphatic System

The lymphatic system is a second system of vessels that differ anatomically from the blood vessels. The function of the lymphatic vessels is to return whatever fluid and plasma protein has leaked out of the capillaries into the interstitial space back to the circulating blood. Despite the selectively permeable nature of the capillary wall and the balance of Starling's forces, there is a net loss of protein and fluid from the blood to the interstitial space. The extent of this loss increases significantly during exercise and with certain diseases. The normal operation of the lymphatics prevents the formation of edema that would otherwise occur.

The Structure of Lymphatic Vessels

The lymphatic vessels, like the capillaries, are made up of a single layer of epithelial cells. As shown in Figure 19–24, the lymphatic system begins as a series of sacs that have a low internal hydrostatic pressure. These terminal sacs have relatively large endothelial pores, allowing for the movement of both fluid and protein into the lymphatic system. The lymph capillaries drain into larger vessels, which themselves drain into the subclavian veins.

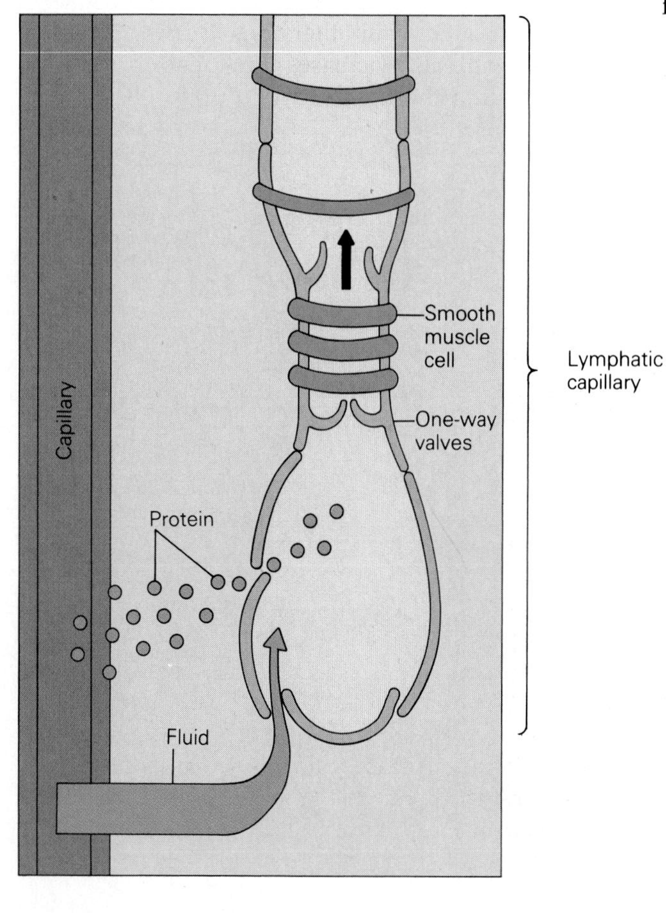

Figure 19–24 Diagram of a lymphatic capillary. The flow of capillary fluid back to the systemic circulation is promoted by one-way valves, contraction of smooth muscle, and compression by skeletal muscle contraction.

The Flow of Lymph

Lymph flow is driven by much the same mechanism that propels venous blood back to the heart. The lymphatic vessels have an extensive series of one-way valves as do the veins. The pumping action comes from phasic contractions of the smooth muscle associated with lymphatic vessels as well as compression of the vessels by skeletal muscle activity. In 24 hours, the lymphatics will return a volume of fluid back to the heart equal to the total circulating blood volume (5 L). In addition, approximately one quarter to one half of the total plasma protein is returned to the blood during this period. The lymphatic system represents the only means by which plasma proteins can be returned to the blood.

The Filtration of Lymph

The lymphatic vessels perform several additional functions. The lymphatics also transport substances absorbed from the gastrointestinal tract, primarily fat, to the circulating blood. Periodic swellings of the lymphatic vessels, called lymph nodes, provide for the removal of microorganisms and other foreign material from the lymph.

Failure of the Lymphatic System

The lymphatic system fails to operate normally in certain diseases. As a result, plasma proteins accumulate in the interstitial space. This extra protein decreases the water concentration of the interstitial fluid and promotes the loss of water from the circulating blood via osmosis. This failure to return protein to the blood ultimately results in edema, or the accumulation of interstitial fluid.

Regulation of Systemic Arterial Pressure

Recall from our previous discussions that the body maintains normal tissue flow by monitoring and maintaining mean arterial blood pressure (MAP). As pointed out in the water system analogy, this approach simplifies the regulatory process and at the same time allows for a great deal of local regulation of blood flow by individual tissues. You will also recall that under conditions in which the demand for flow exceeded the capacity of the pump, it was necessary to selectively decrease the flow to certain users in order to maintain flow to others. Based on our accumulated knowledge of how the heart and vasculature function, we are now in a position to investigate the mechanisms by which the body achieves the necessary coordination of the heart and vasculature to maintain arterial pressure.

As we discuss how the body operates to maintain blood pressure, it will be easy to lose sight of the fact that the ultimate goal is to maintain blood flow, not arterial pressure. If the goal were only to maintain arterial pressure, this could be done with minimal effort by the heart if blood flow to all the tissues were reduced to a trickle; this, of course, would not be compatible with life. Maintaining arterial pressure provides a simple method for adjusting cardiac output to meet the blood flow requirements of the tissues.

As previously noted, the relationship between cardiac output (CO), mean arterial pressure (MAP), and total peripheral resistance (TPR), is described by the equation:

$$CO = MAP/TPR$$

Algebraic rearrangement gives us the equation:

$$MAP = CO \cdot TPR$$

We can see from this equation that two factors determine the magnitude of the mean arterial pressure. One is determined by the operational status of the heart (CO) and the other by the operational status of all the resistance vessels in the body (TPR). Figure 19–25 summarizes all the factors that we have discussed that can change either cardiac output or total peripheral resistance. This figure indicates only where relationships exist, without delineating how a change in one component alters related components. The ways in which these various parameters are integrated to achieve normal cardiovascular function are the subject of the next several sections.

Before proceeding, let us look briefly at the meaning of total peripheral resistance. The total peripheral resistance is the total resistance to the flow of blood from the aorta back to the right atrium. Keep in mind, however, that the major portion of the resistance to blood flow is posed by the arterioles and precapillary sphincters. More importantly, changes in peripheral resistance are entirely due to changes in the diameters of these vessels. Therefore, when we talk about changes in total peripheral resistance, we will be referring to changes in the overall resistance posed by all the resistance vessels in the body. Although other factors such as hematocrit can alter the viscosity of the blood—and consequently, the resistance to flow—these changes do not constitute a mechanism for the homeostatic maintenance of arterial pressure. We will therefore consider the viscosity of the blood to be constant under most conditions.

Cardiac Output, Venous Return, and End-Diastolic Volume

These terms must be defined clearly if we are to understand how these factors interact. The terms **cardiac output** and **venous return** obviously refer to blood flow from the heart to the capillaries and from the capillaries to the heart, respectively. What is not obvious is whether they refer to the average flow over a period of time or to the instantaneous flow at any point in time.

If we use these terms to describe average flow over a period of many minutes, then cardiac output and venous return must be equal since this is a closed system. Frequently, however, we must consider transient changes in flow that occur in selected portions of the system. Consider for a moment the changes that occur following an increase in venomotor tone. Prior to this event, cardiac output and venous return were equal. As the veins contract and shift blood volume from the peripheral veins to the heart, venous return will for a short time be slightly

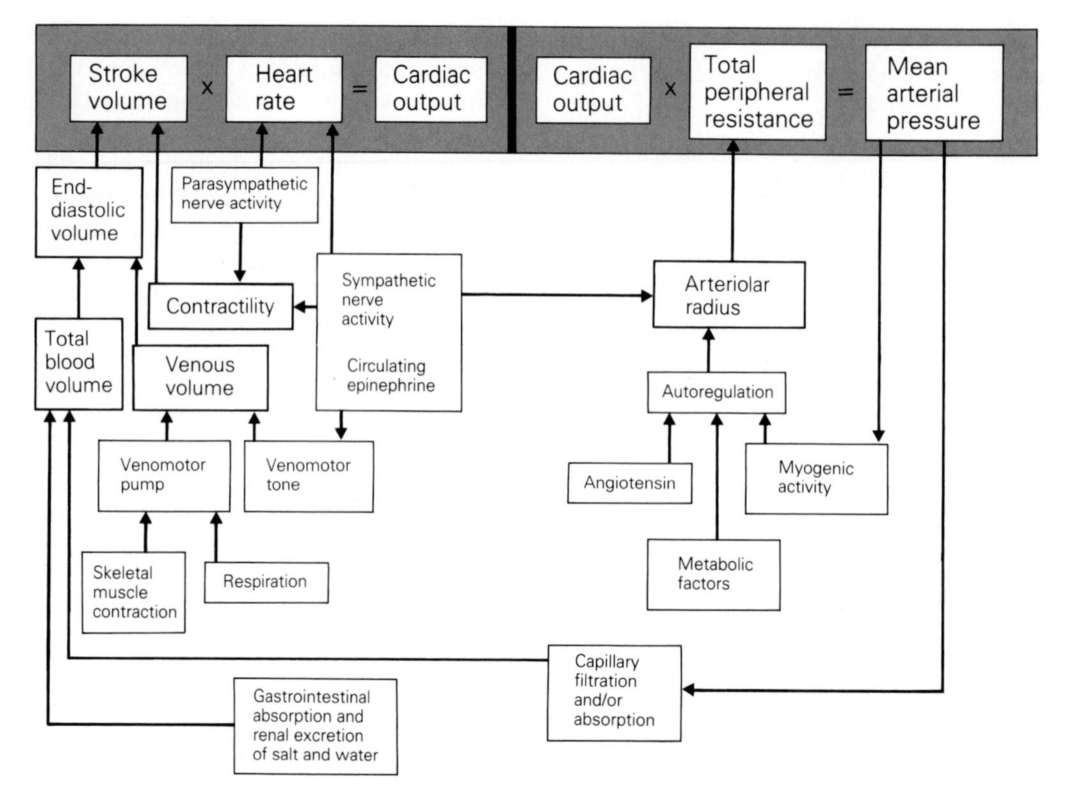

Figure 19–25 Summary of the factors that can alter cardiac output and arterial pressure.

higher than cardiac output because of the extra volume of blood flowing back to the heart from the peripheral veins. As long as venous return to the heart exceeds cardiac output, the volume of blood in the heart will be increasing. The increase in end-diastolic volume will then lead to an increase in stroke volume and cardiac output (according to the Frank-Starling law). After the venoconstriction, cardiac output and venous return will again be equal (but greater than before the venoconstriction occurred). One might then state that an increase in venous return (temporarily) produced an increase in cardiac output, leading to an increase in venous return (average).

This sounds like a circular argument if it is not made clear that in one case the term *venous return* refers to instantaneous flow and in the other it refers to average flow. A clearer description of the same sequence of events is that the redistribution of blood volume from the peripheral veins to the heart produces an increase in end-diastolic volume, leading to an increase in cardiac output and venous return (average). The important point is that cardiac output is very sensitive to changes in end-diastolic volume, which in turn depends on the total blood volume and the distribution of blood within the vascular system.

Consider the following analogy. You are sitting on a stool with a glass in your right hand. Two buckets are placed on the floor, one to either side of you, with a short piece of tubing connecting the bottoms of the two buckets. The two buckets are each half full of water, and you can just barely reach down far enough to fill your glass with water. Your goal is to fill the glass with water from the bucket on your right and pour it into the bucket on the left; you are pumping water from one bucket to another just as the heart pumps blood from the veins to the arteries. If you work hard enough, you will begin to accumulate more water in the bucket to your left because you are transferring water faster than it can flow back through the tubing.

But here is where you get into trouble; as you transfer water to the left bucket, the water level in the right bucket goes down. The further down the water level goes, the less water you can get in your glass, and you reach a point where the water level in the bucket sets an absolute limit on your performance. Your capabilities have not diminished, there has been no loss of water, the buckets holding the water are unchanged, but you are limited as a pump because you can no longer fill your glass.

There are two ways you can remedy the problem: (1) you can put more water in the buckets to raise the water level (increased blood volume), or (2) you can raise the level of the buckets so you can reach the water in the bottom (increased venomotor tone). Whichever action you take, the results will be

The conducting units (trachea, bronchi, and bronchioles) have three important functions: (1) to warm and humidify the air, (2) to distribute air to the gas exchange surface of the lung, and (3) to serve as part of the body defense system. Since the inside of the lung is literally exposed to the external environment, the conducting unit contains a **mucociliary transport system** that keeps microorganisms, dust particles, and noxious gases from entering the alveoli (Fig. 20–6). The conducting units are lined with cilia that beat synchronously, moving foreign material and microorganisms up the trachea. The cilia are buried in a carpet of sticky mucus. At every bend in the airway, many of the heavy particles in the inspired air fail to make the turn and stick to a film of mucus. The mucus is coughed up and either excreted or swallowed.

The cilia are greatly affected by tobacco smoke. When the cilia come in contact with tobacco smoke, they become partially paralyzed, slowing the mucociliary transport system. As a result, microorganisms and excess mucus accumulate in the lower airways (e.g., the bronchioles). Often, the excess mucus plugs the lower airways, and a cough is required to dislodge and remove it. If excess mucus accumulates over a long period of time, bacterial infection will often be present. When this occurs, the mucus is no longer a light yellow but green in color as a result of the infiltration and rupturing of white blood cells.

Lung Volumes and Ventilation

Air Flow

The process of moving air in and out of the lungs is known as **ventilation.** When you take a breath of air into your lungs, you can hear the sound of air rushing in, especially if you breathe rapidly. The sound comes from air flow in the airway tree. There are two basic types of air flow.

Turbulent Air Flow

One type, occurring in the trachea and large bronchi, is **turbulent flow** (Fig. 20–7*a*). Turbulent flow occurs at high flow rates and consists of completely

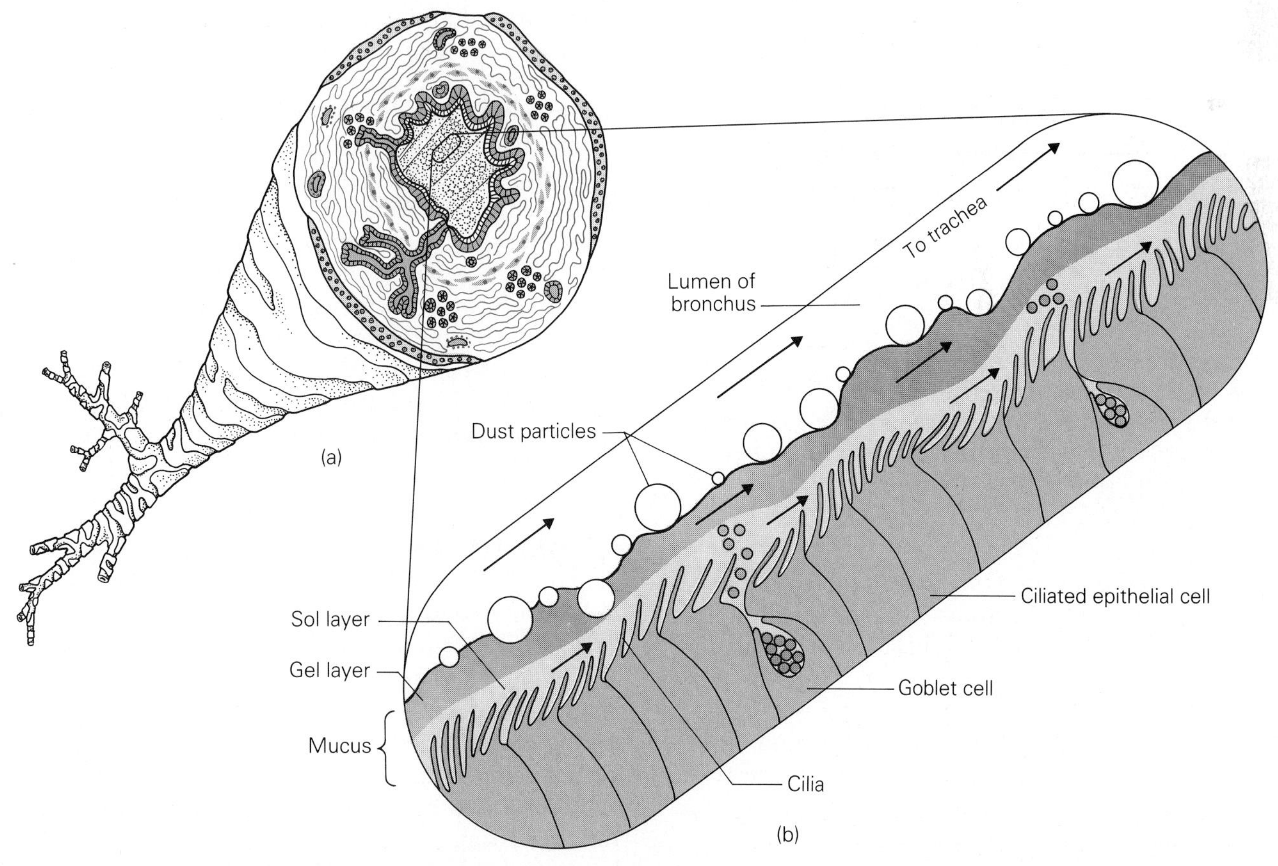

Figure 20–6 Mucociliary transport system. Mucus layer traps inhaled particles. The mucus layer is propelled by cilia toward the trachea to remove bacteria and particulate matter.

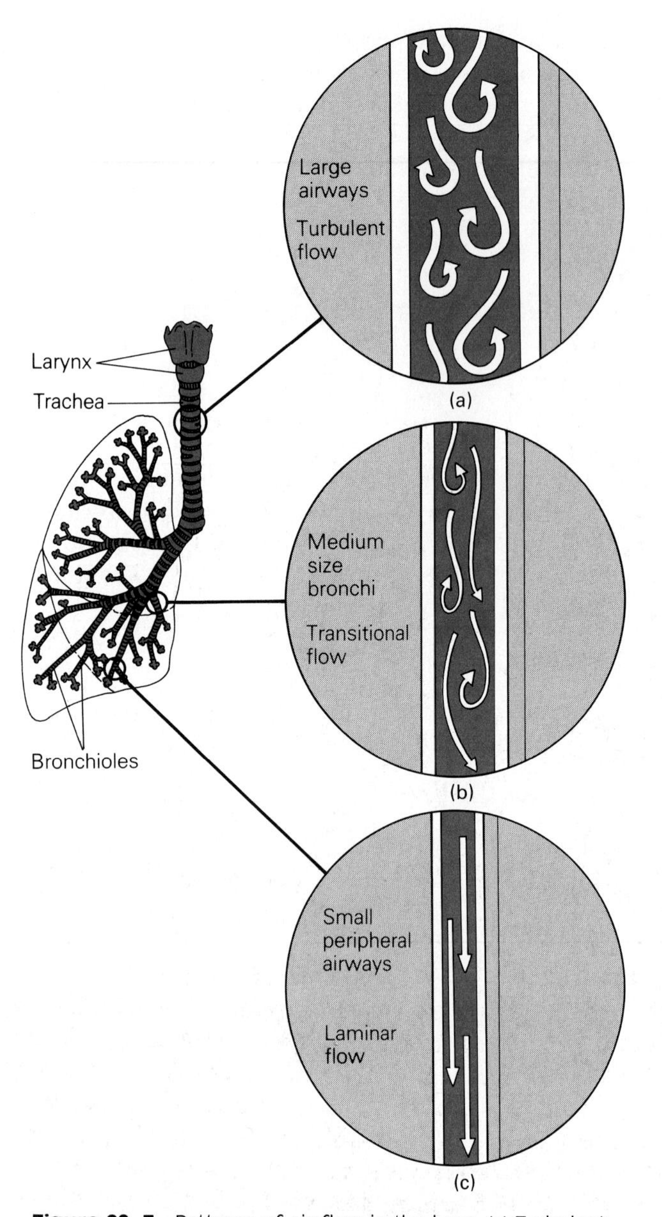

Figure 20–7 Patterns of air flow in the lung. (*a*) Turbulent flow occurs at high flow rates in large airways. (*b*) Transitional flow is a combination of laminar and turbulent flow and occurs in medium-sized bronchi. (*c*) Laminar flow occurs at low flow rates in small peripheral airways.

disorganized patterns of air flow in which the molecules literally collide with each other, resulting in a noise that can be heard on inhalation or exhalation. Thus, the faster you inhale or exhale air, the more turbulent the noise that is produced.

An important consideration regarding turbulent flow is that turbulence in the large airways is the primary source of resistance to breathing. Turbulent flow is also greatly affected by gas density. If one breathes a helium-oxygen mixture, which is a much lighter gas than air (consisting of nitrogen and oxygen), less turbulence is produced, meaning that less resistance and less work are involved in breathing. Gas density also affects the vocal cords in the larynx by changing the pitch of our voices, making us sound like Donald Duck when we talk after breathing a helium-oxygen mixture.

Laminar Air Flow

The second type of air flow in the lungs is **laminar air flow,** which is characterized by streamlined flow that parallels the side of the airways (Fig. 20–7*c*). Laminar flow is silent, because air molecules slide over each other. This type of air flow occurs mainly in the small peripheral airways where air flow is extremely slow.

Laminar flow is greatly affected by airway diameter. Specifically, laminar flow is directly related to the radius raised to the fourth power (Flow $\propto$ radius^4). This means, for example, if a peripheral airway constricts because of asthma such that the radius is halved, then airflow in that tube decreases 16-fold.

$$F = 1^4 = 1$$

$$F = \left(\frac{1}{2}\right)^4 = \frac{1}{16}$$

This relationship between airflow and radius in the small airways explains why a person who suffers from asthma (a condition caused by spasmodic contraction of the bronchiole smooth muscles) has great difficulty in getting air in and out of the alveoli. During asthma attacks, many asthmatics take epinephrine, a hormone that acts on the beta receptor to cause the smooth muscle to relax and in turn dilate the airways. At lower flow rates during expiration—particularly at branches of the airway tree where flow from separate tubes comes together and empties into a single airway—the air flow pattern is a mix. This is referred to as **transitional flow,** a combination of turbulent and laminar flow (Fig. 20–7*b*).

Spirometry and Lung Volumes

During inspiration, the lungs are inflated, and air is drawn in. The volume of air that is breathed in and out of the lung is measured by a **spirometer** (Fig. 20–8). This device is a simple volume recorder consisting of a doubled-walled cylinder in which an inverted bell is immersed in water to form a seal. The bell is attached by a pulley to a pen that writes on a rotating drum. As air enters the spirometer from the lungs, the bell rises. Because of the pulley arrangement, the pen is lowered. Thus, a downward pen deflection represents expiration, and an upward

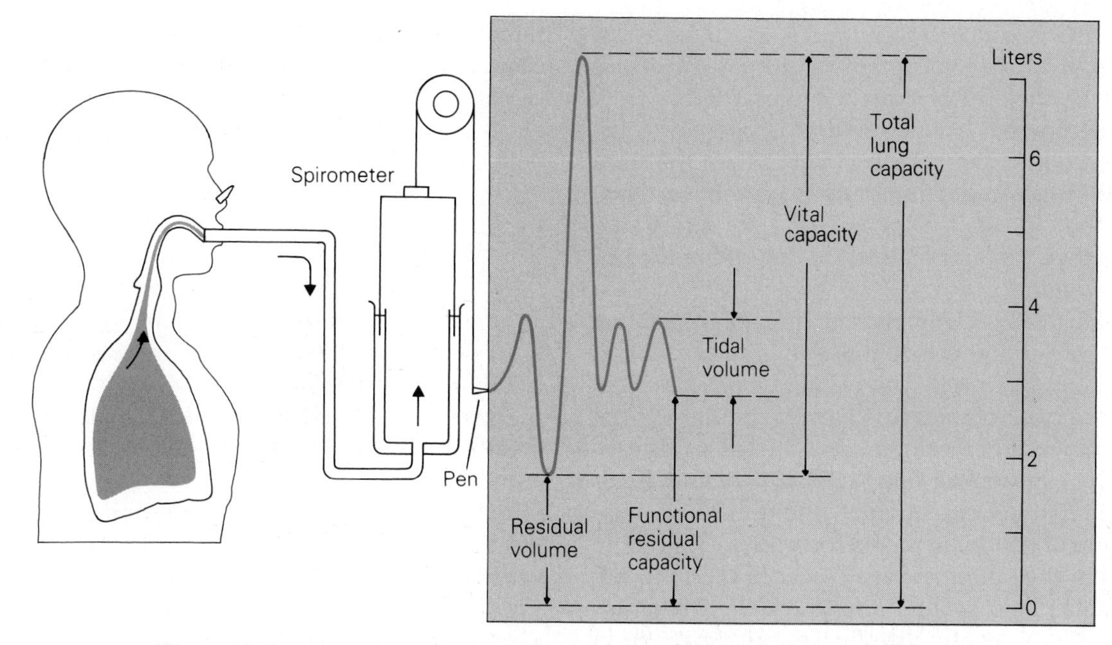

Figure 20–8 Lung volumes and capacities measured by a spirometer. With inspiration the pen shows an upward deflection on the spirograph and, with expiration, a downward deflection. Note that the functional residual capacity and residual volume cannot be measured directly with a spirometer.

pen deflection represents inspiration. The recording is known as a **spirogram,** which measures two factors: (1) how fast air moves in and out of the lungs (rate of air flow) and (2) the volume of air moved.

The volume of air entering or leaving the lungs during a single breath is called **tidal volume** (see Fig. 20–8). Under resting conditions, this volume is approximately 500 ml. The resting tidal volume represents only a fraction of the total air that the lungs can take in or out.

Forced Vital Capacity

The maximum amount of air that can be forcefully and rapidly exhaled after a deep breath (maximal inspiration) is called **forced vital capacity** and measures approximately 5 liters for an adult. Forced vital capacity is one of the most useful tests to assess the overall ability to move air in and out of the lungs (ventilation). For example, a well-trained athlete can have a forced vital capacity of more than 7.0 L, whereas an asthmatic may have a vital capacity no greater than 3 L. The constricted bronchioles of asthmatics can collapse during forced expiration, trapping air in the lungs, leading to a decrease in forced vital capacity. A similar problem exists with individuals who smoke or have **bronchitis,** a condition in which the lining of the bronchioles becomes inflamed and swollen, thereby reducing airway diameter.

Forced vital capacity is determined primarily by three factors: (1) strength of chest and abdominal muscles, (2) airway resistance, and (3) lung volume. Any condition that decreases the strength of the respiratory muscles (e.g., poliomyelitis), decreases lung volume (e.g., tuberculosis), or increases airway resistance (e.g., bronchitis) will lead to a significant decrease in forced vital capacity.

Residual Volume and Functional Residual Capacity

Even after an individual expires as forcefully and completely as possible (forced vital capacity), some air, approximately 1.0 L, remains in the lungs and is called **residual volume.** When air becomes trapped in the lungs during expiration because of airway plugging or airway closure, the residual volume becomes abnormally high.

Another lung capacity that is often measured is **functional residual capacity.** This is the volume of air remaining in the lungs after a normal expiration and amounts to about 2.5 L.

Because the lungs cannot be emptied completely following forced expiration, neither residual volume nor functional residual capacity can be measured directly by simple spirometry. These two volumes are measured indirectly by a dilution technique based on the following relationship:

$$C_1 \times V_1 = C_2 \times V_2$$

in which C is concentration and V is volume. For example, if a 1-L box contains eight gas molecules (Fig. 20–9), then the box can be represented by $C_1 \times V_1$. If the box has a window that is opened to another box with an unknown volume, then the eight gas molecules can equilibrate between the two boxes, and $C_1 \times V_1 = C_2 \times V_2$. If C_1, V_1, and C_2 are known, then V_2 can be solved to determine the unknown volume.

Since residual volume and functional residual capacity are unknown volumes, a similar dilution technique can be used. The technique used to measure these two volumes involves the dilution of helium, an inert and insoluble gas. The subject is connected to a spirometer filled with 10% helium in air (see Fig. 20–9). Consequently, the initial 10% concentration of helium in the spirometer is C_1, and the volume in the spirometer is V_1. Since the lung initially contains no helium, after the subject breathes the helium-air mixture, the helium concentration in the lungs becomes the same as in the spirometer after equilibration. Because helium is insoluble and is not taken up by the blood, the concentration of helium in the lung after equilibration is C_2, and the unknown volume in the lung is $V_2 - V_1$.

It is important to start the test at precisely the right time. If the test begins at the end of a normal tidal volume (end of expiration), the volume of air remaining in the lung represents functional residual capacity. If the test begins at the end of a forced vital capacity and the subject starts breathing the helium-air mixture, then the test will measure residual volume.

Ventilation

So far, the lung volumes that have been discussed are primarily static volumes. However, ventilation is a dynamic process. If 500 ml of air is brought into the lungs with each breath (tidal volume) and the breathing rate is 14 times a minute, then the total volume of inspired air that enters the lungs each minute is 500 × 14 = 7,000 ml per minute or 7 liters per minute. This volume of air is known as **minute ventilation.**

Dead Space Volume

However, it is important to realize that not all of the minute ventilation enters the alveoli for gas exchange. The tidal volume is distributed between the conducting airways and alveoli (Fig. 20–10). Since gas exchange occurs only in the alveoli and not in the conducting airway (trachea, bronchi, bronchioles), part of the minute ventilation becomes wasted

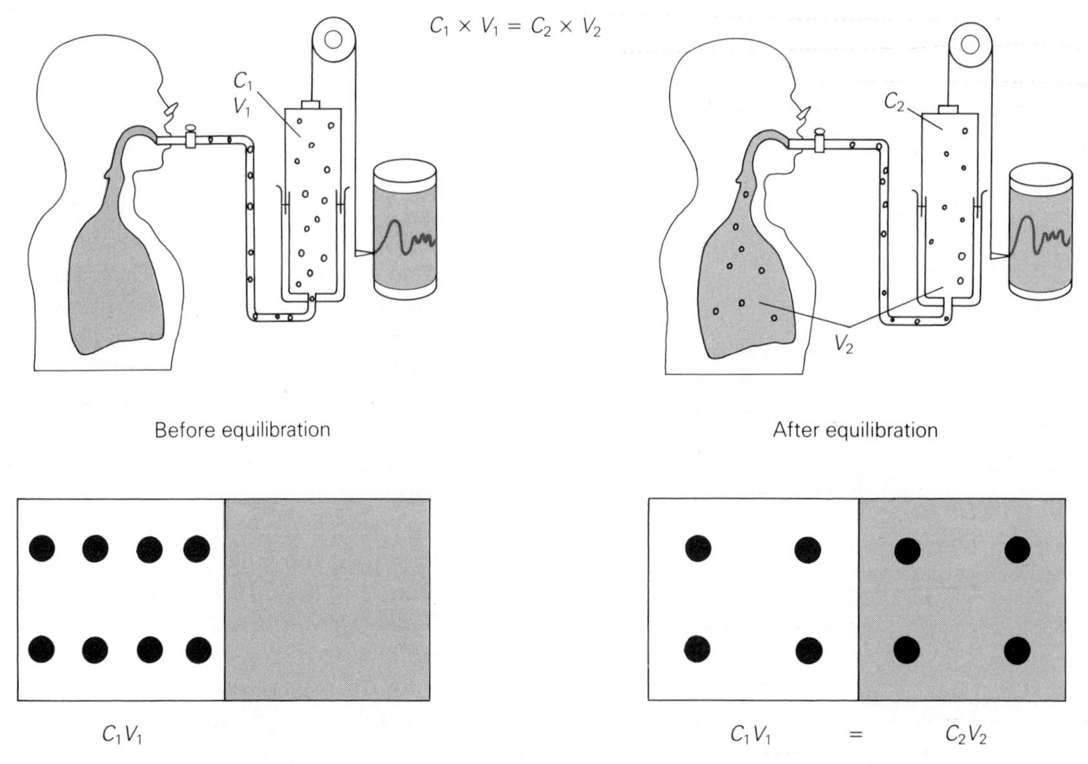

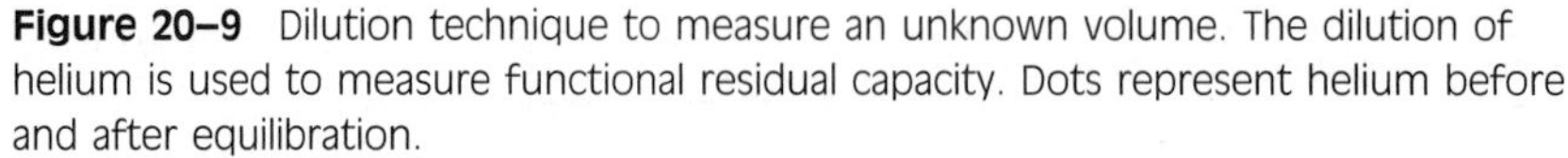

Figure 20–9 Dilution technique to measure an unknown volume. The dilution of helium is used to measure functional residual capacity. Dots represent helium before and after equilibration.

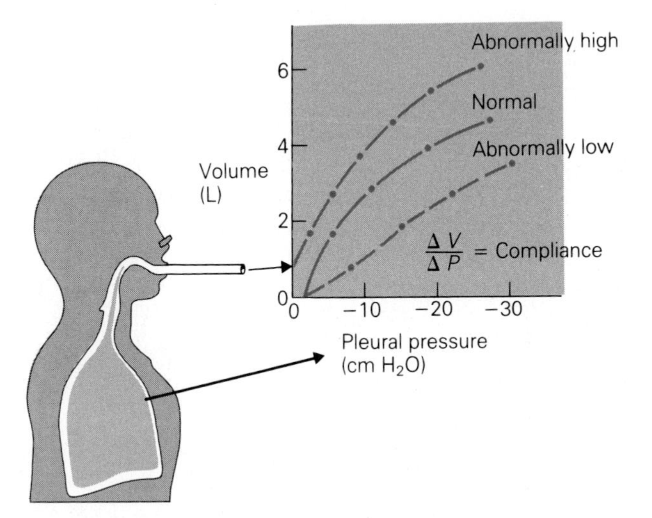

Figure 20–17 Lung pressure-volume curve and compliance. A lung with a slope greater than normal would have an abnormally high compliance. A lung with a slope smaller than normal would have an abnormally low compliance.

Δ pressure). This value is known as **compliance** (C_L = Δ volume/Δ pressure).

We can generate a similar pressure-volume curve for the human lung by measuring a change in lung volume with a spirometer and a change in transpulmonic pressure with a pressure gauge (Fig. 20–17). In practice, a pressure-volume curve is produced by first having the subject inspire maximally to total lung capacity and then expire slowly while recording lung volume and transpulmonic pressure measurements. Compliance (the slope of the pressure–volume curve) provides a measure of the distensibility (i.e., ability to stretch) of the lung. For example, a steep slope (high compliance) indicates that the lung is very distensible and easy to inflate. This means the lung can be inflated to a higher volume with a given amount of effort than a less distensible lung. In other words, with the same transpulmonic pressure, the lung with the higher compliance will inflate to a higher volume because it is more distensible. A lung with a low compliance (decreased slope) is stiffer or not as distensible.

What is the significance of having an abnormally high or abnormally low compliance? Low compliance, indicating a stiff lung, means more work is required to bring in a normal volume of air. Any time the lung becomes injured (by infection or by toxic environmental insults), the elastic properties are often altered, and the lung loses its distensibility (becomes stiffer). A very high compliance, however, is just as detrimental as a low compliance. Consider the disease **emphysema,** in which the lung has been overstretched from chronic coughing and congested airways. Lungs of patients with emphysema (which is strongly linked with smoking) have a high compliance (high distensibility) and are extremely easy to inflate. However, getting air out again is another matter! Lungs with abnormally high compliance (see Fig. 20–17) have poor **elastic recoil,** the property of stretched tissue that allows it to return back to normal. Thus, a lung affected by emphysema is easily distended but will not recoil back during expiration. Consequently, a lot of effort is required to get air out of the lungs. A simple analogy is a sock or stocking that has become overstretched. It can be easily stretched out but does not recoil to fit the contour of the legs and appears to sag, or look "baggy." Similarly, smoking-induced emphysema can lead to "baggy lungs."

Surface Properties of the Lung: Surface Tension

Another property that markedly affects lung compliance is **surface tension** occurring within the inner surface of the alveoli at the gas-liquid interface. Surface tension is a molecular force created at the gas-liquid interface, specifically, the attractive forces between the liquid and air molecules. These forces create a surface film. Since the surface of the alveolar membrane is moist and is in contact with air, a large gas–liquid interface is produced. Surface tension in alveoli produces a force that pulls inwardly. This tension tends to cause alveoli to become unstable by increasing alveolar pressure proportionately more in smaller alveoli than in larger ones at low lung volumes (Fig. 20–18).

To counteract the high surface tension, specialized epithelial cells lining the alveoli secrete a deter-

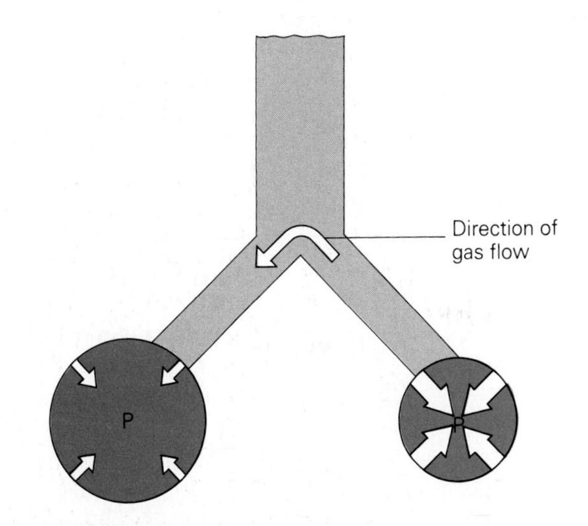

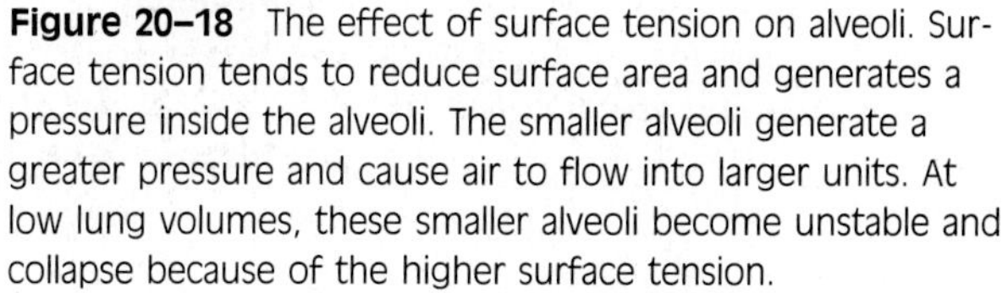

Figure 20–18 The effect of surface tension on alveoli. Surface tension tends to reduce surface area and generates a pressure inside the alveoli. The smaller alveoli generate a greater pressure and cause air to flow into larger units. At low lung volumes, these smaller alveoli become unstable and collapse because of the higher surface tension.

Focus on Asthma

One of the airway diseases that many of us have encountered is asthma. Asthma is a breathing disorder brought on by narrowing of the small airways, namely the bronchi. The asthmatic person experiences labored breathing, shortness of breath, and a distressing tightness in the chest. Shortness of breath can occur suddenly and without warning. During the asthmatic attack, the person feels as if he or she is suffocating, and must sit up or stand and devote all of their energy to breathing. The duration of the attack can last several minutes to several hours. At the end of an attack, a cough may occur, which at first is slight and dry but often becomes more marked with considerable sputum. Following an asthmatic attack, there can be considerable soreness in the chest.

The cellular mechanism of asthma, which causes the shortness of breath, wheezing, and tightness in the chest, is thought to be the release of vasoactive mediators (histamine, prostaglandins, and leukotrienes). These agents cause the bronchial smooth muscles to constrict the airways and also cause an excess secretion of mucus, which plugs the terminal airways. The mucus is usually coughed up at the end of an attack.

One form of asthma is triggered by an allergic reaction and is often called allergic asthma. This type of asthma is most prevalent in children and adolescents. In most cases, these allergic reactions occur in the bronchial tree and are triggered by a foreign protein such as pollen, animal dander, mites, mold, spores, feathers, and, less commonly, food (e.g., milk, nuts, chocolate, and seafood). Certain synthetic drugs, such as aspirin and penicillin, although not proteins, can also trigger asthma. Such agents or allergens react with antibodies and induce the release of the potent vasoactive mediators, most importantly histamine. Underlying allergic asthma is an inherited tendency to develop a hypersensitivity to these allergenic agents.

A second form of asthma is idiopathic asthma, in which a specific cause cannot be identified. This form usually occurs in adults over 30 and can be triggered by exercise, emotions, inhalation of cold air, and infections.

About 1% of the entire population (1 in 100) are estimated to have asthma. Asthma in children under the age of 15 are estimated to be between 5% and 15%. Asthma is more frequent among boys and in men 30 years and older. An important aspect of the treatment of asthma is to identify, when possible, the agent that triggers the attack. Medications include oral drugs, which will relax airway smooth muscle. Aerosol drugs are also used in the medications if the attack is severe.

gentlike material that coats the inner surface of the alveoli and drastically reduces surface tension. This material is known as **pulmonary surfactant** and is a complex mixture of lipids (fat) and protein. The main component of pulmonary surfactant that is responsible for reducing surface tension is the phospholipid **phosphatidylcholine.** Thus, pulmonary surfactant lowers surface tension, prevents alveolar collapse, and makes the lung more distensible (more compliant).

This is extremely important because the effects of surface tension on lung compliance are just as great as the effects of elastic forces. A striking example of what happens when inadequate amounts of surfactant are not present is seen with a disorder known as **respiratory distress syndrome.** This disorder frequently afflicts premature neonates who are born with insufficient amounts of surfactant. These infants have extremely labored breathing because surface tension is high and makes it difficult to inflate the lungs. Because of the high surface tension, these infants also have alveolar collapse, a phenomenon known as **atelectasis.** Forty to fifty thousand babies are afflicted each year with infant respiratory distress syndrome. These infants are at high risk until the lung becomes mature enough to secrete surfactant. Consequently, infant respiratory distress syndrome is still the major cause of the deaths among newborns in the United States.

The Work of Breathing

During inspiration, muscular work is involved in expanding the thoracic cavity, inflating the lungs, and overcoming airway resistance. Since work can be measured as force multiplied by distance, the amount of work involved in breathing can be expressed as a change in lung volume (distance) multiplied by the change in transpulmonic pressure (force). Thus, energy is expended during muscular contraction to create a force (transpulmonic pressure) to inflate the lungs. When a greater transpulmonic pressure is required to bring the same volume of air into the lungs, more muscular work and greater energy will be required.

What factors affect the amount of work involved in ventilating the lungs? First is lung elasticity. As discussed earlier, lung inflation requires energy to overcome elastic forces. Thus, the more compliant the lung, the more easily it is stretched and the less energy is required in breathing. When the lung becomes stiff (less compliant), more mustinually leaking from the capillaries into the interstitial space. Much of the fluid is reabsorbed by the capillaries, but part of it remains behind and must be drained away through the lymphatics. If the fluid accumulates, the distance between the alveolar gas and the capillary blood will increase and oxygen can no longer be readily absorbed. Accumulation of fluid in the lung is called **pulmonary edema.** This is a serious condition that can, in extreme cases, result in death. When edema forms, the lymphatics go into full operation in order to keep the alveoli dry. The lymphatic fluid drains into systemic veins.

Metabolic Functions in the Pulmonary Circulation

Chapter 20 showed that the alveoli have a large surface area. Since the capillary bed takes up about 90% of the alveolar wall, the surface area of the capillaries is impressively large as well. Owing to the central location of the pulmonary circulation, all of the blood is exposed to the capillary endothelium each time it is pumped through the body. The inside of the capillary wall is completely covered by metabolically active endothelial cells. These cells act upon a number of substances. For example, **angiotensin,** a potent vasoconstrictor, is an important hormone in the regulation of blood flow, as mentioned in Chapter 19. This material is made by the kidneys and released into the blood in an inactive form, **angiotensin I.** During its passage through the pulmonary circulation, angiotensin I is converted to active **angiotensin II** by an enzyme located on the surface of the capillary endothelium. Angiotensin II is delivered by the blood to the resistance vessels (arterioles) in the systemic circulation to cause constriction and raise blood pressure. Many other vasoactive substances, such as histamine, serotonin, and the prostaglandins, are both released and inactivated by the endothelium. In this way, the pulmonary circulation serves important metabolic as well as gas exchange functions.

Vascular Pressure and Blood Flow in the Pulmonary Circulation

Cardiac Catheterization

Blood pressures in the pulmonary circulation are much more difficult to measure than those in the systemic circulation where arteries and veins are conveniently located near the surface of arms and legs. The pulmonary arteries and veins lie deep within the thoracic cavity and are therefore inaccessible. **Cardiac catheterization,** a remarkable technique developed around the middle of this century, permits measurement of the blood pressures in all four chambers of the heart.

Cardiac catheterization is routine today. For example, the least accessible chamber of the heart, the left atrium, can be catheterized by passing a meter-long needle with a curved tip along the femoral vein and inferior vena cava until the needle reaches the right atrium. There, the wall between the right and left atria is punctured, and left atrial pressure is measured and blood samples taken. Because there are no somatic sensory nerves in the heart, the patient does not feel the catheter. Therefore, only a local anesthetic is needed where the skin is cut to isolate a superficial vein. The cardiac catheterization procedure can be performed rapidly and safely. Measurements made with this technique, along with other important technological advances such as the cardiac by-pass pump, have enabled surgeons to save countless lives.

Blood Pressures in the Pulmonary Circulation

Pressure in the pulmonary artery is much lower than in the rest of the body. The right ventricle generates a pressure of about 25 mm Hg, which becomes the pulmonary arterial systolic pressure as the ejected blood passes through the pulmonic valve. Once the pulmonic valve closes, the pressure falls to diastolic pressure, just as it does in the systemic circulation. Pulmonary arterial diastolic pressure is about 10 mm Hg. Mean pressure for a normal adult is 15 mm Hg. These numbers are stated the same way as is done for systemic arterial pressure: 25/10, with a mean of 15 mm Hg (Table 21–1). Pul-

Table 21–1
Pressures in the Pulmonary Artery During Various Conditions

Condition	Systolic/ Diastolic (mm Hg)	Mean (mm Hg)
Rest	25/10	15
Moderate exercise	30/15	20
Maximal exercise	50/25	33
At very high altitude	50/25	33
At very high altitude and maximal exercise	75/45	55
Severe pulmonary hypertension	175/100	125

From Groves, B.M., et al. *J. Appl. Physiol.* 63:521–530, 1987.

monary arterial pressure is usually reported as a mean pressure. These pressures are shown in Fig. 21–4, which is a pressure tracing obtained as the tip of a catheter is passed through the right atrium and right ventricle and into the pulmonary artery. If the catheter is advanced farther and farther out along the pulmonary arteries, it will reach vessels that are too small to permit further passage. The catheter will become mechanically wedged in the vessel and pressure will drop to a "wedge" pressure. In this position, the vessels downstream from the catheter act like an extension of the catheter. The wedge pressure is near left atrial pressure.

The low pulmonary arterial pressure is just enough to pump blood to the top of the lung.* Although all of the lungs receive some perfusion even during resting conditions, the large majority of the blood flows through the lower lung (Fig. 21–5). This creates a gradient of blood flow in the lung with not much of the flow going to the top of the lung and most of the flow going to the bottom of the lung. Almost all of the capillaries in the lower lung are open and have blood flowing rapidly through them; under these circumstances, the capillary bed is said to be highly "recruited." In the upper lung, only a few capillary pathways are open (recruited), and the blood cells flowing through them travel slowly.

*To calculate the height to which the right ventricle can pump the blood, consider the following. Blood has a specific gravity near 1, whereas mercury has a specific gravity of 13.6—that is, mercury is 13.6 times heavier than water. Multiply the mean pulmonary arterial pressure of 15 mm Hg by 13.6 = 204 mm H_2O. Thus, mean pulmonary pressure is sufficient to pump the blood 200 mm high, which is nearly the height of the top of the lung above the heart.

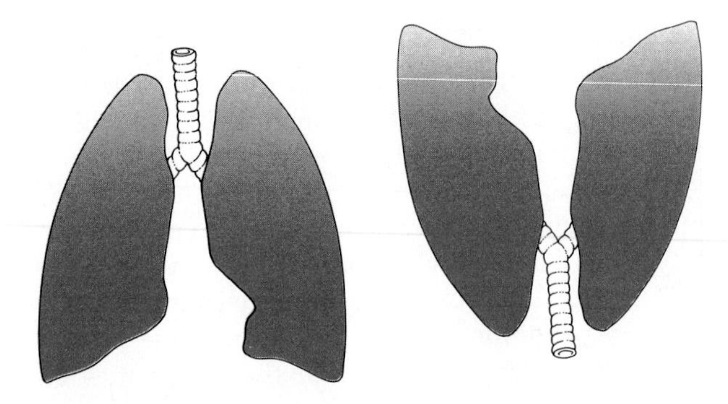

Figure 21–5 Because of the low values of both pulmonary arterial pressure and vascular resistance, most of the blood flows to the lower part of the lung during rest. If a person stands on his or her head, blood will flow to the apexes of the lung.

Ventilation-Perfusion Balance

As long as inspired air is distributed evenly through the lung, venous blood (the blue blood in Fig. 21–6*a*) can flow to any alveolar region and pick up oxygen from the alveolar gas; as the blood flows past the alveolar gas, it becomes **oxygenated.** Inspired air, however, is not always evenly distributed to all alveoli. This causes regional **hypoxia** (low oxygen)

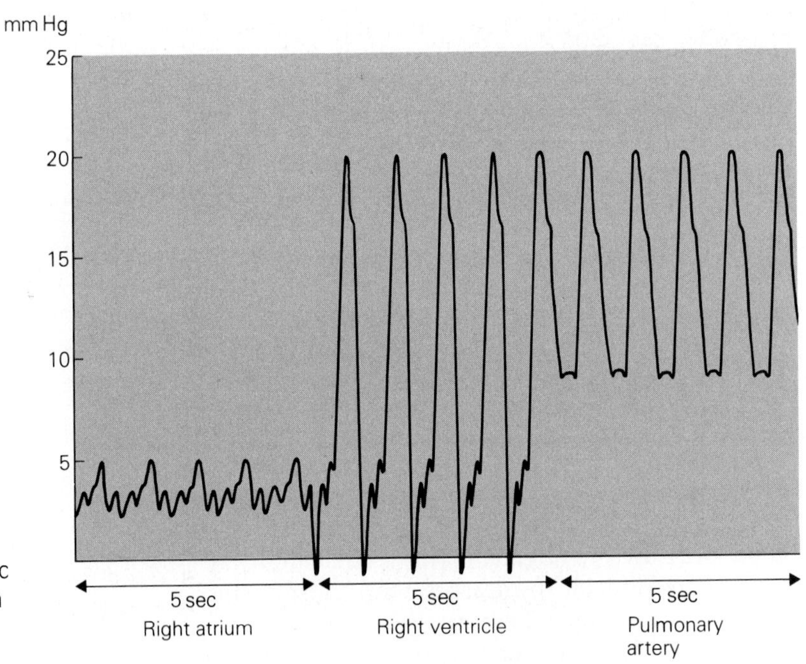

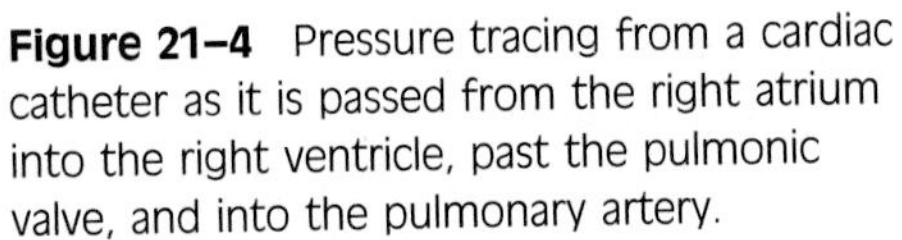

Figure 21–4 Pressure tracing from a cardiac catheter as it is passed from the right atrium into the right ventricle, past the pulmonic valve, and into the pulmonary artery.

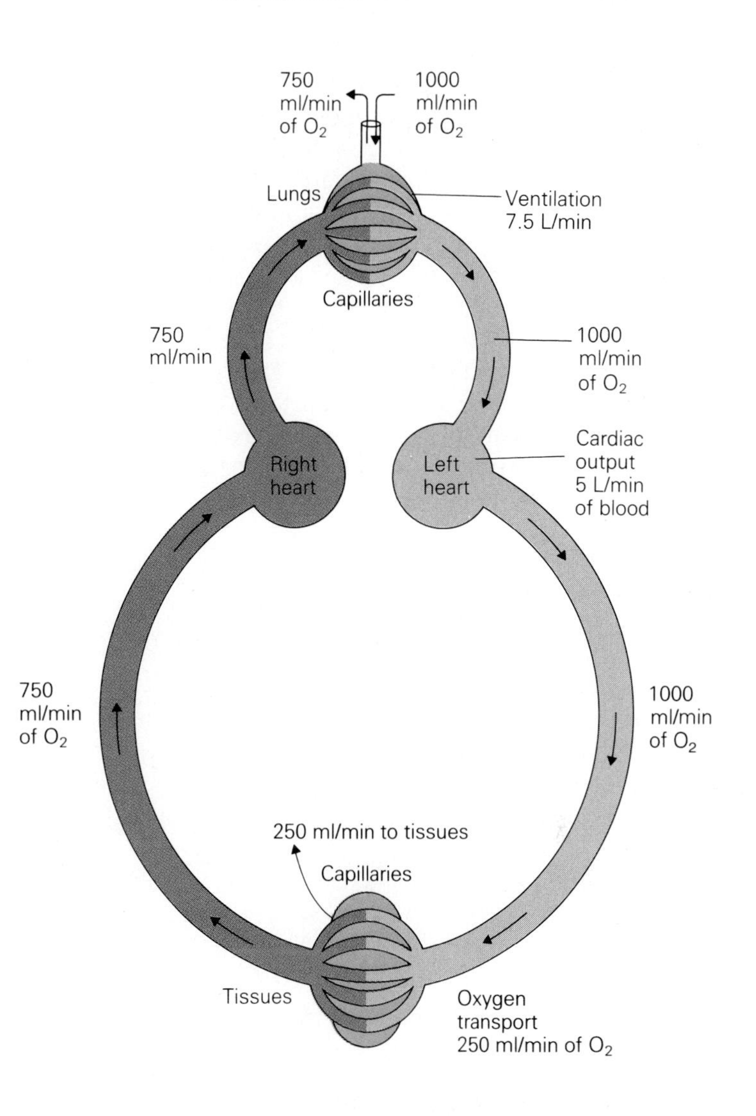

Figure 21–9 Schema of the circulation showing oxygen deliveries.

be transported by the blood, oxygen had to dissolve in blood. If such were the case, the cardiac output would have to be more than 1000 liters/min. That flow rate would completely fill an automobile gasoline tank in 3 seconds! Such a delivery rate would require a huge heart. Each beat of the heart would cause the blood to exceed the sound barrier.

Hemoglobin in Oxygen Exchange

A remarkable molecule, hemoglobin, solves the problems of oxygen demand. Hemoglobin is a chemical that permits large amounts of oxygen to be carried to the tissues in an efficient way. The structure of hemoglobin was presented in Chapter 17. Each hemoglobin molecule contains four heme molecules, each of which contains an iron atom. Each iron atom is capable of binding with a molecule of oxygen. When oxygen combines loosely with the heme portion of hemoglobin in the lung, where the pressure of oxygen is high, it forms **oxyhemoglobin.** When the oxyhemoglobin reaches the tissue where oxygen pressure is low, the hemoglobin releases the oxygen. Because of the loose way in which O_2 binds with the iron atoms, the oxygen is delivered to the tissue fluids not as ionic oxygen but as dissolved molecular oxygen.

The capacity of hemoglobin for oxygen is much higher than that of plasma. About 98% of the oxygen in the blood is carried by the hemoglobin. Every 100 ml of blood can combine with 20 ml of oxygen. This is written as 20 volumes percent (vol.%).

The Oxyhemoglobin Dissociation Curve

Now we can see how hemoglobin loads with oxygen in the lung. Fig. 21–10*a* shows an important physiological curve that is "S" shaped and is called the **oxyhemoglobin dissociation curve.** If the blood is 100% saturated, then all of the binding sites on the hemoglobin contain an oxygen molecule. If the blood is 50% saturated, then half of the hemoglobin binding sites contain oxygen molecules. Saturation gives an indication of how completely the binding sites are filled and how many are left in reserve. Another way to measure the amount of oxygen in the blood is to measure its content of oxygen—that is, how many milliliters of oxygen are carried by each 100 ml of blood. As we have seen before, the unit for content is vol.%. That measure is found on the right hand axis of Figure 21–10*a*. Under normal resting conditions, arterial blood has an oxygen tension of 100 mm Hg and venous blood an oxygen tension of 40 mm Hg (see Table 21–2).

Oxygen Delivery During Rest

Under normal conditions when alveolar oxygen tension is 100 mm Hg, the blood saturation is 97% or has 20 vol.% in it. Normal venous blood with its oxygen tension of 40 mm Hg is 75% saturated or has 15 vol.% oxygen in it. Thus, the arteriovenous oxygen difference is 5 vol.%, meaning that 5 ml of oxygen is delivered for each 100 ml of blood pumped by the heart.

Oxygen Delivery During Exercise

During exercise, the oxygen tension in the muscles drops to lower values than at rest because oxygen is

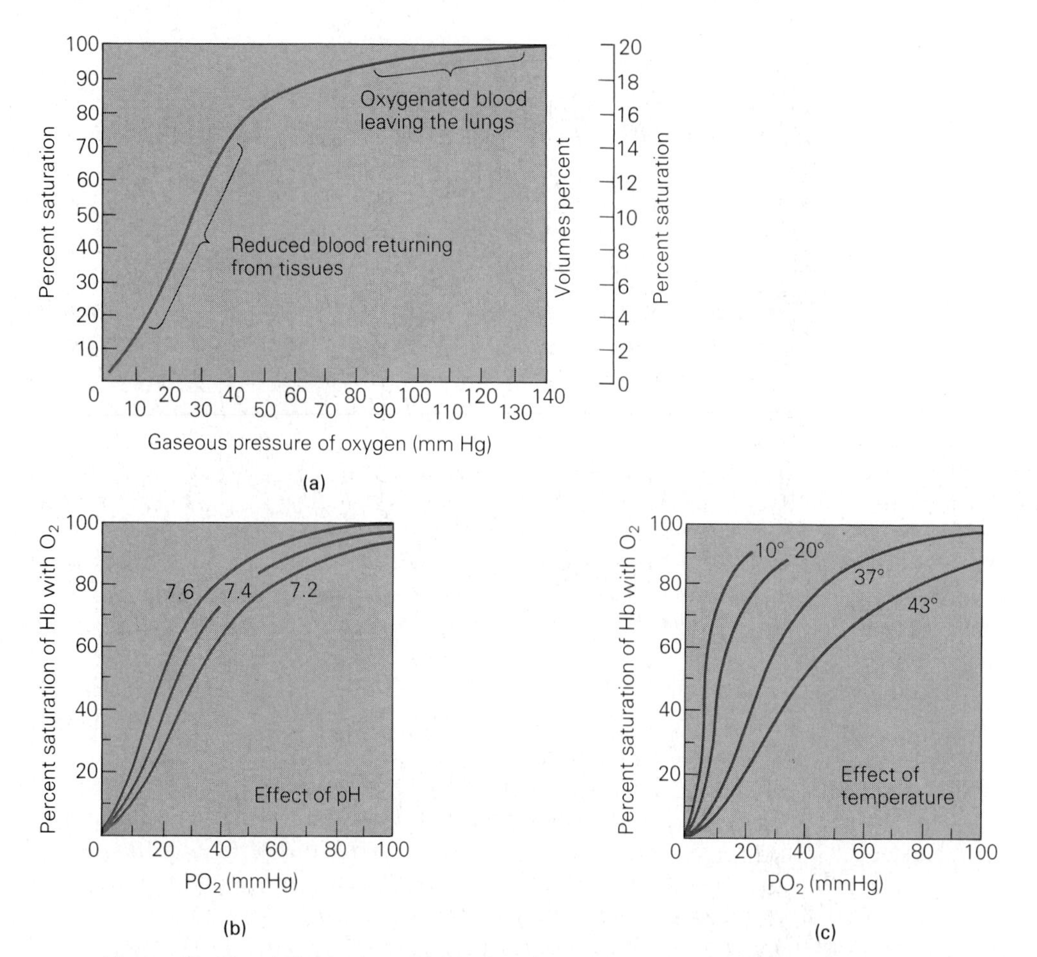

Figure 21–10 (*a*) The oxyhemoglobin dissociation curve, showing the way in which oxygen binds to hemoglobin in the lung and unloads oxygen in the tissues. Note that as oxygen tension increases in the mid-range, the hemoglobin loads rapidly with oxygen. As the hemoglobin molecules become saturated with oxygen, loading slows as oxygen tension increases. (*b* and *c*) Both pH and temperature cause the oxyhemoglobin curve to shift. A shift in the curve causes the hemoglobin to load or unload more easily. In the lungs, where the temperatures are relatively cool and the pH is relatively alkaline, loading is facilitated by the leftward shift of the curve. In an exercising muscle where the environment is relatively warm and acidotic, unloading is facilitated.

being used so rapidly by the working muscle. Heavy exercise can reduce muscle oxygen tension to less than 15 mm Hg. This enlarges the gradient that favors the movement of oxygen from the capillary blood and into the tissue. The delivery of oxygen is increased because of greater extraction.

Some interesting characteristics of hemoglobin facilitate oxygen transport. The effect of blood acidity (see Fig. 21–10*b*) is to shift the position of the oxyhemoglobin dissociation curve to the right. For a given oxygen tension, the amount of oxygen (oxygen content) is lower. This means that for a given oxygen tension in the tissue, the hemoglobin will unload more oxygen. Exercising muscle has greater acidity owing to increased carbon dioxide and lactic acid production. The unloading of oxygen from hemoglobin is therefore facilitated. Similarly, increased temperature also causes a rightward shift of the curve (see Fig. 21–10*c*). Blood temperature rises when it reaches a hot exercising muscle, and oxygen unloading is further enhanced.

A leftward shift in the oxyhemoglobin dissociation curve facilitates oxygen loading. At a given oxygen tension, the content and saturation will be higher. As the blood enters the alveolar capillaries, carbon dioxide leaves the blood, the pH becomes more basic, and oxygen loading is enhanced. Similarly, cooler temperatures cause a leftward shift of the curve. Since the lungs are cooler than exercising muscle is, a leftward shift is also promoted. Thus, the oxyhemoglobin curve shifts in a remarkable way from right to left and back to the right with each passage around the circulation in order to favor oxygen loading in the lungs and unloading in the tissues.

Transport of Carbon Dioxide by the Blood

Carbon dioxide is transported in the circulation more readily than is oxygen, because carbon dioxide—being a nonpolar molecule—is highly soluble in lipid and thereby moves easily across plasma membranes. Even in many abnormal conditions, blood carbon dioxide tension tends to remain more nearly normal than oxygen. Because the maintenance of pH in the body is so crucial to many chemical reactions, it is important that carbon dioxide remain normal, since P_{CO_2} in the blood is intimately linked to acid-base balance as described in Chapter 26.

In a normal resting adult, 4 vol.% of carbon dioxide are transported to the lungs to be exhaled each minute. Carbon dioxide is transported in the blood in three ways. All three begin with CO_2 being dissolved in the plasma (Fig. 21–11) after it has diffused from the tissue into systemic capillaries.

Bicarbonate Ions in CO_2 Transport

The largest fraction of carbon dioxide (60% to 70%) is transported as bicarbonate ions. These ions are formed by an important series of reactions that take place inside the red blood cell.

First, carbon dioxide and water in the red blood cell combine to form carbonic acid as shown in this equation:

$$CO_2 + H_2O \xrightarrow{\text{carbonic anhydrase}} \underset{\text{carbonic acid}}{H_2CO_3}$$

Ordinarily, this is a slow reaction requiring many seconds for completion, but when catalyzed by the enzyme **carbonic anhydrase,** the reaction is accelerated by a factor of 5000 and is complete in a fraction of a second.

In the next chemical step, the carbonic acid dissociates into a bicarbonate ion (HCO_3^-) and a hydrogen ion (H^+):

$$\underset{\text{carbonic acid}}{H_2CO_3} \rightharpoonup \underset{\text{bicarbonate}}{HCO_3^-} + H^+$$

The bicarbonate ion moves from the red cell into the plasma, where it is dissolved much more readily than carbon dioxide. This leaves a hydrogen ion free in the red cell, where it readily combines with hemoglobin in the following way.

$$H^+ + Hb^- \rightharpoonup HHb$$

These reactions are summarized in Figure 21–11.

Carbaminohemoglobin in CO_2 Transport

Carbon dioxide can combine directly with hemoglobin to form carbaminohemoglobin. The carbon dioxide does not bind at the same site on the iron atom as does oxygen but instead binds loosely by a direct reaction with some of the amino groups that form the hemoglobin molecule. Between 15% and 30% of carbon dioxide transport is accomplished in this form. The hemoglobin molecule has a higher affinity for carbon dioxide when it is oxygen-desaturated—exactly the condition that exists in venous blood, in which oxygen is low and carbon dioxide levels are high. The bond between hemoglobin and carbon dioxide is loose enough to be reversible in

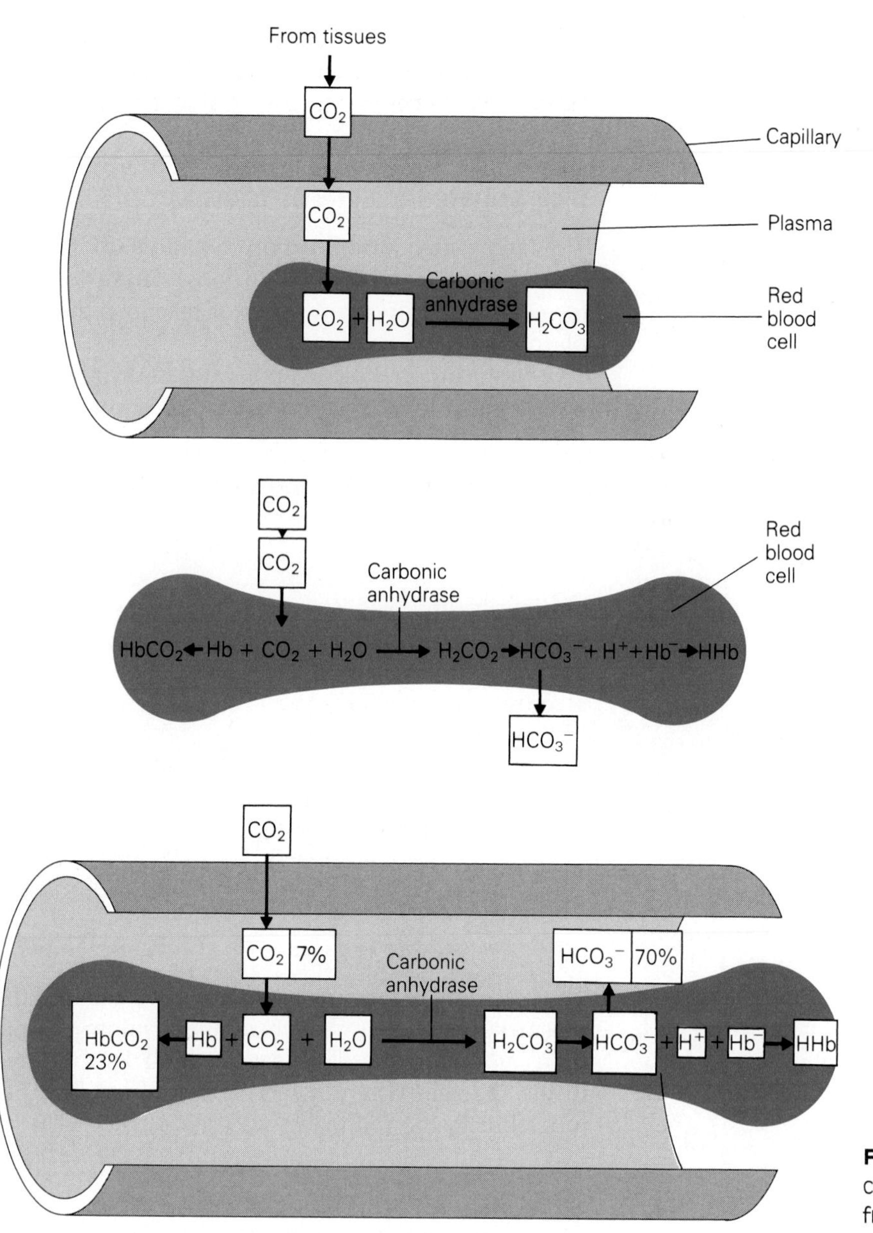

Figure 21–11 The ways in which carbon dioxide is carried by the blood from the tissues to the lungs.

the lung. The carbaminohemoglobin reaction is much slower than the enzyme-catalyzed reaction that forms bicarbonate.

Dissolved Carbon Dioxide in CO_2 Transport

Between 7% and 10% of the carbon dioxide is transported in a dissolved state in the plasma and red blood cells. This accounts for only about 0.3 vol.% of the total carbon dioxide transport.

All three of these methods of transport are readily reversible in the pulmonary capillaries, where the concentration gradient favors the movement of carbon dioxide from the blood into the alveolar gas. As the equations read in the reverse as well as the forward direction, when carbon dioxide is given off in the lung, the transport cycle is complete.

It is important to remember that the blood does not empty itself of carbon dioxide in the lung or of oxygen in the tissues. Even during the most strenuous exercise—when oxygen extraction is high in the tissues and ventilation is high in the lungs—oxygen and carbon dioxide always remain in the blood. The quantities of these gases in various regions of the circulation are shown in Figure 21–12.

resistance is lower and more oxygen is available.

D. The balancing system is effective even when an individual is exposed to whole lung hypoxia such as at high altitudes. The ventilation–perfusion balancing system senses the low oxygen and responds by triggering vasoconstriction of the arteries, increasing pulmonary artery pressure and routing blood to the upper regions of the lung, where more oxygen is available.

E. During exercise, cardiac output increases, pulmonary arterial pressure rises because of increased quantity of blood flowing through the lungs, and the lung routes more blood to its upper regions, resulting in a more even perfusion of the lungs. The body further compensates for increased demand by enhancing the oxygen uptake from the alveolar gas in two ways: (1) by increasing the number of capillaries through which the blood flows, thus increasing gas exchange surface area, and (2) by shortening the capillary transit time.

III. A. A convenient method of measuring gas concentration is to calculate the partial pressures of a gas.

B. In the process of gas exchange, oxygen and carbon dioxide cross the alveolar–capillary membrane in response to their respective concentration gradients.

IV. A. At rest, the blood delivers 5 ml of oxygen for every 100 ml of blood that circulates through the tissues. Resting cardiac output is 5 liters per minute; therefore, 250 ml/min of oxygen are delivered to the tissues. With exercise, this delivery can be increased 15 to 20 times. Hemoglobin greatly facilitates oxygen delivery to the blood by loosely binding oxygen to its four iron atoms and then releasing the oxygen in the tissues. The oxyhemoglobin dissociation curve describes how hemoglobin is loaded and unloaded with oxygen at a given oxygen tension.

B. During rest in a normal adult, 4 vol.% of carbon dioxide are transported to the lungs to be exhaled each minute. There are three ways in which carbon dioxide is transported in the blood: bicarbonate ions, carbaminohemoglobin, and dissolved carbon dioxide.

V. In pulmonary edema, fluid leaks from the pulmonary capillaries into the interstitial space and, under severe conditions, into the alveolar space, making gas exchange very difficult.

VI. A. Before birth, the lungs of the fetus do not serve any respiratory function. Gas exchange takes place between the mother and the fetus through the placenta. In the fetus, blood is shunted past the lung by the foramen ovale and the ductus arteriosus.

B. After birth, the blood flow through the lungs increases, systemic arterial pressure rises, and the foramen ovale and ductus arteriosus close.

C. Sometimes, the ductus remains open, causing patent ductus arteriosus. Ventricular septal defect is another congenital heart defect.

Review Questions

1. What are the important features of pulmonary capillary anatomy and function?
2. What is the function of the pulmonary lymphatic vessels?
3. What are the metabolic functions of the pulmonary circulation?
4. What is cardiac catheterization?
5. How does pressure change as a cardiac catheter is advanced from the right atrium into a pulmonary artery?
6. How is ventilation–perfusion balance maintained through hypoxic vasoconstriction?
7. What is the normal distribution of pulmonary blood flow? How is it altered by whole lung hypoxia and exercise?
8. How does the pulmonary capillary bed respond to exercise to increase gas uptake?
9. What are normal P_{O_2} values in the following places: the atmosphere at sea level, in Denver (barometric pressure =

630 mm Hg), in the alveolus, in venous blood, in arterial blood?

10. What are oxygen uptake values during rest? During exercise?

11. What is the oxygen-carrying capacity of normal blood?

12. What is an oxyhemoglobin dissociation curve? What are the causes of both right and left shifts in the curve? Describe how these shifts facilitate gas exchange during exercise.

13. What are the three major ways in which CO_2 is carried by blood?

14. What are two causes of pulmonary edema?

15. How does the pulmonary circulation protect the body from emboli?

16. What is the circulatory anatomy in the fetus, and how is cardiac output distributed?

17. What changes occur in the fetal circulation at birth?

Suggested Readings

Altman, L. K. *Who Goes First?* New York: Random House, 1987.

Forster, R. E., DuBois, A. B., Briscoe, W. A., and Fischer, A. B. *The Lung.* Chicago: Year Book Medical Publishers, 1986.

Guyton, A. C. *Textbook of Medical Physiology.* Philadelphia: W. B. Saunders, 1985.

Mines, A. H. *Respiratory Physiology.* New York: Raven Press, 1981.

Simon, T. *The Heart Explorers.* New York: Basic Books, 1966.

West, J. B. *Respiratory Physiology.* Baltimore: Williams & Wilkins, 1985.

22

Respiration in Unusual Environments

For much of our history, humans have been restricted to breathing atmospheric air. The largest variations in what we breathed were encountered on ascents to high altitude, where alveolar oxygen tension dropped as barometric pressure fell. There were also some cases, beginning in Grecian times, of underwater activities in primitive diving bells, in which divers would have encountered air pressures that were greater than one atmosphere (1 atm). In the eighteenth century, the chemistry of gases blossomed, and individual gases such as oxygen, nitrogen, and carbon dioxide were isolated. Then, the study of peculiar gases began in earnest. Joseph Priestley said of pure oxygen, "Only two mice and myself have had the privilege of breathing it." Today we breathe in high-oxygen environments, low-oxygen environments, polluted environments, and artificial environments such as in space capsules and in underwater devices. Various anesthetic gases relieve pain during surgery, and we have contrived all sorts of artificial breathing devices to keep us alive in hostile environments and during times when our natural ability to breathe is inadequate. This chapter deals with this interesting area of pulmonary physiology.

Respiration in Low-Oxygen Environments

The condition of low oxygen **(hypoxia)** is most commonly encountered during travel to high altitudes. As the altitude increases, the barometric pressure decreases. The percentage of oxygen in the air, however, remains constant at near 21%. The decreasing barometric pressure causes the partial pressure of oxygen to fall (Table 22–1). This fact was not appreciated until the last century when the French physiologist Paul Bert was shocked by the deaths of balloonists who ascended to altitudes of more than 15,000 feet (Fig. 22–1). His lengthy scientific work prompted by this tragedy was published in the classic book *La Pression Barometricque,* in which he emphasized, "Oxygen tension is everything; barometric pressure in itself is nothing or almost nothing." Bert has been acclaimed as the founder of aerospace medicine; he also did important work on the effects of high pressure experienced by divers.

Immediate Effects of Low Oxygen

The detrimental effects of hypoxia are exaggerated if the exposure is sudden. For example, a pilot who loses his or her oxygen supply at 30,000 feet has just over one minute to descend to lower altitude before losing consciousness. In contrast, after careful acclimation, several climbers have been able to reach the summit of Mount Everest, nearly 30,000 feet, without the use of supplemental oxygen. This amazing achievement emphasizes that the body is much more able to tolerate hypoxia if the ascent is slow. The ideal of successful **acclimatization** through slow ascent is highlighted repeatedly in the literature on high altitude and mountain climbing.

Figure 22–1 The first successful aerial devices were hot air balloons. Although dangerous, these balloons issued in the age of aviation.

Since millions of people travel to high altitudes annually for recreation, the acute effects of hypoxia are of both physiologic and clinical interest. Some effects are present at even moderate elevations of 5000 feet, an altitude at which major cities are found throughout the world. There are noticeable changes in ventilation, especially during exercise, even at these relatively low elevations. Acclimatization occurs over a matter of weeks to months (Fig. 22–2). Most residents of elevations of 5000 feet soon do not notice any effect of the altitude.

As the altitude increases, however, the body makes noticeable efforts to deliver normal amounts of oxygen to the tissues. Chief among these responses to altitude is **hyperventilation**—that is, deeper breaths taken rapidly. The sequence of events for hyperventilation is as follows: (1) a fall in inspired oxygen tension; (2) decreased alveolar oxy-

Table 22–1
Effects on Alveolar Gas Concentrations and on Arterial Oxygen Saturation of Acute Exposure to Low Atmospheric Pressure

			Breathing Air		Breathing Pure Oxygen	
Altitude (ft)	Barometric Pressure (mm Hg)	P_{O_2} in Air (mm Hg)	P_{O_2} in Alveoli (mm Hg)	Arterial Oxygen Saturation (%)	P_{O_2} in Alveoli (mm Hg)	Arterial Oxygen Saturation (%)
0	760	159	104	97	673	100
10,000	523	110	67	90	436	100
20,000	349	73	40	70	262	100
30,000	226	47	21	20	139	99
40,000	141	29	8	5	58	87
50,000	87	18	1	1	16	15

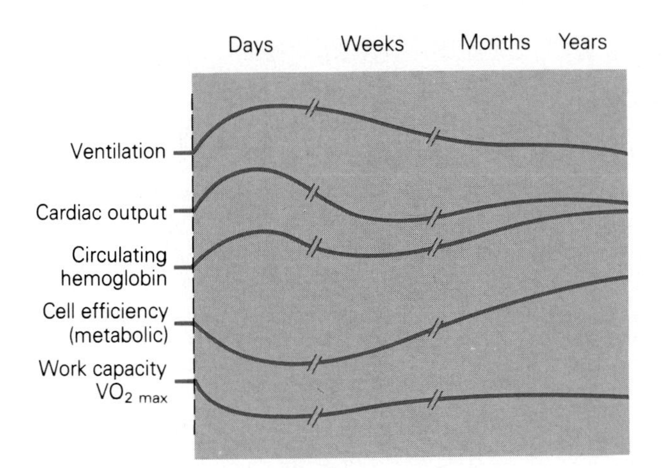

Figure 22–2 Approximation of the acclimatization changes that take place during a stay at a moderate altitude of 14,000 to 16,000 feet. (From Houston, C. S. *Hypoxia: Man at Altitude.* New York, Thieme-Stratton, 1982.)

gen tension; (3) decreased arterial oxygen tension; (4) increased firing of peripheral chemoreceptors (the carotid bodies); (5) hyperventilation; and (6) increased alveolar and arterial oxygen tension. The loop is regulated in part by a negative feedback limb: (1) hyperventilation; (2) elimination of alveolar carbon dioxide; (3) decreased arterial carbon dioxide tension; (4) decreased arterial and cerebrospinal fluid (CSF) pH; (5) decreased firing of both central and peripheral chemoreceptors; and (6) decreased ventilation. Ventilation increases immediately upon ascent and stays elevated for many weeks (see Fig. 22–2).

Sometimes, too much carbon dioxide is blown off during hyperventilation, resulting in **acute mountain sickness.** This disorder is most common after rapid ascent to altitudes above 7000 or 8000 feet and consists of headache, nausea, vomiting, sleep disturbances, and breathlessness (dyspnea). A drug called acetazolamide (Diamox) can reduce hyperventilatory-induced alkalemia and lessen the symptoms of acute mountain sickness. This drug is thought to work by interfering with the catalytic effect of carbonic anhydrase on the reaction of carbon dioxide and water in the blood (see Chapter 21). This helps overcome the alkalemia (basic conditions) of hyperventilation and ameliorates acute mountain sickness. Although this drug is still in the experimental stages, it does offer hope for avoiding the symptoms of acute mountain sickness.

At higher altitudes (9000 to 10,000 feet), a rarer disorder called **high-altitude pulmonary edema** occurs. The cause of the edema is not clear, but fluid is known to leak from the pulmonary microvessels, forming edema and causing dyspnea (breathing difficulty), cough, weakness, headache, stupor, and (rarely) death. The disorder is treated with supplemental oxygen; even more important is a rapid descent to low altitude. The disease clears rapidly (in hours or days) at lower altitudes.

A more serious disorder is **high-altitude cerebral edema,** which occurs above altitudes of 10,000 to 12,000 feet. This disease is characterized by severe headache, hallucinations, problems with muscular coordination (ataxia), weakness, impaired mental ability, stupor, and death. Immediate descent is mandatory. Fortunately, high-altitude cerebral edema is rare. Altitude illnesses are summarized in Table 22–2.

Long-Term Effects of Low Oxygen

The body undergoes other physiological changes to adjust to low environmental oxygen. One important adaptation is that more red blood cells are manufactured, which causes elevated hematocrit. Up to a certain point, this change enables the blood to carry more oxygen to the tissues (Fig. 22–3). However, the higher the hematocrit, the more viscous (thickened) the blood becomes, a change that makes it more difficult for the heart to pump. Normal hematocrit is near 45%. In some cases, high-altitude residents develop a condition called **polycythemia** in which hematocrits reach levels of over 70%. These individuals must have blood withdrawn periodically to reduce their hematocrit to a range that permits the heart to pump effectively.

The body increases ventilation, cardiac output, and circulating hemoglobin (see Fig. 22–2) during

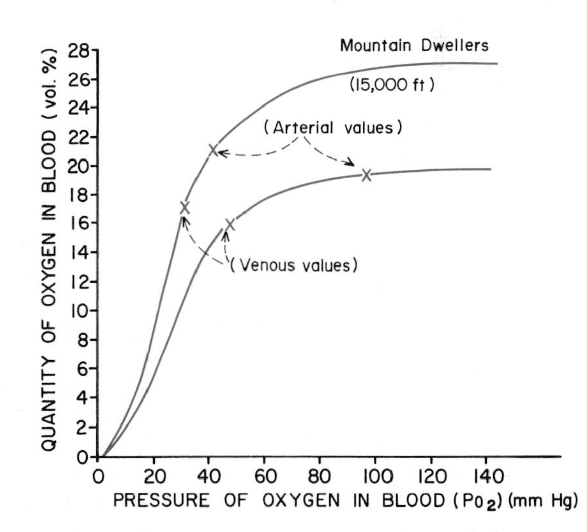

Figure 22–3 The oxygen-carrying capacity of blood of mountain dwellers is greater than that of sea level natives, due in part to an increase in hematocrit. (From "Oxygen-dissociation curves for bloods of high-altitude and sea-level residents." PAHO Scientific Publication No 140, Life at High Altitudes, 1966.)

Table 22–2
Altitude Illness

Acute hypoxia	Mental impairment and usually collapse after *rapid* exposure above 18,000 feet. Rare on mountains.
Acute mountain sickness	Headache, nausea, vomiting, sleep disturbance, dyspnea. Above 7000 to 8000 feet. Self-limited. Common.
High altitude pulmonary edema	Dyspnea, cough, weakness, headache, stupor, and rarely death. Above 9000 to 10,000 feet. Requires rapid descent or treatment early. Uncommon.
High altitude cerebral edema	Severe headache, hallucinations, ataxia, weakness, impaired mentation, stupor, and death. Above 10,000 to 12,000 feet. Uncommon. Descent mandatory.
Subacute and chronic mountain sickness	Failure to recover from acute mountain sickness may necessitate descent. Some develop dyspnea, fatigue, plethora and heart failure after years of asymptomatic residence at altitude. Rare.
Altitude-related problems	Retinal hemorrhage, edema, thrombophlebitis and embolism, and cold injury.
Chronic conditions made worse at altitude	Sickle trait, chronic cardiac or pulmonary disease.
High altitude deterioration	Long periods spent above 18,000 to 19,000 feet cause insomnia, fatigue, weight loss, and general deterioration. Permanent residence is not possible above this altitude. Deterioration is more rapid at even higher elevations.

From Houston, C. S. In: Hypoxia in Man at Altitude. Ed. by Sutton, J. R., Jones, N. L., and Houston, C. S. New York: Thieme-Stratton, 1982.

the first few weeks to months at high altitude. Over a long period, cardiac output and ventilation tend to return to low-altitude values, while hematocrits continue to climb. The tissue cells also change to become metabolically more efficient. For example, the mitochondria increase in number, and there are alterations of metabolic pathways. Even with these changes, the maximum work capacity never reaches sea-level values. Nevertheless, the adaptation of long-term residents can be impressive. In the Andes, the natives return from the high mines in the evenings to their "low" altitude homes for a rousing soccer game played at 15,000 feet; their team has an impressive home field advantage.

The best acclimatized of all high-altitude residents are the Sherpas in Nepal, whose climbing feats are legendary (Fig. 22–4).

Other people—for example, the residents of high altitudes (10,000 feet) in Colorado—are often not as well adapted to low oxygen (Fig. 22–5). The percentage of Coloradans who are aged 60 years and older and live at high altitude is much lower than the percentage of older people who live at low altitude. The other older high-altitude residents tend to leave for health reasons, usually because of heart and lung disease (Fig. 22–6). The former residents report that their health is improved after they settle at low altitude (Fig. 22–7).

Figure 22–4 Sherpas are the people most highly adapted to high altitudes. Here a Sherpa evacuates a climber from the extreme altitudes of the Himalayas. (Credit: John Cleare/Mountain Camera)

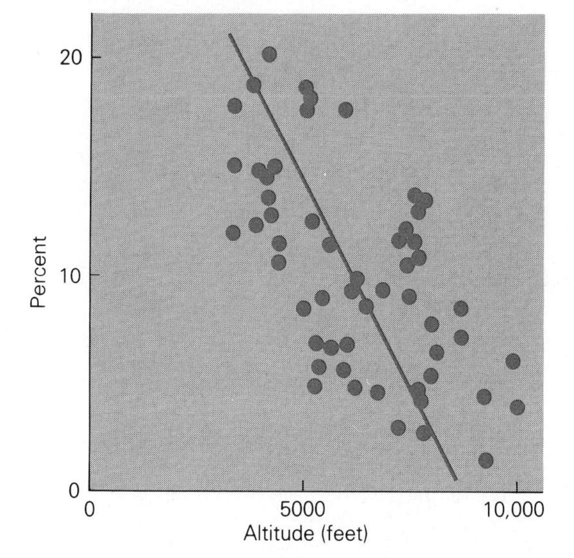

Figure 22–5 In Colorado, there is a progressive decrease in the percentage of elderly people living at high altitude. (Adapted from Regensteiner, J. G. and Moore, L. G. Migration of the elderly from high altitudes in Colorado. *JAMA* 253: 3124, 1985.)

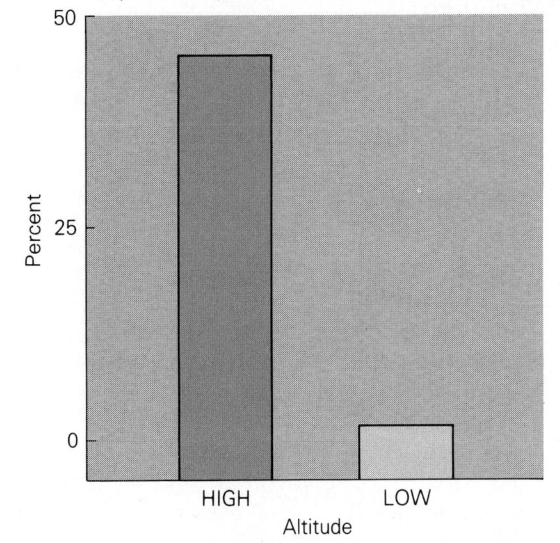

Figure 22–7 Of the high altitude migrants, about half reported improved health when they reached low altitude. (Adapted from Regensteiner, J. G. and Moore, L. G. Migration of the elderly from high altitudes in Colorado. *JAMA* 253: 3124, 1985.)

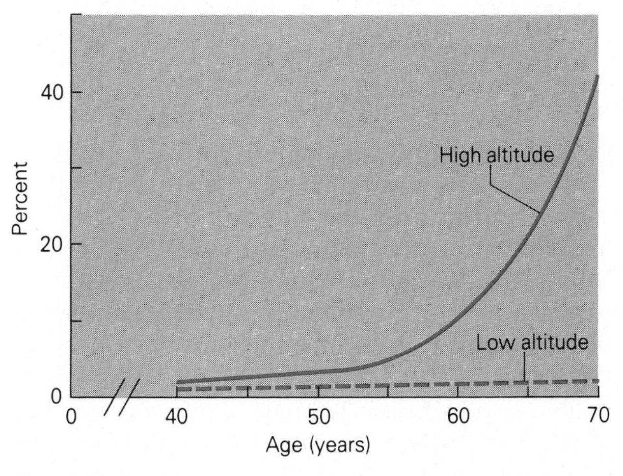

Figure 22–6 The chief reason for migration of the elderly from high altitudes is heart and/or lung disease. This study compared the migration patterns of a high-altitude and a low-altitude town of equal size and economic status. (Adapted from Regensteiner, J. G. and Moore, L. G. Migration of the elderly from high altitudes in Colorado. *JAMA* 253: 3124, 1985.)

Respiration in Deep Sea Diving

Gas Pressures Under Water

As a diver submerges under water, the pressure surrounding his or her body rises because of the weight of the water. To visualize how the pressure rise takes place, imagine diving with a balloon that is filled with air at a pressure of 1 atm. The deeper the balloon is taken, the greater is the water pressure pushing in on it. This increased pressure causes the volume of the balloon to decrease. Every 33 feet below the surface, the pressure increases by 1 atm. As the balloon gets smaller and smaller, the air molecules are compressed closer and closer together, increasing the pressure in the balloon. Similarly, as a diver submerges, the air is compressed (Fig. 22–8).

For thousands of years, people have used equipment to help them remain underwater for prolonged periods. The earliest devices were probably snorkels (Fig. 22–9). These devices provide a practical way to breathe to a depth of about 18 inches. Below that depth, two problems occur. First, the water pressure makes it difficult to expand the chest to draw in fresh air. Secondly, the snorkel adds dead space through which the diver must draw air.

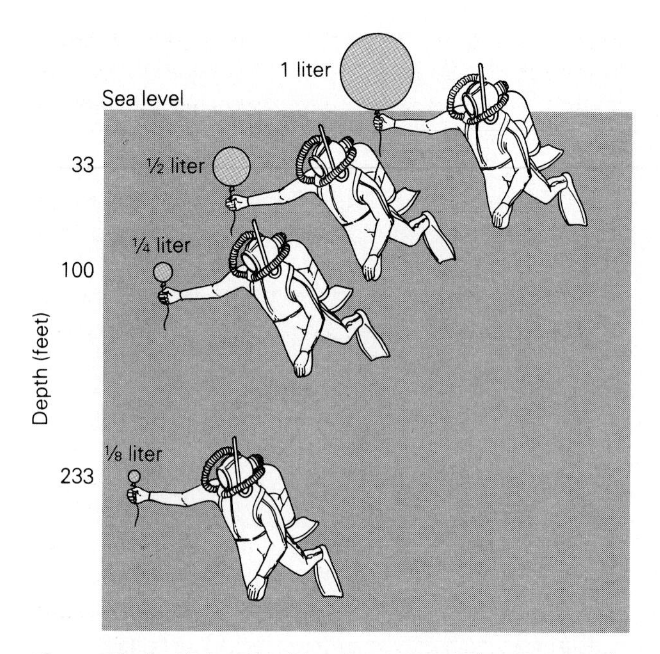

Figure 22–8 Air in a container is progressively compressed as the container is lowered into water, where the surrounding pressure increases.

If the dead space becomes too large, then the diver is simply moving air back and forth in the snorkel and does not get any fresh air.

The next apparatus to be invented was most likely the diving bell. These devices have been used for centuries. The first ones were simple, inverted chambers, open at the bottom, in which the diver stood and breathed the air in the chamber as it submerged (Fig. 22–10). These diving bells had many limitations, including air that became rapidly fouled by carbon dioxide. Also, the air volume was compressed to a smaller and smaller volume as the pressure increased during descent. Alexander the Great is said to have used a diving bell in the fourth century B.C. Modern diving bells are completely enclosed, so compressed air volume is no longer a problem. By the early nineteenth century, the enclosed diving suit had been invented (see Fig. 22–10). Fresh air was fed from the surface into the suit by a pump to sustain the diver for long periods under water.

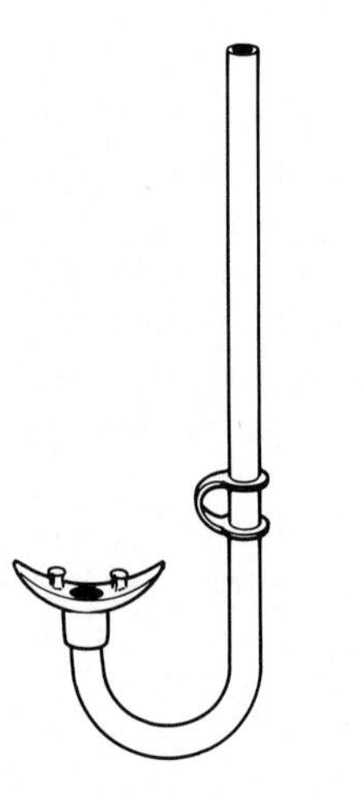

Figure 22–9 Although a swimmer can remain submerged for long periods with a snorkel, the depth of a dive is limited because the snorkel increases respiratory dead space and because the increasing pressure on the chest makes it difficult to draw in fresh air.

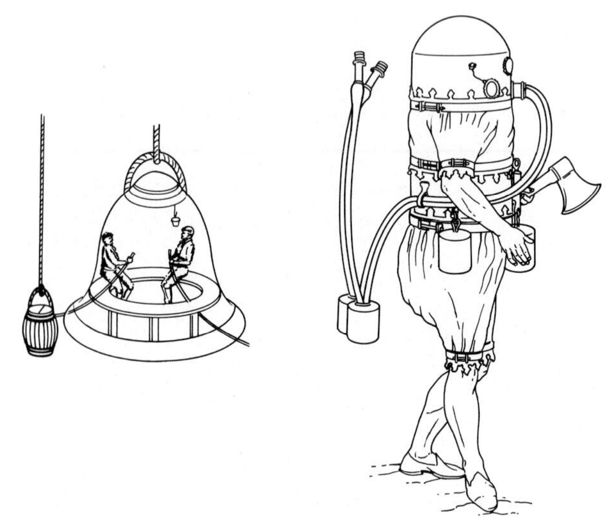

Figure 22–10 Diving bells have been used since the time of the ancient Greeks. During the eighteenth century, the advent of pumps made diving bells and diving suits practical.

Work was begun in earnest during World War II on the Self Contained Underwater Breathing Apparatus, "SCUBA" (Fig. 22–11). This kind of gear has become very popular as its development has progressed for recreation, exploration, and work on underwater oil rigs and pipelines.

Effects of High Pressures on Respiration

A characteristic of all these devices is that air enters the lungs at higher-than-normal pressure. Unfortunately, breathing such high pressures has some potentially serious consequences. The problems come not during the descent, but when the diver ascends. When the pressure becomes high enough, particularly at depths below 100 feet, significant amounts of nitrogen in the compressed air is absorbed into the tissue. The longer the time spent under water, the more nitrogen is absorbed. As the diver rises to the surface, the nitrogen is released from the tissues. If the ascent is slow, the nitrogen can be picked up by the blood and returned to the lungs to be exhaled. If the ascent is rapid, bubbles can form both in the blood and the tissue. When bubbles form near the joints, they cause agonizing pain.

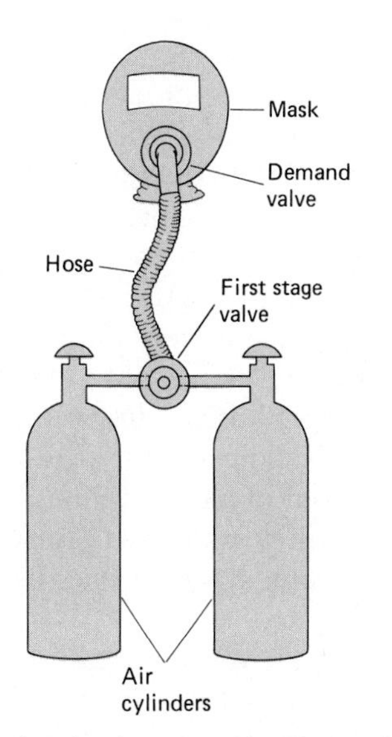

Figure 22–11 The development of Self Contained Underwater Breathing Apparatus (SCUBA) has enabled many people to pursue underwater activities. It is important, however, to be cautious during deep or prolonged dives because gases such as nitrogen can be absorbed into the body and cause problems during both the dive and the ascent.

Since the diver bends over to relieve the pain, this syndrome has been called the "bends." The risk of the bends increases both as the depth of the dive increases and the length of time at depth increases. The treatment for the bends is to spend time in a compression chamber in which the air pressure is raised until the bubbles are compressed and the nitrogen redissolves into the tissue. Pressure in the chamber is gradually lowered, giving the nitrogen time to be exhaled through the respiratory system instead of forming bubbles in the body.

At depths below about 150 feet, the number of nitrogen molecules in the body rise to a point at which they have a narcotic effect. This causes "raptures of the deep." Under these conditions, divers can often make unintelligent, life-threatening decisions. One way they can avoid this is to switch from a nitrogen-oxygen mixture to a helium-oxygen mixture. Helium is less soluble in tissues than nitrogen, and so fewer helium molecules dissolve in the body. This reduces the narcotizing effects as well as the number of molecules that must be removed from the tissue during ascent. Also, helium has a smaller molecular size than nitrogen, so it diffuses more rapidly, which aids in elimination of the gas.

Marine Mammals: Champion Divers

Pearl divers in the Far East can work for one to two minutes at depths of 50 or 60 feet using a single breath. Yet this remarkable achievement pales beside the diving abilities of marine mammals. Seals, dolphins, and whales have developed numerous diving reflexes that assist in conserving oxygen. The blood supply to most tissues is reduced virtually to zero by powerful vasoconstriction (Fig. 22–12). The heart rate slows so that myocardial oxygen consumption decreases and permits a decrease even in coronary blood flow. Only the blood vessels that supply the brain are unaffected by the diving reflex, so that blood flow is maintained at normal levels. The reduced demand for blood flow permits cardiac output to be reduced; at a depth of 300 feet, the seal's heart rate is only 4 beats per minute; on the surface, its heart rate is 120 beats per minute.

Surprisingly, the seals exhale just before they dive. In Weddell seals that were monitored on free dives of up to 23 minutes and depths of 700 feet, nitrogen uptake from the alveoli was found to cease at depths of 90 feet, because the gas-containing alveoli were collapsed shut. The gas was compressed into nonexchanging airways. The seals use this technique to avoid the bends.

Whales have the best diving performance of all marine animals. These animals have been known to spend more than an hour at tremendous depths.

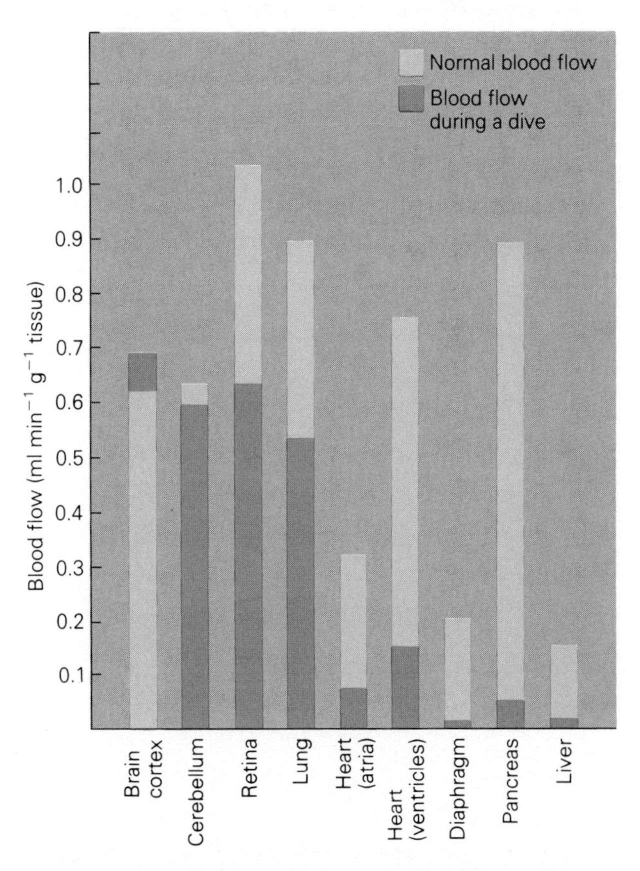

Figure 22–12 When a seal submerges, the diving reflex causes intense vasoconstriction in most of the body and reduces blood flow. (Adapted from K. Schmidt-Nielsen. *Animal Physiology.* Cambridge, Cambridge University Press. 1983.)

The record is 3717 feet recorded by an unfortunate sperm whale that became entangled in a trans-Atlantic cable that was broken at that exact depth. The whale dives so well because its metabolism is low and because it carries a large reserve of oxygen in its muscle. The whale and other diving mammals can tolerate much higher levels of carbon dioxide than nondiving animals.

Respiration in Submarines

The first practical submarines were built in the 1800s, although crude hand-propelled models were fabricated in the 1700s. By the start of World War I, Britain had 77 submarines, France 45, the United States 35, Germany 29, and Russia 28. The submarines in the early 1900s could stay submerged only for several hours and operated exclusively in shallow coastal waters. The effectiveness of submarines in warfare made them popular with every major power. To date, thousands of these vessels have been built.

One of the unpleasant aspects of these machines is that when they sink they usually take all hands aboard and trap them below. That grim image has prompted much work to be done on methods of escape. All sorts of scuba type gear have been developed, but usually to no avail. During World War II, only five men were saved by mechanical lungs. Ironically, a simple, excellent escape technique effective to depths of 300 feet had been known for nearly a century. It was used by Wilhelm Bauer in 1851 when his hand-propelled submarine sank in 60 feet of water (Fig. 22–13). The hull began to collapse from the pressure. Bauer waited until the air pressure in the vessel was equalized with the water pressure and then opened the hatch. The captain and his two crew members by then were breathing air at 3 atm, but had not breathed it long enough to build up nitrogen pressure in the tissue. Thus, they could ascend very rapidly without experiencing the bends. As they rose to the surface, the compressed air in their lungs expanded. This caused them to exhale steadily during the ascent. They never ran out of air even though they exhaled the whole time they ascended. As Captain Bauer explained, "We came to the surface like bubbles in a glass of champagne."

This technique was ignored and forgotten. It was accidentally repeated in several instances during World War II, but was never developed. Instead, with technology gone awry, navies continued to develop impractical mechanical contrivances that did not work, while all the time the body was superbly adapted to escape. In the 1950s, the British and U.S. navies recognized that the best technology for escape was no technology. Water-filled escape training towers have been built, and unaided escapes at depths to 300 feet are a routine part of submarine training.

One step has been added to the natural escape method: sailors now inflate a small life vest at the moment of departure. This extra buoyancy causes the submariner to ascend very rapidly to the surface. The decompression is so rapid under these circumstances that the sailors exhale as fast as they can. The air they breathed while at 300 feet is so

Figure 22–13 Wilhelm Bauer's submarine sank in 60 feet of water, but the captain and crew escaped by swimming easily to the surface. (Photo courtesy of: Submarine Force Library and Museum, Groton, CT)

highly compressed that it expands rapidly in their lungs as they ascend and, even with continual maximal forced expiration, they never run out of air on the way up. In navy jargon, this technique is known as "blow and go."

"Homo Aquaticus"

Two methods that are in the research stage offer promise for underwater use. One is the actual breathing of liquid. If saline (salt solution) or some other fluid such as fluorocarbon is put under several atmospheres of pressure by 100% oxygen, then the oxygen tension of the saline rises to thousands of millimeters of mercury. If an experimental animal such as a rat is held under the surface, it will eventually be forced to breathe the hyperoxygenated saline. Rats can breathe such fluid and swim around for periods of up to 18 hours (Fig. 22–14).

There are two major problems with liquid breathing. The first is that the removal of carbon dioxide is difficult because the respiration depends on the bellows action of the lungs to wash carbon dioxide out. If the lungs are filled with fluid, it is much more difficult for the bellows to work effectively, and carbon dioxide builds up in the blood. The second is that the fluid washes the surfactant out of the lungs, leading to alveolar collapse when the animal attempts to return to air. This makes the transition from fluid back to air difficult. It has been done successfully, however, by a number of rats, mice, and even some dogs.

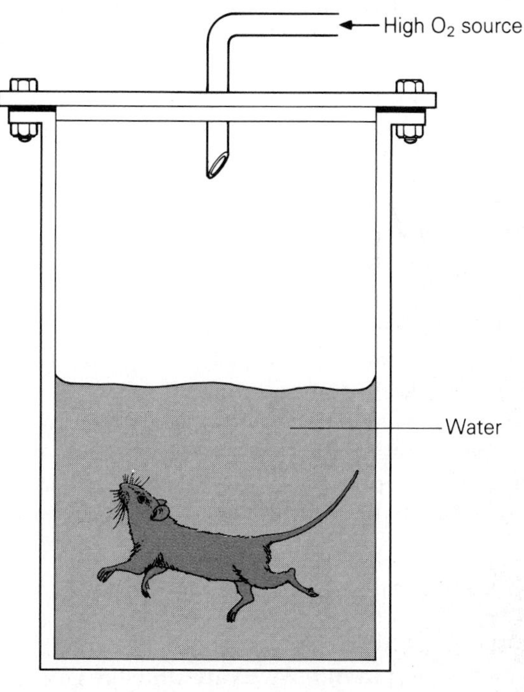

Figure 22–14 Small animals have survived for hours under water if the oxygen tension is raised to high levels.

Another possibility for prolonged undersea activities is the development of artificial gills. This is not a small engineering feat, but it seems possible. If a gill could be developed that a person could hold in his or her mouth and have trail behind, the large surface area of the artificial gill could extract oxygen from the water and excrete carbon dioxide just as fish do. Animals such as the mudpuppy have external gills as well as lungs (Fig. 22–15). They use the gills to stay successfully under water for prolonged periods. Liquid breathing techniques, artificial gills, or some similar method might lead to the development of what Jacques Cousteau describes as *Homo aquaticus,* an underwater man capable of living indefinitely beneath the surface of the sea.

Respiration in High-Oxygen Environments

The usual problem in breathing is getting enough oxygen into the body. However, with modern chemistry and technology it is possible to breathe 100% oxygen and actually get too much. High-oxygen environments are often found in hospitals, where oxygen is administered to patients with oxygen deficiency, and in aviation and space flights, where the pilot or astronaut breathes air with high oxygen concentration because of extremely low oxygen tension.

Atelectasis: Collapse of Alveoli

One of the physical dangers of breathing pure oxygen is **atelectasis** (collapsed alveoli) which can occur in pilots, astronauts, and in hospital patients. This

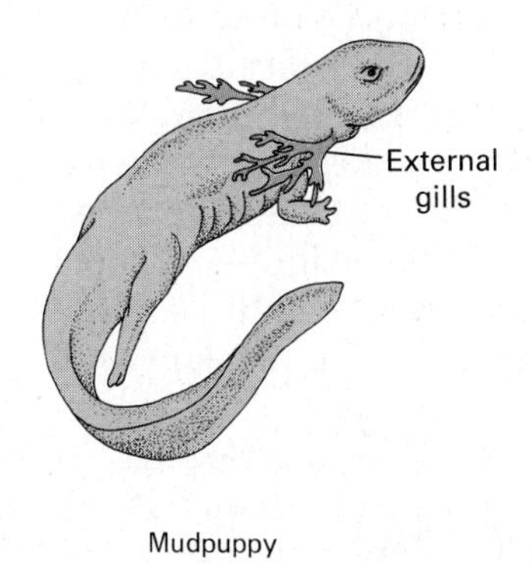

Figure 22–15 Animals like the mudpuppy have external gills.

physical complication occurs when small airways become blocked by mucus; the oxygen in the plugged alveoli gets absorbed, and the alveoli rapidly collapse (Fig. 22–16). The total pressure of the trapped gas is close to 760 mm Hg, but some of the individual gas pressures in the blood are much less than 760 mm Hg. If gas is trapped in the course of breathing air (see Fig 22–16), nitrogen acts to deter collapse because it is not utilized and therefore is not absorbed. However, when one is breathing pure oxygen, there is no nitrogen; hence, oxygen is continuously absorbed, and the alveoli collapse (Fig. 22–16*a*, *b*).

Postoperative atelectasis occurs commonly in patients who are ventilated with high-oxygen mixtures. Atelectasis is also prevalent in fighter pilots who are breathing high oxygen and experiencing large gravitational forces (see the section on space travel). The added gravitational effects augment alveolar collapse. Atelectasis is most likely to occur at the bottom of the lung where alveoli are poorly expanded and airways are partially closed. Reopening atelectic areas is difficult because the surface tension effects are more pronounced on small units. Atelectasis can be reopened by deep breaths.

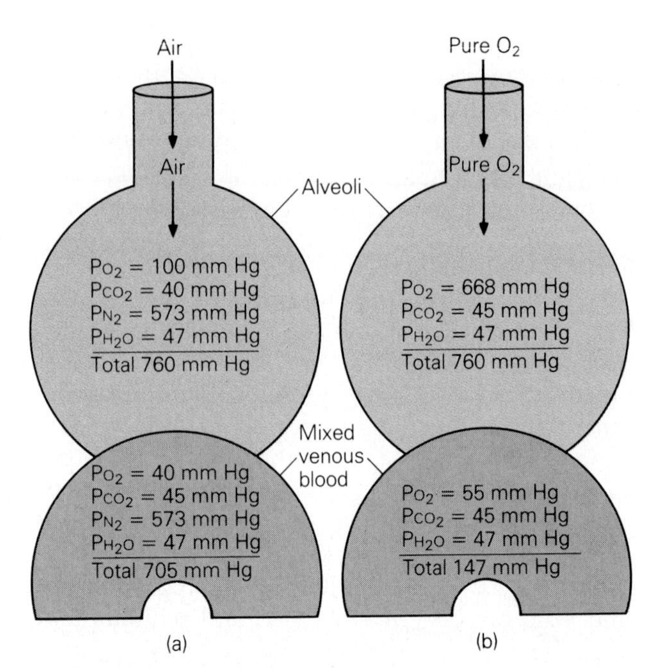

Figure 22–16 Cause of atelectasis as a result of a plugged airway. When (*a*) air is breathed and (*b*) oxygen is breathed, alveolar pressure in both cases is the same, but the sum of the gas pressures in the mixed venous blood is smaller.

Toxicity Associated with Breathing High Oxygen

The second problem associated with high oxygen, which is much more serious than atelectasis, is toxic effects. These occur when a person breathes high concentrations of oxygen for more than one day. The primary target organ for oxygen toxicity is the lung, where damage can occur within 24 hours and can cause pulmonary edema.

Free-Radical-Induced Toxic Effects

The first site of injury is the pulmonary capillary endothelial cells, which result in leaky pulmonary capillaries. The mechanism for tissue damage is the formation of **free radicals.** These highly reactive molecules are formed as byproducts from molecular oxygen. (Before molecular oxygen (O_2) can be used in the oxidation of fuels for energy, it first must be converted to "active forms". These active forms are called **oxidizing free radicals,** and the most important one in the cell is the **superoxide anion radical** ($O_2^{-}\cdot$). These free radicals are continually being produced from dissolved molecular oxygen under normal oxygen tensions (tissue P_{O_2} 40 mm Hg). However, under these normoxic conditions, the free radicals are rapidly removed by protective cellular enzymes—namely, **superoxide dismutase, catalase,** and **peroxidase.** As long as P_{O_2} is normal, the oxidizing free radicals are rapidly removed and no cellular damage occurs.) When tissue P_{O_2} is greatly increased—such as in breathing of high oxygen—excess free radicals are formed and overwhelm the protective enzymes. As a result, these free radicals cause destructive and lethal cellular effects by (1) oxidizing membrane lipids, thus altering essential properties of membranous structures of the cells, (2) oxidizing cellular proteins and DNA, and (3) oxidizing enzymes, thus shutting down cellular respiration.

High oxygen is not the only source of oxidant damage from free radicals. Breathing ozone (O_3) or nitrogen dioxide (NO_2) from polluted air will also cause free radical injury to the lung. Some herbicides, such as paraquat, are extremely toxic because of their free-radical-induced injury. If someone smokes tobacco or marijuana that has been treated with paraquat, the paraquat in the smoke will react with lung cells to form free radicals. Crop dusters and migrant workers are also at risk because of the exposure to paraquat through the lungs or skin.

Retrolental Fibrosis

Another hazard of breathing high concentrations of oxygen is seen in intensive care units in which premature infants develop blindness owing to **retro-**

lental fibrosis. The mechanism for the blindness is deterioration of the retina and inadequate development of retinal blood vessels. The poor vasculation leads to the formation of fibrosis tissue behind the lens (retrolental fibrosis). If the Po_2 is kept below 140 mm Hg in the incubator, retrolental fibrosis can be avoided.

Toxic Effects of High-Pressure Oxygen

A third type of oxygen toxicity occurs when oxygen is breathed at pressures exceeding 760 mm Hg. Oxygen is breathed at pressures greater than 1 atm in deep sea diving or in hyperbaric (high-pressure) chambers. When oxygen is breathed at 2 to 4 atm, it stimulates the central nervous system (CNS), leading to convulsions. The exact cause for the toxic effect on brain tissue is not fully known, but the high oxygen pressure is thought to inactivate dehydrogenase enzymes. The convulsions are often preceded by twitching of the face, ringing in the ears, and nausea. Convulsions can occur when an amateur scuba diver inadvertently uses a tank filled with O_2.

Respiration in Unusual Gaseous Environments

High Concentrations of Carbon Dioxide and Respiration

Breathing high carbon dioxide levels usually does not occur. Normal Pco_2 in the air we breathe has a partial pressure of 0.3 mm Hg (0.04%), which is negligible. However, in certain types of diving gear and closed environments (e.g., submarines and spacecraft), carbon dioxide can build up and be rebreathed because of defects in equipment. Rebreathing higher-than-normal levels of carbon dioxide will stimulate ventilation. A buildup of an alveolar Pco_2 of 80 mm Hg, which is about twice normal, is the maximum that an individual can tolerate. With these high concentrations, hyperventilation is maximum, and the individual becomes severely acidotic. When Pco_2 levels are greater than 80 mm Hg, the situation becomes intolerable, and the respiratory centers no longer are stimulated but are inhibited. Respiration begins to fail, and acidosis becomes more severe. Eventually, the CO_2 acts like a narcotic, and the individual becomes lethargic and eventually unconscious.

Carbon Monoxide and Respiration

Carbon monoxide (CO), unlike carbon dioxide, does not stimulate respiration. The problem with carbon monoxide is that it is an odorless, colorless, and poisonous gas that comes from incomplete combustion of fuels that cannot be detected by the body. Carbon monoxide becomes lethal because it competitively binds at the same site as does oxygen on the hemoglobin molecule to form **carboxyhemoglobin (COHb).** The formation of COHb interferes with the transport of oxygen to the cells. The reaction ($Hb + CO \rightleftharpoons HbCO$) is reversible and is a function of the partial pressure of CO. This means breathing a higher concentration of CO will shift the reaction to the right to form more HbCO, and breathing lower concentrations of CO will shift the reaction to the left to form less HbCO. The problem is further complicated because the binding affinity of CO for hemoglobin is about 210 times that of O_2. Because of this greater tendency for binding to Hb, even trace amounts of CO in the respired air can be extremely dangerous. For example, an alveolar Pco of 0.5 mm Hg (1/210 that of alveolar Po_2) will cause half the hemoglobin in the blood to bind with CO and half to bind with O_2. Under these conditions, the amount of oxygen carried by the blood to the cells is halved. Thus, a Pco of 0.7 mm Hg (0.1% of air) can be lethal because the cells become **anoxic** (deprived of oxygen). The first organ to become sensitive to the lack of oxygen is the brain. Altered vision, reaction times, and coordination are the first signs of CO poisoning and often explain why there are more accidents on the freeways when CO levels rise. The real danger of CO poisoning is that it cannot be detected by the body. With hypoxemia, there are physical signs: the individual will turn blue around fingertips and lips. However, the formation of HbCO makes the blood crimson red and masks the bluish tint of hypoxemia. The second problem is that there are no chemoreceptors to detect CO. The problem is compounded because the chemoreceptors that detect changes in Po_2 are not stimulated. The reason for the latter is that Po_2 in blood is normal with CO poisoning, yet O_2 cannot bind with hemoglobin.

An individual severely poisoned with CO can best be treated by administration of 100% O_2. Administering pure oxygen elevates the blood Po_2, which displaces CO from the hemoglobin. Also beneficial is the simultaneous administration of 5% to 7% carbon dioxide. Breathing the high concentration of carbon dioxide stimulates ventilation and reduces alveolar CO, which concomitantly releases CO from the blood. When the combination of high oxygen and carbon dioxide is breathed, CO can be removed 10 to 15 times faster than when just fresh air is breathed.

Anesthetic Gases and Respiration

Without anesthesia, modern surgical procedures such as appendectomies, cardiac surgery, and neurosurgery could not be done. Until the 1930s, all anesthetics were administered via the lungs.

Prior to the introduction of anesthesia in the nineteenth century, surgical pains were relieved by alcohol or opium. For over 2000 years, opium had been known for its narcotic effects, but was first used as an anesthetic in the ninth century A.D. The patient inhaled the opium vapor, which diffused across the alveolar membrane as would oxygen to cause drowsiness and to deaden the pain. However, many deaths resulted because of overdoses that depressed the respiratory centers and stopped breathing. It was not until the 1800s that effective, safe opium administration became possible thanks to a German pharmacist who isolated the main alkaloid (morphine) from opium.

The first real anesthetic gas, **nitrous oxide,** was discovered by Joseph Priestley in 1772. However, nitrous oxide was not used until 1884, when a Connecticut dentist employed it for a tooth extraction. Nitrous oxide was commonly called "laughing gas" because it literally caused individuals to laugh and to feel happy. Actually, for almost one hundred years after its discovery, nitrous oxide was not taken seriously and was very popular at parties in the nineteenth century because of its pleasurable intoxicating effects. Another dentist in the 1800s suggested the use of a second gas, **ether,** as an anesthetic. Ether was used for the first time at Massachusetts General Hospital in 1846 and subsequently became very popular. Later, **chloroform** was introduced as an anesthetic by a British physician who used it first for women in childbirth.

The discovery of the anesthetic properties of ether, nitrous oxide, and chloroform satisfied the immediate needs of surgery despite their toxic effects and the possibility of explosion in the case of oxygen. However, in the 1950s, the fluorinated and halogenated anesthetics were discovered. These newer anesthetics had the advantage that they were not flammable or toxic, and had a less depressant effect on respiration.

There are three stages of anesthesia with inhaled anesthetics. The first stage is **analgesia,** in which the response to pain is reduced. The second stage is the **loss of consciousness.** The third stage is **surgical anesthesia,** in which all sensations are lost, motor functions are inhibited, and regular automatic breathing commences.

Respiration in Space Travel

With the rapid development of spacecraft and supersonic planes, the human body can be subjected to acceleratory forces owing to sudden changes in velocity or direction. For example, a spacecraft must attain a speed of 25,000 miles/hr to escape the earth's gravitational field and to leave its atmosphere. High acceleration rates are required to attain these speeds. When an aircraft or spacecraft takes off, the human body inside is subjected to **positive acceleration.** (Inertial forces are directed in a direction that may be described as head-to-foot with respect to the pilot.) When an aircraft or spacecraft slows down at the end of a flight, **negative acceleration** occurs (inertial forces are directed in a foot-to-head direction).

High G-Forces and the Human Body

Any time an aircraft turns, centrifugal and acceleration forces come into play as a result of gravitational attraction. The human body's sensation of weight is due to **gravitational attraction** and is referred to as **G-force.** The G-force is a unit used to express the magnitude of the acceleratory force. One G is equal to the force exerted by one times the normal gravitational force. For example, when an individual is sitting in a chair, the force that presses the body against the chair seat is equal to 1 G (one times the pull of gravity). If the force that presses against the seat is four times the normal weight of the body, the force would be 4 G. If force is less than the body weight, then the force is a negative G. Positive and negative G forces are illustrated in Figure 22–17, in which an aircraft pulls out from a dive or goes through an outside loop so that the pilot is held down by the seat belt. In the dive, the pilot experiences positive G force with inertial forces acting in a head-to-foot direction. In the loop, the pilot experiences a negative G force, in which the inertial forces are directed from foot to head.

The physiological effects of these acceleratory forces are due to the fact that some tissues and blood in particular are "loose" with respect to body. When a positive G force is exerted in a head-to-foot direction, blood is pulled toward the lower parts of the body. In Figure 22–17, when the pilot "pulls 4G's," then venous pressure in the leg veins would be four times the normal pressure. The heart cannot pump unless venous blood is returned, and consequently there is considerable pooling of blood in the lower extremities. At four G's, a pilot in the seated position experiences a fall in arterial systemic pres-

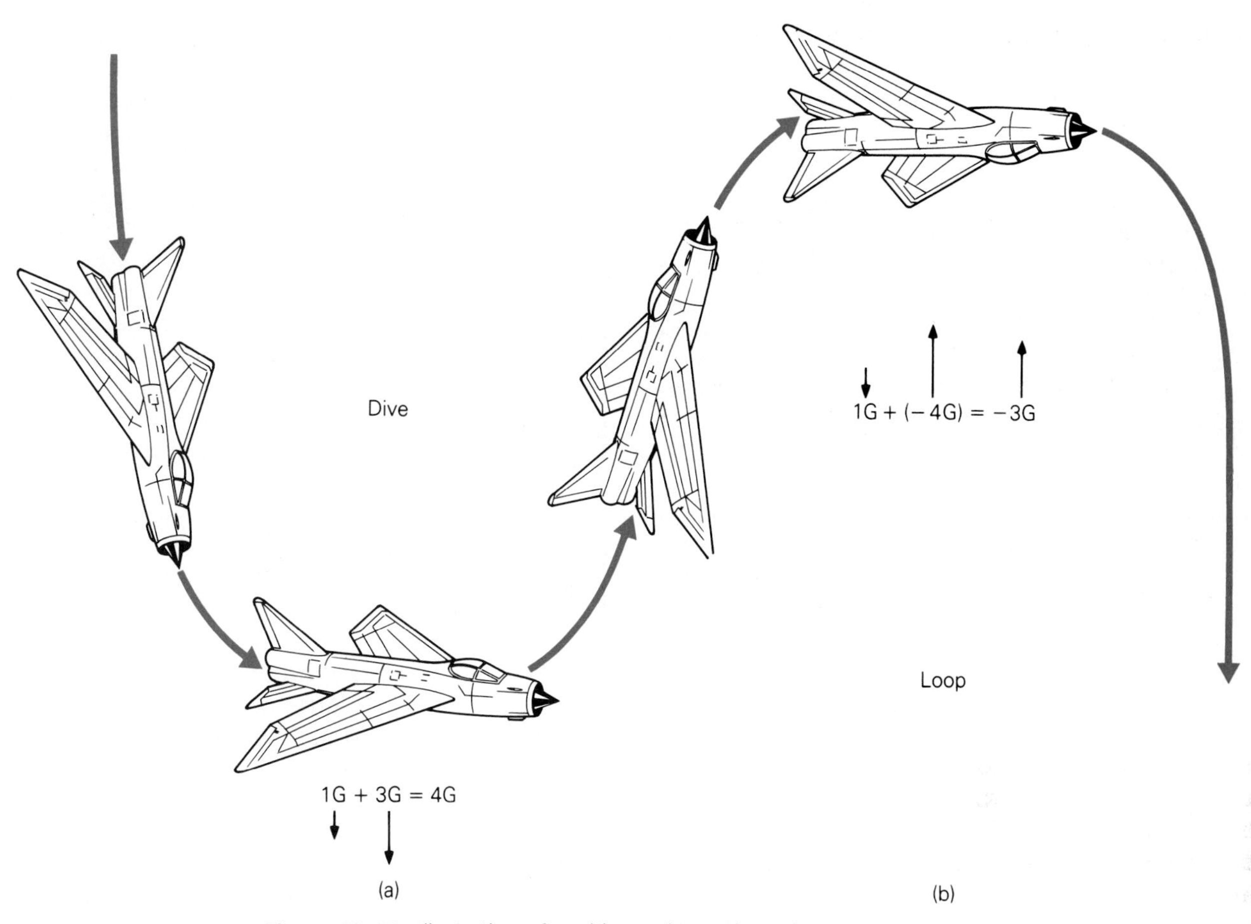

Figure 22–17 Illustration of positive and negative G forces on an airplane.

sure, at the level of the heart, of approximately 40 mm Hg. Under these conditions, blood flow to the brain almost stops. Consequently, loss of vision, referred to as **blackout,** occurs. Consciousness is finally lost because of cerebral hypoxia. Other symptoms are also prevalent. At 3 G to 4 G, a pilot or astronaut experiences difficulty in the use of muscles. Facial tissue sags and eyelids droop. At 5 G, movement of the body is almost impossible, and respiration stops because the respiratory muscles cease to function. Between 5 G and 9 G, legs become congested and calf muscle cramps are prevalent. Pulmonary edema occurs because of venous engorgement in the base of the lung, and vision as well as hearing is lost. Finally, consciousness is lost.

Most of these symptoms quickly vanish when the positive G force is removed. To minimize the G forces, astronauts lie down during take off or reentry and wear special antigravity suits. On acceleration, these suits automatically pump up, and increased pressure exerted on the body prevents pooling of the blood. With negative G-force (foot-to-head) the motion of blood is toward the head. Up to −3 G, there is pressure in the eye sockets. Neck veins are overdistended and the face becomes puffy. Ringing in the ears occurs, and headaches may ensue. Above −3 G, retinal vessels and vision become engorged and vision is impaired. The light filtering through the blood gives a sensation of redness, a phenomenon known as **redout.** Often, at high negative G (greater than −3 G), retinal and cerebral hemorrhages can occur. The physiological effects of negative G are much more lasting than those of positive G, and problems with vision and headache may persist for hours following a negative G exposure.

Weightlessness and Respiration

Once a spacecraft leaves the earth's gravitational forces, astronauts are free of the problem of acceleratory forces, but are faced with the problem of weightlessness. The physiological problems associated with weightlessness do not appear to be severe. One concern is the lack of hydrostatic pressures, which results in decreased blood volume, red cell mass, and cardiac output. The second physiological problem related to weightlessness is diminished physical activity, because very little muscle contraction is required for movement. The decrease in physical activity leads to a decrease in work capacity and loss of calcium from bone.

Artificial Atmospheres in Spacecraft

There is no atmosphere in space, so consequently astronauts must be provided with an artificial atmosphere that is pressurized and equipped with life-sustaining oxygen. The types of atmospheric conditions selected depend, in part, on the sophistication of the engineering. For example, the Russian space program tends to use 21% oxygen in nitrogen at a pressure equivalent to sea level, while the early Apollo missions used 100% oxygen at one third of atmospheric pressure, which gives a Po_2 equivalent of 1 atm. The reason for these two different gas mixtures was based on weight and space. In the earlier Apollo missions, NASA did not have booster rockets sufficient to launch a heavy spacecraft. Using 100% oxygen at ⅓ atm gives a lighter weight load than a thick-shelled spacecraft using air at 1 atm. One of the dangers of 100% oxygen is the likelihood of an explosion or a fire. Such an accident did occur aboard one of the early Apollo crafts on the launch pad, resulting in a fatal fire and forcing NASA to change gas mixtures. Now all U.S. space flights use air as a gas mixture.

Radiation Hazards in Space

The earth's upper atmosphere is continuously being bombarded by cosmic radiation, particularly from the sun. Many of these radiation particles are trapped in the earth's magnetic field, termed the **Van Allen radiation belt.** Two major belts around the earth trap radiation particles: the inner Van Allen belt starts at an altitude of 300 miles and extends to 3000 miles; the outer belt starts at an altitude of about 6000 miles and goes to 20,000 miles. Since these two belts contain high amounts of radiation, spacecraft must not travel within them because astronauts can receive fatal doses of radiation there. At present, all spacecraft orbit below an altitude of 200 miles to avoid any radiation hazard.

Summary

I. A. At high altitudes, the barometric pressure is lower than at sea level. Since the percent of oxygen in air remains the same at near 21%, regardless of the altitude, the amount of oxygen falls as altitude increases. The body's efforts to adapt to high altitude center on compensating for the lack of oxygen (hypoxia). A person who lives at sea level who is suddenly exposed to a very high altitude (over 20,000 feet) will rapidly lose the ability to function. If he or she has time to *acclimatize,* the person can climb to these altitudes without the use of supplemental oxygen. The first response of the body to hypoxia is hyperventilation. Increased respiration raises arterial oxygen tension, but also blows off too much carbon dioxide, which can cause acute mountain sickness, a self-limiting disorder that is not serious, although it can be quite uncomfortable. Rapid ascent to altitudes of 10,000 feet or more can lead to high altitude pulmonary edema, a disease in which fluid collects in the lung and increases the diffusion distance for oxygen and carbon dioxide. Although serious, the edema clears rapidly on descent. Starting at about 12,000 feet, high altitude cerebral edema can occur from rapid ascent. This is an extremely serious condition that can result in death; immediate descent is mandatory.

B. Over the first few weeks to months at high altitude, the body makes a number of adjustments to the hypoxia. The oxygen-carrying capacity of the blood is raised by increased hematocrit. Both ventilation and cardiac output increase to improve oxygen delivery. After a longer time, ventilation and cardiac output return to normal, but hematocrit remains elevated.

The chronic effects of high altitude are pronounced in the elderly. In the United States and other countries where people have lived at high altitudes for only a few generations, there are relatively few old people adapted to high altitudes. With thousands of years to *acclimate,* the Andean natives and the Sherpas have successfully adapted to chronic hypoxia.

II. A. Pressure increases rapidly as a diver submerges. For every 33 feet below the surface, the pressure increases by 1 atm. The most primitive underwater diving apparatus is the snorkel, the length of which is limited by the added *dead space* and by the increasing pressure on the chest wall.

B. The high pressure experienced with modern diving gear causes nitrogen to be absorbed by the blood and tissue. If the diver ascends too rapidly, the nitrogen forms bubbles and causes the painful disorder called the bends. Below 150 feet, nitrogen narcosis occurs and causes raptures of the deep, a dangerous condition that affects the diver's mental capacity.

C. Marine mammals are adapted for diving for long periods to depths of more than 1000 feet. Their diving reflex slows the heart and redistributes blood flow away from all organs except the brain.

D. In the 1950s, the Navy relearned that free ascent could be made from depths of 300 feet by simply floating up to the surface and exhaling rapidly. Since the air in the lungs expands during ascent, navy submariners never ran out of air.

III. A. Breathing high concentrations of oxygen can cause alveolar collapse (atelectasis). This occurs through irritation of the small airways, which leads to excess mucus secretion and subsequent plugging of airways. Once airways are plugged, oxygen diffuses across alveoli and, in the absence of nitrogen, rapid collapse occurs.

B. Breathing pure oxygen leads to pulmonary edema due to damage of pulmonary capillaries. Lung tissue is damaged by the formation of free radicals, which oxidize membrane lipids, enzymes, proteins, and DNA molecules. In addition to causing lung damage, high oxygen will also cause fibrosis in the eye (retrolental fibrosis), which will lead to blindness in the newborn.

IV. A. In artificial environments, carbon dioxide can build up and be rebreathed. High concentrations of CO_2 stimulate respiration, leading to hyperventilation and acidosis. When alveolar carbon dioxide exceeds 80 mm Hg, respiration is no longer stimulated but is depressed and can lead to death.

B. Carbon monoxide is an odorless, colorless, and poisonous gas that cannot be detected by the body. Carbon monoxide becomes lethal because it competitively binds at the same site as does oxygen on hemoglobin molecules. The problem is further complicated by the fact that CO has a much greater binding affinity for hemoglobin than oxygen. Thus, only trace amounts of CO in the blood are necessary to cause death.

C. The first anesthesias used were gases that had to be breathed. The first real anesthetic gas was nitrous oxide, first used in the 1800s by a dentist. Later, ether and chloroform were introduced as anesthetic gases. It was not until the 1950s that the fluorinated and halogenated anesthetics were discovered. These newer anesthetics represented a major improvement over their predecessors since they were neither toxic nor flammable.

V. A. In space flights and supersonic flights, the human body can be subjected to acceleratory forces (G-forces) by a sudden change in velocity or direction. When a positive G-force occurs, blood is pulled toward the lower parts of the body. Consequently, venous pulling occurs, and blood flow to the brain decreases. If the G-force is severe, loss of vision (blackout) and unconsciousness occurs. When a negative G-force is experienced, blood moves toward the head. Neck veins become overdistended, pressure builds up in the eye sockets, and retinal vision becomes blurred owing to engorgement of retinal vessels (redout).

B. Another challenge astronauts face once in space is weightlessness, the lack of gravity. At present, the physiological problems associated with weightlessness are not severe, but with future longer space flights problems of weightlessness may become more serious.

C. During space flights, astronauts must be provided with an artificial atmosphere suitable to sustain life. Currently, the artificial atmosphere is a mixture of air at 1 atm. However, with longer space flights, alternative gas mixtures may be required.

D. In addition to weightlessness and artificial atmospheres, astronauts are also potentially exposed to cosmic radiation. The earth's magnetic field (Van Allen radiation belt) traps radiation particles and prevents them from entering the earth. As long as the space flights are below the Van Allen belt (300 miles), astronauts are fairly safe from radiation exposure. However, once spacecrafts begin to venture beyond the earth's magnetic field, special precautions will need to be taken to protect astronauts.

Review Questions

1. What are the effects of increasing altitude on barometric pressure and oxygen tension?
2. What is the partial pressure of oxygen at sea level, Denver, and Mount Everest (barometric pressure = 760, 630, and 247 mm Hg, respectively).
3. How does the body adapt to high altitude during a short-term trip?
4. What disorders can develop during rapid ascent to various altitudes?
5. What adaptive changes occur during long-term residence at high altitude?
6. What happens to pressure surrounding a diver as he or she submerges?
7. What happens if a diver ascends too rapidly?
8. What happens if a diver goes too deep?
9. How can a person escape from a submarine from a depth of 300 feet without diving equipment?
10. What are two major side effects of breathing high oxygen on the human body?
11. Why does carbon monoxide become poisonous to the body?
12. Who discovered the first anesthetic gas, nitrous oxide?
13. How does the G-force affect the body?

Suggested Readings

Fishman, A. P., and Richards, D. W. *Circulation of the Blood.* New York: Oxford University Press, 1964.

Lambertsen, C. J. *Underwater Physiology.* Baltimore: Williams & Wilkins, 1967.

Stoelting, R. K., and Miller, R. D. *Basics of Anesthesia.* New York: Churchill Livingstone, 1984.

Sebel, P., Stoddart, D. M., Waldhorn, R. E., Waldmann, C. S., and Whitfield, P. *Respiration.* New York: Torstar Books, 1985.

Altman, L. K. *Who Goes First?* New York: Random House, 1987.

Table 23–2
The Digestive Enzymes of the Gastrointestinal System and Their Actions

Source of Enzyme	Site of Action	Enzyme	Substrate	Products of Digestion
Salivary glands	Mouth and gastric lumen	Amylase	Polysaccharides	Oligosaccharides
Stomach	Gastric lumen	Pepsin	Proteins	Peptides
Pancreas	Small intestinal lumen	Amylase	Polysaccharides	Oligosaccharides
		Trypsin, chymotrypsin, and carboxypeptidase	Proteins and peptides	Peptides and amino acids
		Lipase	Triacylglycerols	Monoglycerides and free fatty acids
Small intestinal mucosa	Small intestinal epithelium	Oligosaccharidases Peptidases	Oligosaccharides Peptides	Monosaccharides Amino acids

The luminal membranes of the intestinal epithelial lining contain a number of transport systems specialized for the absorption of specific components of the diet. Some of these transport systems are capable of moving certain digestion products and inorganic ions from lumen to blood by active transport against electrochemical gradients. Other substances move by diffusion, at varying rates depending on their molecular size and lipid solubility.

The Regulation of Gastrointestinal Function

Few mechanisms regulate the process of digestion and absorption. Since this lack of regulation applies to those constituents of the diet that supply calories to the body, the more of these foods that are eaten, the more will be digested and absorbed. Thus, most cases of obesity result from eating too much food too often.

In contrast to digestion and absorption, motility and secretion are highly regulated—in particular, by neural and hormonal mechanisms. These processes are regulated by stimulation of receptor cells, which in turn triggers a chain of events operating through various pathways to produce specific responses of smooth muscle and exocrine gland cells.

Many of the stimuli that initiate the regulatory events are present in the lumen of the digestive tract. The most important stimuli are digestion products themselves, hydrochloric acid (HCl) secreted by the stomach, and distention by the contents of the tract. However, responses of the gastrointestinal system are not limited to stimuli applied to the tract, because stimulation of many other regions of the body can elicit responses of the gastrointestinal system. For example, both the sight and smell of food stimulate salivary, gastric, and pancreatic secretion. Another example is the profound motor activity (i.e., vomiting) that results from stimulation of the vestibular apparatus during motion sickness.

The receptors of the tract include sensory nerve receptors, such as **chemoreceptors** that respond to chemical agents in the contents of the tract, and **mechanoreceptors** that respond to distention of the tract. The reflex activity that results from excitation of these nerve receptors plays an important role in the regulation of gastrointestinal secretory and motor activity. In addition, a number of different types of endocrine cells in the digestive tract serve as receptors. When excited by a variety of stimuli, these cells release hormones into the circulation. These hormones, as do the nerve reflexes, also participate in the regulation of secretion and motility.

Neural Control of Gastrointestinal Processes

The neural pathways for the control of motor and secretory activity of the gastrointestinal system are summarized in Figure 23–4. Two types of neural pathways are involved in the regulation of gastrointestinal motor and secretory activity: **short reflexes,** mediated through nerve cells that lie entirely in the wall of the digestive tract, and **long reflexes,** mediated over nerves outside of the tract.

Short Reflex Regulation. Regulation by short reflexes occurs through the myenteric and submuco-

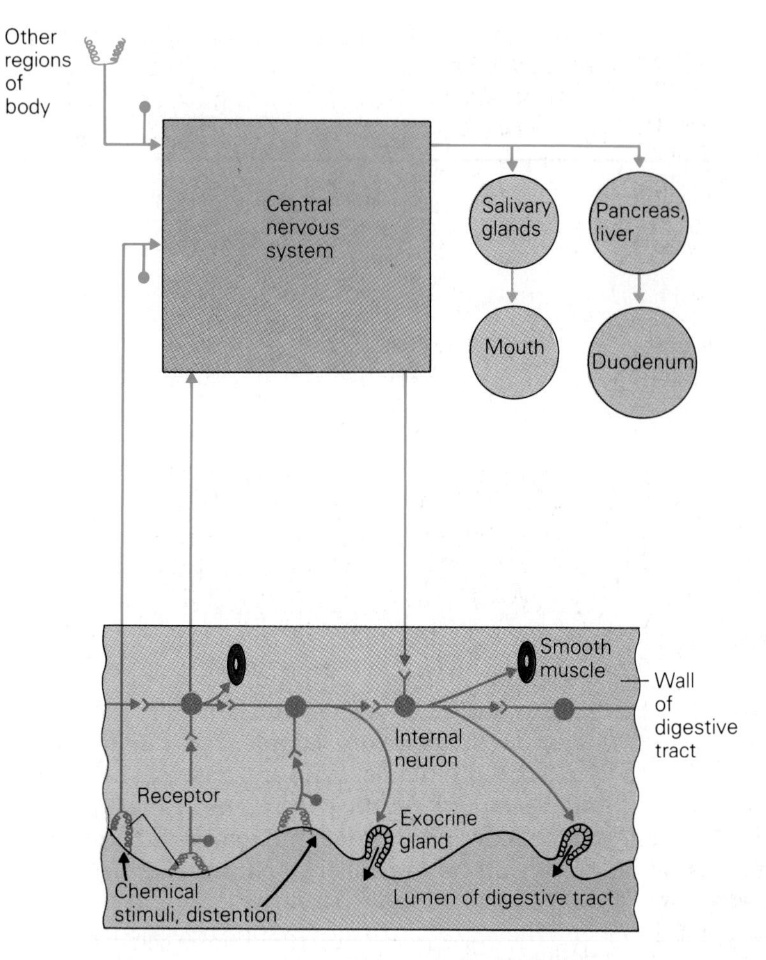

Figure 23–4 Neural pathways for the control of motor and secretory activity of the gastrointestinal system. Green lines show short reflex pathways; blue lines show long reflex pathways.

sal nerve plexuses that are present in the wall of the digestive tract (see Fig. 23–2). These nerve plexuses possess all of the elements required for the local regulation of motor and secretory activity; i.e., they contain (1) sensory neurons that are excited by the intraluminal stimuli mentioned previously, (2) interneurons that conduct impulses up and down the tract, and (3) motor and secretory neurons that innervate the effector smooth muscle and secretory cells. The short reflexes that are mediated over these nerve cells give the digestive tract a degree of autonomy in the sense that muscle contraction and exocrine secretion are to an extent independent of regulatory influences that originate outside the wall of the tract. These short reflexes can be either excitatory or inhibitory depending on the mediator that is released at the effector cell by motor and secretory neurons. If the mediator is acetylcholine, as is usually the case, the result is excitation of exocrine secretion and contraction of visceral smooth muscle.

Long Reflex Regulation. Gastrointestinal activity is also regulated by reflexes mediated through nerves outside of the gastrointestinal system. The major function of these long reflexes is to correlate activities, not only between different regions of the gastrointestinal system, but also between this system and other parts of the body. These long reflex arcs are mediated from either the gastrointestinal system or other regions of the body over afferent sensory nerves to the central nervous system. The central nervous system (CNS) in turn relays impulses over efferent parasympathetic and sympathetic nerves to the gastrointestinal system. Since the efferent autonomic nerves synapse with cell bodies of the internal nerve plexuses of the digestive tract, impulses over external nerves can modify short reflex activity that is initiated in the wall of the tract. This modification can be either excitatory or inhibitory, depending on the type of efferent fibers involved. The vagus nerve (a parasympathetic preganglionic nerve) is the major motor and secretory external nerve of this organ system, and most vagal efferent fibers are cholinergic and excitatory. The splanchnic nerves (sympathetic postganglionic nerves) are adrenergic, and the norepinephrine they liberate depresses excitability of the nerve plexuses, thereby inhibiting short reflexes mediated through these local nerve networks.

Hormonal Control of Gastrointestinal Processes

Four hormones of gastrointestinal origin—**gastrin, cholecystokinin (CCK), secretin,** and **gastric inhibitory peptide (GIP)**—are important in the control of motility and secretion. All of these hormones are polypeptide hormones consisting of chains of amino acids of varying lengths. The four types of endocrine cells that secrete these hormones occur as single cells throughout the epithelial lining of certain regions of the tract. When stimulated, these endocrine cells release their hormones into the blood, which carries them to target organs of the gastrointestinal system, where they produce their motor and secretory effects. All four hormones are released in response to intraluminal stimuli acting directly on the endocrine cells in the luminal lining. Gastrin is also secreted in response to acetylcholine released by nerve impulses—that is, through short reflexes in the intestinal wall that are activated by intraluminal stimuli and through long reflexes, the efferent limbs of which are in the vagus nerve. The location of the gastrointestinal endocrine cells, the stimuli for the release of their hormones, and the major physiological actions of these hormones are summarized in Table 23–3.

Chewing, Salivary Secretion, and Swallowing

Chewing

The process of chewing is accomplished by the combined actions of the skeletal muscles of the jaws, lips, cheeks, and tongue. The contraction and relaxation of these muscles is coordinated by impulses over a number of cranial nerves. Although these muscles can be controlled voluntarily, the act of chewing is partly reflexive in nature. The pressure that results when the jaws close following the intro-

Table 23–3
The Gastrointestinal Hormones

	Gastrin	CCK	Secretin	GIP
Location of Endocrine Cells	Antrum of stomach	Upper small intestine	Upper small intestine	Upper small intestine
Stimuli for Release	Amino acids, peptides, acetylcholine	Amino acids, peptides, fatty acids	Acid	Fatty acids, glucose
Major Physiological Actions				
Stomach				
Acid secretion	Stimulates		Inhibits	Inhibits
Gastric emptying	Stimulates		Inhibits	Inhibits
Mucosal growth	Stimulates			
Pancreas				
Enzyme secretion		Stimulates		
Bicarbonate secretion			Stimulates	
Growth		Stimulates		
Biliary system				
Liver bicarbonate secretion			Stimulates	
Gallbladder contraction		Stimulates		
Sphincter of Oddi		Relaxes		
Small intestine				
Ileal motility	Stimulates			
Ileocecal sphincter	Relaxes			
Mucosal growth	Stimulates			
Large intestine				
Motility	Stimulates			

duction of a bolus of food into the mouth produces a reflex inhibition of the jaw-closing muscles and a reflex contraction of the jaw-opening muscles. The mouth opens, and the resulting reduction in pressure causes a reflex contraction of closing muscles and relaxation of opening muscles. This sequence of muscular activity continues rhythmically until the bolus of food has been chewed.

A major function of chewing is to reduce a mouthful of food to particles of a size convenient for swallowing. Although this varies from one person to another, the extent to which food is chewed has no effect on digestion and absorption. The maximal force that can be exerted during chewing is much greater than that required for the chewing of ordinary food. More important in the grinding of food is the occlusive contact area between the molars and premolars. Another function of chewing is to move food around the oral cavity and in so doing to stimulate receptors for taste and smell. This is important because much of the satisfaction of eating is derived from the perception of these sensations. In addition, stimulation of these receptors promotes secretion of saliva, which softens and lubricates the bolus of food.

Secretion of Saliva

The acinar cells of the parotid, submaxillary, and sublingual glands combine to secrete between 1 to 2 liters of **saliva** every day. The resting rate of secretion is about 0.5 ml per minute. When stimulation is maximal (by acid solutions), the secretory rate increases to about 4 ml per minute. The parotid gland contains mostly **serous cells,** which secrete a watery solution containing inorganic ions and **amylase,** a digestive enzyme that helps break down starch.

In addition to serous cells, the submaxillary and sublingual glands contain **mucous cells,** which secrete **mucin.** When mucin mixes with the watery secretion of the serous cells, a solution of high viscosity called **mucus** is produced, and this gives a thick viscous characteristic to certain types of saliva.

The composition of the saliva that is present in the mouth is variable because it is a mixture of serous and mucous cell secretions. The actual composition of saliva depends on the extent to which each cell type is stimulated to secrete, and this in turn depends on the nature of the stimulus and thus the function that is to be performed. For example, if a dog is given fresh meat, saliva that contains much mucus is secreted. This secretion lubricates the food mass and facilitates its passage into the stomach. If the same dog is given dry meat powder, a watery secretion is produced that washes the powder from the mouth.

The secretion of saliva is regulated by long reflexes. Afferent impulses originate in chemoreceptors in the mouth and nose when these receptors are stimulated by the taste and smell of food. Another sensory pathway is through stimulation of oral mechanoreceptors by food. In all instances, impulses are sent to salivary centers in the medulla, which in turn relay impulses over the two divisions of the autonomic nervous system to the various salivary glands, causing these to secrete their specific juices. Both parasympathetic cholinergic and sympathetic adrenergic stimulation *excite* salivary secretion, in contrast to their usual opposing actions on other effector organs of the body. However, the parasympathetic nerves are the most important regulators of salivary secretion, because the mouth is dry when these nerves are nonfunctional.

Salivary secretion serves a number of functions. Saliva softens and lubricates food, which facilitates swallowing. The moistening properties of saliva facilitate chewing and speech. Saliva contains a high concentration of bicarbonate ion. This alkaline (basic) ion neutralizes acids produced by oral bacteria, thereby preventing these acids from dissolving the enamel of teeth. In addition, the decreased secretion of saliva that occurs during dehydration contributes to the sensation of thirst.

The digestive enzyme salivary amylase is synthesized by the serous cells and stored in these cells as vesicles of enzymes called **zymogen granules.** This enzyme (Fig. 23–5) degrades complex carbohydrates such as starch. Starch is a **polysaccharide** consisting of many branched and straight chains of the monosaccharide glucose. The major end-products of this enzyme are **oligosaccharides,** molecular fragments consisting of two to nine molecules of glucose. The major oligosaccharide that is produced is the disaccharide maltose, which is a combination of two glucose molecules. Salivary amylase requires for optimal activity a pH close to neutrality (7), as is found in the mouth. Although food remains in the mouth for only a short time, and although the contents of the stomach are highly acidic, a bolus of food remains intact for a while after entering the stomach, so that zymogen digestion can continue inside the bolus for a period of time.

Swallowing

Once a mouthful of food is chewed and mixed with saliva, it is propelled rapidly from the mouth through the pharynx and esophagus into the stom-

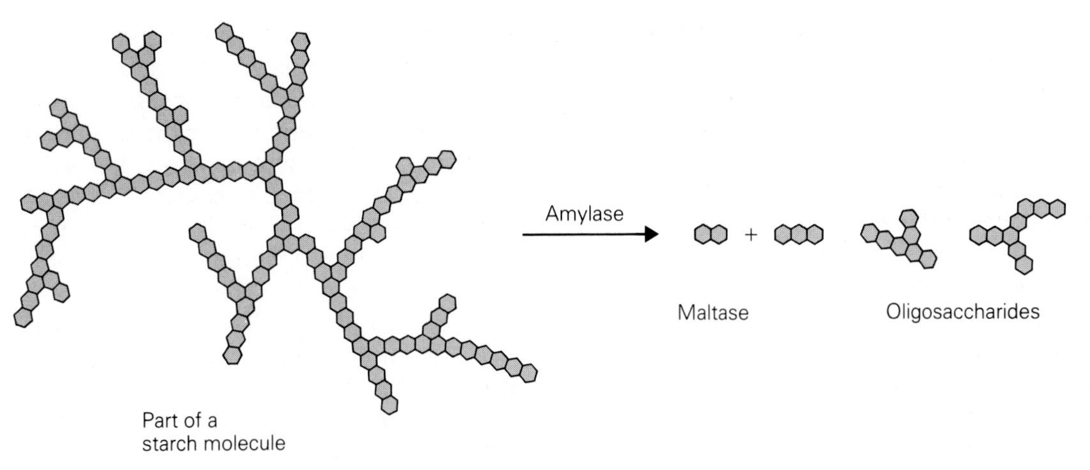

Figure 23–5 Structure of part of a starch molecule and the action of amylase.

ach. The food bolus moves in response to skeletal and smooth muscle contractions that occur sequentially from above to below. These muscular movements are coordinated mainly by long reflexes mediated through a swallowing center in the brain.

Swallowing Movements

Swallowing Movements in the Mouth and Pharynx. A bolus of food moves from the mouth through the pharynx into the esophagus (Fig. 23–6) in about 1 second. The movement is accomplished by the coordinated contractions of a number of skeletal muscles.

The front of the tongue is pushed up against the hard palate, and the food mass, which can be either liquid or solid, is rolled toward the back of the tongue. The bolus is then forced into the pharynx by the contraction of skeletal muscles in the back of the throat. Respiration is inhibited briefly. The soft palate is raised and closes the airways to the nose. Food is prevented from entering the trachea by the elevation of the larynx, the region between the base of the tongue and the trachea, and by the approximation of the vocal cords, whose combined actions close the glottis (the opening of the vocal cords). An auxilliary mechanism to prevent food from entering the respiratory passages is that, as food passes over the epiglottis, this lidlike structure is pressed down over the closed glottis. As these openings close, the constrictor muscles of the pharynx contract and force the bolus into the esophagus.

The Role of the Esophagus in Swallowing. The esophagus, a muscular tube located mostly in the thorax, conducts material from the pharynx to the stomach. The major anatomical features of this organ (Fig. 23–7*a*) are the **upper esophageal sphinc-**

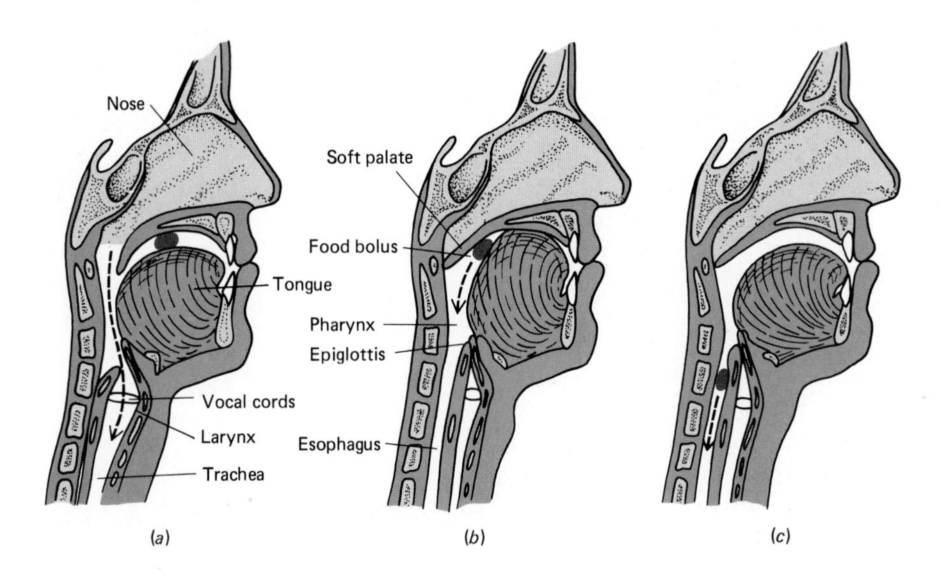

Figure 23–6 Passage of a bolus of food from the mouth through the pharynx into the upper esophagus during a swallow.

ter (a band of skeletal muscle); the **body** of the esophagus (the upper third of which is skeletal muscle and the lower two thirds of which are smooth muscle); and the smooth muscle **lower esophageal sphincter,** the last portion of which lies below the diaphragm in the abdominal cavity.

In order to understand esophageal function, it is necessary to know what the pressures are in the lumen of the different regions of the esophagus both at rest and during a swallow. It is also important to know how these pressures relate to those in the pharynx at the upper end of the esophagus and in the stomach below, because it is the existence and direction of pressure gradients that determine not only whether contents will move through the esophagus but in which direction they will move.

When the esophagus is at rest (see Fig. 23–7*a*), pressure in the lumen of the body of the esophagus is the same as intrathoracic pressure, which is subatmospheric (see Chapter 20). On the other hand, pressures in the upper and lower esophageal sphincters are above atmospheric pressure, indicating that these sphincters are contracted when there is no swallowing activity. Since pressure in the resting pharynx is atmospheric, a pressure gradient would exist between pharynx and body of the esophagus if it were not for the zone of high pressure that results from the contraction of the upper esophageal sphincter. If this pressure gradient were allowed to manifest itself, large volumes of air would flow from pharynx to esophagus during the normal course of breathing, and much of this air would be swallowed. The above-atmospheric pressure produced by closure of the upper sphincter prevents this from happening.

At the lower end, intragastric pressure (the pressure inside the stomach) is greater than atmospheric pressure, but less than that of the contracted lower sphincter. Consequently, gastric (stomach) contents do not ordinarily reflux into the esophagus.

However, reflux is common despite the barrier imposed by the lower sphincter, and because the contents of the stomach are quite frequently acid, the esophagus is irritated, producing the unpleasant sensation known as **heartburn.** In some circum-

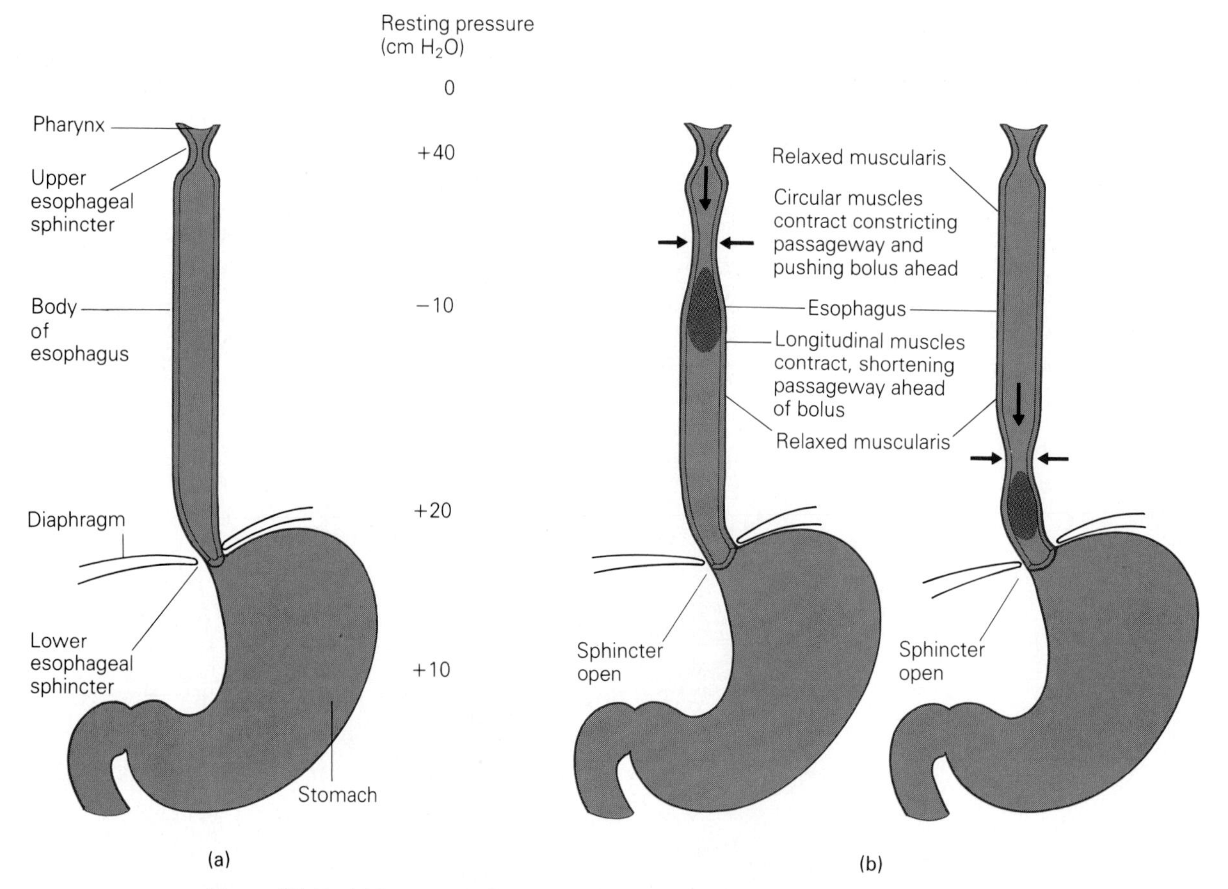

Figure 23–7 (*a*) Representative resting intraluminal pressures in the pharynx, upper esophageal sphincter, body of the esophagus, lower esophageal sphincter, and stomach. (*b*) Movement of a bolus of food through the esophagus by a peristaltic contraction.

stances, such as bending or stooping, when the abdominal muscles contract forcibly, and when the diaphragm descends during deep inspirations, intra-abdominal and thus intragastric pressure increase greatly. However, reflux does not ordinarily occur in these circumstances because the increased intra-abdominal pressure is transmitted to the same extent to the stomach and to that portion of the lower esophageal sphincter that lies below the diaphragm in the abdominal cavity—that is, the same pressure barrier as before is maintained. If for any reason the subdiaphragmatic lower esophageal sphincter (and the top of the stomach) is displaced into the thoracic cavity, a condition known as **hiatus hernia,** the terminal segment of this sphincter can no longer be assisted by changes in intra-abdominal pressure, and reflux can occur.

Soon after a swallow is initiated, pressure in the pharynx increases as a result of the skeletal muscle contraction that occurs here. Also, pressures in the upper and lower sphincters abruptly drop to atmospheric pressure, showing that these sphincters have relaxed. A bolus of food moves from the pharynx through the relaxed upper sphincter into the esophagus in response to the downward pressure gradient that now exists through these regions.

Peristalsis in the Esophagus. After the bolus enters the esophagus, the upper sphincter closes, the glottis opens, and respiration resumes. Then pressure rises sequentially from above to below in the body of the esophagus. This sequence of pressure changes is due to the fact that a wave of contraction moves down the length of the esophagus. The contractile wave propels a semisolid food mass ahead of it (see Fig. 23–7*b*) and into the stomach through the previously relaxed lower sphincter. This type of propulsive activity, which is characteristic of most regions of the digestive tract, is called **peristalsis.** The esophageal peristaltic wave travels at a rate of 2 to 4 cm per second and takes about 9 seconds to traverse the esophagus. If a bolus is liquid, it is shot through the esophagus by the initial force of swallowing and travels by force of gravity to the stomach in about 1 second. Once the bolus enters the stomach, the lower sphincter contracts, preventing reflux of gastric contents into the esophagus.

The Swallowing Reflex

The all-or-none swallowing reflex is activated when a bolus of food stimulates mechanoreceptors in the pharynx. Afferent impulses are sent to a swallowing center in the medulla. This center in turn sends out, in the proper sequence, efferent impulses over somatic nerves to the skeletal muscles and autonomic nerves to the smooth muscles that are involved in the swallowing response (Fig. 23–8). Motor impulses that regulate the activity of the skeletal muscles of the pharynx, larynx, upper esophageal sphincter, and upper body of the esophagus travel over various cranial somatic nerves. The esophagus is innervated primarily by the **vagus nerves.** The vagal efferent nerve fibers that supply the upper esophageal sphincter and upper body of the esophagus terminate at skeletal muscle motor end-plates and, thus, are somatic (and not autonomic) nerves. The vagal efferent nerve fibers that coordinate both the orderly progress of the peristaltic wave through the smooth muscle portion of the lower esophagus and the activity of the lower esophageal sphincter are autonomic nerves. These vagal preganglionic fibers synapse with cholinergic fibers of the internal nerve plexuses in the smooth muscle wall, and it is

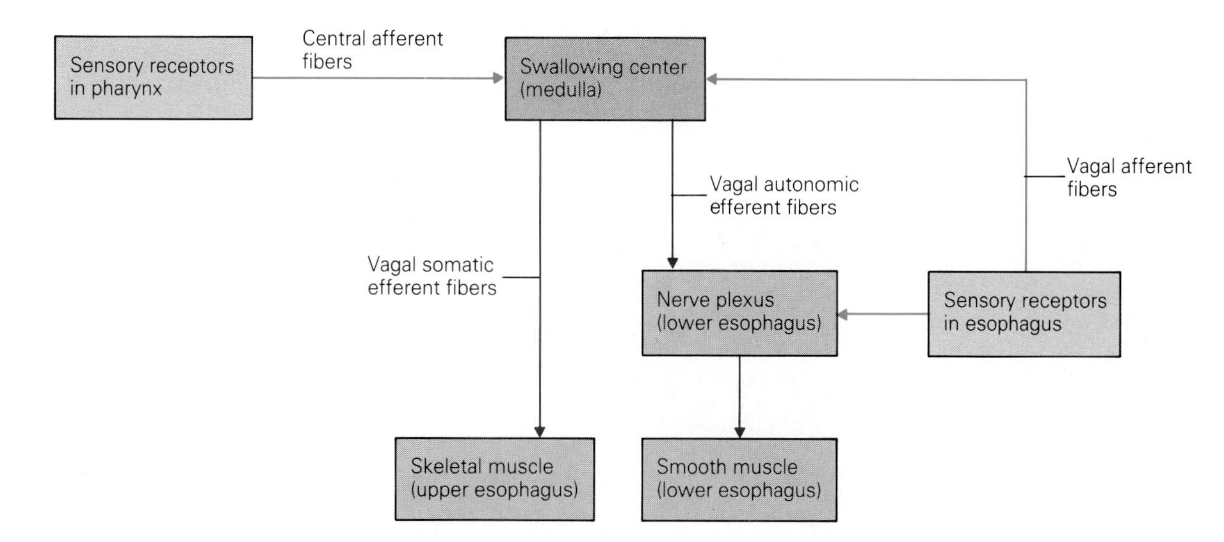

Figure 23–8 Neural pathways for the regulation of esophageal motor activity.

through these nerve networks that the vagus nerves manifest their motor effects. Some people lack the esophageal nerve plexuses. In such cases, the lower esophageal sphincter fails to relax following a swallow, and the resting tone of this sphincter is higher than normal. As a result, food has difficulty entering the stomach and tends to accumulate above the sphincter, causing the esophagus to dilate. This condition is known as **achalasia.**

The peristalsis that follows a conscious effort to swallow is **primary peristalsis. Secondary peristalsis** follows stretch of the esophageal wall without being preceded by a swallowing movement. This type of peristalsis is mediated by long reflexes over the vagus nerves from the esophagus to the swallowing center and back to the esophagus, and also by short reflexes in the esophageal wall (see Fig. 23–8). Secondary peristalsis is important because it facilitates the removal of any food that remains in the esophagus following the passage of a primary peristaltic wave.

Any one of a variety of disorders associated with the nerves and muscles involved in swallowing create difficulty in swallowing, a condition known as **dysphagia.** One example is poliomyelitis, which paralyzes certain cranial nerves that are a part of the swallowing reflex. Another is myasthenia gravis, a condition in which certain skeletal muscles involved in the swallowing response can contract only weakly.

The Processes of Motility and Secretion in the Stomach

The stomach (Fig. 23–9), located between the esophagus and small intestine, functions as a storage organ for food. Without the stomach, it would be necessary to consume many small meals every day rather than a few large meals in order to meet the energy requirements of the body. In addition to storing food, the stomach secretes HCl and the enzyme **pepsin,** both of which contribute to the digestion of food. The digestive activity, along with the muscular mixing movements of the stomach, converts large, solid food particles into a liquid suspension of finely divided particles. The small intestine can best perform its digestive and absorptive functions when presented with contents in this form. Intestinal digestion and absorption is also optimized by regulatory mechanisms that ensure that the musculature of the stomach discharges gastric contents into the small intestine slowly—that is, the small

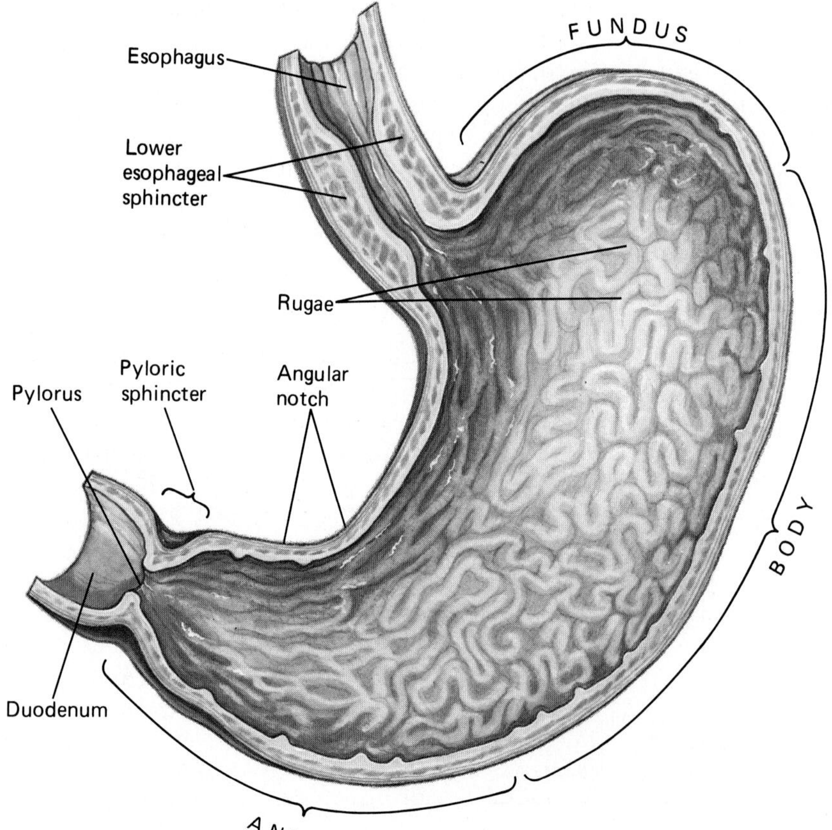

Figure 23–9 Anatomy of the stomach.

intestine is not overwhelmed by a flood of contents from above.

Motility of the Stomach

The motor activity of the stomach serves the following major functions: (1) accommodation of a meal with a relatively small increase in intragastric pressure, (2) mixing of food with gastric (stomach) secretions so that digestion can begin, (3) reduction of the size of food particles, and (4) emptying of gastric contents into the duodenum (the upper region of the small intestine).

Muscular Movements of the Stomach

Plasticity and Receptive Relaxation. The storage function of the stomach is served by the smooth muscle of the **body** and **fundus** of the stomach. These regions of the stomach can adapt to the volume of contents they contain in that, as the stomach fills to capacity, intragastric pressure increases only slightly (Fig. 23–10). A property of smooth muscle that allows the stomach to accommodate large volumes with minimal changes in intragastric pressure is its ability to adjust its resting length over a wide range without large changes in resting tension (see Chapter 16). This property of **plasticity** is passive in the sense that it does not depend on neural or hormonal influences. However, a neural mechanism does contribute to the volume adaptation. As a meal is eaten, with each swallow the body and fundus of the stomach relax slightly. This **receptive relaxation** is mediated by a long reflex over the vagus nerves.

Peristalsis in the Stomach. Peristaltic contractions of the stomach mix food with gastric juice, reduce food particle size, and expel gastric contents into the duodenum. When the stomach contains food, peristaltic contractions travel over the stomach from above to below at a frequency of about three per minute. These contractions are much more vigorous in the **antrum** than in the body and fundus because antral muscle is so much thicker than elsewhere. Consequently, it is primarily the muscular activity of the antrum that generates the forces responsible for emptying the stomach. The driving force behind the emptying process is the pressure differential that develops between the antrum and the upper small intestine as a peristaltic wave travels over the antrum. At this time, pressure in the antrum rises momentarily above that in the duodenum, and a few milliliters of stomach contents move into the upper small intestine through the open **pyloric sphincter,** the thick bundle of circular smooth muscle and connective tissue separating the stomach from the small intestine. As the peristaltic wave passes over the pyloric sphincter, this structure contracts and offers enough resistance to prevent further evacuation.

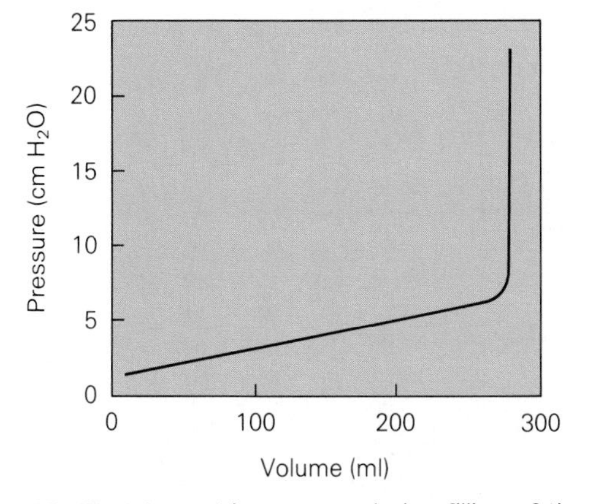

Figure 23–10 Intragastric pressure during filling of the stomach of a rabbit.

Retropulsion. Another consequence of closure of the pyloric sphincter is **retropulsion,** in which stomach contents are forcibly squirted back into the body of the stomach. Retropulsion achieves a very effective mixing of food with gastric juice and also breaks lumps of food into small particles. The process of evacuation and retropulsion continues about three times a minute until the stomach is empty (Fig. 23–11).

Regulation of Stomach Motility

The Basic Electrical Rhythm of the Stomach. Peristalsis occurs at the rate of three contractions per minute because cycles of spontaneous depolarization and repolarization originate at this rate in muscle pacemaker cells at the top of the stomach. Because this activity is continuous, it has been termed the **basic electrical rhythm** of the stomach. These depolarization-repolarization cycles, which have amplitudes of 10 to 15 millivolts and durations of 1 to 4 seconds (Fig. 23–12), travel as **slow waves** through the gastric muscle from above to below as a ring around the stomach. In the absence of excitatory stimuli, the depolarizations are too small to cause the muscle membrane to reach threshold and action potentials to be fired to elicit contraction. When excitatory influences (such as release of acetylcholine by nerve fibers or the release of gastrin from the antral mucosa) are sufficient to produce action potentials, these occur at the peak of the slow waves because muscle excitability is closest to

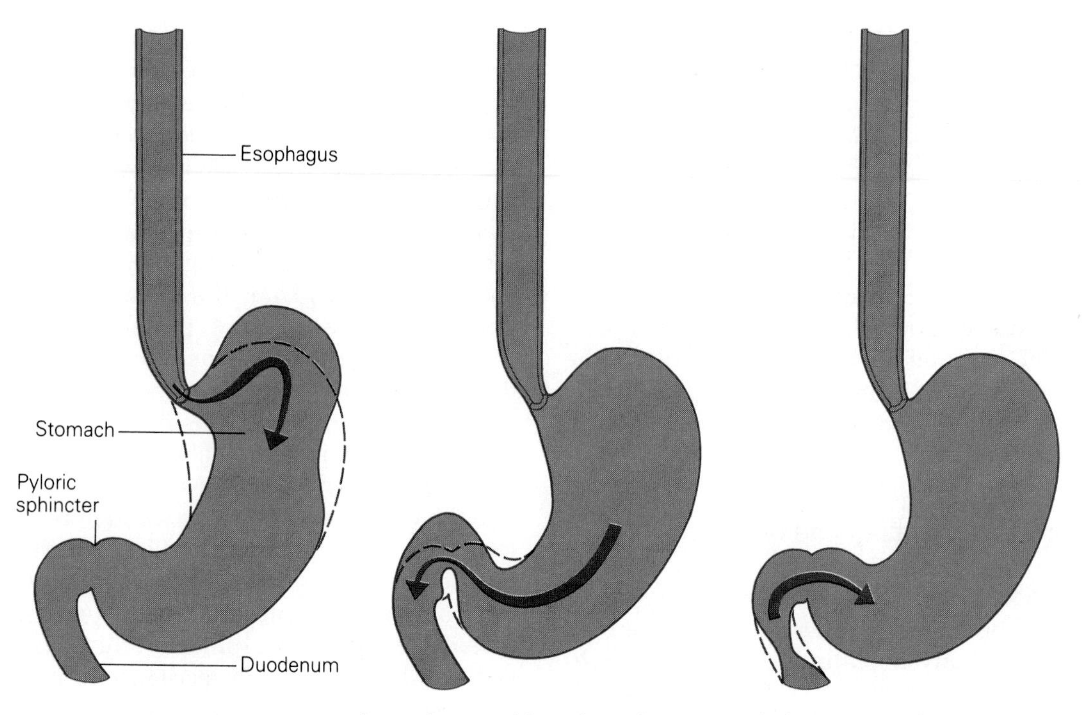

Figure 23–11 Evacuation and retropulsion of gastric contents during passage of a peristaltic wave down the stomach. (Modified from Vander, Human Physiology, McGraw-Hill, 1985, p. 488.)

threshold at this time. The intensities of excitatory stimuli do not alter the frequency of the basic electrical rhythm and thus do not alter the frequency of peristalsis. On the other hand, intensity of stimulation does affect the magnitude of contraction: that is, as intensity of stimulation increases, more action potentials are fired and contraction is more vigorous (see Fig. 23–12).

Neural and Hormonal Control. The rate at which the stomach empties is the result of the interplay between mechanisms that either excite or inhibit contraction of gastric smooth muscle. Both neural and hormonal mechanisms are involved in regulating the rate of gastric emptying.

The mechanisms that increase the vigor of gastric contractions originate in the stomach (Fig. 23–13). When the stomach is distended with food, stimulation of mechanoreceptors in the gastric wall activates short reflexes in the wall and long reflexes over the vagus nerves. Both types of reflex activity result in increased gastric motor activity. The greater the distention, the more impulses are fired and the more vigorous is gastric peristalsis. Peristalsis is also augmented by the hormone **gastrin,** which is released from the antrum by intraluminar

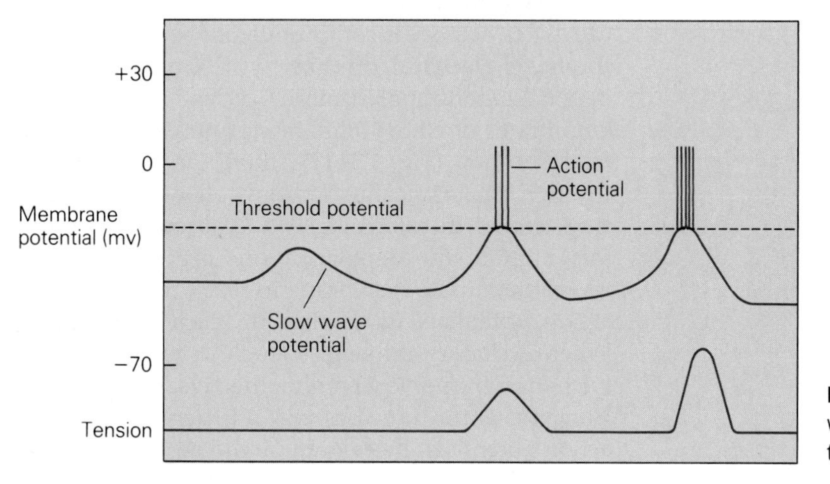

Figure 23–12 Relationship between slow waves, action potentials, and generation of tension in gastric smooth muscle.

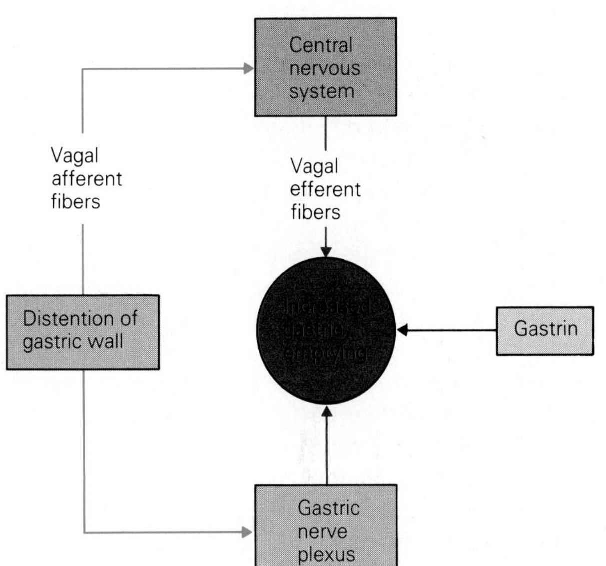

Figure 23–13 Excitation of gastric emptying.

stimuli that are present during the digestion of a meal.

The excitatory influences that originate in the stomach are kept in check by inhibitory mechanisms that originate in the duodenum (Fig. 23–14*a, b*). These duodenal inhibitory mechanisms prevent the upper small intestine from being overwhelmed by material from the stomach. Gastric emptying is retarded through signals from the duodenum when the contents that enter this region are either high in fat, acidic (having a pH less than 3.5), hyperosmotic, or bulky enough to stretch the duodenal wall. The duodenal mucosa, when contacted by acid, releases the hormone **secretin** into the blood. This hormone is carried to the stomach, where it inhibits motor activity. **Gastric inhibitory peptide,** another hormone that inhibits gastric contractions, is released from the duodenal mucosa by the products of fat digestion. In addition, certain endocrine cells may respond to the intraluminal stimuli just listed by releasing as-yet-unidentified hormones that inhibit gastric motility. A variety of duodenal nerve receptors, such as osmoreceptors, mechanoreceptors, and chemoreceptors respond to these intraluminal stimuli to initiate reflex inhibition of gastric motility. These neural mechanisms, collectively called the **enterogastric inhibitory reflex,** include long reflexes over external nerves and short reflexes through the internal nerve plexuses in the duodenal

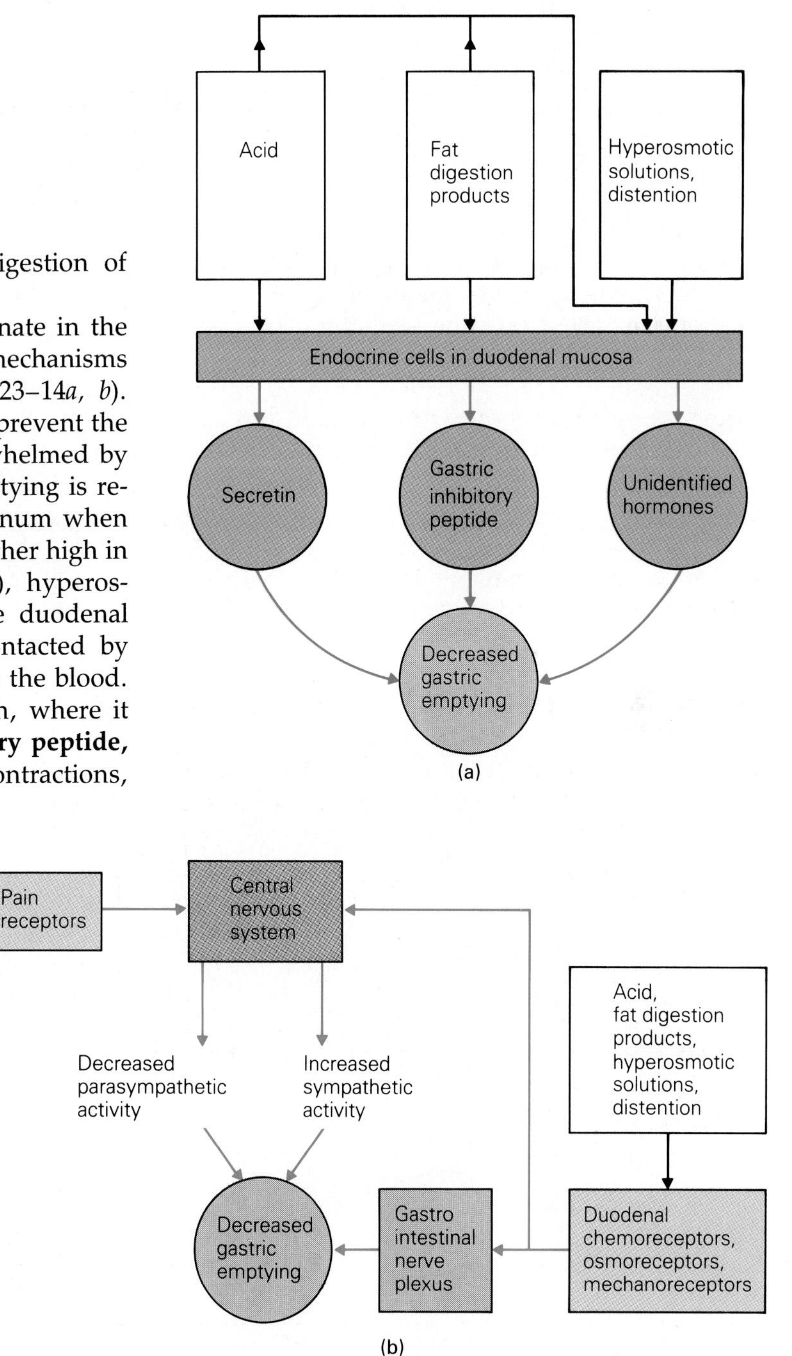

Figure 23–14 Inhibition of gastric emptying. (*a*) Hormonal pathways. (*b*) Neural pathways.

and gastric walls. In general, long reflex inhibition occurs when parasympathetic (vagal) activity to the stomach is decreased or when sympathetic activity is increased. The reverse is true for excitation of gastric motility.

Although control of gastric movements largely centers on mechanisms that originate in the stomach and duodenum, gastric motility may be influenced by long reflexes initiated from any sensory region of the body. For example, stimulation of visceral and somatic pain receptors inhibits gastric motility (Fig. 23–14*b*). In addition, various emotional states, such as anger, fear, and depression, produce changes in gastric motor activity, although the direction of these changes is not always predictable.

Secretion of Gastric Juice

The stomach secretes about 2 to 3 liters of gastric juice every day. Most of this juice is produced by exocrine glands located in the body and fundus of the stomach. These exocrine secretory glands contain three types of secretory cells: chief cells, parietal cells, and mucous cells (Fig. 23–15).

The **chief cells,** which are located in the basal regions of these glands and also in the glands of the antrum, secrete **pepsinogen.** Pepsinogen is the inactive precursor of **pepsin,** an enzyme that initiates the digestion of protein in the digestive tract.

The **parietal cells** secrete the strong (highly dissociated) acid **hydrochloric acid (HCl).** Hydrochloric acid, in addition to activating pepsin, also provides an optimal pH for the activity of this enzyme. This acid contributes to the breakdown of muscle fibers and connective tissue and kills bacteria that enter the digestive tract through the mouth. Hydrochloric acid is the major factor in the formation of ulcers in the upper digestive tract. The parietal cells also secrete **intrinsic factor.** This substance is necessary for the absorption of vitamin B_{12}, which is required for formation of normal red blood cells (Chapter 17).

Mucous cells are located at the necks of these glands, in the glands of the antrum, and scattered throughout the gastric epithelial lining. The alkaline, mucus-containing fluid that these cells secrete

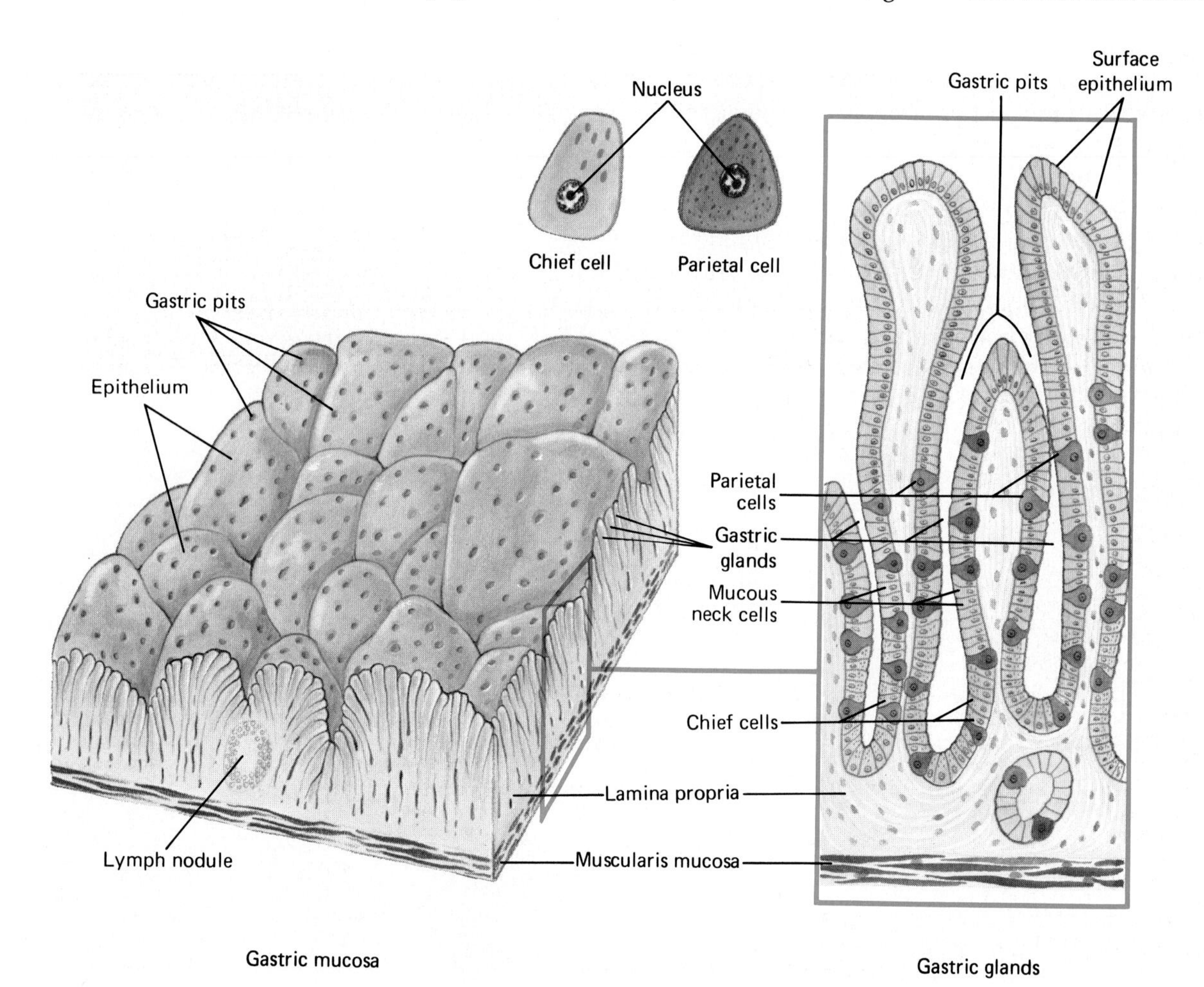

Figure 23–15 Exocrine gland from the body of the stomach.

fluid material or, more quantitatively, an output of fecal water greater than 500 ml per day. This condition results from poor absorption of fluid or increased fluid secretion. Either or both can occur when the intestines are subject to bacterial and viral infections. In such cases, larger-than-normal quantities of digestive contents are present in the large intestine, and the excessive distention—and thus increased motor activity—that results is the primary cause of the diarrhea. Severe or prolonged diarrhea can result in the loss of significant quantities of water and inorganic ions. One consequence is the depletion of blood volume. In addition, since the secretions of the lower digestive tract have a relatively high concentration of sodium bicarbonate, a possible result of the loss of this alkaline ion is **metabolic acidosis** (Chapter 26).

Appendicitis is inflammation of the appendix, a small fingerlike, blind tube that extends from the cecum. This condition is preceded by obstruction of the lumen of the appendix; the infection that usually follows may result in edema, ischemia, gangrene, and perforation. Appendicitis begins with referred pain in the umbilical region, followed by nausea and vomiting. After several hours, the pain becomes severe and is localized to the right lower quadrant of the abdomen. Early removal of the appendix is recommended in all suspected cases of appendicitis.

Summary

I. A. The major function of the gastrointestinal system is to process food and, by doing so, provide nutrients to the cells of the body. A major component of the gastrointestinal system is the digestive tract, which consists of the mouth, pharynx, esophagus, stomach, and small and large intestines. Accessory structures connected to the digestive tract by ducts are the salivary glands, pancreas, liver, and gallbladder.

Blood flows to and from the gastrointestinal system by way of the splanchnic circulation. Blood that leaves the stomach, intestines, and pancreas goes to the liver by way of the portal vein (portal circulation) before entering the central venous system.

B. In the process of motility, coordinated contractions of muscle, both skeletal and smooth, move ingested material through the digestive tract and also mix this material with the juices of the exocrine glands that are secreted en route. In the process of secretion, the various exocrine glands contain a number of types of secretory cells, each of which produces a characteristic secretion. Some of these secretions consist entirely of water and inorganic ions, whereas others also contain organic compounds, including the digestive enzymes and the bile salts. In the process of digestion, the digestive enzymes catalyze the conversion of large organic molecules of the diet to smaller molecules that are readily absorbed. In the process of absorption, substances travel from lumen to blood and lymph mainly from the small intestine by means of a variety of both active and passive transport processes.

C. Neural mechanisms for the control of gastrointestinal function include both short reflexes over nerve cells that lie entirely in the wall of the digestive tract and long reflexes over external nerves, the most important external nerve being the vagus nerve. The polypeptide hormones gastrin, secretin, cholecystokinin, and gastric inhibitory peptide also regulate gastrointestinal function.

II. A. Chewing involves the coordinated contraction and relaxation of a variety of skeletal muscles to reduce the size of food particles and to mix them with saliva.

B. The long reflex secretion of saliva that accompanies chewing provides salivary amylase, an enzyme that initiates the digestion of starch, the major dietary carbohydrate.

C. Between swallows, sphincters at either end of the esophagus impede the entrance of air into the esophagus from above and of gastric contents from below. A swallow moves food rapidly from the mouth through the pharynx and esophagus to the stomach in response to the pressure gradients generated by the muscular contractions that occur sequentially from above to below. The characteristic propulsive muscular activity of the esophagus is peristalsis, a moving wave of contraction.

D. Coordination of swallowing is by long reflexes mediated through a swallowing center in the brain and short reflexes in the wall of the esophagus.

III. A. Receptive relaxation of the body and fundus, the storage regions of the stomach, allows large volumes of ingested material to be accommodated with only small increases in intragastric pressure. Peristaltic contractions of the pyloric antrum generate a pressure difference between antrum and duodenum, which is the driving force that empties the stomach.

The frequency of peristalsis is governed by the basic electrical rhythm of the stomach, a small wave of depolarization that is conducted down the gastric musculature about three times every minute. The strength of gastric peristalsis depends on a balance between the excitatory and inhibitory mechanisms that influence the gastric musculature at any given time. Excitatory mechanisms are mediated primarily through short reflexes in the gastric wall and long reflexes over the vagus nerves, both of which are triggered in response to stomach distention. Chemical and physical stimuli in the material that enters the duodenum from the stomach inhibit gastric motor activity. These stimuli produce inhibition by releasing *secretin, gastric inhibitory peptide,* and unidentified hormones from the duodenal mucosa, and through short and long enterogastric reflexes.

B. The major secretory components of gastric juice are hydrochloric acid and intrinsic factor, secreted by the parietal cells, and pepsinogen, secreted by the chief cells. HCl is secreted by active transport. When food is eaten, short and long reflexes triggered by stimulation of receptors in the head and stomach cause the release of acetylcholine by nerve fibers in the acid-secreting portion of the stomach and the release of gastrin from endocrine cells in the pyloric antrum. Both acetylcholine and gastrin excite the secretion of HCl by the parietal cell. Inhibition of HCl secretion is mediated by suppression of gastrin release by HCl and by neural and hormonal inhibitory mechanisms that originate in the duodenum. Mechanisms that protect the upper digestive tract from the corrosive actions of HCl include the impermeability of gastric cell membranes to HCl, dilution and neutralization of acid, and rapid cellular regeneration. The chief cells secrete inactive pepsinogen, which is activated by HCl to active pepsin, an enzyme that initiates the digestion of protein. The most effective stimulus for pepsinogen secretion is the acetylcholine that is released at the chief cell by the short and long reflexes that occur following the eating of food.

IV. A. The intercalated duct cells of the pancreas secrete a large volume of a water-electrolyte juice with a relatively high concentration of sodium bicarbonate. The sodium bicarbonate of the aqueous juice is transported actively by the duct cells and is largely responsible for the neutralization of the HCl in gastric contents.

B. Pancreatic acinar cells secrete a small volume of juice containing a wide variety of digestive enzymes whose combined actions almost complete the digestion of food.

C. Excitation of secretion of pancreatic bicarbonate is largely by secretin released in response to acid stimulation of the duodenal mucosa. Excitation of secretion of enzymes is largely by cholecystokinin released in response to stimulation of the duodenal mucosa by fat and protein digestion products.

V. A. Between meals, bile secreted by the liver is stored in the gallbaldder where the organic constituents of bile are concentrated by the transport of salt and water from lumen to blood. Contraction of the gallbladder and relaxation of the sphincter of Oddi results in the flow of bile into the upper small intestine in response to the release from the duodenal mucosa of cholecystokinin by fat and protein digestion products.

B. Bile salts, the major secretory components of bile, are synthesized and actively secreted by the cells of the liver. Following transit through the small intestine where they perform their function in the digestion and absorption of fat, most of the bile salts are actively absorbed from the terminal ileum into the portal circulation, which returns them to the liver for resecretion (enterohepatic circulation of bile salts). The bile pigments are degradation products of hemoglobin metabolism and are excreted largely in the feces.

C. Circulating bile salts are the most potent stimulus for secretion of bile by the cells of the liver. The bile ducts respond to secretin stimulation by secreting a water-electrolyte fluid that has a high bicarbonate content.

VI. A. Segmenting contractions mix the contents of the small intestine, thereby facilitating digestion and absorption; peristalsis propels unabsorbed material into the large intestine. Short and long reflexes are largely responsible for regulating small intestinal motility.

B. Fluid enters the lumen of the small intestine as a result of the osmotic shift of fluid from blood to lumen in response to the presence of hyperosmotic contents in the lumen. Mucosal exocrine glands of the small intestine are an additional source of fluid.

C. The major process involved in the absorption of water from the small and large intestines is the active transport of sodium, which carries water along passively from lumen to blood. Iron and calcium absorption are regulated processes mediated by special transport systems.

Digestion of carbohydrate to monosaccharides and protein to amino acids is completed by enzymes in the intestinal brush border membrane. Most monosaccharides and amino acids are absorbed into the portal circulation by sodium-dependent active transport systems.

Water-insoluble monoglycerides and free fatty acids, the end products of pancreatic lipase digestion of triacylglycerol fats, interact with bile salts to form water-soluble aggregates called micelles. Monoglycerides and free fatty acids diffuse passively from monomolecular dispersion into the intestinal epithelial cell, where they are resynthesized to triacylglycerols and passed on to the lymph as chylomicrons.

VII. A. A firm fecal mass is formed by the absorption of water from the large intestine; this dehydrated material is stored for variable periods of time before defecation occurs. Products of the metabolic activities of the bacteria that reside in the lumen of the large intestine include certain vitamins and flatus. Mucus, bicarbonate, and potassium are secreted into the lumen by the colonic mucosa.

B. Segmenting contractions promote absorption of inorganic ions and water, and propulsive *mass movements* move contents into the distal large intestine at infrequent intervals. Distention of the rectum elicits the defecation reflex, which consists of short reflexes and long reflexes through a sacral region of the spinal cord. Although the muscular movements of the distal colon usually suffice to move feces to the outside of the body, contraction of certain skeletal muscles (straining movements) can supply a considerable amount of force in defecation. The accumulation of feces in the large intestine is known as constipation. Diarrhea is the frequent defecation of highly fluid materials.

Review Questions

1. What are the neural and the hormonal pathways that regulate the secretory and motor activity of the gastrointestinal system?
2. What functions are performed by the processes of chewing and salivary secretion?
3. What are the approximate pressures in the pharynx, various regions of the esophagus, and stomach, both at rest and after swallowing? Relate these pressures to muscular activity in each region.
4. How does the stomach adapt to the wide range of volumes it contains with only small increases in intragastric pressure?
5. What gastric and duodenal stimuli are of major importance in regulating the rate at which the stomach empties? Describe the neural and hormonal pathways through which these stimuli manifest their regulatory influences.

6. What is a proposed mechanism by which the gastric parietal cells secrete HCl? Draw a diagram.
7. What are the pathways, including stimuli, involved in the excitation and inhibition of gastric secretion?
8. How is pepsinogen activated to pepsin? What is the digestive action of pepsin?
9. What are the activities of both smooth and skeletal muscles that participate in the act of vomiting?
10. What are the major enzymes of pancreatic juice? Describe the digestive activity of each of these enzymes.
11. What pathways and stimuli are involved in the excitation of secretion of both the aqueous and the enzyme components of pancreatic juice?
12. What are the major components of bile? Explain any differences in the concentrations of the organic components of liver and gallbladder bile.
13. What are the stages in the enterohepatic circulation of bile salts? Draw a digram and indicate the points at which bile salts are synthesized, actively transported, and lost from the body.
14. What are the two major types of motor activity of the small intestine, and what is the function of each?
15. What is the mechanism by which the small and large intestines absorb sodium chloride and water?
16. How is glucose transported across the intestinal epithelial cell?
17. Why are pancreatic lipase and the bile salts essential for the intestinal absorption of triacylglycerols?
18. What are the muscular movements of the distal colon that occur during defecation, and what mechanisms coordinate these movements?

Suggested Readings

Brooks, F. P., ed. *Gastrointestinal Pathophysiology*, 2nd ed. New York: Oxford University Press, 1978.

Davenport, H. W. *A Digest of Digestion*, 2nd ed. Chicago: Year Book Medical Publishers, 1978.

Davenport, H. W. *Physiology of the Digestive Tract*, 5th ed. Chicago: Year Book Medical Publishers, 1982.

Granger, D. N., Barrowman, J. A., and Kvietys, P. R. *Clinical Gastrointestinal Physiology*. Philadelphia: W. B. Saunders Company, 1985.

Johnson, L. R., ed. *Physiology of the Gastrointestinal Tract*. New York: Raven Press, 1981.

Sanford, P. A. *Digestive System Physiology*. Baltimore: University Park Press, 1982.

Sernka, T. J., and Jacobson, E. D. *Gastrointestinal Physiology: The Essentials*, 2nd ed. Baltimore: Williams & Wilkins, 1983.

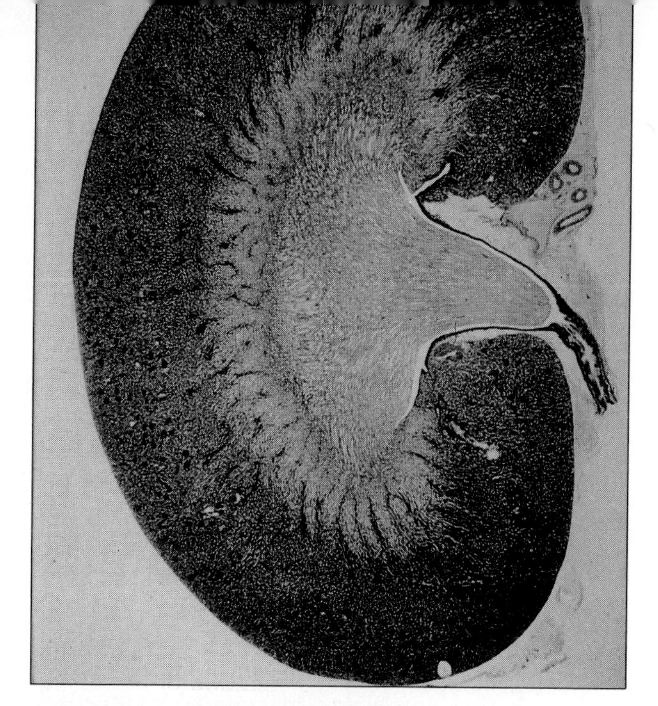

The Kidney

The kidneys play a dominant role in regulating the composition and volume of the extracellular fluids (blood plasma, interstitial fluid, and lymph). They provide a suitable "internal environment" for the body's living cells. The kidneys are commonly thought to simply dispose of waste products in the urine, but they really do much more. This chapter will consider the functional anatomy of the kidneys, renal blood flow, the processes involved in urine formation, and the ways in which the kidneys handle a variety of solutes and water. Chapter 25 will primarily consider how renal (kidney-related) and other mechanisms are controlled and integrated to accomplish the important task of regulating the internal environment. Before considering kidney function in detail, we will outline some of the functions of the kidneys.

1. The kidneys regulate the osmotic pressure of the plasma and other extracellular fluids. Most plasma membranes are highly permeable to water, and so cells have the same osmotic pressure as the extracellular fluid. By stabilizing the osmotic pressure of the extracellular fluid, the kidneys minimize shifts in water between cells and their extracellular environment. In so doing, the kidneys protect the stability of cell volume.
2. By regulating the excretion of sodium and water, the kidneys regulate the volume of the extracellular fluid.
3. The kidneys regulate the individual concentrations of numerous electrolytes in the extra-

cellular fluid, including sodium, potassium, calcium, magnesium, chloride, sulfate, and phosphate ions.

4. The kidneys regulate the plasma bicarbonate concentration and hence the hydrogen ion concentration and so play an important role in acid-base regulation. This topic will be discussed in Chapter 26.
5. The kidneys eliminate metabolic waste products such as urea (an end product of protein metabolism), uric acid (an end product of purine metabolism), and creatinine (an end product of muscle metabolism). They also eliminate many foreign compounds from the body, including drugs such as penicillin.
6. The kidneys produce a number of special substances. These include: **erythropoietin,** a hormone that stimulates the rate of production, maturation, and release of red blood cells from bone marrow; **renin,** a proteolytic enzyme important in the regulation of extracellular fluid volume and blood pressure; **kallikrein,** a proteolytic enzyme that leads to the formation of kinins, which are vasodilators; and various **prostaglandins** and **thromboxane,** fatty acid derivatives that act as local hormones. Prostaglandins E_2 and I_2 have several actions in the kidneys, including vasodilation, enhancement of renal excretion of salt and water, and stimulation of renin release. Thromboxane is a vasoconstrictor that may be responsible for reduced renal blood flow in a number of renal diseases.
7. The kidneys have several special metabolic functions. They are responsible for converting the inactive form of vitamin D to its active form, **1,25-dihydroxy-vitamin D_3.** The latter is a hormone that stimulates intestinal calcium absorption. The kidneys synthesize ammonia from amino acids. This is important in acid-base regulation. The kidneys can synthesize glucose from noncarbohydrate sources (e.g., amino acids), a process called *gluconeogenesis*. During a prolonged fast, the glucose added to the blood by the kidneys helps to maintain the blood sugar concentration. The kidneys are an important site of degradation (hence inactivation) of several polypeptide hormones, including insulin, glucagon, and parathyroid hormone.

Functional Anatomy of the Kidney

The functions of the kidneys can only be understood in terms of their unique anatomy. Figure 24–1 presents an overview of the urinary system and a section through the human kidney. The kidneys are bean-shaped organs that lie in back of the abdominal cavity on either side of the vertebral column. The kidneys are drained by the **ureters,** which carry the urine to the **urinary bladder.** The tube draining the bladder is called the **urethra.** If a kidney is cut (see Fig. 24–1), two parts are easily distinguished: an outer part, called the **cortex,** and an inner part, called the **medulla.** The cortex is reddish-brown and has a granular appearance. The cortex contains **glomeruli, convoluted tubules, cortical collecting ducts,** and associated blood vessels. The medulla has a lighter color and is striated (striped). Striations result from the parallel arrangement of **loops of Henle, medullary collecting ducts,** and blood vessels. The medulla is usually divided into an **outer medulla** (closer to the cortex) and **inner medulla** (further from the cortex).

The human kidney is divided into about a dozen **lobes.** Each consists of a **pyramid** of medullary tissue plus the cortical tissue overlying its base and covering its sides. The apex of a medullary pyramid forms a renal **papilla,** which drains the urine into a **minor calyx.** Minor calices unite to form a **major calyx;** these in turn lead to the expanded **renal pelvis,** which is drained by the ureter. The ureter, renal artery, renal vein, nerves, and lymphatic vessels enter or leave the kidney at the **renal hilum,** a depression of its medial surface.

The Structure of a Nephron

The basic unit of kidney structure and function is the **nephron** (Fig. 24–2). Each human kidney contains about one million nephrons. Each nephron consists of a **renal corpuscle** and its attached **renal tubule.**

The renal corpuscle consists of a tuft of capillaries, the **glomerulus,** surrounded by an expanded, double-walled cup, **Bowman's capsule.** The **urinary space** of Bowman's capsule is continuous with the lumen of the renal tubule. Traditionally, the renal tubule has been divided into several anatomically distinct segments. The first segment is called the **proximal tubule.** This is divided into two parts, the **proximal convoluted tubule,** which coils and twists (convolutes) in the neighborhood of its renal corpuscle, and the **proximal straight tubule,** which plunges toward the renal medulla. The **loop of Henle** is the portion of the nephron between the proximal and distal convoluted tubules. Its first part is the proximal straight tubule. Next is a variable **thin limb** (descending and ascending portions), and finally a **thick ascending limb** (distal straight tubule). This is followed by a short **distal convoluted tubule.** Distal convoluted tubules join **connecting tubules,** which lead to **cortical collecting ducts.**

as serving to clear substances from the blood plasma. Different substances have different clearance rates. The **renal plasma clearance** of a substance is defined as the rate of excretion of a substance divided by its plasma concentration. By convention, the symbol $\mathbf{C_x}$ indicates the clearance of a substance x; $\mathbf{U_x}$ the **urine concentration** of the substance x; $\mathbf{P_x}$ the **plasma concentration** of the substance x; and $\dot{\mathbf{V}}$ the **urine flow rate.** The rate of excretion of a substance is equal to the product of its urine concentration and the urine flow rate, or in symbols, $U_x \times \dot{V}$. From the definition of clearance, we can write the **clearance formula,** $C_x = U_x \times \dot{V}/P_x$. Plasma and urine concentrations of substances are often expressed in terms of milligrams per milliliter, and urine flow is given as milliliters per minute. Substituting these dimensions into the clearance formula, we see that the units of clearance are milliliters of plasma per minute.

$$C_x = \frac{\text{mg } x/\text{ml urine} \times \text{ml urine/minute}}{\text{mg } x/\text{ml plasma}}$$
$$= \text{ml plasma/minute}$$

Renal clearance may be considered as the volume of plasma per unit time that must be completely freed (cleared) of a substance to supply the quantity of a substance excreted in the urine per unit time.

To measure GFR, we need a substance that is cleared from the plasma solely by glomerular filtration. This substance, therefore, should not be reabsorbed or secreted by the kidney tubules. It should not be destroyed, synthesized, or stored by the kidneys. It should pass through the glomerular filtration membrane unhindered; in other words, it should not be too large a molecule or a molecule that is bound to plasma proteins. It should be nontoxic. Finally, it should be possible to measure this substance in plasma and urine using simple analytical methods.

Such an ideal substance is the polysaccharide **inulin,** a fructose polymer, derived from the roots of certain plants; it has an average molecular weight of about 5000. (Do not confuse inulin with the hormone insulin!) Figure 24–9 shows that the amount of inulin (IN) filtered per unit time, $P_{IN} \times \text{GFR}$, equals the amount of inulin excreted per unit time, $U_{IN} \times \dot{V}$. The inulin clearance, C_{IN}, is defined by the equation $U_{IN}\dot{V}/P_{IN}$ and is therefore equal to the GFR.

The way in which inulin is used to measure GFR can be illustrated by an example. An inulin solution is infused intravenously into a person to achieve a constant plasma inulin concentration. A timed urine sample is collected, and average urine flow rate is calculated by dividing the urine volume by the duration of collection. In the middle of the urine collection period, a blood sample is taken. Subsequently, plasma and urine samples are analyzed for inulin. Suppose the following values are found: $P_{IN} = 0.30$ mg/ml, $U_{IN} = 30$ mg/ml, and $\dot{V} = 1.25$ ml/min. Substituting in the inulin clearance equation, $C_{IN} = U_{IN} \times \dot{V}/P_{IN} = \text{GFR}$, we get

$$\text{GFR} = \frac{30 \text{ mg/ml urine} \times 1.25 \text{ ml urine/min}}{0.30 \text{ mg/ml plasma}}$$
$$= 125 \text{ ml plasma/min}$$

You may have noticed that in this example the inulin concentration in the urine is 100 times that in plasma. This is not due to tubular secretion of inulin; rather, it is caused by water reabsorption. As water is reabsorbed by the kidney tubules, filtered inulin is left behind and therefore becomes concentrated in a smaller volume of water. For example, if the inulin contained in 100 ml of filtrate is concentrated in 1 ml of urine, the inulin concentration will be increased 100-fold, and 99 ml of water (or 99% of the filtrate) must have been reabsorbed by the kidney tubules.

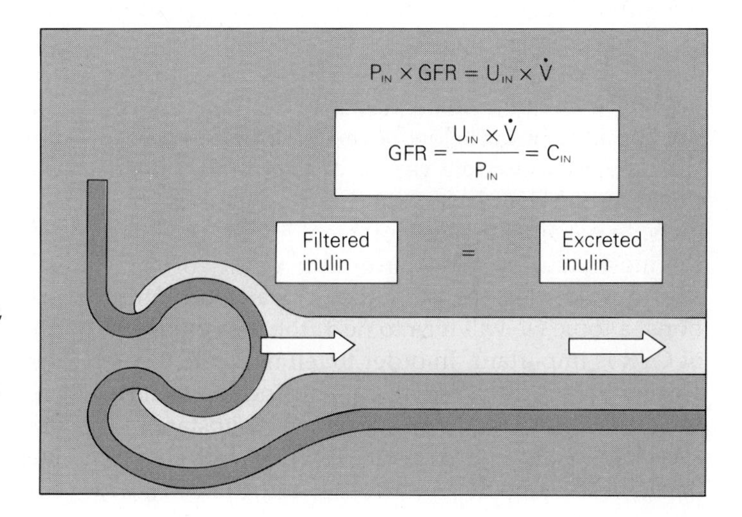

Figure 24–9 Principle behind the measurement of glomerular filtration rate (GFR). Inulin is freely filterable, so the amount filtered per unit time equals its plasma concentration (P_{IN}) multiplied by the GFR. Inulin is neither reabsorbed, secreted, synthesized, destroyed, nor stored by the kidney tubules, so filtered and excreted amounts are equal. Rearranging the equation, we find that GFR is equal to the inulin clearance (C_{IN}).

GFR in people depends on body surface area; it is usually corrected to a standard body surface area of 1.73 square meters. In normal young males, it averages 125 ml/min (180 liters/day), and in normal young females, it averages 115 ml/min (165 liters/day). GFR is very low in the newborn, 20 ml/min per 1.73 square meters of body surface area, and attains adult values (when corrected for body surface area) by about one year of age. In men, GFR is maintained at about 125 ml/min to about 45 to 50 years of age, after which it declines about 13 ml/min per decade.

Inulin is the "gold standard" for measuring GFR, but it is not often used clinically for this purpose. It is somewhat inconvenient to infuse this substance. Instead, the **endogenous creatinine clearance** is used. Creatinine does not have to be infused since it is normally produced in the body. Its plasma levels are normally quite constant, and it can be easily measured in plasma and urine. In people, creatinine is actually secreted in addition to being filtered, but the error introduced by tubular secretion is small, and the endogenous creatinine clearance does provide a clinically valuable estimate of GFR.

If GFR falls, substances produced in the body that depend primarily on glomerular filtration for their elimination from the body (including creatinine and urea) will accumulate, and their plasma concentrations will rise. The extent to which the plasma concentration of creatinine is elevated may serve as an index of the degree of impairment of filtration rate. Likewise, the degree to which blood urea nitrogen (BUN) is elevated may reflect the degree of impairment of filtration rate; however, urea is less well suited for this assessment than is creatinine, because BUN also depends on other factors, such as the rate of protein catabolism in the body and the urine flow rate.

We have devoted considerable attention to the measurement of GFR because it is so important in assessing kidney function. Glomerular filtration is a major process by which the kidneys regulate the volume and composition of the body fluids. A reduction in GFR far below normal will impair a person's ability to excrete various waste products and the proper amounts of water and mineral electrolytes. In patients, the measurement of GFR provides the most valuable indicator of the severity of renal failure. Also, in studying tubule transport functions, a topic we will turn to next, the measurement of GFR is important. In order to tell how much of a substance is reabsorbed, we must first know how much was filtered. To tell whether a substance is secreted by the kidney tubules, we cannot simply measure the amount excreted; we must know how much entered the tubules by glomerular filtration. Measurements of GFR are thus important to the study of transport by the kidney tubules.

Tubular Reabsorption

In the process of glomerular filtration, large quantities of mineral electrolytes, water, and organic compounds are presented to the tubules. Varying amounts of some of these substances will be excreted (see Table 24–1). Most of the glomerular filtrate is not excreted, because the tubules selectively reabsorb the constituents of the glomerular filtrate.

Reabsorption may be either active or passive. **Active reabsorption** (active transport) is a process requiring the local expenditure of metabolic energy by the tubular epithelium and can effect the net movement of a substance against concentration or electrical gradients, or both. Examples of actively reabsorbed substances are sodium, glucose, and phosphate. **Passive reabsorption** (passive transport) does not depend directly on the expenditure of metabolic energy, and movement occurs down concentration, electrical, or osmotic gradients. Passively reabsorbed substances include urea, chloride, and water. In this section, we will discuss the mechanisms for reabsorbing various organic compounds. Later, we will consider the reabsorption of various mineral electrolytes and water.

Reabsorption of Glucose

To study how glucose, or any other substance, is handled by the nephron, scientists have used a technique called **kidney micropuncture.** With this technique, the kidney is exposed in anesthetized animals, and small samples of glomerular filtrate or tubule fluid are collected from the nephron with micropipettes. The collected fluid is analyzed and the site of collection determined by nephron microdissection. From such studies, scientists can localize the site of transport, determine the magnitude of transport, and gain insights into the nature of the transport process. Using this method, physiologists have found that the concentration of glucose in fluid collected from the urinary space of Bowman's capsule is identical to that in plasma. In the early proximal tubule, however, the glucose concentration in tubule fluid falls steeply. The conclusions from these measurements are: that filtration of glucose is not restricted by the glomerular filtration membrane; that glucose is reabsorbed by the proximal tubule; and that glucose reabsorption is active. Glucose is moved out of the tubule fluid and returned to

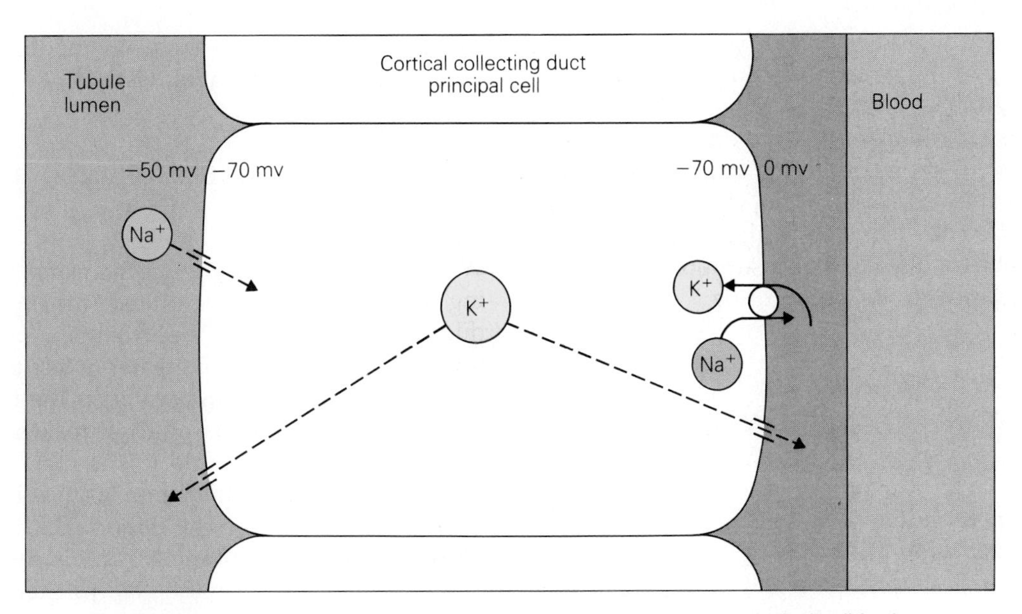

Figure 24–15 A model of potassium and sodium transport by a principal cell in the cortical collecting duct. The two steps in potassium secretion are (1) uptake into the cell via the basolateral membrane sodium-potassium pump and (2) passive diffusion across the luminal cell membrane. Sodium reabsorption involves passive diffusion into the cell through a sodium-selective ion channel in the luminal cell membrane, and active pumping of sodium across the basolateral cell membrane.

of our intake is excreted in the feces. The regulation of plasma calcium involves control of intestinal calcium absorption, renal excretion, and exchanges with bone. The parathyroid hormone and vitamin D play important roles in this regulation. Calcium balance will be discussed in Chapter 27.

Calcium is filtered and reabsorbed by the kidneys. Since about 40% of plasma calcium is bound to plasma proteins, only the nonbound portion (60%) is filtered. About 60% of the filtered calcium is reabsorbed in the proximal convoluted tubule, and most of the rest in the thick ascending limb, distal convoluted tubule, and collecting ducts. The quantity excreted is about 1% to 2% of the filtered load. Regulation of urinary excretion of calcium occurs primarily in the distal portions of the nephron. The parathyroid hormone stimulates calcium reabsorption and so acts to conserve filtered calcium for the body.

Tubular Transport of Magnesium

About two thirds of the magnesium in our diet is normally excreted in the feces, and one third in the urine. Approximately 30% of plasma magnesium is bound to plasma proteins and so is not filterable at the glomeruli. The proximal convoluted tubules reabsorb about 20% to 30% filtered magnesium, and 50% to 60% is reabsorbed in the loop of Henle. Only 3% to 5% of the filtered magnesium is excreted under normal conditions. The kidneys play a major role in regulating the plasma magnesium concentration. Excesses of magnesium are rapidly excreted in the urine; in magnesium-deficient states, magnesium is avidly reabsorbed by the kidney tubules and virtually disappears from the urine.

Tubular Transport of Phosphate

The kidneys play an important role in regulating the plasma concentration of inorganic phosphate. We usually excrete in the urine the amount of phosphate which we ingest in our diet. Filtered phosphate is reabsorbed by sodium-dependent, secondary active transport in the proximal tubule. The reabsorptive mechanism has a limited rate, or *Tm*. The *Tm* is normally exceeded by the quantities of phosphate filtered, and so phosphate is excreted in the urine. The situation here differs from that of glucose, where the *Tm* is reached only when the plasma glucose concentration is raised dramatically. Because the filtered phosphate load and *Tm* are normally so close to one another, the kidneys can participate in regulating the plasma phosphate concentration.

If phosphate intake is increased, plasma phosphate levels will rise, and the filtered phosphate load will be increased. The *Tm* will be exceeded more than usual, and so phosphate excretion will increase, thereby ridding the body of the extra phosphate. Conversely, if phosphate intake is decreased, plasma phosphate and filtered phosphate amounts are decreased, and all of the filtered phosphate will be reabsorbed, thus conserving phosphate for the body. Phosphate excretion is also regulated by changing the *Tm*. For example, the parathyroid hormone inhibits phosphate reabsorption, thus promoting phosphate loss in the urine.

Tubular Reabsorption of Water

The rate of water excretion in the urine depends on the rates of filtration of water (GFR) and tubular water reabsorption. Since the urine volume is mostly water, we can assume that renal water excretion rate is simply equal to the urine flow rate. Urine flow rate can vary widely, depending on conditions. If a person drinks a large quantity of water, urine flow rate may be as high as 20 ml/min. The osmolality of this urine may be as low as 30 to 40 mosm/kg H_2O, much more dilute than plasma (approximately 300 mosm/kg H_2O). On the other hand, in a dehydrated person, urine flow rate may be as low as 0.3 ml/min. The osmolality of this urine may be as high as 1200 to 1400 mosm/kg H_2O, or approximately 4 to 5 times the normal plasma osmolality. A "normal" urine flow rate is about 1 ml/min (1.5 liters/day), and the urine is usually modestly concentrated (600 to 800 mosm/kg H_2O).

Production of Hyperosmotic Urine to Save Water

We will consider now the mechanisms for producing an osmotically concentrated (hyperosmotic) or dilute (hypo-osmotic) urine—that is, a urine with a total solute concentration greater or less than that of plasma. If there is excess water in the body (e.g., due to drinking a lot of liquid), it makes sense that the kidneys should get rid of the extra water and excrete the urinary solutes in a large volume of osmotically dilute urine. The importance of excreting an osmotically concentrated urine is more subtle. Recall that the kidneys are always called upon to excrete solutes. These solutes include waste products of metabolism and various mineral electrolytes that are consumed in the diet. The excretion of these solutes requires the excretion of water, since this is the vehicle in which these substances are dissolved. Suppose someone had to excrete 600 mosm of solutes per day. Suppose also that the person could form only an isosmotic urine—that is, a urine with an osmolality of 300 mosm/kg H_2O. How much water would have to be excreted? The answer is that it would take 2 kg (or 2 liters) of water to excrete 600 mosm. Suppose now that the same amount of

Focus on Kidney Stones

A kidney stone is a hard mass that forms in the urinary tract. At least 1% of Americans develop kidney stones some time during their life. Stone formation occurs more commonly in men than in women and most often affects men between 30 and 60 years of age. A stone lodged in the ureter causes severe pain. The condition causes considerable suffering and loss of time from work, and it may lead to kidney damage. Once stone formation occurs in a person, it often recurs.

Stones form when poorly soluble substances in the urine precipitate out of solution, and crystals form, aggregate, and grow. Most kidney stones (75% to 85%) are made up of calcium salts, calcium oxalate and calcium phosphate. Stones may form from precipitated ammonium magnesium phosphate, uric acid, or cystine. Since a low urine flow rate raises the concentration of poorly soluble salts in the urine, thereby favoring precipitation, an important key to preventing stone formation is to drink plenty of water and maintain a high urine output.

Fortunately, most stones are small enough so that they are passed down the urinary tract and are spontaneously eliminated. Microscopic and chemical examination of the eliminated stones help in identifying the nature of the stone, in considering possible causes of stone formation, and in selecting the appropriate treatment. Sometimes a change in diet is recommended to reduce the amount of potential stone-forming material (e.g., calcium, oxalate, or uric acid) in the urine.

If the stone is not passed, several options are available. The classical treatment has been surgery to remove the stone, but such operations are not without risks. A recently developed treatment for kidney stones that does not require surgery is called **extracorporeal shock-wave lithotripsy.** A device called a **lithotriptor** is used. The patient is placed in a tub of water, and the stone is localized by X-ray imaging. Shock waves generated in the water by electric discharges are focused upon the stone through the body wall. The shock waves fragment the stone so that it can be passed down the urinary tract and eliminated.

solutes could be concentrated in a urine with an osmolality of 1200 mosm/kg H_2O. How much water would have to be excreted? The answer is that 0.5 kg (or 0.5 liter) would contain 600 mosm solute as a 1200 mosm/kg H_2O solution. We can see then that by excreting the urinary solutes in an osmotically concentrated (hyperosmotic) urine, the kidneys save water for the body. In the example given, by excreting the same amount of solutes in urine with an osmolality of 1200 mosm/kg H_2O instead of 300 mosm/kg H_2O, the kidneys, in effect, saved 1.5 liters of pure water for the body.

Since water is often scarce, the ability to form an osmotically concentrated urine is an important adaptation to a terrestrial environment. By getting rid of the urinary solutes in a hyperosmotic urine, the need for water intake is diminished. In the animal kingdom, only birds and mammals can produce a urine that is osmotically more concentrated than the blood. These animals are the only classes that have loops of Henle. Furthermore, in general, those mammals with relatively long loops of Henle are able to produce the most concentrated urines. For example, the kangaroo rat, a small rodent that inhabits the southwestern deserts of the United States, has an exceptionally long loop of Henle and can produce urine with an osmolality of 5500 mosm/kg H_2O, some 18 times that of plasma. These facts from comparative anatomy and physiology suggest an association between loops of Henle and the ability to produce a concentrated urine.

The Countercurrent Hypothesis for the Formation of Osmotically Concentrated Urine

The **countercurrent hypothesis** is now universally accepted as the explanation for the mechanism whereby an osmotically concentrated urine is formed. Scientists are still not convinced about certain aspects of this hypothesis—in particular, how the osmotic gradient is established in the inner medulla. The term *countercurrent* indicates that fluid flows in opposite directions in adjacent limbs of tubules and blood vessels in the kidney medulla.

There are actually two countercurrent mechanisms. The loops of Henle act as **countercurrent multipliers;** they set up a gradient of osmolality that increases from the junction of the cortex and medulla to the tip of the renal papillae. The blood vessels of the medulla (vasa recta) act as passive **countercurrent exchangers;** they help to preserve the osmotic gradient in the medulla. The collecting ducts act as **osmotic equilibrating devices;** it is here that the urine finally becomes osmotically concentrated. Depending on the plasma level of ADH, the fluid in the collecting ducts tends to equilibrate osmotically with the surrounding interstitium.

If kidney tissue from a thirsted animal is taken from various levels of the medulla, a progressively higher osmolality is found with increasing depth of the medulla. The highest osmolalities are found at the papillary tip. All of the structures in the medulla participate in this increasing gradient. Also, in a thirsted animal, with high plasma levels of ADH, the final urine has the same osmolality as tissue fluid at the tip of the kidney papilla.

This gradient is established by countercurrent multiplication in the loops of Henle. Figure 24–16 shows a simplified model for the countercurrent multiplication process. Consider a loop filled with fluid isosmotic to plasma (300 mosm/kg H_2O). Assume that the membrane separating the two loops is impermeable to water. Also assume that the loop is able to establish an osmotic gradient of 200 mosm/kg H_2O between the two limbs of the loop at any level (see Fig. 24–16*b*). This step is often called the **single effect.** This gradient could be produced by transport of salt out of the ascending limb and its deposition in the descending limb. Water is left behind (since the intervening membrane is water-impermeable), and so the osmolality of the ascending limb fluid decreases and that of the descending limb increases by the same amount. Next, we add new fluid to the loop (see Fig. 24–16*c*), as fresh fluid flows in from the proximal convoluted tubule. Again allow the 200 mosm/kg H_2O gradient to be established (Fig. 24–16*d*). Repeat the process several times (Fig. 24–16*e* to *h*). What is the final result? Notice that at any level of the loop a gradient of 200 mosm/kg H_2O exists. Along the length of the loop, however, a larger osmotic gradient (about 400 mosm/kg H_2O) has been established. This is what is meant by countercurrent multiplication; namely, a gradient existing at any level of the loop is multiplied by countercurrent flow so as to produce a larger gradient along the length of the loop.

The model just discussed should not be taken too literally. For example, flow through the loop of Henle is not a discontinuous process. Also, ascending and descending limbs do not share a common membrane; rather, there is an intervening interstitial space. The model is realistic, however, in that salt is transported out of the ascending limb across a water-impermeable membrane. The descending limb is water-permeable, and its osmolality is increased primarily by water removal, not by active deposition of salt into its lumen. Notice that fluid leaving the ascending limb is osmotically dilute,

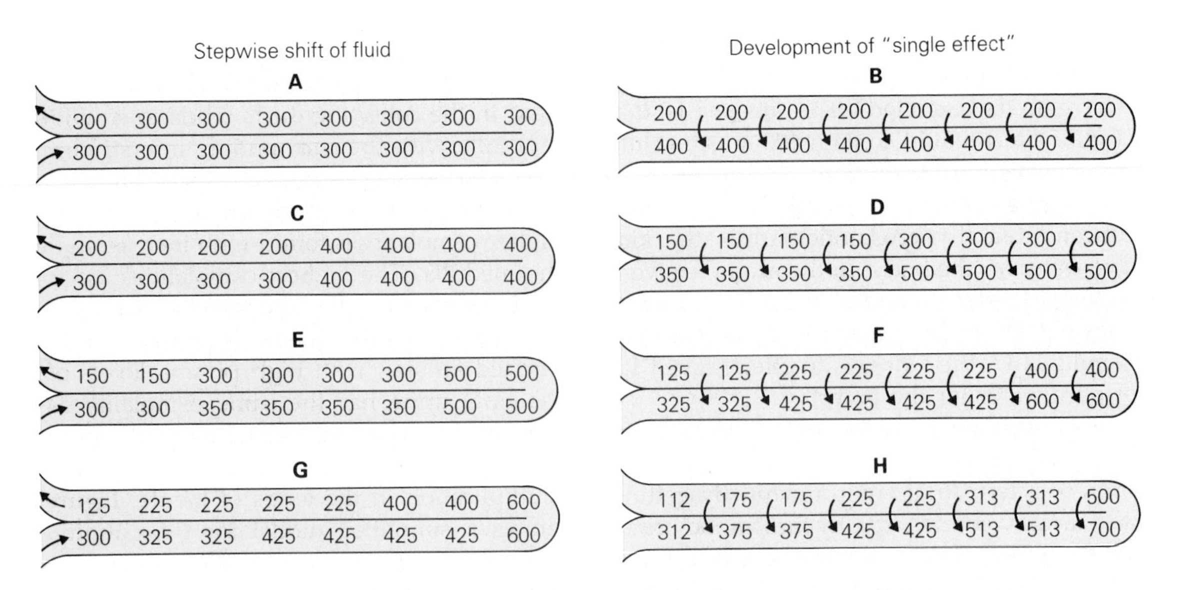

Figure 24–16 The principle of countercurrent multiplication, based on the assumption that, at any level along the loop, an osmotic gradient of 200 mosm/kg H_2O can be established by salt transport between ascending and descending limbs. (Modified from Pitts, R. F. *Physiology of the Kidney and Body Fluids.* 3rd edition. Chicago: Year Book, 1974).

even in a kidney that will be forming a concentrated urine. This is so because salt has been removed along the ascending limb.

Another important feature of countercurrent multiplication is that it is an energy-consuming process. In other words, in order to establish a gradient, there must be an energy source. The energy source for operating the countercurrent multiplier is ultimately active sodium transport, which is linked to ATP hydrolysis. Note that the magnitude of the gradient established depends on the size of the single effect and the length of Henle's loop. If the single effect is reduced, or if the loops are short, then a large gradient along the length of the loop cannot be established.

Figure 24–17 shows a model for the operation of the countercurrent mechanism in the kidney. In this model, we are assuming that a maximally concentrated urine is formed. The numbers in blue give the osmotic concentration in mosm/kg H_2O. The boxed numbers in green give the relative amounts of filtered water present at various points along the nephron. The heavy blue shading along the ascending limb indicates a low water permeability. A juxtamedullary nephron, with a long loop of Henle extending to the tip of the papilla, is depicted.

Seventy percent of the filtered sodium and water is reabsorbed along the proximal convoluted tubule. This reabsorption is essentially isosmotic. Fluid entering the descending limb of Henle's loop has an osmolality of about 300 mosm/kg H_2O. The descending limb is water-permeable, and since the interstitium surrounding the tubules is hyperosmotic, water leaves the descending limb and the osmolality increases. Since NaCl is the main solute in the descending limb fluid, its concentration rises.

In the thick ascending limb of Henle's loop, there is a vigorous sodium-potassium pump whose activity results in sodium chloride deposition in the outer medulla. An osmotically dilute fluid, which contains mainly NaCl and urea, leaves the loop. As the fluid traverses the cortical collecting ducts, in the presence of ADH, water is reabsorbed. This is because water moves along its osmotic gradient from the tubule fluid into the cortical interstitium, where it is carried away by a high blood flow. Before the tubule fluid once again re-enters the outer medulla, it is now isosmotic, and its volume is substantially reduced. The reabsorption of dilute fluid in the cortex is important because if a large volume of water were presented to the medulla, too much water would be added to the medullary interstitium, and the urine could not be maximally concentrated. The cortical collecting ducts are urea-impermeable, so the urea concentration rises.

As the fluid enters the outer medulla, it becomes hyperosmotic, as water is reabsorbed into the medullary interstitium. The urea concentration rises further. Next, when the urine enters the inner medulla, it encounters a urea-permeable segment. Urea diffuses out of the collecting ducts and accumulates in the inner medullary interstitium. The urea in the medulla can be thought of as osmotically balancing urea in the urine, so that other solutes can be osmotically balanced by the salt deposited in the medulla. As fluid moves down the length of the

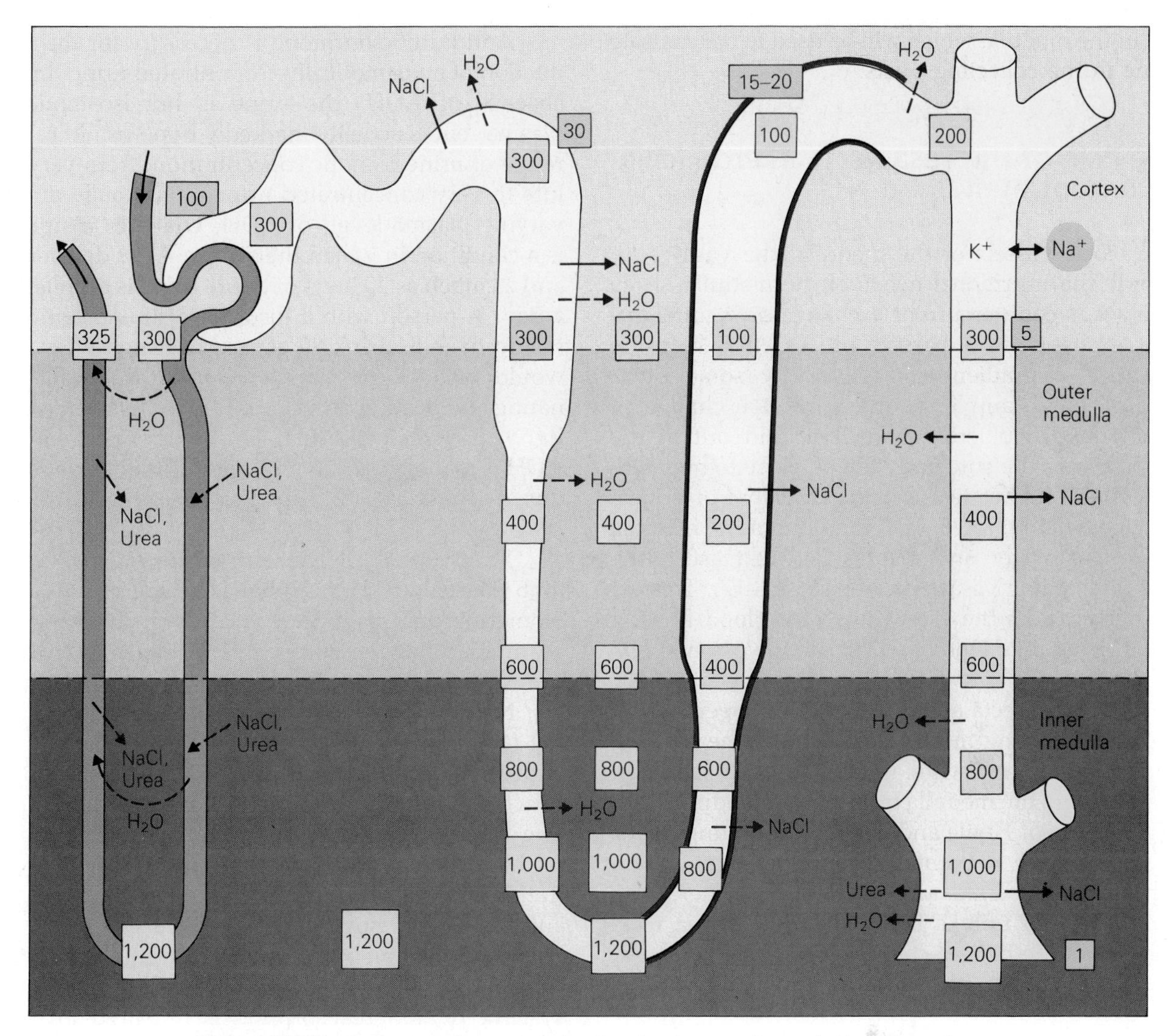

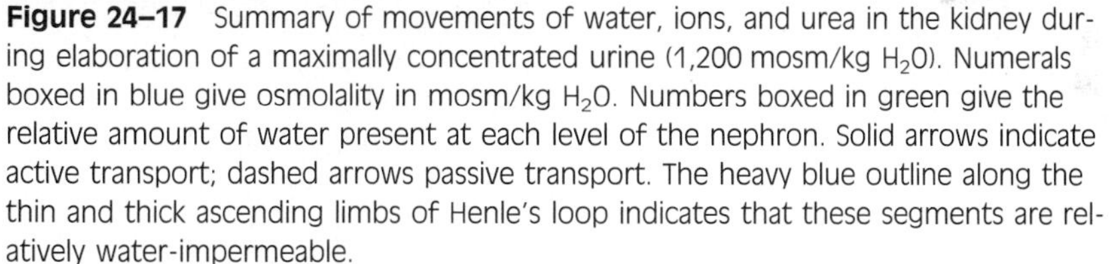

Figure 24–17 Summary of movements of water, ions, and urea in the kidney during elaboration of a maximally concentrated urine (1,200 mosm/kg H_2O). Numerals boxed in blue give osmolality in mosm/kg H_2O. Numbers boxed in green give the relative amount of water present at each level of the nephron. Solid arrows indicate active transport; dashed arrows passive transport. The heavy blue outline along the thin and thick ascending limbs of Henle's loop indicates that these segments are relatively water-impermeable.

inner medullary collecting duct, water is reabsorbed. In the presence of high levels of ADH, fluid in the collecting ducts achieves the same osmolality as in the surrounding interstitium. Urine of high osmolality and low volume is finally excreted.

A Special Role for Urea in Producing a Concentrated Urine

In the model presented, urea plays an important role in the concentrating mechanism in the inner medulla. This is a region of the kidney in which there are thin (but no thick) ascending limbs. The epithelium of the thin ascending limb is flat, with few mitochondria. It does not look like the type of epithelium usually associated with vigorous salt transport. Urea is added to the inner medulla from the collecting ducts and from the calyceal urine which bathes the papilla. The osmotic pressure due to urea causes the removal of water along the descending limb and thereby concentrates NaCl in this descending limb fluid. When the NaCl-enriched fluid enters the NaCl-permeable (but water-impermeable) thin ascending limb, a concentration gradient for outward, passive movement of NaCl is established. This outward movement drives the countercurrent multiplier in the inner medulla. It results in net addition of solute to this region of the kidney, which is important in establishing the gradient in the medulla which will be used to concentrate urine in the collecting ducts.

The Role of the Vasa Recta in Producing a Concentrated Urine

The blood vessels of the medulla, the vasa recta, supply the nutritional needs of the medulla. They also act as countercurrent exchangers. Countercurrent exchange is a passive process that helps to maintain a gradient established by some other means. For example, countercurrent exchange of heat between blood flowing into and out of the limbs allows the core body temperature to be maintained at a higher temperature than that of fingers and toes. Blood flowing into and out of the medulla exchanges water and solutes between ascending and descending vasa recta (see Fig. 24–17). This exchange reduces the extent to which blood flowing into the medulla tends to dissipate the osmotic gradient.

The vasa recta are essential to the operation of the concentrating mechanism because they are responsible for removing water from the medulla. Water enters the medulla from the descending limbs of the loops of Henle and from the collecting ducts; there must be a pathway for removal of water; otherwise, it would accumulate here. There is a force for fluid uptake in the ascending vasa recta because of the plasma colloid osmotic pressure and because of an elevated concentration of small solutes.

Factors Affecting Urinary Concentrating Ability

Many factors influence the ability to form an osmotically concentrated urine (Table 24–3).

Table 24–3 Factors Affecting Urinary Concentrating Ability
Antidiuretic hormone (ADH)
Delivery of NaCl to ascending limb of Henle's loop
Reabsorption of NaCl by ascending limb
Delivery of fluid to medullary collecting ducts
Medullary blood flow
Urea
Length of Henle's loop

Antidiuretic hormone is necessary for the production of an osmotically concentrated urine. In the absence of ADH, the urine is not isosmotic to plasma, but is actually markedly hypo-osmotic. The range of urine osmotic concentration, from very dilute to very concentrated urine, is normally due to varying plasma levels of ADH. **Diabetes insipidus** is a condition in which there is an ADH deficiency, and as much as 20 liters of dilute urine is excreted in a day. A person with a disorder of this magnitude would have to drink 80 glasses of water a day and would spend a large part of the day and night urinating and drinking. In **central (hypothalamic, pituitary) diabetes insipidus,** synthesis or secretion of ADH is inadequate. In **nephrogenic diabetes insipidus,** ADH is present, but the collecting ducts are unresponsive.

An adequate delivery of salt to the ascending limb is necessary for maximal urine concentration. If glomerular filtration rate is abnormally low—for example, following severe hemorrhage—then concentrating ability is impaired.

Not only must delivery of salt be adequate, but the rate of reabsorption of salt along the ascending limb must be maintained. After all, this is the single effect essential for countercurrent multiplication. The "loop" diuretic drugs inhibit salt reabsorption in the thick ascending limb; they decrease concentrating ability and cause a marked increase in salt and water excretion.

The urine can be maximally concentrated only if a trickle of fluid flows through the collecting duct system. If fluid reabsorption is impaired in early portions of the nephron, as occurs during an osmotic diuresis, then a maximal urine osmolality cannot be achieved.

Medullary blood flow can affect the osmotic gradient in the medulla and thus affect concentrating ability as well. An excessive blood flow to this region tends to wash out the gradient and thereby reduce the maximum urine osmolality that can be attained.

Urea plays a critical role in the concentrating mechanism, especially in the establishment of an osmotic gradient in the inner medulla. With a low-protein diet and the resulting decrease in urea production, concentrating ability is impaired. The deficient protein intake of people in drought-stricken areas results in a greater need for water, because of impaired urine concentrating ability caused by less urea in the medulla.

Finally, the length of Henle's loops affect concentrating ability. Some desert-living mammals have very long loops and a correspondingly enhanced ability to concentrate the urine and save water.

Production of an Osmotically Dilute Urine

In principle, the production of an osmotically dilute urine is quite straightforward. All that is needed is the reabsorption of solute across a relatively water-impermeable membrane, with water being left behind. In the thick ascending limb, active reabsorption of sodium (accompanied by chloride) results in considerable dilution of the tubular fluid, as we have already seen. This segment is often called the **diluting segment.** In the absence of antidiuretic hormone, the collecting ducts are very water-impermeable. Even though the medullary interstitium is hyperosmotic (but not as much as in a concentrating kidney), collecting duct water mostly stays in the duct lumen. Continued reabsorption of salt results in further dilution of the urine along the collecting ducts. The end result is the excretion of a large volume of hypo-osmotic urine.

Summary

I. A. The kidneys play a dominant role in regulating the volume and composition of the extracellular fluid, the environment of the living body cells. The basic unit of kidney structure and function is the nephron.

B. Each kidney is usually supplied by a single renal artery. Important blood vessels in the kidney include interlobar arteries, cortical radial arteries, afferent arterioles, glomeruli, efferent arterioles, peritubular capillaries, and vasa recta.

C. The kidneys are richly innervated by sympathetic efferent fibers and by afferent fibers that mediate pain. The kidney cortex contains lymphatic vessels.

D. The juxtaglomerular apparatus consists of specialized cells thought to play a role in regulating glomerular blood flow and filtration rate.

E. The kidneys have a huge blood flow because of the need to sustain a high rate of plasma filtration.

II. A. Urine formation starts with the filtration of plasma in the kidney glomeruli. Glomerular filtration is favored by the high hydrostatic pressure of the blood in the glomerular capillaries and is opposed by the hydrostatic pressure in the urinary space of Bowman's capsule and by the glomerular capillary colloid osmotic pressure. Glomerular filtration is rather nonselective; proteins are mostly retained in the plasma by the glomerular filtration membrane, but all low-molecular-weight substances are freely filtered. Glomerular filtration rate is most accurately measured by determining the inulin clearance.

B. Glucose is reabsorbed by the proximal tubule. Reabsorption is active and sodium-dependent and shows a maximal rate *(Tm)*. Normally, all of the filtered glucose is reabsorbed. Urea is filtered by the kidney glomeruli and is passively reabsorbed by the kidney tubules. Active reabsorption of sodium provides the driving force for tubular reabsorption of water, glucose, amino acids, chloride, and phosphate.

C. Some organic compounds are secreted from the blood surrounding the kidney tubules into the tubular urine.

III. A. Most of the filtered sodium is reabsorbed by the kidney tubules; the quantity of sodium excreted plays an important role in body sodium balance. The proximal convoluted tubule reabsorbs the greatest fraction (70%) of the filtered sodium and water; the tubule fluid stays essentially isosmotic to plasma in this nephron segment. The loop of Henle reabsorbs about 20% of the filtered sodium and 10% of the filtered water. The distal convoluted tubule and collecting ducts reabsorb about 9% of the filtered sodium and 19% of the filtered water. The collecting ducts are the site of final regulation of sodium and water excretion; aldosterone and antidiuretic hormone (ADH) increase sodium and water reabsorption, respectively, by the collecting ducts.

B. Potassium is filtered, reabsorbed, and secreted by the kidney tubules. The cortical collecting duct is an important site determining potassium secretion, and hence, excretion.

C. About 10% of dietary calcium is excreted by the kidneys.

D. The kidneys play a major role in regulating the concentration of magnesium in the plasma.

E. The kidneys are also important in regulating the plasma concentration of inorganic phosphate.

IV. A. The mammalian kidney can form a urine that is osmotically more dilute or concentrated than plasma. By forming a dilute urine, the kidneys get rid of extra water from the body; by forming a concentrated urine, the kidneys save water for the body. In the absence of ADH, an osmotically dilute urine is excreted. ADH results in a more concentrated urine by increasing the water permeability of the kidney collecting ducts, thereby allowing the collecting duct urine to equilibrate with the osmotically concentrated medullary interstitium.

B. The formation of an osmotically concentrated urine depends upon the establishment of an osmotic gradient in the medulla by the loops of Henle. These structures act as countercurrent multipliers.

C. The addition of urea to the inner medulla allows for efficient operation of the urinary concentrating mechanism.

D. The vasa recta act as countercurrent exchangers and remove water from the kidney medulla.

E. Factors affecting urinary concentrating ability are ADH, salt reabsorption by the ascending limb of Henle's loop, tubule fluid and blood flow, urea, and the length of Henle's loops.

F. Production of an osmotically dilute urine involves the reabsorption of solute across a membrane, with water left behind.

Review Questions

1. What are the segments of the nephron and collecting duct? What blood vessels are important for the kidney?
2. What is the juxtaglomerular apparatus, and why is it significant?
3. What is renal autoregulation, and why is it important?
4. Why does a loss of fixed negative charges from the glomerular filtration membrane lead to proteinuria?
5. What factors account for the very high rate of filtration of plasma in the glomeruli of the human kidney?
6. How will the following affect the rate of glomerular filtration: 1) a decrease in K_f, 2) a decrease in glomerular capillary hydrostatic pressure, 3) an increase in hydrostatic pressure in the urinary space of Bowman's capsule, and 4) a decrease in plasma colloid osmotic pressure?
7. What is meant by the term *renal clearance?*
8. To estimate glomerular filtration rate, a patient (50 kg woman, 20 years of age) was instructed to collect a 12-hour urine sample. A blood sample was collected, and urine and plasma were analyzed for creatinine by a standard colorimetric method. The following values were obtained: urine volume, 720 ml; urine creatinine, 0.80 mg/ml; and plasma creatinine, 0.010 mg/ml. What is her GFR in ml/min?
9. Why does a fall in glomerular filtration rate lead to an increase in plasma creatinine concentration?
10. Why is uncontrolled diabetes mellitus often accompanied by glucosuria?
11. What is the difference between diabetes mellitus and diabetes insipidus?
12. Why can the clearance of PAH be used to estimate renal plasma flow?
13. What are the pathways for sodium reabsorption across the proximal tubule epithelium?
14. How does sodium and water reabsorption in the proximal convoluted tubule differ from this process in the collecting duct?
15. How do the following affect potassium excretion: 1) an increase in plasma aldosterone concentration, 2) an increase in plasma potassium concentration, 3) an increase in transtubular potential difference in the cortical collecting duct, and 4) an increase in tubule fluid flow rate in the cortical collecting duct?
16. What is the difference between countercurrent multiplication and countercurrent exchange?
17. In the production of an osmotically concentrated urine, what is the role played

by 1) loops of Henle, 2) vasa recta, 3) collecting ducts, 4) ADH, and 5) urea?

Suggested Readings

Brenner, B., Coe, F. L., and Rector, F. C., Jr. *Renal Physiology in Health and Disease.* Philadelphia: W. B. Saunders Company, 1987.

Kriz, W., and Bankir, L. A standard nomenclature for structures of the kidney. *Am J Physiol* 254:F1–F8, 1988.

Marsh, D. J. *Renal Physiology.* New York: Raven Press, 1983.

Pitts, R. F. *Physiology of the Kidney and Body Fluids,* 3rd ed. Chicago: Year Book Medical Publishers, 1974.

Seldin, D. W., and Giebisch, G., ed. *The Kidney. Physiology and Pathophysiology.* New York: Raven Press, 1985.

Smith, H. W. *From Fish to Philosopher.* Boston: Little, Brown, and Company, 1953.

Sullivan, L. P., and Grantham, J. J. *Physiology of the Kidney,* 2nd ed. Philadelphia: Lea and Febiger, 1982.

Valtin, H. *Renal Function: Mechanisms Preserving Fluid and Solute Balance in Health,* 2nd ed. Boston: Little, Brown, and Company, 1983.

Vander, A. J. *Renal Physiology,* 3rd ed. New York: McGraw-Hill Book Company, 1985.

Regulation of Fluid and Electrolyte Balance

In the last chapter, we discussed how various substances were handled by the kidneys. We considered the processes of glomerular filtration, tubular reabsorption, and tubular secretion. In this chapter, we will first describe in more detail the volume and composition of the various body fluids. Then we will discuss how we keep in balance with respect to water, and sodium and potassium ions. We will indicate some of the body disturbances that occur when our kidneys fail and discuss some of the methods for treating renal failure. Finally, we will briefly consider the functions of the ureters and urinary bladder.

Body Fluids and Fluid Compartments

The fluids of the body can be divided into two categories: **intracellular fluid** (fluid within cells) and **extracellular fluid** (fluid outside of cells). These fluids differ strikingly in composition and in volume. Because most cell membranes are water permeable, however, the osmotic pressure inside and outside of our cells is the same.

Body Water and Its Divisions

The body fluids contain both solutes and water, but in terms of amount or volume occupied, water is the major constituent. Therefore, we will consider the water content of the body first and later cover the various solutes dissolved in the body fluids.

The percentage of total body weight that is water varies from 45% to 75% in different individuals. This range mostly reflects differences in the amount of body fat. The water content of adipose (fat) tissue is about 10% by weight; most other tissues contain 70% to 75% water by weight. An obese individual has a low percentage of body weight made up of water, and a lean individual has a high percentage. Young adult males average about 60% water by weight; young adult females average about 50% water. The difference between the sexes is due to females' greater amount of subcutaneous fat. With aging, the percentage of body weight which is water decreases because water-rich muscle tissue tends to be replaced by water-poor adipose tissue.

In a healthy person, the water content of the body hardly changes from day to day. A simple way to evaluate *changes* in total body water is to weigh a person. An increase in body weight may signal abnormal fluid (water) retention; a loss of weight may indicate abnormal fluid (water) loss.

Total body water is distributed in several divisions or compartments (Fig. 25–1). Approximately two thirds of body water is contained in the **intracellular fluid compartment,** and one third in the **extracellular fluid compartment.**

The intracellular fluid compartment is not one continuous space, but is made up of trillions of cells. The intracellular space is separated from the extracellular space by cell plasma membranes. If we assume an idealized, young adult 70-kg male, in whom total body water is 60% of body weight, we can calculate that total body water is 42 kg (or 42 liters), intracellular water is 28 liters, and extracellular water is 14 liters.

The extracellular fluid can be further subdivided into two major subcompartments, separated from each other by the endothelium of the blood vessels. Blood vessels contain **blood plasma,** the part of the blood exclusive of blood cells and platelets. The plasma is about 93% water (by weight or volume). Plasma water accounts for about one quarter of the extracellular water (3.5 liters in a 70-kg young adult male). Outside of the blood vessels are the **interstitial fluid** and **lymph.** Interstitial fluid is the fluid directly bathing most body cells, and lymph is the fluid in lymphatic vessels. The interstitial fluid and lymph together contain three quarters of the extracellular water (10.5 liters in a 70-kg male). Blood plasma, interstitial fluid, and lymph are nearly identical in chemical composition, except for a higher protein concentration in plasma.

Figure 25–1 does not show an additional small extracellular fluid compartment, the **transcellular fluid.** This fluid compartment includes specialized

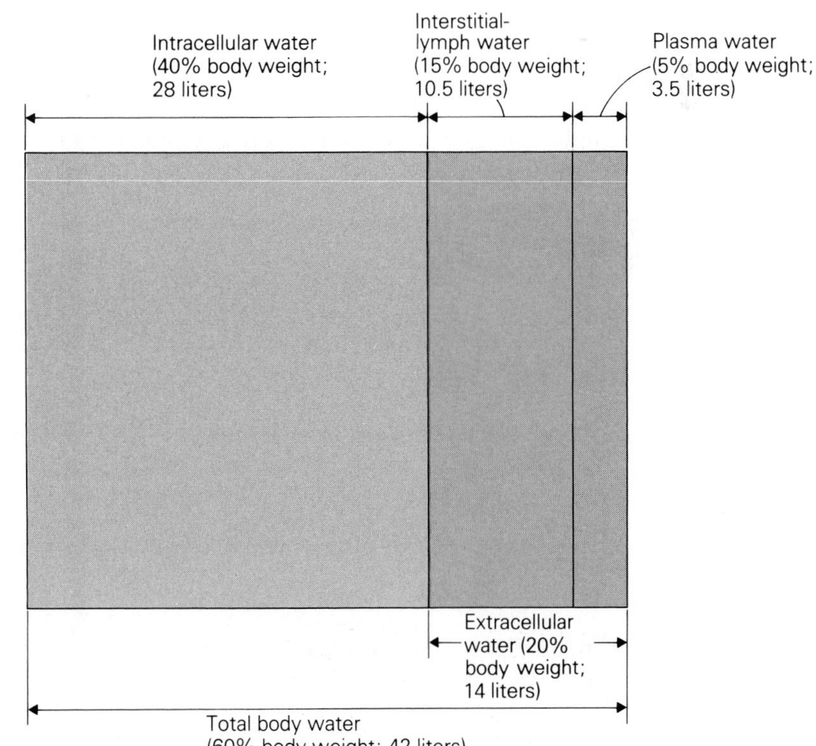

Figure 25–1 Distribution of water in the body of an average young adult 70 kg male.

fluids such as **cerebrospinal fluid,** the **aqueous humor** of the eye, secretions of the digestive glands, sweat, and renal tubular fluid and urine. These fluids are separated from the plasma by an endothelium as well as a continuous epithelial cell layer. This epithelium modifies the chemical composition of the transcellular fluid, so it is not a simple ultrafiltrate of plasma, as is interstitial fluid. Transcellular fluid amounts to only 1% to 3% of body weight, but it is extremely important, physiologically speaking. The transcellular fluids, in general, are continuously formed, and abnormal loss of these fluids or blockage of fluid drainage may have serious consequences for the organism.

Electrolyte Composition of the Body Fluids

The body fluids contain many uncharged organic substances (e.g., glucose, urea), but in terms of concentration, electrolytes are quantitatively more important. Electrolytes contribute most to the osmotic pressure of the body fluids and consequently are crucial in the distribution of body water.

The concentrations of various electrolytes in plasma, interstitial fluid, and intracellular fluid are summarized in Table 25–1. The intracellular fluid values are based on measurements in skeletal muscle cells. These cells were chosen as a representative cell type because they account for about two thirds of the cell mass in the human body. Concentrations are expressed in terms of milliequivalents per liter solution or per kg H_2O. One **equivalent (Eq)** contains Avogadro's number (6×10^{23}) of plus or minus charges; one **milliequivalent (mEq)** is equal to 1/1000 Eq. For singly charged **(univalent)** ions, 1 mEq is the same as 1 millimole (mmole). Thus, we can express plasma sodium concentration as 142 mEq per liter or 142 mmole per liter. For doubly charged **(divalent)** ions, 2 mEq is the same as 1 mmole. For example, a plasma Ca^{2+} concentration of 5 mEq per liter is the same as 2.5 mmole per liter. Some electrolytes, in particular proteins, are **polyvalent,** so there are several milliequivalents per millimole. The virtue of expressing concentration in terms of milliequivalents per liter is that in any fluid compartment the sum of the positive ions (cations) must equal the sum of the negative ions (anions), because of the requirement of electroneutrality in solutions. In Table 25–1, note that in every compartment the total equivalents of cation and anion are equal.

Plasma concentrations are listed in the first column of Table 25–1. Sodium is the major cation in plasma, and chloride and bicarbonate are the major anions. The plasma proteins (mostly serum albumin) bear a net negative charge at physiological pH. The electrolytes are actually dissolved in the watery phase of plasma, so concentrations in plasma water are presented in the second column of Table 25–1. These values were calculated by assuming a plasma water content of 93%. For example, for Na^+, 142 mEq per liter plasma divided by 0.93 liter water per liter plasma equals 153 mEq per liter water. Since a liter of water weighs 1 kg, the concentrations could also be expressed per kg of water.

The interstitial fluid (third column of Table 25–1) is an ultrafiltrate of plasma. It contains essentially the same concentrations of all low-molecular-weight substances as in plasma, but little protein. The differences in concentrations between the small ions in plasma water and interstitial fluid (compare columns two and three in Table 25–1) arise because

Table 25–1
Electrolyte Composition of the Body Fluids

Electrolytes	(1) Plasma (mEq/liter)	(2) Plasma Water (mEq/kg H_2O)	(3) Interstitial Fluid (mEq/kg H_2O)	(4) Intracellular Fluid (Skeletal Muscle) (mEq/kg H_2O)
Cations				
Na^+	142	153	145	10
K^+	4	4.3	4	159
Ca^{2+}	5	5.4	3	1
Mg^{2+}	2	2.2	2	40
Total	153	165	154	210
Anions				
Cl^-	103	111	117	3
HCO_3^-	25	27	28	7
Protein	17	18	—	45
Others	8	9	9	155
Total	153	165	154	210

of differences in protein concentration between these two fluids. Two factors are involved. The first is the so-called **Donnan distribution** effect. Because the large plasma proteins are negatively charged and are kept within the circulation, this leads to an excess of small cations and a deficit of small anions in the circulation. In other words, the concentrations of small cations in interstitial fluid will be lower than in plasma, and the concentrations of small anions in interstitial fluid will be greater than in plasma. The Donnan factor is 0.95 for univalent cations (Na^+, K^+) and is 1.05 for univalent anions (Cl^-, HCO_3^-). For example, if the plasma sodium concentration is 153 mEq per kg H_2O, then the interstitial fluid sodium concentration is 0.95 × 153 mEq per kg H_2O or 145 mEq per kg H_2O. The Donnan factor for divalent cations (Ca^{2+}, Mg^{2+}) is 0.90.

The differences between plasma and interstitial fluid concentrations of Ca^{2+} and Mg^{2+} are larger than can be accounted for by the Donnan effect alone. Approximately 40% of plasma calcium and 30% of plasma magnesium are bound to plasma proteins. It is only the non-protein-bound Ca^{2+} and Mg^{2+} that are in equilibrium across capillary walls. Note that in most instances, the differences between plasma and interstitial fluid are small. For practical purposes, we can assume (except for plasma proteins) that electrolyte concentrations in plasma and interstitial fluid are the same.

If we examine intracellular fluid composition (column four of Table 25–1), major differences between this fluid and the extracellular fluids are apparent. The concentrations of potassium, magnesium, and protein in the cell are much higher than in the surrounding interstitial fluid. The concentrations of sodium, calcium, chloride, and bicarbonate are much lower in the cell. The anions in muscle cells labeled as "Others" are mainly proteins and organic phosphate compounds (e.g., creatine phosphate, ATP), and various other anions to which the cell membrane is mostly impermeable.

The high internal potassium concentration and low internal sodium concentration is maintained by activity of the sodium-potassium pump, which extrudes sodium ions and takes up potassium ions. The lower intracellular concentrations of chloride and bicarbonate can be mostly accounted for by the membrane potential difference, approximately −90 mv, which favors the outward movement of these negatively charged ions.

The distribution of ions within the cell is not uniform; different organelles may have different ion concentrations. For example, the cell nucleus has a higher sodium concentration than the cell as a whole. In muscle, most of the intracellular calcium is in the sarcoplasmic reticulum and mitochondria, and the cytoplasmic calcium concentration is very low.

Do not be misled by the higher number of milliequivalents per kg H_2O in the cell than in the interstitial fluid (compare totals in columns three and four of Table 25–1). A single protein molecule or phosphate compound may bear several negative charges, so the number of milliequivalents in the cell is greater than the number of milliosmoles. Also, magnesium is divalent (40 mEq per kg H_2O corresponds to 20 mmoles per kg H_2O) and is largely protein-bound within the cell, so it is not effective osmotically. It is generally accepted that the total solute concentrations (osmolalities) in the cell and surrounding extracellular fluid are identical. Water moves freely through most plasma membranes, so there is a condition of osmotic equilibrium between cellular and extracellular fluids.

Distribution of Water Between Cellular and Extracellular Fluids

Osmotic pressure is of prime importance in determining the distribution of water between cells and extracellular fluid. Osmotic pressure is directly related to total solute concentration (osmolality). An increased solute concentration lowers the activity (or concentration) of water; water will move from a region of lower osmolality to a region of higher osmolality. Such water movement could be prevented by applying a hydrostatic pressure to the more concentrated solution. Animal cells cannot employ such a mechanism, since a hydrostatic pressure difference of a few centimeters of H_2O will lead to rupture of the plasma membrane. In animals, osmotic pressures inside and outside of cells are equal. If solute or water is added to or lost from the extracellular fluid, osmotic equilibrium will be temporarily upset. Water will move into or out of the cells until a new osmotic equilibrium is achieved.

Most of the volume in any body fluid compartment is occupied by water. At a given osmolality, the total volume of (or amount of water in) a compartment is directly related to the amount of solute in the compartment. This important relationship follows from the equation which defines the term *concentration*—namely, concentration = amount/volume. If solute concentration (osmolality) is maintained constant, then the volume must change directly with the amount of solute.

In the extracellular fluid, the amount of sodium present is a major determinant of the volume of water present. This relationship results because sodium and its accompanying anions, chloride and bicarbonate, are the major osmotically active solutes present in extracellular fluid. If we add up the concentrations of these ions in interstitial fluid (see Table 25–1), we get 145 + 117 + 28 = 290 mmoles per kg H_2O. The corresponding osmolality is about 270 mosm per kg H_2O. (This figure is lower than 290 because of interactions among ions in solution; it is only in infinitely dilute solutions that ions behave as completely independent particles.) Since interstitial fluid (or plasma) osmolality averages 287 mosm per kg H_2O (approximately 300 mosm per kg H_2O), it is apparent that sodium salts account for more than 90% of the osmolality. The osmolality of the extracellular fluid is closely regulated by the thirst sensation and control of renal excretion of water by ADH. Consequently, if osmolality (sodium concentration) is kept constant, it follows that the amount of sodium in the body will determine the amount of water in the body. To express it another way, if a person takes in extra salt (NaCl), thirst results, water will be ingested, and extracellular fluid volume will be increased. If a person loses salt, extracellular fluid volume will be decreased.

Cell volume, also mostly water, is similarly influenced by the amount of contained solute. In the cell compartment, potassium is the major osmotically active solute. Hence, a loss of potassium from the cell results in a decrease in cell volume, and a gain in potassium results in an increase in cell volume. The large amount of impermeable anions (e.g., proteins, organic phosphates) within cells could affect cell volume. In theory, these solutes would cause a redistribution of small, permeable ions so that an excess of solute would be present in the cell (the Donnan distribution effect), and, therefore, water would tend to enter the cell. This tendency is, however, counteracted by the activity of the sodium-potassium pump, which effectively excludes sodium and thereby lowers the total amount of solute in the cell. In this way, the sodium-potassium pump plays a key role in the regulation of cell volume. If sodium-potassium pump activity is decreased (e.g., by low temperature, oxygen lack, metabolic poisons, or cardiac glycosides), cells gain sodium and water, and they swell.

An interesting phenomenon that has been recently described is the following: when certain cells are placed in a hypo-osmotic medium, they initially swell (as expected). Soon, however, they shrink back to their original size. This response is due to loss of potassium, chloride, and water and appears to be caused by activation of membrane ion transport processes. Opposite changes are observed in hyperosmotic media (cells initially shrink and then increase back to normal size). These responses demonstrate that cells have special mechanisms for regulating and preserving their volume.

The normal distribution of water between cellular and extracellular compartments changes in a variety of circumstances. Figure 25–2 provides some examples. Note that the height of the boxes corresponds to solute concentration. The width of the boxes corresponds to volume. The area of each box (height × width) corresponds to the total amount of solute in a compartment (concentration × volume = amount). The dotted lines represent the normal condition and are the same in all four parts of this figure. The solid lines represent conditions after a new osmotic equilibrium has been achieved. Note that the height of the boxes is always identical in intracellular and extracellular fluids, since osmotic equilibrium is attained.

Figure 25–2*a* shows the normal condition. Osmolalities are identical in cellular and extracellular compartments (300 mosm per kg H_2O). The volume of the intracellular compartment (40% of body weight) is twice the volume of the extracellular compartment (20% of body weight).

In Figure 25–2*b*, pure water was added to the extracellular fluid. Ingestion of water would have the same effect. Since the extracellular fluid osmolality was lowered, water moves into the cells until the osmolality in both compartments is lowered to the same level. The entry of water into the cells increases their volume. Since the cells contain two thirds of the body fluid solutes, two thirds of the added water enters the cells, and one third remains in the extracellular space. The decline in plasma osmolality would suppress secretion of ADH, which would permit rapid excretion of the ingested water by the kidneys thereby restoring body fluid osmolality and volume back to normal.

In Figure 25–2*c*, isotonic saline was added to the extracellular fluid. This can be done by intravenous infusion or ingestion of a 0.9% NaCl solution. Note that the final osmolality is not different from normal. This is not surprising, since isotonic saline is also isosmotic. Note also that cell volume does not change. By definition, an isotonic solution will not cause cells to shrink or swell. All of the isotonic saline is retained in the extracellular fluid and accordingly increases only this volume. In a normal person, the expanded extracellular fluid volume (by mechanisms we will discuss later) leads to an increase in renal excretion of salt and water, so normal conditions would eventually be restored.

Although it is not illustrated, a loss of isotonic fluid from the extracellular compartment would have predictable consequences: no change in body fluid osmolality or in cell volume and a decrease in

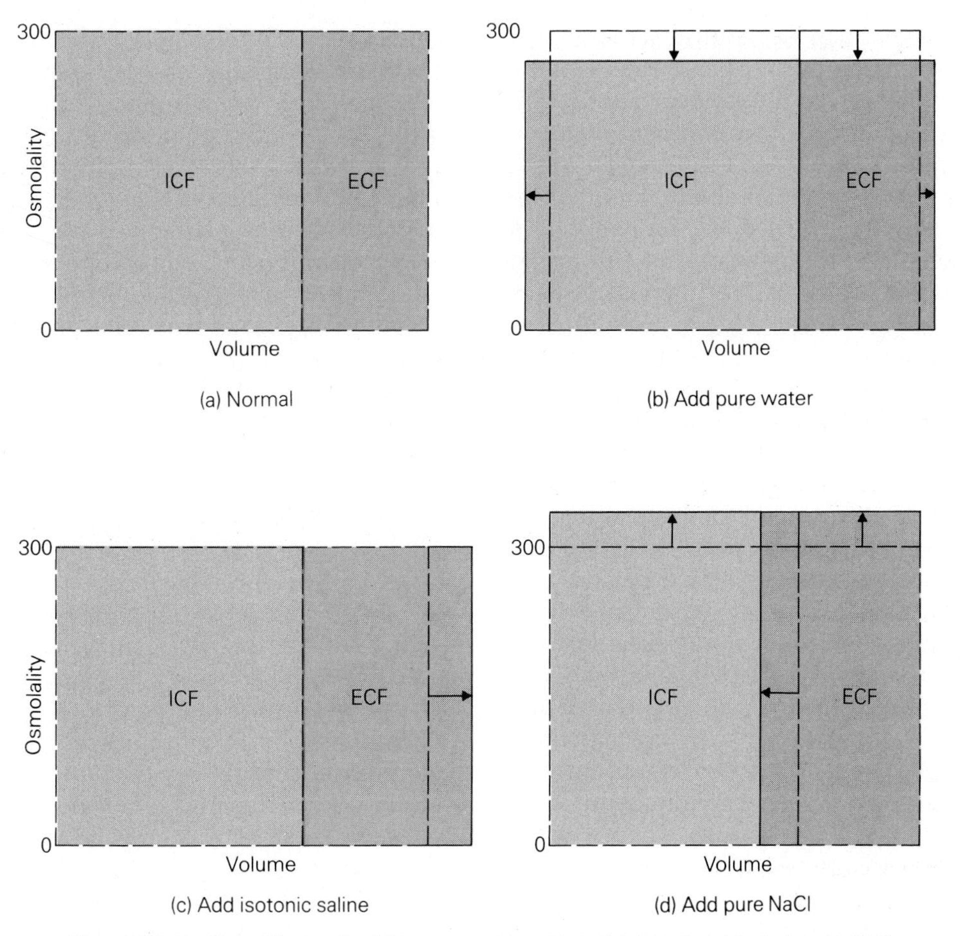

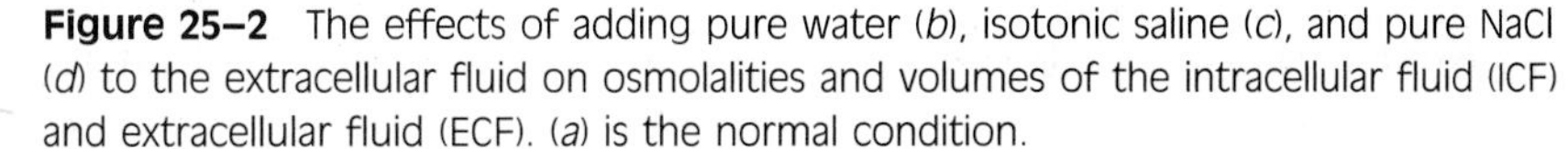

Figure 25–2 The effects of adding pure water (*b*), isotonic saline (*c*), and pure NaCl (*d*) to the extracellular fluid on osmolalities and volumes of the intracellular fluid (ICF) and extracellular fluid (ECF). (*a*) is the normal condition.

extracellular fluid volume. Hemorrhage and diarrhea are two cases where an isotonic fluid is lost from the body. The body acts to restore depleted extracellular fluid by renal retention of salt and water and by increased water intake (due to thirst).

Figure 25–2*d* shows the effect of adding sodium chloride without water to the extracellular compartment. Extra salt intake (e.g., eating salted potato chips) would be a corresponding circumstance. The added salt increases the osmolality of the extracellular fluid, which results in water movement out of the cells. Extracellular fluid volume is increased at the expense of cell volume. At equilibrium, osmolality is uniformly increased. When pure NaCl (or a hypertonic NaCl solution) is added to the extracellular fluid, the NaCl is diluted by the total body water, even though all of the NaCl stays in the extracellular fluid. The body's response to NaCl excess involves (1) thirst (stimulated by cellular dehydration) and increased ingestion of water and (2) increased renal excretion of salt and water (stimulated by the expanded extracellular fluid volume).

Distribution of Water Between Blood Plasma and Interstitial Fluid

The relative volumes of plasma and interstitial fluid are primarily determined by the Starling-Landis forces that operate across capillary walls. As discussed in Chapter 19, movement of fluid across capillary walls is determined by the balance between capillary and tissue hydrostatic and colloid osmotic pressures. The hydrostatic pressure in the capillaries tends to push fluid out of the capillaries; the colloid osmotic pressure due to the plasma proteins acts to retain fluid within the vascular system. An increase in capillary hydrostatic pressure, a decrease in plasma colloid osmotic pressure, or an increase in capillary membrane permeability to proteins would favor net movement of fluid out of the vascular system and would result in a decrease in plasma volume and an increase in interstitial fluid volume (edema). On the other hand, a decrease in capillary hydrostatic pressure or an increase in plasma colloid osmotic pressure would favor movement of interstitial fluid into the circulation. In most vascular beds, the net force favoring outward movement of the fluid at the arterial end of the capillaries is roughly the same as the net force favoring inward movement of fluid at the venous end of the capillaries. So little fluid leaves the circulation. Extra fluid that has left the circulation is returned by the lymphatics. Inadequate lymphatic drainage can lead to an appreciable increase in size of the interstitial space.

The Concept of Fluid and Electrolyte Balance

Despite varying daily intake of food and water, the volume and composition of body fluids remains constant. This suggests that we stay in balance with respect to many substances. A person in a stable balance maintains the same amount of a specified substance in the body over a period of time. The rates of gain or input (due to intake or production in the body) and loss or output (due to excretion or destruction in the body) of a particular substance are exactly equal, so there is no net accumulation or net loss.

If fluids and electrolytes are not in balance, two alternative conditions may occur. If input exceeds output, then accumulation of a substance in the body results, and a **positive balance** exists. For example, if the intake of salt exceeds the rate at which it is excreted, then salt and water will accumulate in the body and extracellular fluid volume and body weight will increase. If output exceeds input, then net loss of a substance from the body results, and a **negative balance** exists. For example, if urinary excretion exceeds the daily intake of potassium, a fall in body potassium and plasma potassium concentration results.

The kidneys play a major role in maintaining balance with respect to water, mineral electrolytes, hydrogen ions, and various organic compounds. They accomplish this by adjusting the output of these substances in the urine in such a way as to match the rate of addition to the body. Renal excretion of water is under the continuous control of the antidiuretic hormone and is normally adjusted to maintain water balance. The kidneys are the primary site of control of electrolyte output. The mineral electrolytes, such as sodium, potassium, and phosphate, are ingested in variable amounts. They are not produced or destroyed in the body, and so, in a balanced system, the amounts taken in are excreted, mostly in the urine. Some substances, such as hydrogen ions or urea, are produced by metabolic reactions in the body. Since these substances should not normally accumulate, they are excreted at a rate matching their production rate.

The Regulation of Water Balance

Table 25–2 presents a balance chart for water for an average 70-kg male. We assume that the person is in water balance—that is, total input and output are equal (2500 ml per day). The input of water is normally derived from the diet. In a hospital setting, intravenous infusions may also be a source of water (and electrolytes). The various beverages we drink contain water. This intake of water is largely conditioned by habit but can also be stimulated by the thirst sensation. Water is also contained in solid food. For example, most fruits and vegetables are 80% to 90% water, and a hamburger is about 55% water. Water is also produced by oxidation of foodstuffs in the body. For example, for each mole of glucose oxidized, six moles of water are produced. (Recall from an earlier discussion the following reaction: $C_6H_{12}O_6 + 6\ O_2 \rightarrow 6\ CO_2 + 6\ H_2O$.)

On the output side, we need to consider loss of water by way of the skin, lungs, gastrointestinal tract, and kidneys. Some water is always lost by way of the lungs and skin. This loss is usually not sensed and so is called **insensible water loss.** On a cold day, however, it is possible to perceive this water loss, because exhaled air is saturated with water vapor, and the water will condense to form visible droplets in cold air. Large quantities of water also can be lost from the skin through sweating. In a hot environment or during heavy exercise, as much

Table 25–2
Daily Water Balance in an Average 70-kg Male

Input		Output	
Water in beverages	1000 ml	Skin and lungs	900 ml
Water in food	1200 ml	Gastrointestinal tract (feces)	100 ml
Water of oxidation	300 ml	Kidneys (urine)	1500 ml
Total	2500 ml	Total	2500 ml

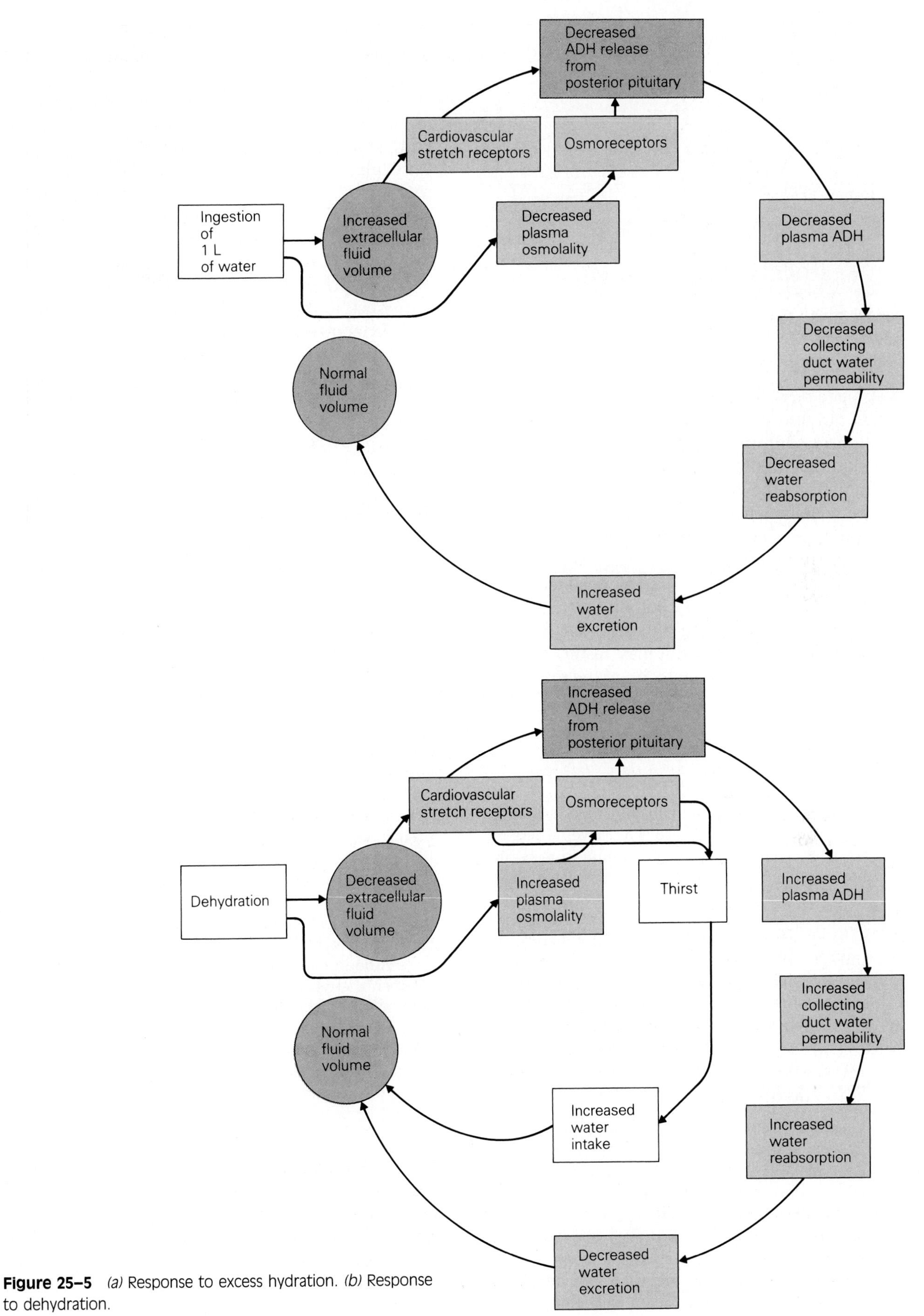

Figure 25–5 *(a)* Response to excess hydration. *(b)* Response to dehydration.

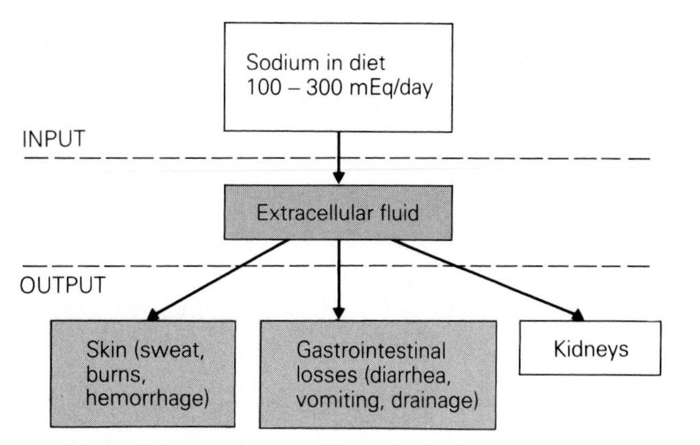

Figure 25–6 Sodium balance. The kidneys are usually responsible for 95% of the sodium output.

NaCl per day). In a hospital setting, intravenous fluids can also be a source of sodium.

On the output side, sodium may be lost from the skin, gastrointestinal tract, or kidneys. Loss of sodium from the skin may occur with sweating, burns, or hemorrhage. Sweat contains 5 to 80 mEq of sodium per liter, and so with heavy exercise or in warm environments, an appreciable amount of sodium may be lost in sweat. Gastrointestinal losses of sodium are normally small, but with diarrhea, vomiting, or external drainage of gastrointestinal secretions, large losses of sodium may result. The urine normally accounts for about 95% of the total output of sodium from the body. The kidneys constitute the major site of control of sodium output and normally adjust excretion so as to maintain balance.

A positive sodium balance results if sodium input exceeds output. Since water is retained along with sodium, and since sodium is primarily an extracellular ion, extracellular fluid volume will be increased. If sufficient salt and water are retained, expansion of the interstitial fluid space leads to generalized edema. This condition is associated with weight gain (due to the extra water) and skin puffiness, especially in the feet and ankles (due to the effects of gravity). A positive sodium balance and generalized edema occur in a number of clinically important conditions, including congestive heart failure, hepatic cirrhosis (a liver disease), and nephrotic syndrome (a kidney disease). In order to reduce the edema, it is common practice to (1) put the patient on a low salt diet and (2) use diuretic drugs to promote sodium excretion by the kidneys. In this way, the physician tries to relieve the edema fluid and bring the patient back into sodium balance.

A negative sodium balance results if sodium output exceeds input. This may occur if there is excessive loss of sodium via the skin, gastrointestinal tract, or kidneys. A negative sodium balance leads to a reduction of extracellular fluid volume. In turn, this produces a decrease in plasma volume and con-

Focus on Sodium Appetite

Mammals generally maintain the amount of sodium in their bodies at a constant level. There are, however, obligatory losses of sodium in the urine and in other body secretions (e.g., sweat, saliva), for which the body must compensate. In carnivores, replacement is not a problem, since meat contains large amounts of sodium. For herbivores, however, there may be a serious lack of sodium in the diet, since most vegetation has a relatively low salt content.

Sodium intake in people is quite variable. Some people automatically salt their food without tasting it, and other people never add salt to their food. A typical American diet contains more than enough sodium.

A craving for salt in response to a real need for sodium may occur in people. Interestingly, many patients with Addison's disease show an unusual desire for salt. In Addison's disease there is deficient adrenal production of aldosterone, so that the patient tends to lose excessive amounts of sodium in the urine. Increased salt intake in this disease would be of obvious value in maintaining a normal extracellular fluid volume. The medical literature contains many descriptions of excessive salt intake in people with Addison's disease. For example, one 34-year-old man with marked Addison's disease is reported to have put approximately a 1/8-inch layer of salt on his steak, used nearly 1/2 a glass of salt for his tomato juice, used salt on oranges and grapefruit, and made lemonade with salt!

Laboratory rats presented with two drinking bottles, one containing tap water and the other a 1% solution of sodium chloride, fail to discriminate between the two solutions and drink about the same amount from both bottles. After removal of both adrenal glands, however, they show a marked preference for the salt-containing solution.

It is well known that herbivores, such as deer and cattle, are attracted to salt-enriched mineral deposits (salt licks). The need for salt is greatly increased during pregnancy or lactation.

The physiological basis for sodium appetite remains incompletely understood. A decreased blood volume, resulting from diminished extracellular fluid sodium, may be a critical internal signal. Some studies have suggested that angiotensin II and mineralocorticoids are involved in sodium appetite, but more research needs to be done on this subject.

sequently may lead to an inadequate circulation, hypotension, and even death.

It should be clear that sodium balance is closely related to regulation of extracellular fluid volume. This is so because sodium salts are the major solutes of the extracellular fluid, as discussed earlier. Since the osmolality of the extracellular fluid is closely regulated by ADH, the kidneys, and thirst, it follows that retention or loss of sodium from the body will be accompanied by parallel changes in the amount of water in (and hence volume of) the extracellular fluid. The regulation of sodium balance falls largely upon the kidneys, although there is some evidence for a **sodium appetite** (which stimulates sodium intake) when there are sodium deficits. Since the kidneys are intimately concerned with control of sodium balance and regulation of extracellular fluid volume, it should not be surprising that extracellular fluid volume is a major determinant of renal sodium excretion. We will elaborate on this later.

There is compelling evidence that hypertension (high blood pressure) may often be due to a disturbance in sodium (salt) balance. Excessive dietary intake of salt or inadequate renal excretion of salt tends to increase intravascular fluid volume; somehow this is translated into an increase in blood pressure. Avoidance of a high salt intake may help to prevent high blood pressure. A reduced salt intake and diuretic drugs are used to keep blood pressure under control in people with hypertension.

The Renal Response to Changes in Dietary Intake of Sodium

Let us consider the renal response to changes in dietary intake of sodium. Figure 25–7 illustrates a balance study. A person was put on a fixed sodium intake of 100 mEq per day. During the first three days of the study (the control period), the person was in a stable balance, as indicated by the fact that urinary excretion matched the sodium intake (in this study, we will neglect extrarenal losses of sodium, which are usually small). On day 4, the sodium intake was increased to 300 mEq per day and was maintained at that level until the end of day 10. What happened? Sodium excretion rose, but for the first few days (days 4 to 7) was less than the sodium input. Consequently, there was a phase of positive sodium balance. During this time, the person retained salt, and since the subject was allowed free access to water, water intake increased, and salt and water were retained in isosmotic proportions. If we measured plasma sodium concentration, we would find that it was not detectably altered throughout

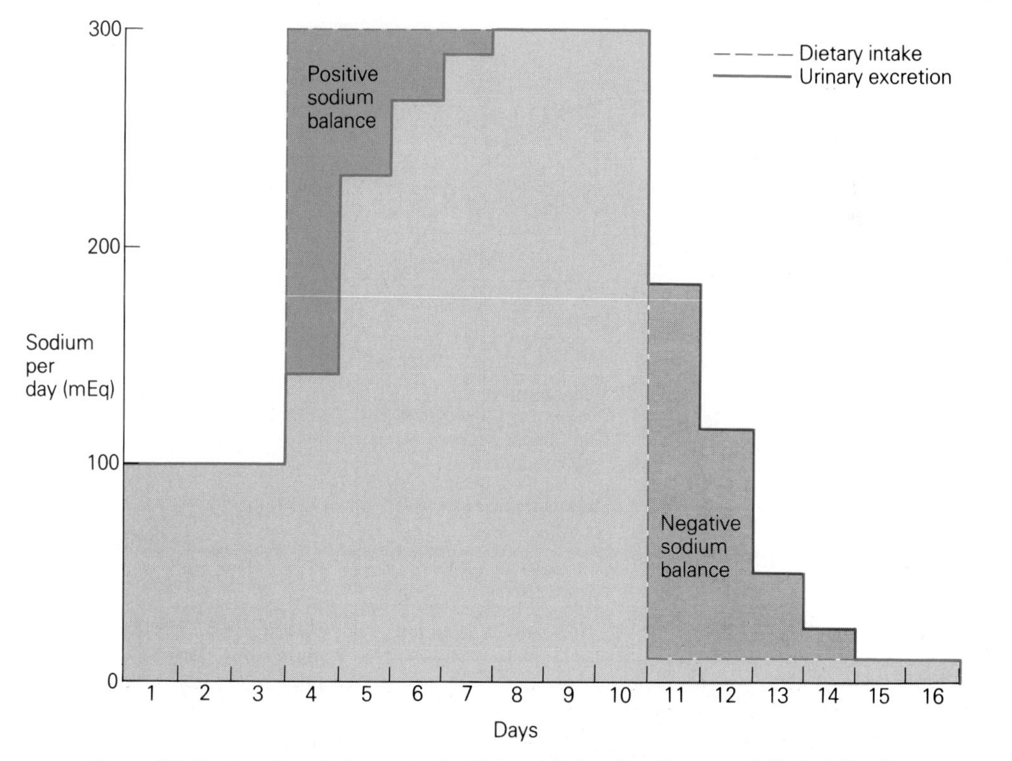

Figure 25–7 A sodium balance study. Dietary intake of sodium was initially 100 mEq per day, was then increased to 300 mEq per day (days 4 to 10), and was finally reduced to 10 mEq per day (days 11 to 16). Daily urinary excretion of sodium (solid orange lines) was measured.

this study. The sodium content of the body, the extracellular fluid volume, and body weight, however, increased during days 4 to 7. During days 8 to 10, the subject was back in a stable balance—that is, input and output of sodium were once again equal. When the person switched to a low sodium diet (10 mEq per day), sodium excretion fell. During days 11 to 14, sodium excretion was higher than sodium intake, and so a negative sodium balance resulted. During this time, extracellular fluid volume and body weight decreased. Finally, a new stable balance on the low salt diet was established (days 15 and 16).

This study demonstrates that changing the dietary sodium intake leads to appropriate changes in sodium excretion by the kidneys. An increase in sodium intake leads to increased urinary sodium excretion, and a decrease in sodium intake results in decreased sodium excretion. The kidneys can adjust sodium excretion over a wide range. Note, however, that the renal response is sluggish, and so it takes a few days to reach the appropriate rate of renal sodium excretion. During the time that renal sodium excretion is less than sodium intake (days 4 to 7), extracellular fluid volume is increasing. Extracellular fluid volume is stable, but higher than the control volume, during days 8 to 10. The increased extracellular fluid volume may be thought of as the stimulus that promotes renal sodium loss. When the sodium intake is decreased, a fall in extracellular fluid volume may be thought of as the cause of the reduced sodium excretion.

Effector Mechanisms Activated by Altered Extracellular Fluid Volume

What mechanisms allow the kidneys to adjust their output of sodium to maintain balance and relative stability of extracellular fluid volume? Figure 25–8 summarizes how sodium excretion may be controlled by a negative feedback system. According to this scheme, the controlled variable is extracellular fluid volume. Changes in extracellular fluid volume are sensed by cardiovascular volume receptors and by the kidneys. The effector mechanisms include changes in (1) glomerular filtration rate, (2) plasma aldosterone levels, (3) peritubular capillary Starling forces, (4) renal sympathetic nerve activity, (5) intrarenal blood flow distribution, and (6) plasma atrial natriuretic factor. There may be additional effector mechanisms. Changes in these mechanisms lead to changes in sodium excretion and so act to bring extracellular fluid volume back to normal.

We will first discuss the six effector mechanisms just listed and then the nature of the sensors and what they may actually be sensing. Sodium excretion represents the difference between filtered and reabsorbed amounts, and so factors that affect so-

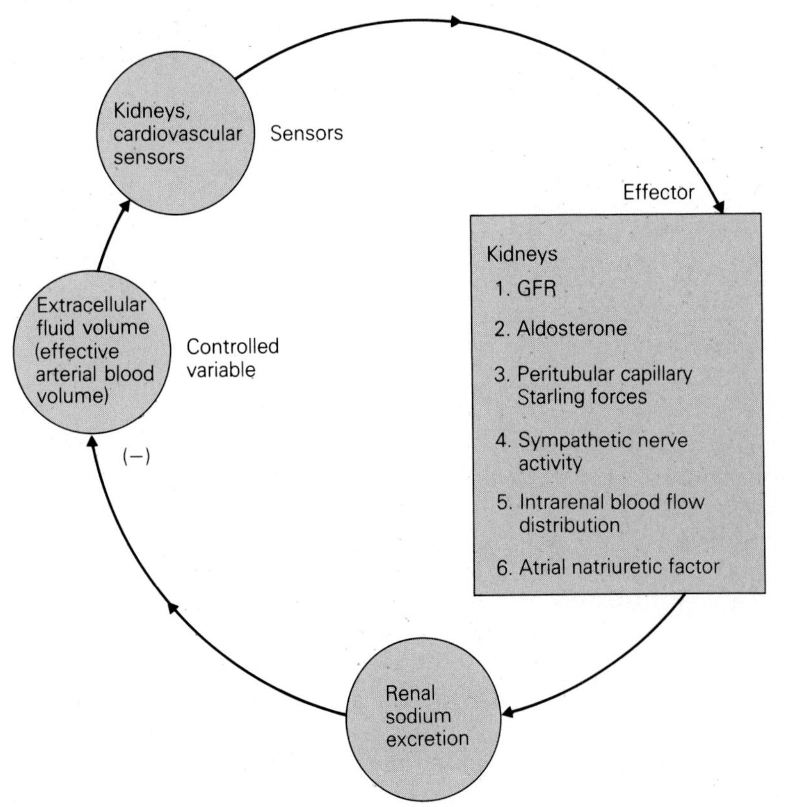

Figure 25–8 Regulation of extracellular fluid volume (or effective arterial blood volume) by a negative feedback control system. Changes in effective arterial blood volume are sensed by cardiovascular and renal sensors, and are translated into changes in renal effector pathways, leading to changes in sodium excretion that bring effective arterial blood volume back to normal.

tabolism. Tissue growth and repair require potassium. Conversely, tissue breakdown or increased protein catabolism lead to release of cell potassium into the extracellular fluid.

The normal range for plasma potassium concentration is 3.5 to 5.5 mEq per liter. A plasma potassium concentration below 3.5 mEq per liter is called **hypokalemia.** This condition may cause skeletal muscle weakness and even paralysis. A plasma potassium concentration above 5.5 mEq per liter is called **hyperkalemia.** Plasma potassium concentrations above 7 mEq per liter demand immediate attention because of possible cardiac irregularities. A plasma potassium concentration greater than 10 to 12 mEq per liter is usually fatal; the cause of death is cardiac arrhythmia or arrest. The changes in skeletal and cardiac muscle function that occur with abnormal plasma potassium levels result from the influence of potassium on cell membrane potentials.

Figure 25–10 summarizes potassium balance. Normally, we ingest about 100 mEq per day of potassium. The amount of potassium we take in can be quite variable, depending on what we eat and drink. About 10% of the daily potassium intake is excreted in the feces, and about 90% in the urine. Loss of potassium in sweat (which contains 5 to 10 mEq potassium per liter) is usually negligible. The kidneys are clearly the major site of potassium loss from the body, and they are the major site of control of potassium balance.

Figure 25–10 also re-emphasizes that most of the body's potassium is inside cells. The amount of potassium in extracellular fluid can be calculated from its concentration (4.5 mEq per liter) and the extracellular fluid volume (about 14 liters) and is equal to about 60 mEq. This amounts to only 2% of the body's potassium.

Several factors affect the distribution of potassium between cells and extracellular fluid. First, activity of the cell membrane Na,K−ATPase, which pumps potassium into cells, is of key importance. Impaired cell metabolism or cardiac glycosides (e.g., digitalis) inhibit the pump, and tend to produce lower cell and higher extracellular potassium levels. Second, acid-base status affects the distribution of potassium between intracellular and extracellular compartments. A fall in plasma pH causes a shift of hydrogen ions into cells in exchange for cell potassium ions. The consequence then is a rise of plasma potassium concentration. Third, the availability of insulin is an important factor affecting potassium distribution. Insulin promotes the movement of potassium into skeletal muscle and liver cells; lack of insulin causes more of the body's potassium to stay outside of cells. Fourth, shifts of potassium from cells to extracellular fluid will occur with cell break-

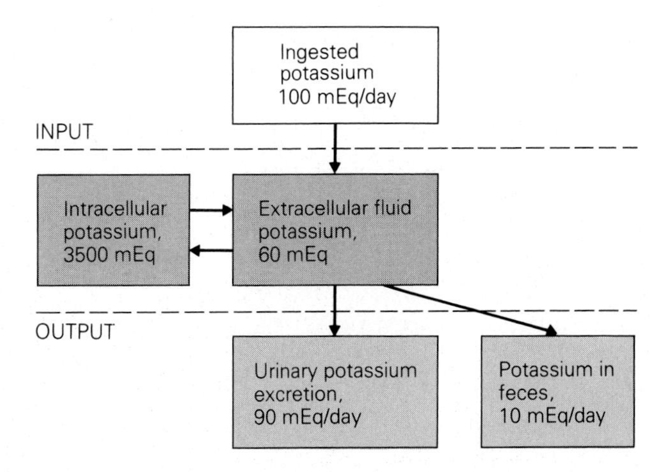

Figure 25–10 Potassium balance (input-output) and the distribution of potassium between the two major body fluid compartments.

down, due to tissue trauma, infection, ischemia (inadequate blood flow), or heavy exercise. Since so much of the body's potassium is within cells, a small loss from or gain by this compartment could profoundly affect the plasma potassium concentration.

Plasma potassium must be closely regulated. Part of this regulation involves the hormones insulin, epinephrine, and aldosterone, which stimulate the uptake of potassium by cells. The kidneys also play an important part in regulating plasma potassium. When increased potassium loads are presented to the body, the kidneys normally can rapidly excrete excess potassium.

In Chapter 24, we discussed some of the factors that affect potassium excretion by the kidneys. Recall that most of the filtered potassium is reabsorbed by early portions of the nephron and that the collecting ducts secrete potassium. Some of the factors that affect potassium excretion include intracellular potassium concentration, aldosterone, excretion of anions, and urine flow rate.

Figure 25–11 shows the mechanisms whereby an increase in potassium intake in the diet leads to increased renal potassium excretion. Increased potassium intake tends to raise the plasma potassium concentration. By a direct effect on the adrenal cortex, an elevated extracellular potassium concentration stimulates the release of aldosterone. Aldosterone then travels in the blood to the kidneys, where it acts on the principal cells of the collecting duct. It increases activity of the Na,K−ATPase in the basolateral cell membrane and potassium permeability of the luminal membrane of these cells, thereby favoring potassium secretion (see Figure 24–15). An increased potassium intake raises the potassium concentration in most body cells, including the principal cells, and so potassium secretion is directly increased in this way.

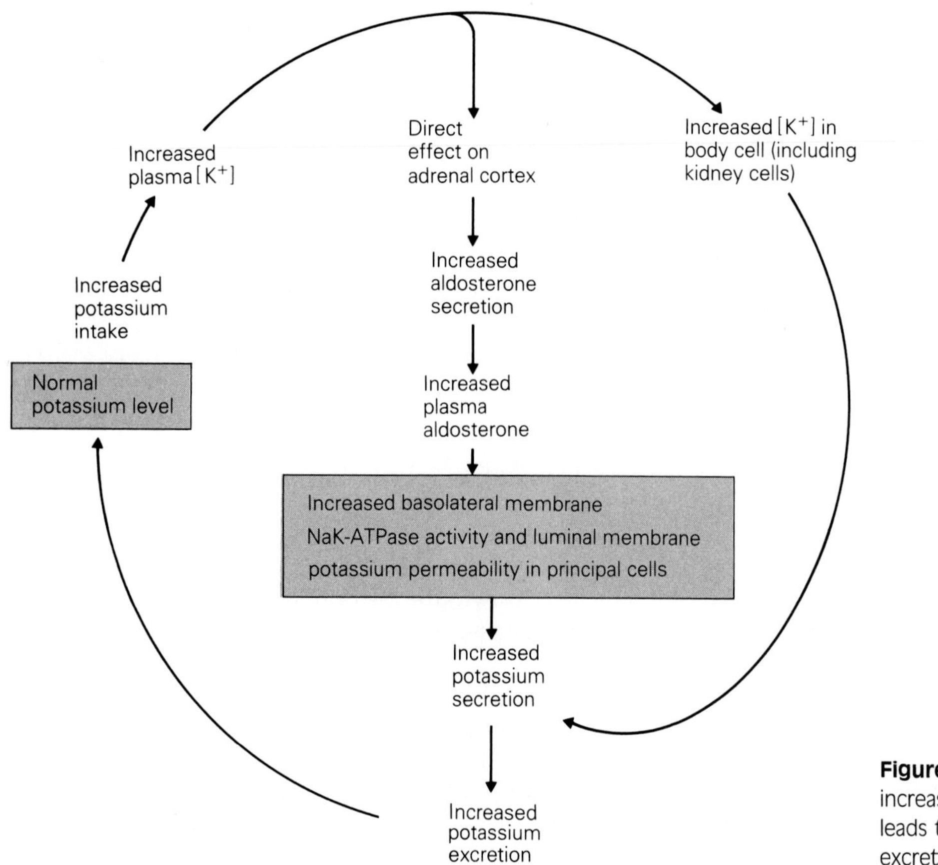

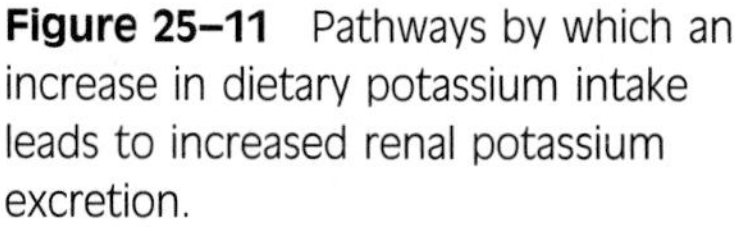
Figure 25–11 Pathways by which an increase in dietary potassium intake leads to increased renal potassium excretion.

The opposite changes occur if potassium intake is reduced. Plasma and cell potassium levels tend to fall, aldosterone secretion is inhibited, and secretion of potassium by the collecting duct is diminished. The result is a decreased rate of potassium excretion.

The kidneys normally do a remarkable job of maintaining potassium balance. But abnormal renal excretion of potassium (either too much or too little) is a common cause of disturbed potassium balance.

Too little renal excretion of potassium causes hyperkalemia. Most foods are rich in potassium, so continued food intake with inadequate renal excretion leads to a positive potassium balance. In acute renal failure—that is, with sudden renal shutdown—a life-threatening hyperkalemia may develop. Often compounding inadequate renal excretion (in patients with acute renal failure) are tissue trauma, infection, and acidosis, all of which tend to raise the plasma potassium level. Chronic renal disease leads to hyperkalemia when the filtration rate falls below 15 to 20 ml/min. Up to that point, hyperkalemia does not develop because each surviving nephron can excrete a greater-than-normal quota of potassium. In adrenocortical insufficiency (Addison's disease), secretion of aldosterone is inadequate. A lowered plasma aldosterone level results in decreased stimulation of renal potassium secretion and excretion, so that ingested potassium tends to be retained in the body.

Too much excretion of potassium by the kidneys leads to hypokalemia. With adrenocortical hyperfunction, aldosterone levels are abnormally high, and potassium secretion is stimulated. Treatment with glucocorticoids can also result in excessive potassium excretion. In a number of kidney diseases, the ability of the tubules to reabsorb filtered potassium is impaired and excessive potassium excretion results. Excessive renal potassium loss occurs in uncontrolled diabetes mellitus. This occurs for several reasons: (1) an osmotic diuretic effect of glucose (increased urine output produces potassium loss), (2) increased production of ketone body acids (excretion of anions results in increased excretion of cations), and (3) increased plasma aldosterone levels (secondary to volume contraction caused by excessive loss of salt and water in the urine). Diuretic drugs (used in the treatment of hypertension and edema) are the most common cause of abnormally high renal potassium loss. This effect is probably due mostly to the increased rate of fluid flow through the collecting ducts that these agents produce. Patients taking diuretics are often advised to eat bananas, a rich source of potassium.

Excessive amounts of potassium can also be lost from the body through the gastrointestinal tract. Diarrheal fluid may have a potassium concentration of 50 to 60 mEq per liter, so a disturbance of the lower gastrointestinal tract may rapidly lead to potassium depletion. Potassium depletion also develops during vomiting, but the reasons are complex. First, gastric juice contains some potassium (about 5 to 10 mEq per liter), which would account for some of the loss. Secondly, food (potassium) intake is reduced. Finally, the kidneys actually increase potassium excretion. This is due in part to the metabolic alkalosis that develops. It is also due to volume depletion, which stimulates aldosterone secretion via renin and angiotensin II. This promotes sodium conservation by the kidneys but can lead to enhanced potassium excretion.

Kidney Failure and Its Treatment

Kidney failure can occur at any age. It may develop suddenly and lead to death in a few days, or it may be due to a chronic renal disease that progresses slowly over decades. In chronic kidney disease, there is a progressive loss of nephrons. Kidney disease has a variety of causes, among which are bacterial infections, congenital defects, immunological injury, impaired renal blood flow, toxins, tumors, and urinary tract obstruction. The specific type of kidney disease can usually be diagnosed from the patient's history and clinical tests. In all forms of renal failure, however, the signs and symptoms are very similar. This symptom complex has been called **uremia** (urine in the blood). Basically, uremia indicates a disturbed internal environment due to renal failure. The disturbed internal environment impairs the function of all body cells.

Disturbances in Renal Failure

Following are some of the disturbances that occur in uremia. This alphabetical list is not comprehensive but should serve to re-emphasize many of the important functions of normal healthy kidneys.

Anemia, with a fall in blood hematocrit to 20% to 35%, may develop in chronic kidney failure. The anemia is mainly due to deficient production of **erythropoietin,** a hormone produced mainly by the kidneys. Erythropoietin stimulates the bone marrow to produce red blood cells (erythropoiesis). Recently, human erythropoietin has been synthesized in the laboratory by recombinant DNA techniques. When administered to renal failure patients, it was found to cure the anemia.

Azotemia, the accumulation of nitrogenous waste products in the blood, develops as a consequence of a markedly decreased GFR. Plasma urea, creatinine, and uric acid levels are elevated.

Calcium and phosphate metabolism are altered with renal failure. In uremia, plasma calcium is typically decreased, and plasma phosphate is increased. Secondary hyperparathyroidism, deficient synthesis of the active form of vitamin D by the kidneys, and acidosis all lead to bone disease.

Hypertension (high blood pressure) is often present in renal failure. Its origin is not always known, but there is evidence for several possible pathways. The diseased kidney may produce excessive amounts of renin. Renin causes increased production of angiotensin II (a potent vasoconstrictor) and aldosterone (which would favor salt retention). The inability of the kidneys to excrete a normal amount of sodium could also lead to hypertension owing to excess volume. Finally, the diseased kidneys may fail to produce vasodilator substances.

Metabolic acidosis typically accompanies severe renal failure. Acid end products of metabolism are not excreted at a normal rate and hence accumulate in the body. Impaired renal synthesis of ammonia contributes to the decreased ability to excrete hydrogen ions.

Potassium balance is usually well maintained until GFR falls below 15 to 20 ml/min. Retention of potassium can cause plasma levels to rise to lethal levels.

Sodium balance is typically impaired in patients with severe renal failure. On a normal sodium diet, excessive salt and water may be retained, leading to generalized edema. On a low-salt diet, patients may become salt-depleted.

Urinary concentrating ability is impaired with renal failure. Urine with an osmolality close to that of plasma is excreted.

Numerous other cardiovascular, gastrointestinal, muscular, and nervous abnormalities occur in the uremic state. In many instances, it is not understood why renal failure causes these disturbances.

Treatment by Dialysis

Most of the signs and symptoms of uremia can be relieved by dialysis. **Dialysis** is the separation of smaller from larger molecules in solution by diffusion of the small molecules through a selectively permeable membrane. In treating uremia, two methods of dialysis may be used.

In **peritoneal dialysis,** about 2 liters of a balanced salt solution is introduced into the abdominal cavity. The peritoneum acts as a dialyzing membrane. Small molecules and ions (e.g., urea, potassium) diffuse into the injected solution. The solution is drained and discarded. This procedure is used by some patients with chronic renal failure on an outpatient basis and is called **continuous ambulatory peritoneal dialysis.** The patient can perform self-dialysis and usually does this several times in a day.

Hemodialysis (Fig. 25–12) is a more efficient process involving an **artificial kidney.** The patient's blood is pumped through cellophanelike tubing

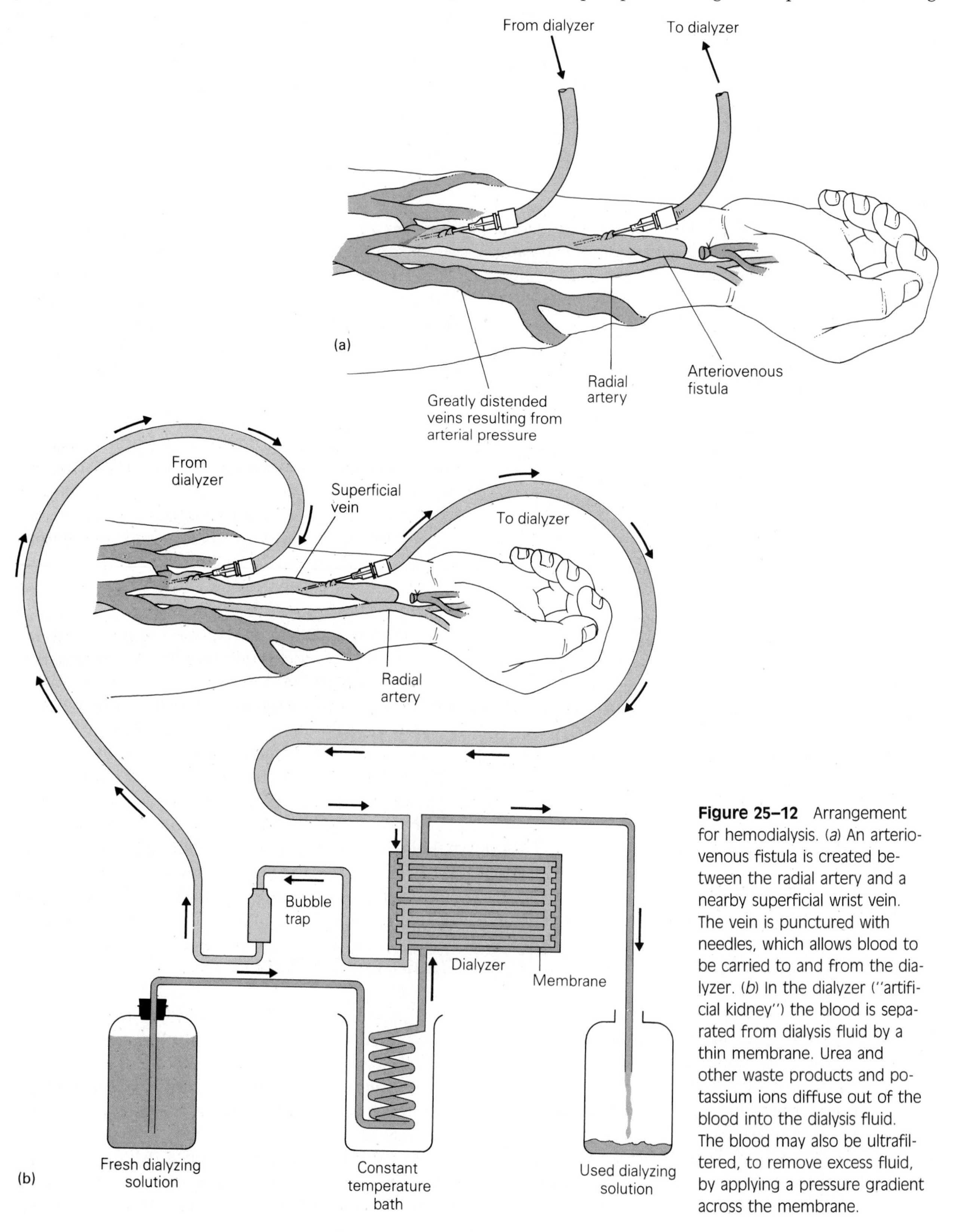

Figure 25–12 Arrangement for hemodialysis. (*a*) An arteriovenous fistula is created between the radial artery and a nearby superficial wrist vein. The vein is punctured with needles, which allows blood to be carried to and from the dialyzer. (*b*) In the dialyzer ("artificial kidney") the blood is separated from dialysis fluid by a thin membrane. Urea and other waste products and potassium ions diffuse out of the blood into the dialysis fluid. The blood may also be ultrafiltered, to remove excess fluid, by applying a pressure gradient across the membrane.

immersed in a bath of balanced salt solution. Urea, potassium, and other substances in excess in the body diffuse through the tubing membrane out of the blood into the bath. Patients are usually dialyzed three times a week, and they can often undergo this procedure at home for 4 to 6 hours while watching television or reading.

Dialysis can allow patients with otherwise fatal chronic renal failure to live useful and productive lives. It has some drawbacks, however. For one thing, it does not correct the anemia. It controls hypertension only with difficulty. Uremia develops between the periods of dialysis. There is a constant risk of infection and, with hemodialysis, hemorrhage. Dialysis does not maintain normal growth and development in children. Finally, it is costly. Dialysis is, however, invaluable in keeping patients alive and functioning until a suitable kidney transplant becomes available.

Transplantation of the Kidney

Renal transplantation is the only "cure" for patients with terminal chronic renal disease. It may restore complete health and function. In 1985, almost 8000 renal transplant operations were performed in the United States. At present, about 90% of kidneys grafted from a living donor related to the patient function for one year; grafts derived from an unrelated person who has just died (cadaver donors) have a one-year survival of about 80%.

The main problem in transplantation is the immunological rejection of the transplanted kidney by the host. A better understanding of immunology should lead to greater success with organ transplantation.

Functions of Ureters and Urinary Bladder

The kidneys form urine constantly. Urine is conveyed to the urinary bladder by the ureters and stored there until the bladder is emptied by way of the urethra.

The Ureter

The two ureters are muscular tubes that carry the urine from the pelvis of each kidney to the urinary bladder. Peristaltic movements originating in the pelvis force the urine along the ureters toward the bladder. These muscular contractions are probably myogenic in origin. They help to overcome the gradually increasing pressure as the urine accumulates in the bladder. The ureters enter the base of the bladder obliquely, thus forming a valvular flap that passively prevents reflux of urine during bladder contraction. The ureters are innervated by sympathetic and parasympathetic nerve fibers. Afferent sensory nerve fibers are present, as evidenced by the excruciating pain felt when a stone is present in the ureter.

The Urinary Bladder

The urinary bladder (see Fig. 10–8) is a distensible hollow organ containing smooth muscle in its wall. The muscle is called the **detrusor,** which means in Latin "that which pushes down." The neck of the bladder **(internal sphincter)** also contains smooth muscle. The body of the bladder and bladder neck are innervated by parasympathetic **pelvic nerves** and sympathetic **hypogastric nerves.** The external sphincter is composed of skeletal muscle and is innervated via somatic nerves, the **pudendal nerves.** Pelvic, hypogastric, and pudendal nerves contain both motor and sensory fibers.

The bladder has two functions: (1) to serve as a distensible reservoir for storage of the urine, and (2) to empty its contents at suitable intervals. As the bladder fills, it adjusts its wall tension to its capacity, so that minimal increases in bladder pressure occur. The external sphincter is kept closed via impulses along the pudendal nerves. The first awareness of bladder filling occurs at a volume of about 100 to 150 ml, and the desire to void is usually experienced when the bladder contains 150 to 250 ml of urine. A person becomes uncomfortably aware of a full bladder when the volume is 350 to 400 ml; at this volume, pressure in the bladder is about 10 cm of water. With further increases in volume, bladder pressure rises steeply, in part due to reflex contractions of the detrusor. An increase in volume to 700 ml creates pain and often loss of control. The sensations of bladder filling, of conscious desire to void, and of painful distension are mediated by afferent fibers in the pelvic nerves.

Micturition

The periodic complete emptying of the urinary bladder is called **micturition** (or **urination**). It is a complex act involving coordinated activity over both autonomic and somatic nerve pathways and involves several reflexes that can either be facilitated or inhibited by higher centers in the brain. The basic reflexes occur at the level of the sacral spinal cord and are modified by centers in the midbrain and cerebral cortex. Distension is sensed by stretch receptors in the bladder wall; these induce reflex contraction of the bladder and relaxation of internal and external sphincters. This reflex is released by removing inhibitory impulses from the cerebral cortex. Fluid flow through the urethra reflexly causes further contraction of the detrusor and relaxation of the external sphincter. Increased parasympathetic nerve activity is responsible for contraction of the detrusor and relaxation of the internal sphincter. During micturition, the perineal and levator ani muscles relax, which shortens the urethra and decreases its resistance. Descent of the diaphragm and contraction of abdominal muscles raises intra-abdominal pressure, thus aiding in the expulsion of urine from the bladder.

Fortunately, micturition is under voluntary control. In young children, however, it is purely reflex and occurs whenever the bladder is sufficiently distended. At about $2\frac{1}{2}$ years of age, it begins to come under cortical control, and complete control is usually achieved by 3 years of age.

Summary

I. A. The human body is about 60% water by weight. Two thirds of body water is in the intracellular fluid compartment, and one third is in the extracellular fluid compartment.

B. Potassium ions are the major osmotically active solutes in cells; potassium influences cell volume, excitability, and metabolism. Sodium and its accompanying anions, chloride and bicarbonate, comprise the major osmotically active solutes in extracellular fluid. The amount of water in, and hence volume of, the extracellular fluid space is determined primarily by the amount of sodium in this compartment.

C. In general, cell membranes are highly permeable to water, and so osmotic equilibrium between intracellular and extracellular fluids results.

D. In the vascular system, the colloid osmotic pressure exerted by the plasma proteins keeps fluid within the circulation.

II. A. The kidneys play a major role in keeping us in a stable balance with respect to water, sodium and potassium ions, and numerous other substances. Abnormal renal excretion is a common cause of imbalances (positive or negative balances).

B. Plasma osmolality is regulated by renal excretion of water, which is controlled by ADH, and by the thirst mechanism. ADH release and thirst are stimulated by cellular dehydration (an increase in effective plasma osmotic pressure) and by extracellular dehydration (hypovolemia, a decrease in effective arterial blood volume). Thirst is an emergency mechanism. Under ordinary circumstances, the main stimulus for ADH release is a rise in plasma osmolality detected by osmoreceptor cells in the hypothalamus.

C. The kidneys are the major site of sodium output and regulation of extracellular fluid volume. An increase in extracellular fluid volume (or effective arterial blood volume) leads to increased renal sodium excretion and vice versa. Renal excretion of sodium represents the difference between filtered and reabsorbed amounts. The following factors influence sodium excretion: (1) glomerular filtration rate, (2) mineralocorticoids (aldosterone), (3) peritubular capillary Starling forces, (4) renal sympathetic nerve activity, (5) intrarenal blood flow distribution, (6) atrial natriuretic factor, (7) glucocorticoids, (8) estrogens, (9) osmotic diuretics, (10) poorly reabsorbed anions, and (11) diuretic drugs.

D. Most of the body's potassium is within cells. The kidneys normally maintain potassium balance by excreting most of the ingested potassium. Changes in dietary potassium intake cause changes in plasma aldosterone levels and kidney cell potassium concentration, and in this way change secretion and excretion of potassium by the kidneys.

III. A. No matter what the cause, with renal failure, the symptom complex (uremia) that develops is much the same and reflects an abnormal internal environment.

B. Renal failure may be treated with dialysis, but dialysis is not a cure.

C. Successful renal transplantation is the best hope for patients with chronic renal failure.

IV. A. The two ureters are muscular tubes that carry the urine from the kidneys to the bladder.

B. The urinary bladder functions as a reservoir for urine and is periodically emptied (micturition).

C. Micturition is a complex act involving autonomic and somatic nerves, spinal reflexes, and higher brain centers.

Review Questions

1. For an average 70-kg young adult male, what are the approximate volumes (in liters) of the intracellular fluid, extracellular fluid, interstitial fluid-lymph, and plasma compartments?
2. How does plasma differ from interstitial fluid in chemical composition? How does transcellular fluid differ from interstitial fluid?
3. Which of following is present at a higher concentration in cells than in extracellular fluid: Na^+, K^+, Ca^{2+}, Mg^{2+}, Cl^-, HCO_3^-, proteins?
4. What effect would ingestion of a liter of 150 mM NaCl (isotonic saline) have on: (1) intracellular fluid volume, (2) extracellular fluid volume, (3) plasma osmolality, (4) plasma ADH concentration, (5) plasma renin activity, and (6) plasma aldosterone concentration?
5. Why is thirst (like shivering) considered to be mainly an emergency mechanism?
6. Drinking a liter of water results in a large increase in urine output. What are the intermediate steps responsible for this response?
7. A person with severe congestive heart failure has the following signs and symptoms: (1) thirst, (2) high plasma ADH levels, (3) low plasma osmolality (hyponatremia), (4) reduced renal sodium excretion, and (5) massive edema. What are the probable causes for these abnormalities?
8. How is ADH thought to produce its effects on the collecting duct epithelium?
9. What are six factors that *might* contribute to the enhanced sodium excretion that results from an increased intake of sodium?
10. How do renin and angiotensin II participate in the regulation of blood pressure and extracellular fluid volume?
11. What two factors contribute to decreased renal potassium excretion following a decrease in dietary potassium intake?
12. What are two functions of the urinary bladder, and how are these functions accomplished?

Suggested Readings

Brenner, B., Coe, F.L., and Rector, F.C., Jr. *Renal Physiology in Health and Disease.* Philadelphia: W.B. Saunders Company, 1987.

Pitts, R.F. *Physiology of the Kidney and Body Fluids,* 3rd ed. Chicago: Year Book Medical Publishers, 1974.

Valtin, H. *Renal Function: Mechanisms Preserving Fluid and Solute Balance in Health,* 2nd ed. Boston: Little, Brown and Company, 1983.

26

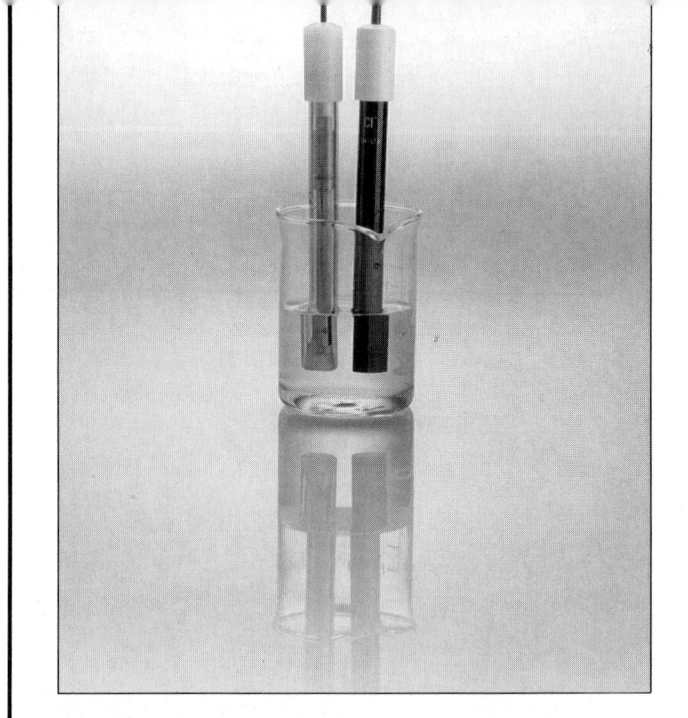

Regulation of Acid–Base Balance

The **hydrogen ion concentration, [H^+],** of the extracellular fluid is one of the most important of all homeostatic states. Many abnormalities in the functioning of cells, tissues, and organs result if the extracellular fluid is too acidic or too alkaline (basic). This fact is not surprising, since hydrogen ion concentration affects the electrical charge and hence configuration and binding properties of proteins. Hydrogen ion concentration, typically expressed in terms of pH, influences (1) the speed of chemical reactions in the body (catalyzed by protein enzymes), (2) cell permeability properties, and (3) cell structure.

This chapter will focus upon the regulation of the pH of the extracellular fluid. It is obviously important for a multicellular organism to provide a suitable "internal environment," a controlled environment for its cells. As you know, the extracellular fluid is this internal environment, and pH is one of the most important conditions in that environment.

In clinical medicine as well as in physiology, **systemic arterial blood** is used to evaluate acid–base status. The pH of whole blood, measured with a pH meter, is actually the pH of the plasma (not the cells) and so is a measurement of extracellular fluid pH.

In terms of the functioning of individual cells, although extracellular pH is important, the pH within cells is probably even more important. Usually, though not always, pH within cells and pH outside of cells change in the same direction. By maintaining a normal extracellular pH, the body provides for some stability of intracellular pH.

We will first review some fundamental definitions of acid–base chemistry, expanding on the introduction provided in Chapter 2. Then we will consider how changes in hydrogen ion concentration are minimized by buffers in the body. These buffers include chemical buffers in extracellular and intracellular fluids and in bone, and the physiological buffers, the lungs and kidneys. We will discuss briefly how the lungs participate in acid–base regulation by controlling the partial pressure of CO_2 (P_{CO_2}) in arterial blood. Then we will consider how the kidneys eliminate extra acid or base from the body. We will describe the four fundamental types of acid–base disturbances, their characteristics and causes, and ways in which the hydrogen body attempts to compensate for disturbances of pH. Finally, we will comment briefly on regulation of intracellular pH.

Principles of Acid–Base Physiology

Definition of Acid and Base

As first presented in Chapter 2, an **acid** is a substance that can donate hydrogen ions. A hydrogen ion (H^+) consists of a bare proton, a hydrogen atom without its orbiting electron. Examples of acids include **hydrochloric acid (HCl), sulfuric acid (H_2SO_4), nitric acid (HNO_3), phosphoric acid (H_3PO_4), ammonium ion (NH_4^+), lactic acid, acetic acid,** and **carbonic acid (H_2CO_3).** An acid will donate its hydrogen ion to a base.

Accordingly, a **base** is a substance that can accept or bind hydrogen ions. Examples of bases include **sodium hydroxide (NaOH), potassium hydroxide (KOH), ammonia (NH_3),** and **lactate, acetate,** and **bicarbonate (HCO_3^-)** ions.

Not all acids have the word *acid* in their names. For example, the ammonium ion (NH_4^+) is an acid. If we mix aqueous solutions of the salts ammonium chloride (NH_4Cl) and sodium bicarbonate ($NaHCO_3$), the following reversible reaction takes place:

$$NH_4^+Cl^- + Na^+HCO_3^- \rightleftharpoons NH_3 + H_2CO_3 + Na^+Cl^-$$

In this reaction (left to right), the ammonium ion is an acid; it donates a hydrogen ion to the base bicarbonate.

The Neutralization Reaction

In a **neutralization reaction,** an acid reacts with a base to form a salt and water. For example, in the reaction

$$HCl + NaOH \longrightarrow NaCl + H_2O$$

the acid HCl donates its hydrogen ion to the hydroxide ion of NaOH to form water and sodium chloride (a salt).

Quantifying Acids and Bases: Milliequivalents

The amount of an acid or base is often expressed in terms of **equivalents** or **milliequivalents (mEq)** (1 mEq is 1/1000 of an equivalent). One equivalent of acid will neutralize one equivalent of base. One mole of HCl contains one equivalent of acid. One mole of H_2SO_4 contains two equivalents of acid because sulfuric acid can donate two hydrogen ions. Therefore, two moles (two equivalents) of NaOH are required to neutralize one mole (two equivalents) of sulfuric acid, according to the following two steps:

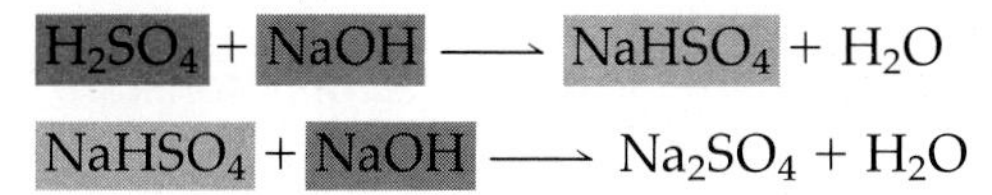

$$H_2SO_4 + NaOH \longrightarrow NaHSO_4 + H_2O$$

$$NaHSO_4 + NaOH \longrightarrow Na_2SO_4 + H_2O$$

Amphoteric Substances

Some chemical substances can function as both an acid and a base and are referred to as **amphoteric substances** (*amphoteros* in Greek means "pertaining to both"). These include amino acids and proteins. For example, the amino acid **glycine,** $^+H_3N{-}CH_2{-}COO^-$, acts as an acid in the following reaction:

$$NaOH + {}^+H_3N{-}CH_2{-}COO^- \longrightarrow H_2N{-}CH_2{-}COO^-Na^+ + H_2O$$

The ammonium group of glycine donates a hydrogen ion to the hydroxide ion of NaOH. Glycine acts as a base in the following reaction:

$$HCl + {}^+H_3N{-}CH_2{-}COO^- \longrightarrow Cl^{-+}H_3N{-}CH_2{-}COOH$$

The carboxylate group of glycine accepts a hydrogen ion from HCl. Proteins contain many different side groups that can act as acids or bases.

The Acid Dissociation Constant

When an acid (generically indicated as **HA**) is added to water, the following reversible reaction takes place:

$$HA + H_2O \rightleftharpoons H_3O^+ + A^-$$

The species H_3O^+ is a hydrated hydrogen ion and is called the **hydronium ion.** For simplicity, it is common practice to ignore water in this reaction and to write

$$HA \rightleftharpoons H^+ + A^-$$

The reaction going from left to right is called the **dissociation reaction,** and the reaction going from right to left is the **association reaction.** The rate of the dissociation reaction is equal to the product of the concentration of HA and the **dissociation rate constant k_1,** a specific value for this reaction. The rate of the association reaction is equal to the product of the concentrations of H^+ and A^- and the **association rate constant k_2,** a specific value for this reaction. At equilibrium, the rates of dissociation and association reactions are equal. We can therefore write:

$$k_1 \times [HA] = k_2 \times [H^+] \times [A^-]$$

Rearranging, we get $[H^+] \times [A^-]/[HA] = k_1/k_2$. We define a new constant $\mathbf{K_a}$ as equal to this ratio k_1/k_2. K_a is the **equilibrium constant** for this reaction and is called the **dissociation constant** or **ionization constant** for the acid.

The higher the acid dissociation constant, the more completely an acid is dissociated or ionized and, consequently, the more free (unbound) hydrogen ions are present in solution. Acids with high dissociation constants are **strong acids.** These include acids such as hydrochloric, sulfuric, nitric, and phosphoric acids. Hydrochloric acid, for example, is essentially completely dissociated into hydrogen ions and chloride ions in dilute aqueous solution. There is very little undissociated HCl. In a 0.1 M HCl solution, the free hydrogen ion concentration is nearly 0.1 M.

A low acid dissociation constant is characteristic of **weak acids.** In other words, a weak acid is incompletely dissociated in solution. For example, the acid dissociation constant of acetic acid is equal to 1.8×10^{-5}. If we have a 0.100 M solution of acetic acid in water, the concentration of undissociated acid (CH_3COOH) is 0.0987 M, and the concentrations of free hydrogen and acetate (CH_3COO^-) ions are both equal to 0.0013 M. In other words, nearly 99% of the acetic acid molecules are not dissociated in this solution. Note especially that the concentration of free hydrogen ions is low for a solution of a weak acid.

Acid dissociation constants are usually small and awkward to manipulate mathematically, so they are often presented in a logarithmic form. The $\mathbf{pK_a}$ is defined as the **negative logarithm** to the base 10 of K_a or

$$pK_a = -\log K_a$$

For example, for acetic acid, $pK_a = -\log 1.8 \times 10^{-5} = 4.7$. The negative logarithm of a number is equal to the logarithm of the inverse of the number. So the equation can be written as

$$pK_a = \log (1/K_a)$$

This equation shows more clearly that pK_a and K_a are inversely related. A low pK_a corresponds to a high dissociation constant (strong acid), and a high pK_a corresponds to a low dissociation constant (weak acid).

pH Values of Aqueous Solutions

As discussed in Chapter 2, we can also define a term called **pH** as the negative logarithm to the base 10 of the molar activity (concentration) of hydrogen ions. In equation form:

$$pH = -\log [H^+] = \log (1/[H^+])$$

The pH of a solution is inversely related to the free hydrogen ion concentration. A low pH indicates a high hydrogen ion concentration, and a high pH indicates a low hydrogen ion concentration. The pH scale is useful because it compresses the wide range of hydrogen ion concentrations possible in aqueous solutions into a more manageable scale. The pH scale usually goes from 0 to 14. A 1 M solution of the strong acid HCl has a pH of about 0 ($\log 1 = 0$). A 1 M solution of the strong base NaOH has a pH of about 14 ($[H^+] = 10^{-14}$M). A pH of 7 corresponds to a $[H^+]$ of 10^{-7} M. If the pH of a solution is below 7, it is considered to be **acidic;** if equal to 7, **neutral;** and if greater than 7, **alkaline** or **basic.**

The Henderson-Hasselbalch Equation

The equilibrium equation for dissociation of an acid is often expressed in logarithmic form. We can rearrange the equation

$$[H^+] \times [A^-]/[HA] = K_a$$

and solve for $[H^+]$:

$$[H^+] = K_a \times [HA]/[A^-]$$

Taking the logarithms of this equation, we get:

$$\log [H^+] = \log K_a + \log ([HA]/[A^-])$$

Multiplying both sides of the equation by -1, we get:

$$-\log [H^+] = -\log K_a + \log ([A^-]/[HA])$$

Since $pH = -\log [H^+]$ and $pK_a = -\log K_a$, we can write the equation

$$pH = pK_a + \log ([A^-]/[HA])$$

This logarithmic form of the acid dissociation equation is known as the **Henderson-Hasselbalch equation** and is very useful, as we will see later.

The Ion Product of Water

Water reversibly dissociates into hydrogen and hydroxide ions according to the reaction:

$$H_2O + H_2O \rightleftharpoons H_3O^+ + OH^-$$

In this reaction, water is acting as both an acid (proton donor) and base (proton acceptor). It is customary to simplify this reaction and write:

$$H_2O \rightleftharpoons H^+ + OH^-$$

The dissociation constant (K) for this reaction is equal to

$$K = [H^+] \times [OH^-]/[H_2O]$$

The tendency of water to dissociate is very small. The concentration of water ($[H_2O]$) is large and is essentially constant. We can therefore define a new constant, $\mathbf{K_w}$, as equal to $K \times [H_2O]$, and write

$$K_w = [H^+] \times [OH^-]$$

This equation indicates that in any aqueous solution, the product of the hydrogen and hydroxide ion concentrations is constant. This means that $[H^+]$ and $[OH^-]$ are inversely related. $\mathbf{K_w}$, called the **ion product of water,** has a value of 1×10^{-14} M^2 at room temperature (25°C). In *pure* water, $[H^+] = [OH^-]$. Therefore, $[H^+]^2 = 1 \times 10^{-14}$ M^2 or $[H^+] = 10^{-7}$ M (pH 7). Hence, the pH of pure water is 7. If we add a strong base (NaOH) to water, OH^- ions will combine hydrogen ions supplied by water, thereby shifting the equilibrium toward undissociated water and lowering the free hydrogen ion concentration. For example, in a 1 M solution of NaOH, $[OH^-] = 1$ M, and so $[H^+] = 10^{-14}$ $M^2/1$ M $= 10^{-14}$ M (pH 14). In a 0.1 M solution of HCl, $[H^+]$ is essentially 0.1 M, so $[OH^-] = 1 \times 10^{-14}$ $M^2/0.1$ M or 1×10^{-13} M.

Table 26–1
pH Values of Aqueous Solutions of Chemicals, Body Fluids, and Beverages

Chemicals	
0.100 M hydrochloric acid	1.1
0.010 M hydrochloric acid	2.0
0.001 M hydrochloric acid	3.0
0.100 M lactic acid ($pK_a = 3.9$)	2.5
0.100 M acetic acid ($pK_a = 4.7$)	2.9
pure water	7.0
0.100 M sodium bicarbonate	8.4
0.100 M ammonia	11.1
0.100 M sodium hydroxide	13.0
Body fluids	
Arterial blood	7.40
Venous blood	7.35
Cerebrospinal fluid	7.35
Skeletal muscle cell cytoplasm	6.9
Saliva	5.8–7.1
Gastric juice	0.7–3.8
Gallbladder bile	5.6–8.0
Pancreatic juice	7.5–8.8
Intestinal juice	7.0–8.0
Feces	5.9–8.5
Urine	4.5–8.0
Beverages	
Lemon juice	2.2–2.4
Sodas	2.8–3.7
Orange juice	3.7
Beer	4.4
Coffee	5.0
Milk	6.7
Tea	6.9

pH Values for Solutions of Acids and Bases, Body Fluids, and Beverages

Table 26–1 presents pH values for some aqueous solutions of chemicals, body fluids, and beverages. Notice that the pH of various solutions of acids depends not only on the concentration of acid, but also on the dissociation constant of the acid. Strong acids will produce solutions with a lower pH than will the same concentrations of weak acids because the stronger acid is dissociated (ionized) more, so it yields a greater concentration of free hydrogen ions in solution.

The pH of arterial blood averages 7.40, which corresponds to a $[H^+]$ of 4×10^{-8} M or 40 nanomolar. The pH of venous blood is slightly lower (average 7.35) owing to its higher content of carbon dioxide (hence the acid, carbonic acid). Most body fluids, with the exception of gastric juice, have pH values in the range 5 to 8. Cell cytoplasm is more acidic than extracellular fluid; the pH of skeletal muscle cell cytoplasm averages 6.9. pH may differ strikingly from one organelle to another within a cell; for example, in lysosomes, which contain enzymes that operate best in an acidic environment, pH is about 4.5. Gastric juice is quite acidic owing to its high concentration of hydrochloric acid. Pancreatic juice is slightly more alkaline than blood because of its high bicarbonate concentration. The pH of urine can vary between 4.5 and 8.0.

Many foods are acidic or alkaline. When we consume food, the load of acid or base depends not only on the free hydrogen ion concentration, but also on how much undissociated acid (or base) is present. An undissociated acid may liberate hydrogen ions (or a base may bind hydrogen ions) at the pH of body fluids. The effect of a chemical substance on acid–base balance also depends on how the substance is metabolized in the body, a subject we will discuss later.

Buffer Systems: General Principles

A buffer, by definition, is something that minimizes a shock and thereby promotes relative stability. A **pH buffer,** as described in Chapter 2, minimizes a change in pH when either acid or base is added to a solution.

A pH buffer *does not prevent* a pH change. Whenever an acid is added, pH will fall. If a base is added, pH will rise. The extent of change in pH depends on the amount and nature of added acid or base and on the amount and nature of the pH buffer.

Previously, we wrote the equation for dissociation of an acid: $HA \rightleftharpoons H^+ + A^-$. In this reaction, HA is the acid and A^- is the **conjugate base.** *Conjugate* means "joined in a pair." A **chemical buffer** consists of a weak acid and its conjugate base (or a weak base and its conjugate acid). Following are some examples of conjugate acid–base buffer pairs:

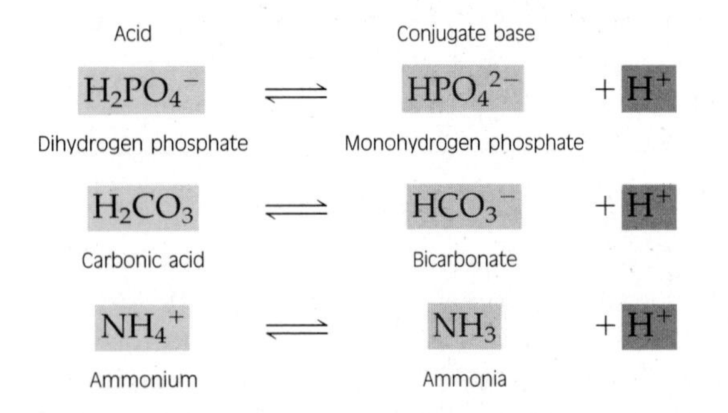

The equilibrium expression for dissociation of an acid can be written in the Henderson-Hasselbalch equation form:

$$pH = pK_a + \log ([\text{conjugate base}]/[\text{acid}])$$

For example, for a mixture of monohydrogen and dihydrogen phosphate,

$$pH = 6.8 + \log ([HPO_4^{2-}]/[H_2PO_4^-])$$

This equation demonstrates that if the ratio of conjugate base to acid is defined for a buffer of known pK, then the pH is automatically determined. This principle is useful in making up pH buffer solutions.

What, for example, is the pH of an aqueous solution containing a mixture of 0.1 M Na_2HPO_4 (sodium monohydrogen phosphate) and 0.1 M NaH_2PO_4 (sodium dihydrogen phosphate)? Since the ratio $HPO_4^{2-}/H_2PO_4^-$ is 1, and the logarithm of 1 is zero, the pH is $6.8 + 0 = 6.8$. If the ratio of $HPO_4^{2-}/H_2PO_4^-$ is 4, then the pH of this solution is $6.8 + \log 4 = 6.8 + 0.6 = 7.4$.

Chemical buffers stabilize the pH of aqueous solutions. For example, if we add strong acid (HCl) to a phosphate buffer solution, the following reaction takes place:

$$HCl + Na_2HPO_4 \longrightarrow NaH_2PO_4 + NaCl$$

In this reaction, the base component of the buffer pair (HPO_4^{2-}) binds hydrogen ions. The strong acid (HCl) is, in effect, converted into a weak acid ($H_2PO_4^-$), thereby minimizing the increase in free hydrogen ion concentration (fall in pH). If we add strong base (NaOH) to the phosphate buffer solution, the following reaction takes place:

$$NaOH + NaH_2PO_4 \longrightarrow Na_2HPO_4 + H_2O$$

The acid component of the buffer pair ($H_2PO_4^-$) liberates hydrogen ions, thereby diminishing the rise in pH caused by addition of strong base. The strong base (NaOH) is, in effect, converted into a weak base (HPO_4^{2-}).

Figure 26–1 shows a titration curve for a phosphate buffer solution. The *y* axis gives the amount of strong acid or base added to the solution. The *x* axis gives the pH. As we move from left to right along the curve, $H_2PO_4^-$ is converted to HPO_4^{2-} by addition of strong base. As we move from right to left along the curve, HPO_4^{2-} is converted to $H_2PO_4^-$ by the addition of strong acid. The slope of the titration curve is an index of the effectiveness of the buffer in resisting a change in pH. The slope of this curve is steepest when the pH is equal to the pK_a of the phosphate buffer. This means that for a given amount of added acid or base, the pH change will be least near the pK_a of the buffer. As we move away from the pK_a, the effectiveness of the buffer declines. In choosing a buffer to stabilize a certain pH, it is always best to select a buffer pair with a pK_a close to the desired pH.

The effectiveness of a buffer in minimizing pH changes depends on two factors. The first factor is the pK_a of the buffer in relation to the desired pH. The pK_a of the imidazole group of histidine (an amino acid found in hemoglobin) is close to 7.4, and hence this is an ideal buffer in blood. The pK_a of

($CO_2(d)$ = 1.2 mmoles per liter). The pH of the blood can be calculated as follows: pH = 6.10 + log ($[HCO_3^-]/0.03\ Pco_2$) = 6.10 + log (24/0.03 × 40) = 6.10 + log 20 = 6.10 + 1.30 = 7.40. Suppose we now add 10 mmoles of strong acid to each liter of blood. For simplicity, we will ignore the fact that bicarbonate is not the only buffer base that combines with the added hydrogen ions. According to the following reactions, 10 mmoles of dissolved CO_2 will be formed:

$$H^+ + HCO_3^- \longrightarrow H_2CO_3 \longrightarrow H_2O + CO_2$$

If we had a closed container (Fig. 26–4*b*), from which none of the CO_2 formed could escape, the resulting pH would be equal to 6.10 + log ([24 − 10]/[1.2 + 10]) = 6.20. If we had an open system (see Fig. 26–4*c*), the CO_2 produced could be blown off as it is formed and Pco_2 could be maintained constant at 40 mm Hg ($CO_2(d)$ = 1.2 mmoles per liter). In this case, the pH is equal to 6.10 + log ([14]/[1.2]) = 7.17, which is much better than a pH of 6.20 (a lethal pH). A fall in blood pH causes hyperventilation so that the Pco_2 is not maintained constant, but actually falls—for example, to 30 mm Hg ($CO_2(d)$ = 0.90 mmoles per liter) (see Fig. 26–4*d*). In this case, the pH is equal to 6.10 + log ([14]/[0.90]) = 7.29. We can see then that the ability to blow off CO_2 and to lower its level in the blood by hyperventilation diminishes the fall in pH produced by adding strong acid.

The bicarbonate-CO_2 buffer system is "open" in other respects, too. There is a continuous source of CO_2 from metabolism, which can replace CO_2 that is consumed when strong base is added to the body. The kidneys can change the amount of bicarbonate in the extracellular fluid by excreting bicarbonate in the urine (when there is excess base in the body) or by synthesizing new bicarbonate and adding it to the blood (when there is excess acid in the body). The ability of the body to change the amounts of components of this particular buffer pair makes the bicarbonate-CO_2 system a remarkably effective buffer.

The Isohydric Principle

We have discussed the various buffer systems separately, but in reality they all work together. In fact, in any fluid compartment, all are in equilibrium with the same hydrogen ion concentration. This idea is called the **isohydric principle.** The term *isohydric* means "same hydrogen ion." We can therefore write:

$$\begin{aligned} pH &= 6.8 + \log([HPO_4^{2-}]/[H_2PO_4^-]) \\ &= pK_{protein} + \log([proteinate^-]/[H\text{-}proteinate]) \\ &= 6.1 + \log([HCO_3^-]/0.03\ Pco_2) \end{aligned}$$

When an acid or base is added to the body, all buffer pairs participate in buffering. The importance of each depends on its pK, amount, and accessibility. The expression just given re-emphasizes the point that the ratio of conjugate base to acid for any buffer pair of known pK defines the pH. For the remainder of this chapter, we will stress the bicarbonate-CO_2 system and its effects on blood pH. You should recognize, however, that other buffers are also important and interact with the bicarbonate-CO_2 buffer pair. Changes in the ratio HCO_3^-/CO_2 indicate that other body buffers are changing too. The reason for emphasizing the bicarbonate-CO_2 system is mainly because the physiological buffers act on its components. The respiratory system controls arterial blood pH by regulating the Pco_2, and the kidneys control blood pH by regulating plasma bicarbonate concentration.

Physiological Buffers

The Respiratory System as a Buffer System: Control of the Pco_2

The respiratory system controls the partial pressure of CO_2 in arterial blood. Recall from Chapter 20 that the level of alveolar ventilation is a determinant of the Pco_2 in the alveolar spaces of the lung. CO_2 pressures in alveoli and arterial blood are equal because of equilibration of CO_2 between alveolar gas and pulmonary capillary blood. For a constant rate of CO_2 production, alveolar ventilation and alveolar Pco_2 are reciprocally related. The higher the rate of alveolar ventilation, the lower will be the alveolar Pco_2, and vice versa. Normally, CO_2 is expired at the same rate as it is produced from metabolism, and the Pco_2 remains near 40 mm Hg.

In a state of hyperventilation, CO_2 will be flushed out of the alveolar spaces at a rate greater than the rate at which it is produced in the body and added to the alveolar spaces. Consequently, the Pco_2 of alveolar gas and arterial blood will fall to a new lower-than-normal level, and the reactions

$$CO_2 + H_2O \rightleftharpoons H_2CO_3 \rightleftharpoons H^+ + HCO_3^-$$

will be pulled to the left. The hydrogen ion concentration of the blood will fall; in other words, the blood will become more alkaline. During hypoventilation, CO_2 added to the alveolar spaces will not be removed at an adequate rate. CO_2 pressure in the alveoli and arterial blood will rise. The above reactions will be pushed to the right, so that a more acidic blood pH results. From these two examples, it should be clear that abnormal respiratory system function—that is, hyperventilation or hypoventilation, produces disturbances in blood pH. We will return to this topic toward the end of this chapter.

The respiratory system normally acts in such a way as to minimize pH changes in the blood. Such changes might be produced by adding a "fixed" acid or a base to the blood or by adding or removing too much CO_2.

By the term *fixed* acid, we mean an acid other than carbonic acid. In contrast to the concentration of carbonic acid, the concentration of a fixed acid in blood is not affected by lung function. If, for example, hydrochloric acid is added to or produced in the body, it will be neutralized by chemical buffers, and then the hydrogen ions (combined with urinary buffers) and chloride will be excreted by the kidneys. A fixed acid cannot be blown off by the lungs.

If the blood is made more acidic by addition of fixed acid, pulmonary ventilation is increased. The receptors that sense the fall in pH and reflexly increase ventilation are mainly the **peripheral chemoreceptors,** the carotid and aortic bodies. A large fall in blood pH may also stimulate central chemoreceptors or the medullary respiratory center itself. Increased ventilation serves to lower the arterial blood CO_2 pressure and carbonic acid concentration, and so reduces the acidic shift in blood pH. Figure 26–4*d* illustrates how hyperventilation in response to added acid minimizes a fall in blood pH.

If the blood is made more alkaline by infusing base, the opposite changes occur. In this case, hypoventilation results, which leads to retention of CO_2, a higher carbonic acid concentration in the blood, and a less alkaline shift in blood pH. Hypoventilation is limited, because it causes retention of CO_2 and a fall in arterial oxygen tension, and both of these changes stimulate ventilation.

If excess CO_2 accumulates in the body, because of inadequate alveolar ventilation or breathing of a CO_2 mixture, ventilation will be stimulated. CO_2 is a very powerful stimulus to ventilation; an increase in arterial blood P_{CO_2} of 2 to 3 mm Hg causes the rate of ventilation to double. The receptors mediating this increase in ventilation are chiefly **central chemoreceptors** in the medulla of the brain. Carbon dioxide diffuses from the blood into brain interstitial and cerebrospinal fluids, and there it forms carbonic acid, which liberates hydrogen ions. It is generally believed that chemoreceptors respond to hydrogen ion concentration (or pH) and not to molecular CO_2 directly. The peripheral chemoreceptors are also stimulated by an increase in arterial blood P_{CO_2}, but are less important than are the central chemoreceptors in the ventilatory response to CO_2. The increase in ventilation caused by CO_2 is useful in bringing about removal of CO_2, thereby minimizing carbonic acid accumulation in the blood and an acidic shift in blood pH.

The Kidneys as a Buffer System: Acidification of the Urine

The kidneys play a major role in acid–base regulation by getting rid of acid or base in the urine when there is an excess or deficit of hydrogen ions in the body. As mentioned, the usual condition is one of excess acid, since metabolism of food produces about 50 to 100 mEq of strong acid in a day. This acid is first neutralized by bicarbonate (and other chemical buffer bases) in the body. Then the kidneys get rid of the acid in the urine. In the process of excreting acid, the kidneys add new bicarbonate to the blood and so bring the blood pH back to normal.

Most of the hydrogen ions excreted in the urine are combined with urinary buffers. The limits of urine pH are from a maximum of 8.0 to a minimum of 4.5. Suppose we excrete 1.5 liters of pH 4.5 urine in a day. How many *free* hydrogen ions are excreted? This can be calculated by multiplying the urine volume (1.5 liters) by the hydrogen ion concentration ($10^{-4.5}$ mEq per liter) and amounts to about 0.05 mEq. This amount is clearly inadequate to take care of the amount of acid that needs to be excreted. Most of the hydrogen ions excreted in the urine are not free, but are combined with buffers. These buffers include titratable acids (e.g., inorganic phosphate) and ammonia.

To **titrate** an acid is to determine the amount of acid present by adding base until the acid is neutralized. **Titratable acid** is measured in the clinical laboratory by determining how many milliequivalents of strong base (NaOH) are needed to bring a urine sample back to the pH of arterial blood (usually 7.4). Most of the titratable acid is normally inorganic phosphate buffer. Depending on the circumstances, other substances in the urine (creatinine, various organic acids) may also be included. In the test tube, the following reaction takes place:

$$H_2PO_4^- + OH^- \longrightarrow HPO_4^{2-} + H_2O$$

This reverses the reaction that took place along the kidney tubules when hydrogen ions were added to the tubular urine:

$$H^+ + HPO_4^{2-} \longrightarrow H_2PO_4^-$$

Ammonia also acts to buffer secreted hydrogen ions in the urine, according to the following reaction:

$$H^+ + NH_3 \rightleftharpoons NH_4^+$$

The pK_a for the ammonium ion (about 9.3) is so high that ammonia is not included when titratable acid is measured. In blood plasma and urine, most ammonia is in the form of ammonium ion and not as the free base NH_3. Of the hydrogen ions excreted in urine, normally nearly one third are in the form of titratable acid, close to two thirds are present as

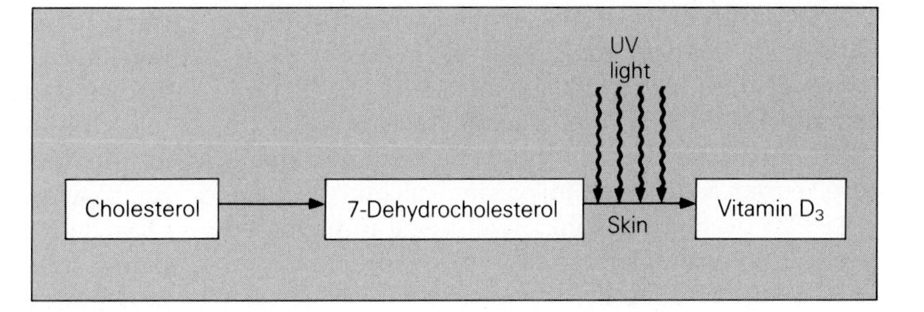

Figure 27–14 Illustration of the pathway of formation of vitamin D_3, beginning with cholesterol.

skin, is first carried to the liver via the blood. In the liver, a specific enzyme catalyzes the addition of a hydroxyl group to carbon number 25 to form 25-hydroxy-vitamin D_3. The 25-hydroxy-vitamin D_3 then leaves the liver and travels via the blood to the kidneys. In the proximal tubule cells of the kidney, a specific enzyme known as **1α-hydroxylase** catalyzes the conversion of 25-hydroxy-vitamin D_3 to 1,25-dihydroxy-vitamin D_3 (1,25-OH-vitamin D_3). It is this form of the vitamin that functions as a hormone and is involved in the regulation of plasma calcium concentrations.

The formation of 1,25-OH-vitamin D_3 is a highly regulated process. The primary site of regulation is the kidney. The activity of 1α-hydroxylase is markedly increased by parathyroid hormone (see Fig. 27–15). Thus, the activity of the enzyme will change indirectly as a result of changes in the plasma calcium concentration. (As mentioned, if the plasma calcium concentration decreases below normal, parathyroid hormone secretion is stimulated. This results in an increase in 1α-hydroxylase activity, leading to increased formation of 1,25-OH-vitamin D_3.)

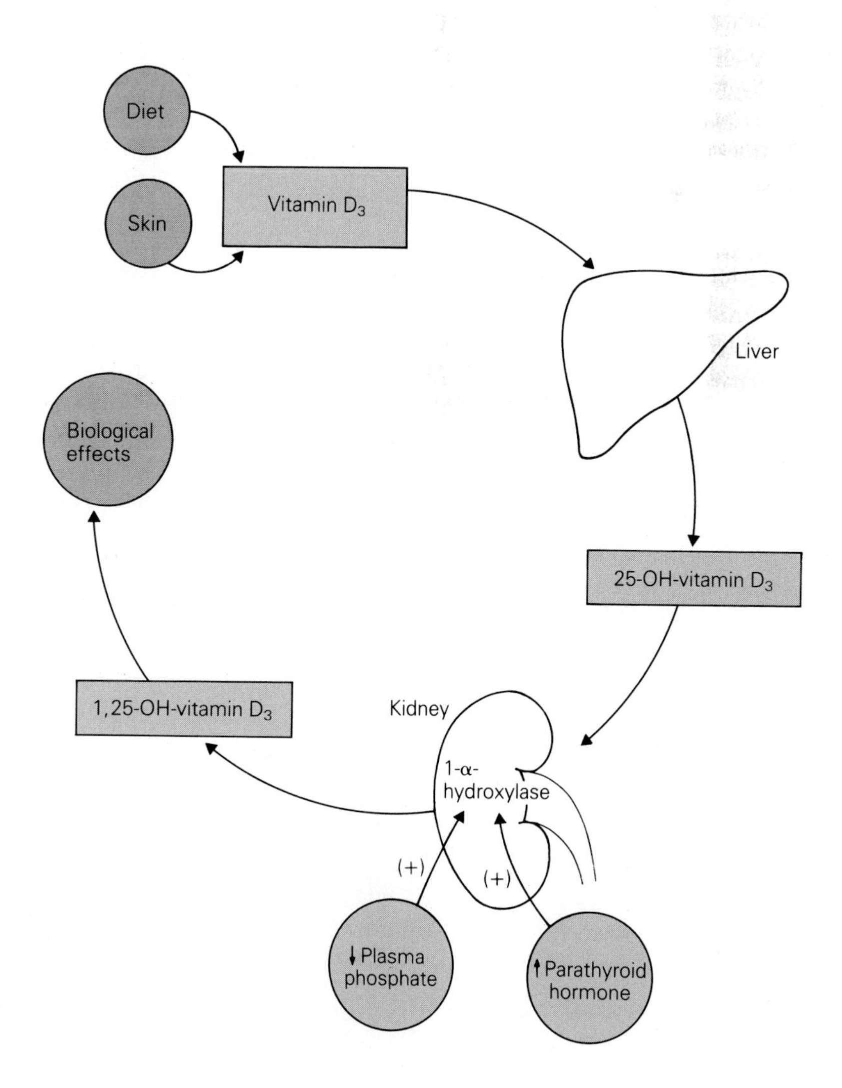

Figure 27–15 Diagram of the pathway for conversion of vitamin D_3 into 1,25-dihydroxy-vitamin D_3. Hydroxylation of vitamin D_3 on the 25 position occurs first in the liver, followed by hydroxylation on the 1 position in the kidney. Parathyroid hormone stimulates the kidney enzyme, which catalyzes the addition of a hydroxyl group to carbon number 1.

In addition, a decrease in plasma phosphate causes an increase in activity of 1α-hydroxylase, which also leads to increased formation of 1,25-OH-vitamin D_3. In this case, phosphate appears to directly modulate the activity of the enzyme. Therefore, either a decrease in plasma calcium or a decrease in plasma phosphate will result in increased production of the hormonally active form of vitamin D (see Fig. 27–15.)

Actions of Hormones Involved in Regulating Calcium and Phosphate Homeostasis

The actions of parathyroid hormone, calcitonin, and 1,25-OH-vitamin D_3 are summarized in Table 27–6. The specific effects of these hormones will be described in the following sections.

Actions of Parathyroid Hormone in Calcium and Phosphate Homeostasis. As a result of its actions in regulating plasma calcium concentrations, parathyroid hormone is *acutely* required for life. Removal of the parathyroid glands without any subsequent intervention will usually lead to death owing to hypocalcemic tetany within 48 to 72 hours. Figure 27–1 shows the initial signs of hypocalcemic tetany, involving spasm of muscles of the hand and forearm.

As indicated in Table 27–6, the overall effect of parathyroid hormone is to increase the plasma calcium concentration and decrease the plasma phosphate concentration. With these effects in mind, what specific responses does the hormone elicit that lead to these changes?

One important effect of parathyroid hormone in the kidney is to increase the activity of 1α-hydroxylase, which catalyzes the formation of 1,25-OH-vitamin D_3, the active metabolite of vitamin D. Parathyroid hormone also stimulates the active transport mechanism for calcium reabsorption in the distal tubules, leading to an increase in calcium retention and a decrease in urinary excretion of calcium. Parathyroid hormone also decreases the proximal tubular reabsorption of phosphate, resulting in **phosphaturia** (increased phosphate excretion in the urine). This phosphaturic effect of parathyroid hormone is an important one, since other effects of parathyroid hormone tend to increase the flow of *both* calcium and phosphate into the blood. Except for the increased phosphate excretion in the urine, both calcium and phosphate would accumulate in the plasma and then simply recrystalize in bone mineral. This would counteract the desired effect, which is to raise the plasma calcium concentration.

In bone, parathyroid hormone has several effects, all of which lead to increased net resorption of bone. The primary effect is to stimulate the dissolution of bone by increasing the activity of osteoclasts, the large, multinucleated cells involved in bone resorption. In addition to increasing the activity of existing osteoclasts, parathyroid hormone also stimulates the maturation of immature osteoclasts, resulting in a greater number of active osteoclasts.

Table 27–6
Summary of the Effects of Hormones Involved in Regulating Calcium and Phosphate Metabolism

Parameter or Target Tissue	Parathyroid Hormone	Calcitonin	1,25-OH-vitamin D_3
Plasma calcium concentration	↑	↓	↑
Plasma phosphate concentration	↓	↓	↑
Kidney	↑ Reabsorption of calcium ↓ Reabsorption of phosphate ↑ Activity of 1α-hydroxylase	↓ Reabsorption of calcium ↓ Reabsorption of phosphate	↑ Reabsorption of calcium ↑ Reabsorption of phosphate
Bone	↑ Resorption of bone	↓ Resorption of bone	Promote PTH actions
Gastrointestinal tract	Indirect effects by increasing 1,25-OH-vitamin D_3 formation	No effect	↑ Absorption of calcium and phosphate

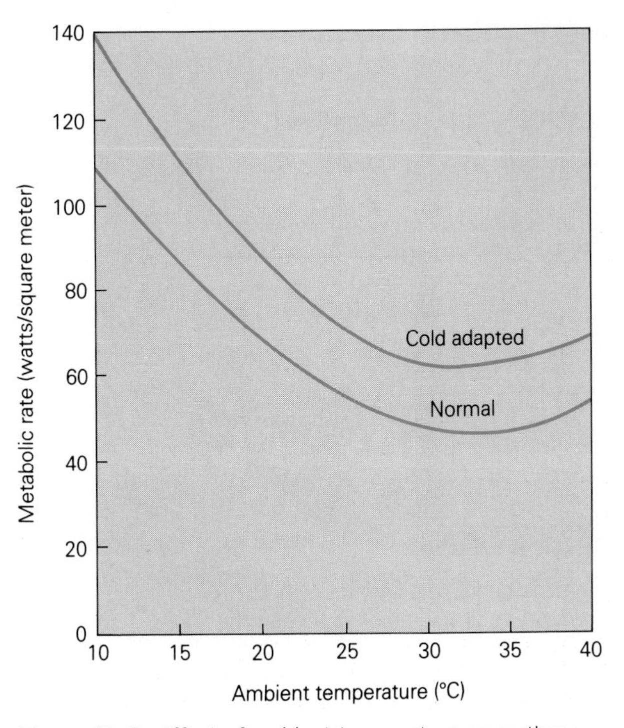

Figure 28–9 Effect of ambient temperature on resting metabolic rate before and after chronic cold exposure.

long-term exposure to cold produces a 20% to 30% increase in metabolic rate (Fig. 28–9). The increase is caused, in part, by adaptation of the thyroid gland. Thyroxin secretion increases in cold climates and decreases in hot climates.

The Basal Metabolic Rate

The concept of **basal metabolic rate (BMR)** has evolved bacause of our increasing knowledge of the many factors that can influence a person's metabolic rate. BMR is a standard method for determining metabolic rate that allows comparison between individuals. To record BMR, a set of standard conditions has been established, which usually measures the so-called basal conditions. Actually, the BMR is not truly "basal," since metabolic rate during sleep is lower than BMR. It is more appropriate to describe BMR as the "metabolic cost of living." The following four conditions usually are specified for measurements of BMR:

1. The person must have fasted for 12 hours.
2. The measurement usually is made in the morning after a night of restful sleep.
3. The person must remain at complete rest for 30 to 60 minutes before the test.
4. The atmospheric temperature must be comfortable, about 25°C.

Measurements of Metabolic Rate

The metabolic rate is a measure of the total energy used by the body per unit time. In a person completely at rest sitting in a comfortable temperature, all of the metabolic energy is transformed to body heat. Furthermore, a normal person under these conditions keeps body temperature constant by releasing an amount of body heat into the environment that is equal to the heat produced inside the body. A person's metabolic rate, therefore, can be determined by measuring the heat given off from the body in a known period of time.

Direct Calorimetry. A person's metabolic rate can be measured by direct or indirect methods. In the direct method for measuring metabolic rate, a person is placed in a large insulated chamber called a **human calorimeter** (Fig. 28–10). The temperature of the calorimeter is kept constant by a water cooling radiator system. The heat lost from a person's body is picked up by the cooling water system. The rate of heat gain by the radiator system can be accurately measured with a thermometer and is equal to the rate of heat liberated from the person's body. Although direct calorimetry is a very accurate technique since it measures that heat output of the body directly, it is a time-consuming, expensive, and cumbersome method. It is therefore used most commonly for research purposes and to validate less costly and more convenient methods for measuring metabolic rate.

Indirect Calorimetry. A commonly used indirect method for measuring a person's metabolic rate is based on the determination of a person's **oxygen consumption** per unit of time. This method is based on the concept that since 95% to 100% of the energy available to the body is derived from cellular reactions with oxygen, a person's metabolic rate can be accurately estimated from oxygen use. Studies comparing metabolic rates of humans using both indirect and direct methods have shown good agreement, with differences being less than 5%. The errors in the indirect method are not large enough to outweigh its advantages.

The Energy Equivalent of Oxygen. In practice, the indirect method for measuring metabolic rate is dependent on the concept of the energy equivalent of oxygen and a special instrument for the accurate measurement of oxygen consumption called a **respirometer** (Fig. 28–11). The energy equivalent of oxygen is the quantity of energy liberated when the body uses one liter of oxygen to produce metabolic

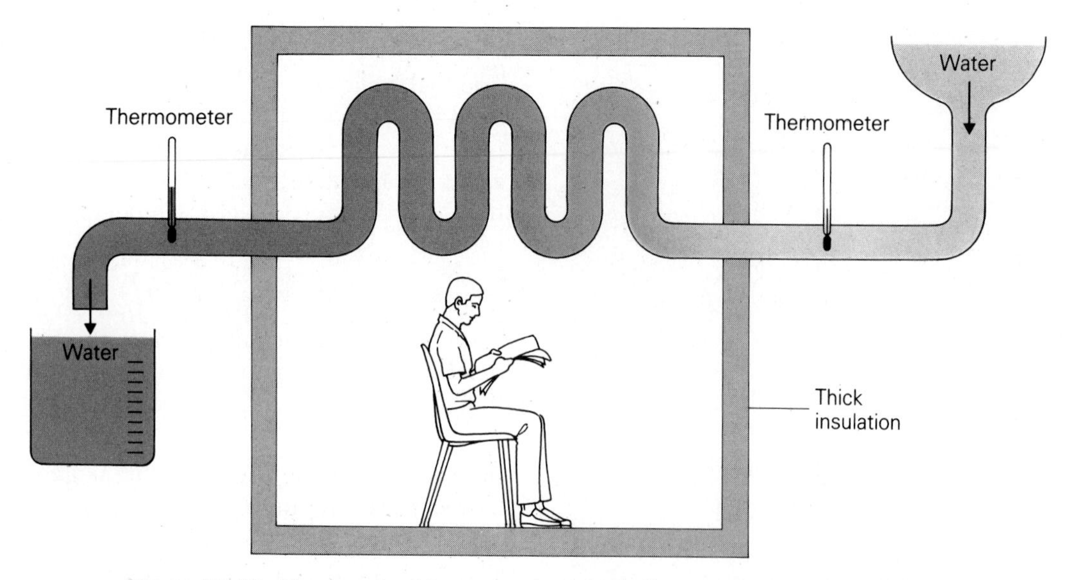

Figure 28–10 Direct method for measuring metabolic rate. The amount of heat produced by the person's body is calculated from the total volume of water and the difference between inflow and outflow temperature.

energy. For example, the amount of energy released when the body uses one liter of oxygen to oxidize **carbohydrates** is 5.0 calories. When the body uses one liter of oxygen to burn **fats,** the energy released is 4.7 calories. The corresponding value for **protein** is 4.6 calories. Since the energy equivalent for all three food types is about the same, an approximate average of these values is commonly used and is equal to *4.825* calories of heat produced per liter of oxygen consumed. Metabolic rate, therefore, is commonly determined by the following equation:

$$\text{Metabolic rate (Cal/m}^2\text{/hr)} = \text{liters of } O_2 \text{ consumed} \times 4{,}825 \text{ calories per liter of } O_2 \div \text{surface area (meters}^2\text{)}$$

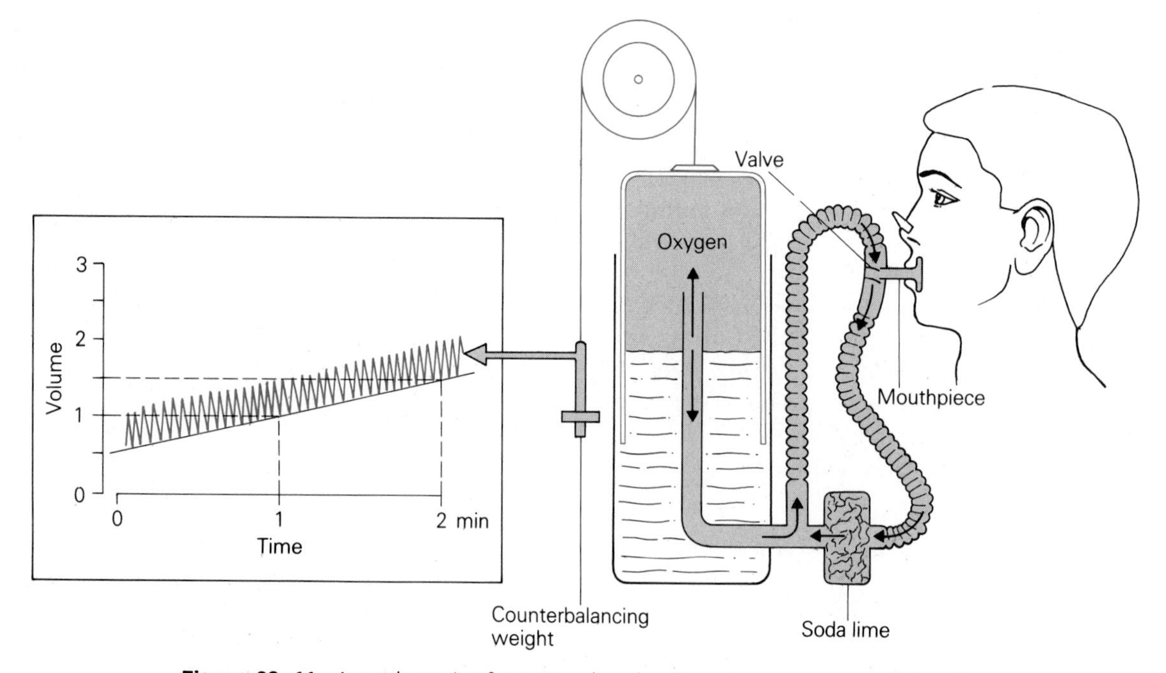

Figure 28–11 A respirometer for measuring the rate of oxygen consumption. The subject inspires oxygen from a bell gasometer through a mouthpiece and valve. CO_2 is removed from the expired air by soda lime before it returns to the bell.

severity of a person's illness. Some pathophysiological conditions produce either **hypothermia** or **hyperthermia,** which are **unregulated** decreases or increases in body temperature, respectively.

Fever: A Regulated Increase in Body Temperature

Fever is usually caused by chemical agents called **pyrogens** released into the body during disease processes. The exact way that these agents work is not known, but they appear to **reset** the controlled level of the hypothalamic thermostat at a higher temperature level (Fig. 28–18). The hypothalamic integrating center compares existing peripheral and core temperature thermoreceptor information with the new elevated reference setting, activating heat production and conservation responses and inhibiting heat loss responses. Simultaneously, the person feels cold and may react behaviorally by turning up the heat or putting on more clothing. After several hours, the body temperature will rise to the new reference temperature, at which point the hypothalamus will equalize heat loss and heat conservation responses. The body temperature continues to be regulated at the elevated level as long as the pyrogen is present. When the pyrogen is no longer effective or present, the so-called **crisis** occurs, and the reference temperature returns to the lower, normal range. When this happens, the hypothalamic integrator activates heat loss and inhibits heat gain processes that gradually return the body temperature to normal levels.

Hyperthermia: Unregulated Increases in Body Temperature

Heat Stroke

Heat stroke is a form of an unregulated hyperthermia or heat illness associated with a breakdown of the body's temperature regulating system. Heat stroke usually occurs during heavy work or exposure to hot, humid conditions. In these situations, the rate of heat gain can exceed the maximal rate of heat loss so that the body temperature continues to rise unless the level of work is decreased or the person moves to a cooler, less humid atmosphere.

If, however, the body temperature rises beyond a critical temperature in the range of 41 to 42°C, the person is likely to develop heat stroke. The symptoms include dizziness, abdominal distress, absence of sweating, sometimes delirium, and eventually loss of consciousness and death if the body temperature is not decreased quickly. Some of the symptoms result from circulatory shock or insufficiency brought about by fluid and electrolyte loss before the onset of the symptoms. The high temperature itself may damage body tissues, particularly in the brain, leading to loss of sweating, and a vicious cycle is started in which core temperature is raised continually toward lethal temperatures.

Heat Exhaustion

Heat exhaustion, a less serious heat illness caused by **circulatory insufficiency** and **hypotension,** often takes the form of fainting due to decreased blood pressure. This is brought about by depletion of

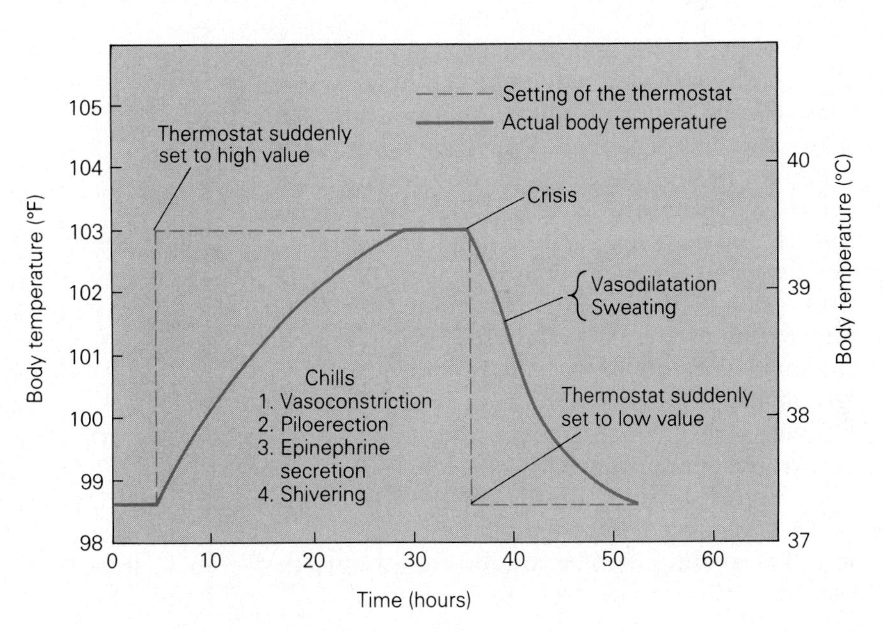

Figure 28–18 Time course of typical febrile episode. Effects of changing the setting of the "hypothalamic thermostat."

Focus on Malignant Hyperthermia

Malignant hyperthermia is the result of an inherited defect in skeletal muscle metabolism. The disorder is inherited as an autosomal dominant trait. In genetically disposed persons, several other disorders are seen, such as hypoxia, hypotension, and acidosis. Susceptibility to the condition appears to involve a defect in the uptake of calcium by the sarcoplasmic reticulum, and the event is produced by anesthetic agents. The probability that the condition will develop in susceptible individuals increases with successive exposure to anesthetics.

In malignant hyperthermia, the calcium concentration of the sarcoplasm becomes abnormally high, and the skeletal muscles undergo prolonged contractions. The high metabolic rate of the continuously contracting muscles increases metabolic heat production and produces the hyperthermia. The hyperthermia is not of CNS origin, and there are no signs of coordinated rhythmic shivering. The resulting high body temperature is a major contributor to the mortality of this disorder. Temperatures in excess of 45°C are often reported.

The preferred treatment consists of immediate removal of heat with a cooling blanket and removal of the anesthetic agents. Respiratory support is also important to provide for the high consumption of oxygen and production of carbon dioxide. Bicarbonate helps to offset the metabolic acidosis caused by the high rate of lactate production through glycolysis. The intravenous infusion of iced lactated Ringer's solution also helps to remove body heat and supports renal function. Once a person is known to be susceptible to malignant hyperthermia, general anesthesia should be avoided.

plasma volume secondary to sweating and by vasodilatation of skin blood vessels, which is a normal vasomotor response to heat stress. Unlike heat stroke, heat exhaustion can occur under relatively mild heat stress and results from overactivity of heat loss mechanisms. People, such as the elderly, with deficiencies in blood pressure regulation are particularly susceptible to heat exhaustion.

Hypothermia: Unregulated Decreases in Body Temperature

Cold illnesses or disorders in humans usually involve peripheral circulatory or cardiac defects. Hypersensitivity of the arteries in the skin to cold (Raynaud's disease) can reduce blood flow to such critical levels that pain and local asphyxia may occur if the stress is continued. When the body mechanisms for heat production and conservation are exceeded by exposure to severe cold, **hypothermia** results, and body temperature drops below the normal range. At body temperature of about 26 to 28°C, death can occur due to **myocardial fibrillation.** Older persons are particularly susceptible to hypothermia, since in some elderly individuals, the hypothalamic reference temperature can be reset to a lower level. Some elderly people can maintain a core temperature of 35°C or even less without the onset of shivering. In the elderly, this appears to be a regulated hypothermia, and to some extent this condition can be considered a counterpart of fever.

Summary

I. A. Humans are homeotherms. They usually regulate their deep body temperature within a narrow range when exposed to a wide range of climatic conditions. The regulation of body temperature near the upper limit of heat tolerance ensures both rapid metabolism and a near-constant and optimal internal thermal environment.

The body has both longitudinal and radial temperature gradients. Normal body temperature should be conceived as a range of body temperatures. The most widely accepted measures of body core temperature are rectal, oral, and axillary temperatures.

B. The mean skin temperature is the accepted measure of the temperature of the body shell.

C. The mean body temperature is obtained by weighing mean temperatures of the core and the skin.

D. Many factors can affect the body temperature of an individual including exercise, diurnal rhythms, the reproductive cycle, age, and disease.

II. A. Heat is produced continuously in the body as a by-product of catabolism.

Both physiological and physical processes are continuously operating to balance heat production with heat loss, thereby maintaining thermal homeostasis and a relatively constant body temperature. The metabolic rate is the amount of food energy converted to heat per unit of time. Many factors can affect metabolic rate, including exercise, food, hormones, age, sex, size, and temperature. Basal metabolic rate (BMR) is a measure of the metabolic cost of living measured under standard conditions. It allows comparisons between individuals. A person's metabolic rate can be measured by direct or indirect calorimetry.

B. The metabolic heat produced must be continually removed from the body to maintain thermal balance. The body can exchange heat with its surroundings through radiation, conduction, convection, and evaporation. The evaporation of *sweat* is the only process for heat loss from a body exposed to hot environments. Sweating is the active secretion of mostly water by specialized glands in the skin. Sweat must be rapidly evaporated from the skin for cooling to occur.

C. To keep the body temperature relatively constant, the heat that is produced in the core must be transferred to the body shell, where it can be exchanged with the environment. Heat is transferred by internal conduction and circulatory convection. The rate of heat transfer due to internal conduction is too slow to maintain a balance between heat production and heat loss. Circulatory convection by the blood, therefore, is of prime importance for thermal balance. The vasomotor responses that control peripheral blood flow determine the effective insulation of the body and the rate of heat transfer by circulatory convection from the body core to the skin.

III. A. A negative feedback system is activated by temperature receptors that coordinate the responses for effective regulation of body temperature. Peripheral temperature receptors measure skin and shell temperature and are located just beneath the skin. They are nerve endings classified as cold or warm receptors. Nerve impulses from peripheral receptors enter the spinal cord at all levels and ascend to the brain.

B. Central thermoreceptors measure the temperature of the body core. They are located in deep body areas and include both cold and warm receptors. They send impulses into the brain, where integration with peripheral thermal information takes place.

C. The hypothalamus—through interactions with other brain areas, as well as with central and peripheral temperature receptors—acts as a central integrator and thermostat to prevent excessive body heating or cooling. Stimulation of the hypothalamus by overheating initiates antirise responses such as vasodilatation and sweating. Stimulation of the hypothalamus by excessive cooling stimulates antidrop responses such as vasoconstriction and shivering. The regulation of body temperature is a classic example of a precise negative feedback control system that has evolved to maintain thermal homeostasis.

IV. A. Fever is a regulated increase in body temperature usually caused by chemical agents called pyrogens released into the body during disease processes. Pyrogens appear to reset the control level of the hypothalamic thermostat at a higher body temperature level. Fever, therefore, is not a loss of effective regulation of body temperature.

B. Heat stroke is a form of an unregulated heat illness associated with a failure of the body's temperature regulating system. Heat stroke usually occurs during heavy work and/or exposure to hot humid conditions in which the rate of heat gain can exceed the maximal rate of heat loss. Under these conditions, the body temperature continues to rise beyond a critical temperature of 41 to 42°C typically associated with heat stroke. The high temperature itself may damage brain tissue, leading to failure of the temperature regulating system and death. Heat exhaustion is caused by circulatory insufficiency and often results in fainting due to decreased blood pressure. It occurs as a result of overactivity of heat loss mechanisms, including fluid loss due to excessive sweating and peripheral vasodilatation. People with deficiencies in blood pressure regulation such as the elderly are particularly susceptible to heat exhaustion.

C. Cold illnesses in humans usually involve peripheral circulatory or cardiac defects. Hypersensitivity of the arteries in the skin to cold can reduce blood flow to levels that induce local asphyxia and pain. In severe cold, heat loss may exceed the body's maximal ability for heat production, and body temperature will continue to fall, resulting in hypothermia. During severe hypothermia, in which body temperature drops to about 26 to 28°C, death can occur owing to myocardial fibrillation.

Review Questions

1. How does the regulation of body temperature in homeotherms differ from that of poikilotherms?
2. What are some of the benefits of homeothermy?
3. What are the location and thermal characteristics of the body shell and the core?
4. What is the most widely accepted measure of "normal" core temperature? What is its normal range of variance?
5. What is the importance of the concept of "mean body temperature" in temperature regulation?
6. What factors affect an individual's body temperature?
7. What is the relationship between heat production, heat loss, and thermal homeostasis?
8. Distinguish between metabolic rate and basal metabolic rate.
9. What factors affect metabolic rate?
10. What are the physical processes for heat exchange between the body and its surroundings?
11. Why is sweating so important in body temperature regulation?
12. What is the role of the cutaneous vasomotor system in the regulation of body heat transfer?
13. Describe the roles of warm- and cold-sensitive central and peripheral temperature receptors in the regulation of body temperature.
14. What is the role of the hypothalamus in the regulation of body temperature?
15. Discuss the functional differences between fever and heat stroke.

Suggested Readings

Bligh, J. Regulation of body temperature in man and other mammals. In Shitzer, A., and Eberhart, R., Eds. *Heat Transfer in Medicine and Biology,* Vol. 1. New York: Plenum Press, 1985, p. 15.

Crawshaw, L.I. "Temperature regulation in vertebrates." *Annual Review of Physiology,* 42:473, 1980.

Elizondo, R.S. Temperature regulation in primates. In Robertshaw, D., Ed. *International Review of Physiology: Environmental Physiology II,* Vol. 15. Baltimore: University Park Press, 1977, p. 71.

Elizondo, R.S., and Johnson, G.S. Peripheral effector mechanism of temperature regulation. In Szelenyi, Z., and Szekely, M., Eds. *Advances in Physiological Sciences,* Vol. 32. New York: Pergamon Press, 1980, p. 397.

Gordan, C.J. and Heath, J.E. "Integration and central processing in temperature regulation." *American Review of Physiology,* 48:595, 1986.

Heller, H.C., et al. "The thermostat of vertebrate animals." *Scientific American,* 239 (2):102, 1978.

Hensel, H. "Neural processes in thermoregulation." *Physiology Review,* 53:948, 1973.

Veale, W.L. Ruwe, W.D., and Cooper, K.E. Mechanism of fever and antipyresis. In Khogali, M., and Hales, J.R.S., Eds. *Heat Stroke and Temperature Regulation.* New York: Academic Press, 1983, p. 79.

29

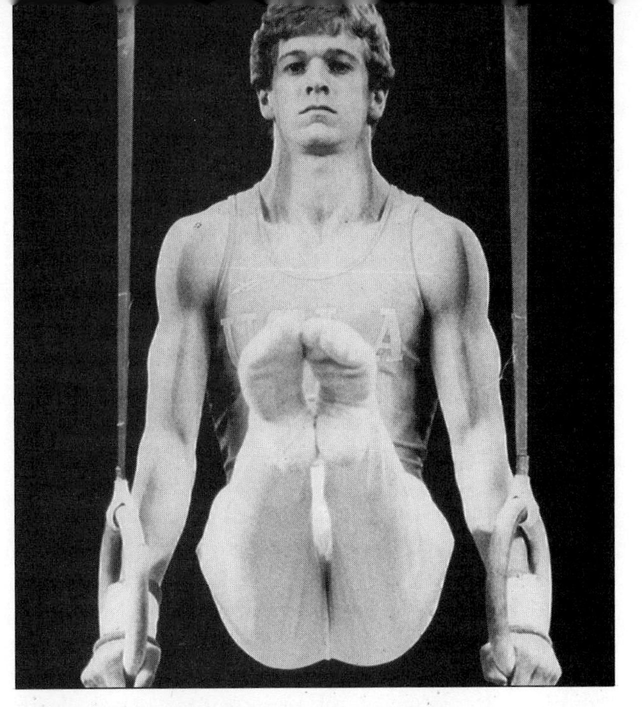

Physiology of Exercise

In all types of exercise, the final common **physiological event** is the **contraction** of skeletal muscle. But exercise also places demands on various other body systems that play a supportive role. The basic biochemical alterations at the level of the muscle cell that convert chemical energy into mechanical energy are accompanied by increases in the integrated activity of many organ systems, including the **central nervous system** (CNS), which initiates coordinated skeletal muscle activity. The **respiratory system** and **circulatory system** increase their activity during exercise to provide increased oxygen and nutrients to active muscle cells and to remove CO_2, other metabolites, and heat. Contracting muscle cells may increase heat production by 20 times that of resting levels; thus, exercise can also place extra demands on the **temperature regulating system.** The extent of the physiological adjustments in organ systems in response to exercise is often a direct measure of the **strain** or **stress** of exercise on the body.

The basic concepts of exercise physiology, which this chapter will discuss, have been derived mostly from studies on young adult male subjects because relatively complete measurements of the physiology of exercise have traditionally been made only on this group of athletes. Recent studies, however, in which measurements have been made on females, indicate that the same basic principles apply. However, some **quantitative differences** are found, such as differences in body size, body composition, and sex hormones. In general, most quantitative values for women, such as ventilation, cardiac output, and muscle strength, which are related to muscle mass, will vary between 65% to 75% of the values recorded in males. Differences in athletic performance between the sexes, on the other hand, are smaller because the female body is on average smaller, requiring smaller muscles to perform the same activities.

Exercise is often classified as either **dynamic exercise** or **static exercise.** Dynamic exercise is performed anytime muscle length changes. In a physical sense, resistances are overcome along a certain distance. In this type of exercise, which includes such activities as bicycling, running, stair climbing or mountain climbing, exercise is quantified in physical units such as watts. In **positive dynamic exercise,** the muscles act as a "motor" such as in ascending a mountain. In **negative dynamic exercise,** muscles act as "brakes" such as in descending a mountain. Static exercise is performed during **isometric muscle contraction.** During isometric exercise, shortening of the muscle does not occur and no distance is traversed, so it cannot be defined as work in the physical sense. Nevertheless, static exercise is associated with increased metabolic demands. The **elevated metabolic demand** is the common feature of all types of exercise, and, in most instances, this is more important for physiological adjustments than any other factor.

Metabolic Aspects of Exercise

Energy Requirements of Muscular Contraction

The primary energy source for skeletal muscle contraction, as for other types of cellular work, is high-energy phosphate compounds such as **adenosine triphosphate (ATP).** These are produced by both anaerobic and aerobic metabolic processes (Fig. 29–1). Three important metabolic systems that supply energy for muscle contraction are (1) the phosphagen system, (2) the glycogen-lactic acid system, and (3) the aerobic system **(the citric acid cycle).**

The Phosphagen Energy System

The term *phosphagen* is another name for a high-energy phosphate molecule, one that serves as a reservoir of phosphate bond energy. The **phosphagen energy system** of muscle includes ATP and **phosphocreatine.** When one phosphate group is removed from ATP (producing ADP), approximately 11,000 calories of energy become available for the contractile process. When a second phosphate is removed, leaving AMP, an additional 11,000 calories become available. The amount of energy present as ATP in the muscle is enough to maintain maximal exercise for only 5 to 6 seconds.

Phosphocreatine, however, has slightly more energy than the bond energy of ATP and can reconstitute ATP by donating phosphate groups to AMP and ADP. Muscle cells have two to three times more phosphocreatine than ATP, and together the phosphagens ATP and phosphocreatine can maintain maximal muscle contraction for only 10 to 15 seconds. Thus, the energy supply from the phosphagen system is used for short bouts of maximal exercise.

The Glycogen-Lactic Acid System

The glycogen-lactic acid system in muscle also can supply energy for contraction. The stored muscle glycogen can be split into glucose **(glycogenolysis)** and then used for energy. Recall from Chapter 6 that glycolysis occurs in the absence of oxygen and is therefore **anaerobic.** During glycolysis, each glucose molecule is converted into two molecules of pyruvic acid, and energy is released to form two molecules of ATP. If oxygen is available, pyruvic acid enters the citric acid cycle, and much more ATP is formed. If there is insufficient oxygen to support the oxidative stage of glucose metabolism, however, the pyruvic acid is converted to lactic acid. The glycogen-lactic acid system can produce ATP molecules

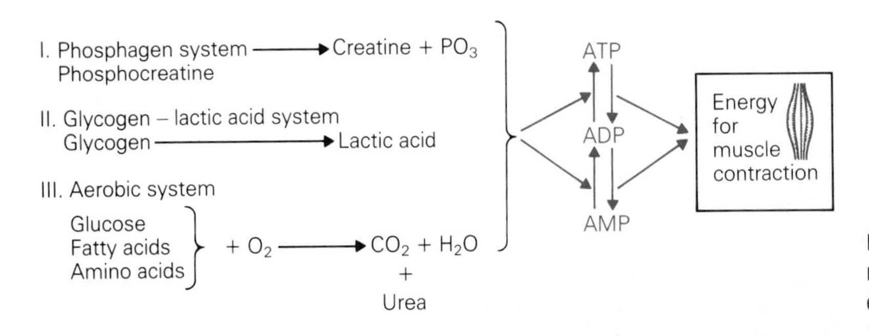

Figure 29–1 Summary of the three primary metabolic pathways that supply energy for muscle contraction.

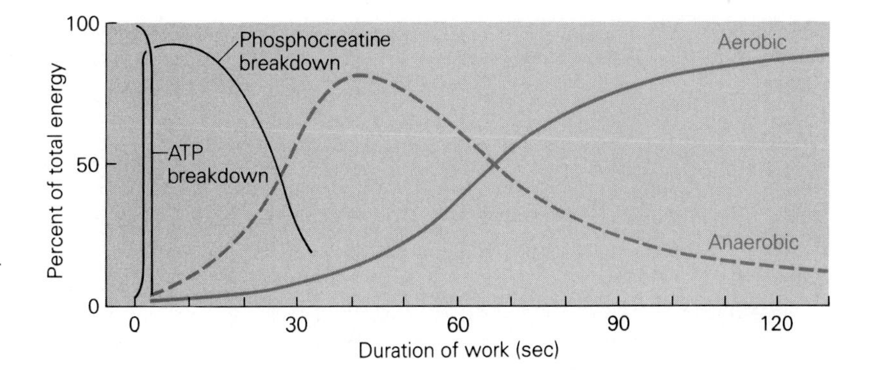

Figure 29–2 Relative contribution of various energy sources to the total energy used by the muscle at the beginning of light exercise.

more rapidly than the oxidative mechanisms in the mitochondria and is used as a rapid source of energy for muscle contraction. The glycogen-lactic acid system can provide the energy for maximal exercise for only 30 to 40 seconds.

The Aerobic Energy System

Recall from Chapter 6 that the aerobic energy system involves the use of oxygen within the mitochondria for the metabolism of glucose, lipids, and proteins in the citric acid cycle and electron-transport chain to produce 36 ATP molecules for every molecule of glucose. Only 20% of the energy released, however, is in the form of the mechanical energy of contraction; 80% is released as heat energy. The heat released during exercise places additional demands on the cardiovascular system and the sweat response, which play a major role in temperature regulation and the transfer of heat from the body to the environment (see Chaper 28).

Aerobic metabolism is normally the source of energy for prolonged light work. During light exercise, energy is obtained anaerobically only during the first few minutes when muscle blood flow is gradually increasing (Fig. 29–2). Thereafter, metabolism is entirely aerobic, utilizing glucose, fatty acids, and amino acids. During heavy exercise, by contrast, part of the energy is always obtained anaerobically, and if large amounts of lactic acid are produced, muscle fatigue and cramps result.

Oxygen Consumption During Exercise

During exercise, oxygen consumption increases by an amount corresponding to the intensity of the exercise. As illustrated in Figure 29–3, oxygen consumption increases during the first few minutes of exercise and then reaches a "steady state" in which the oxygen uptake corresponds to the demands of the tissues. The average person in the basal state consumes about 0.25 liters (L) of oxygen per minute. An untrained person can increase O_2 consumption threefold (0.75 L/min) during light work and eightfold to tenfold (2–3 L/min) during heavy exercise. Well-trained athletes can increase their O_2 consumption over 20 times (5–6 L/min) above basal level. Oxygen consumption is often standardized in units of body weight (O_2 mL/kg/min), and averages 44 to 51 mL/kg/min for untrained males. Values between 50 to 60 mL/kg/min are often obtained in young, well-trained males. Untrained females normally have O_2 consumption during exercise that is about 20% less than that of untrained males.

Oxygen consumption increases linearly with increasing levels of exercise up to a point where the maximum for oxygen transport and uptake in the tissues appears to be reached and a plateau, or the so-called **maximal oxygen consumption ($\dot{V}O_2$max),** is attained (Fig. 29–4). The $\dot{V}O_2$max is generally as-

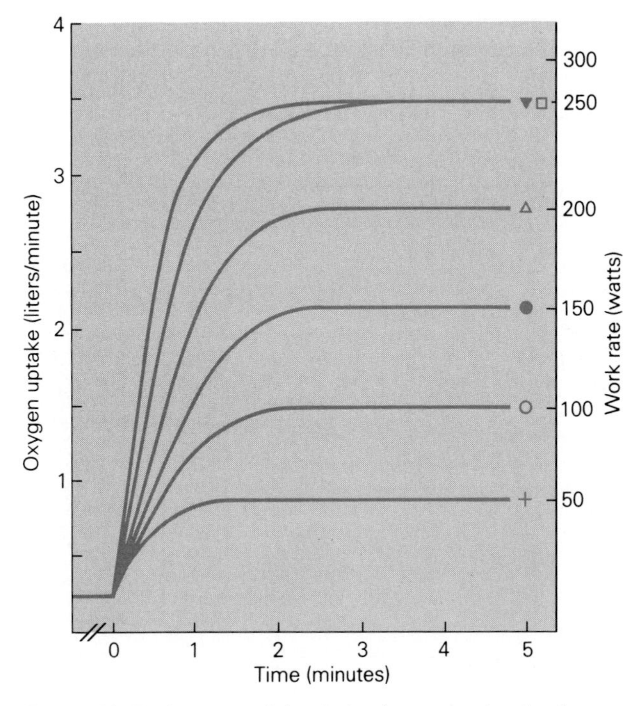

Figure 29–3 Oxygen uptake during increasing levels of steady state exercise.

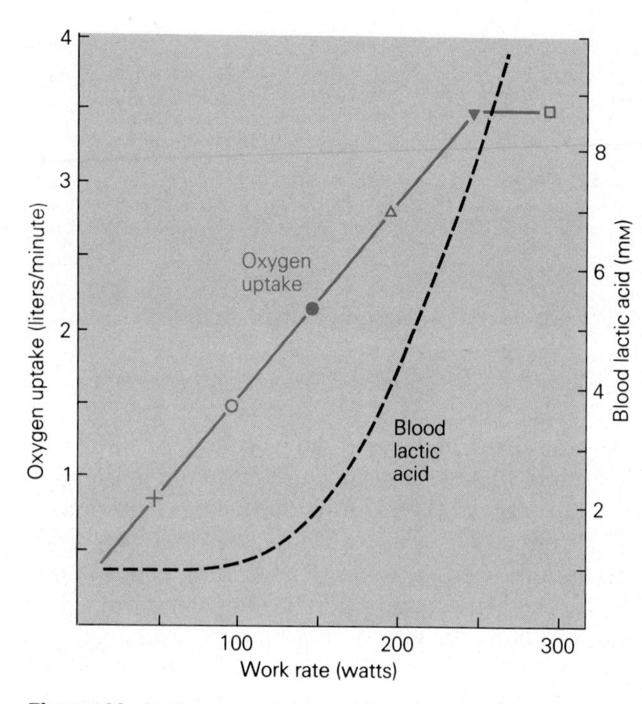

Figure 29–4 Oxygen uptake and blood lactic acid during increasing levels of exercise.

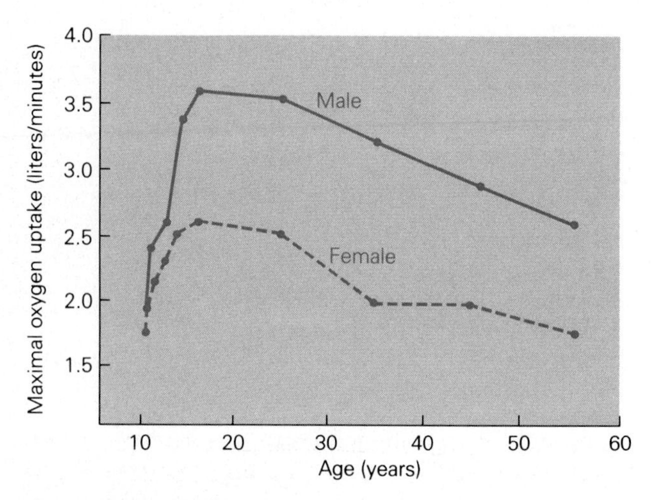

Figure 29–6 Maximum oxygen consumption as a function of age in males and females.

sumed to represent a person's **maximal aerobic power** and thus is an important factor in determining an individual's ability to sustain high-intensity exercise. Three main criteria have been established to certify that a measurement of $\dot{V}O_2$max actually represents a person's maximal aerobic power: (1) there is no further increase in oxygen consumption despite further increases in the rate of exercise; (2) the blood lactate concentration exceeds 8 mM; and (3) large muscle groups are exercised for at least 3 mintues. Figure 29–5 illustrates that the $\dot{V}O_2$max peak obtained during maximal exercise is maximal only when a relatively large muscle mass (i.e., two legs) is being exercised.

Numerous factors can affect an individual's $\dot{V}O_2$max. With age, $\dot{V}O_2$max decreases in both males and females. As shown in Figure 29–6, maximal uptake in liters per minute increases during the growth stages of the early years and reaches a peak between 18 and 25 years of age. This early increase in oxygen consumption is directly related to the increase in body mass that occurs at this age. Thus, the apparent increase in aerobic capacity is essentially eliminated when the $\dot{V}O_2$max is expressed relative to body weight as mL/kg/min. After age 25, the $\dot{V}O_2$max in terms of liters per minute, as well as relative to body weight, steadily declines so that at age 55 it is about 72% below values obtained for 20-year-olds. While active adults tend to retain relatively higher $\dot{V}O_2$max than inactive adults, a progressive decline in this physiological capacity also occurs with advancing age (Fig. 29–7).

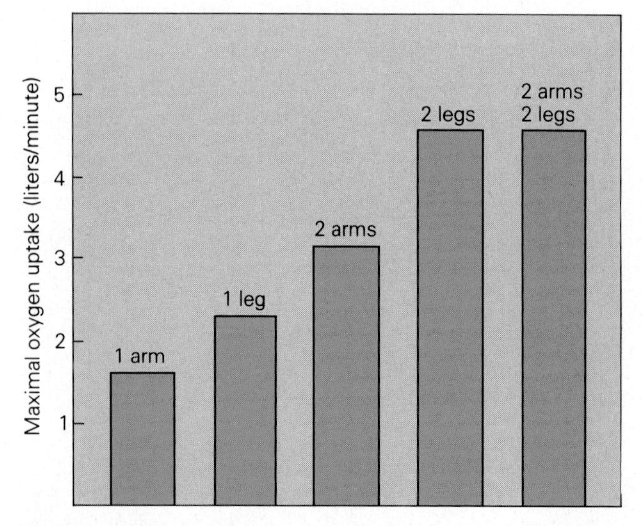

Figure 29–5 Peak oxygen consumption during maximum exercise using different size muscle mass.

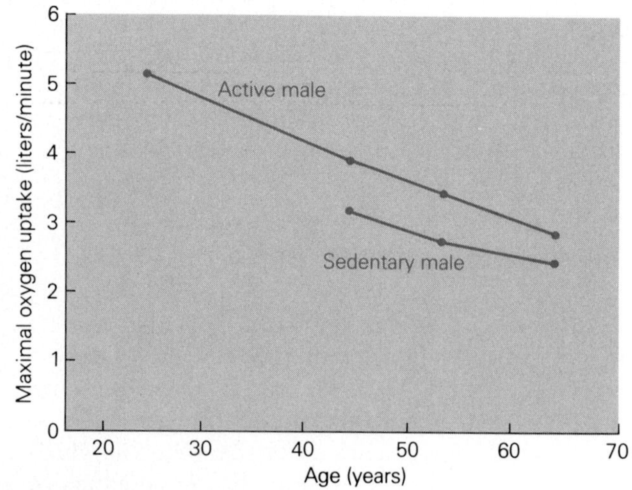

Figure 29–7 Maximum oxygen consumption as a function of age and level of activity.

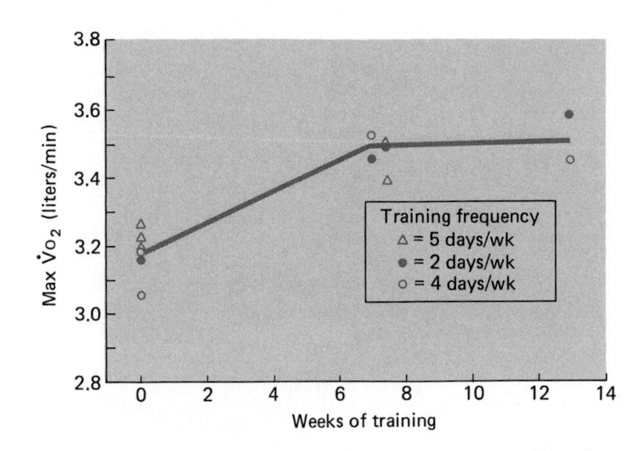

Figure 29–8 Change in maximal oxygen consumption during 7 to 13 weeks of athletic training. (Reprinted from Fox: *Sports Physiology*. Philadelphia, Saunders College Publishing, 1979.)

Training increases $\dot{V}O_2$max and can partially counteract the effects of aging (Fig. 29–8). For example, the mean $\dot{V}O_2$max for nine men on an initial exposure to running 7 mph (9% grade) on a treadmill averaged 53 mL/kg/min. After 2 months of training by running on a treadmill at a rate of 8 mph (zero incline) for 10 to 15 minutes, their $\dot{V}O_2$max was increased to 56 mL/kg/min. The physiological factors that may limit human O_2 consumption include (1) the rate of cardiovascular transport to active tissue, (2) the rate of O_2 use by active tissue, and (3) the O_2 diffusion capacity of the lungs. There appears to be a dual basis for an increased $\dot{V}O_2$max with training, including (1) an increase in maximal cardiac output, and (2) an increase in blood arterial-venous oxygen difference. Training, on the other hand, has a relatively small effect on pulmonary function.

Oxygen Debt During Exercise

When exercise begins the demand for oxygen increases immediately. However, tissue blood flow and aerobic oxidative metabolism require some time to adjust to the new demands and an **oxygen debt** is incurred as oxygen is used from existing body stores (Fig. 29–9). During light work, the oxygen debt remains constant after a steady state is reached in which oxygen uptake and consumption are in balance. During heavy exercise, the oxygen debt continues to build up until the exercise ends. At the end of exercise, although the demand for oxygen by contracting tissue decreases abruptly, the rate of oxygen uptake remains above the resting level for several minutes as the oxygen debt is being repaid. The oxygen debt is often defined as the extra amount of oxygen that must be taken into the body after exercise above the pre-exercise or resting level. The oxygen debt repaid after the end of exercise is often greater than that incurred during the work itself and is affected by other factors such as increases in body temperature and hydrogen ion concentration. After light work, the oxygen debt amounts to as much as 4 liters, and after heavy exercise it can be as high as 20 liters.

The repayment of the oxygen debt occurs in two stages (Fig. 29–10). The first stage is called the **alactacid oxygen debt,** the term referring to the oxygen needed to replenish the oxygen stores of the body as well as the phosphagen system. The alactacid debt is usually repaid within 2 to 3 minutes after exercise. The second stage is called the **lactic acid oxygen debt** and requires the removal of lactic acid from the body. This stage is repaid very slowly and requires at least 1 hour. During this time, some of the lactic acid is converted to pyruvic acid and then metabolized by body tissues. Also, much of the lactic acid is converted by the liver to glucose, which is used to replenish the glycogen stores in the muscle.

The Use of Nutrients During Exercise

Carbohydrates, fatty acids, and (to a lesser degree) amino acids can be utilized as fuel during exercise in proportions that vary depending on the intensity

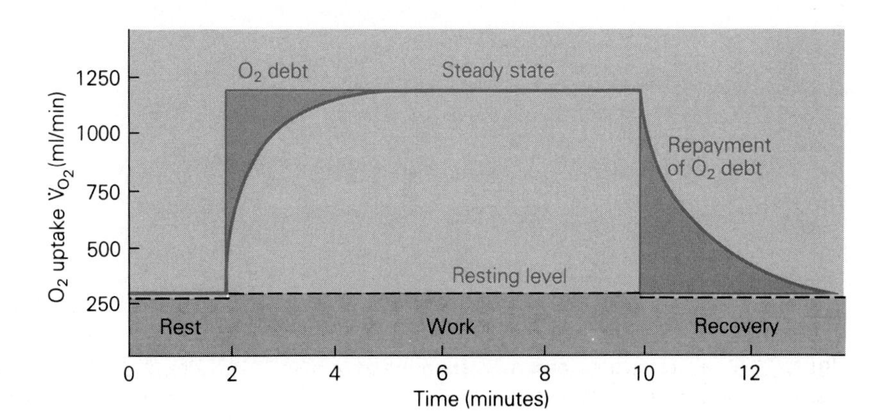

Figure 29–9 Oxygen consumption during light steady state exercise.

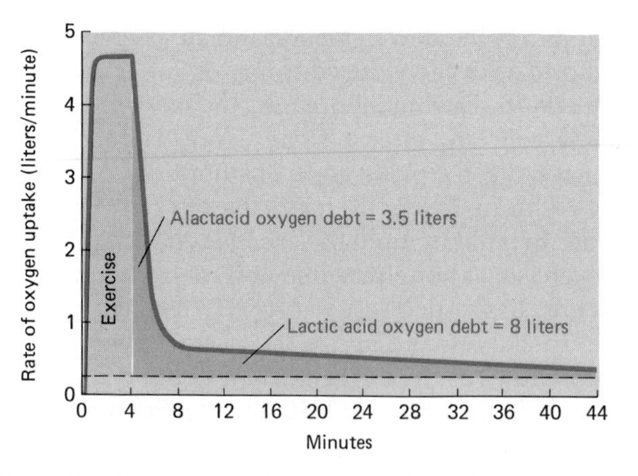

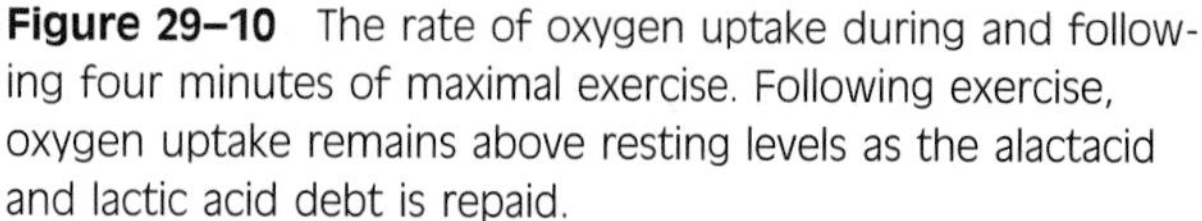

Figure 29–10 The rate of oxygen uptake during and following four minutes of maximal exercise. Following exercise, oxygen uptake remains above resting levels as the alactacid and lactic acid debt is repaid.

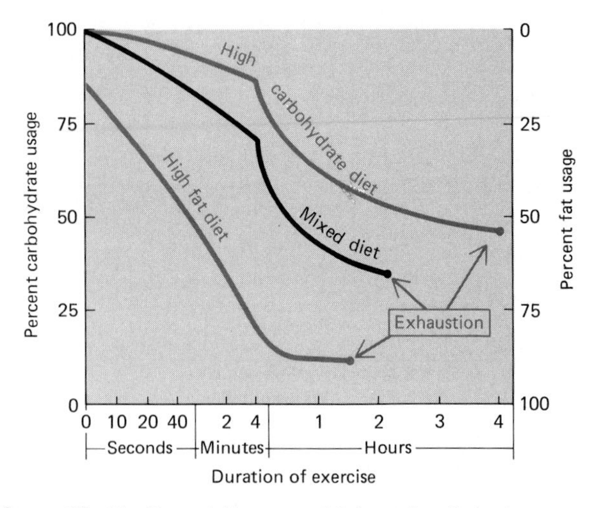

Figure 29–11 The relative use of fats and carbohydrates as a function of diet and duration of exercise.

and duration of the exercise and the availability of stored fuels. At low rates of exercise of short duration such as a short walk, lipids supply a large portion of the required energy. With increasing intensity and duration of exercise, there is a gradual change toward a proportionally greater share of energy yield from carbohydrates. However, if moderately heavy exercise continues for 4 to 5 hours, the glycogen stores of the muscle become depleted. The muscle now depends upon the limited supply of glucose that can be absorbed from the gut, or upon energy from other sources, mainly fats, and (to a much lesser extent) protein. Clearly, the choice of fuel for exercising muscle is mainly limited to carbohydrates and fat.

The relative usage of carbohydrates and fat for energy yield during prolonged exhaustive exercise is illustrated in Figure 29–11. Note that in an individual on a mixed diet, most of the energy at this level of exercise is derived from carbohydrates during the first few minutes, but as the exercise continues there is a gradual shift toward fat use, and at exhaustion as much as 60% to 70% of the energy is being derived from fats rather than carbohydrates. It is also clear that diet can influence the relative proportion of stored fuels and thus influence the relative usage of fat and carbohydrates. A high-carbohydrate diet is associated with a relatively greater share of the energy yield from carbohydrates and longer endurance, whereas a high fat diet results in a proportionally greater share of the energy yield from fat and relatively shorter endurance.

Diet can also affect glycogen recovery following exhaustive exercise. Figure 29–12 shows the rate of muscle glycogen replenishment in three different situations, namely: a high-carbohydrate diet, a high-fat/high-protein diet, and starvation. Note that even in individuals on high-carbohydrate diets,

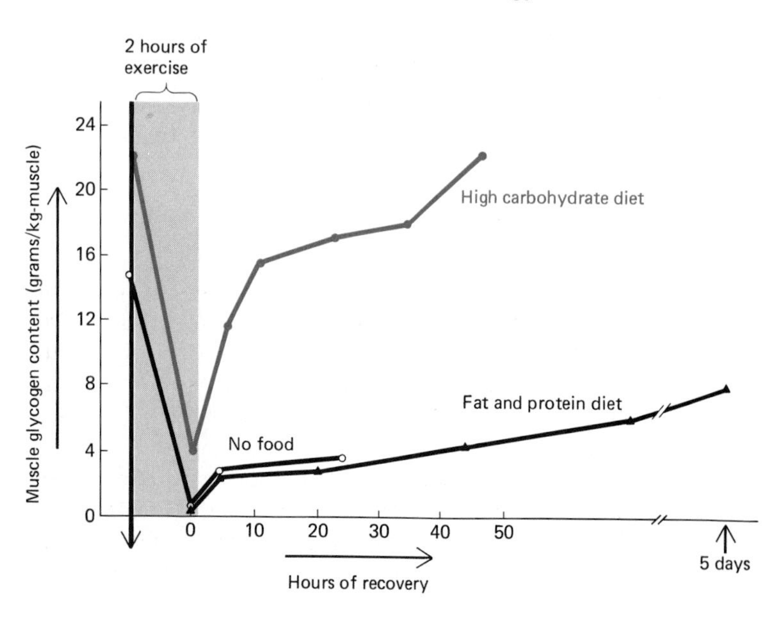

Figure 29–12 Rate of replenishment following prolonged exercise with different diets. (Reprinted from Fox: *Sports Physiology*. Philadelphia, Saunders College Publishing, 1979.)

muscle recovery is a relatively slow process, lasting approximately two days. Individuals on high-fat and high-protein diets, or persons eating no food at all, showed very little recovery even after 5 days. It seems clear that to achieve optimal performance approximately 48 hours of recovery after exhausting exercise are essential.

Respiration During Exercise

Pulmonary Ventilation During Exercise

The increased aerobic metabolism in muscle during exercise places extra demands on the respiratory system. The increased oxygen demands of an exercising individual are nicely paralleled by increased **pulmonary ventilation.** During exercise, the increase in ventilation is proportional to the metabolic demands (O_2 consumption), as illustrated in Figure 29–13. Pulmonary ventilation, which is normally 5 to 6 liters per minute, may approach 150 liters per minute in severe, short-term exercise. The change in total ventilation is associated with a fourfold increase in respiratory rate and a sixfold increase in tidal volume. These changes require greater work by the respiratory muscles, and the work of breathing is increased by one-hundredfold (Fig. 29–14). Note, however, that maximal breathing capacity in a healthy young individual is approximately 170 liters per minute and is therefore 50% greater than the pulmonary ventilation achieved during maximal exercise. The point is that the respiratory system is not normally the most important factor limiting the delivery of oxygen to the muscles during maximal exercise.

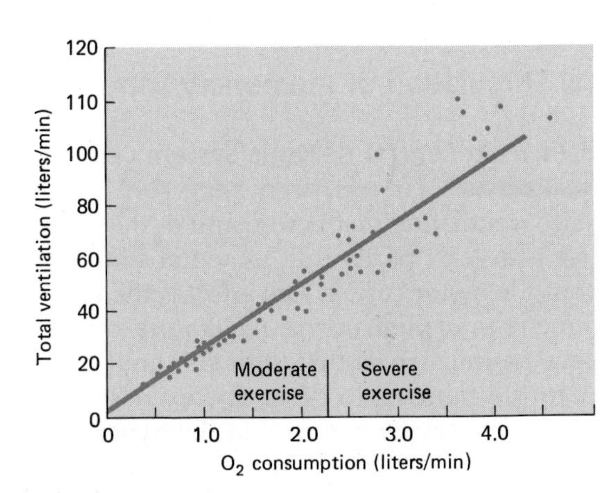

Figure 29–13 Effect of exercise on oxygen uptake and total ventilation.

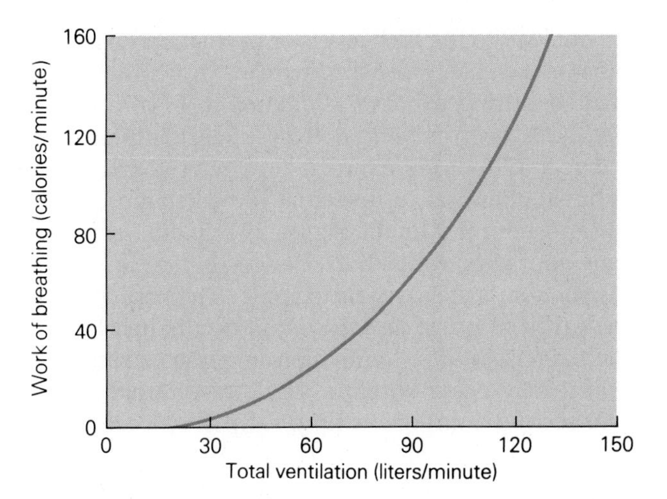

Figure 29–14 Work of breathing as a function of minute volume (pulmonary ventilation).

The Diffusing Capacity of the Lung During Exercise

The **diffusing capacity of the lung** is defined as the milliliters of a gas diffusing across the pulmonary membrane per minute per millimeter of mercury pressure difference between the alveolar air and the pulmonary blood. The diffusing capacity of the lung for the respiratory gases provides an index of the dimensions of the pulmonary capillary bed, the pulmonary membrane, and the overall efficiency of the system in the exchange of respiratory gases. The diffusion capacity of oxygen and carbon dioxide increases during exercise. Oxygen diffusion capacity increases from an average of 23 mL/mm Hg at rest to about 80 mL/mm Hg at maximum exercise in a highly trained athlete. The increased diffusing capacity in the lung is largely the result of the increased pulmonary blood flow that accompanies exercise and ensures maximum perfusion of pulmonary capillaries and thus provides a greater surface area for gas diffusion.

Control of Pulmonary Ventilation During Exercise

Pulmonary ventilation at rest exhibits inherent rhythmicity. Its control involves interactions between CNS inspiratory and expiratory centers, pulmonary stretch receptors, and peripheral and central chemoreceptors sensitive to blood gas pressure and H^+ concentrations.

The mechanisms that control ventilation at rest cannot entirely explain the increased ventilation or **hyperpnea** during exercise. For example, artificial changes in Po_2, Pco_2, and acidity in the body at rest do not lead to the high levels of ventilation observed during exercise. How this increase in ventilation is elicited during exercise is unknown. Several theories have been advanced that suggest interactions between numerous stimuli falling into two main categories, **chemical,** or **humoral** (fluid-related), **stimuli** and **neural stimuli.** Figure 29–15 illustrates the time course of ventilation changes before and during moderate steady state exercise. The first component (Phase I) is a rapid rise that occurs just before the exercise starts. Ventilation increases so rapidly that it may occur within a single respiratory cycle. The second component (Phase II) is a slower rise, and the third component (Phase III) is associated with the development of steady state ventilation. The fast component of the response is believed to be associated with neural stimuli, since they occur sooner than the metabolic changes. The slow components are believed to be related to the slower humoral stimuli that alter CNS respiratory center activity. While the exact nature of the neural and chemical mechanisms that control ventilation during exercise is not known, it is clear that they interact to provide the proper increase in pulmonary ventilation to meet the metabolic demands of the exercise and thus keep blood respiratory gases almost normal. Some of the factors that have been studied as possible stimuli for the exercise hyperpnea will now be briefly discussed.

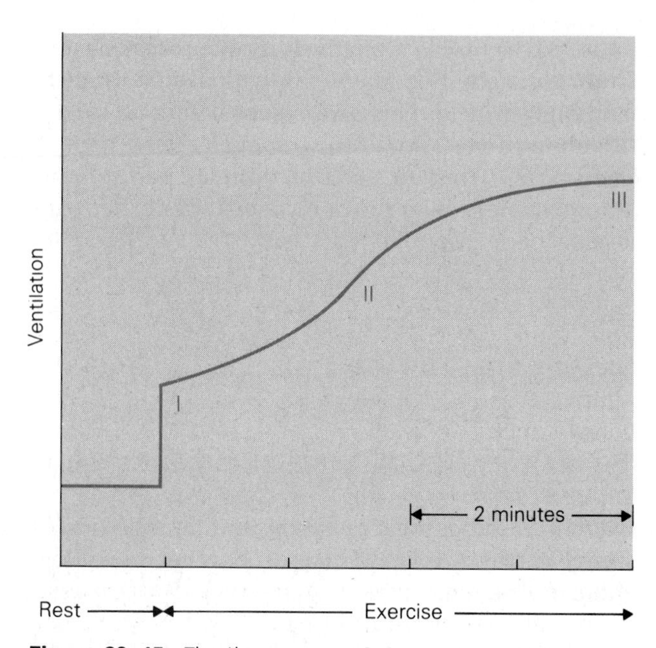

Figure 29–15 The time course of change in ventilation before and during moderate exercise.

Chemical Stimulation of Pulmonary Ventilation

Increased Arterial Carbon Dioxide Concentration During Exercise. The close correlation between ventilation rate and metabolic rate during steady state exercise would suggest that the ventilatory drive may be associated with a chemical constituent that is altered at rates proportional to metabolism. Carbon dioxide production is greatly increased in muscle during exercise and could theoretically increase arterial P_{CO_2} sufficient to stimulate CNS respiratory centers. Actual measurements of arterial P_{CO_2} indicate that the increases are nonexistent during moderate exercise and that a sizable decrease can occur during very heavy exercise. Much larger increases in arterial P_{CO_2} than have been measured would have to occur if P_{CO_2} alone were the primary stimulus.

Increased Arterial Hydrogen Ion Concentration During Exercise. Lactic acid production is greatly increased in muscles during exercise and could lead to an increase in arterial hydrogen concentrations sufficient to stimulate CNS respiratory centers. Actual measurements have shown that significant changes in arterial lactic acid concentration occur only during very strenuous exercise. Furthermore, at the end of exercise, lactic acid levels may increase slightly while ventilation rate rapidly decreases. It is therefore unlikely that the blood hydrogen concentration alone provides the major stimulus for exercise hyperpnea.

Decreased Arterial Oxygen Concentration During Exercise. Arterial **hypoxemia** can stimulate respiration by eliciting chemoreceptor activity from the carotid and aortic bodies. The threshold for these receptors, however, is an arterial blood oxygen pressure approximately 60 mm Hg. Even during strenuous exercise, arterial oxygen pressure usually remains absolutely constant, and this mechanism alone cannot provide the primary stimulus for the increased ventilation during exercise.

Neural Stimulation of Pulmonary Ventilation

Impulses from Central Nervous System Centers During Exercise. It has been suggested that the increased ventilation just before and at the onset of exercise is at least partly due to neural stimuli arising from the motor cortex. The effect is the result of the direct stimulation of the respiratory center by the same neural signals that radiate from the motor cortex to the muscles to cause the exercise. While neural signals may be involved in the hyperpnea of exercise, it is difficult to envision how purely neural activity can be so finely tuned to the metabolic demand of the exercise.

Impulses from Peripheral Receptors During Exercise. Passive movement of the limbs is often associated with an increase in ventilation. The ventilatory response is closely related to the number of joints and muscles involved in the movement. Researchers have suggested that muscle contraction and movement of the extremities may stimulate peripheral proprioreceptors in the joints and in muscle spindles. These receptors, in turn, are believed to send afferent neural signals to the respiratory center and stimulate ventilation. The role of peripheral neural receptors in fine tuning ventilation to metabolic demands is unknown.

Impulses from Lung Stretch Receptors During Exercise. Lung stretch receptors are known to be stimulated by filling of the lungs and to send afferent neural signals to the respiratory center, resulting in an inhibition of inspiration and initiation of expiration. Lung stretch receptors may play a role in the hyperpnea of exercise by increasing respiratory rate. This mechanism, however, is not considered the primary stimulus for exercise hyperpnea, since voluntary increases in respiratory rate are not self-perpetuated, as is required during exercise.

Increased Sensitivity of the Respiratory Centers During Exercise. An increased sensitivity of the respiratory centers to arterial P_{CO_2} could account for the hyperpnea of exercise. The change in sensitivity, however, would have to be very large since even in strenuous exercise arterial O_2 and P_{CO_2} change very little. The fact remains, however, that when breathholding time is measured during moderate exercise, rebreathing occurs at a lower alveolar CO_2 and higher O_2 than at rest. It is not possible to ascribe this effect only to increased receptor center sensitivity as the neural factors described previously may be operating to increase the activity of the respiratory center.

Increased Body Temperature During Exercise. An increase in body temperature increases ventilation and could play a role in exercise hyperpnea. The increase in ventilation, however, in exercise is much greater and occurs before any significant rise in body temperature. While an increase in body temperature is not the primary stimulus for exercise hyperpnea, the possibility remains that the hyperthermia of exercise may have a direct effect on the activity of joint and muscle receptors as well as the respiratory center during exercise.

We must conclude that neither neural nor chemical stimuli alone can explain the hyperventilation of exercise. Furthermore, while neurogenic factors cannot be considered as a regulating stimulus, they may act as an activator of the response. The humoral factors, on the other hand, appear to be critical for the finer adjustments of ventilation to keep the blood respiratory gases almost normal during exercise.

Circulation During Exercise

Muscle Blood Flow During Exercise

One of the most important functions of the cardiovascular system during exercise is to deliver oxygen and other nutrients to the working muscles and remove CO_2 and other metabolic by-products. As illustrated in Figure 29–16, this function is facilitated by a marked increase in muscle blood flow during dynamic exercise. Muscle blood flow can increase approximately twenty-fivefold during maximal exercise. Approximately 50% of the increase in flow is the result of local vasodilatation caused by the direct effects of the increased muscle metabolism. The rest of the increase in muscle blood flow during exercise is the result of several factors, an important one being a moderate increase in arterial blood pressure of about 30%. Figure 29–16 also shows that during contraction muscle blood flow temporarily decreases as the intramuscular vessels are compressed. Strong tonic contractions are known to cause rapid muscle fatigue owing to the inadequate delivery of oxygen and nutrients during the continuous contraction.

Cardiac Output During Exercise

Cardiovascular regulation during exercise is designed to provide an adequate cardiac output to ensure an adequate oxygen and nutrient supply to the

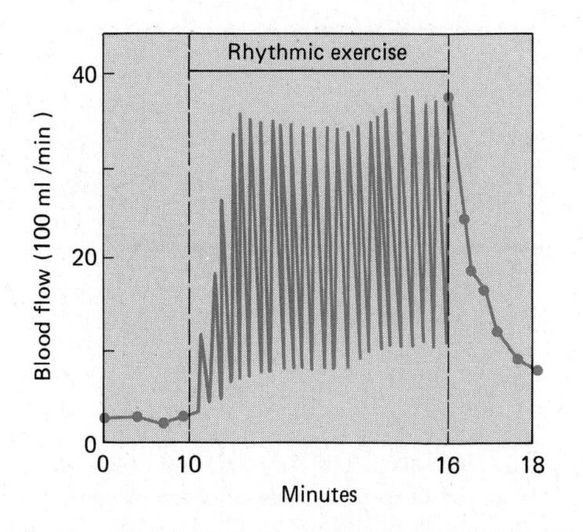

Figure 29–16 Effect of rhythmic muscle contraction on human calf muscle blood flow. (From Bancroft and Dornhurst; *J. Physiol., 109:*402, 1949.)

Focus on Exercise and Immunity

A relationship between exercise and immunity was first demonstrated in 1920, when it was shown that intense exercise caused leukocytosis. That is an increase in white blood cell (WBC) count from between 5000 and 7000 WBC per milliliter of blood to between 25,000 and 30,000 WBC per milliliter. This increase in WBC count cannot be explained by hemoconcentration alone. It is also due to the washing out of WBCs from their storage places in the bone marrow, spleen, liver, and lungs, which receive increased blood flow as a result of the increased cardiac output with exercise.

Recent studies have shown that there are many similarities in the response of the immune system to the stress of exercise and its response to infection. During both exercise and infection there is an increase in body temperature, which increases the rate of all chemical reactions, including those associated with the immune system. Other specific immune responses to exercise also seen in infection include (1) leukocytosis, (2) increased natural killer cell activity, (3) increased plasma levels of acute phase reactants such as C-reactive protein, and (4) increased levels of plasma interferon. These responses to exercise are mediated by the secretion by macrophages of the polypeptide hormone interleukin 1 (IL1). IL1 is a prominent member of a group of polypeptide mediators now called cytokines. IL1 is one of the key mediators of the body's response to microbial invasion, inflammation, immunological reactions, and tissue injury. The specific stimulus for the secretion of IL1 during intense exercise is unknown but may be related to local muscle inflammation or tissue injury.

working muscles. As illustrated in Figure 29–17, there is a remarkable consistency between cardiac output, oxygen consumption, and work output. During light and moderate exercise, as work output increases, there is a proportional increase in oxygen consumption, which dilates the muscle blood vessels. This dilation increases venous return, which results in a linear increase in cardiac output. There is a deviation from linearity as oxygen consumption approaches VO_2max related to thermoregulation at high work rates. During strenuous exercise, as body temperature rises, skin blood flow must be increased so that the high internal heat loads can be transferred to the skin and dissipated to the environment as required for effective body temperature regulation.

The increased blood flow to the skin under these conditions places an additional demand on the available cardiac output. Typical cardiac outputs for nontrained and highly trained athletes are illustrated in Table 29–1. The table shows that an untrained individual during maximal exercise can increase cardiac output approximately four times the resting level while the well-trained person can increase cardiac output about six times the resting level. The greater cardiac output in trained individuals is mainly due to **hypertrophy** (enlargement) of the heart, which results in an increase in both the size of the chambers and the mass of the heart. Hypertrophy of the heart, however, occurs only in endurance-type training and has no important clinical significance.

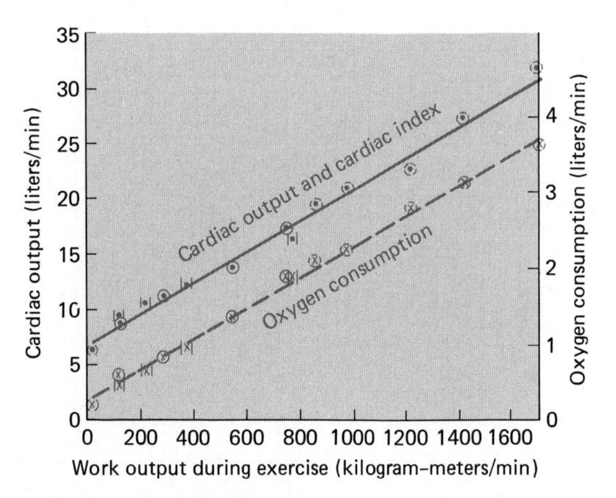

Figure 29–17 Relationship between cardiac output, oxygen consumption, and work output during different levels of exercise.

Stroke Volume and Heart Rate During Exercise

Table 29–1 also compares stroke volume and heart rate in nonathletic and endurance trained athletes at several levels of exercise. It shows that both at rest

Table 29–1 Comparison of Stroke Volume and Heart Rate in a Trained Athlete and a Nonathlete

State	Stroke Volume (mL)	Heart Rate (beats/min)
Resting		
Nonathlete	75	75
Athlete	105	50
Maximum exercise		
Nonathlete	110	195
Athlete	162	185

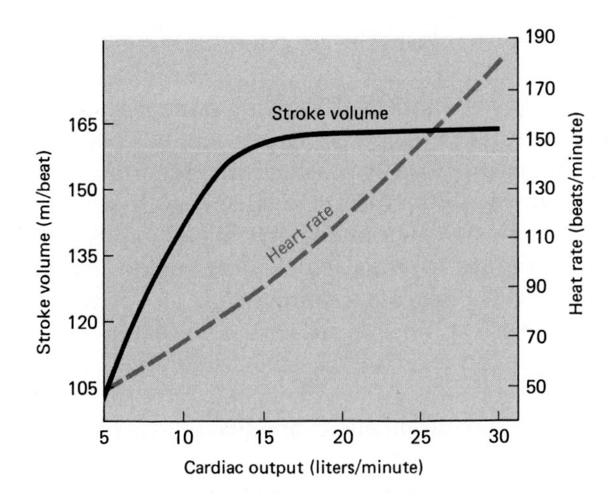

Figure 29–18 Relationship between stroke volume, heart rate, and different levels of cardiac output during exercise in an endurance trained athlete.

and during maximum exercise, the trained individual has a significantly greater stroke volume and a lower heart rate than the untrained person. Because of the greater pumping effectiveness of the heart, the trained individual can achieve the same cardiac output both at rest and during exercise with a decreased heart rate. Figure 29–18 illustrates the interrelationships between stroke volume, heart rate, and cardiac output in a trained person with a resting cardiac output of 5.5 liters per minute and a maximum exercise output of 30 liters per minute. The heart rate increases more than 250%, while stroke volume increases about 50% and reaches a maximum when the cardiac output approaches 15 liters per minute. Further increases in cardiac output are the result of increases in heart rate. Therefore, in strenuous exercise, the increase in heart rate accounts for the major increase in cardiac output.

During maximum exercise, cardiac output increases to about 90% of the maximum that the person can achieve. The capacity of the cardiovascular system therefore is normally a limiting factor in terms of maximal oxygen consumption and exercise performance. The maximum cardiac output of a person decreases wth age and significantly contributes to the decrease in the maximal exercise performance in the elderly.

Local Blood Flow to Organs During Exercise

The active muscles, including the respiratory muscles, receive the bulk of the increased cardiac output during exercise. The increase in muscle blood flow is associated with drastic alterations in the blood flow distribution to other organs. Blood flows and the percentage of the cardiac output perfusing various tissues during rest and strenuous exercise are illustrated in Table 29–2. As the table shows, brain blood flow is maintained in heavy exercise. Splanchnic and renal blood flow, on the other hand, are decreased while skeletal muscle, heart, and skin blood flow are increased. Exercise has a dual effect on skin blood flow. At the start of exercise, there is a decrease in cutaneous blood flow that facilitates increased muscle blood flow. However, as exercise continues, skin blood flow increases above resting levels to permit effective transfer and dissipation of body heat. Clearly, during heavy exercise some of the large vascular beds in the viscera have reduced blood flow so that blood flow through more vital and active areas can be maintained or increased.

Body Temperature in Exercise

Strenuous prolonged exercise markedly increases body temperature. Since the mechanical efficiency of the human body is less than 25%, more than 75% of the metabolic energy of exercise is converted to

Table 29–2
Blood Flow Distribution During Rest and Exercise in an Athlete

Area	Rest		Heavy Exercise	
	mL/min	%	mL/min	%
Splanchnic	1400	24.0	300	1.0
Renal	1100	19.0	900	4.0
Brain	750	13.0	750	3.0
Coronary	250	4.0	1,000	4.0
Skeletal muscle	1200	21.0	22,000	85.5
Skin	500	9.0	600	2.0
Others	600	10.0	100	0.5
Cardiac output	5800	100.0	25,650	100.0

heat. The greater the exercise intensity, the greater the total amount of heat produced, and at maximal exercise it can be 10 to 20 times greater than resting levels. The excessive heat must be removed and dissipated from the body to prevent overheating and tissue damage from excessive hyperthermia.

Figure 29–19 illustrates the relationship between muscle and rectal temperature as a function of the percentage of a person's maximal oxygen uptake. As expected, the highest temperatures occur in the exercising muscles, where most of the increased heat is being produced. It should be noted that the temperatures do not depend on the absolute magnitude of the energy output, but on the level of metabolism relative to the individual's maximal oxygen consumption. Rectal temperatures may reach 40 to 41°C in highly trained athletes during prolonged exertion. Body temperature increases during exercise are independent of environmental temperature except at very extreme levels. Normally, internal body temperature levels off at a few degrees above normal regardless of the atmospheric temperature. The increase in body temperature during exercise is not due to a failure in the mechanisms for effective temperature regulation. On the contrary, it is a well-regulated response that occurs even during exercise in a cold environment. The primary stimulus for the increase in body temperature during exercise is not known, but it may involve a change in the set point for body temperature regulation.

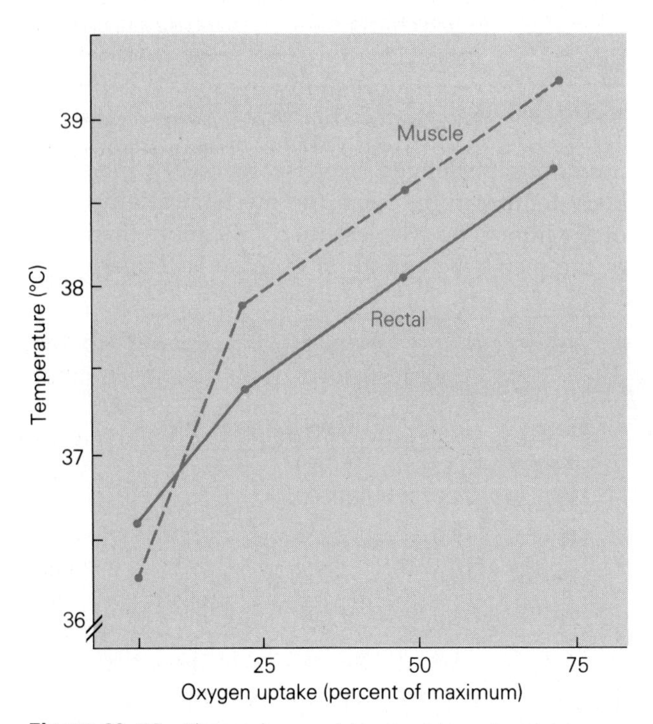

Figure 29–19 Change in exercising muscle and rectal temperature in relation to the oxygen uptake as a percent of a person's maximum.

A key advantage of allowing body temperature to rise during exercise is that less strain is placed on the thermoregulatory mechanisms. Maintenance of a resting body temperature during exercise would necessitate a greater increase in skin blood flow and could reduce essential blood flow to the working muscles. The rate of sweating would also have to be increased resulting in possible dehydration earlier in prolonged exercise. These considerations do not mean that heat is always beneficial to the athlete. In very hot and humid environments, or great excesses in athletic clothing, nonregulated increases in body temperature exceeding 42°C can easily occur. When this happens, multiple symptoms begin to appear, including extreme weakness, dizziness, decreased sweating, headache, nausea, confusion, collapse, and unconsciousness. This whole complex is called **heat stroke** (see Chapter 28), and failure to treat by appropriate body cooling can lead to death.

Hormonal Responses to Exercise

Exercise is known to stimulate the secretion of numerous hormones. With few exceptions, however, their specific role in the physiology of exercise is unknown. Two hormonal systems deserve mention in this regard: (1) secretion of catecholamines by the adrenal medulla, and (2) secretion of adrenocortical-stimulating hormone by the anterior pituitary.

During exercise, catecholamine secretion by the adrenal medulla system increases in response to stimulation by the sympathetic nervous system. Frequently, the secretion of catecholamines (epinephrine and norepinephrine) begins before the onset of exercise as an anticipatory response and then catecholamine secretion coincides with the onset of work. The total amount of catecholamines secreted is related to the intensity of the exercise. Among other actions, catecholamines mobilize the fat and glycogen deposits of the body and thus play an important role in regulating the availability of nutrients during exercise. Catecholamines also enhance cardiac activity and play an important role in the cardiovascular adjustments in exercise.

During exercise, there is also an increase in pituitary secretion of hormones that stimulate the adrenal cortex. Specifically, after exercise starts and with a latency of about 2 minutes, the anterior pituitary increases its secretion of ACTH. This increased ACTH then stimulates the release of corticosteroids from the adrenal cortex. The significance of the increase in corticosteroids during exercise is not fully surrounding the sperm is reabsorbed, and hence the sperm entering the vas deferens are highly concentrated.

Storage of Mature Sperm in the Vas Deferens

A small number of sperm are stored in the epididymis, but most of the sperm travel to the vas deferens where they can be stored for several months in the absence of ejaculation. The **vas deferens (ductus deferens),** components of the spermatic cord, arise from the tail of the epididymis. The vas deferens are composed of long slender tubes, or ducts, which run from the scrotum up into the abdominal cavity and behind the bladder, where they empty into the **ejaculatory duct** (see Fig. 30–8). The ejaculatory duct runs through the center of the **prostate gland,** which is located at the base of the bladder surrounding the urethra (see Fig. 30–1). The thick wall of the vas deferens is composed of smooth muscle and innervated by sympathetic nerves. (The importance of this innervation for ejaculation will be discussed in Chapter 32.) As the vas deferens enter the ejaculatory duct, they are joined by the secretory ducts of the seminal vesicles.

Addition of Nutrient Fluid by the Seminal Vesicles

The **seminal vesicles** are composed of two sacs, one on either side of the bladder. The seminal vesicles secrete a viscous material rich in **prostaglandins,** fructose, and other nutrients. The prostaglandins in seminal vesicle fluid stimulate contractions of the uterus in the female; these contractions play a role in moving the sperm through the female reproductive tract. Nutrients in seminal vesicle fluid serve as an energy source for the sperm following ejaculation. Seminal vesicle fluid accounts for approximately 60% of the total volume of ejaculated semen. During ejaculation, the contents of the vas deferens are emptied into the ejaculatory duct immediately before the contents of the seminal vesicles. Seminal vesicle fluid washes sperm out of the ejaculatory duct.

Addition of Alkaline Fluid by the Prostate Gland

The **prostate gland** secretes an alkaline liquid that contains citric acid, calcium, and various enzymes. Fluid from the prostate gland adds to the volume of the semen, and the alkaline nature of prostatic fluid may be very important in protecting sperm fertility. Secretions of the vagina are acidic, and sperm mobility is inhibited in acidic fluids. **Prostatic fluid** may neutralize vaginal acidity and increase sperm mobility and fertility.

The prostate gland contains an elaborate system of valves. During ejaculation, a valve in the prostate opens, and the semen is ejected into the urethra. A sphincter in the prostate contracts during ejaculation and seals off the entrance to the bladder, which prevents the passage of urine from the bladder into the urethra. Thus, although urine and semen share a common anatomical pathway to the outside of the body, the valves in the prostate prevent these two fluids from mixing. During ejaculation, semen passes from the ejaculatory duct into the urethra, where it is joined by secretions from the bulbo-urethral glands, also called Cowper's glands (see Fig. 30–8).

Addition of Lubricant Fluid by the Bulbo-Urethral Glands

The **bulbo-urethral glands** are located on each side of the urethra. They secrete a clear viscous liquid that adds to the volume of the semen. During sexual arousal, a small amount of fluid from the bulbo-urethral glands may appear at the tip of the penis before ejaculation. This fluid, termed **pre-ejaculatory fluid,** may act as a lubricant during sexual contact. Pre-ejaculatory fluid occasionally contains a small number of live sperm cells, which makes the practice of withdrawal prior to ejaculation an unreliable method of contraception.

The **urethra** transports the semen through the shaft of the penis to the outside of the body during ejaculation (Fig. 30–8). Stimulation of the sympathetic nerves of the epididymis, vas deferens, prostate, and seminal vesicles during sexual arousal results in smooth muscle contractions in these organs and expulsion of the fluids that compose the semen.

The average volume of semen ejaculated is 3 to 5 milliliters (5 milliliters = about one teaspoonful). The normal concentration of sperm in the ejaculate is highly variable (40 to 120 million per milliliter) and depends in part on frequency of ejaculation. Once sperm are ejaculated in the semen, their maximum life span is only 24 to 72 hours at body temperature.

The major organs involved in sperm production, maturation, and transport are described in Table 30–2 (see page 864).

Neuroendocrine Control of Reproduction in the Male

The **pituitary gland,** which lies at the base of the brain, was described in Chapter 13. Two decades ago, it was thought to be the source of the hormonal signals that control reproduction. We now know that the pituitary gland acts as a relay station, receiving neural and hormonal input from the brain and relaying this information, in the form of hormonal messages, to other organs of the body.

Table 30–2
Major Organs Involved in Sperm Production, Maturation, and Transport

Testes: Sperm production: mitosis, meiosis, and differentiation
Epididymis: Sperm transport and maturation: motility and fertility
Vas deferens: Sperm storage
Seminal vesicles: Production of seminal fluid containing nutrients, fructose, and prostaglandins
Prostate: Production of prostatic fluid which is alkaline and contains calcium and citric acid
Bulbo-urethral gland: Production of "pre-ejaculatory" fluid
Penis: Erection and ejaculation

Neuroendocrine Control of Spermatogenesis

As described in Chapter 13, the hypothalamus secretes a releasing factor called **gonadotropin releasing hormone,** or GnRH (Fig. 30–9). GnRH is transported to the anterior pituitary gland via a complex set of blood vessels known as the **hypothalamic-pituitary portal vessels** (see Fig. 13–3, Fig. 13–4). When GnRH reaches the anterior pituitary in the male, it stimulates the release of two major reproductive hormones, termed the **gonadotropins.** The first gonadotropin, **luteinizing hormone (LH),** is secreted by the anterior pituitary into the blood, and is transported to the testes. There it stimulates the Leydig cells to synthesize and secrete testosterone.

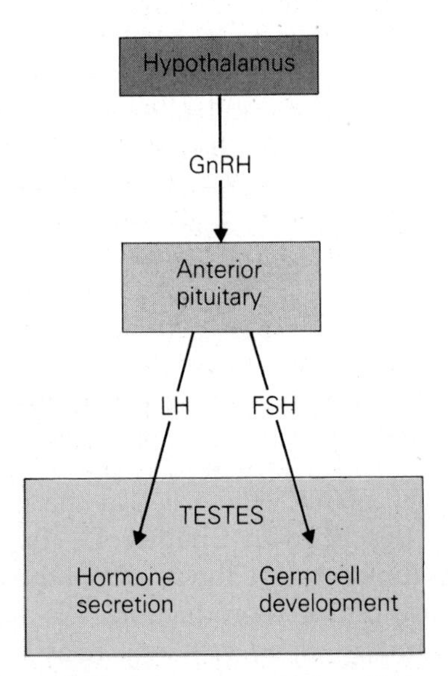

Figure 30–9 Relationship among the hypothalamus, pituitary gland, and testes.

The second gonadotropin, **follicle-stimulating hormone (FSH),** is secreted by the anterior pituitary into the blood, and is also transported to the testes, where it binds to specific receptors on the plasma membranes of Sertoli cells. FSH stimulates the conversion of spermatogonia into spermatocytes in the seminiferous tubules of the testes. Both LH and FSH are required for spermatogenesis (see Fig. 30–9).

Feedback Regulation of Neuroendocrine Function

The secretion of GnRH, LH, and FSH is fairly constant from day to day in the male, in marked contrast to the cyclic secretion of these hormones in the female. The constancy of hormone secretion in the male is due to the presence of a **continuous feedback loop** between the hypothalamus, pituitary gland, and testes (Fig. 30–10).

The hypothalamus is sensitive to changes in the circulating levels of sex hormones. It monitors the level of these hormones and responds by increasing or decreasing the production of its releasing factor,

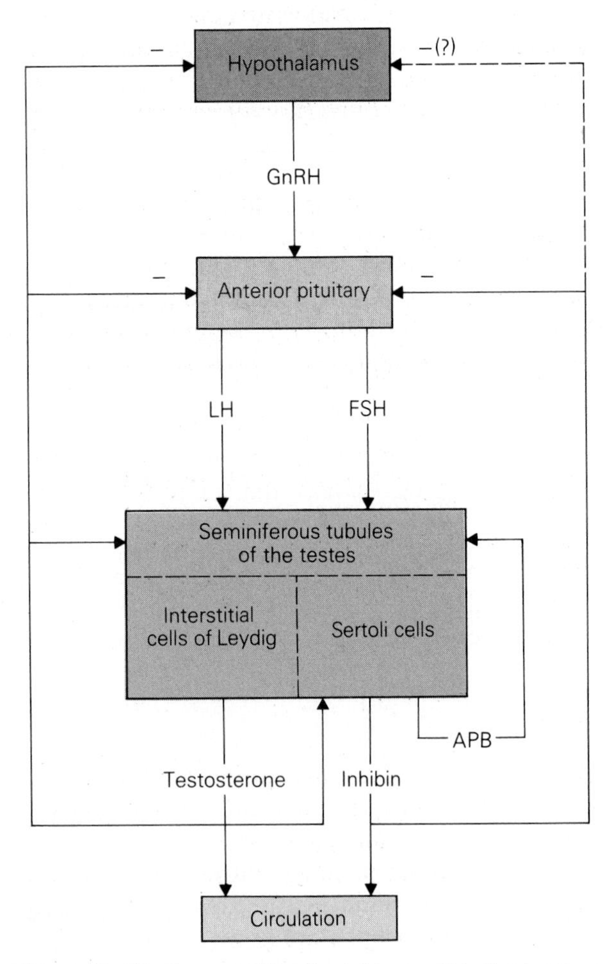

Figure 30–10 Hormonal feedback loops within the hypothalamic-pituitary-gonadal system.

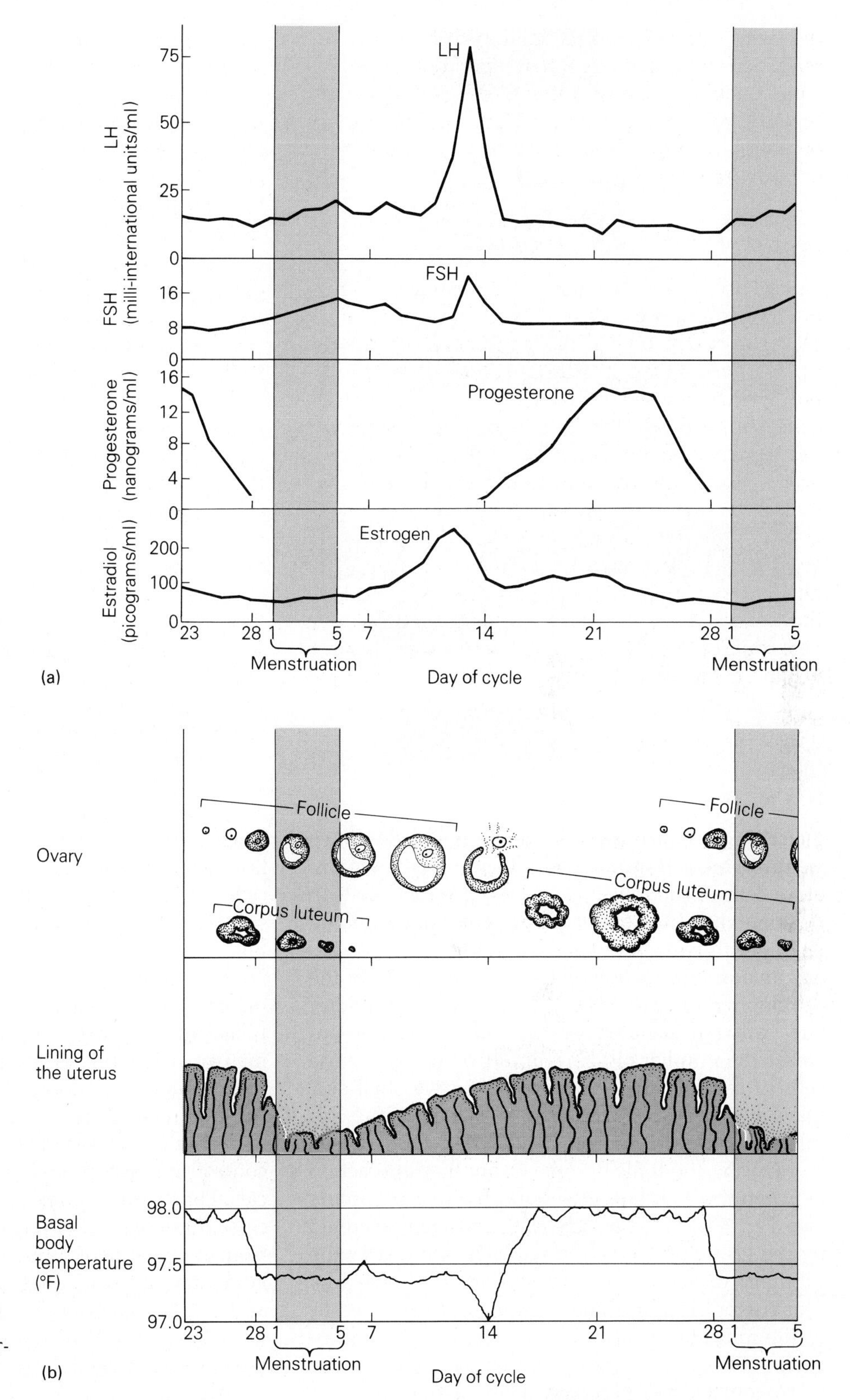

Figure 30–17 Hormonal, ovarian, and uterine changes that occur during the ovarian cycle. (Courtesy of Dr. Resco, Oregon Health Sciences University)

local action of estrogen, which results in an increase in the number of FSH receptors on the granulosa cells, augments the growth-promoting effect of FSH on the follicular granulosa cells. As the follicle grows, it secretes more estrogen, which continues to augment the action of FSH in the follicle.

Follicular growth and development in the ovary can be viewed as a self-perpetuating mechanism that involves a complex interaction between the pituitary gland and the ovary that results in a steady rise in follicular growth and estrogen secretion. As discussed in Chapter 12, when a hormone acts on

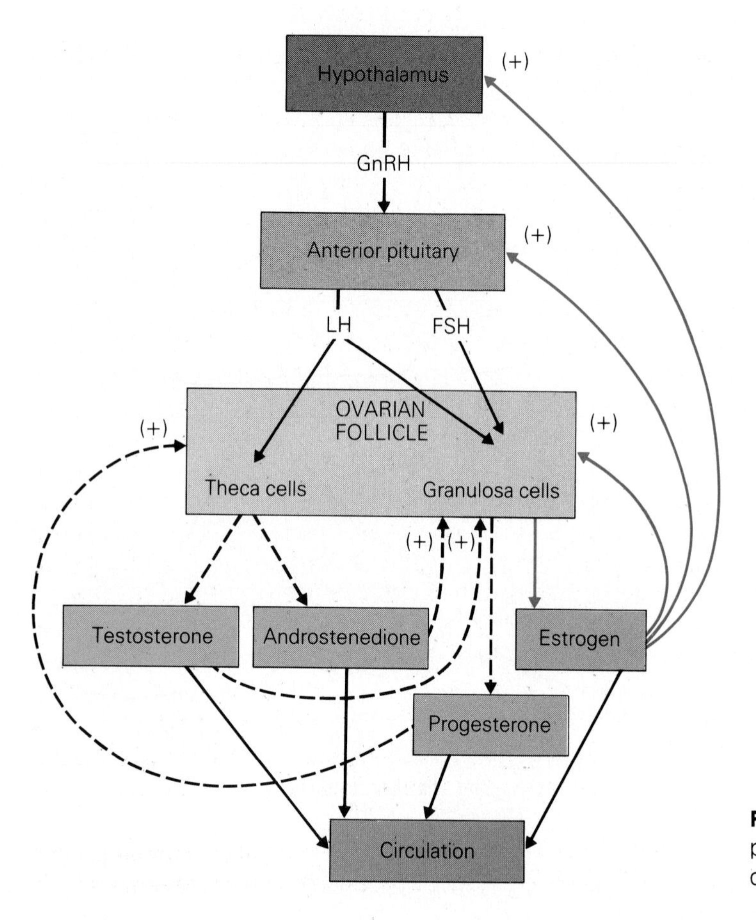

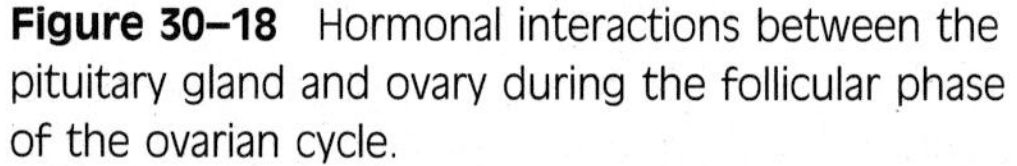

Figure 30–18 Hormonal interactions between the pituitary gland and ovary during the follicular phase of the ovarian cycle.

the cell from which it was secreted, it is said to have an **autocrine action.** Estrogen, secreted by the granulosa cells of the follicle, acts on these same cells to augment the further secretion of estrogen and to augment the action of FSH. Therefore, estrogen exerts an autocrine action on the granulosa cells of the developing follicle. FSH secreted by the pituitary and estrogen secreted by the follicle together increase the number of LH receptors on the granulosa cells of the follicle. This increase, as we shall see, plays an important role in the formation of the corpus luteum.

During the follicular phase, another interaction between the pituitary gland and the ovaries occurs (see Fig. 30–18). The increasing estrogen, secreted by the granulosa cells of the follicle, stimulates the pituitary to synthesize and secrete additional LH. This rise of LH in the blood stimulates the theca cells of the follicle to synthesize and secrete two androgens: testosterone and **androstenedione.** These androgens exert a **paracrine** action on neighboring cells in the follicle. The androgens diffuse into the follicular granulosa cells and serve as precursors for estrogen production in the granulosa cells.

The rising blood levels of LH stimulate the granulosa cells of the follicle to secrete progesterone. Progesterone enters the circulation, and blood levels of progesterone rise (see Fig. 30–17). In addition, some of the progesterone from the granulosa cells diffuses back into the theca cells and acts as a substrate for further androgen synthesis by the theca cells (see Fig. 30–18).

As we have seen, as many as 20 follicles may increase in size each month during the follicular phase in the ovary, but only one follicle reaches maturity. The others undergo atresia. The mechanisms that control the selection of a single follicle for full maturation are not completely understood. It appears that complex endocrine events in the ovary control the growth and atresia of the various follicles. For example, atresia may result from an insufficient number of FSH receptors in some follicles, or from an inability of the granulosa cells of certain follicles to produce adequate amounts of estrogen from androgen precursors. Fluid from atretic follicles contains lower levels of both FSH and estrogen than fluid from mature follicles.

Neuroendocrine Control of Ovulation

During the follicular phase of the ovarian cycle, increased release of estrogen from the ovary stimulates LH and FSH release from the anterior pituitary, which results in progressive follicular growth for approximately 10 days (see Fig. 30–17). On day 14, just prior to ovulation, blood levels of LH and FSH are higher than at any other time in the ovarian cycle. These peaks in blood levels of LH and FSH are referred to as the **preovulatory surges** of LH and FSH. The surge of LH induces changes in follicular structure that result in ovulation (see Fig. 30–17). As we have discussed, these LH-induced changes include an increase in the production of antral fluid, compression of the ovum toward the wall of the follicle, protrusion of the follicle against the wall of the ovary, and, finally, enzymatic degeneration of the ovarian wall and release of the antral fluid and ovum into the abdominal cavity during ovulation. One or two days before ovulation, the follicle stops secreting estrogen, and blood levels of estrogen begin to fall (see Fig. 30–17). The mechanism that controls the cessation of estrogen secretion by the follicle is not known. Ovulation marks the end of the follicular phase of the ovarian cycle in the female.

In many women ovulation is accompanied by a slight upward shift of approximately 0.5 to 1.0 degree in basal body temperature, which persists throughout the luteal phase (see Fig. 30–17). The exact cause of this rise in temperature is not known. However, since ovulation can occur in the absence of a detectable change in body temperature, measurements of body temperature are not reliable indicators of ovulation.

Neuroendocrine Control of the Luteal (Postovulatory) Phase

Immediately following the release of the ovum from the follicle on day 14 of the ovarian cycle, the remaining granulosa cells of the follicle increase in size and number and undergo chemical changes to form the corpus luteum (see Fig. 30–17). This process is called **luteinization.** Differentiation and growth of the corpus luteum are stimulated by LH from the anterior pituitary. In fact, LH derives its name, "luteinizing hormone," from its action on the corpus luteum. The role of FSH in the luteal phase is unknown. Although the presence of LH is necessary for the formation of the corpus luteum, the ovary is thought to secrete an additional substance that inhibits the formation of the corpus luteum in the mature follicle prior to ovulation as well as in follicles that do not ovulate.

During the luteal phase, the corpus luteum secretes a large amount of **progesterone** and a smaller amount of **estrogen,** increasing blood levels of these two hormones (see Fig. 30–17). The estrogen secreted by the corpus luteum is structurally identical to the estrogen secreted by the follicle during the follicular phase. However, while estrogen stimulates release of GnRH, LH, and FSH during the follicular phase, the estrogen and progesterone that are secreted during the luteal phase *inhibit* GnRH secretion from the hypothalamus and LH and FSH secretion from the pituitary (Fig. 30–19). During the luteal phase, blood levels of LH and FSH fall (see Fig. 30–17). The opposite effect of estrogen and progesterone on GnRH, LH, and FSH release during the ovarian cycle is due to a change in hypothalamic sensitivity to estrogen and progesterone during the follicular and luteal phases of the cycle.

As LH and FSH blood levels fall during the luteal phase, the signal for follicular growth diminishes, and no new follicles begin to grow in the ovary. The corpus luteum increases in size for about 7 to 8 days following ovulation and then begins to degenerate. The mechanism responsible for degeneration of the corpus luteum is not completely known. Falling blood levels of LH may not provide an adequate signal for luteal growth (see Fig. 30–17). In addition, the corpus luteum itself is thought to secrete hormones, known as **prostaglandins,** which induce further degeneration of the corpus luteum.

As the corpus luteum degenerates, it secretes less progesterone and estrogen, so that blood levels

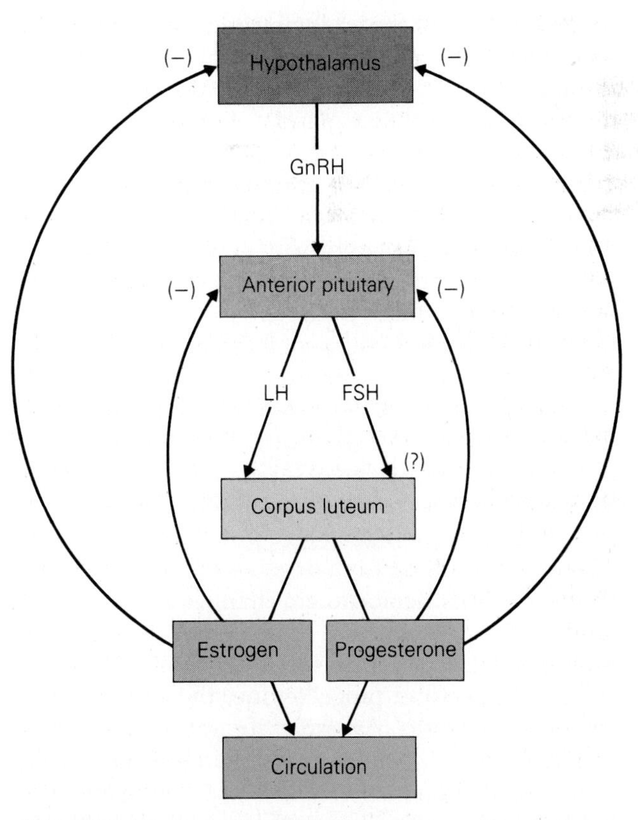

Figure 30–19 Hormonal interactions between the pituitary gland and ovary during the luteal phase of the ovarian cycle.

of estrogen and progesterone begin to fall (see Fig. 30–17). The inhibition of GnRH, LH, and FSH release, which was maintained by progesterone and estrogen from the corpus luteum, is now removed, so that blood levels of LH and FSH begin to rise again. Rising levels of LH and FSH in the blood stimulate growth of another set of follicles in the ovary, which marks the beginning of a new follicular cycle (see Fig. 30–17).

Neuroendocrine Control of the Menstrual Phase

Dramatic changes in the structure and secretory capacity of the endometrium of the uterus occur during the ovarian cycle in the female. These changes are induced by both estrogen and progesterone. As we have seen, during the follicular phase of the ovarian cycle, the growing follicle secretes large amounts of estrogen. This estrogen stimulates proliferation of the endometrium of the uterus (see Fig. 30–17). In addition, estrogen increases the number of progesterone receptors in the uterus. At the end of the follicular phase in the ovary, just before ovulation, the follicle stops secreting estrogen, resulting in a drop in circulating levels of estrogen (see Fig. 30–17).

During the luteal phase of the ovarian cycle, which follows ovulation, estrogen and progesterone, secreted by the corpus luteum, stimulate continued growth of the endometrium (see Fig. 30–17). Estrogen and progesterone stimulate growth and proliferation of the coiled blood vessels of the endometrium. The tubular glands of the endometrium fill with glycogen, and enzymes accumulate in the connective tissue of the endometrium. During the luteal phase of the ovarian cycle, the endometrium swells and becomes approximately 4 to 6 mm thick owing to the progressive growth of glands, blood vessels, and connective tissue.

The function of the proliferation of endometrial cells is to provide a continuing source of nutrients for the ovum if it is fertilized and implants in the uterus. If the ovum is not fertilized, the corpus luteum begins to degenerate on day 8 following ovulation and stops secreting progesterone and estrogen.

A drop in circulating levels of estrogen and progesterone resulting from degeneration of the corpus luteum causes **hemorrhagic changes** in the uterine endometrium that result in the onset of menstruation (see Fig. 30–17). In the absence of stimulation from estrogen and progesterone, the endometrium becomes **necrotic.** Arterioles of the endometrium constrict, which results in a slowing of the circulation. Blood enters the vascular layer of the endometrium, resulting in blood pooling. Gradually, the outer layers of the endometrium separate from the uterus at the site of hemorrhage.

Prostaglandins, produced by the uterus, stimulate the uterine smooth muscles to contract rhythmically. These uterine contractions result in the discharge of blood and endometrial cells that constitutes the **menstrual flow.** The menstrual flow lasts for approximately 5 to 7 days (see Fig. 30–17), during which the uterus expels an average of 20 to 200 milliliters of blood. An overproduction of prostaglandins in the uterus during the menstrual phase of the ovarian cycle can lead to excessive uterine contractions in many women, which result in menstrual cramps, a condition called **dysmenorrhea.** Prostaglandin action on smooth muscles in other areas of the body can also result in nausea, vomiting, and headache, which can also accompany the menstrual phase.

A summary of the hormonal, ovarian, and uterine changes that occur during the ovarian cycle are contained in Table 30–4.

Table 30–4
Summary of the Hormonal, Ovarian, and Uterine Changes That Occur During the Ovarian Cycle

1. During the early part of the follicular phase, blood levels of LH and FSH begin to rise.
2. This rise in LH and FSH stimulates growth of 6 to 12 primary follicles and the full maturation of a single Graafian follicle.
3. The growing follicles secrete large amounts of estrogen, lesser amounts of progesterone, and small amounts of testosterone and androstenedione.
4. Rising blood estrogen stimulates growth of the endometrium of the uterus.
5. Rising blood estrogen also stimulates further secretion of GnRH, which results in the pre-ovulatory surges of LH and FSH.
6. Just prior to ovulation, the follicle stops secreting estrogen and blood estrogen levels fall.
7. The surges of LH and FSH at midcycle induce ovulation.
8. Following ovulation, the remaining follicular cells in the ovary become the corpus luteum.
9. The corpus luteum secretes large amounts of progesterone and lesser amounts of estrogen. Blood levels of these hormones begin to rise.
10. Progesterone and estrogen from the corpus luteum stimulate further growth of the endometrium of the uterus.

11. Progesterone and estrogen also inhibit secretion of LH and FSH, and blood levels of these pituitary hormones begin to fall.
12. As blood levels of LH and FSH fall, the signal for follicular growth diminishes and no new follicles begin to grow.
13. On day 8 following ovulation, the corpus luteum begins to degenerate and secretes less progesterone and estrogen.
14. Falling blood levels of progesterone and estrogen result in hemorrhagic changes in the uterine endometrium, which culminate with the onset of menstruation.
15. As blood levels of progesterone and estrogen fall at the end of the luteal phase, the inhibition of GnRH, LH, and FSH, which was exerted by progesterone and estrogen, is removed. Blood levels of LH and FSH begin to rise, thus marking the beginning of another follicular phase (see point 1).

Effects of Estrogen and Progesterone in the Female

The major sex hormones in the female are **estrogen** and **progesterone.** As described in Chapter 12, both of these hormones are steroids. In addition to their roles in regulating the ovarian cycle, estrogen and progesterone have widespread effects on many other physiological and metabolic processes. The effects of estrogen are summarized in Table 30–5.

Effects of Estrogen at Puberty

Estrogen stimulates the development of the sex organs and the appearance of the secondary sex characteristics in the female. Therefore, the effects of estrogen in the female are analogous to the effects of testosterone in the male. At puberty, estrogen stimulates growth of the vagina, uterus, and fallopian tubes as well as the external genitalia, breasts, and body hair. Estrogen stimulates linear growth but also accelerates closure of the epiphyses of the long bones, which results in a shorter height in women than in men. Stimulation by estrogen results in the particular pattern of distribution of body fat around the hips, buttocks, and abdomen that characterizes the female form. When the ovary is absent or when it secretes inadequate amounts of estrogen, puberty either fails to occur or progresses very slowly.

Metabolic Effects of Estrogen

Estrogen interacts with a variety of other hormones that control metabolic processes. For instance, estrogen decreases the sensitivity of peripheral tissues to insulin, which can result in decreased glucose tolerance. Estrogen increases the levels of plasma proteins that bind steroids such as thyroxine, cortisol, and progesterone. In addition, estrogen increases circulating levels of renin and angiotensin II, which can result in hypertension.

Evidence is accumulating that estrogen protects against bone loss. For instance, removal of the ovaries, or inadequate ovarian secretion of estrogen, results in the acceleration of bone loss known as **osteoporosis.** In such cases, replacement of estrogen can dramatically slow the rate of bone loss. Postmenopausal women, in whom ovarian secretion of estrogen is very low, exhibit osteoporosis most often. Note that, although men have very low circulating levels of estrogen, they are not particularly prone to osteoporosis. However, impairment of gonadal function in the male, a condition called **hypogonadism,** is often accompanied by low bone mass. It is possible that androgens may protect against bone loss in the male.

Effects of Progesterone

Progesterone is the most active naturally occurring antagonist of androgens. Progesterone antagonizes the effects of both testosterone and dihydrotestosterone by competing for the androgen-binding sites present in adrogen-dependent tissues. Progesterone prevents target tissues from responding to the normally low levels of circulating androgens in the female. Progesterone and its derivatives are very useful in treating abnormally heavy hair growth in

Table 30–5
Effects of Estrogen

Puberty
Stimulates growth of internal reproductive organs, external genitalia, and breasts
Stimulates growth of body hair
Stimulates growth of long bones and early closure of the epiphyses
Stimulates "female" pattern of fat distribution
Hormonal
Stimulates secretion of GnRH, LH, and FSH during the follicular phase and inhibits secretion of these same hormones during the luteal phase
Decreases sensitivity of peripheral tissues to insulin
Increases circulating levels of renin and angiotensin II
Metabolic
Protects against bone loss

the female, a condition termed **hirsutism,** which is thought to be due to an overproduction of androgens.

Progesterone, secreted by the corpus luteum, may also promote edema (salt and fluid retention), which occurs in many women at the end of the luteal phase of the ovarian cycle prior to the onset of menstruation. Progesterone is thought to increase circulating levels of renin, which stimulates aldosterone secretion and results in salt and fluid retention. However, the severity of premenstrual fluid retention does not appear to correlate with circulating levels of renin or aldosterone. The role of ovarian hormones in controlling fluid and electrolyte levels in the female is not completely understood.

Menopause

Menopause, the cessation of menstruation in the female, occurs on the average at age 50. For several years before the onset of menopause, menstruation occurs less frequently and at variable intervals. As previously discussed, the number of follicles in the ovary declines with age, and hence estrogen secretion declines as well. When all of the follicles have completely disappeared, the ovary stops secreting estrogen. The low circulating levels of estrogen found in postmenopausal women are due to secretion of estrogen precursors by the adrenal glands, which are then converted to estrogen in the periphery.

During menopause, the remaining follicles in the ovary become less sensitive to LH and FSH, and to compensate, blood levels of these pituitary hormones begin to rise. The decline in circulating estrogen during menopause results in a loss of vaginal epithelium, a decrease in breast mass, and an increase in bone loss. Other symptoms that accompany menopause include vascular flushing ("hot flashes"), rapid shifts in mood and emotion, and an increase in coronary vascular disease.

Focus on Menopausal Osteoporosis

Menopause is accompanied by decreased ovarian production of estrogen. One of the long-term metabolic effects of decreased estrogen is a reduction of total bone mass or bone density. Following menopause, women lose approximately 1% to 2% of their bone mass per year. This reduction of bone mass is caused by more bone matrix eroding than being replaced. Consequently, bones become thin and brittle, and fractures occur easily. Bone pain is often the first symptom indicating the presence of osteoporosis. Intense pain may develop suddenly because of compression fractures of the vertebrae or fractures of the long bones. In x-ray images, the bones appear abnormally lucent (transparent), owing to thinness, and the spine is curved because of compression. Such curvature of the spine may result in decreased height over a period of years.

Treatment of menopausal osteoporosis is difficult, because no pharmacologic agent has been discovered that is both safe and effective in inducing the formation of new bone. At present, the best treatment of menopausal osteoporosis involves maintaining estrogen levels through hormone replacement therapy (HRT). When begun at the onset of menopause, estrogen replacement decreases the rate of bone loss that normally occurs following menopause. Estrogen is usually administered in the lowest dose that is found to relieve symptoms, since high doses are associated with increased risk for certain types of cancer.

In treatment of menopausal osteoporosis, replacement of estrogen is accompanied by maintenance of an adequate dietary intake of calcium. However, excessive intake of calcium is of limited value since absorption of calcium from the gastrointestinal tract is never complete. Dietary calcium is derived from dairy products such as milk, cheese, and eggs, as well as white bread. Adults on normal diets consume approximately 1000 mg of calcium daily. Of this amount, less than half is absorbed from the gastrointestinal tract, and the rest is lost in the feces and urine without having been absorbed. Most women with menopausal osteoporosis excrete normal amounts of calcium into the urine.

Although a cure for menopausal osteoporosis remains elusive, prevention involves ensuring that excessive bone loss does not occur before the onset of menopause. Prevention also involves slowing the rate of bone loss following menopause. Maintenance of physical activity and an adequate diet throughout life, together with estrogen replacement at the onset of menopause, will decrease bone loss in later years.

Summary

I. A. The external reproductive organs in the male are the penis and scrotum. The internal reproductive organs include the testes, epididymis, vas deferens, seminal vesicles, and the prostate gland.

B. Spermatogenesis takes place in the seminiferous tubules of the testes. Spermatogenesis begins at puberty and is constantly repeated throughout the reproductive years of the male. During spermatogenesis, germ cells undergo a series of divisions and differentiate from spermatogonia into mature spermatozoa. The Sertoli cells, which form a protective blood-testes barrier around the inside of the seminiferous tubules, ensure that the sperm develop in a protected environment.

C. Sperm mature in the epididymis and are stored in the vas deferens. The seminal vesicles, prostate gland, and bulbo-urethral gland contribute important fluids that nourish sperm and add volume to the ejaculated semen.

D. Neuroendocrine control of reproduction in the male involves the secretion of gonadotropin releasing hormone from the hypothalamus, which in turn stimulates the release of luteinizing hormone and follicle-stimulating hormone from the pituitary. LH stimulates the Leydig cells of the testes to secrete testosterone. FSH stimulates spermatogenesis in the testes. Release of GnRH, LH, and FSH is controlled through feedback loops involving both testosterone, secreted by the Leydig cells, and inhibin, secreted by the Sertoli cells within the seminiferous tubules of the testes.

E. Testosterone stimulates the onset of puberty and promotes spermatogenesis in the testes.

II. A. The external reproductive organs in the female are the labia majora, the labia minora, and the clitoris. The internal reproductive organs include the vagina, uterus, ovaries, and fallopian tubes. In contrast to the male, the urinary and reproductive tracts in the female are anatomically separate.

B. Oogenesis takes place in the ovaries. In contrast to the male, germ cell division in the female begins during embryonic development but is then arrested until puberty. Oogenesis begins with the growth of primary follicles in the ovary. Once each month beginning at puberty, one of the primary follicles in the ovaries grows at a rapid rate and differentiates into a mature Grafian follicle containing an ovum.

C. At midcycle, this mature follicle releases its ovum in a process termed ovulation. Following ovulation, the ovarian follicle becomes the corpus luteum.

D. Neuroendocrine control of reproduction in the female involves the secretion of GnRH from the hypothalamus and LH and FSH from the pituitary. During the follicular phase of the ovarian cycle, FSH stimulates the growth of ovarian follicles, which in turn secrete estrogen. The luteal phase, which begins following ovulation, is marked by growth of the corpus luteum in response to LH stimulation. The corpus luteum secretes both estrogen and progesterone, which in turn stimulate growth and proliferation of the endometrium of the uterus. As the corpus luteum begins to degenerate at the end of the luteal phase, secretion of estrogen and progesterone diminishes, resulting in necrotic changes in the endometrium, culminating in the onset of menstruation.

E. Estrogen, secreted first by the developing follicles and later by the corpus luteum, stimulates the onset of puberty in the female and has a wide variety of effects on other endocrine systems, such as the pancreas, and on many tissues of the body, including bone.

F. Menopause, the cessation of menstruation, marks the end of the reproductive years in the female. As ovarian follicles begin to degenerate during menopause, follicular secretion of estrogen diminishes. The decline in circulating levels of estrogen during menopause results in partial atrophy of tissues that depend on estrogen stimulation, including the vaginal epithelium and breasts.

Review Questions

1. What will happen if the testes are removed prior to puberty? Why?
2. What is the hormonal feedback relationship among the hypothalamus, the pituitary, and the follicle during the luteal phase of the ovarian cycle? Draw a diagram to explain.
3. Why do spermatocytes differ from each other in terms of their genetic composition?
4. What are the actions of estrogen at puberty?
5. Which structures contribute fluid to the volume of ejaculated semen? Why are these fluids important?
6. What is the corpus luteum? How is it formed, and what role does it play in follicular growth?
7. Draw a diagram of the internal structure of a seminiferous tubule to show the arrangement of the Sertoli and spermatogenic cells. Explain the importance of this arrangement.
8. What is the principal difference between germ cell production in the male and female?
9. Which hormonal changes result in the onset of menstruation?
10. Why don't urine and semen mix with one another?
11. List three physiological changes that accompany menopause. Why do these changes occur?

Suggested Readings

Hafez, E.S.E. *Human Reproduction.* New York: Harper & Row, 1973.

Hedge, G.A. *Clinical Endocrine Physiology.* Philadelphia: W.B. Saunders Co., 1987.

Shearman, R.P. *Clinical Reproductive Endocrinology.* Churchill Livingstone, 1985.

Warshaw, J.B. *The Biological Basis of Reproductive and Developmental Medicine.* New York: Elsevier Science Publishing Co., 1983.

Williams, R.H. *Textbook of Endocrinology.* Philadelphia: W.B. Saunders Co., 1981.

31

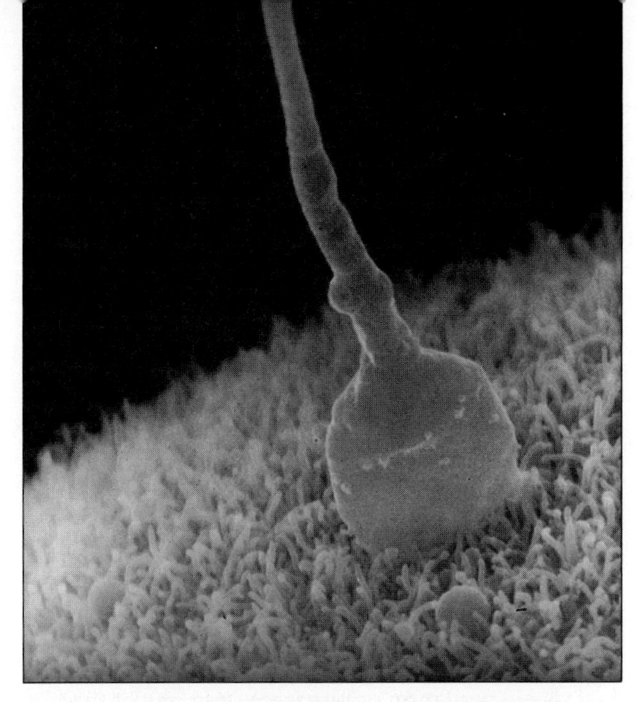

Pregnancy, Fetal Development, and Lactation

In Chapter 30, we investigated the cyclic hormonal and physiological changes that occur in the female, culminating in the release of an ovum once each month during the years between puberty and menopause. We also examined the processes that control production and expulsion of sperm in the male. We will now investigate the processes of fertilization of the ovum and maturation and development of the fetus, as well as the physiological changes that occur in the female during pregnancy, parturition (childbirth), and lactation (milk production and release).

The Process of Fertilization

Because both sperm and ova are viable only for very short periods of time, fertilization of an ovum will occur only when sperm are deposited in the vagina close to the time of ovulation. Following ejaculation, sperm normally retain their capacity to fertilize an ovum for approximately 24 to 72 hours in the female reproductive tract, although longer periods of sperm viability have been reported. Following ovulation, the ovum is "receptive" to fertilization for approximately 10 to 15 hours. Therefore, in order for the ovum to be fertilized, sperm must be present in the female reproductive tract during the 72 hours before or during the 15 hours after ovulation. This restricted time interval may make the odds against fertilization appear to be rather high. However, one out of every four women becomes pregnant after only one month of repeated intercourse without contraception.

Transport of the Ovum to the Fallopian Tube and Uterus

As described in Chapter 30, ovulation occurs in response to stimulation from high blood levels of LH and FSH approximately 14 days after the termination of the previous menstrual cycle. Except in rare instances, ovulation occurs from only one of the two ovaries each month.

During ovulation, a single ovum is released from an ovary into the abdominal cavity at a point directly below the open end of the fallopian tube (Fig. 31–1). The open end of the fallopian tube is composed of long fingerlike projections called **fimbriae.** The fimbria and fallopian tubes are lined with **cilia,** which move or sway in a rhythmic pattern toward the uterus and induce wavelike motions on the inner surface of the fallopian tube. The uterus and cervix are also lined with cilia, but not as many as are found in the fallopian tubes. Following ovulation, the ovum is drawn from the abdominal cavity into the fallopian tube by the sweeping motion of the cilia (see Fig. 31–1).

After entering the fallopian tube, the ovum is transported to the uterus by the wavelike motion of the cilia. The ovum takes approximately 3 to 5 days to travel from the ovary, through the fallopian tube, to the uterus. Because the ovum is receptive to fertilization only for approximately 10 to 15 hours following ovulation, fertilization normally takes place in the fallopian tube before the ovum has reached the uterus.

Transport of Sperm to the Fallopian Tube

Following ejaculation into the vagina, sperm are transported by way of the vagina, cervix, and uterus into the fallopian tubes (Fig. 31–2). The time required for sperm to reach the fallopian tubes is not precisely known, but it is thought to be a few hours, with some sperm reaching the fallopian tubes within minutes following ejaculation.

Sperm are propelled through the female reproductive tract toward the fimbrial ends of the fallopian tubes by wavelike movements in the tail portion of the sperm and by contractions of the muscles of the uterus and fallopian tubes. In the female, vaginal and cervical stimulation during intercourse results in the release of **oxytocin** from the pituitary, which in turn may stimulate uterine contractions during intercourse. In addition, seminal fluid contains a high concentration of **prostaglandins,** which also act to stimulate contractions of the uterus.

Cervical Mucus and Sperm Transport

The ease with which sperm can migrate through the cervix into the uterus is partially determined by secretions of the cervix. The cervix contains glands that secrete a mucus composed of glycoproteins, salt, and water. The volume and consistency of the cervical mucus change during the ovarian cycle in

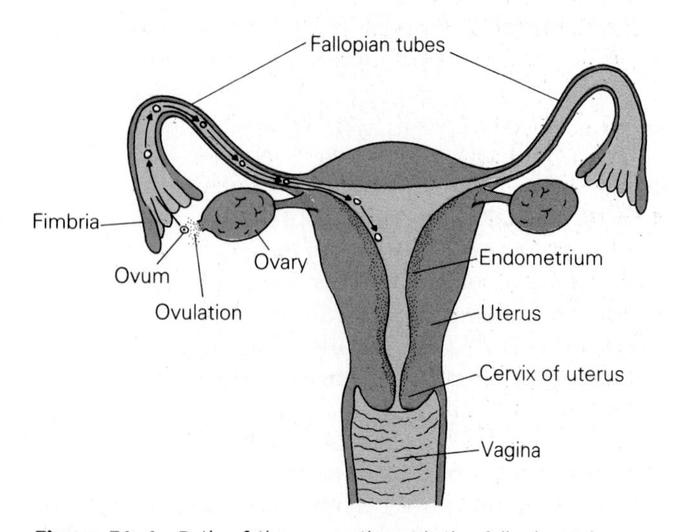

Figure 31–1 Path of the ovum through the fallopian tube and into the uterus following ovulation.

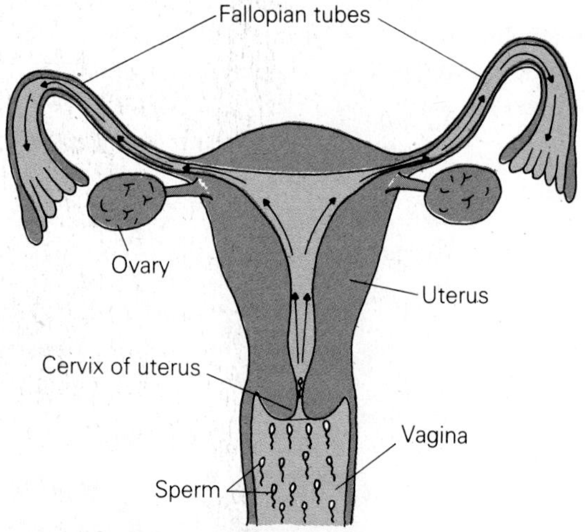

Figure 31–2 Pathway of sperm through the vagina, cervix, and uterus into the fallopian tubes.

response to changes in the circulating levels of estrogen and progesterone.

Cervical Mucus Secretion at Midcycle. As was discussed in Chapter 30, estrogen secretion by the ovary increases at midcycle just prior to ovulation. Estrogen stimulates the cervix to secrete large amounts of mucus, which is clear, watery, and nonviscous. Upon estrogen stimulation, the glycoproteins in the cervical mucus assemble to form elongated fibers arranged in channels that allow sperm to penetrate the mucus and pass upward through the cervix (Fig. 31–3). These changes in the fluidity and glycoprotein structure of cervical mucus at midcycle facilitate sperm migration through the cervix into the uterus at a time that coincides with ovulation. Note that many structurally abnormal sperm do not penetrate the cervical mucus at midcycle. Whether this cervical exclusion of abnormal sperm is due to decreased motility of the sperm or to immunological factors present in the cervical mucus is not known.

Cervical Mucus Secretion before and after Midcycle. Prior to and following ovulation, circulating levels of progesterone are higher than at midcycle. **Progesterone** stimulates the cervix to secrete a thick, viscous, sticky mucus that lacks glycoprotein channels. This mucus plugs the cervical opening into the uterus and acts as a barrier that impedes sperm migration into the uterus at times other than at midcycle. As we shall see in Chapter 32, the efficacy of many oral contraceptives that contain steroids is due in part to steroid-induced alterations of the physical and chemical properties of cervical mucus.

Sperm Loss in the Female

Although the number of sperm found in ejaculated semen depends on many factors—including the quantity of the ejaculate and the time since the previous ejaculation—an average of 100 to 500 million sperm are deposited in the vagina during intercourse. Since only one sperm can fertilize the ovum, this number may seem much larger than is necessary to ensure fertilization. However, of this enormous number, only a few thousand sperm reach the fallopian tubes, and only a few hundred reach the vicinity of the ovum.

Sperm loss is high for many reasons. Up to 50% of the sperm in a normal ejaculate may be incapable of fertilization because of abnormal shape or reduced motility. When deposited into the vagina, some sperm are destroyed by the acidic secretions of the vagina. Upon reaching the cervix, sperm entry

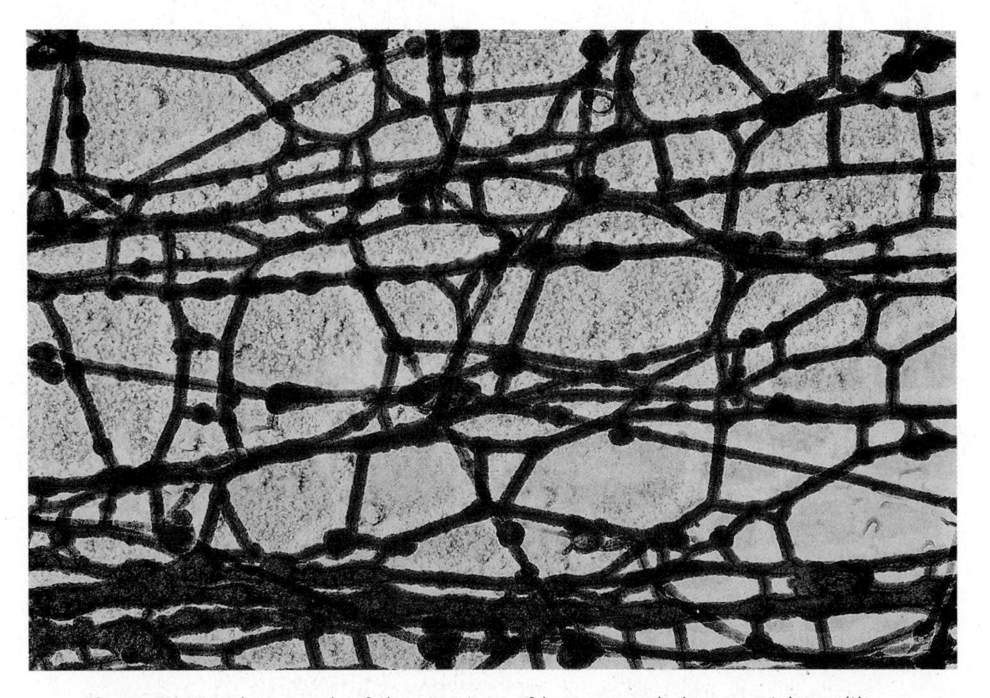

Figure 31–3 Photograph of the structure of human cervical mucus, taken with a scanning electron microscope. At midpoint of the ovarian cycle, coinciding with the time of ovulation, large channels are formed that permit sperm migration through the cervix. (Visuals Unlimited/David Phillips)

into the vagina may be blocked by the cervical mucus at certain times of the ovarian cycle, as we have previously seen. Of the sperm that manage to reach the uterus, about half will enter the fallopian tube that does not contain an ovum.

Although these factors result in a large reduction in the number of sperm that reach the ovum, only one sperm is required for fertilization of the ovum.

Capacitation and Activation of Sperm

As stated in Chapter 30, sperm are not mobile when they leave the seminiferous tubules of the testes and enter the epididymis. It is in the epididymis that sperm mature and acquire the ability to move. However, mature sperm from the epididymis are not capable of fertilizing an ovum. Sperm must first undergo a process termed **capacitation.** Although capacitation is not well understood, it appears to involve at least two steps. First, a physiological change in the plasma membrane of the sperm takes place that allows the sperm to penetrate the surface membrane of the ovum. Secondly, the movements of the tail of the sperm become faster and more pronounced, resulting in an increase in sperm motility. Capacitation normally takes place when the sperm are in the female reproductive tract before the fusion of sperm and ovum.

The Fusion of Sperm and Ovum

Before we examine the process of sperm–ovum fusion, let us review the anatomy of the mature sperm and ovum. Figure 31–4 illustrates the general structure of a mature sperm (also see Fig. 30–6). The head of the sperm is composed of a dense, compact nucleus, an acrosome that caps the nucleus, and a surface membrane. The **acrosomal portion** of the sperm head contains enzymes that play an important role in the process of fertilization. The midpiece of the sperm contains spirals of mitochondria, while the tail of the sperm is composed of microtubules surrounded by a fibrous sheath.

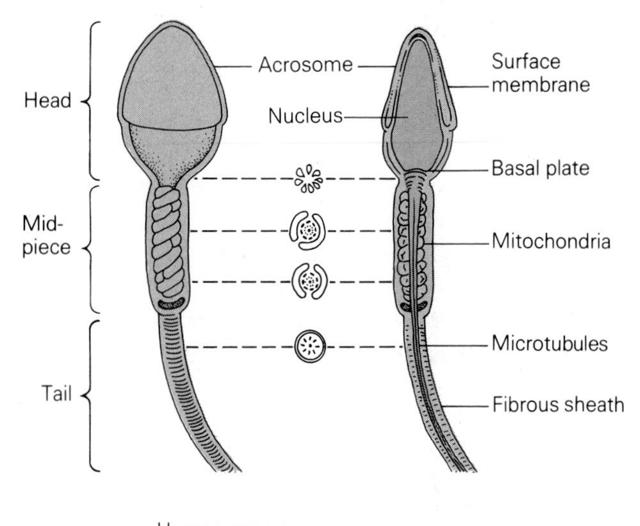

Figure 31–4 Structure of a human sperm. (Redrawn from Warsaw, J.B., Ed., *The Biological Basis of Reproductive and Developmental Medicine.* 1983.)

The organization of the mature human ovum prior to fertilization is illustrated in Figure 31–5. The mature ovum is surrounded by the **vitelline membrane.** On top of the vitelline membrane is the vitelline space, which separates the vitelline membrane from the next cell layer, the zona pellucida. The **zona pellucida** is composed of a layer of cells that form a jellylike coating around the ovum. The zona pellucida itself is surrounded by several layers of granulosa cells called the **cumulus oophorus.**

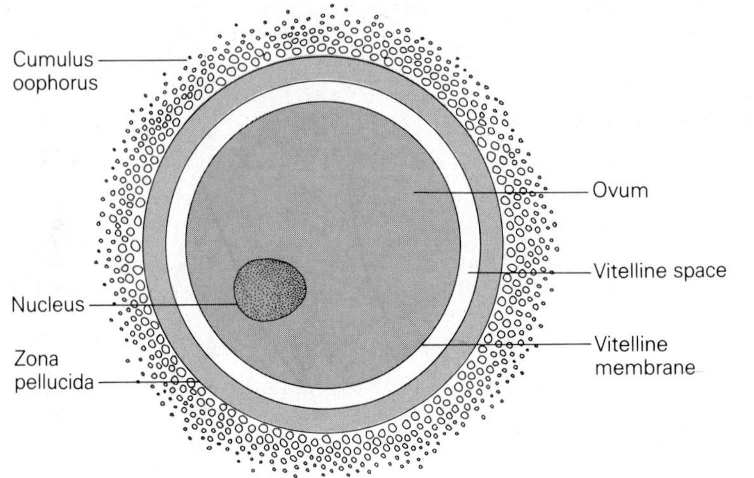

Figure 31–5 Organization of the mature human ovum prior to fertilization.

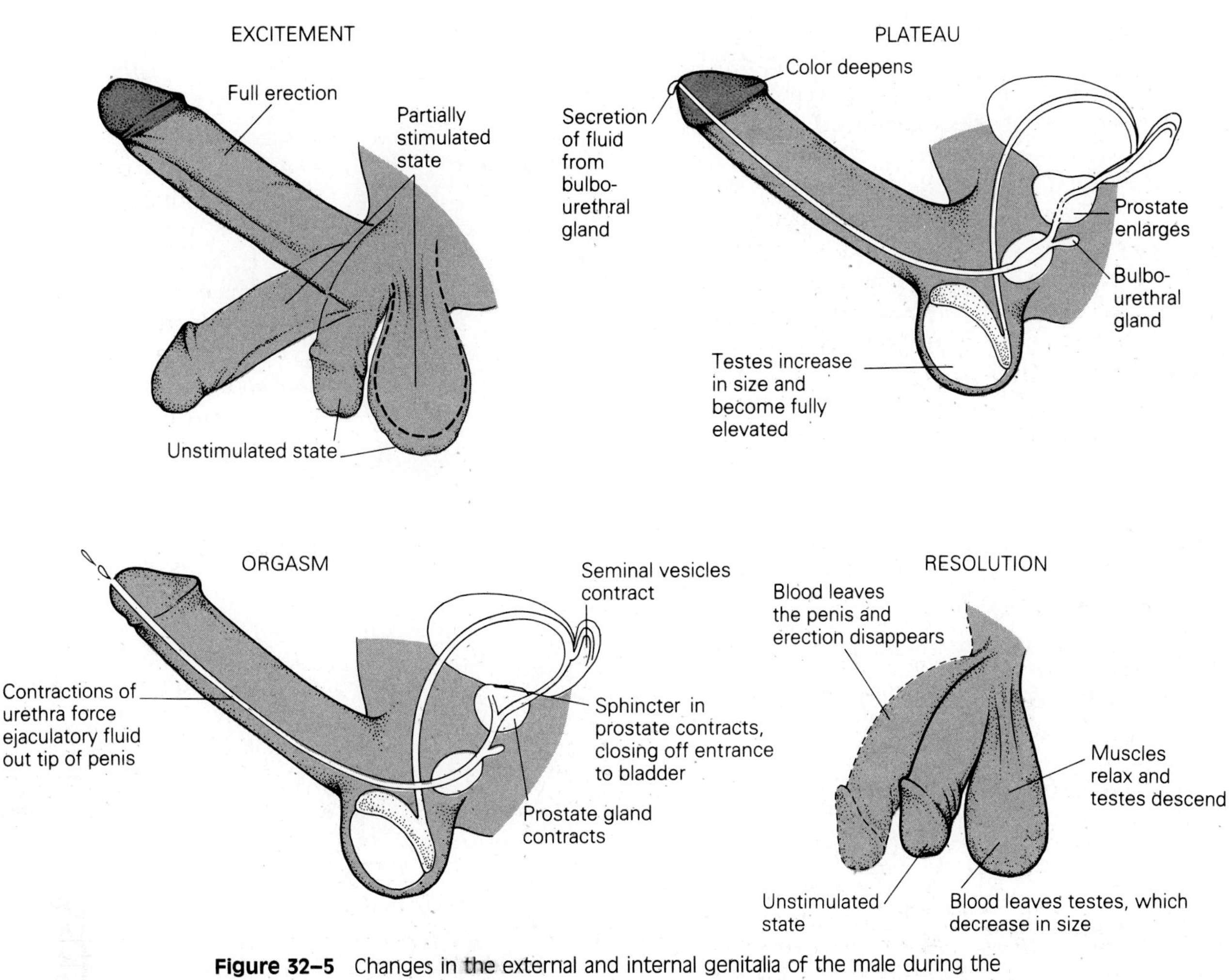

Figure 32–5 Changes in the external and internal genitalia of the male during the sexual response cycle. (Redrawn from Masters, Johnson, and Kolodny, *Human Sexuality*, Little, Brown, 1985.)

animals contains a bone, human males do not have a penile bone. The human penis does contain a thin skeletal muscle, but this muscle does not contribute significantly to penile erection as evidenced by the fact that normal erection occurs in the absence of penile muscle activity following certain spinal cord injuries. Instead, penile erection depends upon increased blood flow to the penis and increased intrapenile pressure. Penile vasoconstriction during sexual arousal in the male is not constant. Decreased blood flow to the penis can result in a reduction in penile size and firmness even though sexual arousal and neuromuscular tension have not diminished. Such fluctuations in blood flow to the penis are usually short-lived.

Sexual arousal also results in increased blood flow to the testes, causing an increase in testicular size (see Fig. 32–5). During the excitement phase, the muscles surrounding the vas deferens contract, lifting the testes closer to the body and smoothing the scrotal skin (see Fig. 32–5). An increase in general muscle tone or tension, termed **myotonia,** also accompanies sexual arousal. Erection of the nipples of the breast occurs in some (but not all) men.

Excitement Phase in the Female

One of the physiological responses to sexual arousal during the excitment phase in the female is **vaginal lubrication** (Fig. 32–6). Both penile erection in the male, and vaginal lubrication in the female, occur in response to vasocongestion. Sexual arousal results in increased blood flow to the vagina, and vasocongestion of the vagina leads to release of moisture from the vaginal lining, a process termed **transudation.** As vaginal secretion of moisture increases, some moisture may reach the vaginal opening and the surrounding labia. Vaginal secretion of moisture lubricates the vagina and facilitates insertion of the penis without discomfort. Note that there are great individual variations in the amount of moisture secreted by the vagina during sexual arousal, and the

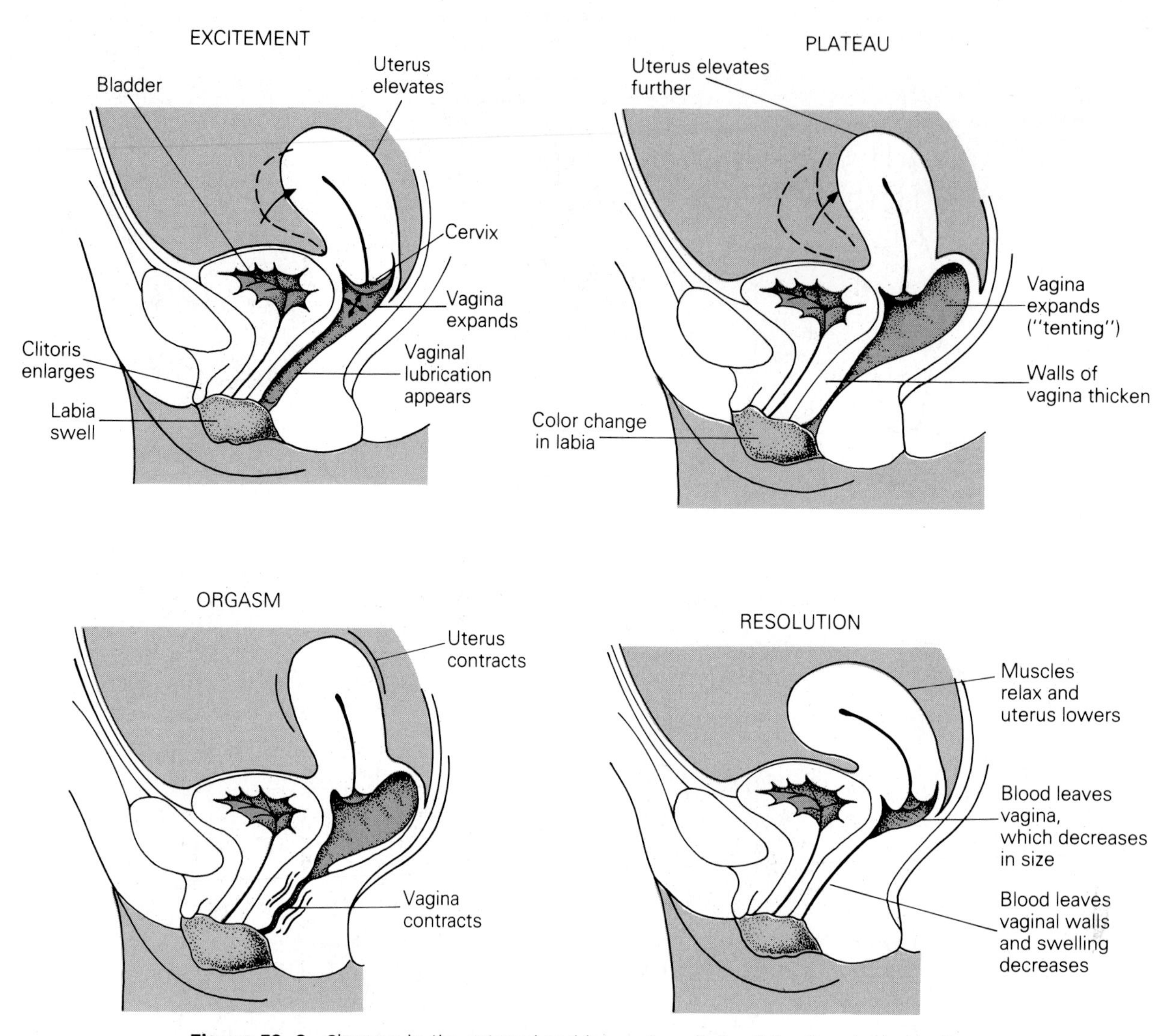

Figure 32–6 Changes in the external and internal genitalia of the female during the sexual response cycle. (Redrawn from Masters, Johnson, and Kolodny, *Human Sexuality*, Little, Brown, 1985.)

amount of moisture present cannot be used as an index of sexual arousal. Heightened sexual arousal may be accompanied by sparse vaginal lubrication, and vaginal secretion of moisture can occur in the absence of sexual arousal.

During the excitement phase, the vagina expands, and the cervix and uterus are lifted upward away from the vagina (see Fig. 32–6). Increased blood flow to the external genitalia results in increased size of the clitoris, labia minora, and labia majora (Fig. 32–7*a*). Muscle contractions in the breast tissue often result in nipple erection, and increased blood flow to the breast causes breast enlargement (see Fig. 32–7*b*). As in the male, there is an increase in general muscle tension, or myotonia, in the female during the excitement phase.

Plateau Phase in the Male

The plateau phase of sexual arousal precedes orgasm. Time spent in the plateau phase varies widely within and among individuals. Men who ejaculate quickly spend little time in this phase. During the plateau phase, blood pooling in the head of the penis causes this area to enlarge and to become darker in color (see Fig. 32–5). The testes continue to increase in size because of increased blood flow, and muscle contractions in the scrotum lift the testes up and back toward the anus (see Fig. 32–5). As was discussed in Chapter 30, pre-ejaculatory fluid from the bulbo-urethral glands, which contains viable sperm, is often secreted at the tip of the penis during the plateau phase of sexual arousal (see Fig. 32–5). Heart rate, blood pressure, respiration rate,

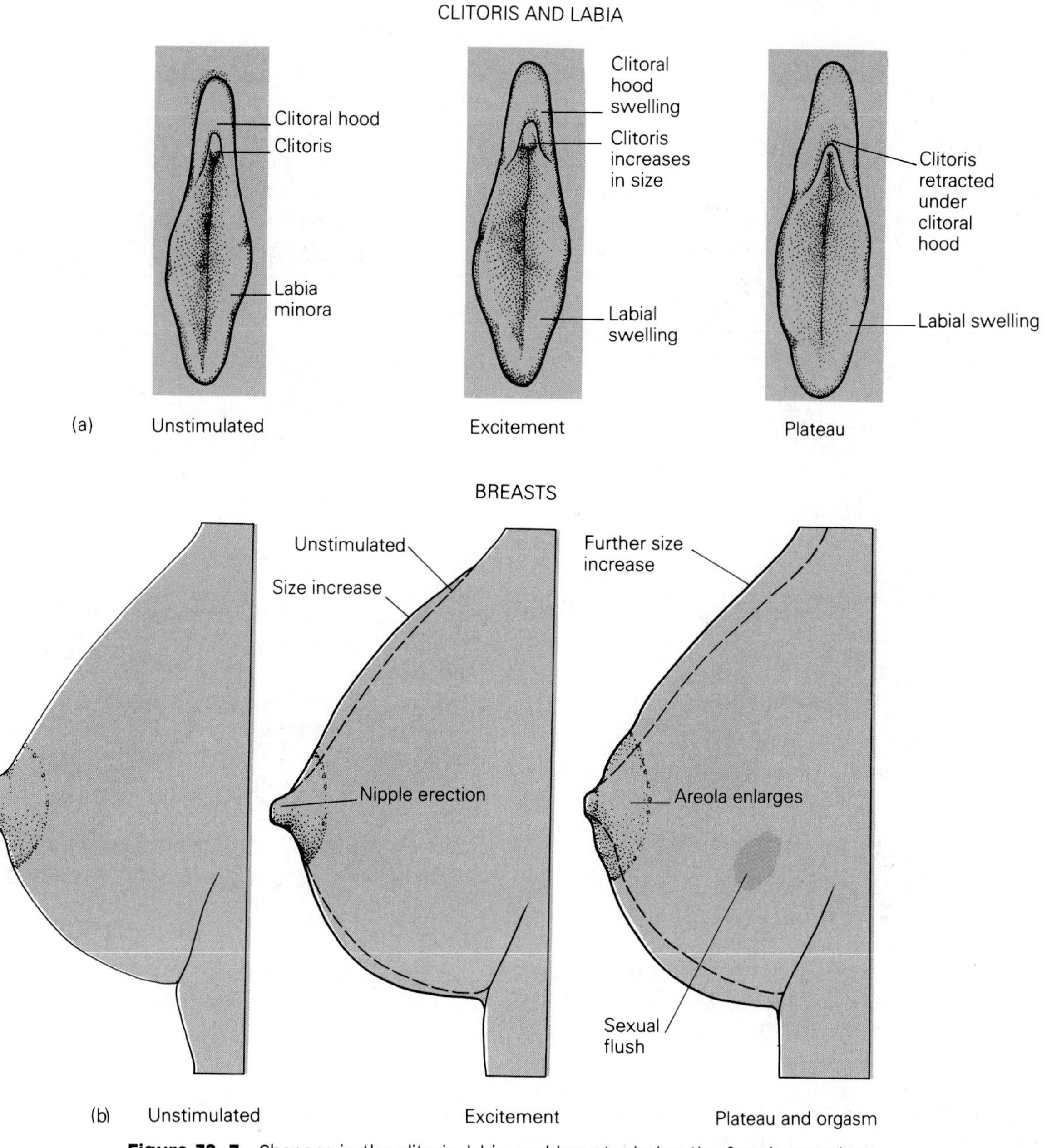

Figure 32–7 Changes in the clitoris, labia, and breasts during the female sexual response cycle. (Redrawn from Masters, Johnson, and Kolodny, *Human Sexuality,* Little, Brown, 1985.)

and muscle tension all increase significantly during this phase.

Plateau Phase in the Female

In the female, blood flow to the vagina increases, resulting in swelling of the walls of the vagina, which reduces the size of the vaginal opening (see Fig. 32–6). This increases the vaginal surface area that contacts the penis during intercourse. The uterus is elevated away from the vagina in a process termed *tenting* (see Fig. 32–6). The clitoris, which is highly sensitive to touch during this phase, is retracted against the pubic bone and becomes partially covered by the skin of the clitoral hood, which reduces direct physical contact of the clitoris (see Fig. 32–7). Increased blood flow to the labia produces a deepening in the color of the labia minora and labia majora. Blood flow to the breasts results in increased size of the areoli and darkening of the skin color in areas on the breasts, referred to as the "sexual flush" (see Fig. 32–7). Darkening of the skin on other areas of the body may occur as well. As is seen in the male, heart rate, blood pressure, respiration rate, and muscle tension all increase during the plateau phase in the female.

Orgasmic Phase in the Male

Ejaculation in the male involves two processes: the **emission** and the **expulsion** of ejaculatory fluid (see Fig. 32–5). During emission, increased sympathetic nerve activity stimulates rhythmic muscular contractions of the vas deferens, seminal vesicles, and prostate, forcing ejaculatory fluid into the urethra. Once these contractions have begun, they cannot be stopped until ejaculation is complete. During the expulsion phase, rhythmic muscular contractions of the urethra force ejaculatory fluid out of the tip of the penis. These two phases of ejaculation usually last only a few seconds. As was discussed in Chapter 30, a sphincter in the prostate contracts during ejaculation, sealing off the entrance to the bladder and preventing urine from mixing with ejaculatory fluid (see Fig. 32–5).

Orgasmic Phase in the Female

Orgasm in the female is characterized by rhythmic muscular contractions of the uterus and vagina that may last for as long as 15 seconds (see Fig. 32–6). As in the male, muscles in various parts of the body involuntarily contract during female orgasm, resulting in muscular rigidity and spasms.

Some controversy surrounds the importance of clitoral stimulation in inducing orgasm in the female, and whether females ejaculate. It appears that most women, while capable of achieving orgasm in response to vaginal stimulation, experience more intense orgasms when clitoral and vaginal stimulation are combined. Although most females do not ejaculate fluid from the urethra during orgasm, ejaculationlike responses have been reported by some females. At present, it is not known whether the fluid expelled from the urethra during orgasm in these women resembles prostatic fluid or urine.

Note that female orgasm is not essential for reproduction, since insemination of the female can occur in the absence of female sexual arousal.

Resolution Phase in the Male

Following ejaculation, most males enter a refractory period during which no further orgasm or ejaculation is possible. The length of the refractory period is highly variable (from a few minutes to many hours) and is prolonged by fatigue, time since the last orgasm, and age. During the refractory period, increased sympathetic nerve activity constricts the arteries in the penis, and blood leaves the penis through the veins, resulting in a loss of penile erection (see Fig. 32–5). Decreased blood flow to the testes also results in a decrease in testicular size. As muscles relax, the testes are lowered away from the body (see Fig. 32–5). Heart rate, blood pressure, respiration, and muscle tension all return to normal during the resolution phase. When intense sexual arousal is not followed by orgasm, the length of the resolution phase is prolonged and can be accompanied by aching sensations in the testes due to continued vasoconstriction.

Resolution Phase in the Female

Many females, unlike most males, are **multiorgasmic** (capable of experiencing more than one orgasm before entering the resolution phase) if intense sexual arousal is maintained. During the resolution phase, muscular relaxation results in a lowering of the uterus and cervix toward the vagina, which increases the probability of cervical contact with any ejaculatory fluid present in the vagina (see Fig. 32–6). Blood flow away from the breasts and genitals results in a reduction in the size of the breasts, clitoris, labia, and vagina. As in the male, the female's heart rate, blood pressure, respiration, muscle tension, and skin flushing all return to normal during the resolution phase. Breast and genital tissues are extremely sensitive to touch following orgasm in both sexes, and continued physical stimulation of these tissues during the resolution phase can be irritating or painful.

Sexual Reponse During Infancy

Sexual arousal in response to physical stimulation occurs even before birth. Penile erections have been detected by ultrasound in the male fetus during the last few months of gestation. At birth, many male infants exhibit penile erections, while female infants manifest vaginal lubrication and clitoral enlargement. Following birth, certain forms of physical contact, including breastfeeding, can result in penile erection in male infants and vaginal lubrication and enlargement of the clitoris in female infants. Although both males and females are capable of achieving orgasm from birth, the ability of the male to ejaculate does not occur until the internal and external genitalia have matured at puberty.

Sexual Response During Aging

Although physiological changes in the sexual response occur in both males and females during aging, the ability to become sexually aroused does not normally disappear.

Aging and the Female Sexual Response

In the female, the breasts, clitoris, and vagina remain sensitive to stimulation in old age. However, after menopause, the ovaries stop secreting estrogen, which leads to changes in the vagina. A low level of circulating estrogen results in diminished blood flow to the vagina during sexual arousal, which in turn leads to decreased vaginal lubrication. Use of an artificial lubricant can circumvent the pain that might result from intercourse in the absence of sufficient vaginal lubrication, a condition termed **dyspareunia.** Some evidence suggests that regular sexual activity following menopause slows the decline in vaginal lubrication that occurs during aging. A decrease in the elasticity of the vaginal walls in aging women leads to reduced vaginal expansion during sexual arousal. In addition, enlargement of the breast, development of the sexual flush, and increased muscle tension during arousal are significantly reduced in older women. Changes in the female sexual response during aging are summarized in Table 32–4.

Table 32–4
Changes in the Sexual Response During Aging

Female
1. A decrease in circulating estrogen following menopause results in decreased blood flow to the vagina which, in turn, results in diminished production of vaginal lubrication. Decreased vaginal lubrication can result in painful intercourse, or dyspareunia.
2. The vaginal walls lose elasticity and sexual arousal produces less vaginal expansion.
3. Many responses to sexual arousal—including enlargement of the breast, deepening of skin color (sexual flush), and muscle tension—are significantly reduced.
Male
1. Penile erection occurs more slowly and may require more prolonged stimulation.
2. The penis is less firm when fully erect.
3. Time spent in the refractory period following orgasm, during which no erection or ejaculation is possible, is prolonged.
4. The volume of ejaculatory fluid expelled at orgasm is decreased.
5. Testicular size and elevation during sexual arousal are reduced.
6. Many responses to sexual arousal—including deepening of skin color (sexual flush) and muscle tension—are reduced.

Aging and the Male Sexual Response

Although circulating testosterone levels begin to decline gradually after age 55 to 60 (see Chapter 30), there is no dramatic drop in testosterone levels during aging in the male that parallels the rapid decline in circulating estrogen levels in the female following menopause. Healthy males can maintain the capacity for penile erection throughout their entire life span, and sperm production continues till age 80 or 90 although the production rate slows down after age 40.

However, physiological changes in the male sexual response do occur with aging. The sensitivity of penile sensory nerves to tactile stimulation decreases with age, and hence more prolonged and direct penile stimulation may be needed to induce erection. Blood flow to the penis during sexual arousal may also be reduced with age, and so the penis may be less firm when fully erect. Testicular elevation during sexual arousal is slower and less complete in the older male. The amount of fluid ejaculated at orgasm is often reduced, and the refractory period following orgasm is prolonged. As in the female, muscle tension during sexual arousal in the older male is also reduced. A summary of the changes in the male sexual response during aging is contained in Table 32–4.

Sexual Dysfunction

The term *sexual dysfunction* refers to any condition in which the normal physical responses to sexual stimulation are impaired. Sexual dysfunction can be a source of anxiety, frustration, and distress, disrupting personal relationships and family life.

Endocrine Factors in Sexual Dysfunction

Estrogen Deficiencies in the Female

Estrogen plays an important role in maintaining vaginal lubrication. Estrogen stimulates proliferation of the vascular bed beneath the vaginal epithelium and increases blood flow to the vagina, which results in secretion of moisture from the vaginal walls. Inadequate estrogen stimulation can result in **atrophic vaginitis,** which is characterized by dryness and thinning of the vaginal mucosal membrane and decreased elasticity and muscle tone of the vagina. Following menopause, when circulating estrogen levels are decreased, vaginal elasticity and vascularization are reduced, and blood flow to the vagina can be decreased by as much as 80%. Estro-

Focus on Amenorrhea

Amenorrhea, or the absence of menstruation, is classified as either primary or secondary. Primary amenorrhea is the absence of menstruation in females who have never menstruated. Secondary amenorrhea is the cessation of menstruation in females who have previously menstruated. Primary amenorrhea is rare and usually results from developmental abnormalities of the ovaries or the reproductive tract, or from the formation of scar tissue in these structures in response to physical injury or infections occurring before the first menstrual cycle would normally begin.

Secondary amenorrhea is more common. Diagnosis can be problematic, because the length of the menstrual cycle varies greatly among women. A general rule that can be used to diagnose this disorder is the absence of menstruation for an interval greater than three times the individual's normal cycle length.

Secondary amenorrhea can result from malfunctions at the level of the ovary, uterus, pituitary, or hypothalamus. During adulthood, autoimmune diseases, radiotherapy or chemotherapy used as cancer therapy, and surgery or infections that cause scar tissue to form can result in ovarian failure and a reduction in the number of viable follicles. At the level of the pituitary, the most common cause of secondary amenorrhea is the presence of a prolactin-secreting tumor, which results in hyperprolactinemia and consequent suppression of GnRH release (see text). At the level of the hypothalamus, many factors can contribute to a suppression of GnRH release and the onset of secondary amenorrhea. Stress, which often accompanies intense programs of physical exercise, can suppress GnRH release and result in secondary amenorrhea. Extreme weight loss, which can occur during dieting; in response to drugs such as alcohol, narcotics, or amphetamines; and in the condition termed "anorexia nervosa," can also suppress GnRH release and result in secondary amenorrhea.

Extreme weight gain can result in secondary amenorrhea through a different mechanism. An increase in adipose tissue results in increased conversion of adrenal androgens to estrogen, and this increase in circulating estrogen disrupts the normal feedback signals to the hypothalamus (see text). Psychological stresses of many types also result in amenorrhea.

Secondary amenorrhea can be seen following discontinuation of birth control pills in approximately 2% of the women who have taken the pills for two or more months, a condition termed "post-pill amenorrhea." It is not clear whether this type of amenorrhea is caused by a hypothalamic disturbance in response to the steroid content of the pills, or whether the women displaying post-pill amenorrhea have a history of menstrual irregularities prior to administration of the pills.

Secondary amenorrhea is often reversible and tends to disappear with prolonged discontinuation of oral contraceptives, when normal dietary patterns are resumed, or when the source of stress is removed. It should be noted that the most common nonpathological condition that results in secondary amenorrhea is pregnancy.

gen replacement therapy following menopause can increase blood flow to the vagina by fourfold.

Premenopausal women who have very low levels of circulating estrogen (for instance, following ovarian disease or surgical removal of the ovaries) can also experience atrophic vaginitis with accompanying loss of vaginal lubrication and elasticity. Inadequate vaginal lubrication can produce discomfort or pain during sexual intercourse (dyspareunia). Women with very low levels of circulating estrogen can also experience symptoms of peripheral neural dysfunctions, which can include slowed response to tactile stimulation or loss of clitoral sensation. Estrogen replacement therapy can reduce the severity of these symptoms.

Testosterone Deficiencies in the Male

Some evidence suggests that testosterone is necessary for normal sexual activity in the male. Although males with low circulating levels of testosterone can attain penile erection, the capacity to ejaculate during orgasm may be delayed, diminished, or absent. In addition, low levels of testosterone are often associated with reduced sexual desire or sexual libido. Androgen replacement therapy often increases both **ejaculatory capacity** and **sexual desire.** The exact mechanism by which testosterone facilitates ejaculation or sexual desire is not known. It has been proposed that androgens are involved in sensory perception and the capacity for development of muscle tension within the pelvis.

Neural Disorders and Sexual Dysfunction

Afferent sensory impulses from the internal and external genitalia enter the spinal cord and travel to the brain. Efferent nerve impulses from the brain leave the spinal cord and activate muscles and blood vessels of the internal and external genitalia. The excitement phase of sexual arousal in both males and females involves increased afferent neural input from sensory nerves in the genitals and efferent neural induction of vascular changes in the internal and external genitalia. Orgasm involves increased efferent motor nerve activity, which induces muscular contractions in the internal and external genitalia. Therefore, both mechanical and disease-related injuries to the spinal cord or brain can result in impairment of sexual arousal and orgasm in both males and females. Interruption of afferent sensory pathways can interfere with sexual arousal. Interruption of efferent pathways may disrupt increased blood flow to the genitals during sexual arousal and hence may reduce penile erection in the male and vaginal swelling and lubrication in the female, as well as muscle contractions during orgasm in both sexes.

Multiple sclerosis, a disease characterized by patches of demyelinization in the spinal cord and brain, can be accompanied by a reduction or elimination of sexual arousal and orgasmic capacity in both males and females because of lesions in the spinal cord that disrupt afferent and efferent nerve transmission.

Severe **diabetes mellitus** can be accompanied by dysfunctions of the sensory fibers that innervate the penis and clitoris, resulting in a reduction or absence of the genital sensations necessary for sexual arousal and orgasm. Diabetic neural dysfuction can also result in decreased blood flow to the genitals, which can result in insufficient penile rigidity to achieve vaginal penetration.

Drugs and Sexual Dysfunction

Although small amounts of **alcohol** are often used to reduce sexual inhibitions, large amounts of alcohol have a depressant effect on the central nervous system, which can interfere with sexual arousal. In addition, chronic and excessive consumption of alcohol, as seen in alcoholics, can lead to damage of the CNS, which results in alcoholic neuropathy in males and females. Alcohol-induced injury to the somatic and autonomic nerves in the periphery, spinal cord, or brain can result in a reduction or absence of sexual arousal and orgasmic capacity.

Sedatives (e.g., barbiturates, chloral hydrate), **narcotics** (e.g., codeine, morphine), and **tranquilizers** (e.g., diazepam, meprobamate), when taken in high doses, can retard or inhibit sexual arousal and orgasm in both males and females because of a depression of nerve conductance in the central nervous system.

Some drugs used to treat hypertension, such as reserpine and alpha-methyldopa, can alter neural control of blood flow to the internal and external genitalia in both males and females, resulting in decreased vaginal swelling and lubrication in the female and difficulty in achieving erection in the male.

Some antihistamines that contain ephedrine decrease blood flow to the vagina and vaginal lubrication, resulting in dyspareunia.

Other Sources of Sexual Dysfunction

Anorgasmia, Infection, and Vaginismus

Difficulty in reaching orgasm, which is referred to as **anorgasmia,** represents the largest category of sexual dysfunction in the female. However, organic disorders that can interfere with orgasm, such as hormone deficiencies, alcoholism, diabetes, or neurological disorders, account for less than 5% of the cases of anorgasmia. Anorgasmia appears to be largely due to psychological factors, which may include sexual inhibitions acquired during early training, unrealistic performance expectations for oneself or one's partner, difficulty with sexual communication, and negative self-image. These issues are beyond the scope of this chapter. For further information, consult the volumes on human sexuality listed at the end of this chapter.

Vaginal infections, allergic reactions, or chemical irritations can produce a condition termed **nonatrophic vaginitis.** This condition, which is quite common and usually of short duration, can result in vaginal dryness or tenderness.

Endometriosis (see "Infertility") and **pelvic inflammatory disease** can both cause pain during sexual arousal and intercourse. This is due to the fact that vasocongestion and pelvic muscle contractions during sexual arousal place increased pressure on diseased tissues, in the case of endometriosis, or on chronically inflamed tissues in the case of pelvic inflammatory disease. Further pressure due to intercourse can result in deep pelvic pain. Pain associated with sexual arousal and intercourse can result in decreased vaginal lubrication, loss of orgasmic capacity, and a loss of sexual desire, or sexual aversion.

Vaginismus is defined as involuntary contractions and spasms of the vaginal (or pubococcygeal) muscles, which makes vaginal penetration very difficult or impossible (Fig. 32–8). Vaginismus may stem from emotional problems or traumatic or painful sexual experiences; or it may arise as a protective reflex to avoid pain in the case of pelvic disease. Prolonged dyspareunia, or painful intercourse, can sometimes also lead to vaginismus. Note that dyspareunia and/or vaginismus in the female can cause sexual problems in male partners when they feel

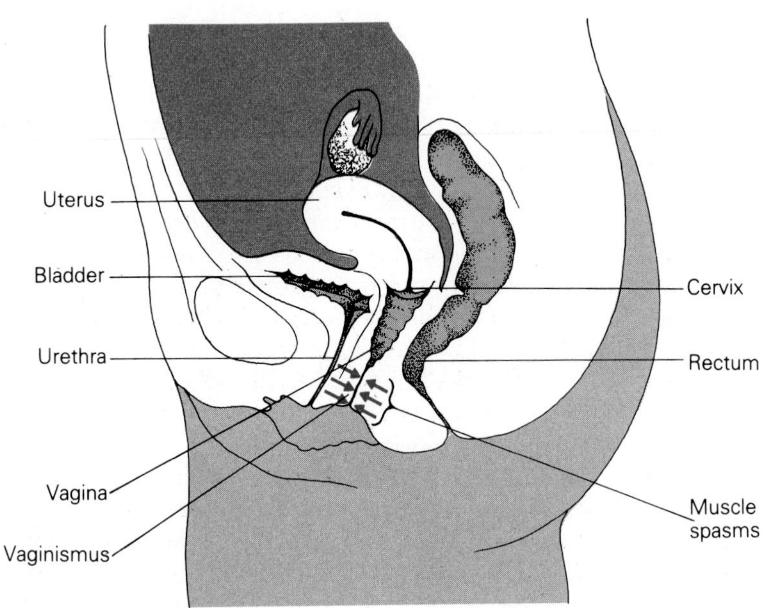

Figure 32–8 The involuntary muscle spasms of vaginismus. (Redrawn from Masters, Johnson, and Kolodny, *Human Sexuality*, Little, Brown, 1985.)

responsible for unsuccessful attempts at intercourse. Feelings of inadequacy and guilt can result in erectile difficulty and loss of sexual desire.

Erectile Dysfunction, Premature Ejaculation, and Dyspareunia

Erectile dysfunction, or **impotence,** is defined as the inability to achieve penile erection sufficient for vaginal penetration. Erectile dysfunction is classified as either primary or secondary. **Primary erectile dysfunction,** which is very rare, defines the male who has never been able to achieve erection sufficient for vaginal penetration. **Secondary erectile dysfunction** is defined as periods of inability to achieve or maintain erection. Secondary erectile dysfunction can occur in response to anxiety, stress, illness, fatigue, lack of privacy, alcohol consumption, or change of sexual partner. Episodes of secondary erectile dysfunction are usually short-lived, and isolated episodes occur in most men.

Premature ejaculation can be defined as consistent, unintentional ejaculation prior to vaginal penetration. When defined in this way, very few men experience premature ejaculation. More typically, loss of ejaculatory control results in ejaculation soon after vaginal penetration. The probability of premature ejaculation increases with time since last orgasm, novelty of the sexual partner, decreased age, and prolonged stimulation prior to vaginal penetration. Manual application of pressure to the penis for a few seconds, from front to back just below the coronal ridge or at the base of the penis, can reduce the urge to ejaculate and help to control premature ejaculation (Fig. 32–9). Application of pressure to

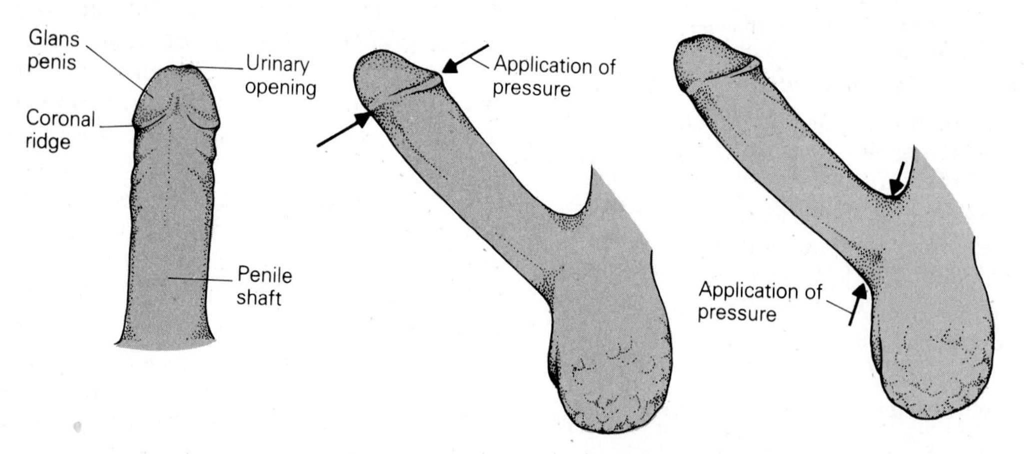

Figure 32–9 Application of manual pressure to the penis, used to prevent premature ejaculation. (Redrawn from Masters, Johnson, and Kolodny, *Human Sexuality*, Little, Brown, 1985.)

the penis may also cause a partial loss of erection that is only temporary.

Dyspareunia in the male is classified as painful erection, intromission, or ejaculation. Painful erection and/or pain on intromission can result from many factors, including inflammation of the foreskin of the penis (termed **balanitis**), reduced elasticity of the skin covering the penis, and penile scarring. These conditions can result from physical trauma to, or infection of, penile tissue. Painful ejaculation can result from physical, infectious, or chemical irritation of the urethra, termed **urethritis.**

Sources of sexual dysfunction in the male and female are summarized in Table 32–5.

Table 32–5
Origins of Sexual Dysfunction

Male
1. Neurological disorders resulting from physical or disease-related injury (multiple sclerosis, diabetes mellitus) to the brain, spinal cord, or peripheral nerves.
2. Testosterone deficiency characterized by decreased penile erection, capacity to ejaculate, or sexual desire.
3. Excessive use of drugs including alcohol, sedatives, narcotics, tranquilizers, antihistamines, and antihypertensive agents.
4. Primary or secondary erectile dysfunction resulting from physiological or psychological causes.
5. Premature ejaculation.
6. Dyspareunia due to physical, chemical, or infectious damage to the penis or urethra.
Female
1. Neurological disorders resulting from physical or disease-related injury (multiple sclerosis, diabetes mellitus) to the brain, spinal cord, or peripheral nerves.
2. Chronic estrogen deficiency that results in atrophic vaginitis characterized by vaginal dryness, thinning, and decreased elasticity.
3. Acute estrogen deficiency that results in nonatrophic vaginitis characterized by vaginal dryness.
4. Excessive use of drugs including alcohol, sedatives, narcotics, tranquilizers, antihistamines, and antihypertensive agents.
5. Dyspareunia due to physical, chemical, or infectious damage to the vagina, clitoris, or labia.
6. Endometriosis and pelvic inflammatory disease.
7. Involuntary contractions of the vaginal muscles (vaginismus).

Infertility

Infertility, or the inability to reproduce, can be due to anatomical, hormonal, chromosomal, immunological, or psychological factors.

Fertility in the male depends on several factors: adequate secretion of the gonadotropins that act to stimulate spermatogenesis; testes that are capable of responding to gonadotropins by producing sperm; glandular production of seminal fluid; an intact ductal system for sperm delivery; and an intact nervous system for the control of penile erection and ejaculation. A disruption of normal function at any of these levels can result in infertility.

In addition, three aspects of the sperm themselves are important for sperm penetration and fertilization of the ovum: sperm number, sperm morphology (shape and structure), and sperm motility. Although it is always difficult to establish "normal" ranges for physiological processes, it is generally accepted that 15 to 250 million sperm per milliliter of semen represents a normal sperm count. A further definition of sperm normality requires that at least 60% of the sperm have a normal appearance and that at least 60% are active or motile.

A low sperm count and/or decreased sperm motility can result from gonadotropin deficiencies or from damage to the testes. Testicular damage can result from infections, inflammation, impaired circulation, and certain diseases such as mumps. In addition, a variety of environmental factors such as fever, prolonged heat exposure, chemical agents, and irradiation can result in testicular damage and impaired spermatogenesis. The extent of testicular damage produced by these agents is determined by the intensity and duration of testicular exposure. Abnormal sperm shape or structure, which includes a nonoval head section, a broken midpiece, or a bent or coiled tail, can result in infertility by preventing sperm migration through the cervical mucus or penetration of the ovum. Infection and inflammation of the epididymis, vas deferens, or ejaculatory ducts can reduce or prevent sperm transport, which can also result in infertility.

As in the male, infertility in the female can be due to a number of factors. One of the most common causes of female infertility is failure to ovulate. Ovulatory failure can result from inadequate secretion of gonadotropins, failure of the ovary to respond to gonadotropins, variations in the anatomical structure of the ovaries, and ovarian damage resulting from infection or inflammation. Other factors that can cause female infertility include overproduction of thick cervical mucus, which can block sperm entry into the uterus, and infection or disease of the uterus or fallopian tubes. For instance, **endo-**

metriosis is a disease characterized by the presence of endometrial-like tissue in locations other than the uterus, such as the fallopian tubes or ovaries. Tissue formation, which may appear as cysts or nodules, can reduce the diameter of the fallopian tubes and block ovum transport to the uterus.

Immunological factors may also contribute to infertility. Antisperm antibodies have been found in both males and females. Antibodies, produced by the male, coat the surface of the sperm and reduce the ability of sperm to penetrate the cervical mucus. Antibodies produced by the female can induce **agglutination,** or aggregation of sperm into clumps, which also reduces the ability of sperm to penetrate the cervical mucus. Whether immunological factors play an important role in infertility remains to be determined.

Contraception

Being able to block conception ("contra-ception") has resulted in increased independence of human sexuality from reproduction. Various methods are used to interfere with conception, but all of them fall within four basic categories: **barrier methods, chemical methods, surgical methods,** and **abstinence.** In addition, all methods of contraception work by disrupting one of three major steps in the reproductive process: ovulation, sperm and ovum transport, or implantation of the fertilized ovum in the uterus. The sites of action of various methods of contraception are illustrated in Figure 32–10.

Barrier Methods

All barrier methods of contraception prevent sperm transport. Barrier methods include condoms, diaphragms, cervical caps, and contraceptive sponges. Condoms are thin sheaths of rubber that fit over the penis and prevent sperm from entering the vagina. Diaphragms, cervical caps, and sponges are rubber or polyurethane devices that cover the entrance to the cervix and prevent sperm from entering the uterus. Condoms, diaphragms, and sponges are used only once and must be replaced prior to each sexual encounter. Cervical caps, which work similarly to the diaphragm, can be worn for longer periods prior to removal; they are held in place by suction.

Barrier methods of contraception vary in effectiveness and are often combined with chemical methods to increase their effectiveness. For instance, the contraceptive sponge contains **spermicides,** chemicals that kill sperm. Spermicides, contained in cream or jelly, are also applied to the inside of the diaphragm to kill sperm that may pass the rim of the diaphragm.

Barrier methods of contraception have minimal side effects, which include the possibility of an allergic reaction to the spermicide or to the material used in construction, and a chance of introducing infection into the vagina.

Chemical Methods

The most popular method of contraception, the **birth control pill,** is taken orally and acts to prevent contraception by blocking ovulation (see Fig. 32–10). There are many types of birth control pills, but they all contain synthetic variations of the gonadal hormones estrogen and/or progesterone, and they act to inhibit the release of follicle stimulating hormone (FSH) and luteinizing hormone (LH) from the pituitary. As we learned in Chapter 30 (see Fig. 30–17), a rise in circulating levels of LH and FSH at the midpoint of the ovarian cycle stimulates maturation of the ovarian follicle, resulting in ovulation. When LH and FSH release are suppressed by oral contraceptives, as illustrated in Figure 32–11, follicular maturation does not occur, and ovulation is inhibited. Continuous use of oral contraceptives will inhibit follicular maturation, ovulation, and the onset of menstruation. Discontinuation of oral contraceptives for one week each month results in a drop in circulating levels of estrogen and progesterone, which stimulates the onset of menstruation. The contraceptive efficacy of birth control pills does not require monthly discontinuation of use.

Progesterone, contained in oral contraceptives, also interferes with conception in two additional ways. First, progesterone increases the viscosity of cervical mucus, which, as we have seen in Chapter 31 (see Fig. 31–3), blocks the entry of sperm into the uterus. Secondly, progesterone interferes with implantation of the ovum in the uterus by inhibiting proliferation of the endometrial lining of the uterus, which is required for implantation.

Birth control pills can produce a variety of side effects, including increased risk of nausea, constipation, elevations of blood pressure, skin rashes, and salt retention with accompanying weight gain. However, the most serious side effect is increased risk of cardiovascular disease, which becomes significant in women over 35 who use both oral contraceptives and tobacco.

Another type of oral contraceptive, termed the **morning-after pill,** contains a very high level of **estrogen,** which alters the endometrial lining of the

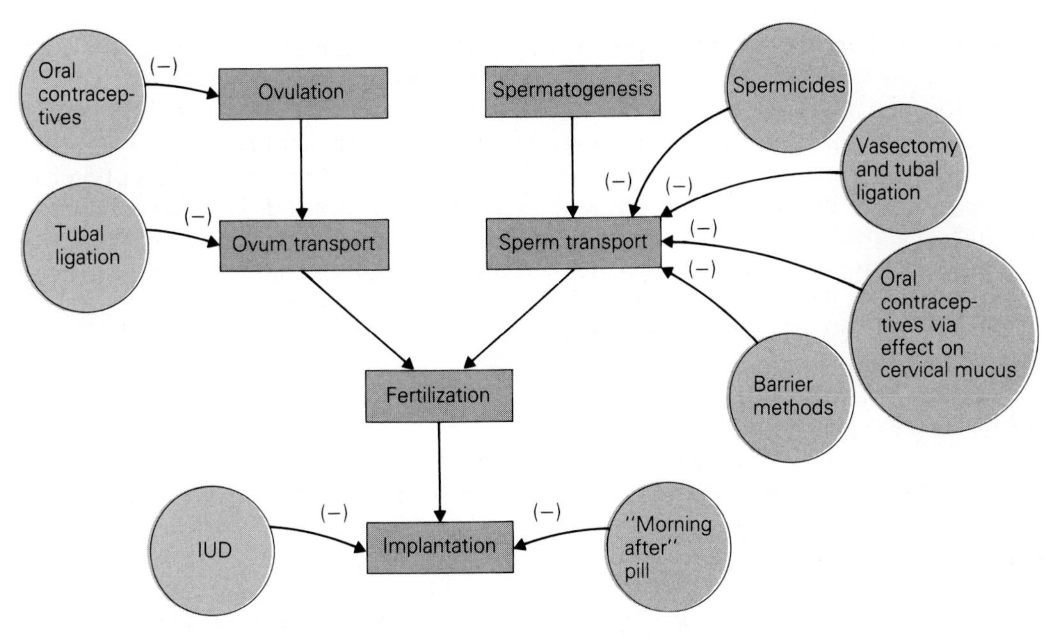

Figure 32–10 Sites of action of various methods of contraception.

uterus and prevents implantation of the fertilized ovum. However, the high levels of estrogen contained in the morning-after pill often induce severe nausea and vomiting, which have limited the use of this type of oral contraceptive.

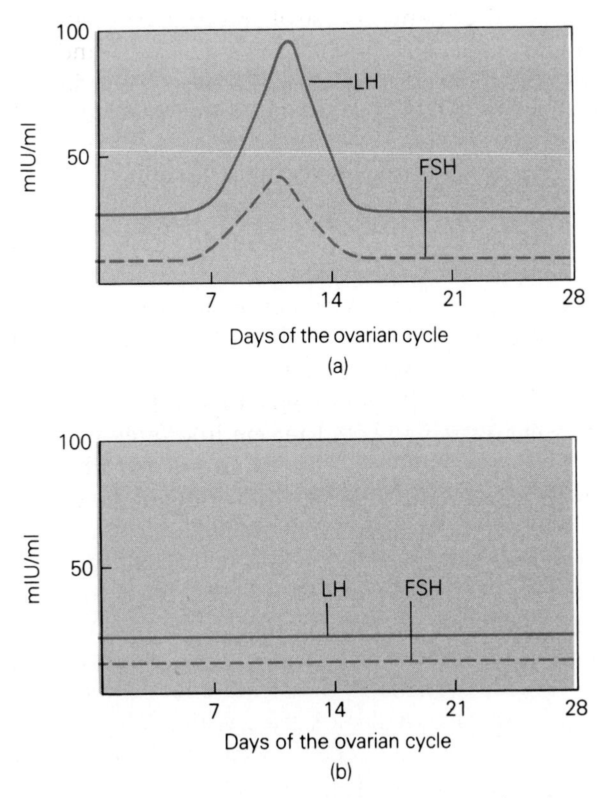

Figure 32–11 Plasma levels of gonadotropins during the ovarian cycle in women *(a)* not using oral contraceptives and *(b)* using oral contraceptives.

Another type of contraceptive method is the **intrauterine device (IUD),** which can be worn for extended periods. The IUD is made of plastic or metal and is inserted by a physician into the uterus through the vagina and cervix. Although the mechanism of action of the IUD is not completely understood, it is thought to induce a local inflammatory reaction in the endometrium of the uterus, which prevents implantation of the ovum. Side effects of the IUD include uterine bleeding, spasms and cramping of uterine muscles, pelvic infection, and perforation of the uterine wall.

Surgical Methods

The most effective form of birth control, with the exception of abstinence, is surgical sterilization. Sterilization procedures, which include vasectomy in the male and tubal ligation in the female, have enjoyed a recent increase in popularity because they are safe, effective, and permanent.

Vasectomy

Vasectomy is a simple surgical procedure that is performed within 15 to 20 minutes under local anesthesia. As illustrated in Figure 32–12, vasectomy consists of a small surgical incision on each side of the scrotum to expose a small section of the vas deferens. The vas deferens are severed, and the ends are tied, clamped, or cauterized to prevent sperm transport through the vas deferens.

Vasectomy does not interfere with sperm production. Following vasectomy, sperm accumulate in

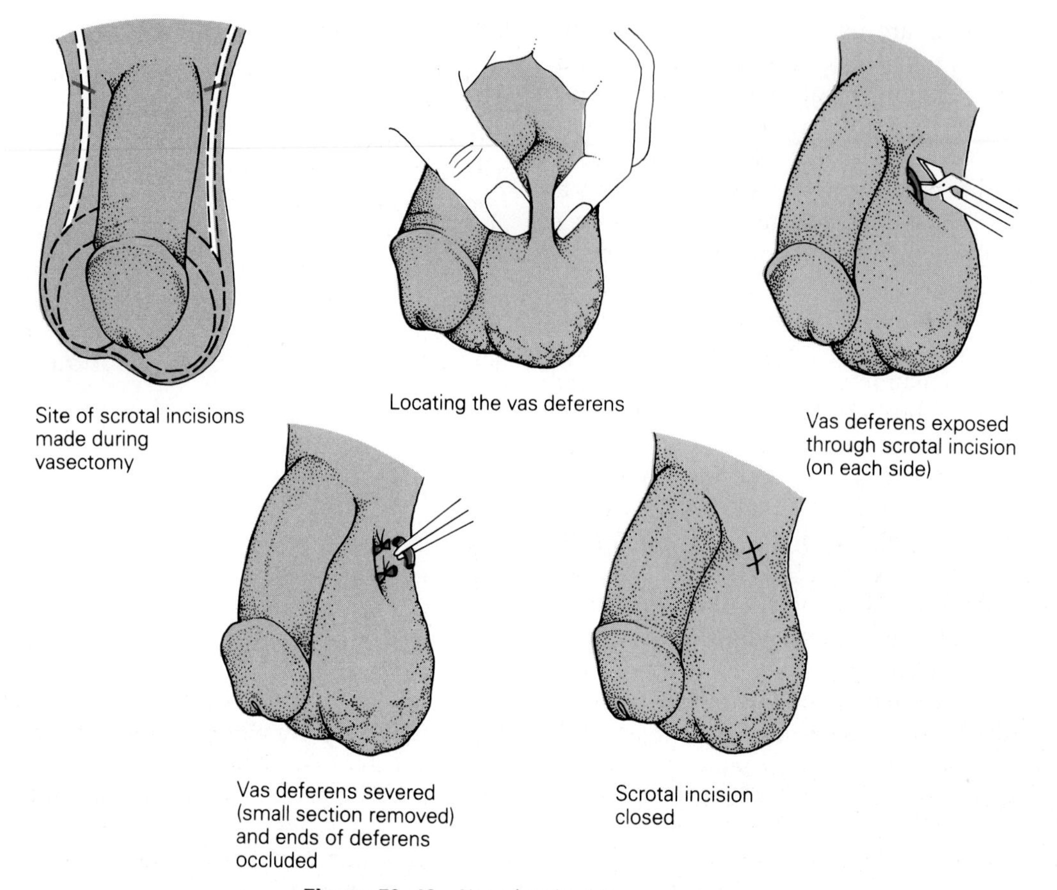

Figure 32–12 Steps involved in a vasectomy.

the epididymis of the testes, where they are engulfed and destroyed by phagocytosis. Vasectomy does not interfere with the synthesis or release of testosterone from the testes and does not affect erection, ejaculation, or orgasm in any way. Vasectomy is a very effective method of contraception, with failure rates below 0.15%. Postsurgical complications, which can include bleeding, infection, and swelling of scrotal tissue due to sperm leakage, are rare. No long-range health risks appear to be associated with vasectomy.

Tubal Ligation

Tubal ligation consists of an abdominal incision to permit insertion of an instrument called a **laparoscope,** which is used to locate and sever the fallopian tubes on each side of the body. The cut ends of the fallopian tubes are occluded with rings, clips, or cautery, blocking both ovum transport to the uterus and sperm transport through the fallopian tubes. When performed after childbirth, tubal ligation often consists of placing two ligatures around the midsection of each fallopian tube and removing the intervening segment, as illustrated in Figure 32–13. Tubal ligation is a very effective method of preventing pregnancy, and postsurgical complications, which can include bleeding and infection, are rare. As with vasectomy, no long-range health risks appear to be associated with tubal ligation.

Abstinence

Complete abstinence, which is the only completely effective form of birth control, is unpopular for obvious reasons. However, many people practice **periodic abstinence** during the time of ovulation when fertility is the greatest (termed the **rhythm method**). The efficacy of this method of preventing pregnancy depends on the accuracy of predicting the time of ovulation.

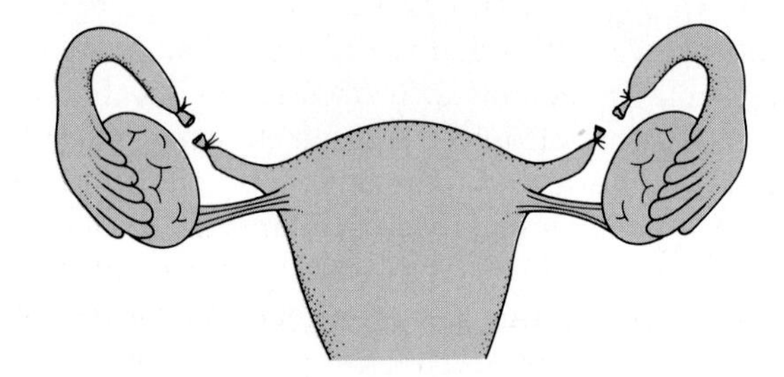

Figure 32–13 Illustration of a tubal ligation.

Three methods are used to predict ovulation. The first involves monitoring the ovarian cycle and length of menstruation for at least 6 months and avoiding intercourse on the days preceding and following ovulation (at midcycle). This method has a high failure rate because many women exhibit a large variation in length of monthly cycles. The second method involves daily monitoring of basal body temperature and avoiding intercourse on the days preceding and following ovulation. As we discussed in Chapter 31 (Fig. 31–17), body temperature decreases prior to ovulation and rises again following ovulation. This method also has a high failure rate because the temperature shifts that occur during the ovarian cycle are small (usually less than 1 degree) and are difficult to identify accurately. The third method of identifying the time of ovulation requires monitoring the consistency of cervical mucus. As described in Chapter 31, cervical mucus becomes clear, thin, and stringy around the time of ovulation. Predicting ovulation on the basis of consistency of cervical mucus has a high failure rate because of the difficulty of accurately identifying these changes.

Future Trends

Several new approaches to contraception are currently being developed and clinically tested. The important variables that should be identified for all new methods of contraception are efficacy, duration of action, reversibility, ease of use, and incidence of both short- and long-term side effects. In the female, research efforts are being directed toward blocking ovulation and implantation. In the male, the focus is on blocking production, maturation, motility, and/or transport of sperm.

Female Contraception

Long-acting forms of estrogen and progesterone are being administered by subcutaneous implantation and by injection, which can prolong the effectiveness of these substances for months. Other hormonal analogs that suppress LH and FSH release are also being investigated, as are substances that antagonize the actions of progesterone and human chorionic gonadotropin (hCG) on the endometrium and hence interfere with implantation of the ovum. Synthetic analogs of the prostaglandins, which induce uterine contractions, are being investigated for their efficacy in inducing the onset of menstruation, which would evacuate the contents of the uterus whether or not a fertilized ovum is present.

Male Contraception

Combinations of estrogen, progesterone, and testosterone are currently being examined for their ability to transiently block sperm production. Other naturally occurring substances, such as an extract from cottonseed oil (termed *gossypol*), that are capable of immobilizing or killing sperm, are also being investigated. Efforts are focused as well on identifying and testing substances, such as inhibin, that inhibit FSH release from the pituitary and hence remove the hormonal signal for sperm production in the testes.

Summary

I. A. The genetic information contributed by the parents determines the chromosomal or genotypic sex of the zygote. The chromosomal sex of the zygote determines the gonadal sex, and the gonads in turn determine the phenotypic sex of the individual. The presence or absence of a Y chromosome determines the sex of the zygote. The Y chromosome codes for, or regulates the expression of, the H-Y antigen, a protein that binds to receptors in the gonads. The H-Y antigen induces gonadal cells to develop as testes. In the absence of a Y chromosome (XX), the gonads develop as ovaries.

B. During the first six weeks of embryonic development, the gonads in all embryos, whether genetically male or female, are anatomically and morphologically identical. Hormones secreted by the developing gonads direct gonadal differentiation and the development of the internal genitalia. In the male embryo, the developing testes secrete Mullerian Inhibiting Substance (MIS), which induces regression of the female internal genitalia. Testosterone secreted by the developing testes stimulates development of the male internal genitalia (the Wolffian duct system). In the absence of the H-Y antigen, the gonads of the female embryo develop as ovaries rather than as testes. The developing ovaries do not secrete testosterone or MIS. In the absence of MIS, the Mullerian duct system develops in the female embryo.

C. The external genitalia in the male and female differentiate in response to the presence and absence of testosterone, respectively.

D. Exposure of the female embryo to androgens during development will not alter differentiation of the female internal genitalia, which occurs normally in the absence of Mullerian inhibiting factor. However, exposure to testosterone during development may produce masculinization of the female external genitalia.

E. Abnormal patterns of sexual differentiation can occur at four levels: chromosomal sex (genotype), gonadal sex, internal genital sex, and external genital sex. Chromosomal abnormalities can alter gonadal differentiation as well as enzyme activity and hormone receptor activity, resulting in altered hormone secretion from the gonads or altered action in hormone-sensitive tissues.

II. A. The ability to become sexually aroused is present before birth and remains throughout the life span of both males and females. Sexual arousal in the adult, which can result from physical or mental stimulation, is divided into four stages: excitement, plateau, orgasm and resolution. The excitement phase of sexual arousal is marked by increased blood flow to the genitals, which results in penile erection in the male and transudation in the female as well as increased muscle tension or myotonia in both sexes. During the plateau phase of sexual arousal, heart rate, blood pressure, respiration rate, muscle tension, and genital vasocongestion increase in both sexes. The orgasmic phase is marked by muscular contractions of the internal and external genitalia of both sexes which, in the male, results in emission and expulsion of ejaculatory fluid from the penis. During the refractory period, increased sympathetic nerve activity in both sexes reduces blood flow to the internal and external genitalia, and blood leaves the genitals through the genital veins. The dimensions and position of the penis, vagina, and cervix return to those of the unaroused state. Heart rate, blood pressure, respiration, and muscle tension all return to normal as well. The external genitalia of both sexes are extremely sensitive to touch during the resolution phase.

III. A. Endocrine deficiencies can result in reduced sexual arousal and decreased capacity for orgasm. Insufficient estrogen secretion in the female results in reduced vaginal lubrication and painful intercourse, or dyspareunia. Chronic estrogen deficiencies can lead to atrophic vaginitis. Testosterone deficiencies in the male can reduce ejaculatory capacity and sexual desire.

B. Mechanical or disease-related damage to the peripheral nerves, spinal cord, or brain, can result in impaired sexual arousal and/or orgasm in both sexes. Both multiple sclerosis and diabetes mellitus can produce neuropathies that can compromise sexual arousal and orgasm.

C. Alcohol, antihistamines, sedatives, narcotics, and tranquilizers can retard or inhibit sexual arousal and orgasm by altering blood flow to the genitals or the neural transmission of sexually arousing stimuli.

D. Other conditions that interfere with sexual arousal and/or orgasm include anorgasmia, endometriosis, and vaginismus in the female and erectile dysfunction, urethritis, and premature ejaculation in the male.

E. Infertility can result from hormonal deficiencies, destruction of gonadal tissue resulting from inflammation or disease, and the production of antisperm antibodies.

IV. A. Barrier methods of contraception, including diaphragms, cervical caps, contraceptive sponges, and condoms, interfere with sperm transport. The efficacy of barrier methods in preventing pregnancy is increased by spermicides.

B. Chemical methods of contraception include hormones, such as estrogen and progesterone, which are taken orally in the form of birth control pills. Estrogen and progesterone inhibit the secretion of pituitary hormones that control ovulation. Progesterone also increases the viscosity of cervical mucus, impeding sperm entry into the uterus. In addition, progesterone inhibits proliferation of the endometrial lining of the uterus, interfering with implantation of the ovum in the uterus. The morning-after pill contains high levels of estrogen, which prevent implantation of the ovum in the uterus.

C. Surgical methods of contraception include vasectomy in the male, which interferes with sperm transport, and tubal ligation in the female, which interferes with both sperm and ovum transport.

D. The rhythm method of contraception refers to periodic abstinence from intercourse at midcycle around the time of ovulation. The efficacy of this contraceptive method depends on the accuracy of predicting the time of ovulation in the female. Changes in body temperature, consistency of cervical mucus, and length of the ovarian cycle are all used to predict time of ovulation, with variable accuracy.

E. Future trends in contraception include the development of long-acting natural and synthetic chemicals that inhibit ovulation, spermatogenesis, sperm transport, ovum transport, and/or implantation of the ovum in the uterus.

Review Questions

1. Why is the sex of the offspring determined by the father?
2. How does a single X chromosome in the female affect development of the gonads and the internal and external genitalia?
3. What are the genetic factors that control testicular development in the male?
4. What are the effects of exposing a genetically female embryo to high levels of androgens (1) during the first 12 weeks of gestation and (2) from week 15 to term?
5. How do hormonal secretions from the testes in the male influence differentiation of the internal and external genitalia?
6. How are Klinefelter's syndrome and the syndrome of testicular feminization different?
7. What physiological changes occur in the human sexual response as a function of aging in the male and female?
8. What are the four stages of sexual arousal in the male and female?
9. Do males develop the capacity for both orgasm and ejaculation at the same age?
10. What do the terms *dyspareunia* and *vaginismus* mean?
11. What are three drugs that can induce sexual dysfunction? Explain how they work and describe the symptoms they produce.
12. What are the four basic categories of contraceptive methods, and how do they work?
13. Why do birth control pills that contain progesterone prevent pregnancy?
14. What are four new approaches being developed to prevent pregnancy?

Suggested Readings

Allgeier, E.R., and Allgeier, A.R., Eds. *Sexual Interactions.* Lexington, Mass.: D. C. Heath and Co., 1984.

DeCherney, A.H., Ed. *Reproductive Failure.* Churchill Livingstone, 1986.

Geer, J., Heiman, J., and Leitenberg, H., Eds. *Human Sexuality.* Englewood Cliffs, N.J.: Prentice-Hall Inc., 1984.

Kaplan, H.S., Ed. *The Evaluation of Sexual Disorders.* Brunner/Mazel, 1983.

Masters, W.H., Johnson, V.E., and Kolodny, R.C., Eds. *Human Sexuality.* Boston: Little, Brown and Co., 1985.

Shearman, R.P., Ed. *Clinical Reproductive Endocrinology.* Churchill Livingstone, 1985.

Williams, R.H., Ed. *Textbook of Endocrinology.* Philadelphia: W. B. Saunders Co., 1981.

Index

Section and Chapter Opening Photo Captions

Chapter 1, p. 1: Many body systems are extremely active in these athletes, whereas other systems are relatively inactive. Physiology is the science of studying the functioning of bodily systems. (Bruce Berg/Visuals Unlimited.)

Section I, pp. 26–27. Large photo courtesy of Manfred Kage/Peter Arnold, Inc.; small photo by Lennart Nilsson from Sheila Kitzinger, *Being Born,* New York, Grosset & Dunlap, 1986, p. 14.

Chapter 2, p. 29: The body is composed of both inorganic and organic compounds. Among the latter are the important vitamins, such as these vitamin C molecules. (G. Musil/Visuals Unlimited.)

Chapter 3, p. 80: Some organisms are composed of single cells but the human body contains many billions of these tiny structures. Life is a phenomenon that occurs only in cells, the structural and functional units of all organisms. (Harold Sweetman/Visuals Unlimited.)

Chapter 4, p. 111: Living processes depend on bilayer membranes, which include the plasma membrane of the cell and the membranes in the various organelles. In many instances, the membrane surfaces are folded to facilitate function by increasing surface area. (David M. Phillips/Visuals Unlimited.)

Chapter 5, p. 136: The control center of the cell is its nucleus, which contains the genetic blueprint that guides all life processes. (K.G. Murti/Visuals Unlimited.)

Chapter 6, p. 179: The "powerhouse" of the cell is the mitochondrion, where the processes of cellular respiration are coordinated to provide energy for metabolic processes. (K.G. Murti/Visuals Unlimited.)

Section II, pp. 208–209. Large photo courtesy of Manfred Kage/ Peter Arnold, Inc.; small photo © 1985 SIV/Peter Arnold, Inc.

Chapter 7, p. 211: The structural and functional unit of the nervous system is the neuron. While all neurons are integrated functionally, they are not connected physically, being separated from one another by small synaptic spaces. (Stan Elems/Visuals Unlimited.)

Chapter 8, p. 254: Some sensory structures inform us of internal conditions, while others receive information about the external world. In the latter category, the most important sense to most people is vision. (A.L. Blum/Visuals Unlimited.)

Chapter 9, p. 304: Muscles are among our most important effectors, and their actions involve the nervous system, as is clear in this view of a neuromuscular junction. (Fred Hossler/Visuals Unlimited.)

Chapter 10, p. 333: The pathways for messages between the central and the autonomic nervous systems are nerves, such as these myelinated nerves. (David M. Phillips/Visuals Unlimited.)

Chapter 11, p. 354: The spinal cord and brain perform most integrative functions in our nervous system, with the latter being of greatest importance. (John D. Cunningham/Visuals Unlimited.)

Chapter 12, p. 374: The vehicles by which endocrine glands communicate with one another and with other organs are hormones, such as this crystallized molecule. (David M. Phillips/Visuals Unlimited.)

Chapter 13, p. 403: For its size, the pituitary gland is one of the body's most influential structures. Among its cells are produced an array of diverse hormones. (John D. Cunningham/Visuals Unlimited.)

Chapter 14, p. 427: Sitting like a cap atop the kidneys are the adrenal glands. Although their outer cortex and their inner medullary cells are physically connected, they produce quite different hormones that influence the body in diverse and important ways. (J.A. Carney/Visuals Unlimited.)

Chapter 15, p. 445: The exocrine activities of the pancreas are important in digestion. The central cells seen here are the islets of Langerhans, which produce many of the pancreatic hormones. (John D. Cunningham/Visuals Unlimited.)

Section III, pp. 470–471. Large and small photos courtesy of Manfred Kage/Peter Arnold, Inc.

Chapter 16, p. 473: The muscles that we control voluntarily are the striated or skeletal muscles seen here. Cardiac muscle also appears striated, but unlike striated muscle (and similar to smooth muscle), it doesn't tire as quickly. (David M. Phillips/Visuals Unlimited.)

Chapter 17, p. 514: Macrophages or "big eater cells" are among our important arsenal of defense mechanisms. This macrophage is seen binding to three cancer cells. (W. Johnson/Visuals Unlimited.)

Chapter 18, p. 549: Cardiac muscle starts beating a mere three weeks after conception and may continue to beat after a person is declared legally dead. Its structure distinguishes it in major ways from striated and smooth muscles. (John D. Cunningham/Visuals Unlimited.)

Chapter 19, p. 576: The two principal types of blood cells are the erythrocytes and the leukocytes. The biconcave shape of the red blood cells is important in providing a large surface area for oxygen transport. The white blood cells are important in many aspects of defense. (David M. Phillips/Visuals Unlimited.)

Chapter 20, p. 618: Although the respiratory system is one of the body's most compact organ systems, being restricted to the thoracic cavity, its functioning is important to the entire body. (O. Averback/Visuals Unlimited.)

Chapter 21, p. 645: Millions of microscopic alveoli like these, each surrounded by a network of capillaries, make up the lungs and provide us with an enormous surface area for gas exchange. (David M. Phillips/Visuals Unlimited.)

Chapter 22, p. 665: Humans explore and work in almost every environment, including underwater. Diving provides many physiological challenges. Breathing gases must be matched carefully to the depth and to the degree of activity in order to avoid problems such as the bends developing in the diver's body. (W.H. Hughes/Visuals Unlimited.)

Chapter 23, p. 681: Our tube-within-a-tube gastrointestinal tract includes a large number of diverse structures. Their combined functions would be of little value, however, if we lacked an efficient mechanism to absorb the chemically simplified nutrients. That important function is accomplished largely by the villi in the small intestine, and their very large surface area assists in this effort. (Fred Hossler/Visuals Unlimited.)

Chapter 24, p. 723: Although various organs excrete wastes, our prime organ of excretion is the kidney, seen here in cross-sectional view along with the upper portion of the ureter. (John D. Cunningham/Visuals Unlimited.)

Chapter 25, p. 754: The maintenance of the proper quantity and composition of our body fluids depends upon the kidney. The microscopic units that perform the important tasks of secreting wastes and conserving useful substances are the nephrons. (David M. Phillips/Visuals Unlimited.)

Chapter 26, p. 780: The measurement of the acidity or alkalinity of body and other solutions is easy in the clinical laboratory, often using ion-specific electrodes such as those shown here. Far more miniaturized equipment is necessary for acid-base measurements inside the body. (Science Vu/Visuals Unlimited.)

Chapter 27, p. 800: The prime reservoir for many of the body's critical minerals, such as calcium and phosphorus, is bone, seen here in cross-sectional view. (John D. Cunningham/Visuals Unlimited.)

Chapter 28, p. 823: When we're hot, our major physiological mechanism for maintaining our body temperature homeostatically is perspiration, and the resulting evaporative cooling that occurs when fluid is released from the sweat glands. (Veronika Burmeister/Visuals Unlimited.)

Chapter 29, p. 841: Every organ system contributes to the skill of these athletes. In addition to obvious strength of the skeletal and muscular systems, the digestive, respiratory, circulatory, and urinary systems interact to supply the body with needed nutrients, energy, and gases, and to excrete metabolic wastes. All these functions are under the joint control of the nervous and endocrine systems. (H. Oscar/Visuals Unlimited.)

Chapter 30, p. 856: This egg and its associated cells is seen near the time of ovulation. Eggs are produced regularly during menstrual cycles but only if they are fertilized is there a possibility of a genetic blueprint being passed to another generation. (D.M. Phillips/Visuals Unlimited.)

Chapter 31, p. 879: It all starts when the sperm fertilizes the egg to produce a zygote that contains the combined genetic blueprints of the parents. Within hours, the zygote divides to become two cells, and by the time of birth we consist of billions of somatic cells with diverse functions but the same basic blueprint. (David M. Phillips/ Visuals Unlimited.)

Chapter 32, p. 905: Sexual activities involve every organ system, and although there are numerous measurable physiological and psychological changes during the sexual response cycle, there are also many unexplained aspects of it. (Bruce Berg/Visuals Unlimited.)